Integral Ecology

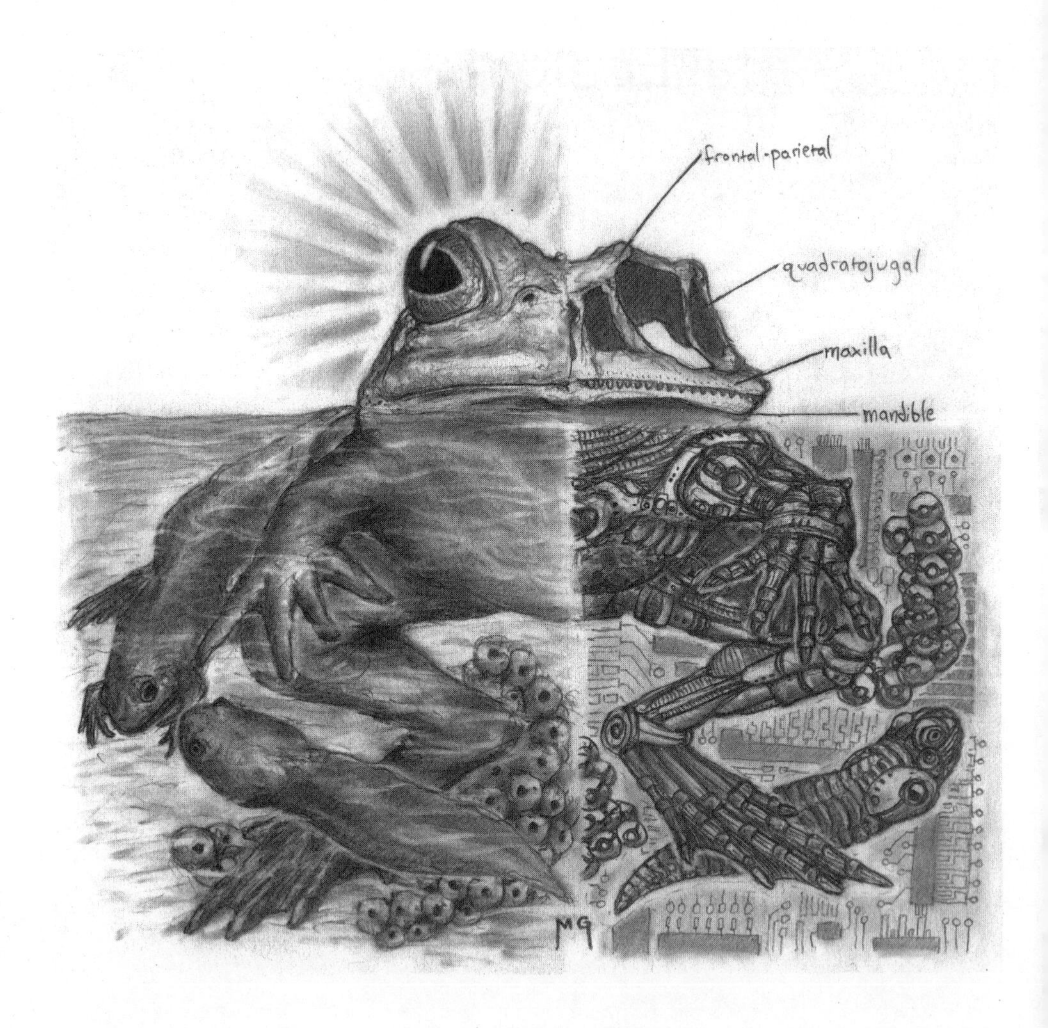
frontal-parietal
quadratojugal
maxilla
mandible
MG

Integral Ecology

Uniting Multiple Perspectives on the Natural World

Sean Esbjörn-Hargens, PhD, *and*
Michael E. Zimmerman, PhD

With case studies by Gail Hochachka, Brian N. Tissot, and Darcy Riddell

Foreword by Marc Bekoff, PhD

INTEGRAL BOOKS
Boston & London
2009

INTEGRAL BOOKS
An imprint of Shambhala Publications, Inc.
Horticultural Hall
300 Massachusetts Avenue
Boston, Massachusetts 02115
www.shambhala.com

Frontispiece by Michael Garfield

9 8 7 6 5 4 3 2 1

First Edition
Printed in the United States of America
Designed by DEDE CUMMINGS DESIGNS
♾ This edition is printed on acid-free paper that meets
the American National Standards Institute z39.48 Standard.
♻ This book was printed on 30% postconsumer recycled paper.
For more information, please visit www.shambhala.com.
Distributed in the United States by Random House, Inc.,
and in Canada by Random House of Canada Ltd

Library of Congress Cataloging-in-Publication Data

Esbjörn-Hargens, Sean.
Integral ecology: uniting multiple perspectives on the natural world /
Sean Esbjörn-Hargens, and Michael E. Zimmerman; with case studies by
Gail Hochachka, Brian Tissot, and Darcy Riddell; foreword by Marc Bekoff.—
1st ed.
p. cm.
Includes bibliographical references and index.
ISBN 978-1-59030-466-2 (hardcover: alk. paper)
1. Human ecology—Philosophy. 2. Nature. 3. Human beings.
I. Zimmerman, Michael E., 1946– II. Title.
GF21.E75 2009
304.201—dc22
2008032370

To Tatiana Rose, the queen of all those unseen at Sea Frog Haven

—SEAN ESBJÖRN-HARGENS

To my wife, Teresa, and my daughter, Lizzie

—MICHAEL E. ZIMMERMAN

We do not easily know nature, or even know ourselves. Whatever it actually is, it will not fulfill our conceptions or assumptions. It will dodge our expectations and theoretical models. There is no single or set "nature" either as "the natural world" or "the nature of things." The greatest respect we can pay to nature is not to trap it, but to acknowledge that it eludes us and that our own nature is also fluid, open, and conditional.

Hakuin Zenji put it "self-nature that is no nature / . . . far beyond mere doctrine." An open space to move in, with the whole body, the whole mind.

—GARY SNYDER,

No Nature: New and Selected Poems

CONTENTS

PART THREE

PART FOUR

ILLUSTRATIONS

PREFACE

SEAN ESBJÖRN-HARGENS

In hindsight, it now seems inevitable that our paths would not only cross but that we could become companions in and explorers of Integral Ecology. As Michael published *Contesting Earth's Future* (1994), the first environmental philosophy text to draw heavily on Ken Wilber's analysis, I was heading to Africa, where I would encounter Wilber's writings. Separately, and at the same time, we were developing much of what is presented in this book. Thus, for over five years both of us were independently connecting Wilber's integral analysis to ecological issues and environmental thought. As a result of our efforts we had both become friends with Ken Wilber. Wilber brought us together and introduced us in the summer of 2000 by inviting us (along with several others applying Integral Theory to ecology) to Boulder, Colorado, for the first Integral Ecology Center meeting, hosted by the recently formed Integral Institute.[1] Soon after this meeting, Wilber's *A Theory of Everything* was published. Within its pages occurs his first published usage of the phrase "integral ecology."[2]

Over the next several years, Michael and I stayed in touch and compared notes as we both taught academic courses in Integral Ecology and wrote articles connecting the Integral Model to ecological issues.[3] In April 2002 the Integral Institute hosted another Integral Ecology meeting. This time the focus was to discuss the possibility of writing a textbook on Integral Ecology.[4] Over the next few months, the two of us began sketching out the contents of such a book and starting writing articles that would eventually become chapters.[5]

It is our hope that this book supports a new kind of ecology, one that is informed by the strengths of many approaches and methods, while at the same time exposing the limits and blindspots of various perspectives. May this book serve a flourishing of mutual understanding between differing perspectives.

ACKNOWLEDGMENTS

As with any long-term project with a scope as large as this one, there are many to thank and acknowledge. To begin with, we want to recognize Ken Wilber for his brilliant mind and open heart, his friendship, and his guidance in manifesting this book through countless hours of review and conversation. Annie McQuade has done the heavy lifting, placing our word clay into the editorial fire and creating a durable text with a nice glaze. We are deeply indebted to her intellectual labor, sense of humor, and much-needed encouragement all the while being patient and persistent with our big life transitions (I became a father, and Michael moved his family across the country). We were blessed to have a strong seasoned team at Shambhala with Kendra Crossen, Liz Shaw, Chloe Foster, Lora Zorian, and Hazel and Sara Bercholz sharing legs in our publishing triathlon and ultimately carrying the project across the finish line. Matt "Wrench" Rentschler had the laborious job of tending to endnote editing details that would drive most people to drink. Not only did he do a superb job on this task, but he also provided some important content research for chapter 9. Jon Geselle provided his embodied presence and sharp mind while reading the entire manuscript and made many important suggestions that greatly improved the text. For many years Jon has been a sounding board for many of the ideas presented here and has consistently helped refine the theory and praxis of Integral Ecology. Throughout the project Stan Salthe provided valuable conversation and feedback on the content of the manuscript. Marc Bekoff's support and generous foreword are greatly valued. Brad Reynolds did a wonderful job providing all the tables and figures, of which there are many. His skill and aesthetic sense has greatly enhanced the book. Likewise, Michael Garfield's work on the 4-quadrant frog frontispiece and cover art is superb. If only all authors could have such skill and vision tending to their book covers. Clint Fuhs supported us in unpacking Integral Calculus—lighting up the path where few have gone. Nick Hedlund and Carrissa Wieler did important research locating key quotes and texts for chapter 9. Yotam Schacter did a meticulous job of scanning the case studies in part four. Gail Hochachka, Brian N. Tissot, and Darcy Riddell have provided valuable illustrative case studies of Integral Ecology in action for part four. These case studies do the important work of connecting our theoretical musings with the concrete details of application. Similarly, we are appreciative of all those who have provided us with examples of Integral Theory applied to ecological and environmental issues: Cameron Owens, Wade Prpich, Kevin Feinstein, Brad Arkell, Brian Eddy, Barrett Brown, Cynthia McEwen, Will Varey, Mark DeKay, Ian Wight, Chris Reidy, Gail Hochachka, Tim Winton, Nick Wilding, Marilyn Hamilton, Kevin Snorf, Joel Kreisberg, John Dupuy, Chris and Ilsa Preist, David Johnston, and Stephan Martineau. Integral practitioners Ingrid Bamberg and Dan Wheeler have provided much support and camaraderie in various forms.

In addition, I want to acknowledge that a profound debt is owed the wild places, in particular the rocky shores of the Pacific Northwest and the redwoods of Northern California (especially those at Sea Frog Haven, which held me in the final stages of this process connecting me to what in many respects is all that matters!).

An equal recipient of my appreciation is Michael Zimmerman. Michael has not only served as mentor, colleague, and friend, but he has also been a fellow explorer seeking more Integral frameworks and articulations of our relationship with the natural world. I consider it a rare opportunity to be able to work so closely with someone as philosophically astute and openhearted as Michael. It has been a total joy to collaborate on this project with him.

Thanks goes out to my mom, Rochelle "Rody" Hargens, for her unmatched mother's love and all the support she has given me from the beginning. She has truly made this book possible. And I bow to her for allowing me to explore, unaccompanied, the woods around our house in Shelton—it was there that the seeds of Integral Ecology were planted. Deep gratitude goes to my dad, Gary Hargens, for all the times he placed us in the wilderness: hunting blacktail deer, stalking the elusive pine mushroom and the tasty "yellow trumpet" chanterelles, fishing for cutthroat trout and Chinook salmon, and sitting in freezing water at the break of dawn, shotguns in hand, listening to mallards and canvas backs cut the air above us.

Gratitude also goes out to Joe Yuska at Lewis & Clark College Outdoors for all his mentoring in loving the outdoors; Stefan Aumack, John Barrett, and Matt Couch, who accompanied me on many camping trips in the Pacific Northwest and the Southwest; Bill Rotschaefer; my crews with Northwest Youth Corps; faculty at the California Institute of Integral Studies: Sean Kelly, Brian Swimme, Robert McDermott, Jorge Ferrer, Steven Goodman, Agana Chatterji, and Richard Shapiro; my biological father, Jerry Frye; the Bay Area Integral Community; my students at John F. Kennedy University, the California Institute of Integral Studies, Fielding Graduate University, and the Institute of Transpersonal Psychology; the people of Bhutan; my integral interlocutors Kevin Snorf, Forest Jackson, Andre Marquis, Barrett Brown, and Frank Poletti; and especially to my beloved wife and dharma companion, Vipassana Esbjörn-Hargens; she *more than anyone* has been an anchor and an inspiration—I simply cannot imagine having done this book without her daily soul illumnation, which has continually revealed deeper layers of Being to me.

SEAN ESBJÖRN-HARGENS
Sebastopol, Calif.
August 2008

I want to recognize Sean for being such an outstanding partner in this enterprise. Never have I worked with someone so closely on a writing project of this magnitude and complexity. Sean always lived up to his agreements in all areas. I am honored to have worked with him and to have learned so much in the process. Likewise, I am deeply indebted to Ken Wilber, whose visionary and integrative thinking has played such an important role in my continuing academic and personal development. Sean has already acknowledged many of the people whom I would like to thank, including all those associated with the Integral Ecology branch of the Integral Institute.

I would like particularly to thank Stan Salthe for engaging in a most stimulating dialogue with Sean and me during the writing of this book. Stan is a formidable integrative thinker.

I want also to thank my students, who allowed me to explore with them

Integral themes in several different courses, and who forced me to go back to the drawing boards on more than one occasion through persistent and insightful criticism of some of the positions that I put forth. In my view, Integral thinkers must above all be self-critical, and thus must always be willing to learn from others.

Finally, I would like to thank my wife, Teresa, and my daughter, Lizzie, who supported me in the years of traveling and writing required to complete this volume. I dedicate this book to them.

We are, I hope, about to see a significant increase in integral thinking on many different ecological fronts. Yet, I have learned not to be attached to the consequences of my efforts, which may or may not bear fruit. Gratifying it is to take a stand with Sean and many others on behalf of this glorious planet and the life forms that thrive upon it.

May my intentions be honorable!
May all beings be happy!
May Divine blessings be on all Creation!

MICHAEL E. ZIMMERMAN
Boulder, Colo.
August 2008

FOREWORD

MARC BEKOFF, PhD

The Public Lives of Animals: Bringing Interiority to the Surface

Integral Ecology is a wonderful book—but actually, it's many books in one. Its scope is enormous, and the authors are to be congratulated for getting so much important and timely information between two reasonably spaced covers (considering the scope of their project—to integrate 200 perspectives on the natural world—even 800 pages is reasonable).

Our relationship with animate and inanimate nature is a complex, ambiguous, challenging, and frustrating affair. Many people claim to love nature and to love other animals, and then, with little forethought, concern, or regret, go on to abuse them in egregious ways, far too numerous to count. I often say when someone tells me that they love various landscapes or animals and then partake, either directly or indirectly, in subjecting them to intentional pain and suffering or wanton destruction, that I'm glad that they don't love me!

The messages in the book with which I agree are far too numerous to mention, but among them is a strong call for integrated holism, compassion, and respect for our one and only planet. The authors also make the important work of Ken Wilber more accessible to a new audience in ecology and environmental studies who might not have previously known about it. This book is a wonderful example of interdisciplinary lateral thinking grounded in a multiple methodological approach. The authors also note that science isn't value-free. That science is value-free is a myth, and I think that when we realize that this is the case, we do better science. Science isn't the only show in town. We need to blend scientific data (what I call "science sense") with intuition, common sense, indigenous knowledge, and qualitative research, as we try to comprehend the world in which we're immersed.

As I read this book I often thought of Thomas Berry's claim that each and every individual is a member of a communion of subjects. The authors insist on returning "interiority" to the conversation about humankind and nature. Any ecological view that doesn't consider the depths and interconnections among all components of nature is only a "partial ecology" rather than an "integrated ecology" and misrepresents the magnificent webs of nature that abound all over the place. By providing a framework for ecologists to include the interiors of organisms and environmentalists to include the interiors of the general public (e.g., by taking into consideration the three major worldviews of traditional, modern, and postmodern), the authors provide a great service to the field of ecology and environmental studies. *Integral Ecology* articulates a way of doing ecological science that investigates the behavior, experience, systems, and cultures of organisms.

The science of ecology has excluded the interiors of ecosystem members for too long! The experiences and lives of animals and all sorts of vegetation are all bound into a social nexus, and all members are indispensable for maintaining the integrated whole of nature. Let's not forget that vegetation can also have social lives. Suffice it to say, and to risk being trite, the deep reciprocal interconnections among members of the earth community are such that we're all in this together, and we all need others we can lean on. A view of nature that sanitizes, reduces, and simplifies the complex interrelationships that exist is bound to tarnish and diminish the appreciation that people have for what's out there and to result in a feeling of alienation. This feeling of alienation can then feed back and produce more alienation.

Many scientists are also control freaks, but once they realize that full control isn't a reality and that reductionism misrepresents nature, a clearer view of the magnificence of nature emerges, along with greater appreciation for her splendor. All branches of science are burdened by uncertainty, but this doesn't mean that we don't have enough information to offer solid explanations with strong predictive value. We need to be able to live with uncertainty and give up control. This is not to be antiscience but to accept the world as it is.

Animal Passions as Public Interiorities

Concerning my own work, the authors generously note that I've been working for decades on trying to get others to realize that other animals have rich and active minds and emotional lives. In the authors' words, I've been studying the "interiors" of organisms despite the fact that most of my colleagues ignore them or write them off as nonexistent. Toward this end, I've been arguing for an interdisciplinary holistic approach and methodological pluralism that counters strong tendencies toward narrow research endeavors that ignore cross-disciplinary collaboration, foster reductionism, and stimulate territorial behavior among different disciplines among many of my colleagues.

I've also been arguing that we need more compassion and subjectivity in science and that the emotional and inner lives of other animals aren't all that private. We often know more than we give ourselves credit for about interiority; the passionate lives and feelings of animals are public matters, and the privacy of mind argument frequently is overblown. We can indeed factor in subjectivity and intersubjectivity with solid science. I maintain that we know that some nonhuman animals feel something some of the time, just as do human animals. It's nonsense to claim that we don't know if dogs, pigs, cows, or chickens feel pain or have a point of view about whether they like or don't like being exposed to certain treatments. Who are we kidding? Frankly, I think we're kidding ourselves. While there always will be mysteries about other minds and what's happening in them, animals tell us how they feel rather freely and clearly. And even if we're wrong some of the time, we surely aren't wrong all of the time. Attributions of emotions help us to explain and understand various aspects of animal behavior, and allow us to make very accurate predictions about what individuals are likely to do in specific situations. In the majority of cases, it is apparent that differences among species are differences in degree rather than differences in kind, à la Charles Darwin's ideas about evolutionary continuity.

To sum up briefly, many animals are actually experiencing the lives they're living and the emotion they're expressing. Of this there is no doubt.[1] Thus, ecology must recognize animal subjectivity and the enormous ethical implications of the fact that many animals are thinking and feeling beings. This book takes many important steps toward creating such an ecology: an integral ecology that includes animal interiors alongside behavioral and environmental analysis. Such an ecology is long overdue and entirely possible.

One of the most exciting aspects of research in animal cognition and animal emotions is that scientists are collecting data on the nitty-gritty details of the behavior patterns in which they're interested. As we collect more detailed data, we discover much that was overlooked in previous research. A good example of this is my research on the question of whether animals can be moral beings: do they know right from wrong? I call this "wild justice."[2] It took years of analyzing single frames of ongoing play interactions for me to see the link between how animals play, how they negotiate the cooperation necessary for them to play with one another, and the likelihood that they know what's right and wrong in the context of social play. I discovered that cooperation, empathy, forgiveness, trust, and apology all play a significant role in allowing individuals to play with one another without play escalating into fighting. Now other researchers are also discovering that species other than domestic dogs and their wild relatives show moral behavior.

Cognitive ethology (the study of animal minds) is the unifying science for understanding the subjective, emotional, empathic, and moral lives of animals because it is essential to know what animals do, think, and feel as they go about their daily routines in the company of their friends and when they are alone.[3] The more detailed the analyses, the more richness emerges concerning the emotional lives of animals. It's important to learn why both the similarities and differences in cognitive capacities and sentience between humans and other animals have evolved. The more we come to understand other animals, the more we will appreciate them as the amazing beings they are, and the more we will come to understand ourselves. We must pay close attention to what animals do in their worlds and recognize other animals have a perspective and a "way of knowing."

Nowadays it's getting more difficult to find skeptics concerning the passionate lives of animals, and it's about time that skeptics who question the existence of animal emotions have to "prove" their case rather than vice versa. The line that animals merely act "as if" they're feeling pleasure or pain never has been very strong, and now it's essentially dead. Indeed, many prestigious scientific journals, including *Science*, *Nature*, and the *Proceedings of the National Academy of Science* are now publishing essays on animal emotions including research on joy and laughter in rodents.

Anthropomorphism and Interiority: The Abundant Evidence of Our Own Senses

Many of my colleagues dismiss my views and those of the authors by exclaiming, "Oh, you're just being anthropomorphic!" They're right. There are many different ways of describing and explaining what animals do. How researchers choose to summarize what they see, hear, or smell when studying animals depends on the questions in which they're interested. There isn't only one correct way to describe or to explain what animals do or feel. Using words that describe such human characteristics as thinking, joy, grief, embarrassment, and jealousy is being anthropomorphic, a practice that irks many of my colleagues although they freely engage in it. Anthropomorphism is the attribution of human characteristics to nonhuman creatures and inanimate objects. Intersubjectvity and interiority clearly are involved. Such intersubjectivity, between humans and animals, serves as one way of legitimately accessing and experiencing what animals are feeling, providing us with the capacity to accurately describe the inner worlds of animals.

As humans who study other animals, we can only describe and explain their behavior using words with which we're familiar from a human point of view. Anthropomorphizing can be viewed as an adaptation, so when we anthropomorphize, we're doing what comes naturally.[4] So, when I tried to figure out what was happening in my late dog Jethro's head, I had to be anthropomorphic, but I tried to do it from a dog-centered—"dogocentric"—point of view. Just because I tell you Jethro was happy or jealous, this doesn't mean he was happy or jealous like humans or, for that matter, like other dogs. Being anthropomorphic, in a reflective way, is a tool to make the thoughts and feelings of other animals accessible to humans. While we surely make errors from time to time, we're pretty good about making accurate predictions in the realm of the mental.[5] Not to mention that the validity of our anthropomorphic descriptions become all the more strengthened when we collaborate them with findings from other methodological investigations. One of the major contributions this book offers is a sophisticated way to make use of multiple modes of research in order to discuss in a meaningful and empirical way the interior lives of animals.

Robert Sapolsky is clearly anthropomorphic in his writing about baboons and places much of the debate about anthropomorphism in a very useful perspective. He wrote: "Do I get grief for the fact that in communicating, say, about the baboons I'm doing so much anthropomorphizing? One hopes that the parts that are blatantly ridiculous will be perceived as such. I've nonetheless been stunned by some of my more humorless colleagues—to see that they were not capable of recognizing that. The broader answer, though, is I'm not anthropomorphizing. Part of the challenge in understanding the behavior of a species is that they look like us for a reason. That's not projecting human values. That's primatizing the generalities that we share with them."[6]

There's no need to think of anthropomorphism as something for which its practitioners should be embarrassed or punished. We're just doing what comes naturally, we don't lose important information that the animals are sharing, and it's very useful for learning more about beastly passions. There's no reason to discount or dismiss the abundant evidence of our own senses.

If we decide against anthropomorphism, we have no alternatives other than to talk about animals as a bunch of hormones, neurons, or muscles absent any context for what they're doing and why. Gordon Burghardt notes that denying our own intuitions about an animal's experience is "sterile and dull."[7] If we don't anthropomorphize, we lose important information. The renowned psychologist Donald Hebb also made some important observations about anthropomorphism. Hebb noted that the keepers' anthropomorphic accounts proved "an intuitive and practical guide to behavior," enabling them to best interact with the captive chimpanzees for whom they cared.[8] In fact, many who favor mechanistic explanations haven't spent much time watching free-ranging animals or have little familiarity with the subtleties of animal behavior.[9] I agree with Henry Pollack when he says that "unfamiliarity breeds uncertainty."[10]

Surely, given the complexity and flexibility of behavior, no explanatory scheme will be correct all of the time. But, more important, reductionists ignore the fact that the utility and accuracy of various sorts of explanations haven't been assessed empirically. So, we really don't know if their flavor of explanations is better for understanding and predicting behavior than those they eschew. Until the data are in we all must be careful in claiming that one sort of explanation is always better than others. It's poor scholarship and sets a poor example for students to take a narrow-minded approach in the absence of supportive data.

Elephant Interiority: It's Okay to Say an Elephant Is "Doing Well" but It's Not Okay to Say She "Isn't Doing Well"

Our strong and irresistible anthropomorphic tendencies also reflect strong intersubjectivity. An illustrative example of what I call "anthropomorphic double-talk" concerns the fate of Ruby, an elephant who, on occasion, was shipped from zoo to zoo in the United States as if she were a piece of furniture. Some people argue against the use of the A word without seeming to know they're using it or hope that others don't see that they are. For example, the then director of conservation and science for the American Zoo and Aquarium Association (AZA), Michael Hutchins, recently claimed that we mustn't be anthropomorphic and that it's bad science to attribute humanlike feelings to animals.[11] Hutchins said: "Animals can't talk to us so they can't tell us how they feel." He was critical of people who claimed that Ruby, a 43-year-old female African elephant at the Los Angeles Zoo, wasn't "doing well" in captivity, that she was unhappy. How could Hutchins be critical of those who claim Ruby is unhappy unless he believes that she is happy, and isn't he then being anthropomorphic? What Hutchins means is that he and others in their camp can be anthropomorphic but their foes can't. They can say that an animal is "doing well," but others can't say the elephant is not doing well. Indeed, the director of the Los Angeles Zoo, John Lewis, said about Ruby, "She seems to be fine." The bottom line is that one can't talk about how animals are doing—well or not so well—without being anthropomorphic, so critics ought to fess up and stop being hypocrites.

Most people who live with and work closely with animals take it as a matter of fact that many animals have emotions, and they know and admit they're being anthropomorphic. For example, mahouts, people who work with elephants, know that one ignores an elephant's "mood" at one's own peril. If we ignore what other animals are feeling, we lose not only much about who these amazing beings are, but perhaps also our own lives.

Deep Ethology

My notion of deep ethology has much in common with integral ecology. Deep ethology comes into play because we need to recognize that we are all integral members of a single earth community and that as moral agents with enormous brains and the capacity to do whatever we want, we have unique responsibilities to nature. We bring a lot to the party along with something special, namely the moral capacity to restrain our exploitive practices regarding other animals and a plethora of landscapes. A deep ethological perspective, like integral ecology, allows us to be sensitive to the lives and experiences of animals and implores us to try to take their point of view in our interactions with them. We need to observe animals carefully, listen to their stories attentively, inhale their odors with zeal and some caution, and never forget that they, like us, have likes and dislikes, preferences, goals, beliefs, and feelings. My views, and those expressed by the authors, on animal use are indeed restrictive. We're not nature's keepers if this means that we "keep nature" by dominating other animals using a narrow, anthropocentric agenda in which other animals are objectified—referred to by numbers and not by names, transformed into points on a graph—and their worldviews discounted.

Deep ethology and integral ecology also highlight that if we allow ourselves to do so, we can feel the feelings of other beings. While this might sound a touch "fluffy" to some of my colleagues, research on "mirror neurons" provides support for there being a neural basis for sharing feelings and intentions.[12] Surely, the sharing of feelings should make us deeply reflect on how we cause so much enduring pain and suffering to other beings, human and other than human. Allowing ourselves, if we dare, to occupy the paws, heads, and hearts of other animals should, I hope, reduce the gap between "them" and "us."

That many animals have subjective and intersubjective communal lives—other animals are in their thoughts and feelings—and a personal point of view on the world that they share with other individuals seems beyond question. In his development of an "anthro-harmonic" perspective on human-nonhuman relationships, Scharper notes that "intersubjectivity is a fundamental reality of all human existence." *Harmonic* means "of an integrated nature" that "acknowledges the importance of the human and makes the human fundamental but not exclusively focal."[13] Working toward an anthro-harmonic understanding of human-nonhuman relationships in the future is a good road to travel. This book provides a wonderful road map for this exciting, challenging, and much-needed journey.

Let me tell a few short stories to highlight some of the above points more succinctly.

Nasty Nick, Reuben, and Ruth: Pissy Baboons

Emotional baboons make a case for strong interiority. Nick, an olive baboon living in the Forest Troop in the southeast corner of Masai Mara National Reserve in Kenya, joined the group when he was an adolescent. According to Robert Sapolsky, you could almost see contempt on his face.[14] Being unashamedly anthropomorphic, Sapolsky noted that Nick dominated his age group and ". . . he was confident, unflinching, and played dirty." During a fight with Reuben, another adolescent male, Reuben "stuck his ass up in the air," a sign of submission, but Nick went over as if to examine his bottom and then slashed Reuben's ass with his canines. To quote Sapolsky, "The guy simply wasn't nice. . . . He harassed the females, swatted at kids, bullied ancient Gums and Limp. On one memorable day, he took exception to something that poor nervous Ruth had done and chased her up a tree. Typically, at this point, the female takes advantage of one of those rare instances when it pays to be smaller than the males—she goes to the farthest end of a flimsy branch and hangs on for dear life, depending on the fact that the heavier male can't crawl out to where she is and bite her. And, typically, the male, thwarted, positions himself to at least trap the female, keeping her screaming on the precarious branch until he gets bored. So Ruth gallops up the tree, Nick after her, and Ruth leaps out to a safe edge. Nick promptly climbs onto a stronger, thicker branch directly above her. And then urinates on her head."[15]

Angry Elephants

In a landmark paper in *Nature* magazine titled "Elephant Breakdown," Gay Bradshaw and her colleagues showed that elephants are very sensitive beings and that social trauma can affect their physiology, behavior, and culture over generations.[16] Indeed, some elephants might suffer from posttraumatic stress disorder (PTSD) that leads to chronic stress and undermines socialization. And these severely traumatized mammoths might also get pissed off, hold grudges, and get even with humans who abuse them. In addition, an article in *New Scientist* magazine followed up on what might be happening to elephants when their families are broken up, noting that unprovoked attacks are getting more common.[17]

Improbable Partners: Intersubjectivity and Cross-Species Interiority

While it's clear that individuals of many species form close social bonds with other members of their species, there are examples of improbable between-species relationships that yield to explanations encompassing intersubjectivity and interiority. Here are two compelling stories.

A Gracious Behemoth: A Winking Whale

In December 2005, a 50-foot, 50-ton female humpback whale got tangled up in crab lines, the weight of which was making it difficult for her to keep her blowhole above the water.[18] After being freed by a courageous team of divers, the whale nuzzled each of her saviors in turn and flapped around in what whale experts said was "a rare and remarkable encounter." James Moskito, one of the rescuers, recalled, "It felt to me like it was thanking us, knowing it was free and that we had helped it." He went on, "It stopped about a foot away from me, pushed me around a little bit, and had some fun." But this wasn't all he had to say. "When I was cutting the line going through the mouth, its eye was there winking at me, watching me . . . it was an epic moment of my life."

A Hamster and a Snake

At Tokyo's Mutsugoro Okoku Zoo a dwarf hamster named Gohan (a tasty rice dish in Japanese) was offered as a meal to a rat snake named Aochan.[19] Aochan had refused to eat frozen mice, and zookeepers figured that Gohan would be more appetizing. However, rather than eating Gohan, Aochan made friends with her and they have shared a cage and remained friends ever since. Gohan even naps on Aochan's back. Even after Aochan began eating frozen rodents, he shows no interest in eating his friend. Kazuya Yamamoto, a keep at the zoo, said, "Aochan seems to enjoy Gohan's company very much."

Where to from Here? Embracing Interiorities

We need a compassionate ethic of caring and sharing our planet. Sensitivity and humility are essential components of our guiding ethic. Expanding our circle of respect and understanding can help bring us all together. We are other animals' guardians, and we owe them unconditional compassion, respect, support, and love. We may have control and dominion over other animals, but this doesn't mean that we have to exploit and dominate them.

The worlds of other animals are laden with magic and wonder. Just as we exclaim "Wow" when we marvel over the mysterious lives of other animals, I wouldn't be surprised if they say "Wow" in their own ways as they experience the ups and downs of their daily lives and the grandeur and magic of the environs in which they live. We need a world where sensing is feeling.

Our starting point should be that we will not intrude on other animals' lives unless we can show that we have a right to override this maxim and that our actions are in the best interests of the animals irrespective of our desires. When unsure about how we influence the lives of other animals, we should give them the benefit of the doubt and err on the side of the animals. It's better to be safe than sorry.

First and foremost in any deliberations about other animals must be deep concern and respect for their lives and the worlds within which they live—respect for who they are in their worlds, and not respect motivated by who we want them to be in our anthropocentric scheme of things. Can we really believe that we are the only species with feelings, beliefs, desires, goals, expectations, the ability to think, the ability to think about things, the ability to feel pain, or the capacity to suffer?

Surprises are always forthcoming concerning the cognitive skills, emotional lives, and sentience of nonhuman animals, and it's essential that people who write about animal issues be cognizant of these findings. I don't see how any coherent thoughts about moral and ethical aspects of animal use could be put forth without using biological/evolutionary, ethological, and philosophical information.

Ethics, compassion, humility, respect, coexistence, and sustainability are among the principles that should guide us when we interact with other animals. In most cases there are more humane alternatives than the methods that we use to intrude into animals' lives and less invasive alternatives when we violate landscapes. When we harm other animals and Earth, we harm ourselves. We're really that connected. We need to keep a worldwide community–based ethic in our hearts. A sense of oneness must be our goal. When you're in pain, I'm in pain; when a tree falls, I fall too. As we come to live more in harmony with nature, we can restore, rekindle, and re-create ourselves, along with our psyches that have been fragmented because of our alienation from nature.

We can always do better in our interactions with other animal beings and Earth. Always. We must always work to reduce harm and cruelty. Always.

Integral Ecology is a forward-looking book that invites compassionate proactive activism when dealing with the messes we've made. I could go on and on extolling the virtues of this wonderful book. I believe I've touched on many of its important messages, so now I invite you to read what the authors have to say and to incorporate their messages into your own interiorities so that we can truly make this world a better place for all beings and all landscapes. We need to open our hearts as we journey here, there, and everywhere.

We can all do better in developing and implementing a broad "worldcentric" ethic for taking multiple perspectives and thereby dealing more effectively with the environmental problems that we face. And individual animals are an integral part of the environment. Time isn't on our side, but my optimism leads me to believe that if we embrace Sean Esbjörn-Hargens's and Michael Zimmerman's messages and put them into action using humility, compassion, heart, and love, we still have a chance to pull ourselves out of the many deep holes we're digging for ourselves, other animals, and ecosystems.

MARC BEKOFF
Boulder, Colo.

NOTES

1. Bekoff, *Animal Passions and Beastly Virtues*; Bekoff, "Animal Emotions and Animal Sentience and Why They Matter"; Bekoff, "The Public Lives of Animals"; Bekoff, *The Emotional Lives of Animals*.
2. Bekoff, "Wild Justice and Fair Play."
3. Bekoff, *Animal Passions and Beastly Virtues*.
4. Horowitz and Bekoff, "Naturalizing Anthropomorphism."
5. Bekoff, *Animal Passions and Beastly Virtues*; Bekoff, "Animal Emotions and Animal Sentience and Why They Matter."
6. Sapolsky, *A Primate's Memoir*.
7. Burghardt, "Animal Awareness."
8. Hebb, "Emotion in Man and Animal."
9. Bekoff, *Animal Passions and Beastly Virtues*.
10. Pollack, *Uncertain Science . . . Uncertain World*.
11. Biederman, "Soft Heart Under Her Thick Skin?"
12. For reviews see Gallese and Goldman, "Mirror Neurons and the Simulation Theory of Mind-Reading"; Allen, "Macaque Mirror Neurons"; Blakeslee, "Cells That Read Minds"; Dobbs, "A Revealing Reflection."

13. Scharper, *Redeeming the Time*.
14. Sapolsky, A *Primate's Memoir*.
15. Ibid.
16. Bradshaw et al., "Elephant Breakdown."
17. Williams, "Elephants on the Edge Fight Back."
18. See Fimrite, "Daring Rescue of Whale off Farallones."
19. See MSNBC, "Hamster, Snake Best Friends at Tokyo Zoo," www.msnbc.msn.com/id/10903211.

REFERENCES

Allen, C. 2006. "Macaque Mirror Neurons: Detecting Intentions Intentionally." In *Naturalizing Intention in Action*, edited by F. Grammont. Cambridge, MA: MIT Press.

Bekoff, M. 2004a. "The Great Divide." *American Scientist Online*. September/October. www.americanscientist.org/template/BookReviewTypeDetail/assetid/35495;jsessionid=aaad8JW-NflEm5.

———. 2004b. "Wild Justice and Fair Play: Cooperation, Forgiveness, and Morality in Animals." *Biology and Philosophy* 19: 489–520.

———. 2006a. "Animal Emotions and Animal Sentience and Why They Matter: Blending 'Science Sense' with Common Sense, Compassion and Heart." In *Animals, Ethics, and Trade*, edited by J. Turner and J. D'Silva. London: Earthscan, 27–40.

———. 2006b. *Animal Passions and Beastly Virtues: Reflections on Redecorating Nature*. Philadelphia: Temple University Press.

———. 2006c. "The Public Lives of Animals: A Troubled Scientist, Pissy Baboons, Angry Elephants, and Happy Hounds." *Journal of Consciousness Studies* 13 (5): 115–31.

———. 2007. *The Emotional Lives of Animals: A Leading Scientist Explores Animal Joy, Sorrow, and Empathy and Why They Matter*. Novato, CA: New World Library.

Biederman, P. W. 2004. "Soft Heart Under Her Thick Skin?" *Los Angeles Times*, November 16.

Blakeslee, S. 2006. "Cells That Read Minds." *New York Times*, January 10.

Bradshaw, I. G. A., A. N. Schore, J. L. Brown, J. H. Poole, and C. J. Moss. 2005. "Elephant Breakdown: Social Trauma; Early Disruption of Attachment Can Affect the Physiology, Behaviour, and Culture of Animals and Humans Over Generations." *Nature* 433: 807.

Burghardt, G. M. 1985. "Animal Awareness: Current Perceptions and Historical Perspective." *American Psychologist* 40: 905–19.

Dobbs, D. 2006. "A Revealing Reflection." *Scientific American Mind*, April/May.

Fimrite, P. 2005. "Daring Rescue of Whale off Farallones." *San Francisco Chronicle*, December 14. www.sfgate.com/cgi-bin/article.cgi?f=/c/a/2005/12/14/MNGNKG7QoV1.DTL.

Gallese, V., and A. Goldman. 1998. "Mirror Neurons and the Simulation Theory of Mind-Reading." *Trends in Cognitive Science* 2: 493–501.

Hebb, D. O. 1946. "Emotion in Man and Animal: An Analysis of the Intuitive Process of Recognition." *Psychological Review* 53: 88–106.

Horowitz, A. C., and M. Bekoff. 2007. "Naturalizing Anthropomorphism: Behavioral Prompts to Our Humanizing of Animals." *Anthrozoös* 20: 23–36.

Pollack, H. 2003. *Uncertain Science . . . Uncertain World.* New York: Cambridge University Press.

Sapolsky, R. M. 2002. A *Primate's Memoir*. New York: Touchstone Books.

Scharper, S. 1997. *Redeeming the Time*. New York: Continuum.

Williams, C. 2006. "Elephants on the Edge Fight Back." *New Scientist*, February 18. www.newscientist.com/channel/life/mg18925391.400.html.

Integral Ecology

Introduction: Whose Environment Is It?

> The "'key-log'" which must be moved to release the evolutionary process for an [environmental] ethic is simply this: quit thinking about decent land-use as solely an economic problem. Examine each question in terms of what is ethically and esthetically right, as well as what is economically expedient. A thing is right when it tends to preserve the integrity, stability, and beauty of the biotic community. It is wrong when it tends otherwise.
>
> —ALDO LEOPOLD[1]

> Are you an environmentalist, or do you work for a living?
>
> —BUMPER STICKER POPULAR WITH OREGON LOGGERS

Whose Environment Is It?

Digging in with its forelegs, a beetle extends its maxillary palps to grasp a tasty bit of decaying wood a few centimeters inside a 200-year-old Douglas fir, weakened by age and felled by a windstorm years earlier. A ravenous woodpecker, hopping along the tree's carcass in search of insects, begins its rapid-fire hammering, which resounds for half a mile or more. Perking up its ears at the arboreal anvil chorus, a bear saunters through stands of cedar and spruce along the way to its favorite valley stream, now roiling with salmon journeying toward their spawning ponds.

Flying in a helicopter over the same steep valley, formed by glaciers thousands of years ago, a British Columbia forester conducts a survey of conditions in the Mid and North Coast Timber Supply Area. At midday, a daring photographer moves in close to capture the image of a grizzly one-handedly spearing a salmon that leaps toward the top of the falls.

Meanwhile, in a side canyon, a gang uses chain saws and bulldozers to establish the road needed to haul out trees that will be cut for faraway markets. Some environmental activists, staging a sit-in to halt the road building, engage in a shouting match with the road crew and loggers. The loggers tell them to go back to the city so that local folks can make a living by doing real work.

A local council of First Nation (indigenous) people, who have long made a living by salmon fishing, announces yet another legal strategy to regain control of their ancestral land—mountains and streams, plants and animals, burial grounds and ritual sites—the future of which is being contested primarily by descendants of European settlers.

A politician in Victoria gets an earful from constituents who differ sharply about logging the coastal rainforests. That evening, on a prime-time news program broadcast across the United States, viewers are informed about tense confrontations building over the fate of the Great Bear Rainforest, otherwise known as the Mid and North Coast Timber Supply Area. Those watching easily make the connection between this troubled region and other disappearing rainforests. In another program, an ecosystem scientist uses satellite imagery and GIS (geographic information system) to show that the rainforest in question is the size of a small country, that significant portions of it have already been clear-cut, and that within 20 years—at current logging rates—all the old growth will be lumber and plywood for voracious consumers. Although he is speaking as a supposedly impartial scientist, the ecologist's tone of voice and facial expression give away his deep concern about the future of the rainforest.[2]

Which is the "real" rainforest? The beetle's? The woodpecker's? The bear's? The forester's? The photographer's? The salmon's? The road worker's? The environmentalist's? The logger's? The First Nation member's? The politician's? The television viewer's? The sawmill worker's? The plywood manufacturer's? Or the ecologist's? We maintain that the rainforest is composed of all these perspectives, and many others. This book is about how to organize and integrate all of these perspectives.

Being able to understand multiple perspectives is essential to sustainable solutions, as Darcy Riddell discovered in the campaign to preserve the Great Bear Rainforest in British Columbia. Along with many other people, she was involved in negotiating the historic April 2001 treaty, in which the provincial government and logging industry agreed to protect significant portions of the Great Bear Rainforest; to continue good-faith discussions to protect other large segments of the rainforest; and to undertake ecologically informed logging in still other parts of the rainforest. After five more years of negotiation, on February 7, 2006, a comprehensive protection package was announced for the Great Bear Rainforest.

The package has four key elements: rainforest protection, improved logging practices, First Nation involvement in decision-making, and conservation financing to enable economic diversification. In total, 5 million acres of forest is to be permanently protected from logging, including new parks (3.3 million acres), previous parks (1 million acres), and new no-logging zones (736,000 acres). Stakeholders agreed to conservation-oriented land management practices to be guided by an Ecosystem Based Management approach by 2009. The overall framework was developed and approved by each First Nation, and grants them greater stewardship and decision-making power over resource development in their traditional territories. Finally, U.S. and Canadian foundations, and the BC government, raised $90 million toward a financing package to fund conservation management projects and ecologically sustainable business ventures in First Nation territories.[3]

According to Riddell, whose account of this historic campaign appears as a case study in part four of this book, environmental activists at first exclusively identified with their own perspective (one of ecological science), according to which clear cutting was seriously degrading the coastal arboreal ecosystem. With these facts in hand, most environmentalists called for a complete logging ban. In so doing, however, they ignored or denied the possibility that well-intended individuals and communities could propose and defend different assessments of the very same facts. Recognizing that to create a sustainable, regional solution, everyone—including environmentalists—needed to understand the Great Bear Rainforest in light of at least some of those other perspectives, Riddell and a number of her colleagues learned about the array of economic, political, cultural, and social factors that drove current logging practices. Because she understood that many well-intentioned people had ties to the forest, Riddell and some of her colleagues began personal transformational practices aimed at reducing their "subtle superiority" based on their previous assumptions that only their ecological perspective was worth adopting. In this respect, Riddell and others practiced some important elements of Integral Ecology, even though most of them had never heard of Integral Theory.[4]

Soon environmentalists realized that they had to get serious about economics, rather than regard it exclusively as the human domain most responsible for destroying the rainforest. Hence, they asked large North American retailers to purchase lumber solely from companies who agreed not to clear cut temperate rainforests. This economic strategy engendered a more flexible attitude on the part of timber officials and British Columbia government representatives at the bargaining table because now loggers might be deprived of their usual markets. Loggers stopped their stalling tactics known as "talk and chop," and they engaged environmentalists and First Nations in increasingly good-faith negotiations. The retail campaign against clear cutting did more than give economic leverage to those opposed to timber company and government intentions to log the entire Great Bear Rainforest; it also focused international attention and criticism on the logging practices employed by powerful groups.

With millions of concerned individuals now looking over their shoulders, those involved in the controversy listened more seriously and sympathetically to opposing views and interests. For instance, an increasing number of environmental activists saw the need to address the pressing economic and social circumstances of the region's human inhabitants, whose ways of life were tied up in the rainforest. Clearly, a viable solution would have to include economic alternatives to unsustainable logging practices. As environmentalists stopped identifying exclusively with their "polarizing identities," they transformed from "being outside agitators to solution-builders."[5] Unfortunately, this shift led some environmentalists to accuse others of selling out to the logging establishment. Despite such accusations, Riddell reports, "Negotiations also enabled opposite sides to engage one another with humanity and mutual respect, fostering [Integral] capacities of mutual understanding."[6] As we will see, integral capacities refer in part to the ability to cease exclusive identification with a particular position, such as modern (industrial logger) or postmodern (green environmentalist), and start sympathizing with multiple perspectives and realities. Riddell writes:

> When [Integral] capacities emerge, complex issues and diverse perspectives can be more readily integrated into holistic, long-term solutions. Leaders acting from Integral capacities act as cultural empathizers and transformers who operate dynamically across multiple worldviews motivating people with diverse interests toward common ecological,

economic, cultural, political, and social goals. Leaders with Integral perspectives can foster healthy ecological worldviews, enabling mutual understanding, and fueling individual and cultural transformations of increasing scope and depth.[7]

The Need for an Integral Ecology

Growing recognition of the complexity of environmental problems has led leaders in environmental organizations, regulatory agencies, corporate offices, and academia to call for greater interdisciplinary, multidisciplinary, and even transdisciplinary models to describe, address, and resolve environmental problems. We agree—we need a more comprehensive map to understand and solve our most intransigent problems. Riddell's application of one version of integral ecology demonstrates just how successful a comprehensive integration of multiple perspectives and disciplines can be. Yet until now, people have not had access to a robust theoretical model that organizes and integrates various disciplines and methods, and generates the most comprehensive solutions. We maintain that Integral Ecology is that theoretical model, built upon the distinctions of Integral Theory.

Integral Theory is a content-free framework developed by Ken Wilber and colleagues. According to Wilber, "the word *integral* means comprehensive, inclusive, non-marginalizing, embracing. Integral approaches to any field attempt to be exactly that: to include as many perspectives, styles, and methodologies as possible within a coherent view of the topic. In a certain sense, integral approaches are 'meta-paradigms,' or ways to draw together an already existing number of separate paradigms into an interrelated network of approaches that are mutually enriching."[8]

As a result of its applicability within, across, and between disciplinary boundaries, Integral Theory has been widely embraced by individuals in many different fields.[9] Applied in the context of environmental problems, Integral Theory organizes insights from more than 200 distinct perspectives to contribute to a more comprehensive understanding of the eco-socio-cultural dimensions involved. Surely there is the need for a model capable of such organization and integration, and surely the field of ecology could make use of such a model.

The Integral Model maintains that there are at least four irreducible perspectives, two of which have been almost entirely excluded from academic and popular ecological discourse. If we exclude any one of these perspectives, we arrive at partial understandings and, unfortunately, partial solutions. We must include objective, interobjective, subjective, and intersubjective perspectives. The objective perspective examines the composition and exterior behavior of individual phenomena, including humans, bears, salmon, and beetles. The interobjective perspective examines the structure and exterior behavior of collective phenomena, ranging from ecosystems to political and economic systems. The data generated by these two perspectives are valuable, yet such data alone do not exhaust the "reality" of the phenomena under investigation, nor do they provide motivation for action. Motivation arises when we experience the phenomena in question through two additional perspectives—subjective (1st-person—I, me) and intersubjective (2nd-person—you, we). These perspectives constitute the interior aspects of phenomena, are traditionally associated with aesthetic experience and cultural values, and have largely been excluded from academic ecological discourse. We cannot understand our complex interiors through natural or social scientific methods, nor can we understand the natural world solely through our interior experience. We need both.

Integral Theory refers to these irreducible perspectives as *quadrants*, and we summarize them as experience (subjective, 1st-person), culture (intersubjective, 2nd-person/1st-person plural), behavior (objective, 3rd-person singular), and systems (interobjective, 3rd-person plural).[10] We cannot understand any one of these perspectives through methods suitable for analyzing the realities of another. Hence, Integral Theory avoids reductionism, especially "gross reductionism," or the reduction of all of reality to individual, objective phenomena (reducing all interiors and systems to atoms—individual "its"); and *subtle reductionism,* or the reduction of all interiors to interobjective phenomena (reducing the "I" and "we" perspectives to interwoven systems—"its"). The science of ecology has typically exemplified the latter form of reductionism, and this subtle reductionism has generated partial understandings of the natural world and continues to generate partial solutions to some of our toughest problems.

Clearly there is a need for subjective and intersubjective perspectives, because they show up at the bargaining table (we don't just have ecological difficulties, we have human difficulties!). Intersubjectivity (2nd-person) arises between two subjects: I and thou, me and you.[11] Different people will experience and assess the same data in different ways. If the subjects involved do not consider the cultural matrices—beliefs, values, norms, religious traditions, ethnic self-identification—of the other subjects, it is difficult to create common ground and understanding. Without understanding and flexibility, it is difficult to agree upon a sustainable solution. Understanding the presuppositions and beliefs that shape your opponent's experience, and discerning how your own experience may be distorted by unyielding adherence to a particular position, are vital to creating common ground and successful, inclusive negotiations.

Genuine mutual respect is difficult to attain, even among experts from different fields, because experts often think that their particular method or perspective is the only correct or most valuable one. There is a need for an integral ecology that resists this method hegemony, the supposition that one or a few perspective(s) can provide the only useful and pertinent truth claims about a complex environmental problem. In resisting method hegemony, Integral Ecology creates a meta-framework that contextualizes and includes the partial truths of all traditions. Indeed, as one commentator stated, "Integral theory carries out a *demythologizing* [i.e., deabsolutizing] mission, in which it undermines the sacred reductionisms and absolutisms practiced by many different methodologies."[12] Instead, it coordinates and organizes all these partial perspectives into a more coherent whole.

Not only does Integral Ecology study interiors in addition to exteriors, but it also studies how those interiors develop within organisms in general and humans in particular. Integral Ecology acknowledges that all organisms have subjective and intersubjective dimensions and describes how interior development in humans determines in profound ways our relationship to the natural world. Until now, ecologists and ecological discourse have mostly excluded an explicit recognition of interiors and their development—and make no mistake, there is a need to understand our interior individual and collective relationship to the natural world, for it is within our interiors that motivation to treat the natural world in healthier ways resides.

To conceptualize an ecosystem, for example, requires a highly developed level of cognition, a level unavailable to children (a level that was even unavailable to most adults many centuries ago). Different kinds of phenomena can manifest—and in that sense be—only within an adequate perspective, clearing, or worldspace (we will discuss this in great detail in chapter 5). If the worldspace needed for a phenomenon to appear is lacking, it cannot show up. In some sense ecosystems subsisted long before ecologists conceptualized them, but in another sense ecosystems, as specifiable phenomena, came into being only when we established the necessary cognitive worldspace. Don't be misled; Integral Theory is not a subjective idealism. Things really do exist, but they manifest only within a worldspace capable of allowing for them.

Based upon decades of research in philosophy and social science, Integral Theory asserts that mind is not a mirror that reflects a pregiven reality. Instead, mind both enables and limits the ways in which things appear. Hence, the worldspace that a child can hold open is clearly more complex than a frog's, but less complex than a mature adult's. During maturation, the human worldspace expands and deepens enormously in many different ways. Because a more expansive and inclusive interior allows a more comprehensive worldspace to emerge, some assertions made about a given phenomenon are more comprehensive, and thus have greater validity, than other claims. Hence, integral perspectivalism is not equivalent to relativism. We do not assert that all perspectives are equal. Some truths are more comprehensive than others. Integral perspectivalism maintains that partial worldviews and partial perspectives reveal partial truths. These partial truths are accurate and essential, yet they must be integrated into a larger, more comprehensive picture. Without an Integral framework, we currently have no framework capable of integrating and organizing these partial perspectives and partial worldviews. Clearly, such a framework is needed.

This ever more comprehensive pattern arises in all 4 quadrants—experience (subjective), behavior (objective), culture (intersubjective), and systems (interobjective). Just as interiors develop (as when a child's worldspace evolves into a more complex, adult worldspace), so, too, exteriors develop (as when an acorn develops into a tree). Integral Ecology recognizes levels of complexity in all 4 quadrants, or throughout all four dimensions or perspectives: systems, behavior, experience, and culture:

- Ecosystems are composed of and influenced by natural and social systems.
- Ecosystems involve the individual behaviors of organisms, at all scales (including microbes and humans). These organisms are understood as being members (not parts) of ecosystems.
- Members of ecosystems have various degrees of interiority (perception, experience, intentionality, and awareness).
- Members of ecosystems interact within and across species to create horizons of shared meaning and understanding.

Integral Ecology creates a framework that allows all aspects of reality to connect with what has traditionally been associated with the scientific study of ecology. But instead of collapsing all connections into an "everything is ecology" position, Integral Ecology highlights the factors that differentiate interrelated phenomena. Thus, while everything can be viewed as (inter)connected, not everything is connected in the same way nor to the same degree! The cliché "Everything is interconnected" becomes "Everything is interconnected, but some things are more connected than others." In other words, there are spectrums of interconnection between variables both in terms of depth and span. As a result, depending on the perspective one is taking, some "parts" are actually not very connected to other "parts."

The four dimensions of any phenomenon co-arise and mutually influence one another in complex ways; none of them has ontological priority. Hence, when we address an environmental problem, we must do more than assess its ecosystemic aspects, such as whether an environmental toxin has altered the food chain. We must also inquire how the pollution affects (or is interpreted by) the aesthetic, recreational, economic, and cultural aspects of communities and organisms that depend upon it.

In short, Integral Ecology advances the development and application of a comprehensive approach to environmental issues. This approach organizes insights from various eco-approaches into an all-inclusive framework. This new framework has promising applications in many contexts: outdoor schools, urban planning, wilderness trips, policy development, restoration projects, environmental impact assessments, community development, and green business, to name a few. Integral Ecology transcends many of the problems that have assailed contemporary partial approaches to the environment and moves toward a developmentally informed understanding of individuals, communities, and systems. As a result, Integral Ecology draws on the expertise of many disciplines and offers extremely comprehensive, far-sighted, and flexible solutions for the environment—solutions that honor the interiors of animals and people and that can carry us into right relationship, at multiple scales, with the Earth.

Aldo Leopold's Inclusion of Interiors

Aldo Leopold, the dean of Anglo-American environmentalism, was a scientist, naturalist, writer, environmentalist, hunter, and farmer.[13] His book, *A Sand County Almanac*, written 60 years ago, contains elements of an integral ecology.[14] As such he was a pioneer of perspectivalism. Over the years Leopold has consistently been cited by ecologists and environmentalists as one of the most important figures in their fields. In fact, *The Environmentalist's Bookshelf* identifies Leopold's *A Sand County Almanac* as the number one most influential book among environmentalists, based on a survey of over 200 experts in ecology and environmental studies.

Leopold defined the land as the Earth's many different habitats and their associated life forms. He recognized that the objective and interobjective (3rd-person singular and plural) methods used by natural science provided important insights into the land and land use. Yet he thought that insights afforded by these perspectives were often insufficient to prevent short- and long-term damage to the land. He believed that other perspectives were clearly needed, and so he employed other, equally valid, subjective and intersubjective perspectives, which he referred to as aesthetic (subjective) and ethical and cultural (intersubjective).

Leopold's land ethic anticipated the need to include consciousness, culture, and nature in the study of ecology in order to achieve a more comprehensive understanding. Leopold reports that his own objectifying, instrumentalist attitude (3rd-person singular) toward nonhuman life changed when, as a young man, he was hunting deer with some friends. Spotting a pack of wolves, he shot a wolf and one of her cubs, members of a species that was then regarded as a worthless and dangerous predator. As he approached the dying mother wolf, he observed "a fierce green fire dying in her eyes."[15] At that moment, Leopold acknowledged that the wolf had a wolfish kind of subjective sentience, and an intersubjective relationship to him. Far from being merely a behavioral mechanism, the wolf exhibited something akin to, yet different from, the yearning, desiring, and fearing that Leopold himself experienced. The wolf had a life of its own. To understand the wolf required more than weighing and measuring it, analyzing the working of its organs, studying its behavior, and comprehending its function as one of the top predators in mountain country. It also required a subjective and intersubjective understanding and appreciation of what it must be like to be a wolf!

Having spent years in the regulatory trenches, and having trained in the natural sciences, Leopold was aware that introducing aesthetics and ethics into land-use policy would not be taken seriously by his colleagues, who were influenced by reductionistic materialism and behaviorism. In fact, he postulated that nothing short of an evolutionary advance—an advance in which humans learned to take multiple perspectives and thus grew to recognize the interiority of other beings—would move society beyond modernity's instrumentalist view that the land is merely raw material for human ends.

Still, Leopold was unable to articulate adequately or defend his intuition that behaviorism (objectivism) and systems theory (interobjectivism) could only partially account for animal life and the land. He recognized the fundamental paradox of environmentalism: Environmentalists value the natural world but typically subscribe to a conception of nature that either excludes value (subjective and intersubjective perspectives) or regards it as a conventional fiction useful for enhancing human survival. In modern cosmology, as Kant feared, there is no place for aesthetic experience, morality, consciousness, and subjectivity. Environmentalists often speak of nature as a complex dynamic system in which humans, like other animals and plants, are merely strands in a cosmic web that lacks any hierarchy or direction. Yet, if humans are merely strands in a complex state of affairs—the *is*—they are in no way capable of calling for alternative actions based on moral obligation—the *ought*.[16]

With this book, we are building on the work of those like Leopold who recognized the need for including human and animal interiority in our understanding of the natural world and humanity's relationship to it.[17] As a result, building on classical definitions, we define ecology as *the mixed methods study of the subjective and objective aspects of organisms in relationship to their intersubjective and interobjective environments at all levels of depth and complexity.*

Interiors and Anthropocentrism

Some critics (mostly scientific ecologists) complained that Leopold was anthropomorphic because he personalized accounts of animals in the wild.[18] As far as positivism, behaviorism, and eliminative materialism are concerned, people fall prey to anthropomorphism even when they ascribe awareness, interiority, and personality to human beings! In addition to being criticized for being anthropomorphic, environmentalists have also often charged him with the crime of anthropocentrism.[19]

Even so, Integral Ecology transcends the anthropocentrism versus anti-anthropocentrism duality that characterizes so many environmental debates in the English-speaking world. Just as it is misguided for anthropocentrists to treat nonhuman life as if it had no intrinsic value, it is also misguided for anti-anthropocentrists to ignore that humans are a remarkable development in terrestrial evolution.

Integral Ecology may seem anthropocentric, because in one of three values (i.e., *intrinsic value*) we maintain that humans are special, in part because humans are endowed with an interior depth that allows us to appreciate the intrinsic value of nature! However, in *extrinsic value* humans are less significant, and within *ground value* humans are of equal value with all life forms.[20] As Leopold remarks, "For one species to mourn the death of another is a new thing under the sun. . . . we, who have lost our [passenger] pigeons, mourn the loss. Had the funeral been ours, the pigeons would hardly have mourned us."[21]

Although humans have richly developed interiors and an astonishing capacity for language, interiority is not restricted to humans, which was Leopold's remarkable revelation. Indeed, Integral Ecology is radically non-anthropocentric insofar as it maintains that interiority goes "all the way down" (i.e., interiority is a basic feature of the universe).[22] The capacity for experience, however meager, is found throughout nature. A deer and a human do not have the same interior experience. Clearly the human's is deeper in important ways, but they both have an experience and they both are of value and must be considered in an integral ecology. Ironically, only humans can have an ecological realization of "oneness" with nature—and even then the amount of humans who do is incredibly small, and the ones that stablize this even smaller. Thus, ecocentric realization is an anthropocentric experience!

Darwin maintained that humans are not a special act of creation, but rather they descend from other animals as the result of chance mutations that proved adaptive, or advantageous. According to Darwinian naturalism, humans are intelligent animals that have evolved accidentally. Many environmentalists often seize on evolutionary theory (and other scientific claims) to reduce arrogant human self-importance, which in fact has at times led to mistreatment of nonhuman nature. In a noble effort to protect animals and habitats from anthropogenic destruction, many environmentalists strike anti-anthropocentric and even misanthropic poses. Indeed, some radical environmentalists would prefer that humankind disappear altogether, thereby removing an alleged cancer from the web of life.

This position is confused. The capacity for significant moral evaluation (even the capacity to evaluate human behavior as self-centered) differentiates humans from nonhumans.[23] In fact, environmentalists have an interior depth that allows them to encourage humans to do the morally right thing and limit their rate of reproduction, preserve habitats, and protect nonhuman species. Yet, if humans are merely another animal species, there is nothing morally wrong with displacing other species as a human expression of the universe's drive to maximize reproduction (certainly, neither a biologist nor an environmentalist would morally critique a nonhuman species that maximized its reproduction). Of course, a population crash and even extinction may result if a species overshoots the carrying capacity of its habitat, but there is no moral failing involved.[24] If we assume a naturalistic conception of humankind, all we can recommend is that humans follow the prudential ought: we ought to alter our behavior toward the nonhuman domains so as to promote long-term human survival and well-being. Many environmentalists insist, however, that a *moral* ought also to apply here: we ought to limit our behavior, including our reproductive drive, so that other life forms can survive and prosper. This recommendation is confused because it cannot be reconciled with the naturalistic view that humans are merely one species among others, bound by the same laws that bind other species. We depict no other species as immoral when it seeks to maximize its own fitness. If we depict similar human behavior as immoral, we do so because we regard human beings as significantly different from all other known species. The (often tacit) presupposition that only humans are morally responsible for their behavior is a reminder that with the emergence of the human species, something novel, extraordinary, and dangerous occurred on Earth.[25] We will unpack the Integral Model and crucial arguments like this, ones that define or obscure the relationship between humans and nature, throughout this book.

Plan of the Book

This book can be characterized as an advanced introduction to Integral Theory in general and Integral Ecology in particular. "Introduction" in the sense that it can be read by someone new to either Integral Theory or ecology and environmental studies. However, the more familiar one is with either domain of discourse, the more one will be able to grapple with the finer details of our presentation. "Advanced" in the sense that it can also be read by an advanced student of either field. An advanced student of Integral

Theory will benefit from seeing an exhaustive presentation of the Integral Model applied. An advanced student of ecology or environmental studies will gain much from seeing how an Integral framework can allow them to unite the many insights within their field. In fact, our endnotes are often geared toward the dedicated scholar-practitioners of Integral Theory. As a result, we feel this book serves those who are just becoming familiar with integral applications and to those who have been long-time scholar-practitioners of integrative efforts.

The book is divided into four parts. Part one introduces Integral Theory, the conceptual framework of Integral Ecology. Part two defines and explores the Who, How, and What of ecological phenomena. Part three applies the Who, How, and What framework to explore a variety of issues and present a number of real-life applications of Integral Ecology. Part four provides three detailed case studies where the Integral Model has been used for ecological and environmental purposes. In general the book moves from a focus on laying out the theoretical basis of Integral Ecology to giving concrete examples of it applied. Hence, each part is increasingly concerned with applications of Integral Ecology.

Part one introduces the theoretical framework of Integral Ecology and is divided into four chapters. This part focuses on theory, as we are summarizing thousands of pages of Wilber's writings and introducing his model in the context of ecology. However, the resulting foundation is capable of supporting a wide range of ecological endeavors. Chapter 1 asserts that Integral Ecology assesses environmental problems through many different perspectives, including objective and interobjective perspectives most commonly associated with the natural and social sciences and the subjective and intersubjective perspectives, which have largely been excluded from ecological discourse. In chapter 2 we begin our presentation of Integral Theory, and we specifically emphasize the quadrants. In the Integral Model, quadrants are both the four major dimensions of holons (from atoms to organisms) and the four major perspectives we can adopt in our research. In chapters 3 and 4 we discuss Integral Theory's contention that evolutionary development occurs in all quadrants. Chapter 3 examines how development has occurred in the exterior objective quadrants studied by the natural and social sciences. In chapter 4 we discuss development as it occurs in individuals and communities, or the quadrants representing the subjective and intersubjective domains.

Part two has four chapters, which explain how Integral Ecology embraces ontological, epistemological, and methodological pluralism. Chapter 5 defines Integral Ecology in a way that allows it to honor the 200+ distinct schools of ecology and environmental thought. In the next three chapters (6, 7, and 8), we clarify that what someone calls "reality" depends on *What* part of reality one is examining, *Who* is doing the examining, and *How* they examine (or which methods they use). Chapter 6 defines and explores the What and introduces the 4 terrains and their 12 niches. Chapter 7 examines the Who and introduces the 8 ecological selves. The 8 eco-selves represent the major developmental worldviews that currently exist and that profoundly influence how we relate to the natural world. Each of these worldviews has an important yet partial piece of the whole picture, and their contribution must be honored if we want to generate mutual understanding among various stakeholders. Chapter 8 presents the How, or the 8 modes of ecological inquiry. These modes of inquiry are the established methodologies (subjective, objective, intersubjective, interobjective) that reveal reliable and verifiable ecological data.

Part three is divided into three chapters, which introduce readers to a variety of applications of the Who, How, and What of Integral Ecology. Chapter 9 examines how Integral Ecology informs our understanding of living in and out of balance with natural systems. In particular we explore an all-quadrant and all-level approach to "being one" with the planet, and an all-quadrant, all-level approach to our understanding of our current ecological crisis. Chapter 10 introduces over a dozen practices we have developed to honor nature as a transformative path. Chapter 11 introduces 15 principles for applying the framework and provides project summaries, case studies, and applications of Integral Ecology in action.

Part four showcases three illustrative case studies by various Integral scholar-practitioners. Case study I presents an example of community development in El Salvador that includes human interiority toward environmental ends. Case study II presents an instance of using the Integral approach to address conflicts occurring over natural resources in Hawai'i. Case study III focuses on environmental activism in Canada's Great Bear Rainforest. In the conclusion, we detail over a dozen distinct advantages of the Integral Ecology framework, present a platform that summarizes these strengths, and then explore the limitations of and road ahead for Integral Ecology.

In spite of its 800+ pages this book still is only the briefest sketch of Integral Ecology. Throughout this volume we cover a lot of ground in order to stretch the Integral canvas as philosophically (part one), theoretically (part two), and pragmatically (parts three and four) as wide as possible. We do this by touching on key debates and controversies, offering powerful Integral distinctions, providing a new framework (i.e., Who x How x What), detailing eighteen personal practices of transformation, describing and drawing on over two hundred unique perspectives on the natural world, incorporating over 1,750 sources from the ecological and environmental literature, and pointing to twenty-three examples and three in-depth case studies of Integral Ecology in action.

We have provided this far reaching overview of Integral Ecology as a means of generating a creative space for exploration of the key ecological issues facing our communities and the planet. Much color, texture, and illumination remains to be added by you and others. Thus, what we present in this book is less about the details, arguments, and examples we give than it is about a new way of thinking about and responding to the ecological and environmental issues confronting us as we head into the 21st century. The two of us can afford—and expect—to be partial in the perspectives we present here but we as a planetary community can no longer afford to approach the natural world with anything less than some form of an integral approach to ecology. We offer this book up as an important step in that direction. May our clarity engender action, may our confusions invite inquiry.

PART ONE

The Historical Context and Conceptual Framework of Integral Ecology

1

The Return of Interiority: Redefining the Humanity-Nature Relationship

> The approach to ecology set forth in *Sex, Ecology, Spirituality* [combines] ecological unity, systems theory, and nondual spirituality, but without privileging the biosphere and without using the Web-of-Life notion, which I maintain is a reductionistic, flatland conception. Rather, an all-quadrant, all-level approach to ecology allows us to situate the physiosphere, the biosphere, the noosphere, and the theosphere in their appropriate relationships in the Kosmos at large, and thus we can emphasize the crucial importance of the biosphere without having to reduce everything to the biosphere.
>
> —KEN WILBER[1]

Where Have Our Interiors Gone?

Many environmental problems are notoriously difficult to define, much less resolve. Natural processes are multifaceted, multileveled, complex adaptive systems, such that uncertainty surrounds efforts to characterize them, much less make reliable predictions about interventions. Indeed, as we all know, some interventions make matters worse. Further, even well-intentioned, thoughtful people often legitimately disagree in various situations as to whether there even is an environmental problem, as different individuals and cultures have different ways of evaluating the same issue. Because environmental problems and solutions manifest differently depending on your perspective, we must include different perspectives and methods—economic, ethical, cultural, scientific, phenomenological, and epistemological—to understand and ameliorate them. Further, environmental problems include features that can only be disclosed by way of subjective and intersubjective perspectives, but these perspectives are rarely considered. Without a full understanding, a full picture of the problem, how can we create a full solution? Why have some perspectives been excluded?

We're not the only ones who recognize the multifaceted nature of environmental problems. In 2004 the Ecological Society of America (ESA) organized a plenary session at its annual meeting. At that meeting Forest Service Chief Dale Bosworth said that he was "humbled" by the daunting intricacy of environmental problems, and called for an expanded ecology: "We need more than technical solutions to problems. We need to focus on the problem in its full dimension—its social and its regulatory and its political and its economic and its ecological dimensions."[2]

Even though Bosworth wants to focus on the "full dimensions" of environmental problems, he fails to mention the ethical, cultural, interpersonal, psychological, or aesthetic (subjective and intersubjective) dimensions of such problems. So even his "expanded approach" is still just an objective and interobjective (3rd-person singular and plural) perspective: socially, politically, economically, and ecologically! He is still inadvertently ignoring our individual and collective interiors.

In his Pulitzer Prize–winning book, *Consilience*, biologist E. O. Wilson attempts the "jumping together" (con-silience) of multiple perspectives, including the sciences and humanities, but he uses only physical, biological, and social methodologies (objective and interobjective methodolgies). He thus reduces the interiority characteristic of subjective (1st-person experience) and intersubjective (2nd-person intersubjective) values to objective and interobjective phenomena (3rd-person singular and plural).[3] We agree with the intention of *Consilience* but insist that we must study subjectivity and intersubjectivity on their own terms, rather than as epiphenomena of an allegedly more fundamental material or physical process. We cannot reduce our subjective and intersubjective dimensions to exterior objects. Interiors must be interpreted on their own terms.

So, while ecologists, policy-makers, and concerned citizens occasionally recognize the need for an expanded ecology, they usually propose a vision confined to an objective and interobjective (3rd-person) perspective. Why have our subjective (1st-person) and intersubjective (2nd-person) perspectives been excluded?

The answer is quite complex and involves many hundreds of years of history. In this chapter, we will discuss how and when our interiors were reduced to our exteriors and why we need to include them to create the most comprehensive understanding and the most comprehensive solutions.

Interiors—Reduced in the Historical Shuffle: Ascent versus Descent

For centuries, European civilization ignored the material world, what St. Augustine called the City of Man. Instead, Christians focused on the world to come, the higher world, the City of God. The Christian tradition urged followers to ascend to the eternal spiritual plane rather than attach themselves to momentary sensory pleasures or temporary earthly offices.

Beginning in early modern times, however, people began demanding greater independence from the Church. Instead of praying for an otherworldly New Jerusalem, moderns asserted that paradise could be achieved in this world through human ingenuity and achievement. Increasingly, religion was regarded as a meddling, corrupt, superstitious institution.

Before the rise of modernity, the Church held the power over all of these domains (hypothesize about gravity—a legitimate scientific endeavor—and go to prison!). Modernity's proponents accomplished the monumental and essential task of differentiating the various domains of truth, power, and authority. Yet it seems moderns took their faith in the material world a bit too far. Reacting to the call to ascend above this earthly domain, moderns descended. They descended so far that they came to believe only in this earthly plane, in what could be encountered with the senses—material objects. The Descenders either denied the possibility of Ascent altogether or else reconceived of it as the endless horizontal "progress" of economic-scientific development. In attempting to create a free space for science, economics, politics, philosophy, and art, moderns ended up throwing out the spiritual baby with the premodern bathwater, and with it went interior perspectives.

Wilber argues that the battle between Ascenders and Descenders is "the central and defining conflict in the Western mind."[4] Ascenders sought to rise above the corrupt and material plane in order to unite with the eternal One. For Descenders, in contrast, God was not the One but the Many. Worshiping the incredibly diverse, visible, sensible, sensual God/Goddess, Descenders "delighted in a creation-centered spirituality that saw each sunrise, each moonrise, as the visible blessing of the Divine."[5]

> Salvation in the modern world—whether offered by politics, or science, or revivals of earth religion, or Marxism, or industrialization, or retribalism, or sexuality, or horticultural revivals, or scientific materialism, or earth goddess embrace, or ecophilosophies, you name it—salvation can be found only on earth, only in the phenomena, only in manifestations, only in the world of Form, only in pure immanence, only in the Descended grid. There is no higher truth, no Ascending current, nothing transcendental whatsoever. In fact, anything "higher" or "transcendental" is now the Devil. . . . And all of modernity and postmodernity moves fundamentally and almost entirely within this Descended grid, the grid of flatland.[6]

Following Max Weber and Jürgen Habermas, Wilber notes that modernity differentiated three domains (art, morals, and science), which he also calls the Big Three (I, We, It): (1) consciousness, subjectivity, self, self-expression (including art), whose mode of truth involves truthfulness and sincerity; (2) ethics, morality, worldview, culture, intersubjective meaning, whose mode of truth involves justice; (3) science, technology, objective nature, whose mode of truth involves correct propositions.[7] Thus, Integral Ecology unites the art of ecology, the Beautiful (environmental aesthetics); the morals of ecology, the Good (environmental ethics); and the science of ecology, the True (environmental science) at multiple levels of complexity.

The differentiation of the Big Three on a large scale created the social and cognitive space necessary for free scientific inquiry, new art forms, market economies, and democratic politics. At last, one could explore the natural universe without ecclesiastic constraints. In such a differentiated worldspace, the individual could develop his or her own views.

Unfortunately, modernity did not adequately integrate the Big Three that it had differentiated. Embracing Descent liberated humanity from religious despotism, and contributed to great scientific, technological, and economic development. Yet exclusive faith in the material world also produced untoward consequences, including the loss of cultural meaning (intersubjective interiors) that followed when humankind was viewed solely through the lens of mechanistic materialism. Wilber depicts the mood of modernity as irony, "the bitter aftertaste of a world that cannot tell the truth about the substantive depth of the Kosmos. . . ."[8] Descenders embodied materialism and thus solidified objective and interobjective (3rd-person singular and plural) perspectives—other perspectives fell by the wayside.

Because personal-artistic and cultural-moral truth claims are more complex and contentious than those made by empirical scientific research, and because scientific knowledge brought such important (and needed) material gains, scientific modes of knowledge (3rd-person perspectives) marginalized the other two modes. Natural science could not acknowledge, much less study, selfhood, interiority, culture, and morality, since sensory-empirical inquiry is suited for material phenomena, not for personal (subjective) and cultural (intersubjective) phenomena.

Modernity's Conception of Nature

Of course, these inner developments had an impact on the conception of nature. Modern science represented nature as "a perfectly harmonious and interrelated system, a great it-system, and knowledge consisted in patiently and empirically mapping this it-system."[9] It conceived of the cosmos in terms of the "great 'web of life,' a great interlocking order of beings, each mutually interwoven with all others," but wholly lacking interiority and intersubjectivity.[10]

The modern, rational ego sought to disenchant nature and achieve complete rational objectivity. So long as one's reasoning is influenced by emotions, so long as one's judgment is tainted by personal, familial, tribal, or racial factors, so moderns argued, one is not truly rational and impartial. Not only must one see the world through a 3rd-person perspective; one must also in effect *become* a 3rd-person perspective, one must become an "it." Following Kant, the modern ego sought to overcome the domain of particularity and corporeality, in order to attain universality and impartiality. Kant, to his credit, still attempted to differentiate the interior domain from the exterior physical-biological domain, and make room for all of them. Other moderns became increasingly skeptical about the existence of interiority, which they lumped into the same category as other discredited ideas, such as the soul. There were significant problems, however, in defending an exclusively material world.

First, moderns could not concede that there was a domain other than the material plane. Interiors, including reasoning and moral judgment, lacked simple material location; hence, interiors simply could not exist. Well, then, exactly where was the location of reason? Furthermore, scientists value their rationality and their research—namely, truth arrived at by rigorous 3rd-person methodologies—but we cannot study the value of such truth using their methodologies, because value is not a material object of study. The value that a physicist assigns to his research can never become an object of his research as a physicist. Third-person, materialistic science cannot justify itself on its own terms, as positivists learned to their dismay when they realized that their own principle of verification (all valid truth claims have to be verified by experience) could not be verified.

> Every great scientific synthesis stimulates efforts to view the whole of reality in its terms. . . . But the views of reality that originate in this way are not themselves scientific, nor are they subject to scientific verification. They attempt to make sense not only of the facts "out there," held at arm's length by the [3rd-person] observer, but also of the facts "in here," facts such as our [1st-person] awareness of our own act of existence, our sense of [2nd-person] moral accountability, our communion with the source of being. Facts of the latter kind lie close to the heart of reality, but they do not lend themselves to scientific formulation. Attempts to explain them scientifically end by explaining them away. But science itself then becomes unintelligible.[11]

Eventually the ego was left in a kind of transcendental limbo that was made increasingly untenable by the relentlessly reductive processes of scientific materialism.[12] Efforts were made to explain interiors objectively in the late 18th century. At that time egoic rationality turned its objectifying gaze back upon itself, treating interiors and rationality as a special kind of object. Objectifying rationality and reason, as Heidegger, Horkheimer, Adorno, Foucault, and Habermas noted, transformed ego into an object suitable for domination, just as it had rendered nature into such an object. The "human sciences" were born. Yet Foucault and other postmodern thinkers (the postmoderns will come later) called these so-called human sciences "dehumanizing humanism." In this view, humans were regarded not as "subjects of communication," but rather as "objects of information."

According to Habermas's reading of Foucault:

> A gaze that objectifies and examines, that takes things apart analytically, that monitors and penetrates everything, gains a power that is structurally formative for these [modern] institutions. It is the gaze of the rational subject who has lost all merely intuitive bonds with his environment and torn down all the bridges built up of intersubjective [and

dialogical] agreement, and for whom in his monological isolation, other subjects are accessible only as the objects of nonparticipant [3rd-person] observation.[13]

To make up for its conceptual erasure, Wilber hypothesizes that the modern ego engages in extraordinary, nature-dominating agency. To demonstrate its own existence, the ego set out to subjugate the material domain (the only domain that supposedly exists). Martin Heidegger, Max Horkheimer, and Theodor Adorno claimed that the striving of modern "man" for world domination showed that he had become an animal seeking power and security.[14] An alternative view is that this world-domination agency represents, at least in part, an effort at self-assertion for those who intuit their own interior and interpersonal reality, but who cannot find any adequate personal or cultural expression for it.

The second problem with the modern quest to overcome particularity and corporeality was that the justifiable differentiation between mind and body became unjustifiable dissociation:

> The rational ego wanted to rise above nature and its own bodily impulses, so as to achieve a more universal compassion found nowhere in nature, but it often simply repressed these natural impulses instead: repressed its own biosphere; repressed its own life juices; repressed its own vital roots. The Ego tended to repress both external nature and its own internal nature (the id). And this repression, no doubt, would have something to do with the emergence of a Sigmund Freud, sent exactly at this time (and never before this time) to doctor the dissociations of modernity.[15]

Before we continue, we think it is necessary to clarify the different ways in which we define the term "nature." For our purposes, we offer three definitions, which we label as NATURE, Nature, and nature.[16] NATURE includes the whole Kosmos in all its dimensions, including interiors and exteriors: the Great Nest of Being. Nature (with a capital *N*) refers to the exterior domains of the Kosmos, the domains that are studied by the natural and some of the social sciences: the Great Web of Life. Finally, nature (lowercase *n*) means the empirical-sensory world in two different but related uses: the exterior world disclosed by the five senses (and their extensions), and the interior world disclosed by feelings, emotional-sexual impulses, somatic experiences as contrasted with rational mind and with culture: the Great Biosphere.[17] Thus, getting in touch with your prerational feelings is the same as getting in touch with the natural world *sans* culture. In other words, by going out into pristine nature (i.e., no signs of human culture or society) you can more easily connect with those nonmental aspects of yourself (e.g., prerational feelings and postrational spiritual experiences—which is why the two get confused and nature becomes equated with NATURE). One of the reasons *Nature* and *nature* get confused is that both refer to the empirical-sensory world. Nature is all the exteriors of the natural and cultural worlds, whereas nature includes the empirical-sensory awareness that knows the noncultural levels of Nature. Each of these "natures" has its champions: the Great Biosphere is worshiped by prerational environmentalists, the Great Web of Life is studied by rational materialists, and the Great Nest of Being is exalted by postrational romantics.[18] In sum, nature is

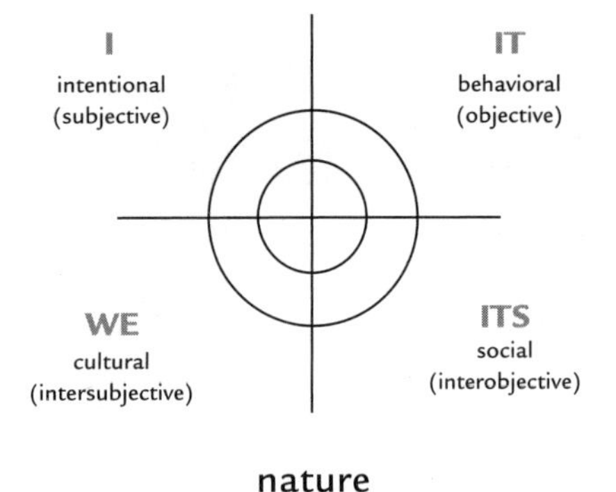

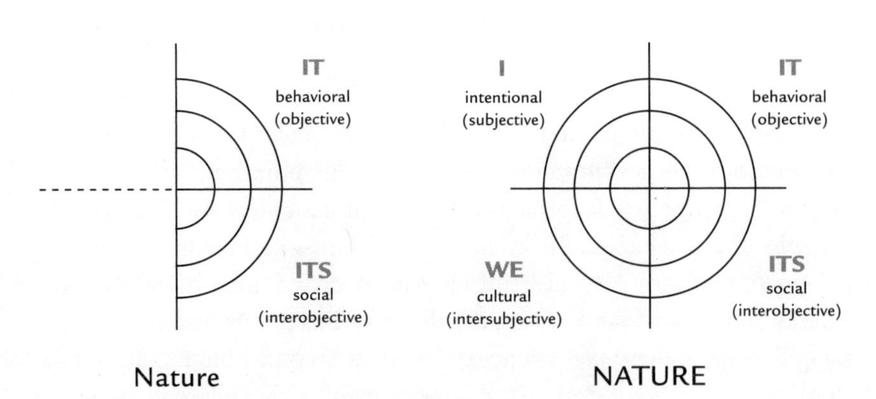

Figure 1.1. nature, Nature, NATURE.

the interior and exterior aspects of the biosphere—e.g., bodily feelings as opposed to mental thoughts (UL), objects of sensory perception as opposed to rational conceptualization (UR), natural systems as opposed to social ones (LR), and somatic connections as opposed to cultural relationships (LL). Capital-*N* Nature is all the exterior levels of the physiosphere, biosphere, and noosphere. And NATURE is all the interior and exterior dimensions of all the levels. See figure 1.1, where the first concentric ring signifies the physiosphere, the second signifies the biosphere, the third the noosphere, and beyond that the pneumasphere.

All three kinds of "nature"—NATURE, Nature, nature—refer to actual dimensions of reality, but problems frequently arise when these three realms are confused with one another.[19] For example, when we are hiking up a steep trail and have a profound intersubjective encounter with a mountain lion that puts us in touch with the sentience that moves through all of manifestation, and say out loud, "Wow, nature is amazing!" we are referring to NATURE: everything in the Kosmos. When we come to a vista on that hike and in response to everything we see, we say, "Isn't nature beautiful?" we are referring to Nature: all the surfaces we can see, the entire sensory-empirical world. When on that same hike we turn to our friend and say, "I love hiking—it really allows me to get out of my head and in touch with nature; being outside makes me feel peaceful," we are referring to nature: our nonrational experience of being away from our typical social-cultural environment. Thus, throughout the rest of the book, whenever we use the term "nature," we will be specifying one of these uses through our spelling or the context.

Another way of representing the relationship between these three natures is in figure 1.2, which shows how they were integrated before the proponents of the rational "ego-camp" and those of the empirical "eco-camp" collapsed NATURE into Nature and nature. The ego-camps absolutized the noosphere while the eco-camps absolutized the biosphere.

Modernity tended to eliminate the interior dimension of reality in favor of exterior dimensions, or material systems. Hence, scientific materialism reduced NATURE to Nature (the mind was reduced to the brain), which is itself part of an organism, which is part of the biosphere. According to this conception, the mind (the noosphere) was nothing more than a part of the biosphere, when in fact the somatic body (UL) is in the mind (UL) and the brain (UR) is in the physical body (UR). Likewise, in terms of interiors,

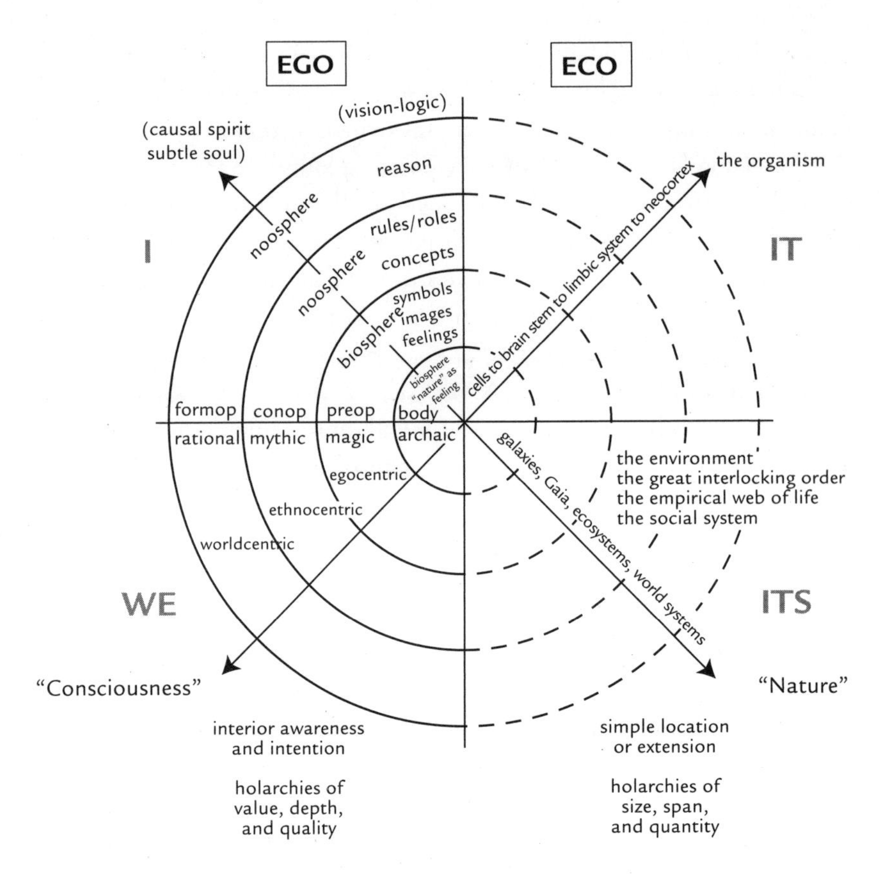

Figure 1.2. The ego and the eco before the collapse.

the *biosphere is in the noosphere*, and at the same time in terms of exteriors the *noosphere is in the biosphere*. Or, as Wilber puts it, "nature is in the mind, but the brain is in Nature."[20] This complex relationship between biosphere and noosphere all too often gets reduced to the latter exterior perspective. "When all interiors are reduced to exteriors, one can no longer recognize degrees of interior depth, and thus everything becomes equally a strand in the great interlocking web of valueless its. Everything is part of Nature. . . ."[21] For modernity, the status of the cognizing, egoic subject was always uncertain, given that it was difficult to depict undeniable subjective (1st-person) experiences in purely objective (3rd-person) terms. In contrast, Integral Ecology, as we will discuss in detail in chapter 2, views the physiosphere (the realm of matter), the biosphere (the realm of life), the noosphere (the realm of mind), and a possible pneumasphere (the realm of Spirit) as four levels of Nature. Thus, nothing transcends Nature, but there are things interior to Nature. Even the pneumasphere is not beyond Nature but an aspect of it. In other words, interiors (emotions, thoughts, spiritual experience) are not beyond Nature but inside it (i.e., NATURE). Integral Ecology integrates monological nature into dialogical culture and then takes both into translogical Spirit: nature into Nature into NATURE.

Many environmentalists today rely exclusively on objective, scientific (3rd-person) accounts of Nature (habitat and organisms). Such a conception, however valuable in its own right, provides no account for the interiority of natural phenomena, much less for the interiority of human beings. Such interiority has no "place" in a world composed solely of material particles and complex systems.[22] These conceptions reinforce what Wilber calls *industrial ontology*: "It is industrialization that holds flatland in place, that holds the objective world of simple location as the primary reality, that colonizes and dominates the interiors and reduces them to instrumental strands in the great web of observable surfaces. That 'Nature alone is real'—that is the voice of the industrial grid."[23]

According to Wilber, "The religion of Gaia, the worship of Nature, is simply one of the main forms of industrial religion, of industrial spirituality, and it perpetuates the industrial paradigm."[24] It only worships objects out there, and ignores the subjectivity in here. Wilber calls it *flatland*, a world of objects and surfaces.

The Romantic Response: It Is Still Reductionism

The Romantics reacted against modernity's conception of nature, but their reaction was confused. The Romantic reaction against rational modernity's materialism, and against the repression (of interiority) that followed from it, was justified, for something serious was amiss. Nevertheless, the Romantics' response to modernity went astray because their efforts involved two competing conceptions of "nature."

The first conception, as we have discussed, was the modernist view that Nature is the all-encompassing, interrelated exterior web of life. According to the Romantics, modernity had lost its connection with the biosphere (despite the modern assertion that everything is enclosed in and flows within the web of life). In positing that culture had deviated or split off from the natural world, the Romantics had to posit a second conception of "nature," a nature from which humankind can deviate. Wilber asks: "what is the relation of this [NATURE] that embraces everything, versus this nature that is different from culture because it is getting ruined by culture?"[25] Romanticism foundered because it could not reconcile these conflicting views of "nature." Great Romantics sought to reconcile this conflict by suggesting that nature is Spirit (NATURE), "because all-embracing Spirit does indeed *transcend* and *include* both culture and nature."[26] Most Romantics, however, were so committed to the Descent path, that "they simply identified [NATURE] with nature. They identified Spirit with sensory nature. And here they went up in smoke, a spectacularly narcissistic, egocentric, flamboyant explosion—because the closer you get to nature, the more egocentric you become. And in search of NATURE, the Romantics headed back to nature, and disappeared into a black hole of their own selfhood, while claiming to speak for the ultimately Divine—divine egoism, it sadly turned out."[27]

Romantic Ecological Thought: Getting Back to Which Nature?

Against the dissociated and alienated rationality that tears up NATURE and represses human interiority (a modernity gone astray), Romantic-influenced environmental resistance movements arose.[28] The Romantic movement was and is often regressive—confusions piled upon confusions. Instead of reflecting upon nature as disinterested scientists, Romantics—including the great German idealist, Schelling—sought to participate in and express NATURE. But too often these Romantics focused on their feelings of awe and wonder inspired by their encounters with the material aspect of nature's sublime power and beauty. Romantics confusedly celebrated what amounted to egocentric feelings and attitudes consistent not with genuine maturity, but with regressive states associated with childhood feelings—thus completing the slide into nature. As Wilber aptly puts it: "At their best, the Eco-Romantics desire a union with NATURE, a union with the Kosmos. . . . But for all the reasons we have discussed, the Eco camps drifted into an equation of NATURE with the great Web of Life, with the vast sensory-empirical world in all its organic richness and vitality. Under the intense gravitational pull of flatland, they reduced NATURE to Nature. And to Nature they pledged their undying allegiance [and from there to nature]."[29]

The Romantics merely merged, communed, and became enmeshed with the material world of sensory nature (versus mind and culture). Intuiting correctly that NATURE can be apprehended only by an awareness that transcends the limits of the rational ego, for the most part, Eco-Romantics confused genuinely transrational awareness—awareness that has included but transcended egoic rationality—with nonrational modes of awareness, including prerational states of awareness. In this way, many Eco-Romantics to this day commit what Wilber calls the pre/post fallacy. Confusing NATURE with nature, and confusing post-egoic with prerational, Eco-Romantics frequently long to get "back to nature," where Nature is the Great Web of Life, and nature includes the feelings of unity that belong to this Web. They want to be one with the biosphere (nature), not one with only the exteriors of the biosphere and noosphere (Nature). While the exteriors of the biosphere (e.g., mountains and streams) are good, the exteriors of the noosphere (e.g., trains and skyscrapers) are bad. Likewise, globalization, climate change, and anthropogenic species extinction are part of Nature, but they are not part of nature. Eco-Romantics reject the upper levels of Nature because they are produced by the noosphere. Instead they focus on the "purer" aspects of Nature—its lower exterior noncultural levels (as well as the interior correlates associated with nature)—and equate that with NATURE. But, unfortunately, going back to nature prevents going forward to NATURE, which transcends and includes rationality, not denies and regresses beneath it. "The Romantics embraced Nature in general, but nature in particular. They aimed for NATURE, got Nature, and ended up glorifying nature."[30] Attempting to overcome alienating dissociation, they ended up in a suffocating, merged, enmeshed identification with nature that left no room for what differentiates humans (and the noosphere) from the biosphere. Keep in mind, however, that while the noosphere and biosphere are differentiated, they are both apects of Nature.

One of the things that makes the distinction between Nature and nature so confusing is that they overlap a little: all of nature is in the noosphere, but only the exteriors of the noosphere are in Nature. In other words, the Right-Hand aspects of the biosphere (nature) are the first few levels of Nature. So aspects of the noosphere are in Nature but not in nature. Hence, you can have two New Yorkers talking in a café about how much they love "nature" in reference to Central Park, and the first one can mean nature (e.g., how the park offers an escape from the city: an oasis surrounded by the concrete desert) and the second one can mean Nature (e.g., how the famous park is almost entirely manmade and provides a beautiful scene of trees and skyscrapers, park benches and runners with iPods, wildflowers and artistic sculptures).

If the Rational-Egoic (modern) camp approached Nature with reason, the Eco-Romantic camp approached it with preconventional feelings. For Romantics, nature amounted to their emotional responses to what we now call the biosphere. However powerful those emotional resonances may be, they omit what can be revealed about Nature through the noosphere, which includes human cognition. Referring to the work of Charles Taylor, Wilber argues that Eco-Romantics join moderns in denying that Nature/nature is a manifestation of Spirit, but instead maintain that Spirit *is itself* the Great Web of Life.[31] That is a materialistic view. In this scenario, attunement with this Web comes not from attaining the highly developed stages of consciousness described by ancient shamanic and contemplative traditions, but instead from regressing and reducing this great NATURE to one's personal feelings. The Romantic phase in which Nature/nature is appraised in terms of the feelings that it evokes in me, as Taylor observes, "presupposes the triumph of the new identity of disengaged reason over the premodern one embedded in an ontic logos [interwoven world order]."[32]

The danger is that this regression to prerational states or preconventional stages leads Romantics to abandon the important and legitimate achievements of rational-egoic selfhood and its associated modern institutions, including constitutional democracy, all of which must be in place in order to move to post-egoic stages of awareness and associated institutions. Although many early-19th-century Romantics were aware of the perils of regression and celebrated the widespread democratic impulse, many Romantics tended to slide into "divine egoism," which celebrated and emphasized what is peculiarly, uniquely my own, where nature becomes the source for egoic sentiment. In other words, Romanticism too easily becomes a form of narcissism in which "nature" is equivalent to preconventional emotional states.

If Wilber is correct, the "back to nature" fantasies are a reprise of this failed Romantic effort to overcome the humanity-nature split: "instead of moving forward in evolution to the emergence of [NATURE] or Spirit (or World Soul) that would indeed unify the differentiated mind and nature, [Romantics] simply recommend 'back to nature.'"[33] Such a view invites psychological and social regression. If nature as biosphere is the "fundamental reality," then that which deviates from nature threatens it. If nature "is the ultimately Real, then culture must be the original Crime."[34] The goal, then, must be to dismantle culture, in order to achieve a lost paradise involving unconscious unity with pristine nature.[35] Cure repression with regression, as some Earth First!ers have done in their call for a return to the Pleistocene age. Such a yearning for primal unity with divine nature is tempting, but potentially disastrous both individually and culturally.[36] Moreover, this regression will not halt ecological destruction. Only a major change in socioeconomic and political institutions will help halt ecological destruction, and such change requires the growth of "mutual understanding and mutual agreement based upon a worldcentric moral perspective concerning the global commons."[37] Such an achievement involves interior growth and development. To escape ecological destruction, a genuinely postmodern humanity must overcome its fear of authentic psychological development and interiority, since these alone can integrate what modernity has dissociated.

While we have already established that interiors are not outside of NATURE but rather are inside of it, there is a long history within ecology and the environmental movement of denying interiors and the development of those interiors (i.e., transcendence). Most environmentalists abjure talk of transcendence and Spirit because they are moderns at heart—they agree that all being is material being. And because transcendence seems to be a uniquely human capacity, environmentalists fear that acknowledging transcendence will only encourage an anthropocentrism that justifies heedless destruction of natural phenomena. But genuine transcendence is not anthropocentric—animals have interiors that grow and complexify—nor is it otherworldly, as interiors are inside of NATURE. An Integral approach always integrates Ascent with Descent.[38]

The Rise of Perspectivalism (Postmodernism): Interiors Making a Slow Comeback

A century after Friedrich Nietzsche linked truth, perspectives, and power, postmodern theorists such as Michel Foucault argued that all truth claims (presumably even his own!) are influenced by power interests that are often unconscious or concealed from those making the claims.[39] Political perspectivalism arose in post-1945 North America, Europe, and colonized nations, when social critics demonstrated that the established elites promulgated truth claims not as perspectives but as objective facts. These "facts," critics pointed out, were often used to legitimate harmful actions and the subjugation of less powerful people.[40] The economically and politically powerful executed such harmful acts and frequently justified them by calling on scientific and technical experts. In the 1960s and 1970s, marginalized and minority voices demanded power and called into question the "facts." Minority groups publicly challenged the representational paradigm. This brought to light the notion that there is not One True Way or One True Perspective, and claimed that we need a host of perspectives to make judgments about complex situations.

Perspectivalism helped recover the importance of the interior domains of reality (particularly intersubjective domains). It maintained that inquiry is inevitably influenced by countless factors, including presuppositions of which the inquirer is not aware. Yet the extreme postmodernists went overboard with their anti-hierarchical assertions, according to which anyone's view is as good as anyone else's. We maintain that some views are better (because they are more comprehensive, more inclusive, and more far-seeing). And even as we affirm the contributions of multiculturalism, an important expression of perspectivalism, we do not share its deep suspicion of all forms of hierarchy, universalism, cross-cultural norms, developmental concepts, and modernity in general. Power interests often influence truth, although truth cannot be reduced to only such interests.

Perspectivalism in Ecological Science

Leading ecological scientists have been directly or indirectly influenced by postmodern perspectivalism, which has at least two major strands: epistemological and political. Perspectivalism is an epistemological challenge to the representational paradigm. The representational paradigm believes that the unbiased, rational mind is capable of mirroring or representing things in the One True Way (have you met that unbiased mind?). Natural science once presented its method as the exemplary and even as the only way for obtaining the Truth. Clearly, times have changed. As ecologist Stanley Dodson remarks: "Whatever perspective is used to view it, ecology is often assumed to be something that exists 'out there.' While it is true that things such as birds and bees and trees and mountains do exist, ecology exists only in our language. *Ecology is an interpretation of our perceptions of organisms and the environment.* As with any interpretation, ecology depends completely on the history and culture of the people making the interpretation."[41]

We can take many perspectives. The very act of choosing a perspective and applying its methods actually helps generate the phenomena in question. Many contemporary ecologists explicitly acknowledge that the phenomena they study arise within and are partially constructed (delimited, defined) by the very act of framing them.[42]

Perspectivalists maintain that mind—far from being a mirror that passively receives independent phenomena—plays an active role in co-constructing phenomena. Methodologies not only reveal, but also in some respects constitute the phenomena under investigation. What we call "facts," in other words, are not ready-made but emerge in a complex process of perceptual, emotional, and cognitive negotiation between knower and known. According to perspectivalism every assertion is made by a person occupying a viewpoint within a cultural worldspace.[43] In fact, Integral Theory asserts that consciousness is embodied in flesh, embedded in culture, and enmeshed in eco-social systems. Assertions about environmental problems are created and interpreted by people with different perspectives and interests.

An environmental problem arises as a problem ("heavy metals in tuna" or "habitat loss") within a shared worldspace, amid particular social structures, and is articulated by linguistic distinctions. Words are not, however, more real than things; but many things can arise only by virtue of language. Hence, a "perturbation" in a complex system is determined as such by scientists with a methodology, informed by a hypothesis, which is influenced by a larger theoretical perspective.[44]

Some ecological scientists explicitly depict the phenomena they study not as preexisting things-in-themselves, but rather as phenomena that arise partly due to the act of "framing." According to proponents of hierarchy theory, even what counts as an entity depends in part on how investigators position it in a selected hierarchical framework, and how they characterize it in terms of time and space. In effect, many scientific ecologists agree that every phenomenon—whether objects of scientific investigation or contents of mystical states—arises within and is interpreted in a shared linguistic-cultural worldspace. Phenomena are perceived or experienced by someone who occupies a particular perspective and applies a particular methodology.[45]

Researchers have attempted to organize these approaches (e.g., Allen and Hoekstra), but an obstacle to unifying ecology as a branch of natural science has been widespread disagreement on principles or foundations (what a surprise).[46] It may come as a shock that there is no single unified school, method, or perspective called "ecology." Instead, there are an overwhelming number of perspectives, concepts, questions, and techniques. To organize this methodological tangle, Stanley Dodson proposed organizing ecological science according to the major perspectives utilized by practitioners.[47] But he restricts the term "ecology" to natural scientific (objective and interobjective) methodology (a 3rd-person perspective). As a result of that restriction, Dodson concludes that there are only six major ecological schools.

But there are now actually over 100 schools of ecology, not to mention schools of environmental studies and schools of ecological thought, and the number continues to grow (see chapter 5 and the appendix). Many of these schools find their homes not in the natural sciences, but rather in the social sciences, the humanities, the arts, or still other domains. The world of ecology has grown so large that it no longer knows itself: practitioners of landscape ecology have never heard of eco-phenomenology; environmental philosophers often do not know the difference between population ecology and community ecology; individuals working in the field of acoustic ecology do not know about ecological hermeneutics. A major advantage of Integral Ecology is that it provides a comprehensive meta-theoretical framework for classifying and coordinating these manifold perspectives.

Ecological Perspectivalism Slowly Begins to Include Interiority

Perspectivalism has been a major impetus in the rise of inter- and multidisciplinary approaches to knowledge formation and problem-solving in recent ecological science. Ecological perspectivalism, however, has been slow to include interior domains, including those pertaining to value and culture. Consider the Ecological Society of America's 2004 publication, "Ecological Science and Sustainability for a Crowded Planet" (ESSCP):

> Ecology is by its very nature an interdisciplinary science, making it impossible for any single ecologist to be well versed in the details of every relevant discipline, method, or instrument. Yet, it is increasingly obvious that ecologists must come together to help understand, solve, and anticipate the environmental issues facing our world. To do so, ecologists may need to think of themselves as entrepreneurs in a shifting and pressure-driven marketplace, where strategic collaborations and rapid responses are keys to scientific success. Our best chance to succeed in those efforts is to have a broadly inclusive approach to ecological research. This ap-

proach must include actively recruiting expertise beyond our discipline, as well as changing our culture to best foster the innovations we need.

. . . If successfully implemented, this new depth and breadth of ecological understanding, including its improved communication beyond the discipline, would allow ecologists to play an influential and eminently helpful role in decisions made at all levels that affect the sustainability of the biosphere.[48]

Although the document's title page encircles a drawing of planet Earth with three phrases, "Anticipatory Research, Informed Decisions, Cultural Change," ESSCP accords only a negligible role to culture and value (interior perspectives). They seek to foster greater collaboration only with other natural scientists, social scientists, businesspeople, and government officials. In our word search, we found no instances of terms such as "norms," "moral development," "interiority," "consciousness," or "subjectivity." "Ethics" appeared only once, in connection with moral rules for using data generated by others. The term "values" occurs in a paragraph encouraging scientists to provide "rigorous ecological knowledge" to religious groups that "have responded to emerging environmental concerns by linking values to an ethos of environmental stewardship."[49]

To bring to fruition ESSCP's important vision of sustainability, scientists must attend to personal and cultural interiors—including values, worldviews, and religious beliefs—because they all play a role both in creating and resolving environmental issues. Joseph Tainter, Timothy Allen, and Thomas Hoekstra call this "post-normal science": "In post-normal science . . . data are insufficient, time is short, and because the stakes are high there is keen public interest and conflicting values. The findings of post-normal sciences are embedded in a larger social framework, in which the audiences consist of contending interest groups, and in which issues more have more than one plausible solution."[50] Post-normal environmental science, "with its focus on stakeholders and cognizance of alternative points of view," has a good deal in common with Integral Ecology's desire to integrate findings from the domains of consciousness, culture, and nature. Others are interested in the integration of interior domains also.

Lance H. Gunderson and C. S. Holling took another step toward integrating these three domains in *Panarchy: Understanding Transformations in Human and Natural Systems*.[51] *Panarchy* calls on the rich conceptual model of resilience in complex adaptive systems to show how natural and social sciences can and must cooperate to address environmental problems.[52] In one essay from *Panarchy*, Frances Westley et al. argue that failure to understand the difference between ecological and social systems "helps to explain the fundamental lack of responsiveness or adaptability to environmental signals that characterize much of natural resource management."[53] Whereas space and time are key categories for understanding ecosystemic structures and patterns, "For social systems, we need to add a third dimension, which is symbolic construction of meaning."[54]

In another essay in *Panarchy*, "A Future of Surprises," Marco A. Jannsen discusses the interrelation of cultural domains, socioeconomic systems, and ecosystems, and outlines a developmental model of culture (as worldviews) that has a lot in common with Integral Ecology's interior developmental approach. According to Jannsen, the most prevalent U.S. worldviews are *hierarchalism* (or conservatism), held by those who defer to authority in defining and solving environmental problems; *individualism*, affirmed by those who put faith in the power of unhindered markets to solve those problems; and *egalitarianism*, adhered to by those (including Greens) who claim that environmental problems can be solved primarily by reducing inequity. Hierarchalism, individualism, and egalitarianism correspond in many important respects to traditional, modern, and postmodern developmental perspectives, which we discuss in more detail in chapters 4 and 7.[55]

These three worldviews—traditional, modern, and postmodern—all manifest themselves within the interior domain of culture. In our view, failure to differentiate among these three worldviews prevents environmentalists from addressing adherents to each in rhetorically effective ways. This is one reason that environmentalism is now widely viewed merely as an interest group, despite the fact that large percentages of individuals from all three worldviews uphold environmental values.[56] Certainly members of the environmental movement do little good when they speak to people who hold differing worldviews as though they were stupid, bad, and wrong. Furthermore, doing so is inaccurate. We hope to recover the study of interiors and the development of those interiors (chapter 4) so that environmentalists will instead acknowledge that environmental issues "arise" differently for people depending on their worldview (traditional, modern, or postmodern). We hope this recognition will encourage environmental leaders to learn how to understand and communicate with these different worldviews, so as to achieve greater participation in solving some of our greatest environmental threats.

Integral Ecologists need to enter into and appreciate the personal and cultural worldspace of the major stakeholders. Everyone has a partial truth to contribute and everyone has a stake in the environment, not just the environmentalists. It is crucial to understand how any given proposal sounds to people operating from different perspectives so that you can collaborate with them in effective ways. It doesn't mean that partial views should dominate our understanding or solution, but they must play a role. Integral Ecology asserts that we are in this together, so we had better start acting like it.

There are real-world reasons we need to revive interior perspectives. By neglecting to study the interiors, including major differences between how "nature" and "ecosystems" are valued at different developmental levels (traditional, modern, and postmodern), ecologists often fail to marshal public support for correcting human activities that may threaten human and nonhuman life.

Many ecologists—not only natural scientists, but also environmental activists informed by ecological science—assume that personal and cultural (interior) changes will automatically arise if we educate people about environmental problems and ecosystem dynamics. Clearly, this is not the case. We argue that the very same facts we use to educate people trigger different value assessments depending on their developmental worldview. Hence, effective environmental advocacy requires rhetorical strategies that consider the values of those you are educating. A conservative Christian might be willing to defend a species against destruction because it is a creature of God, whereas an agnostic modern might defend the same species because it is potentially useful to medical research. Both are acceptable motives regardless of your personal perspective.[57] Facts motivate people differently, depending on their value system or worldview. Only by taking human interiors seriously can we acknowledge the extent to which such interiority pervades the rest of the Kosmos.

A Planetcentric Reach of Interiors: Restoring Depth to the Cosmos

In addition to ascribing interiority to humans, a number of current researchers ascribe some measure of interiority to animals and even plants, although not a self-reflexive interiority.[58] A leading philosopher of mind,

David Chalmers, asserts that just as matter energy is the basic ingredient of material phenomena, so too there must be a basic ingredient—proto-mind—that makes complex interior phenomena such as human consciousness possible.[59] The human brain, then, is closely correlated with the kind of consciousness required for subjective human experience. Chalmers also concluded that the capacity for experience, however meager, goes all the way down, and offers a sophisticated, postmodern version of pan-experientialism that is consistent in many important respects with Integral Theory's pan-perspectivalism.

Experiential capability is ascribable to one-celled organisms. Even though they lack a central nervous system, they are able to recognize food and engulf it. If pan-experientialism is correct, recognizing and engulfing food are not merely mechanical processes but include low-level experiential correlates. Cells that compose organisms engage in elaborate signaling processes, both internally and with other cells. Signaling involves the exchange of information in the form of signs or codes. Natural scientists now conclude that the universe is saturated with information that makes organization possible.[60] DNA, to take the most famous example, is information in the form of a complex code that enables life forms to reproduce themselves.[61] Many now hypothesize that there must be a meta-code, whose location and character are as yet unspecified, which is responsible for organizing the signaling carried out by DNA.

Signaling is also involved in social life. Ferdinand de Saussure defined semiotics as the science "which studies the role of signs as a part of social life."[62] Social life evidently goes all the way down also, as biosemiotics, the study of signaling among animals and plants, asserts.[63] The universality of signaling indicates that the Kosmos is intelligible from the bottom up and that interiority is present from the bottom up. This is not to say that every individual is "conscious" but that all life seems to take a perspective—everything is at least capable of noticing or prehending. Every individual constitutes a finite clearing or perspective, no matter how constricted. This position is a form of pansemiotics as described by Charles S. Peirce, which sees the universe as permeated with meaning at all levels of organization.[64] The Danish theoretical biologist Claus Emmeche summarizes the pansemiotic thesis as follows: "The universe is perfused with signs, semiosis is not only a process found in all living nature among beings which are organic, functional wholes (organisms as interpreters, or interpretants). The sign, its object and its interpretant are universal categories, which existed (eventually in degenerate form) even before the origin of life."[65] Integral Ecology embraces pansemiotics as an intersubjective correlate to subjective pan-interiors.[66]

Although interiority—the capacity for opening a perspective or clearing—is not an exclusive human capacity, humans are endowed with a distinctively rich, linguistically articulated mode of interiority. Humans can even become aware that they are aware, thereby enabling them to deliberately alter how they think and act, and to question their origin, constitution, and purpose. This is a special evolutionary advancement. In virtually all world religions, humans are accorded special status. Even Buddhism, often contrasted by environmentalists with the three allegedly anthropocentric Western versions of monotheism (Judaism, Christianity, Islam), maintains that being born a human being is necessary to be freed from the wheel of suffering (i.e., while nonhuman organisms have buddha nature, they cannot become aware of it and are thereby trapped in their suffering).[67] The Kosmos is not just material objects, but includes interior depth at many different levels.

This interiority makes it possible to use technological power to exploit nonhuman beings, but it also makes it possible to develop a morality that calls for an end to destructive behaviors. The technological domination of nature could only have arisen in the modern era. To significantly limit or transform this destructive domination, in which humans heedlessly exploit nature, we need to do more than develop new technologies or change the social system. We need to facilitate healthy expression of each level of development and to encourage interior development in many different capacities.

Corresponding to any exterior sociopolitical, economic, or technological dominator hierarchy is an interior dominator hierarchy. Although humans have used modern science and technology to dominate others, the same Eros that allowed modern consciousness to emerge also allowed for a worldcentric ethics to emerge, which affirms the universal rights of humankind. Environmentalists have explicitly called upon such rights to work against the untoward consequences of industrial technology. Yet an interior dominator hierarchy remains in place for most modern people with regard to nonhuman beings. Until a critical mass of people evolve to postmodern levels of interiority, in which heedless domination of human and nonhuman beings becomes unacceptable and immoral, environmentalism will remain a reform movement within technological modernity. When this developmental transformation occurs, but not before, we will see the widespread adoption of what Hans Jonas called "the imperative of responsibility."[68] Then and only then will we feel responsible to this world and other species.

It is clear to us that one key to our future as a species lies in reaffirming and exploring interiority. Unlike so many philosophical movements that have come before, we do not believe that enhancing only our interiors will help. We need interior-individual, interior-collective, exterior-individual, and exterior-collective perspectives and methodologies. We need them all. The Kosmos is multifaceted. We cannot expect to understand the Kosmos through partial perspectives and singular methodologies; we must have a multifaceted, kaleidoscopic view.

Integral Ecology coordinates perspectives that are pertinent to environmental concerns. Although human knowledge is finite and perspectival, we assume that it is possible to create ever more encompassing and inclusive models and understandings. Such models provide people with a greater capacity for comprehending and intervening in any complex problem, including environmental problems. Integral Theory provides such a model. To be sure, there are sometimes formidable difficulties involved in coordinating frameworks, which may focus on very different kinds of phenomena, at different temporal and physical scales. Hence, Integral Ecologists will need to devise effective meta-discourses that both recognize such differences and find useful ways of connecting them.[69]

Resisting method hegemony, according to which there is only one or a few reliable ways to interpret any given phenomenon, we adhere to Integral Methodological Pluralism (IMP), which we discuss in chapters 2 and 8. Given the upset such pluralism may provoke on the part of those convinced that their method alone offers the "one" or "most valid" way to truth, IMP (i.e., a mischievous fairy who upsets people) is a particularly apt acronym. IMP is a collection of practices and injunctions guided by the intuitions that "everyone is partially right!" and that everyone brings forth and discloses a different and partial facet of reality.

IMP, according to Wilber, contains three principles: *nonexclusion* (acceptance of truth claims that pass the validity tests for their own paradigms in their respective fields), *enfoldment* (some practices are more inclusive, holistic, and comprehensive than others), and *enactment* (phenomena disclosed by various types of inquiry depend in large part on a host of factors that influence the researcher who is disclosing the phenomena). These three principles guard the validity of different methodologies. Here Wilber describes his commitment to a trans-methodological or Integral approach:

> The whole point about any truly integral approach is that it touches bases with as many important areas of research as possible before returning very quickly to the specific issues and applications of a given practice. . . . An integral approach . . . is a panoramic look at the modes of inquiry (or the tools of knowledge acquisition) that human beings use, and have used, for decades and sometimes centuries. An Integral approach is based on one basic idea: no human mind can be 100% wrong. . . . when it comes to deciding which approaches, methodologies, epistemologies, or ways of knowing are "correct," the answer can only be, "All of them." That is, all of the numerous practices or paradigms of human inquiry—including physics, chemistry, hermeneutics, collaborative inquiry, meditation, neuroscience, vision quest, phenomenology, structuralism, subtle energy research, systems theory, shamanic voyaging, chaos theory, developmental psychology—all of those modes of inquiry have an important piece of the overall puzzle. . . .[70]

Inclusion does not mean we abandon rigor. Instead, it requires that many kinds of rigorous inquiry be brought to bear on complex problems. Often we mistake rigor for the method of inquiry applied by the natural sciences and to a lesser extent by the social sciences (measuring and counting objective data). What English-speaking people call the natural sciences and the humanities are both called sciences in German: *Naturwissenschaften* and *Geisteswissenschaften*. The stems of the word *Wissenschaft* translate as "knowing" and "making." Science, then, is a rigorous inquiry to create knowledge. Each perspective has its own methods, practices, injunctions, and community standards to create knowledge. There are many sciences! There are sciences dedicated to subjectivity, intersubjectivity, objectivity, and interobjectivity. We maintain that they all have something important to say.

Likewise, environmental problems cannot be fully understood solely in terms of ecological science. We need insight from many other perspectives. In recent years, as demonstrated by the Ecological Society of America's plenary session, scientific ecologists have sought to include other perspectives.

Thus, Integral Ecology calls for an exploration of how human interiors such as psychological identities, emotional dynamics, cultural values, and somatic experiences interface with ecological problems. We also call for an acknowledgment of the existence of interiors of all members (including nonhumans) of ecosystems.

In this chapter we have described how we lost interior perspectives and why we need them (as well as other perspectives), and have begun to propose how to recover them. We think the Integral Model is one model capable of organizing all the perspectives necessary to understand and solve the environmental problems we face today. Integral Ecology is a multiperspectival approach to characterizing and solving environmental problems. Such an approach requires that interior perspectives—both subjective (1st-person) and intersubjective (2nd-person), human and nonhuman—be taken as seriously as exterior objective and interobjective (3rd-person singular and plural) perspectives. In the following chapter, we examine the 1st-, 2nd-, and 3rd-person perspectives at work in the quadrant aspect of the Integral Model.

2

It's All About Perspectives: The AQAL Model

The key to protecting oneself against the seductions of confusing maps and territories is to become familiar with how any map is constructed. . . . Recurrent problems with maps arise because their users are unfamiliar with how they are made, or they have forgotten that any map, physical or spiritual, is constructed from an infinite variety of sensual details with a great deal of physical effort, involving endless cycles of trial and error.

. . . Before using any instruments, one has to develop an overall sense of the land to be mapped, as well as a feeling for its peculiarities, how the contours are arranged, where they are broken by swales and outcroppings. Only then can one learn how to use to advantage the transit and tape to track the shape and dimension of each contour.

—DON HANLON JOHNSON[1]

The Value of Multiple Perspectives

A neoclassical economist concludes that an industrial plant emits toxins due to a market failure. He understands this phenomenon in terms of prices, markets, taxes, incentives, marginal costs, discount rates, and so on, and justifiably so. He recommends that government agencies institute an emissions market that forces the company (and others like it) to pay for the air they are polluting.

A Greenpeace activist, who believes that unchecked capitalism is the cause of many environmental problems, is suspicious of this economist's recommendations. To this Greenpeace activist, assigning cost to pollution or placing a price on land seems like yet another example of how capitalism reduces everything to raw materials whose worth is determined solely by market forces.

Despite outer appearances, the economist and the Greenpeace activist have a good deal in common. There is no doubt that both of their perspectives are useful; in fact, any useful perspective sheds light on certain aspects of an environmental problem with varying degrees of clarity. And both of their perspectives are partial. Additionally, both are probably baffled that others simply cannot understand the truth of their recommendations. Both have a difficult time conceiving that their worldview is a perspective, or one way of interpreting the same phenomenon.

We could easily substitute the economist for a theoretical biologist, toxicologist, Sierra Club lobbyist, Earth First! activist, or eco-ethicist. Each would tend toward perspective absolutism. Each would emphasize the truths generated by their own perspective and grant limited validity (and sometimes no validity) to truths generated by alternative perspectives.

Here it is worth emphasizing that our discussion of perspectives has thus far tended to emphasize the interobjective or "its" quadrant, which predominates in the social and natural sciences. Talk of "mapping" and "models" is 3rd-person discourse that objectifies the phenomena under discussion. Whenever possible, Integral Ecologists must remind themselves of the power and limits of such 3rd-person, past tense, objectifying discourse, in order to make room for alternative perspectives, characterized by different tenses, scales, and situatedness.

An Integral Ecologist must bring into dialogue all of the partial and pertinent perspectives of this environmental problem, even while recognizing that some of these perspectives will be incommensurate with one another. For instance, the perspective afforded by a microbiologist interested in how a toxin affects a particular kind of organism is very different from the perspective of a parent whose child has been sickened by that toxin carried in drinking water. Integral Ecologists seek to weave a comprehensive, multitemporal, and multiscalar tapestry that includes truth claims drawn from as many relevant perspectives, methods, practices, and paradigms as possible.[2] Integral Ecologists do not unjustifiably privilege any one perspective at the expense of other legitimate ones—scientific, experiential, social, or cultural—but rather, they attempt to take into account the claims of many fruitful perspectives.

Perspectives, Perspectives, Perspectives

Integral Theory enables us to organize and integrate many different perspectives. This meta-theory organizes proven practices and paradigms, while taking into account as much as possible differences in terms of phenomena being investigated, and at what temporal and spatial scales.[3] Integral Theory has a lot in common with interdisciplinary trends in environmental studies and practices, but it is not an interdisciplinary model. Some have described Integral Theory as a transdisciplinary model, which is described by Julie Klein, drawing on Erich Jantsch's work. She writes: "Whereas 'interdisciplinary' signifies the synthesis of two or more disciplines, establishing a new metalevel of discourse, 'transdisciplinarity' signifies the interconnectedness of all aspects of reality, transcending the dynamics of a dialectical synthesis to grasp the total dynamics of reality as a whole."[4]

While Integral Theory can be used in a transdiciplinary context, it is not only a transdiciplinary approach. In fact, Integral Theory is *postdisciplinary* in that it can be used successfully in the context of *disciplinary* (e.g., helping to integrate various schools of psychology into Integral Psychology), *multidisciplinary* (e.g., helping to investigate ecological phenomena from multiple disciplines), *interdisciplinary* (e.g., helping to apply methods from political science to psychological investigation), and *transdisciplinary* (e.g., helping numerous disciplines and their methodologies interface through a content-free framework) approaches.[5]

Thus, we have a more comprehensive approach: Integral Ecology is a postdisciplinary approach. The Integral Model does not have to generate new data or hypotheses within existing disciplines. Rather it starts out by marshaling and coordinating the data and hypotheses generated by competent practitioners in their respective disciplines. Next it generates new data through combining and juxtapositioning injunctions from various disciplines as well as creating new injunctions. Integral Ecology's postdisciplinary status allows it to be an integral discipline, an integral multidisciplinary approach, an integral interdisciplinary approach, and an integral transdisciplinary approach.

The Integral framework is not just about saying that the arts, morality, and science are important but instead is actually about putting into practice the methods or injunctions associated with different disciplines in order to bring forth the phenomena for which those practices are needed to show up. Thus, the Integral approach is not just about describing more accurately what is "out there" but is about changing our own awareness by following a variety of injunctions. For example, if Integral Ecologists conclude that they need to understand a new aspect of an environmental problem, they can respond to this need by remaining within their discipline, drawing on a number of different disciplines, or using a transdisciplinary approach, but in all cases they will be engaged in 1st-, 2nd-, and 3rd-person injunctions.

In this and the subsequent two chapters, we introduce Integral Theory, which is the theoretical framework for Integral Ecology. Central to this framework is perspectivalism, which includes two related claims. First, sentient beings are capable of taking a perspective, or opening a clearing that allows phenomena to present and reveal themselves. The capacity for perspective taking is ontologically foundational for all individual holons.[6] How far "down" such perspective taking goes is subject to debate, although we posit that even atoms are perspectival insofar as they take into account other atoms.[7] Wilber writes: "If the manifest world is indeed panpsychic—or built of sentient beings (all the way up, all the way down)—then the manifest world is built of perspectives, not perception. . . . Subjects do not prehend objects anywhere in the universe; rather, first persons prehend second persons or third persons; perceptions are always within actual perspectives."[8]

Wilber privileges perspective over perception, in order to clarify the misguided, abstract subject-object scheme that has dominated modern thought. The misunderstanding is this: perception always occurs within a perspective, and the perspective is either 1st-person, 2nd-person, or 3rd-person. All three terms in the phrase "a subject perceives an object" are abstractions, because they do not acknowledge the perspectives involved in the act of perceiving. Instead, it is more accurate to say, "I (1st person) study organisms from a 3rd-person perspective. Those organisms have their own 1st-person perspective. They can encounter other organisms in a 2nd- and 3rd-person perspective."[9] There is no "subject" perceiving an "object." Rather, there are only situated, 1st-person perspectives encountering and experiencing other sentient beings and phenomena.

For example, when Tom plays Frisbee with his dog, they each experience simultaneously arising perspectives. Each experiences the joy of watching the arcing flight of a frisbee (1st person). Each experiences the shared connection through the game (2nd person). Each observes pertinent attributes of the Frisbee (3rd person), including its speed and direction. Along with his 1st-person experience (his own joy or relaxation), Tom encounters his dog within a 2nd-person perspective (I-Thou, or "We"). His dog responds within a 2nd-person perspective. It is mutual resonance, and that is why it is called a "We."[10] The dog's experience is not as complex as Tom's, but it does have an interior experience, and it is capable of feeling their shared connection. And if Tom's girlfriend joins the game, she would be a 3rd person, who has a 1st-, 2nd-, and 3rd-person perspective. All of these perspectives happen at once; it is just a matter of which one(s) you notice or emphasize.

The truth is we are participating in an incomprehensively vast set of 1st-, 2nd-, and 3rd-person relationships. If a fox notices a nearby mouse, we can describe it from a 3rd-person perspective as an organism reacting to a stimulus (which is one essential yet partial view), but in that description the abstract world of subjects perceiving objects remains. We replace this abstract world with a more concrete one when we say: the fox holds a 1st-person perspective within which it encounters other beings from a 2nd- or 3rd-person perspective. "In short," writes Wilber, "a world containing sentient beings [all the way down and up] is a world composed of perspectives—not feelings, not consciousness, not awareness, not processes, not events—for all those are perspectives before they are anything else. . . . If all holons [i.e., individuals] are sentient beings, then all perceptions are actually embedded in perspectives of, from, and between sentient beings, simplified as first-person, second-person, and third-person perspectives."[11]

Given that modernity affirms that the world is composed wholly of inert, insensate, depthless objects (and is unable to account for human consciousness), to assert that there are no objects because there are no subjects is a revolutionary statement. When we dissolve the abstract concept of subjects and objects, we dissolve the mind-body dualism (while retaining the distinction between various definitions of "body" and "mind").

Our second claim is that all knowing is perspectival. Knowing occurs at many different levels, and all behavior has a corresponding experiential or interior state. These experiential, interior states are far more complex in humans by virtue of abstract language.[12] The power of language is so extensive and invisible that people often confuse the conceptual world of language with the phenomenon they are describing. Alfred North Whitehead refers to this process as the fallacy of misplaced concreteness. Hence, our motto for perspectivalism: Don't confuse the map with the territory. Don't confuse your conceptual framework with the scheme it describes. All experience is interpreted within the limits of your conceptual scheme, typically the one consistent with your perspective and developmental center of gravity.[13]

Perspectives Organized: The AQAL Model

Let us now introduce a schematic version of Integral Theory's AQAL diagram (fig. 2.1). AQAL (pronounced *ah-qwul*) is an acronym for "all quadrants, all levels, all lines, all states, and all types." These five terms refer to the intrinsic perspectives that occur at all scales and in all contexts, and to the intrinsic features of all individual holons.[14] By considering these five basic elements, Integral Ecologists can cover all the major facets, dimensions, and aspects of any phenomena. We do not assign ontological or epistemological priority to any one aspect, as these perspectives co-arise in the seamless fabric of reality in every moment and as a result are equally primordial.

Quadrants refer to the four dimensions—intentional, cultural, behavioral, and social (aka systems)—as well as to the four perspective sets by

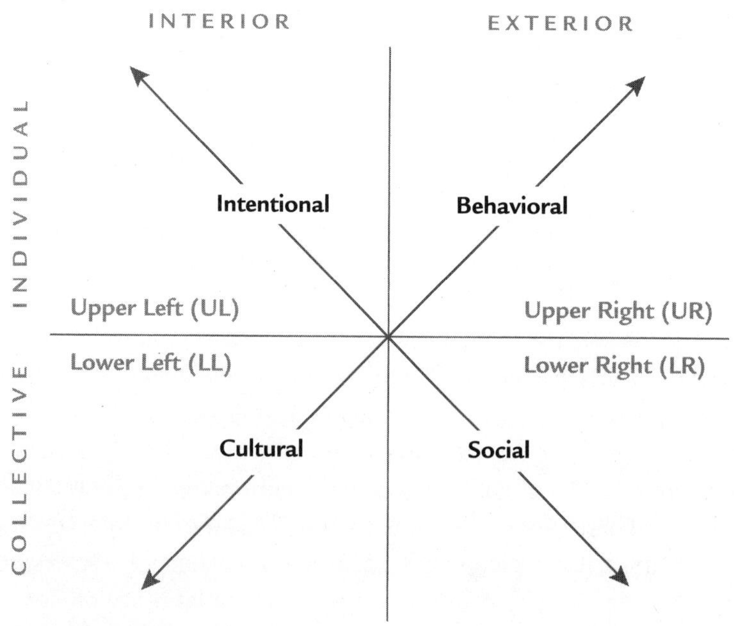

Figure 2.1. AQAL diagram.

which we can gain access to these domains.[15] There is, at any given moment, always an individual and a collective dimension. Alongside each of these dimensions, there is always an interior and an exterior dimension. These four domains—the interior and exterior of individuals and collectives—are also described as the perspectives of (1) experience (individual-interior); (2) culture (collective-interior); (3) behavior (individual-exterior), and; (4) systems (collective-exterior). Figure 2.1 shows how each quadrant represents a different dimension-perspective. The quadrants are designated in Figure 2.1 as Upper Left (UL), Lower Left (LL), Upper Right (UR), and Lower Right (LR). So, for example, in the context of Integral Ecology, if you study animal behavior (UR) you also need to study animal consciousness (UL), animal communication (LL), and animal environments (LR).[16] The remaining four elements discussed below all occur within the quadrants.

Before moving on to these other four elements, let's look at the perspectives offered by the 4 quadrants. To understand the character and consequences of environmental problems, Integral Ecology encourages researchers to solicit perspectives from all domains. We would solicit 1st-person (UL) accounts (including testimony, diaries, letters, documentaries, self-reports, and phenomenological descriptions) from those affected. Hearing the fear, anger, and suffering of people whose health, families, livelihood, or way of life is (or may be) threatened by environmentally destructive practices can have a profound effect on our understanding.[17]

But we cannot deny that people also have beliefs, attitudes, and norms, all of which comprise culture, so we must include 2nd-person (LL) truth claims for a more comprehensive picture. Hermeneutics studies the interiority of cultures. In characterizing an environmental problem, we recognize the need to integrate competing cultural perspectives on what constitutes "nature," as well as what constitutes beauty, goodness, justice, fairness, and compassion. Indeed, what constitutes a problem differs from culture to culture. A conservative culture that emphasizes hierarchy, authority, tradition, and absolute values will usually have a different attitude toward natural phenomena than an atheistic culture that depicts humankind as the source of value and regards nature merely as raw material for serving human needs.

Complex forms of intersubjectivity occur at all levels, not just in human cultures. Biosemiotics and zoosemiotics have contributed a lot to the study of animal culture.[18] Wilber contends that "there is intersubjectivity woven into the fabric of the Kosmos at all levels."[19] Of course, hermeneutics, the analysis of such intersubjectivity, becomes more difficult when we study animal culture, and even more so when we study plant culture, one-celled organisms, and molecules. The farther down the Kosmic scale we travel, the simpler the form of intersubjectivity. But a simple interior does not mean there is an absence of interior. "Certainly when it comes to any integral methodological pluralism worth its name, to dismiss hermeneutics is to dismiss the entire within of the Kosmos—as we said, to completely kill consciousness and culture."[20]

But so far, we only have half of the picture. If we want a truly integral understanding we cannot only study interiors. We must also study exteriors. This requires examining social and systemic phenomena from a 3rd-person perspective (LR). Marx's idea that the economic and technological base determines the cultural and personal superstructure has proven a powerful presupposition for social science (despite the fact that Marx got many other things wrong). Individual behavior is partly a function of social roles determined by socioeconomic factors, which in turn are profoundly influenced by technological innovation. Thousands of years ago the introduction of large-scale agriculture techniques gave rise to urban life, which made a host of social roles possible that were previously unavailable to horticultural and gatherer-hunter societies. Individual behavior may vary, but generally it does so within the constraints of overriding social, political, and economic structures.

Similarly, many ecosystem biologists maintain that individual organisms are primarily functions of their species, which in turn are partially shaped by their prevailing environmental circumstances.[21] For that very reason, habitat protection—not protection of individual organisms—is a top priority for many environmental activists. Here the emphasis is on functional fit: how the many parts of the complex system interact in dynamic but ultimately mechanical ways. The last few decades have seen an explosion in the development and application of the complexity sciences. These emerging fields have revolutionized ecological science, allowing for very sophisticated mapping of multiple nonlinear dynamics. Despite the obvious power of social structural analysis and systems theory, Integral Ecology insists that in and of itself such complex interobjective analysis cannot provide a complete picture. In fact, it is only one-quarter of the picture. According to Integral Theory, all quadrants arise simultaneously ("tetra-mesh") and "tetra-evolve" in interconnected and highly complex ways.

Finally, sciences associated with the UR quadrant, such as chemistry, biology, and zoology, provide powerful, 3rd-person, objective truth claims. Here the focus is on physiological and behavioral dimensions of organisms: for instance, how toxins introduced into a watershed can impact fertility rates of minnows in the local streams and as a result have a domino effect throughout the entire region on organism health and reproduction. Such objective evaluations are the reliable foundation of so much of the ecological sciences and environmental assessments. Many important policies and regulations that protect the natural world and human health are based on these facts.

However, the ideal of investigative objectivity has been successfully challenged by the obvious subjective and intersubjective factors of gender, race, class, and social structure, which inevitably influence the formation of knowledge. Objectivity without presuppositions does not exist. We all have a particular perspective and use a particular method. Nevertheless, scientists following well-established injunctions achieve valid knowledge.

If we examine environmental problems from all of these valid yet partial perspectives, we envision the most current, comprehensive understanding of our environmental problems. We understand that more comprehensive paradigms will emerge, but warn against "new paradigm" enthusiasts who discount or ignore the valid insights of more traditional and modern paradigms. Many people pit the new paradigm against the old, and in the process become polarized against what is perceived as the problem resulting in positions that are anti-Christian, anti-science, anti-Republican, anti-intellectual, anti-corporation, anti-globalization. Integral Ecology explicitly examines the dignity and disaster, the contributions and limitations, of both status quo and new paradigm approaches.

Here is another simple and practical example of the quadrants at work. Let's say I decide to buy flowers for a dinner party. I think: "I need to go to the flower shop." This thought has at least four dimensions, none of which can be reduced to the other. There is my thought and how I experience it (calculating travel time, the joy of shopping, or worrying about the cost). This experience is represented by the psychological structure of formal operational thinking and somatic feelings in the UL quadrant. At the same time, there is the combination of neuronal activity, brain chemistry, behaviors, and bodily states that accompany my thought and my behavior (raising my eyebrows, putting on a coat, getting in the car). All of these are associated with the UR quadrant. Likewise, there are ecological, economic, political, and social systems that co-construct my behavior. These systems supply the flower shop with flowers and determine their price. These are represented by the systems connected to the nation-state and biosphere in the LR quadrant. There is also a cultural context that determines whether I associate "flower shop" with an open-air market, a big shopping mall, or a small store, as well as the various meanings and culturally appropriate exchanges that are associated with giving flowers. This context is represented by the modern worldview in the LL quadrant.

To fully understand and appreciate the thought "I need to go to the flower shop," we cannot explain it in terms of only psychology, or only neurobiology and physiology, or exclusively through social and economic dynamics or cultural meaning. Instead we must consider all of these domains (and their respective levels). Speaking of levels . . .

Levels or waves of development are represented by the arrows bisecting the quadrants.[22] They refer to the stages through which phenomena in all quadrants evolve and complexify. We often represent this evolution with concentric circles overlaid on the quadrants. For example, a dog is physically more complex than an amoeba, and thus located at a higher level than an amoeba in the UR quadrant. We write extensively about levels of development in chapter 4.[23]

Lines of development are another way to describe the distinct capacities and phenomena that develop through levels. In our diagram, lines bisects each quadrant. For example, in the individual-interior quadrant of experience, the capacities or lines that develop include, but are not limited to, cognitive, emotional, interpersonal, and moral capacities. Each of these lines has correlates in the other quadrants. For example, as the cognitive line develops, there are corresponding behavioral and neurophysiological developments in the UR quadrant, and corresponding intersubjective capacities in the LL quadrant and grammatical structures in the LR quadrant. In evolutionary theory they discuss clades (i.e., lines) of evolutionary development (e.g., beetles → wasps → moths → flies).

States are temporary occurrences of aspects of reality. For example, stormy weather is a state that arises in the collective-exterior quadrant (LR) of systems, while joy is a state that occurs in the individual-interior quadrant (UL). There are also brain states (alpha, beta, theta, and delta) and hormonal states in the UR and collective (LL) states such as mob mentality, group-think, and intercorporeal somatic states. States are often used to describe the ways natural phenomena morph (e.g., water turning from ice to liquid to gas).

Types are the variety of styles that arise in various domains. In the UL there are types of personality (Enneagram, Myers-Briggs, Keirsey). In the UR there are blood types (A, B, AB, O) and Sheldon's body types (ectomorph, endomorph, mesomorph). In the LR there are types of ecological biomes (steppe, tundra, and islands), governments (communist, democratic, dictatorial, monarchic, republican, totalitarian), and types of communication exchanges (signifier exchanges, gross and subtle mass-energy exchanges, body language). In the LL there are types of religions and kinship systems (e.g., Eskimo, Crow, Hawaiian, Iroquois, Omaha, Sudanese).

The Integral approach is a 1st-, 2nd-, and 3rd-person set of injunctions and understandings, whereas the Integral AQAL model is a 3rd-person description of interrelated, complex ontological and perspectival dimensions.[24] We encourage you not to confuse the map with the territory—in fact, the map is a performance of the territory, and in important ways they co-arise and are enacted through methodological practices.[25] The model is a useful way to effectively distinguish different aspects of a complex whole. To fully understand the AQAL diagram is to operationalize the distinctions it contains and to recognize these 3rd-person distinctions in your 1st-person awareness. In chapter 10 we offer a variety of practices to support this process.

In fact, the AQAL model is multivarient and can be understood in a number of ways. AQAL is a *map* because it is a series of 3rd-person symbols and abstractions that can guide a person through the contours of their own awareness, as well as through some of the most important aspects of any occasion. It is a *framework* because it creates a space where people can organize and index their own and others' current activities in a clear and coherent manner. It is a *theory* because it offers an explanation for how the most time-tested methodologies and the data they enact can fit together. It is a *practice* because it is not just a theory about inclusion but a practice of inclusion. It involves the meta-paradigm of simultracking humanity's most fundamental methodologies. AQAL can also be practiced in a more personal setting, which results in what is called Integral Life Practice (ILP). It is a set of *perspectives* because it brings together 1st-, 2nd-, and 3rd-person perspectives of 1st-, 2nd-, and 3rd-person realities. It is a *catalyst* because it "psychoactively" scans your entire bodymind and activates or "lights up" any potentials (quadrant, level, line, state, or type) that are not being used at present. Lastly, it is a *matrix* because it combines all quadrants, levels, lines, states, and types in a way that generates a space of potential out of which all of reality can manifest and be accounted. In short, the AQAL model is a 3rd-person map, a postdisciplinary framework for 2nd-person shared language, and a set of 1st-person practices.

The Big Three

Wilber sometimes condenses the UR and LR (domains of It and Its) into the domain of It. The quadrants then become the Big Three (see figure 2.2).[26] The content of these three domains can be characterized as self or consciousness (I, 1st-person), culture (We, 2nd-person), and nature and society (It/s, 3rd-person), which can be analyzed in terms of perspectives such as aesthetics, cultural anthropology, and ecosystem biology or urban sociology, respectively. In brief, the Big Three concern the Good (LL), the True (UR, LR), and the Beautiful (UL).[27]

Wilber posits this tripartite division in part to illustrate that all natural languages contain 1st- (I), 2nd- (We or You, which arises when 1st- and 2nd-person interrelate), and 3rd-person perspectives (It, Its, or They). Knowing

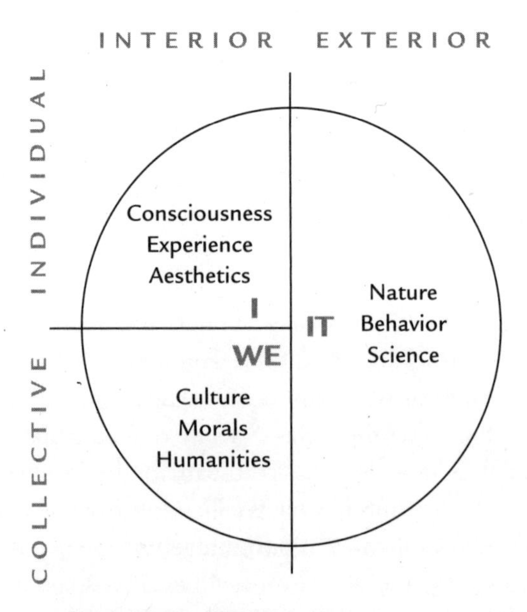

Figure 2.2. The Big Three.

and knowledge claims always occur within the clearing opened by one of these perspectives.

We can understand the Big Three or its expanded version, the quadrant model, both from an ontological standpoint (what do we want to know?) and an epistemological one (what method do we use to know it?). Ontologically, the quadrants indicate that any individual exhibits four different dimensions: (1) an exterior individual facet, (2) an exterior collective facet, (3) an interior individual facet, and (4) an interior collective facet. So a mule deer has 4 quadrants (four distinct dimensions of being) through which it perceives its surroundings (i.e., through 1st- and 3rd- person perspectives in the singular and plural). The mule deer can perceive another deer from the 1st-person perspective (e.g., anxiety or relaxation), and simultaneously within the 2nd-person intersubjective space that it shares with other deer (e.g., recognition) and simultaneously within their 3rd-person perspective (e.g., registering its size, shape, movement, sounds). We can also study the mule deer using methodologies grounded in each quadrant, such as the sciences of animal behavior (UR), cognitive and emotional ethnology (UL), zoosemiotics (i.e., animal communication) (LL), and population ecology (LR).

Quadrants and Quadrivia

There are at least two ways to depict and use the quadrant model: as dimensions or as perspectives. The first, a quadradic approach, depicts an individual situated in the center of the quadrants (see figure 2.3). The arrows point from the individual toward the various phenomena that he can perceive as a result of his own being-in-the-world. Through his use of different aspects of his own awareness or through formal methods based on these dimensions of awareness, he is able to encounter these different phenomena in a direct and knowable fashion. As we will discuss in chapter 8, certain methods are associated with each of the quadrants.

Another way to represent the quadrants is as a quadrivia. "Quadrivia" refers to four ways of seeing (*quadrivium* is the singular). In this approach the different perspectives associated with each quadrant are directed at the phenomenon placed in the center of the diagram. Let's say hundreds of fish are dying in a lake. These fish become the object of investigation, with expertise from each of the quadradic domains evaluating the situation. The arrows pointing toward the center are the methodologies that different investigators use to study the dying fish (see figure 2.4).[28]

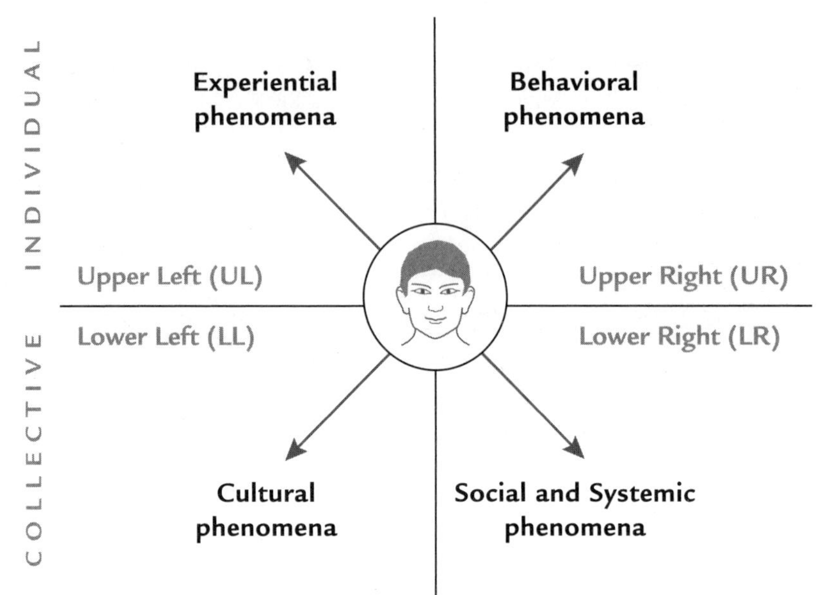

Figure 2.3. The 4 quadrants of an Individual.

In sum, quadrants highlight the four irreducible ontological dimensions that all organisms have, and quadrivia are the four fundamental epistemological perspectives that can be taken on any phenomena. In either case, the 4 quadrants or quadrivia are co-nascent. The result of this is that subjects do not perceive objects. "There is no pure perception in which one entity sees another entity, for that is *already* a first-person perspective on a second or third person. In other words, there is no real space that is not always already a space-arising-as-a-perspective. . . ."[29] Thus, it doesn't even make sense to say that organisms exist and then they see each other, because these are inseparable acts: ontology and epistemology are joined at the perspectival hip, opposite arcs in the same circle. Wilber importantly explains: "This does not mean 'to be is to be perceived,' for that implies there is being per se that can be perceived; nor is this to say that perception creates being, for that implies that perception itself exists apart from something perceived. This is rather to say that being and knowing are the same event within the set of perspectives arising as the event."[30] The only reason we are inclined to think that ontology and epistemology (being and knowing) are separate is the way we shift our awareness from one dimension-perspective to another.

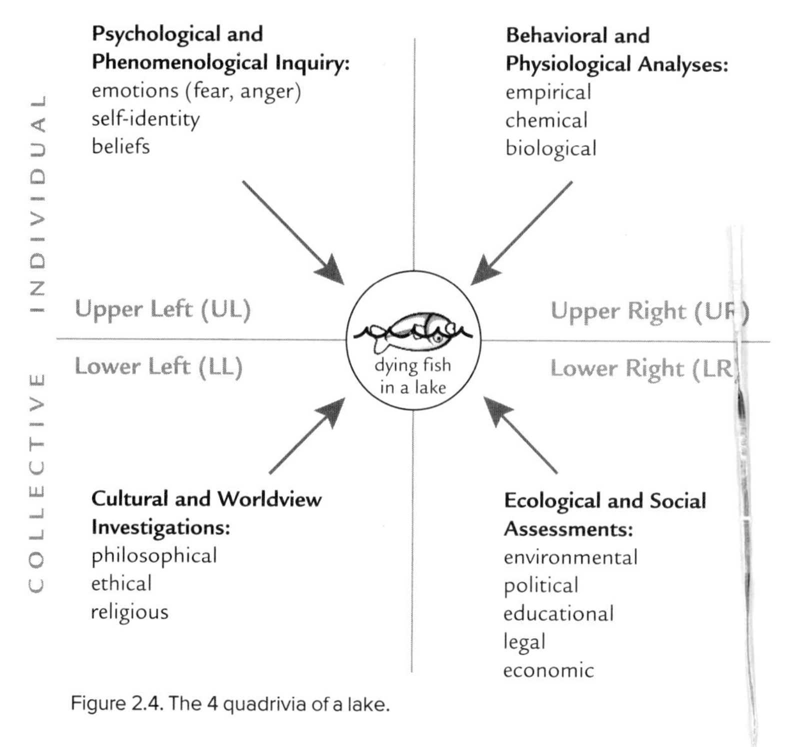

Figure 2.4. The 4 quadrivia of a lake.

"There is simply no perception that is not also a perspective, and therefore no appearance of being that exists other than a phenomenal perspective."[31] In chapter 6 we provide several examples of quadrants and quadrivia that further illustrate their simultaneity.

Beyond Flatland: Integral Theory's Recovery of Forgotten Depth

As discussed in our introduction and chapter 1, Integral Theory arose in part because of the widespread demand that our interior domains—experience and culture—receive commensurate acknowledgment. The addition of interiority transforms the two-dimensional plane, "flatland," into a three-dimensional, full-bodied one. Likewise the quadrant graph is not a Cartesian grid but rather a set of signifiers that represent depth and complexity in each domain. One way to understand this depth is to use a Buddhist mandala. A mandala is a psychocosmogram that represents the interconnectedness of all phenomena.

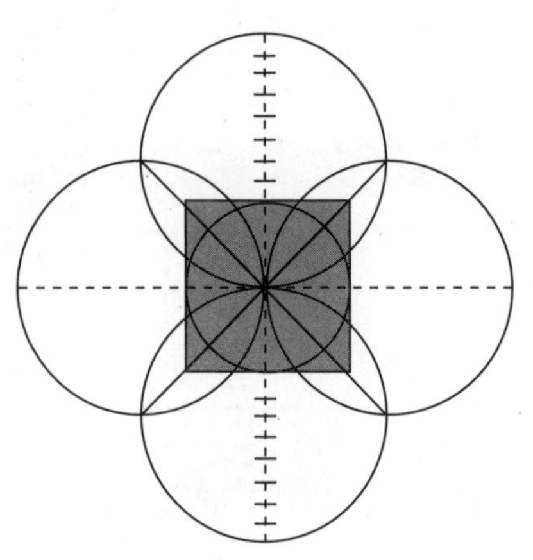

Figure 2.5. Basic outline of a mandala.

Figure 2.5 shows the mandala's two-dimensional outline. The mandala is divided into quadrants, which represent the four directions. These quadrants are bisected by lines and contain levels demarcated by short lines. Figure 2.6 is the same mandala with additional symbolic features. It is more intricate, maybe even more beautiful, but it is still flat. It is still "flat" because it is two-dimensional. Figure 2.7 represents the same mandala but with color and complex shapes. This magnificent and enormously detailed stage of the mandala captures our attention.

Many people conclude that figure 2.7 exhausts the content of the mandala. In fact, however, this mandala is the architectural floor plan of a three-dimensional palace and its surrounding cosmos. Figure 2.8 shows this palace. We must tip this two-dimensional mandala on its side in order to see this palace arise from the flat plane. There is a corresponding interior to each external feature hitherto unnoticed. The nested squares actually represent the walls of the palace, and the circle embracing the palace represents the full extent of the cosmos structured and sustained by the internal order of the palace. Discovering that this mandala is actually a palace is akin to rediscovering our depth, our interiors, in a flatland world.

Aspiring Integral Ecologists may inadvertently reduce the Integral Model with flatland thinking because phenomena arise according to the perspectives afforded by each quadrant, and those perspectives change according to the developmental level of the person interpreting the phenomena in question.[32] As a result, it is important to remember that the quadrants are multidimensional, like the palace, and there are many floors in a multistory

Figure 2.6. Symbolic features of a mandala.

Figure 2.7. Detailed mandala.

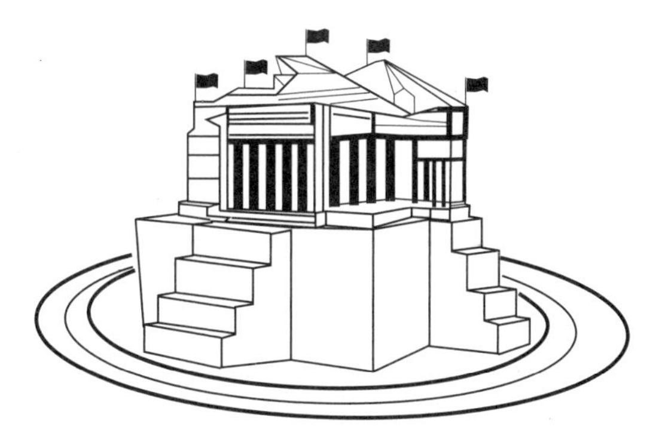

Figure 2.8. Three-dimensional mandala.

perspectival holarchy of quadrants. Reading, studying, and even memorizing Integral Theory's maps and diagrams will not in and of itself generate integral thinking or integral awareness, actions, or solutions. Such awareness requires considerable effort and developmental achievement; you must engage in appropriate practices in all quadrants in order to facilitate the emergence of vision-logic. We will discuss interior development at great length in chapter 4.[33] We provide numerous practices in chapter 10.

Where Is Truth Located?

We discover different kinds of truth claims from different domains, by using different methodolgies. These methodologies and the claims they produce are at work in each quadrant. There are at least two primary methodological families associated with each quadrant and different criteria for truth claims associated with each quadrant. For example, "objective" truths, assertions made from the 3rd-person perspective, correspond to the UR quadrants. But in the UL quadrant, the criteria are sincerity and integrity. (See figure 2.9.)

The quadrants—as content and as perspectives—are illustrated in the four ways in which many North American universities organize their faculties: fine arts (UL), humanities (LL), natural sciences (UR), and social and systemic natural sciences (LR). Figure 2.10 shows the ways that many uni-

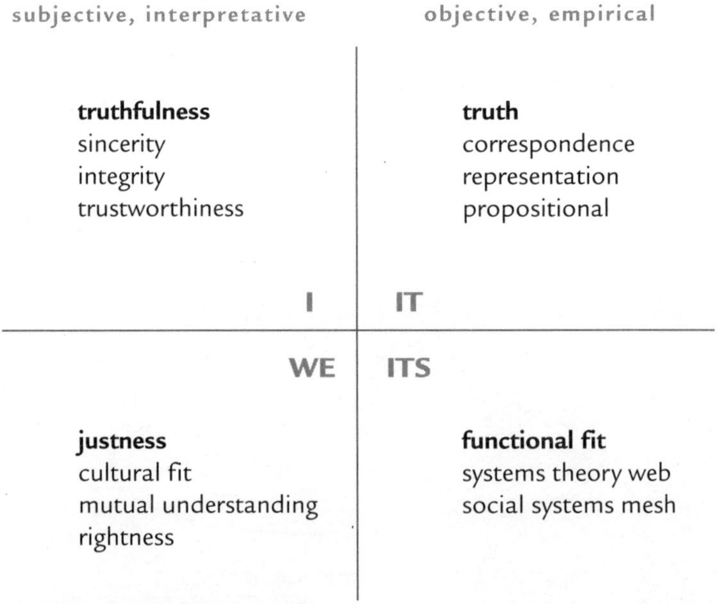

subjective, interpretative	objective, empirical
truthfulness sincerity integrity trustworthiness **I**	**truth** correspondence representation propositional **IT**
WE **justness** cultural fit mutual understanding rightness	**ITS** **functional fit** systems theory web social systems mesh

Figure 2.9. Validity claims.

versity departments are distributed across the quadrants. Even if they cannot teach students every useful perspective, at least some in academia—especially humanists and many social scientists influenced by constructive perspectivalism—have acknowledged modes of inquiry that are not included in university course catalogues.[34] Moreover, humanists, artists, and social scientists have reasserted the validity of their methods of inquiry against the tendency of natural scientists to depict their research methods as incapable of generating warranted truth claims. Many positivists have depicted interior individual truth claims (UL) and interior cultural truth claims (LL), as little better than emotive, and thus nonrational.[35]

Integral Perspectivalism Is Not Relativism

As we noted earlier, we do not think that all perspectives and all truth claims are equal, or that scientific methodology is simply a different kind of narrative, or that ranking knowledge claims hierarchically necessarily involves domination and repression. Postmodern theorists assume that their

	INTERIOR	EXTERIOR
INDIVIDUAL	**Arts (studio and performance)** English (creative writing) Screenwriting Music composition and performance Studio art (painting, sculpting, photography, printmaking) Theater and Dance I	**Natural Sciences** Biochemistry Biomedical engineering Cell and molecular biology Chemical engineering Mathematics Mechanical engineering Psychology Physics IT
COLLECTIVE	WE **Humanities** Anthropology Classical studies Cultural anthropology Literature and languages History Philosophy	ITS **Social and Systemic Natural Sciences** Business Ecology and evolutionary biology Economics Civil/environmental engineering Astrophysics and Astronomy Political science Sociology

Figure 2.10. The departments at a typical university.

perspectivalism and multiculturalism are more inclusive than—and hence superior to—the perspectives utilized by moderns. But strangely, postmodernists deny the superiority of their own perspective, so they end up in a performative contradiction. Integral Ecology resolves this contradiction by reaffirming the reality and importance of holarchy: some truth claims are better—more inclusive, more comprehensive, more insightful, more generative—than others. Integral Ecology affirms that we must take truth claims arising from the natural sciences very seriously in regard to environmental problems. Natural science is not a kind of poetry. Science and poetry involve very different methodologies. Nevertheless, we cannot allow natural scientific truth claims to trump the truths of different methodologies.[36] We can contest objective claims with objective methodologies, and judge subjective claims with subjective methodologies. Within each domain there are claims that are better than others. But you cannot judge a subjective claim with an objective methodology, because the criteria for truth claims are domain dependent.

The Quadrants Complexified

Two concepts significantly influenced Wilber's original account of the quadrants. First, there is no One True Way of seeing; instead, life arises differently depending upon your perspective. The quadrants represent four irreducible perspectives that you need to understand the interior and exterior of individual and collective phenomena. Second, the universe is made up of holons. Holons are parts that form parts of ever-greater wholes. They are parts of larger wholes, and are wholes containing subservient parts. The structure of holons has a lot in common with the quadrants. They both have individual and communal aspects; and they both have interiors and exteriors.[37]

A few years later, however, Wilber concluded that the universe is composed not so much of holons as of the perspectives belonging to holons. To exist is to be a perspective. Wilber's new emphasis on perspectives led him to conclude that the perspective represented by each quadrant can be studied by other perspectives. Each of the quadrants can be viewed by the inside or the outside within the quadrant. Hence, there are at least eight native (or indigenous) perspectives associated with the quadrants.

Consider the 1st-person, experiential perspective represented by the UL quadrant. The 1st-person human perspective is the immediacy of what "I" experience: my feelings, memories, thoughts, etc., which I can reflect upon. If I want to understand the structural features of my 1st-person experience, I can examine these very same thoughts and feelings from a 3rd-person perspective (I can study developmental models). My interior exhibits an exterior, which Wilber calls "the look of a feeling." I can evaluate the structure of my experience—including my developmental level in any number of lines (cognitive, moral, aesthetic, psychosexual, interpersonal).

In other words, phenomenology studies direct 1st-person experience. Structuralism studies the 3rd-person structures and patterns of individual direct experience. Within the quadrants there are 8 perspectives and methodologies. The inside and outside of the four major perspectives actually accommodates all the major methodological families that humans have developed over the last 2,000 years.

You can access the inside or outside view of a holon in each quadrant through a particular methodological family. Each methodology discloses an aspect of reality that other methods cannot. Therefore, the findings of any one method are not accountable to the criteria of other methods. To leave out any of these perspectives (or methods of inquiry) is to fall short of a truly integral understanding. For Wilber, the 8 indigenous perspectives are consistent with his post-metaphysical approach, which avoids positing realities independent of the viewer.[38] Wilber's participatory approach and his epistemological and ontological pluralism are consistent with his view that the Kosmos is constituted of interacting perspectives.

First let's define the 8 methodological families that disclose the inside and outside view of each quadrant, and then revisit the principles behind Integral Methodological Pluralism, which is the framework that guides the discriminating use of these 8 perspectives and their methodological families. *Phenomenology* explores direct experience (the insides of individual interiors). *Structuralism* explores patterns of direct experience (the outsides of individual interiors). *Hermeneutics* explores mutual understanding (the insides of collective interiors). *Ethnomethodology* explores patterns of mutual understanding (the outsides of collective interiors). *Autopoiesis theory* explores self-regulating behavior (the insides of individual exteriors). *Empiricism* explores measurable behaviors (the outsides of individual exteriors). *Social autopoiesis theory* explores self-regulating dynamics in systems (the insides of collective exteriors). *Systems theory* explores the functional fit of parts within a whole (the outsides of collective exteriors). These methods and perspectives seem primarily suitable for human holons, but they are applicable to holons at all levels.

These perspectives exist on a continuum from folk methods to more formal ones. Folk methods are the ways that we are always using 1st-, 2nd-, and 3rd-person methods. For example, when you decide whether to go to bed, you use phenomenological inquiry to assess whether you are tired. When you wonder if it's raining outside, you use empirical observation and look out your window.[39] The more formal methods consist of the established quantitative and qualitative traditions listed above. Unlike the folk methods, which can be performed by anyone, the formal ones require different levels of psychological development. For example, to accurately perform empiricism you need the formal operations associated with a rational worldview. To perform hermeneutics (authentically take the role of "other") you need at least a worldcentric perspective-taking capacity. Likewise to engage in systems theory you must have the post-formal capacities associated with systematic operations.

These 8 primordial perspectives and their accompanying methods all have a partial perspective to offer. This is the first principle of Integral Methodological Pluralism (IMP)—*nonexclusion,* or the acceptance of truth claims that pass the validity tests for their own paradigms in their respective fields. Nonexclusion is guided by the intuition that "everyone is partially right!" and that everyone brings forth and discloses a different and partial facet of reality. This assertion resists method hegemony, according to which there is only one or a few reliable ways to interpret any given phenomenon.

But we are well aware that the assertion of nonexclusion, without any other constraints, could very easily lead to relativism (according to which no one view is better than any other). But some views are better and more inclusive than others! Thus the next principle of IMP—*enfoldment*: some practices are more inclusive, holistic, and comprehensive than others.

Last but not least, IMP adheres to the principle of *enactment* (phenomena disclosed by various types of inquiry depend in large part on a host of factors that influence the researcher). Enactment reaffirms the participatory approach of the Integral approach.

By evaluating, questioning, and probing the findings of many different methodologies, Integral Ecologists have a distinct advantage—compared with those utilizing only one or a few methods—because an integral diagnosis is more comprehensive and thereby produces more comprehensive solutions. For this reason, Integral Ecology encourages teams of experts from different fields to work together. Experts would provide their interpretation according to their investigative methods. Then they would adopt the perspective of other team members. In this way, team members gain insight into the various facets of a problem and may notice something that previously eluded them.

Social Autopoiesis and Communication: A Complex Example

Applying the 8 perspectival families enriches and complexifies our understanding of environmental phenomena. Let's consider a complex example that applies to ecology: social autopoiesis theory and its assertions about the relationship between human society and the natural environment.

Social autopoiesis is the self-maintenance and self-reproduction of collective or communal systems—either human (societies) or nonhuman (ecosystems). These systems are closed and internally self-defining and self-organizing, so they do not enter into direct causal relations with their environment. Instead, they engage in structural coupling. Structural coupling occurs when a social system filters its environment to determine which aspects of the environment the system will resonate with. Certain systems require selectivity to distinguish themselves from otherwise vast and overwhelmingly powerful environments. According to the German social theorist Niklas Luhmann, social theory "has to treat all facts in terms of unity and difference, i.e., in terms of the unity of the ecological interconnection, and the difference of [social] system and environment that breaks this interconnection down. As far as the ecological question is concerned, the theme becomes the unity of the difference of system and environment, not the unity of an encompassing system."[40]

In a remarkable passage in *Ecological Communication*, Luhmann notes that modernity broke with the tradition of viewing the environment as an encompassing body,

> as a living cosmos that assigned the proper place to everything in it. These traditions had in mind the relation of containment of little bodies within a larger one. Delimitation was not viewed as the restriction of possibilities and freedom, but instead as the bestowal of form, support, and protection. This view was reversed only by a theoretical turn that

> began in the nineteenth century when the terms "Umwelt" and "environment" were invented and which has reached its culmination today: [social] systems define their own boundaries. They differentiate themselves and thereby constitute the environment as whatever lies outside the boundary. In this sense, then, the environment is not a system of its own, not even a unified effect. As the totality of external circumstances, it is whatever restricts the randomness of the morphogenesis of the system and exposes it to evolutionary selection. The "unity" of the environment is nothing more than a correlate of the unity of the system since everything that is a unity for the system is defined by it as a unity.[41]

The concept of umwelt, literally "world-around" or "environment," and often used as an "organism's subjective universe," comes from the biologist Jakob von Uexküll.[42] He postulated that organisms create worlds in which things that matter to them can appear. What arises within the umwelt of a flea is radically different from what arises within the umwelt of the animal it feeds on. Umwelt theory often studies the inside of the UL (i.e., the phenomenological experience of the organism) or the inside of the LL (i.e., the hermeneutic meaning of organisms and their worldspace of interpretations). Many biosemioticians use Uexküll's concept of the functional cycle to link umwelt theory to autopoietic processes as associated with Maturana and Varela's research that studies the inside of the UR (i.e., the self-enclosed and self-regulating activity of organisms).[43] We use "umwelt" to refer both specifically to an organism's phenomenological experience of itself and broadly to the capacity of organisms to have 1st-, 2nd-, and 3rd-person awareness of their world. Thus, they simultaneously have a phenomenological interiority, interpret their surroundings through horizons of meaning-making, and use their senses to perceive the environment.[44] In chapter 6 we explore umwelt theory and its contribution to Integral Ecology at length.

Drawing on umwelt theory and autopoiesis theory, Luhmann devised social autopoiesis, which studies the inside of human social structures (LR). According to Luhmann, modern society is a self-producing and self-absorbed system that differentiates itself from the complex natural environment while drawing upon that environment's resources. Luhmann claims that beginning with agriculture, which destroyed existing vegetation in order to plant human-friendly crops, people have redefined, reorganized, and transformed the "natural" world to meet social needs.[45] In the self-producing and self-enclosed modern social system, "nature" is seen and used primarily in terms of its utility. Abstract concepts of a biosphere arrived late on the scene, partly in response to concerns about how industrial society used or abused the environment.

Faced with this damage, environmentalists overlooked how human society differs from the environment and focused on how humans are unified with it. They focused on how complex organisms need an environment to survive—including breathing clean air, drinking pure water, and eating healthy foods. In this same vein, environmentalists argued that humans and human societies are functional parts of biosphere. Following that argument, humans must change the way they treat the biosphere, or risk self-destruction. This conception of humans as parts of the biosphere is misguided, however, because it ignores humans' (and some animals') noospheric aspect. Humans are different from other organisms. They have developed a unique, complex language. Luhmann admits (as we do) that there are environmental problems that society must deal with, yet he insists that we need an adequate conception of human society to address these problems (one that does not diminish our uniquely human capacities and one that does not conceive of us as mere parts, empty of subjectivity). Unfortunately, most ecological literature shows little interest in theoretical reflection. "As if," Luhmann writes, "caring about the environment could somehow justify carelessness in talking about it."[46]

Luhmann maintains that whereas all systems—human or nonhuman—sustain themselves by highly mediated exchanges with the larger environment, human social autopoiesis is additionally and centrally characterized by communication. Luhmann's theoretical perspective in this context shows little interest in the experience of individuals. Indeed, for him the social system is not composed of individuals, but of self-referential communication at multiple levels.

Modern societies are increasingly sensitive to environmental perturbations, especially those due to human activity. Even though recursively closed and attempting to seal themselves off from environmental perturbations, autopoietic social systems are open in some mediated ways to the environment because they require it as the source for water, air, food, and raw materials for industry. But even when environmental perturbations become insistent, they can occur only in terms of the system's own frequencies. Advanced societies contain many different frequencies, which are constituted by the communication necessary for defining and coordinating myriad social roles. Luhmann's crucial claim is this: physical, chemical, or biological facts "create no social resonance as long as they are not the subject of communication. Fish or humans may die because swimming in the seas and rivers has become unhealthy. . . . As long as this is not the subject of communication, it has no social effect. Society is an environmentally sensitive (open) but operationally closed system."[47]

Polluted water has an effect on the human organism that drinks it, but there is no social effect without communication. However, while there is no *social effect* on the mental noospheric level, there is one on a physical biospheric level (e.g., cells die from the toxins in the water). The LL and LR of nonhuman organisms are "lit up" when sick and dying, but ours are not. In this regard Luhmann omits too much. In many ways Luhmann regards the environment as a social construction of society. But other organisms are aware and signaling/communicating with one another about what is going on—organisms move away from poisons. Thus, we apply the following quote from Luhmann to both human and nonhuman organisms: "Communication is an exclusively social operation. On the level of this exclusively social mode of operation there is neither output nor input. The environment can make itself noticed only by means of communicative irritations or disturbances, and then these have to react to themselves."[48]

Within the context of social systems it is helpful to keep in mind that Integral Theory distinguishes between inside-outside and internal-external.[49] Inside-outside refers to being within a boundary. Internal-external refers to being a constituent element. For instance, in a social system, humans are *inside* it, their communication is *internal* to it, but neither they nor their communication is a *part* of the system. As Wilber points out, the aspects of societies that have systems of exchanged signifiers have no organisms internal to them. Exchanges of artifacts and communications are internal to social systems, not organisms. The aspects of societies that are systems of exchange have no organisms internal to them (parts of them), but societies are not merely systems of exchange. Humans are not internal to social systems or ecoystems (not parts of it). They are inside the system boundary, but as members, not parts. Also, organisms can be inside the same ecosystem and members of different cultures as well as inside different ecosystems and members of the same culture.[50] In sum, ecosystems are composed of the interactions and exchanges of organisms. Organisms are not strands in the web, their communications are. "*Organisms* are *members* of a system, their *transactions* are *components* or parts of the system."[51] Organisms can be physically inside an ecosystem, but only their interactions are part of or internal to that system. "Thus, compound individuals or organisms are *inside* an ecosystem when their exterior intersections are *internal* to the ecosystem—that is, when their interobjective exchanges are governed by the nexus-agency for the ecological network—and they are *outside* an ecosystem when their interobjective exchanges are not governed by that particular system."[52] As a result, the only thing that actually is *inside* and *internal* to an ecosystem is the previous moments of that ecosystem—the prior communications of individual organisms, which are members of it.[53]

In the game of chess, the rules are internal to the game, but the human beings playing the game are not internal to the game. But they are inside the game (that is, inside the boundary of who is playing) as members, even though not as constituent parts. Humans play the game and their communication is internal to it, but they themselves are not internal to it; they are members of the game. If a player is caught breaking an internal rule, that player is kicked out of the game. For these reasons, it is profoundly misguided to conceive of humans as parts of, or strands in, the web of life. Organisms are not strands in a web; rather strands are constituted by the intersections of organisms. Human society communicates and intersects through language, but language is not an element inside the members of the social system. Likewise, organisms are not internal to an ecosystem even though they live within its physical and conceptual boundary. Thus, we are not internal to (i.e., parts of) the ecosystem, even though we are inside it. If an individual holon does not follow the internal rules of social holons, that holon will be excluded, kicked out, put in jail, killed, etc. Thus, if we humans do not follow the deep patterns or structures of Gaia, we will be excluded. Although we are affecting the planet at most levels of organization, we are mainly altering planetary structures that pertain mostly to us. So we will kill ourselves before we kill the planet—though we are currently having a negative impact on a variety of species.

Let us now consider how a city might react to the perturbation or disturbance called "benzene pollution of the drinking water supply." One morning, someone says, "The water smells and tastes funny!" This person calls the city water department and says, "What's wrong with the water?" This communication brings the irritant into social existence. According to social autopoiesis, only a communicated environmental disturbance can matter to society. For more to occur, this communication must now react to itself—it must generate more communication. Someone communicates the disturbance to the water treatment plant: "Test the water!" Later, they announce: "The water is heavily polluted with benzene!" The news reaches the mayor's office. Questions arise: "Where is the benzene coming from? When will the water be safe again?" After an investigation, the city communicates that a factory explosion upstream released benzene into the river, and the state governor's office announces its intention to call up the army reserves to distribute bottled water. Communication begets communications to coordinate all necessary actions to restore the equilibrium. If the perturbation is too great, society may not regain equilibrium, in which case it either collapses or transforms into a different kind of social system.

A complex systems theoretical account of this event would focus on externally observable phenomena, including the behaviors of people and institutions, and how those behaviors influence each other. The complex systems theorist examines these behaviors from a 3rd-person perspective. Luhmann, as a social autopoiesis theorist, also wants to understand behaviors, but from the inside (not the interior) of the social system. For him, the inside is constituted by communication, which he defines not in terms of how it is experienced but in terms of how it *coordinates actions* necessary to sustain the system's equilibrium.

As Luhmann states, autopoietic systems can structurally couple with their natural environments in ways that make the environment unsuitable for that system later on. For example, a society that denudes forested hills in order to construct irrigation trenches may go under once the eroding hillsides fill up those trenches with mud. For Luhmann there is a gap between "ecological consciousness" and effective social communication, which has a significant "resonance threshold." Ecological social communication is about whether current dealings with the natural environment will undermine the social structure that depends on it. Such a communication of possible disaster may or may not resonate with the frequencies that define social subsystems. So far, at least, it has been an inevitable aspect of society that the next step in autopoiesis is more important than concern for the long-term future.[54] This fact helps to explain the limited success enjoyed by movements proposing that modern societies adopt the "precautionary principle" or work toward "sustainability."

Let's consider global climate change, which Western society and its systems have communicated about for at least three decades. Because the effects of climate change have not yet sufficiently perturbed the social system, communication about it does not resonate with many subsystems, which are more attuned to communication about the social value of things like SUVs. When climate change disrupts the equilibrium of subsystems and eventually perturbs our larger social system, it will resonate on many different frequencies.[55] Presumably, at this time we will communicate to address the irritation. Luhmann maintains that a competent intervention depends on two factors. First, there must be sufficient competence for "selective behavior," or relative freedom in the face of the natural environment. Second, there must be sufficient communicative competence to carry out the selected intervention.[56] In chapter 7 we will examine the ecological crisis in light of social autopoiesis theory, insofar as it highlights a very important dimension of this crisis (and an important part of the solution).

Our treatment of Luhmann's communicative autopoietic social theory barely scratches the surface of his wide-ranging work. Nevertheless, we learned some important lessons. First, we must distinguish between individual autopoiesis and systemic autopoiesis. Organisms (including humans) are not merely more complicated kinds of social organisms that are "parts" of the biosphere. In chapter 3 we examine how Integral Theory clarifies part-whole relations to characterize humanity's relation to physiosphere, biosphere, and noosphere more appropriately. Second, environmental problems are not self-announcing. They become problems only by way of communication.

But as a social theorist, Luhmann tends to ignore the experiential and cultural (Left-Hand quadrants) contributions to defining environmental problems.[57] Integral Ecology, in contrast, insists that because recognition, definition, and proposed resolution to such problems tetra-evolve, interiority (UL, LL) must be taken into account. Third, social theory maintains that modern society has developed toward scientific modes grounded in the perspectives of the Right-Hand quadrants. Integral Theory adds that individual and cultural evolution have co-evolved with the Right-Hand domains.

Fourth, Luhmann asserts that society is relatively free to develop "selective behaviors" when intervening in environmental problems. This indicates that humans are capable of choosing to live in harmony with the environment. Hence, environment or nature is not the determining factor. This means, in turn, that environmental science as such cannot give rise to ecological consciousness (UL, LL), which belongs to the noosphere. Environmental science deals with phenomena occurring in the biosphere, not the noosphere. No matter how many ecosystem processes we may come to understand, those processes in and of themselves occur in a domain of reality other than the noospheric domain. Fifth, since ecological interventions require competent ecological communication, Luhmann in effect and perhaps in spite of his social theoretic stance makes a case for Integral Methodological Pluralism, which emphasizes the need for a multiperspectival examination of environmental problems, and for Integral Ecology's understanding of developmental models of individuals, cultures, and societies. Unless Integral Ecology considers different kinds of "frequencies" (levels of development), much of our communication about the environment will not resonate.

Summary

In this chapter we introduced the AQAL model and presented some detailed examples of one of its crucial components, the quadrants, in its ontological and perspectival aspects. Individual holons exhibit features that we can categorize by quadrant, and we can gain access to these features by approaching them from appropriate perspectives and methods. Using multiple perspectives, including those that explore the interior domains of complex holonic realities, helps us avoid the widespread tendency toward a two-dimensional or flatland conceptions of the Kosmos and its constituents. We would like to reemphasize that many environmentalists and ecologists, despite their own passionate feelings and concerns, regard the natural world merely as a complex system of matter and energy, lacking interiority. Even when environmentalists do acknowledge interiority, they do so informally and rarely include these interiors in their problem-solving analysis. In this way, they adopt the flatland conceptual scheme that belongs to industrial ontology, which depicts everything—including human beings and other life forms—primarily as material participants in complex systems.

A truly integral ecology moves beyond flatland, so as to acknowledge the interiority characteristic of human and nonhuman individuals and systems. Our brief description of Luhmann's idea of ecological communication has demonstrated, however, that his profound social theory could provide many insights that are consistent with an integral account of the Kosmos as happens in the LR.

3

A Developing Kosmos

> Wilber has sounded a powerful call for us to awaken to the evolutionary process taking place within us, within the universe, not in some distant future but right now. This evolution is fundamentally open and creative, and therefore, at every turn, incomplete and uncertain. We live in systems within systems, contexts within contexts indefinitely, and the systems are constantly sliding and the contexts shifting. The vision of an open universe unfolding and enfolded upwards and downwards without end effectively removes all bases for certainty and completeness. . . . The evolution that we are all part of excludes nothing, not even the contexts that bound our understanding and awareness.
>
> —KAISA PUHAKKA[1]

The Kosmos Is Developing

Because there is widespread agreement about the need to include alternative perspectives in analyzing complex phenomena, our discussion of perspectivalism is relatively uncontroversial. Some may understandably prefer a different categorization of perspectives, or think that the quadrants (and the other elements of the AQAL model) cannot accommodate all the needed perspectives. Others may conclude that our approach is too radically constructivist; still others will regard it as insufficiently so. Nevertheless, our postdisciplinary approach is consistent in many ways with approaches being developed by others.[2]

In this chapter we introduce an idea that was once controversial: that development is a universal feature of the Kosmos. Here we focus on development in the Right-Hand quadrants. We also tackle difficult questions regarding humanity's relationship to the natural world. Despite the widespread tendency to depict humans as part of nature as biosphere, we maintain that humans are actually "part-ners" or members of the biosphere, not parts. This may seem like a mere quibble, but it has real-world impact. Parts are completely subordinate to their senior holons; members have a measure of autonomy in relation to their community. There are both whole individuals and whole systems: compound individuals and compound networks.[3] Each is made of in-kind parts: individuals are made up of smaller individuals (e.g., atoms, cells) and systems are comprised of smaller systems (e.g., a watershed, a bioregion). Renowned environmental designer John Tillman Lyle points out: "We need to recognize that every ecosystem is a part—or subsystem—of a larger system and that in turn includes a number of smaller subsystems. It also has a number of linkages to both the larger and the smaller units."[4] What links the parts in individuals with parts in systems is membership. For example atoms that are parts of individual holons partner with each other and become galaxies, which are parts of more complex systems like ecosystems. If we do not clarify these philosophical points, we could easily, and perhaps unwittingly, construct a counterproductive recipe for such approaches as ecofascism.

The Kosmos Is Developing in All 4 Quadrants

In Integral Theory holons are defined as wholes that are simultaneously parts. So holons are "whole/parts." There are individual holons such as atoms, cells, and organisms. When individual holons get together they form social holons. Figure 3.1 maps major developmental stages that have taken place in individual and social holons from the Big Bang to now. The arrows that bisect each quadrant represent development over sometimes vast periods of time. Of course this diagram greatly oversimplifies complex developmental lines and the quadrant relations among them.[5] As a heuristic device, the diagram represents the ways different levels of reality develop simultaneously in each quadrant. Because development occurs simultaneously in all quadrants, we say that holons tetra-evolve. Let's explore how development is at work in nature, society, and culture. To do so, we will refer to the diagram and provide a few examples.

Soon after the Big Bang, simple atoms such as hydrogen formed. Atoms have an exterior aspect (UR-level 1), which is studied by physics. They also have a very limited interior aspect (UL-level 1), which Wil-

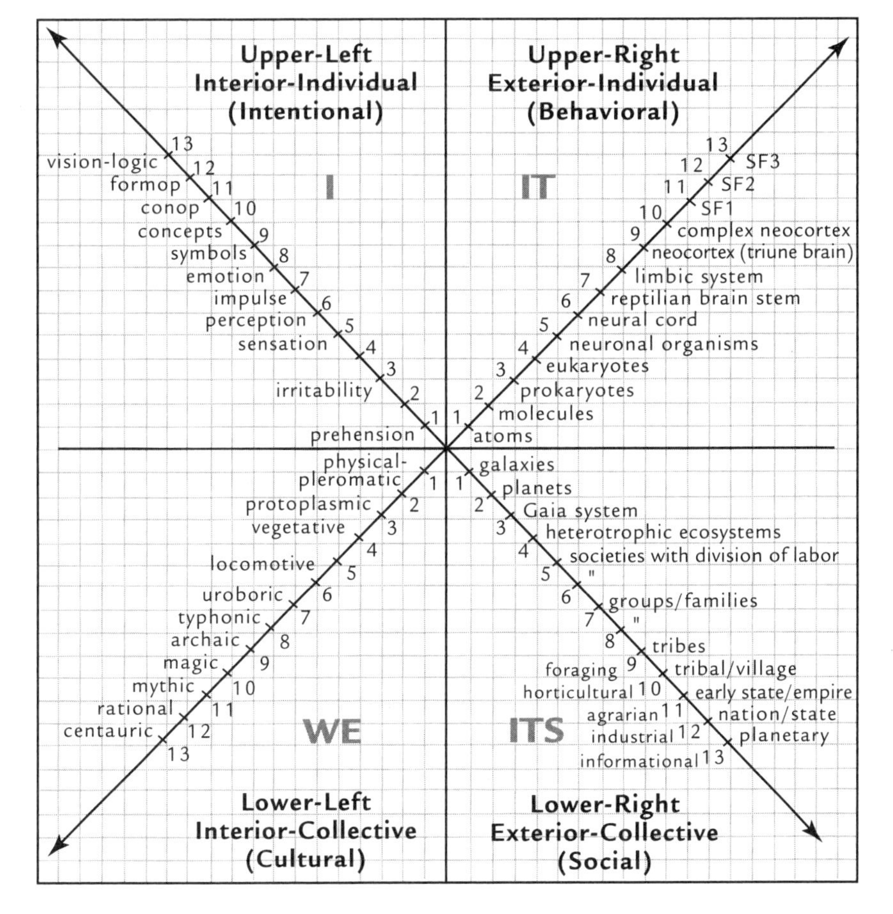

Figure 3.1. The AQAL map.

ber, following Whitehead, calls "prehension." Prehensive atoms are the constituent elements of galaxies (LR). Atoms and galaxies co-evolved. Corresponding at that same level in the LL (interior-collective) is the most primitive kind of intersubjectivity, characteristic of prebiological phenomena—"physical-pleromatic."

A couple of rungs higher, at level 3, prokaryotic cells appeared (UR). They are capable of experiencing irritation (UL). Their intersubjective dimension is still quite limited in complexity (LL), but no more so than their exteriors. Two rungs higher, at level 5, we see neuronal organisms. These organisms co-evolved with societies that could achieve division of labor. These complex organisms can and do experience sensation and an increasingly complex intersubjectivity. At level 8, the limbic system created more complex primates and hominids, and along with them family and tribal groups evolved. Emotions and typhonic culture arose with these Right-Hand developments as humans emerged from their hominid origins. In the final stages the human brain, society, cognitive capacities, and culture developed simultaneously. Although different scholars might include additional developmental stages and omit others, many agree that cosmic history and terrestrial history have evolved in ways consistent with how they are depicted in these four developmental sequences. The relationships among the developmental stages of this diagram are not always self-evident, in part because they omit a great deal in order to be useful as heuristic devices.[6] Nevertheless, the stages in these sequences roughly correspond to one another.

Only recently have large numbers of scholars accepted a developmental conception of the Kosmos. Ancient Greek philosophers assumed that the Kosmos was both eternal and unchanging. Composers of the Hebrew Bible believed that God created the heavens and earth, after which things (including species) remained relatively unchanged. Few seriously considered that the Kosmos, life forms, and human beings had developed and would continue to do so. If anything, the ancients regarded the Kosmos, and particularly human culture, as degenerating when compared with some nobler pristine origin. Today, many environmentalists follow in that same vein, longing for a more pristine past. Many regard modernity not as progress but as a departure from a time when humans were "closer" to nature.[7]

In 18th-century Europe, in part because modern science and industry fueled the promise of improving the human condition, the idea of development took hold. G. L. Buffon spoke of the "common descent" from animals, Georges Cuvier noted that the Earth had experienced several catastrophes that led to widespread species extinctions, and Jean-Baptiste de Lamarck developed an explicit evolutionary theory. Informed by these claims, Charles Darwin and Alfred Wallace devised their own evolutionary theory, which emphasized that speciation occurred because of small, cumulative, accidental changes that contributed reproductive fitness to an emerging species.[8] The idea that life on Earth had evolved, and that humans descended from animal ancestors, raised a firestorm of debate that has not yet ended.

Darwin admitted that his bold hypothesis was undersupported by empirical evidence, including the fossil record, which lacked any "transitional" species. Indeed, few if any such species have yet been found. Nevertheless, the stage was set for neo-Darwinism, which used genetics to explain microevolution so effectively that most scientists assumed it would eventually explain macroevolution as well. This has not happened yet, which gives rise to problems for neo-Darwinism, problems that in no way detract from Darwin's magnificent contributions.[9] One such problem is the existence of DNA, the genetic code responsible for reproducing living beings.[10] No one has any idea how such an enormously complex code, unknown to Darwin, could have arisen by chance encounters of amino acids. So daunting is the task of explaining DNA that the world-famous defender of atheism, Antony Flew, recently became a theist who now believes in the existence of a God similar to Aristotle's conception.[11]

Regardless of how it does so, life does develop. Biological evolution involves ever more complex life forms, even if this complexity is merely an accident. Yet for a long time developmental models were inconsistent with the second law of thermodynamics, according to which all energy moves toward disorganization. How could we be evolving if all energy moves toward disorganization? Entropy, the name for this universal dissipative tendency, led many—already brooding over Darwinism and Nietzsche's proclamation of the "death of God"—to profound personal and cultural despair that characterized *fin de siècle* Europe. Were we dissipating toward disorganization or were we evolving?[12]

The conceptual breakthrough that partly reconciled this confusion came from the Nobel laureate in physics, Erwin Schrödinger, in *What Is Life?*[13] He postulated that although energy moves toward entropy, it can be captured to permit organization and development. For example, plants capture the sun's energy to grow. So long as solar energy is available, other conditions being equal, life can thrive on Earth, even as the universe grows colder. Organisms are entropic or dissipative structures; they consume relatively concentrated energy to fulfill their function. Some researchers—Stanley Salthe and James Kay, for example—even go so far as to argue that life arises and evolves through self-organization that occurs when energy moves through material structures within the appropriate conditions. Even if the universe as a whole is moving toward "heat death," in accordance with the second law of thermodynamics, in our little corner of the universe life utilizes concentrated energy to generate and sustain itself.[14]

Although biology moved toward an evolutionary paradigm by the 1870s, cosmology retained the view that the universe had no beginning in time.[15]

In the 1930s, however, the Big Bang hypothesis challenged this view. After World War II, cosmologists developed evolutionary hypotheses about the Big Bang and the history of the universe.[16] But where did this Big Bang come from? Answering this question may not lie within the province of physics. Subsequent empirical research indicates that the universe's initial conditions were so finely tuned that it is "as if" they were intentionally established. Some cosmologists adhere to a version of the anthropic principle, according to which the existence of life capable of reflecting upon the origins and structure of the universe was built into those initial conditions. It is "as if," some cosmologists remark, the universe were somehow set up in advance to make possible carbon-based life that could eventually ask questions about how cosmic evolution occurred, and even more basically why there is "something rather than nothing."

Around the time that Darwin was writing *The Origin of Species*, Karl Marx was devising his own account of social evolution. Marx drew not on biology but on philosophy and philosophical anthropology, including the work of the German idealist G. W. F. Hegel. In *The Phenomenology of Spirit*, Hegel argued that human consciousness had undergone profound developmental evolution, from naive realism to sophisticated, self-conscious idealism. According to Hegel, human consciousness evolved from the aesthetic stage (as in ancient Athens), to the religious stage (as in Jerusalem and the European Middle Ages), and finally to the philosophical stage (as in German idealism). A crucial feature of this development involved deepening insight into the relationship between nature and mind. Descartes depicted this relationship in dualistic terms, but Descartes made significant contributions to early modern science, which enabled people to recognize that rationality exists not only in mind but also in nature, which Hegel called "petrified spirit." If natural science sets the stage for overcoming humanity-nature dualism, Hegel's metaphysics completed the process by claiming that in modern consciousness Alpha attains Omega—Spirit finally achieves self-recognition as the source of nature (Spirit manifest in space) and human mind (Spirit manifest in time).

Yet Marx interpreted Hegel's thought through a materialistic lens. Marx was influenced by the work of Ludwig Feuerbach, who caused a sensation in mid-19th-century Europe by arguing that the essence of Christianity is human brotherhood and love. Hence, the Christian God is the name we give to a set of human virtues projected onto a nonexistent, otherworldly entity. Marx claimed that anyone who wanted to understand human history and socioeconomic circumstances must purge themselves of prior beliefs by passing through the "fiery brook" (a literal translation of the German name *Feuerbach*). Feuerbach's ideas led Marx to a materialistc interpretation of Hegel. Instead of interpreting Hegel's version of history in terms of the development of consciousness, Marx, the materialist, social revolutionary, interpreted history as a material dialectic between techno-economic conditions and class structure.[17]

Marx assumed that individual consciousness and behavior were determined by social structures, including class, institutions, technology, and economic factors. According to Marx, such factors typically work "behind the back" of individuals, who retain illusions about personal autonomy and freedom. (Marx did understand how techno-economic structures also supported interior development. If the techno-economic conditions were not in place—that is, if greedy capitalists had not exploited workers in ways that made industrial manufacturing possible—then there would not be a revolutionary vanguard nor a class to be led by them.)[18] But Marx seemed to be influenced by an even larger-scale historical process, a greater unseen "behind his own back." As modernity differentiated the domains of inquiry (the Big Three), the Right-Hand quadrants quickly dominated the Left-Hand quadrants. Marx, like many thinkers around him, had already begun to deny the reality of interiors and thus began drawing an incomplete picture of the Kosmos.[19]

As the natural and social sciences moved toward positivism, the denial of human interiority dovetailed with the denial of other invisible domains, including that of Spirit. What began as methodological atheism, in which scientists tried to explain nature without recourse to divinity and dogmatism, soon ended in the conclusion, unwarranted by any empirical evidence, that there is no divinity, spirit, or soul—a dogmatism of its own. Habermas called this the "colonization" of the lifeworld by the natural and social sciences. Wilber calls it "the dominance of the Descenders." Both refer to the same phenomenon—modern scientific, social, and cultural materialism.[20]

Critiques of Development

Despite strong evidence in favor of the hypothesis that development is an intrinsic feature of the Kosmos, the idea of development is not without its critics. Religious conservatives criticize one view of development, evolutionary theory, for ignoring God's role in Creation. Morever, many religious traditions believe that alleged individual, social, and culture human development seems unlikely given the violence and corruption in so-called advanced societies. But it is not only religious conservatives who criticize aspects of the idea of development. Many multiculturalists condemn developmental models as Eurocentric and inherently biased justifications for Western colonialism. Some environmentalists charge that developmental models are anthropocentric and thus legitimate human domination of the biosphere. They argue that the so-called Enlightenment involves a dialectic that leads to global destruction, just as others argue that modernity contained within itself seeds that gave rise to Auschwitz and other horrors.[21]

As we noted earlier, science has not yet accounted for all the conditions necessary to explain either the origin of life or macroevolution. Nevertheless, there are very good reasons to conclude that life on Earth has developed from simple to more complex forms. Such development, we argue, has followed the logic of "transcend and include." Hence, the cell includes the molecules that compose it but transcends the limits of what such molecules can do. Likewise, organisms include cells but transcend the limits of what cells alone can do. Corresponding to the development of life in its exterior aspects is the development of life in its interior aspects. In the case of human beings, interiority includes the capacities that belong to other social mammals but transcends their limitations, thereby making possible complex cognitive acts and moral judgments, among many other things.

As for European colonialism, not to mention ethnocentrism, sexism, and racism, we maintain that such attitudes and practices are inconsistent with the worldcentric ideals of modernity. Hence, we distinguish between modernity, which includes moral development consistent with worldcentric ideals, and modern times, which include premodern, ethnocentric moral views that are at loggerheads with modernity's worldcentrism. Each level of interior moral development enables people to respect more life, human and nonhuman, thereby undermining not only ethnocentrism (as well as sexism, racism, etc.), but also the anthropocentricism that regards nonhuman beings as nothing but resources for enhancing human power and security. It is profoundly misguided to think that we could eliminate anthropocentrism if only we were to accept a radical biocentric egalitarianism that equates human interests with those of all other life forms. Such egalitarianism not only ignores important differences between humans and other organisms, but also fails to see that by calling for what amounts to regression to premodern moral attitudes, biocentric egalitarianism undermines the worldcentric moral domain in which movements for animal rights and environmental protection in fact began to arise.

We maintain that humans are developing—morally, personally, culturally, and socially—even if we can't completely explain the mechanism behind development and even if such development is often erratic and uneven.[22] Now let us focus on developmental factors as they pertain to the Right-Hand quadrants.

Where Is Humanity's Place in Nature? Holons, Emergence, and Holarchy

A major issue for ecologists and environmentalists is determining humanity's place in nature. To what extent are humans part of nature? The answer depends on how we define humans, nature, the logic of parts and wholes, and development in the physiosphere, biosphere, and noosphere. Because appropriately distinguishing among physiosphere, biosphere, and noosphere is so crucial, we make the following initial comments before going further.[23]

By physiosphere, we mean the physical-material circumstances that existed prior to the emergence of the biosphere, by which we mean the domain of a new level of complexity—life—that includes, but transcends the limits of the pregiven physical-material circumstances. However one defines life, and a good definition is not easy, one must make clear that life is not something other than matter-energy (physiosphere), but instead a development that such matter-energy is capable of sustaining (see endnote 13). In other words, rocks can eventually move around, reproduce, and sense the world around them. Life is a level of complexity toward which matter-energy tends to move, if and only if the appropriate conditions are in place. Matter is not "left behind" by life, then, but instead is transformed by drawing from out of itself novel possibilities. The matter-energy in life is not new, but the increasingy complex *organization* of such matter-energy is new. Likewise, when after hundreds of millions of years certain organisms begin to develop neurological systems capable sustaining mental imagery (for instance, mammals and perhaps other organisms as well), a new level of organization of matter-energy is achieved. Animal and human-animal mental activity does not result from a nonmaterial soul but instead is correlated with and a manifestation of developments of the same matter-energy that earlier organized itself into life forms. Matter-energy is always present in and foundational to physiospheric, biospheric, and noospheric beings. In addition to emphasizing that terrestrial development involves the increasing complexification of matter-energy, Integral Ecology also maintains that physiosphere, biosphere, and noosphere tetra-evolve, that is, developments in matter-energy consistent with the emergence of life involve not merely exterior aspects but interior aspects as well. Hence, life forms not only reproduce themselves in ways that physiospheric beings cannot, but also exhibit a more highly developed interiority than physiospheric beings. Living beings have greater possibilities for both 1st-person (subjective) experiences and 2nd-person (intersubjective) encounters. All this must be kept in mind as we describe Integral Ecology's answer to the question of whether humans are "part" of nature.

Rather than merely spell out our answer, however, we prefer to allow it to arise in the course of a discussion of the late Stan Rowe's critique of Ken Wilber's view of humanity's place in nature. Rowe was a noted scientific ecologist who reviewed *A Brief History of Everything* a number of years ago.[24] A detailed response to Rowe's critique is instructive for several reasons.[25] For starters, his anti-anthropocentric, deep ecological views resonate with many ecologists. In critiquing Rowe's position, we hope to engage deep ecologists in a constructive dialogue. Second, answering Rowe's critique provides an opportunity to clarify Integral Ecology's position and common misunderstandings of it. Finally, engaging Rowe provides a venue for complex topics. To respond to Rowe's critical remarks, we reference Wilber's other works, including *Sex, Ecology, Spirituality*.

Rowe concedes that Wilber's multiperspectival approach "is a valuable contribution. It identifies as 'narrow' those prophets and problem-solvers who claim 'my way only' as they charge off in one of the four directions."[26] His criticism is directed at Wilber's holarchical and developmental schemes, which allegedly lead to a distorted understanding of humanity's proper relation to the biosphere. To understand Rowe's critique, we need first to present an account of holons, which is crucial for correctly articulating the complex logic of wholes and parts. In *Sex, Ecology, Spirituality*, Wilber lists 20 tenets of the evolutionary development of holons.[27]

1. Reality as a whole is not composed of things nor of processes, but of holons.
2. Holons display four fundamental capacities:
   2a. self-preservation (wholeness, or agency)
   2b. self-adaptation (partness, or communion)
   2c. self-transcendence (eros)
   2d. self-immanence (agape)[28]
3. Holons emerge.
4. Holons emerge holarchically.
5. Each emergent holon transcends but includes its predecessor.
6. The lower sets the possibilities of the higher; the higher sets probabilities of the lower.
7. "The number of levels which a hierarchy comprises determines whether it is 'shallow' or 'deep'; and the number of holons on any given level we shall call its 'span'" (A. Koestler).
8. Each successive level of evolution produces greater depth and less span.
9. Destroy any type of holon and you will destroy all of the holons above and none of the holons below it.
10. Holarchies coevolve.
11. The micro is in relational exchange with the macro at all levels of depth.
12. Evolution has directionality:
    12a. Increasing complexity
    12b. Increasing differentiation/integration
    12c. Increasing organisation/structuration
    12d. Increasing relative autonomy
    12e. Increasing telos[29]

According to Arthur Koestler, who coined the term "holon," reality is composed of hierarchical and emergent levels. Each level has its own structural uniformities that cannot be reduced to the structures of lower levels. In *The Ghost in the Machine*, Koestler argued that any holon has three different dimensions. It is a whole in its own right. It is composed of parts whose behavior is significantly subordinated. The holon is a part of, and is in important ways controlled by, a more embracing or inclusive whole—a holon at the next hierarchical level.[30] Rowe concurs that the holon "is a whole to its parts below, and it is a part to the whole above. Reality consists of relational holons, not separate 'things.' The concept, a good one, dissolves the antagonism in science between reductionism and holism, for reductionism is a way of understanding that moves downward in hierarchies while holism is the upward view."[31]

Koestler created a pyramidal model of cosmic hierarchy. Vast numbers of holons (subatomic particles) populate the bottom level. Each higher and more complex level—atoms, molecules, organelles, cells, tissues, organs, organ systems, and the organism—has fewer members. Rowe affirmed the usefulness of this nested hierarchy for describing the hierarchical structure of the organism. Yet he and a number of other critics contend that Koestler applied the model beyond its proper limits, by accepting it "as the template for organic development and evolution, for animal locomotion and behavior, for linguistics, and for human societies past and present."[32] Rowe maintains that even though the anatomical structure of an organism is sometimes a helpful analogy for understanding other systems, "organisms are not homologous with all existing systems."[33] Rowe believes that Wilber accepted "Koestler's grab-bag of holons and hierarchies." Wilber's alleged "illogic," according to Rowe, tempts him to "link non-homologous hierarchies" and to "uncritically accept as legitimate for all holarchies" the rules that fit only one kind of holarchy.[34] Further, Rowe thinks that scientists such as Allen and Starr in *Hierarchy: Perspectives for Ecological Complexity* also "prefer to generalize the meaning of holons to include entities of any type" and justify their move by arguing that their common denominator is information.[35] In Rowe's opinion, the spatial arrangement of an organism's nested hierarchy is not applicable to the temporal development of an organism. Rowe believes Koestler glided too quickly over the difference between structures (spatial) and processes (temporal). Informationally triggered feedback occurs within organisms, but it cannot occur in temporal sequences such as organic development. Hence, an adult organism cannot influence its own initial stage of development.[36]

Rowe is correct in saying that organisms cannot influence their own initial developmental stages, but he goes astray in creating an unnecessary dualism between space and time, structures and processes. Structures unfold in temporal sequences. We can study structures in isolation from such sequences, but we fall into serious error if we ignore the temporal aspect not only of their initial sequential appearance, but of their continual maintenance and self-reproduction. The development of a normal phenotype involves material factors that unfold in temporal sequence. The sequence of normal development of a phenotype—which is a kind of holon—embraces a host of junior holons, including atoms, molecules, organelles, and cells. Organisms constitute an emergent level of organization, life, and the specifically emergent properties differ significantly from the properties nonliving phenomena, such as atoms and molecules. As Wilber emphasizes, however, even these emergent properties are not somehow "other" than matter-energy, but developments of capacities inherent within matter-energy. If matter-energy could not readily form itself into the constituents of life forms, life could not have appeared in the first place.

Rowe cites other problems posed by generalizing organic (nested) hierarchies beyond their appropriate application. To do so, he refers to James K. Feibleman's "laws of the levels," which preceded Koestler's book by more than a decade.[37] (Here it is worth remarking that Wilber's conceptual scheme is not reducible to Feibleman's, nor to Koestler's, nor to some combination of the two. Hence, by referring to the work of Feibleman or Koestler, Rowe sometimes sheds little light on Wilber's own view of holons in their spatial and developmental aspects.) Commenting on Feibleman's first law, "Each level organizes the one below it plus one emergent quality," Rowe maintains that this law applies to the nested hierarchy constituting organisms. He adds, however, that "the idea gets hazy when applied to sociological groupings such as family, tribe, ethnic group, societies with division of labor, and nation. . . ."[38] Here, Rowe ignores Wilber's distinction between individual and social holons, which we discuss later on. Development occurs in both individual and social holons, although the 20 tenets do not always apply to social holons, because the latter lack some features possessed by individual holons. According to Rowe's restatement of Feibleman's fourth law, each organic level has some autonomy, but is also an integral part of and thus constrained by the higher level: "The integrative tendency of each holon must overrule its self-assertive tendency if the whole organism is to maintain its health. Such a law is irrelevant to evolutionary sequences. It [overriding the self-assertive tendency] is dangerous when applied to sociological systems for it can be used, as Medawar foresaw, to justify subjugation of the individual to the totalitarian state."[39]

As we will see, Wilber was more attuned than Rowe to the dangers posed by subjugating individuals to the collective. Moreover, some of Rowe's concerns here can be addressed by Wilber's distinction between individual and social holons. Rowe writes that Feibleman's fifth law, "For an organization at any level, its mechanism lies at the level below and its purpose at the level above," is valid for the organic hierarchy but "makes little sense for developmental, evolutionary, sociological, cultural, and mental systems."[40] Wilber, however, does not use Feibleman's terms "mechanism" and "purpose," but instead says that lower-level holons are more fundamental, whereas higher-level holons are more significant. More fundamental holons provide possibility, and more significant holons provide probability.

Rowe observes that Feibleman's law number eight, "The higher the level, the smaller its population of instances," applies to "hierarchies that aggregate upwards, as well as for [most] ecological food pyramids," but Rowe thinks it is dangerous to apply it to cultural evolution "because, for example, it justifies the hegemony of the Western industrial/agricultural system. The fact that the Western cultural system has few variants (small population of instances) . . . does not automatically confer the title 'higher' unless, like Marx and Wilber, one believes in laws of historical necessity."[41] In our view, however, some aspects of traditional society are higher and thus freer than tribal society, and some aspects of modernity are higher and thus freer than traditional society. A worldcentric society is higher and freer than an ethnocentric society. In a highly conformist, traditional, premodern society, campaigns for the rights of women and gays, not to mention for the rights of people with different ethnic, religious, and national backgrounds, simply cannot take place. Such campaigns presuppose the moral space opened up by worldcentrism, according to which equal moral status must be accorded in principle to *all* human beings regardless of factors that tribal and traditional societies use to exclude or to marginalize humans who are somehow "different."

Assuming that moral development does occur, however, we maintain that there are in fact fewer people with highly developed forms of morality than there are people with less developed forms of morality. Moreover, if in fact constitutional democracy does represent an instance of sociocultural development (as we believe it does), one would expect that there are fewer instances of constitutional democracies than other forms of government. Such an expectation is borne out by surveying how contemporary nations govern themselves.

Wilber appeals to Jürgen Habermas's research to argue that in certain respects Western forms of consciousness, rationality, and social systems are more inclusive and comprehensive—and in those respects more "advanced" —than those of other societies. Even though Western countries have often failed to acknowledge and respect non-Western peoples (not to mention nonhuman life forms), it is not an argument against the validity of a world-centric position to which Western democracies theoretically adhere.[42] Wilber has pointed out that individuals, cultures, and societies contain many different lines of development, and every person, culture, and society is somewhat different. Hence, there is no "one" Western system, but many such systems that have what Wittgenstein would call a "family resemblance" to one another.[43]

Later, Rowe states that Wilber ascribes to evolution a drift—even a kind of entelechy—leading toward ever greater complexity. Rowe thinks that Wilber's "sequences are based on a theory of progress onward and upward, like the Marxist faith that historical necessity guides the transition from feudalism to capitalism to communism."[44]

Rowe offers no quotation to support his analogy between Wilber and Marx, because no such quotation exists. Wilber does not adhere to historical determinism. Wilber insists that there is nothing necessary about human social evolution, any more than organic evolution is necessary.[45] Still, we can reasonably engage in retrodiction, that is, we can use historical knowledge to explain how we arrived at the current states of affairs. Hegel remarked that the owl of Minerva paints its gray on gray only at the end of things, that is, wisdom takes flight retrospectively. Looking back over the course of organic evolution, thinkers often claim to discern in it a move toward greater complexity, despite meanderings, tangents, and setbacks. Like Wilber, we see such development in many different areas, including the move from ethnocentrism to worldcentrism.

Rowe believes that Wilber's key law is that "each emergent holon transcends but includes its predecessors, and evolution is a process of transcend and include." Indeed, Wilber does subscribe to this notion. According to Wilber, planetary evolution moves from physiosphere (all quadrant domain of matter-energy) to biosphere (all-quadrant domain that has attained the complexity consistent with life) to noosphere (all-quadrant domain that has attained the complexity consistent with representational mentality).[46] Just as the biosphere is composed of matter-energy that has transcended the limits of prior instances of matter-energy, so too is the noosphere composed of matter-energy that has transcended the previous limitations of the biosphere. If Rowe had understood all this, he would never have misinterpreted (and misquoted) the following passage from *A Brief History of Everything*: "the biosphere is literally internal to us, is part of our being."[47] "Such arguments," writes Rowe, "assume the same kind of structural organization in physical, biological, and mental categories."[48]

This is a curious assertion, especially because Wilber goes to great pains to point out the dramatic differences among the "structural organization" of the three major levels—physiosphere, biosphere, and noosphere—and among the quadrants of each level. For instance, in regard to noosphere: a bear's 1st-person experience lacks physical location, although the brain correlated with such experience does have a physical location. If Rowe's point is that "transcend and include" is common to all three spheres (and that surely is Wilber's point), then Rowe must explain why this general structure does not apply to terrestrial development.

Rowe then builds on his confused argument and asserts that this structural homology, "when teamed up with [Wilber's] Platonic philosophy, provide the bootstraps by which Wilber's system lifts all reality into aspects of consciousness on their way to pure Spirit."[49] Here, Rowe's essay—which hitherto had raised some reasonable reservations about Wilber's position—develops serious problems. First of all, by Platonic philosophy, does Rowe mean an essentialist, ahistorical, eternalist metaphysics, the kind that 19th-century evolutionary theorists overcame to defend their claim that species evolved rather than being instantiations of archetypal forms that exist in the mind of God? Even an uncharitable reader of *A Brief History of Everything* would be hard pressed to impose this view on Wilber! Wilber's post-metaphysical position is even more explicit in later works such as *Integral Psychology* or *Integral Spirituality*. Or by Platonic philosophy does Rowe mean the life-despising, world-negating philosophy that characterizes some versions of Christian Neoplatonism? This criticism does not hold up either. Wilber insists that neither Plato's work nor that of his great follower Plotinus was world-negating. Plotinus castigated the Gnostics precisely because they showed such contempt for material nature. Certainly, Wilber himself does not adhere to *contemptus mundi*. To explain his contention, Rowe states that "the grain of truth" in Wilber's "dogma" is that all "organisms are 'open systems' constantly internalizing energy and materials from the biosphere. . . ."[50] Readers are left to their own devices to understand how this comment pertains to Wilber's claim that the biosphere transcends and includes the physiosphere.

Rowe's statement that Wilber's system lifts reality into consciousness clarifies his intentions. He continues: ". . . the conclusion that such common-sense phenomena as the physiosphere and biosphere—the Earth realities of air-water-landscapes in which humans live, move, and have their being—are interior, structural parts of the mind-noosphere can only ring true for dedicated idealists."[51] Fortunately, Wilber never makes such an outlandish claim. Rowe's reference to "dedicated idealists" explains what he meant when he said that Wilber adhered to Platonism. Yet Plato never adhered to such subjective (or anthropocentric) idealism any more than Wilber does. Rivers and landscapes are not inside and reducible to the human mind. The Mississippi River is not literally a "stream of thought," nor is Mount Everest a figment of the collective (and delusional) human imagination. In suggesting "the biosphere is literally internal to us, is part of our being," Wilber asserts something very different than Rowe thinks. Let us unpack this highly misunderstood position.

Wilber on Parts, Wholes, and Containment

What does Wilber mean by saying that the physiosphere is literally contained in every member of the biosphere? Surely, a single mouse cannot contain the Earth's crust, oceans, and atmosphere! In *Sex, Ecology, Spirituality*, Wilber writes that all physiosphere holons depend on networks of interrelationships with all other holons and with holons at the same level of structural organization (tenet 11).[52] He uses a very simplified analogy for depicting the relation between physiosphere and biosphere holons. Let physiosphere holons be black checkers and biosphere holons be red checkers. These second-level (red, biospheric) checkers add an emergent dimension to the black checkers. Wilber states that level-2 holons embrace and include level-1 holons and then go beyond them. Because all physiosphere holons are wholly interrelated, an organism contains the physiosphere, because the organism contains matter from the physiosphere.

Put in another way, the physiosphere is a part of life because organisms are composed of the constituents of the physiosphere, matter-energy in the form of atoms and molecules. Because an organism is fundamentally composed of nothing but matter-energy, the physiosphere (domain of matter-energy) is a part—that is, a constitutent—of life. Life or biosphere, however, is not a part of pre-organic matter-energy or physiosphere, because life represents a novel phenomenon that builds upon matter-energy but configures it in wholly novel ways.

Let us use a variation of the checkers metaphor to describe the relation between physiosphere and biosphere. Allow a thick black checker to represent the phyisophere, that is, the domain of matter-energy; then imagine that a thin layer of the top of the checker begins turning red. Red represents the organic, biospheric domain, which is fundamentally composed of matter-energy, but which involves an emergent configuration of matter-energy (atoms, molecules) that achieves a much higher level of complexity than is found in pre-organic modes of the physiosphere. The biosphere, then, includes matter-energy (physiosphere) but includes traits missing from matter-energy that is not alive. We can imagine that the entire thin red surface (biosphere) will perish, but its disappearance will not at all affect the physiosphere, which will remain composed of the atoms and molecules left over from perished organisms. Were the physiosphere to disappear, however, the biosphere would disappear along with it. No atoms and molecules, no matter-energy, no organisms or biosphere.

Analogously, the biosphere is a part of humans in the sense that human bodies are organisms related to all other organisms. In this sense, the biosphere is a part of us, it composes us, it constitutes us. Moreover, the physiosphere is also part of us humans, insofar as the biosphere includes as part of itself the physiosphere, and insofar as humans are organisms. But humans are not merely organic beings. Humans are also noospheric beings. To follow our analogy, let us postulate that another, even thinner white layer emerges from the already thin red layer (biosphere), which is in turn a transformation of matter-energy, or physiosphere. The noosphere transcends the biosphere, in the sense that consciousness (including animal consciousness) involves emergent properties that cannot be reduced either to the physiosphere or biosphere. Nevertheless, these properties wholly depend upon physiosphere and biosphere for continued existence. Without the foundation laid by the biosphere, the noosphere could not have emerged. If the biosphere were to vanish, so would the noosphere, because the biosphere is a necessary condition for the noosphere's existence. Hence, conscious reality (noosphere) depends upon the physical and organic domains, but noosphere can be neither reduced to nor wholly explained in terms of the other two more fundamental domains.

This is not "idealism," nor is it an attempt to reduce physical and organic phenomena to mental states. In an important sense, there is *nothing but* matter-energy, or physiosphere. In this respect, humans are "part of" nature, if nature is defined strictly as matter-energy, atoms and molecules, physiosphere. But such a definition of nature is reductionistic, excluding the emergent properties of biosphere and noosphere. In an equally important sense, matter-energy lends itself to processes that bring forth new levels of reality composed of matter-energy, but not restricted to the capabilities of pre-organic, pre-living matter-energy. That rocks cannot get up, walk around, or think is no argument against the importance and even the perfection represented by minerals; and that plants cannot write books is no argument against the importance and even the perfection represented by them. Hence, asserting that noospheric beings (including at least mammals) represent a new level of complexity implies no disrespect for physiospheric realities and biospheric beings. Moreover, even though organisms with neurological systems have a greater degree of subjectivity and intersubjectivity than beings without such systems, this fact does not mean that atoms, molecules, and cells lack subjectivity and intersubjectivity. Instead, all individual physiospheric and biospheric holons have modes of subjectivity and intersubjectivity that are perfect for their own kind, although limited when compared to noospheric beings.

According to Wilber's holarchy, simpler holons are more numerous—that is, have greater span (they are more fundamental)—but have less depth (less significance). There are countless atoms in the universe but far fewer molecules, vastly fewer organic beings, and vastly fewer beings with representational mental capacities. Because the size of the biosphere is much greater than the whole human species, not to mention a single human being, readers understandably resist the assertion that the biosphere is part of the noosphere. Yet humans are not strictly physical beings. If we defined humans strictly as physical beings, then humans are "parts of" of the biosphere. On this topic Wilber states that "the human compound individual is not a part of the biosphere. Rather, a part of the human compound individual [that is, the living physical aspect] is a part of the biosphere, and the biosphere itself is a part of the noosphere. And for just that reason, repression can set in; for just that reason, the noosphere can dissociate the biosphere. . . ."[53]

As merely physical beings, humans are "parts" of the biosphere, which is literally composed of atoms and molecules. As organic or living beings, however, humans are not parts of the biosphere, because individual holons (organisms) relate to same-level social holons (their ecosystem) as members that are in constant exchange with the ecosystem. This is the nugget of truth in the claim made by ecotheorists like Fritjof Capra: humans are strands in the great web.[54] Systems theory environmentalists, however, tend to ignore that humans have interior depth. Instead, Wilber states, they

> reduce the Kosmos to a monological map of the eco-social system—which they usually call Gaia—a flatland map that ignores the six or seven profound interior transformations that got them to the point that they could even conceive of a global system in the first place.
>
> Consequently, this otherwise true and noble intuition of the Eco-Noetic Self [nature mystic] gets collapsed into "we're all strands in the great web." But that is exactly not the experience of the Eco-Noetic Self. In the nature-mystic experience, you are not a strand in the web. You are the entire web. You are doing something no mere strand ever does—you are escaping your "strandness," transcending it, and becoming one with the entire display. To be aware of the whole system shows precisely that you are not merely a strand, which is supposed to be your official [deep ecological] stance.[55]

If we focus on size alone, we will conclude that the noosphere (i.e., human brains) are in the biosphere because the brain's physical location is inside ecosystems. However, if we focus on depth and span, we are hard pressed to conceive of the noosphere as in the biosphere. Let us unpack this further.

Individual and Social Holons

"Depth" refers to increasing interior complexity and individuation; "span" refers to the number of any given kind of holon; and size refers to physical (including volumetric) extension.[56] Wilber uses these distinctions to describe the difference between individual and social holons. Social holons lack the centered nexus-agency that characterizes individual holons, including cells and multicellular complex organisms. Also, individual holons tend to grow larger in size and social holons tend to grow smaller. Both increase in depth. As individual and social holons increase in depth, span almost always decreases. The main theoretical flaw with many holoarchies is that they are based on size alone (the Earth is bigger and humans are

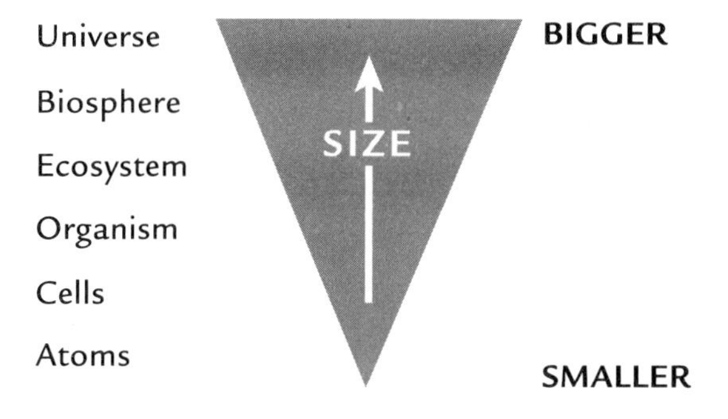

Figure 3.2. Typical ecological hierarchy of span.

smaller, so we must be parts of it). They do not consider depth. In figure 3.2 we see a typical cosmic holarchy. It starts with atoms and ends with the whole universe. The triangle represents the customary (and materialist) view that size (volumetric scale) is the best (and only) way to describe the cosmic holarchy. This holarchy does not account for depth or span, and does not distinguish between social and individual holons. Figure 3.3 represents Wilber's view and divides the Kosmic holarchy into social and individual holons. It shows how size, depth, and span vary.

The evolutionary direction for social holons is from universe to ecosystems, whereas the evolutionary direction for individual holons is from atoms to organisms. According to Wilber, individual holons achieve greater interior depth and individuation as they develop from atoms to organisms.

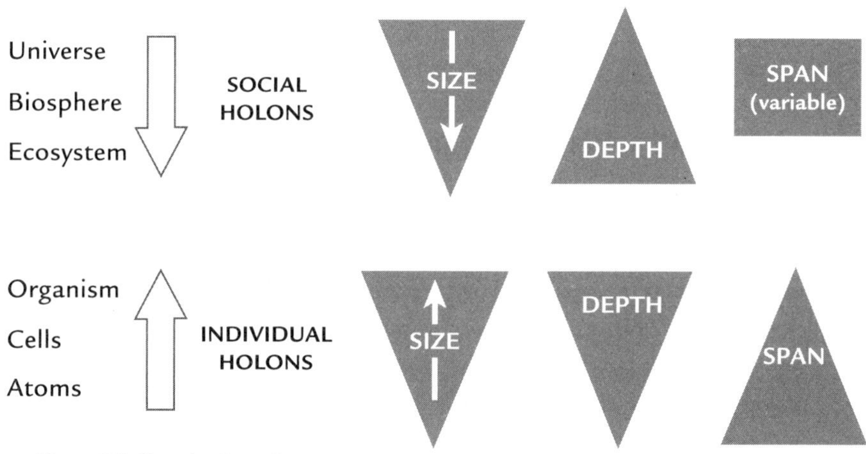

Figure 3.3. Size, depth, and span.

Individual holons tend to grow larger in size. It is true that some molecules are larger than some cells, but any given cell is *always* larger than the molecules composing it. In the evolutionary process that leads from atoms to organisms, size increases, depth increases, but span decreases. There are vastly fewer organisms than atoms, but organisms have far greater depth and thus much greater individuation. Moving from galactic clusters, to galaxies, to solar systems, to planets, to ecosystems (all social holons), size tends to decrease, depth increases, and span tends to increase but is variable. Perhaps Earth started out with one ecosystem but gradually developed more—ecosystem span would have increased, but there would remain only one planet Earth. (We could also assert that pre-biospheric Earth had less depth and thus less significance than today's Earth, which includes ecosystems and organisms.) A holarchy based on increasing size alone is not tenable, because it ignores depth and thus significance, and ignores the important distinctions between individual and social holons.[57]

Building on distinctions made by Erich Jantsch and others, Wilber distinguishes between compound individual holons and social holons, each of which has its own form of part-whole relationship.[58] These distinctions, combined with the distinction between the interiors and exteriors of holons, create four parallel holarchies that avoid the very problems found in most single-axis hierarchies, including the one devised by Rowe. Wilber is true to his own perspectivalism because he insists that assigning a number of holonic levels in any given holarchy is somewhat arbitrary. What we designate as a holon is partly a function of our perspective, although Wilber (like many scientists) insists that a powerful perspective reveals truth. Knowledge is not merely a collection of internally related interpretations but consists of valid interpretations.

Compound individual holons are the ones most consistent with Feibleman's and Koestler's laws. They have a relatively centered agency and autonomy. By comparison, social holons have nexus-agency but not conscious (individual) agency.[59] The distinctions between a social and individual holon are difficult to draw, Wilber concedes, because "it's almost impossible to define what we mean by an individual in the first place."[60] In one moment what reveals itself as a whole is merely a part of a more inclusive whole in the next.[61] Wilber defines an individual holon as an enduring compound individual, composed of junior holons, with its own defining form, wholeness, or deep structure. An individual holon exists inseparably from its social environment, but "to the degree that we can reasonably recognize [its own particular form or pattern], we will refer to an individual holon."[62]

Social holons display a whole-part pattern, are rule-bound, develop (as in stellar or ecosystem evolution), and "function with various degrees of upward and downward causation," depending on their depth.[63] Wilber admits that some social holons, such as ant colonies, behave as if they were superorganisms, but he resists describing "a social holon, such as the State, as being literally a superorganism, because all organisms have priority over their components, and yet with the rise of democratic structures, we like to think that the State is subservient to the people, and to the degree that [the latter] is true, then the social system is not a true organism (it is a social or environmental holon)."[64]

Holons always involve agency-in-communion. The macroscopic is the environment for the microscopic. Jantsch speaks of "the difference between 'vertical organismic and horizontal ecosystemic (symbiotic) organization'—but the point, again, is that [individual and social holons] co-evolve."[65] The micro (organism) is always in relational exchange with the macro (the biome composing the social holon to which the organism is a member). Whereas the organs of an organism (a compound individual holon) are parts of it and thus under its general control, the organisms in a biome are members of it and not under such strict control, because the complexity of organisms confers on them a relatively high degree of autonomy, which we can see in its relative capacity for self-preservation in the midst of environmental fluctuations. Humans achieve such a high degree of complexity and relative autonomy that, as noospheric beings, that they can dissociate from the environment, as in otherworldly religiosity or mind-body dualism that reject consciousness's dependence on the physiosphere and biosphere. Such "pathological agency," however, could land humankind "in ecological hell."[66] In *Sex, Ecology, Spirituality*, Wilber agrees with other hierarchy theorists that there is upward and downward causation in a holarchy. Human behavior can affect the biosphere (downward causation) in a way that threatens it, and the biosphere can then adversely affect humankind (upward causation). Such mutual influence does not mean that humans are merely "parts" of the biosphere.

Humans belong to social groups, including states and towns, which have cultural (intersubjective) principles and social (interobjective) regulations.

Human individuals, however, do not become internal to these social groups the same way that cells become parts of an organism. Sociocultural groups are not composed of the mere bodies of their members, but rather their intersections, especially in the domain of language. Wilber writes:

> . . . in collective or systemic holons, the regnant nexus [that is, the governing sociocultural principle] itself never contains a sentient being as a dominant monad within which other sentient beings become internal subcomponents (as in a giant leviathan), but rather the intersections of the sentient beings become internal to a nexus-agency of which they are members. That regnant nexus itself can be commandeered, taken over, and controlled by dictatorial sentient beings, but never in the sense that other sentient beings then become internal to the dictators, only in the sense that the fascistic holons now have some degree of power over the interactions of the sentient beings inside, not internal to, the network. The types of power exerted in both are dramatically different (which is another reason that the leviathan or Gaia views of ecology often become confused and tend toward fascism, in theory and in practice; in other words, an organism is an "I," an ecology is a "we," and whenever ecology is called an organism, there is a hidden "I," often that of the theorist. . . .[67]

Terrible consequences have ensued from philosophies (such as Stalinist Marxism or National Socialism) that depict individual humans as nothing more than organs of the state. Such philosophies (ecofascism or eco-communism, for example) ignore the agency of individual human holons and overemphasize their communion. Survival of the social collective, according to such philosophies, requires that individuals sacrifice themselves and their interests to the good of the superorganism. Wilber maintains, however, that a social holon, including the state, lacks "a locus of self-prehension, a unity feeling as oneness. More generally, it lacks a locus of individual self-being. . . . the parts in this social system [the State] are conscious, but the 'whole' is not."[68] This serves as a reminder that we should be wary of leaders and philosophers that purport to speak for the whole, be it the state or Gaia. We believe the issue of ecofascism (and fascism in general) is extremely important, not because we fear that it lies around the corner, but because muddled conceptions of part-whole relationships lead to real-world mistakes and can result in the justification of extreme behaviors that discredit legitimate environmental concerns. We discuss this more in the section "Gaia as Super-Organism?"

Kosmic Holarchy, Parts and Wholes

The difficulties involved in sorting out a more accurate description of a Kosmic holarchy are legion. Wilber indicates that many noted thinkers, including Karl Popper and Ervin Laszlo, subscribe to the following version of the hierarchy, which confuses individual and social holons.[69] (We insert "organisms" into the list for clarification.)

Biosphere
Society/Nation
Culture/Subculture
Community
Family
Personal Nervous System
[Organisms]
Organs/Organ Systems
Tissues
Cells
Organelles
Molecules
Atoms
Subatomic Particles

If higher hierarchical levels depend upon lower ones, this hierarchy is deeply problematic. It confuses individual holons with social ones. If this holarchy describes the chronology in which various levels formed, then the biosphere is depicted as having arisen after the emergence of human nations! According to Jantsch, however, the first ecosystem was composed of prokaryotic cells. So the biosphere cannot be the final hierarchical level, because it *already* emerged hundreds of millions of years ago, along with prokaryotic cells. The biosphere, which is a social holon, has co-evolved with organic life forms ever since. Today's biosphere conditions, and is conditioned by, the totality of organisms that are members of it. The biosphere does not depend upon human societies for its existence; human societies depend on the biosphere. If the human species were to became extinct, this event would not destroy the biosphere. But destroying the biosphere would annihilate humankind.

As Wilber puts it, the biosphere is shallower (less complex) but more fundamental than human societies. The distinction between individual (micro) and social (macro) holon should not lead to the conclusion that the macro is on a higher level than the micro. Instead, individual and social are parallel, they "are two aspects of the same thing, not two fundamentally different things (or levels)."[70] Hence, an ecosystem "isn't a particular level among other levels of individual holarchy, but the social environment of each and every level of individuality in the biosphere."[71] Individual organic holons participate in and depend upon environmental or social holons. Hence, "Gaia is not a giant critter that contains individual organisms as cells in its single body. Gaia is the harmonious song sung by a choir of organisms, it is not itself a really big organism."[72]

Following Jantsch, Wilber argues that atoms, early cells, organisms (individual holons), the early universe, the biosphere/Gaia, and terrestrial ecosystems (social holons) co-evolve. (See figure 3.4.) Atoms formed the enormous clumps that characterized the early universe, cells formed the first version of the terrestrial biosphere, and organisms formed ecosystems. Hence, individual holons (e.g., cells) and their social holon counterpart (e.g., biosphere) co-arise and are mutually dependent.

So, is the noosphere part of the biosphere or is the biosphere part of the noosphere? Is there horizontal inclusion; is the noosphere in the biosphere

INDIVIDUAL HOLONS	SOCIAL HOLONS
Organism	Terrestrial ecosystem
Early cells	Biosphere/Gaia
Atoms	Universe

Figure 3.4. Individual and social holons.

as a strand in the web of life? Or is the relationship vertical; is the biosphere in the noosphere as a foundation for the noosphere? If the noosphere were merely a strand in the web of life, rather than an emergent dimension that transcends (and includes) life, then that would mean that the physiosphere gives rise to the biosphere, which includes (contains) all forms of life, including noospheric life. Yet this assumes that interiority is simply a non-emergent function of living matter (bios). Modern science typically accords no interiority (prehension, irritability, perception, cognition) to matter or to life. Hence, the ascription of interiority to noospheric life forms—including human beings—becomes very difficult to defend.[73] Indeed, like mid-20th-century behaviorists, some well-known cognitive scientists continue to deny that "consciousness" is anything other than material brain states. For such scientists, if interiority exists at all, it is a very late (not to mention puzzling) arrival.

Figure 3.5 represents Wilber's approach. It indicates that there are always interior and exterior aspects to the physiosphere, biosphere, and noosphere. Following Alfred North Whitehead, among others, Wilber maintains that noosphere represents a remarkable advance upon the interiority that characterized the biosphere and physiosphere. Holons involve at least four aspects—individual interior (UL), collective interior (LL), individual exterior (UR), and collective exterior (LR)—so a more accurate depiction of the relations among physiosphere, biosphere, and noosphere appears in figure 3.6.

In *A Brief History of Everything*, Wilber occasionally makes assertions that he must later qualify. For example, he writes that even anti-hierarchalists usually subscribe to something similar to the following holarchy: "atoms are part of molecules, which are parts of cells, which are parts of individual

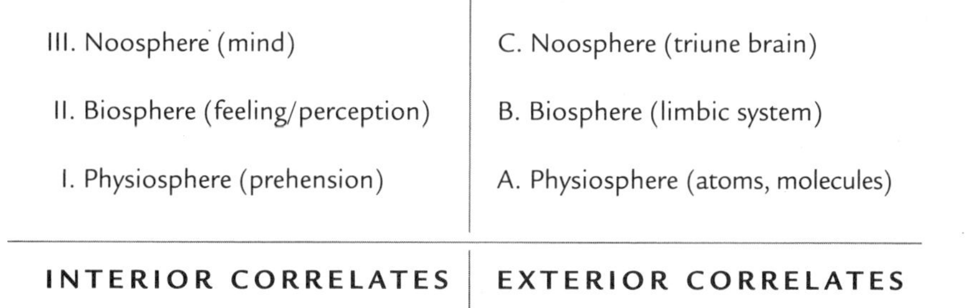

Figure 3.5. Interiors and exteriors of individual holons.

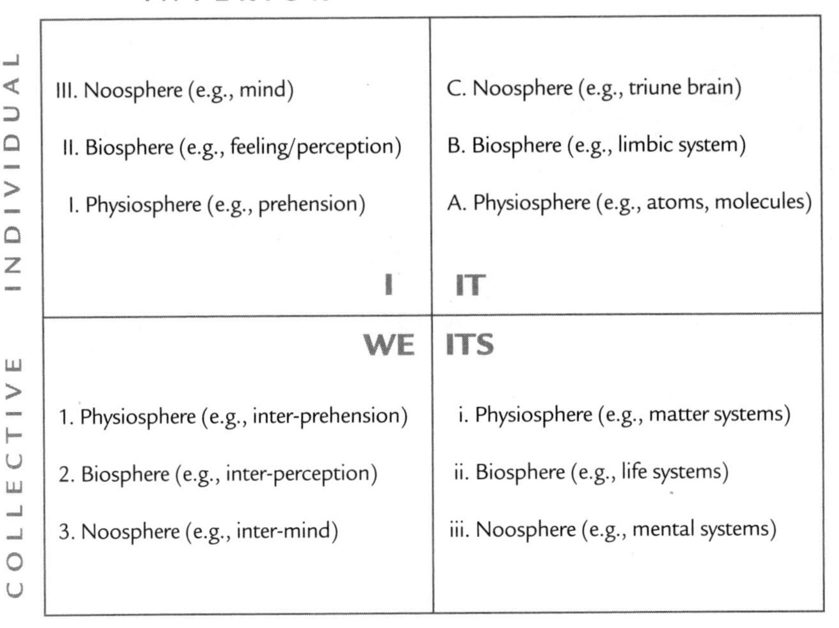

Figure 3.6. The 4 quadrants of the physiophere, biosphere, and noosphere.

organisms, which are parts of families, which are parts of cultures, which are parts of the total biosphere."[74] Wilber remarks that this holarchy is more or less correct, except for the position of biosphere. Rowe comments on this passage, contending that the categories do not mesh with one another (primarily because they do not follow Rowe's logic of volumetric containment). Wilber agrees that such a holarchy has problems, but not the ones Rowe identifies. In the quotation that Rowe cites (which occurs prior to *A Brief History of Everything*'s explicit discussion of social and individual holons), Wilber observes that the placement of the biosphere within the social holon of families, cultures, and biosphere holarchy, is incorrect on two accounts. First, the holarchy confuses individual holons with social ones. Second, biosphere should not be placed last in a sequence of social holons, but instead simultaneously with the development of individual life forms.

According to Rowe, a "logical ecological holarchy follows the principle of containment, viz., each level in the sequence is enveloped as a physical volumetric part by the next higher level."[75] Rowe's holarchy is based on size. Rowe states, "each higher level is the environment of those below." After agreeing with Wilber's nesting hierarchy for compound individual holons (molecules are parts of cells, cells are parts of organs, etc.), Rowe parts company with Wilber by stating that organisms

> are parts of geographic ecosystems, which are parts of the ecosphere. Each higher level is the environment or "field" of the ones below, and each lower level is a functional part of the levels above. Note that in this sequence human organisms appear as one among many species-parts of the sectoral ecosystem that Earth comprises. Humans are made from and sustained by the living Planet. Physically and mentally they are Earthlings. Truly they are marvellous creatures but not the be-all and end-all of creation.[76]

Wilber agrees that humans are one species among many, are sustained by the living Planet, are marvelous, and are not the be-all and end-all of creation. Yet he maintains that humans are not merely organisms but also noospheric or conscious beings. Such consciousness, whether human or animal, has no simple location in the material world. Hence, neither consciousness nor culture can be "contained" within a three-dimensional volumetric framework. True enough, the brain that correlates with consciousness does have a location, and the societies that correlate with cultures (norms, values, philosophies, and so on) do have locations. But because human interiors (both individual and cultural) do not have a location, and because the noosphere (consciousness) both includes and transcends the biosphere, we cannot describe the experience of being a human as merely "part of" the biosphere.

Wilber disagrees with Rowe's placement of the ecosphere as the topmost rung in the terrestrial hierarchy. Virtually all hierarchy ecologists place the biosphere, ecosphere, Gaia, or planetary ecosystem at the top of terrestrial hierarchy. All planets in the solar system have an ecosphere. Jupiter has an awe-inspiring ecosystem but lacks life forms, according to virtually all specialists. No life forms, no biosphere.

Let us look at what we maintain is the correct hierarchical positioning of the biosphere.

Value, Holarchy, and Nature Mysticism

Wilber maintains that there are three ethical value distinctions: *ground*, *extrinsic*, and *intrinsic value*. In terms of ground value, neither the biosphere nor the noosphere is primary. Instead, each is equally valuable as a manifestation of Spirit. Spirit is the ultimate source of all phenomena and ultimate attractor to cosmic development. In terms of extrinsic value, the biosphere is primary because it is more fundamental. If we were to destroy the biosphere, we would also destroy the noosphere. Thus, the biosphere is primary, and this means that the biosphere is part of us. The noosphere is not a part of the biosphere. If it were, the biosphere could not exist without us. But the opposite is true. Many environmentalists intuit this, but they confuse what is most fundamental (Gaia/biosphere), or what has the most extrinsic value, with what is most significant (humans/noosphere), or what has the most intrinsic value.[77] According to Wilber's holarchy, noospheric beings have such enormous depth that they have greater intrinsic value than non-noospheric life. Wilber's views largely overlap with those of Holmes Rolston III, one of the world's leading environmental philosophers.[78]

Of course, many may disagree with a value hierarchy based on increasing depth, but defending most alternatives is no easy task! Wilber sympathizes with the motives of environmentalists who attempt to overcome anthropocentrism by diminishing differences between humans and other organisms and proclaiming that everything is part of nature. This assertion ignores the difficult problem of defining "nature," and denies what differentiates humans from other organisms. Admitting that humans are in some ways unique, however, does not necessarily cause or justify heedless exploitation of the biosphere.

Rowe recognizes the importance of the interior and even spiritual domains when he quotes Fritjof Capra: "When the concept of the human spirit is understood as the mode of consciousness in which the individual feels a sense of belonging, of connectedness, to the cosmos as a whole, it becomes clear that ecological awareness is spiritual in its deepest sense."[79]

Yet, because of his exclusive commitment to a systems theory approach to explaining phenomena, Capra himself cannot provide an adequate description of "interior and spiritual domains." Rowe believes that his statement about spirituality runs counter to the views of Wilber, who, he alleges, "depreciates the physical and natural" and "cannot conceive any other source of values" apart from humankind.[80] In fact, Wilber asserts that everything has some basic ground value, quite apart from any human interest in it. Wilber affirms that all individual holons have a "worldspace" of their own. People should honor the perspectives afforded by such worldspaces. Hardly the words of an anthropocentric philosopher. Movement toward such a dramatically non-anthropocentric view that does not ignore what is unique to humans presupposes the development of worldcentrism, that is, mutual understanding among humans. A genuinely planet-centered perspective, which may emerge in the distant future, would eventually combine ecocentrism with worldcentrism.

In *A Brief History of Everything*, Wilber describes nature mysticism the same way that Capra describes ecological awareness—your separate-self sense temporarily dissolves, and you identify with the entire material world.[81] There is no longer looker and looked at, subject and object; instead, "suddenly there is no looker, just the mountain—and you are the mountain. You are not in here looking at the mountain out there. There is just the mountain, and it seems to see itself, or you seem to be seeing it from within. The mountain is closer to you than your own skin."[82]

As you mature and gain greater moral maturity, you gradually transcend your "merely ethnocentric or sociocentric identity," which can become a source of embarrassment, and expand so as to identify yourself "from a global perspective." You consider the concerns of all people, not just those who share your values and perspectives. Wilber then says: "It's only a small step further to actually experience your central identity, not just with all human beings, but all living beings. Global or worldcentric awareness transcends another notch, escapes its anthropocentric prejudice, and announces itself as all sentient beings. You experience the World Soul."[83]

Wilber's earlier reference to becoming one with the mountain should ward off concerns that nature mysticism limits its identification to plants and animals. Indeed, Wilber adheres to a form of pan-interiorism, according to which all beings have at least some perspective. Nature mystics do not experience nature as "part of" themselves, because it transcends the limits of egoic, ordinary human awareness. Who or what, then, is experiencing this transpersonal awareness? Consistent with the great religious commentators, Wilber answers: A transhuman awareness, World Soul, that includes ordinary human awareness—and in fact, all other forms of organic and inorganic awareness—as an aspect of itself. The nature mystical experience catapults one beyond the ordinary understanding of self and world, and reveals that "consciousness" is profoundly bound up with, and inclusive of (and beyond the limits of), the material realm. Even the capacity to describe the planet as a physiological-organic whole represents an extraordinary feat, one that most systems-theory-oriented ecologists do not mention because they fear that any recognition of unique human capacity reinforces anthropocentrism. Wilber, to his credit, does not believe that interior development is irrelevant to the fact that contemporary environmentalists (or anyone) can appreciate the beauty of, and our dependence upon, this stunning and complex planet. In chapter 9 we discuss nature mysticism in greater detail.

Developmental Holarchy, Volumetric Hierarchy, and Parts and Wholes

As we saw in figure 3.6, Wilber's holarchy involves not one but four hierarchies that correspond to one another. For example, at the individual holonic level, the interior capacity for "irritability" corresponds to the exterior structure of a eukaryotic cell; a cell's irritability corresponds to vegetative capacity at the interior-collective level, and to a heterotrophic ecosystem at the exterior-collective level. According to Wilber, after the development of neuronal organisms, later-evolving levels of complexity centered primarily on brain developments, which co-evolved with interior capacities. The reptilian brain stem corresponds to the interior capacity for impulse, and the limbic system corresponds to the interior capacity for emotion. A major reason that he does not ascribe conscious interiority to social holons, including ecosystems and Gaia, is that they lack the physical structures (e.g., brains) that correlate with such interiority. To be sure, the "interior" status of complex systems remains a very complex and contested area. In any event, Wilber maintains that a comprehensive analysis of a given holon must consider its manifestation in these four dimensions.

Approaching ecological issues from the perspectives provided by "all quadrants and all levels," adjudicates some battles in ecological theory. Take the tension between ecosystem ecology and population ecology, for example.[84] J. Baird Callicott interpreted Eugene Odum's thermodynamic ecosystem ecology to mean that organisms are merely temporary configurations of energy flowing through ecosystems.[85] If an environmental social philosophy is based on the reduction of organisms to functions of systemic processes, then individuals must conform to and support the aims of the systemic whole. In Wilber's view, thermodynamic ecosystem theory has validity but only within the limits imposed by its perspective and methods. Insofar as ecosystem theorists ignore or discount the validity of environmental truths of other legitimate perspectives, such as population dynamics, environmental ethics, 1st-person experience, and so on, ecosystem theorists commit what Wilber calls "quadrant hegemony."

Karen J. Warren and Jim Cheney persuasively argue that ecological hierarchy theory disallows the kind of reductionism that Callicott proposes.[86] Organisms have a degree of autonomy and reality that we cannot easily discount. (Again, organisms are not so much "parts" of an ecosystem as they are members of it.) Life forms and ecosystems mutually condition and influence one another. Despite the many merits of hierarchy theory, it has the same limitations of other instances of general systems theory. Namely, it cannot take into account individual and collective interiority.[87] Hence, ecological hierarchy ignores some crucial dimensions.

Population ecologists, who examine organisms primarily from the Lower-Right (exterior-collective) quadrant, grant reality status to species, demes, populations, communities, and the interactions within and among these various groupings. They often treat ecosystems merely as epiphenomena of, or as resource bases for, organisms in the ongoing struggle for fitness. As in the case of ecosystem theory, however, population ecologists absolutize their perspective if they assume that their perspective provides the most valuable truth about the relationship between populations and ecosystems. Both perspectives are important for observing and interpreting very complex phenomena, and both perspectives are limited. Both approaches offer important truth claims that we must consider in any comprehensive understanding of ecosystem and populations.

However, some part-whole confusion is at work in both perspectives. Ecosystem theorists complain that population dynamics treats the environment merely as resources for organisms, thus failing to consider the allegedly all-embracing character of the biosphere. Arguably, however, organisms/species are not "parts" of the biosphere/ecosphere, but are members of the biosphere that co-evolve with and co-constitute it.[88] Organisms are compound, multilevel individuals that engage in, and interact with, other holons and the environment immediately pertinent at each level. Of course the physiosphere (social holon) and atoms and molecules (individual holons) are foundational to the biosphere.

Population ecologists criticize ecosystem ecologists for reifying a theoretical construct, namely, "ecosystem," as a kind of transcendent structure that is ontologically, temporally, and logically prior to populations. Wilber acknowledges the force of this objection, and argues that from a certain perspective it does make sense to interpret organisms as dissipative structures sustained by the energy flowing through hierarchical levels of an ecosystem. From this perspective, what looks most "real" are energy flows and systemic hierarchy. Other perspectives also reveal important aspects of ecological phenomena.

To look at another perspective, Douglas J. Buege agrees with Rowe and disagrees with Wilber.[89] According to Buege, ecosystems are individuals composed of parts. Buege argues that individual organisms (studied by population ecologists) and ecosystems (studied by ecosystem ecologists) have the interior complexity needed for moral status. He maintains that ecosystems are like baseball teams. Loss of a team (e.g., when the Milwaukee Braves moved to Atlanta) is distinct from and more important than loss of individual players (e.g., when a star pitcher is traded to another team). Buege writes:

> The loss of an entire habitat is a much greater loss than the loss of an equivalent number of individual organisms from various ecosystems because ecosystems are not merely collections of living and non-living beings. Thus, the intuition that higher-level entities such as species and ecosystems are more valuable than the individuals of which they are composed, an intuition shared by many environmental ethicists, may be justified.[90]

The comparison between a baseball team and its members and the constituents of an ecosystem does not hold up to scrutiny. When a major league team moves to another city, the players do not die, whereas loss of an ecosystem entails the death of its constituents. What is the "equivalent number of individual organisms" that can be equated with "an entire habitat"? In what sense are ecosystems "more valuable" than the species and individuals found in them? Ecosystems are more valuable insofar as they are more fundamental (i.e., extrinsic value), but arguably individual organisms have greater intrinsic value because of their relative autonomy and enormous interior and exterior complexity. We also want to keep in mind the importance of comparing intrinsic and extrinsic within either individuals or collectives—comparing each within their own hierarchical organization. In other words, it can be misleading to compare the intrinsic value of an individual with the extrinsic value of a system since both are parts of different hierarchies as individual and social holons.

Despite such problems, Buege raises an important issue—the axiological status of ecosystems (and, by extension, the systems of all such systems, or Gaia). Holmes Rolston III has wrestled with this issue. Although he shares with Wilber the notion that the most intrinsic value is concentrated in complex organisms, Rolston emphasizes both the relative reality and the value-dimension of ecosystems. An ecosystem may lack focused consciousness, but "it has a 'heading' for species diversification, support, and richness. Though not a superorganism, it is a kind of vital 'field.'"[91] For this reason, "Ecosystems are in some respects more to be admired than any of their component organisms because they have generated, continue to support, and integrate tens of thousands of member organisms. The ecosystem is as marvelous as anything it contains. . . . the ecosystem is the satisfactory matrix, the projective source of all it contains. It takes a great world to breed great lives, great minds."[92] The relation between, and the relative value status of, an ecosystem (a social holon) and its organisms (individual holons that are members of it) is an interesting and complex issue that merits further inquiry.

Rowe's holarchy has more in common with Buege's than with Rolston's. Rowe states: "That Nature-as-Earth represents a higher level of integration than the human is a logical extension of the holarchy of containment beyond organisms."[93] Above the organism is the level that Rowe calls the geographic Ecoregion or Bioregion, "a chunk of Earth space resembling a giant terrarium within which humans and other organisms live, move, and have their being."[94] Whereas Wilber describes organisms as members of such a bioregion, thereby recognizing their relative autonomy, Rowe describes organisms as parts of the bioregion, which is part of the more inclusive and integrated whole, the ecosphere. In another essay, Rowe argues that ecosystems and the ecosphere are alive.[95] Restricting life solely to organisms, he argues, invites contempt for the allegedly abiotic constituents of the ecosphere, the terrestrial whole—from Earth core to atmosphere—that is the ultimate source of Earth creativity. Photos of the Earth from space "are intuitively recognized as images of a living 'cell.' Inside that 'cell' cheated by sight, people perceive a particular world separable into important and unimportant parts: the 'organic' and the 'inorganic,' . . . 'living' and 'dead.'"[96] If the Earth were in fact a cell, perhaps it could be regarded as alive, but soon after describing the Earth as an organism, Rowe states that it is not.

Gaia as Super-Organism?

Rowe justifies his claims about the ecosphere by appealing to Feibleman's fourth "law of the levels," according to which the mechanism of any organization lies at the level below, and its purpose at the level above. (This contentious claim seems to deprive any given phenomenon of value or purpose in itself. Instead, something at a given level gains purpose only insofar as it is "part" of something higher or more encompassing. According to such a scheme, we might conclude that individual agency achieves value only insofar as it is part of the communal good. This poses deeply problematic social concepts.) Mechanism, he asserts, can be discerned by analyzing the functional parts at the level below.

> Just so, the function of any given sectoral ecosystem of Earth can be learned by inspecting the interactions of its parts, which are organisms (including people), landforms, soil, air, and water. Ascending the holarchy, the purpose of each holon is revealed in the context of that which encloses it. Thus the role of the heart is to maintain the animal organism in health. The niche of the animal is to play its part in maintaining the ecosystem's integrity. Here is a clue to the role, niche, or purpose of the intelligent human animal in the context of Earth's ecosystems and of Earth itself. Humans, like all holons, ought to act in ways that maintain the health and integrity of the higher-level holons—the regional geographic ecosystems and the ecosphere—in which they are encapsulated.[97]

In "From Shallow to Deep Ecological Philosophy," Rowe reinforces his point: "The purpose of the human being must be found ecologically, in the role played vis-à-vis ecosystems and the ecosphere, not in the narrower roles played vis-à-vis family, ethnic group or society-at-large."[98] Rowe's shorthand conclusion: "Earth before organisms. Ecosystems before people."[99] In "Transcending This Poor Earth," Rowe acknowledges that some will ask whether the "the holarchy that places Earth above people [is] just another path to totalitarianism, to ecofascism."[100] He states that such concerns arise from individualists and humanists who assume that "only people possess high intelligence, are important, and [are] loved by God."[101] Fascism, Rowe correctly points out, is a human institution, not a natural one. Even though it is "ecological reality" that "humans as Earthlings are subservient to the Earth," "Earth's ecosystems express no dictatorial decrees as to human behavior."[102] Humans are free to pursue whatever reckless and self-destructive paths they want. "Earth generally shows humans the folly of their ways slowly, her responses presented as lessons to be learned. Whether Earth is recognized as humanity's body/mind/spirit source and support, and whether or not people act responsibly on that knowledge is their choice."[103]

Let us begin with the notion that humans "ought" to behave as other holons do. Is the "ought" here a prudential ought? If so, then Rowe is surely correct. Humans would be foolish to soil their own nest and destroy the conditions needed for their own survival. But to extend this prudential "ought" to all nonhuman holons is a stretch. Some mammals other than humans engage in prudential behavior, but humans engage in it even more intensively. Nonhuman species sometimes overshoot the carrying capacity of their own niches. Biologists have documented many cases in which the population of a species grows dramatically, due to a temporary abundance of resources, only to collapse or even move to extinction when the resources become scarcer. Most nonhuman holons do not operate under the aegis of a prudential ought. Prudence emerges with organisms capable of reflecting on their situation and intentionally changing their behavior. Surely we would be misusing the prudential meaning of the term "ought" were we to say that a niche "ought" to play its role in maintaining the ecosystem's health and integrity.

Some readers might agree that Rowe also uses "ought" as an ethical imperative. After all, he asserts that Earth is "humanity's body/mind/spirit and support." If Earth is endowed with such honorific qualities, then ought humans to treat the Earth as if "she" were a person, an integrated agency endowed with creative powers and even intentions? Indeed, Rowe asserts that the Earth can be conceived as "one integrated entity," to which we can most properly attribute "the creative synthesizing quintessence called 'life.'. . ."[104] By regarding humans as parts of the ever-higher levels of bioregion and ecosphere, Rowe states that we "shift from navel-gazing homocentrism to Earth-venerating ecocentrism. Matched with Earth's beauty, this [Earth's creative capacity?] is a transcendence that Camus . . . would approve."[105] He personifies Earth as an integrated, transcendently creative female and then adds an aesthetic justification to his claim that humans have an ethical obligation of "ministering to the health of the more creative Being [Earth]" that envelops us: Earth is beautiful.

In presenting truth claims pertaining to Earth's highest position on ontological, moral, and aesthetic hierarchies, Rowe fails to ask what kind of interior development is necessary for him to make such claims. Wilber answers that such cognitive, ethical, and aesthetic judgments presuppose an extraordinarily complex, multileveled interiority that has taken many millions of years to evolve. Yet Rowe's hierarchy cannot acknowledge such interiority. So he concludes that greater external complexity combined with greater size and greater systemic inclusiveness justifies his assertion that Earth, the ecosphere, includes humans as component parts. (This is a typical "bigger is better" formula, which we believe has contributed in important ways to our urgent ecological situation.)

Who exactly is asserting that humans are component parts of the ecosphere? Is it Earth herself, acting as the ventriloquist for whom Rowe serves as the lower-order puppet? If Rowe purports to speak on behalf of Earth, by what criteria should we assess his legitimacy? In "From Shallow to Deep Ecological Thinking," Rowe criticizes what he regards as a misguided hierarchical scheme, "All such hierarchies are abstract conceptual schemes devised by humans and imposed on nature, and clear thinking demands that the different levels be coherent and congruous."[106] Here, at least, Rowe concedes that he is not offering a map of reality, but instead is interpreting the structure of terrestrial reality. Yet elsewhere Rowe displays such strong convictions about the veracity of his far-reaching claims about Earth and humanity's place within it, that he evidently confuses his abstraction with the living Earth. Thus, after indicating that we "intuit" the Earth as a cell, Rowe concludes that in fact, "Earth is not an organism, nor is it a super organism as Lovelock has proposed, any more than organisms are Earth or mini-Earth. The planetary ecosphere and its sectoral volumetric ecosystems are SUPRA-organismic, higher levels of integration than mere organisms. Essential to the ecocentric idea is assignment of highest value to the ecosphere and to the ecosystems that it comprises."[107]

Capitalizing "supra" does not replace the explanation required to demonstrate that Earth has a higher level of organization than organisms. Considering how tightly coupled organs are to organisms, many biologists have resisted the idea that ecosystems—with their more loosely coupled communities—are super-organisms.[108] Still, some support the idea that ecosystems, and even the total terrestrial biosphere, should be considered individuals, because they are self-organizing phenomena.[109] Even Wilber admits that a social holon such as an ecosystem shares many traits with individual holons, but lacks an individual's dominant monad or nexus-agency, because agency is distributed throughout the system's complex matrices.[110] Ecosystems do not transcend and include individuals; they transcend and include smaller ecosystems. Wilber feels that most ecotheorists run into trouble because they fail to understand the relationship between an expanding self-identity and the planet:

> The relationship of an expanding "I" (which transcends and subsumes *its own* lesser identities until it realizes an I-identity with Spirit) leads to an expanding circle of "we" (a "we" that can include *inside* its circle *all* sentient beings), such that individual horizons become fused (or become intersections inside ever-wider circles of care), but individuals themselves don't become "one mass" (or *internal* to a really big organism). As an individual "I" becomes higher/deeper, the circle of "we" becomes larger/wider—but at no point does a particular I subsume other I's, nor at any point does a we swallow individual I's (at no point does a Gaia subsume individuals in an imperium agency. "One Taste" does not mean "one mass" or "one organism" or "one leviathan," but a direct realization that my I is Spirit, my We is all sentient beings, and my It is the entire manifest universe.[111]

Defending the possibility of the individuality of ecosystems, Stanley Salthe argues that "if our observations had the same scale relations to an organism as they have with respect to most ecosystems of biome size, we would not suppose an organism to be an individual either."[112] Elsewhere, Salthe states that "Planet Earth is itself a dissipative structure (Gaia), and it becomes obvious that its degree of control over generally occurring natural forces is, at its own scale, no less elaborate than that in a cell in the sense of requiring highly specified descriptions at appropriate scales. . . ."[113] Although "not advocating ecosystems as 'superorganisms' [as did F. E. Clements in particular]," Salthe maintains that from a very general perspective "organisms could be taken for what one might call 'superecosystems.' . . ."[114] Salthe then concludes that we model individuals on ecosystems, rather than the other way around. In making this move, Salthe seems to make social holons (rather than individual holons, as in Wilber's case) the paradigm of self-organizing hierarchical systems. Nevertheless, Salthe and Wilber would agree that organisms are characterized by internal cohesiveness and a measure of interiority that is at least not yet discernible (not to mention explicable!) in ecosystems.

Despite Rowe's claims to the contrary, the very tight containment/coupling scheme that he proposes, when combined with his ethico-aesthetic imperatives, can in fact justify something like ecofascism.[115] Rowe is correct that ecofascism is a human institution (as are his holarchical conceptions). It is based on a belief system, such as the one he describes, according to which humans are merely parts of a more integrated, transcendent, and valuable whole. If Rowe is correct, our real purpose and our moral obligations are not toward our families and societies, but to Earth itself. Hominids are obligated to be subservient to Earth's own requirements. Rowe, however, neither poses nor answers the following important questions. Who will interpret requirements? Who shall speak for the Earth? And how will disagreement be dealt with? Who will help organize the ecologically ignorant masses, concerned primarily about themselves, their families, and their societies? Who will decide what behaviors help and which hurt the Earth? Are we to believe that there are people endowed with special intuition about what Earth really needs? Will those who prefer to pursue their own narrow purposes be judged as selfish? We will pursue this scenario no further. It may be that Rowe verges on committing what Koestler called the "eighth deadly sin": self-transcendence through misplaced devotion to a powerful figure or a higher cause. Such self-transcendence amounts to personal and social regression.[116] According to Koestler, whose negative experience with Soviet-style communism influenced his views, misguided and unbalanced integrative tendencies (in the form of totalizing social movements, for example) are far more dangerous than self-assertive tendencies.[117]

Bottom Up or Top Down?

In his 1992 essay, "The Integration of Ecological Studies," Rowe indicates his preference for top-down hierarchies—from the most inclusive (biosphere) to the least inclusive (individual organism).[118] Bottom-up approaches, he asserts, emphasize the usual hierarchy of molecules, cells, organelles, and organisms, and only then move to biomes, ecosystems, and ecosphere. The bottom-up approach "invites inconsistencies because, like all 'bottom up' taxonomies that set themselves the task of synthesizing wholes from parts, it lacks an integrating holistic framework from the beginning. Watershed ecosystems cannot be discovered by contemplating populations of trees or forest communities. In short, patterns cannot be discovered 'from below.'"[119]

In Rowe's view, those who follow the bottom-up approach are "unconcerned with ontology (the reality of nature)" and rely instead "only on epistemology (science's analytic mode of knowing)."[120] By contrast, Rowe's top-down approach, according to his own accounts, "imposes internal consistency, because each lower level is derived by subdivision of the one above, as a component of it, and each level constitutes a whole for the levels below."[121] He believes the bottom-up approach (physiology) is related to the lamentable inside-out view adopted by most biologists, who start with the constituents of organisms. By so doing they can never piece together the holistic (ecospheric) puzzle. The alternative approach, which is outside-in (ecology) and top down, reveals "that 'life' is a function of the ecosphere and its sectoral ecosystems, not of organisms per se. Organisms are parts of the supra-organismic systems from which . . . they came. Their roles, purposes, and niches are to be found not so much by reference to others of their kind, or to related kinds, as by reference to the enveloping ecological systems of which they are parts."[122]

Stanley Salthe supports aspects of these claims, when he writes that Gaia's aims "ought never to be conflated with those of people, who (like red blood cells) are at 'her' disposal, not the reverse. We are parts of systems larger in scale than ourselves."[123] Wilber thinks these assertions neglect the fact that humans are not merely physical or organic beings, but are also noospheric ones, and our noospheric aspect is not merely a function or part of an all-encompassing physio-organic system. Materialism is materialism, no matter its guise. Here we can see the subtle ways that materialism has affected virtually all ecological thought.

The hierarchy theorists Allen and Hoekstra question the validity of Rowe's holarchy.[124] They believe that he overemphasizes nesting hierarchies, and thereby neglects the clarity provided by describing scalar phenomena in terms of non-nesting hierarchies (e.g., food chains). Moreover, they find "Rowe's distinction between looking in towards mechanism (physiology) as opposed to outwards towards role (ecology) overstated. . . . By such an important cleaving of the two protocols one misses the important point of

Koestler (1967), that a given structure is the interface between processes driven from below and structural constraints imposed from above."[125]

Rowe's emphasis on top-down hierarchies also runs counter to Koestler's holonic scheme. Bottom-up approaches show the limits of possibility (Wilber writes of constraints, communion, agape, responsibilities), whereas top-down approaches "focus on the special cases that upper level structures allow" (Wilber writes of possibilities, agency, eros, and rights).[126] Of all the arrangements that physical forces make possible, only certain ecological structures actually occur. Rowe correctly asserts that we cannot predict senior-level structures from the more junior-level dynamics that they constrain. But he omits that it is also not possible to tell from senior-level structures alone what limits are fundamental (from below) and which are lonely local structural restrictions. Both approaches, bottom up and top down, are needed to reap the conceptual harvest made possible by Koestler's idea of holonic hierarchy. Finally, Allen and Hoekstra contend, "Rowe is reifying 'ecosystem' even though it is after all only a conceptual device."[127] Because "levels of organization arise from" decisions made by observers, "it makes no sense to us to assert that ecosystems are material, or somehow fundamental, whereas communities and populations are abstract contrivances."[128] By reifying ecosystems (not to mention the SUPRA-individual ecosphere), Rowe engages in a "matter of faith not scientific investigation. Naive realism is the right of anyone who chooses to embrace that philosophy, but it is not verifiable, and so is no position from which to organize a data-driven scientific inquiry."[129]

Wilber emphasizes that his holarchy is a tentative description, a map, drawn from the perspectives devised by hundreds of investigators. He draws on the findings of many different disciplines in order to develop "orienting generalizations" that can help integrate many different cognitive, cultural, and social domains. He emphasizes that his "broad orienting map is nowhere near fixed and final. In addition to being composed of broad orienting generalizations, I would say [*Sex, Ecology, Spirituality*] is a book of a thousand hypotheses. I will be telling the story as if it were simply the case (because telling it that way makes for much better reading), but not a sentence to follow is not open to confirmation or rejection by a community of the adequate. I suppose many readers will insist on calling what I am doing 'metaphysics,' but if 'metaphysics' means thought without evidence, there is not a metaphysical sentence in this entire book."[130]

Although we have criticized Rowe's position, we acknowledge his attempt to make sense of humanity's relation to the wider world, and his desire for humankind to treat the physiosphere and biosphere with respect. But by thinking that what has more span is more fundamental, and more significant, than what has greater depth, and by reducing humans and other organisms to parts of the ecosphere, Rowe aligns himself with the regressive tendency of some deep ecologists. (Zimmerman has argued elsewhere that deep ecology can be interpreted in a progressive way that is generally consistent with Wilber's view.)[131]

A Developing Noosphere

Now that we understand how a developmental conception of Kosmic history can help clarify the complicated issue of humanity's relation to nature, let us study how a developmental conception of interiority (of the noospheric domain) can empower Integral Ecologists to promote solutions in terms that make sense to people at different waves of development.

4

Developing Interiors

It seems a strange thing, when one comes to ponder over it, that a sign should leave its interpreter to supply a part of its meaning; but the explanation of the phenomenon lies in the fact that the entire universe—not merely the universe of existents, but all that wider universe, embracing the universe of existents as a part, the universe which we are all accustomed to refer to as "the truth"—that all this universe is perfused with signs, if it is not composed exclusively of signs.

—CHARLES S. PEIRCE[1]

Look Within: "I" and "We"

In previous chapters, we explored some major developmental concepts in the Right-Hand quadrants. These quadrants comprise Nature, that is, the Kosmos as observed from 3rd-person, monological, objectifying perspectives and methods. We discovered the following about humanity's place in Nature. First, humans—as physiological beings—are composed of the elemental building blocks of cells that are constituent *parts* of themselves and are members the biosphere. In addition, the atoms within a human body are parts of that individual holon and are members of the physiosphere, which is a constituent part of the biosphere. Second, humans—as biological beings (i.e., as animals)—are *members* of the biosphere. Third, humans—as mental or noospheric beings—*contain* the biosphere as constitutive parts without which noospheric beings cannot exist. Thus, humans can simultaneously be parts of, be members of, and contain various developmental levels of Nature, depending on which level of Nature one is talking about: physiosphere, biosphere, or noosphere. Each of these three developmental aspects of Nature has evolved over millions of years. Each aspect transcends but also includes the previous aspect. For instance, the biospheric aspect of an earthworm transcends and includes the physiospheric level developed by Nature. The earthworm thus contains the prior developmental level, physiosphere, but is a member of the social holon that belongs to its own developmental level, in this case the earthworm's ecosystem. Analogously, an atom is a part of an organism (higher level) but is a member of its galaxy (social holon at the same level as atoms). A horse, however, unlike an earthworm, includes a noospheric aspect, because the horse's neocortex makes image generation possible. As in the case of humans, the horse's noosphere transcends and includes both the biosphere and physiosphere. The latter two are "parts" of the noosphere, which depends on them for its foundation. Organisms are constantly involved in exchange relations with physiospheric phenomena, including substances such as water, food, and oxygen. Exchange relations also exist among organisms that are members of the ecosystem, for example, when one organism consumes another. Physiosphere, biosphere, and noosphere are all interdependently related, developmental features of Nature.

In this chapter, we discuss interior development, which develops from prehension at the atomic level all the way through increasingly more inclusive waves of human interiority.[2] Of course, interior growth is inextricably connected to cultural development (LL), and profoundly influenced by Right-Hand developments, including techno-economic development and urbanization (it is always a 4-quadrant affair).[3]

It is difficult to completely understand how our complex noosphere, which is an aspect of Nature, contributes to our definition of "Earth" and "nature."[4] The capacity to conceive of Earth as one planet among many is one that transcends and includes the physiosphere and biosphere. Vast interior development is required to conceive of Earth. Sometimes people talk nostalgically about Earth as it must have been before human beings arrived on the scene. Others imagine how fair Earth would become once again, if only humans were gracious enough to bow out of—or were forcibly removed from—the scene! Yet great interior depth is required to conceive of an Earth without people. This conception of Earth still arises within human cognitive, emotional, and aesthetic capacities, because the human imagining it secretly remains behind after everyone else is imagined away. It is difficult indeed to know how to subtract the ways in which our complex noospheric capacities contribute to what we call "Earth" and "nature."

Human noospheric development—which represents a level of development of Nature—is such that it allows phenomena to appear (including conceptions of Earth) that other beings at other levels of complexity within Nature cannot conceive of. Depending on personal and cultural development, however, people experience, define, and interact with "nature" differently. Hence, there is no single nature, Nature, or even NATURE; rather, "nature" reveals itself differently depending on your perspective and level of development. Yet nature is not entirely socially constructed either. Make no mistake; there are phenomena that reliably show up within the phenomenal worldspace of humans. There is grounded consensual agreement that such things as oceans and trees, unicellular organisms and DNA all exist.

However, we often disagree about the value-status of the natural environment. Is it valuable in itself, or is it merely instrumentally valuable for humankind? How you answer that question depends upon the norms of your culture and your individual moral development—it depends upon your individual and collective interiors. We believe that one of the biggest breakthroughs for dealing with environmental problems would be to acknowledge that individual and collective interiors exist, that they develop, that there are multiple ways of conceiving of nature, and that a successful ecology must learn to communicate in accordance with such an understanding.[5]

Development: How Does It Work?

Before we discuss interior levels and lines of subjective and intersubjective development, we want to explain how the self navigates development. Wilber speaks of the *ladder*, the *climber*, and the *view*. The ladder refers to the developmental waves, levels, or centers of gravity along developmental lines. Wilber is uncomfortable with the ladder metaphor and insists that his use of a ladder in this model is a simplified metaphor and should not be taken to suggest a rigid, step-by-step, linear model of development. Instead, development is complex with moments of progress and backsliding, breakthroughs and stasis. Hence, preferable terms are waves not rungs, streams not lines. In fact, the "ladder" is best thought of as the result of taking a cross-section of the concentric waves of development that enfold each other. As the climber moves up the ladder, the rungs stay in existence; they endure, but the view changes. Thus the climber can see the previous rungs, but the climber can no longer see how things look from those rungs.[6]

The climber mentioned above is the self-system, which is ever-evolving and adjusting to development along its various lines or streams. If we misunderstand development as the self merely acquiring a more comprehensive view of what is already there, we are simply modernists in integral garb, as modernists believe that mind (the viewer) is the mirror of nature. Nothing could be further from our understanding. Wilber agrees in important ways with constructive postmodern theory, and therefore insists that each rung in the ladder opens the self-system to a new clearing in which phenomena can arise that simply are not visible and cannot be apprehended from earlier developmental levels. Wilber writes, "A 'level of consciousness' is simply a measure of the types of things and events that can arise in the first place; a measure of the spaciousness in which a world can appear; a degree of openness to the possibilities of the Kosmos. . . ."[7] Consciousness is not a thing, then, but instead the opening or clearing necessary for things, events, and other phenomena to appear. We can deduce the level of consciousness at which people operate by observing their behavior and what sorts of phenomena they encounter.

At each transition from one developmental stage or wave to another, greater integration can arise—or pathologies can emerge. If a given stage is not integrated, we leave bits of ourselves behind, and become increasingly fragmented. Thus, we might have subpersonalities at a lower rung, which retain the view of that rung. Because these unconscious aspects of ourselves are not integrated, we can project them onto others. Hence, our relationship with the natural world is mediated in part by object-relation dynamics.[8]

In healthy development, we identify with a given center of gravity. When we exhaust the limits of that center, we look in other directions. Ideally, we differentiate from the old center, rather than dissociate from it, as we establish a new center. Yet dissociation from the previous stage occurs all too often.

For example, Michael Zimmerman was the son of a chemical engineer who specialized in producing industrial plastics. Michael began to dissociate from his conventional Catholic upbringing in the late postconventional age of 1960s. He became an Eco-Romantic individualist, increasingly judgmental toward those he thought were still too blinded by religious dogmatism or corporate rationality to see the evils of urban industrialism. The antimodernist sentiments of Nietzsche and Heidegger fed his Eco-Romanticism, but simultaneously conflicted with his postconventional worldcentrism. After publishing essays that portrayed Heidegger as a forerunner of deep ecology, Michael had to confront Heidegger's support for National Socialism. If Heidegger was an antecedent of deep ecology, what exactly did Michael's Eco-Romantic, modernity-bashing deep ecology represent? What kind of worldcentrism refused to embrace hard-working, sincere, and intelligent people like his own father who were part of corporate-industrial society? What kind of worldcentrism critically judged churchgoing people who had not become freethinkers? When Michael encountered Wilber's book *Up from Eden* in 1982, he discovered a way to integrate the noble contributions of modernity and traditional religion without denying their limitations. As a reflection of this understanding, in the mid-1980s, when he was working to end the nuclear arms race, he shaved his beard and cut his long hair, so that businesspeople, attorneys, homemakers, and ordinary citizens would take his concerns seriously. Environmentalism became a way for him to defend the natural world, integrate his own development, and knit together the fabric of human life.

Many researchers assert that all humans pass through at least three stages of moral development: preconventional, conventional, and postconventional.[9] The sequence of these stages is fixed: egocentric to ethnocentric to worldcentric to planetcentric. You cannot operate from a postconventional level without first learning to operate from a conventional level. An infant or toddler is totally preconventional: self-oriented, incapable of encountering the "other" as someone with needs, feelings, and views of their own. (Those who envision enlightenment as returning to the moral and spiritual condition of children are misguided, unless they believe that enlightenment is an undifferentiated, egocentric state. A fully enlightened individual is the very opposite of egocentric and self-absorbed.)

The next generally agreed-upon stage is conventional, characterized as ethnocentric. Here, you strongly identify with the moral beliefs, customs, and principles of your own people, and disapprove or are skeptical of the mores held by others. Those who inhabit the postconventional stage understand that your cultural mores are not absolute but instead are partly particular to your nation or people. This disintegrative and disorienting discovery can result in pathology, unless you integrate the moral insight that people from other countries, religious backgrounds, and ethnic groups are all worthy of respect and proper treatment.

One way of construing the levels or waves of individual and cultural development is shown in figure 4.1, another version of the AQAL diagram.

Developmental Lines: A View of the Interior

As mentioned, these stages tetra-evolve. Hence, at level 4 of figure 4.1, the self's center of gravity is at the mythic level, which corresponds to the mythic order of culture and is consistent with premodern countries, and which requires corresponding brain structure. Of course, this diagram is unavoidably abstract and limited. In *Integral Psychology* and elsewhere, Wilber emphasizes that individuals are constituted of many different lines, each of which develops at its own pace, although in interaction with other lines. If we were to include all of those lines in this diagram, it would lose its usefulness as a

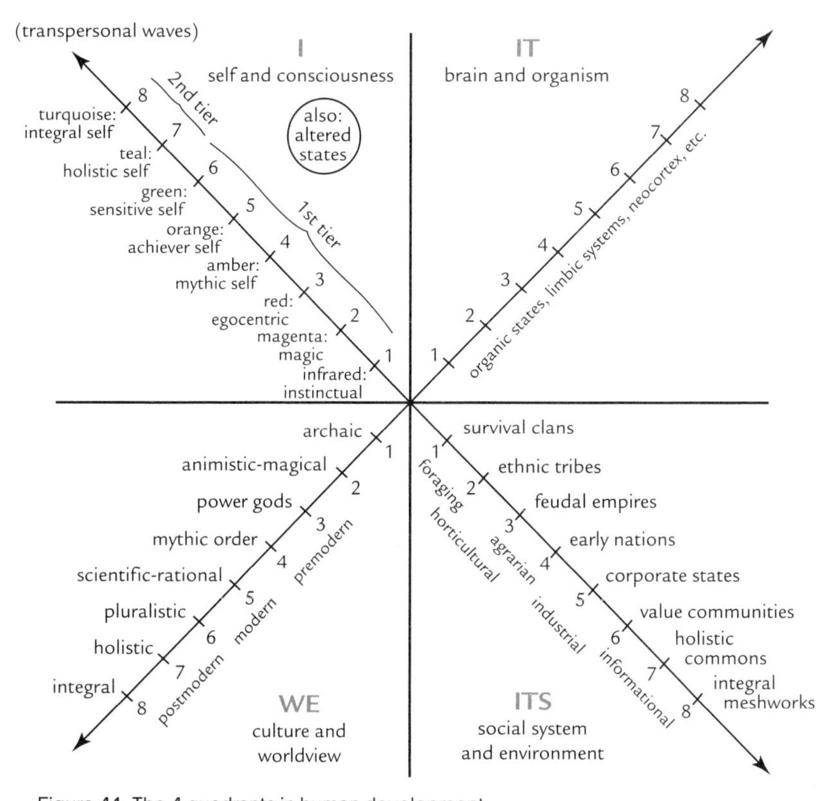

Figure 4.1. The 4 quadrants in human development.

heuristic device. So each of the four lines in the diagram actually represent multiple other lines in each quadrant.

In *Frames of Mind* (1983), Howard Gardner introduced his theory of "multiple intelligences." He found that I.Q. tests measure only verbal and logical-mathematical intelligence, but omit other important kinds of intelligences, such as visual/spatial, bodily/kinesthetic, musical, interpersonal, and intrapersonal.[10] Wilber expands upon the model of multiple intelligences to include still other developmental lines and levels along those lines.

> There is moderate to strong evidence for the existence of the following developmental lines: cognition, morals, affects, motivation/needs, ideas of the good, psychosexuality, kinesthetic intelligence, self-identity (ego), role-taking, logico-mathematical competence, linguistic competence, socio-emotional capacity, worldviews, values, several lines that might be called "spiritual" (care, openness, concern, religious faith, meditative stages), musical skill, altruism, communicative competence, creativity, modes of space and time perception, death-fear, gender identity, and empathy.[11]

Integrating these lines is a life-long endeavor, because we develop along many different lines and integrate or dissociate earlier levels of development in any given line. Just because someone is highly developed in the cognitive line, that does not mean that they are highly developed in others. Recall the bright Nazi doctors, including Joseph Mengele, who tortured enslaved human subjects. Any attempt, then, to categorize an individual based on their level of development in one or two lines is clearly illegitimate.

Figure 4.2 is a psychograph, a simple and elegant diagram that depicts various lines of a fictitious person and the relationships among them. Each column represents a different line. The cognitive line refers to one's capacity to take perspectives, not one's intellectual capacity. Research indicates that it is necessary, though not sufficient, for development in other lines or streams. It creates the clearing in which some other important lines can manifest. For instance, you have to be able to see a perspective (cognitive line) before you can talk to that perspective (interpersonal line); likewise you have to be able to talk to a perspective (interpersonal line) before you can be that perspective (self line).[12]

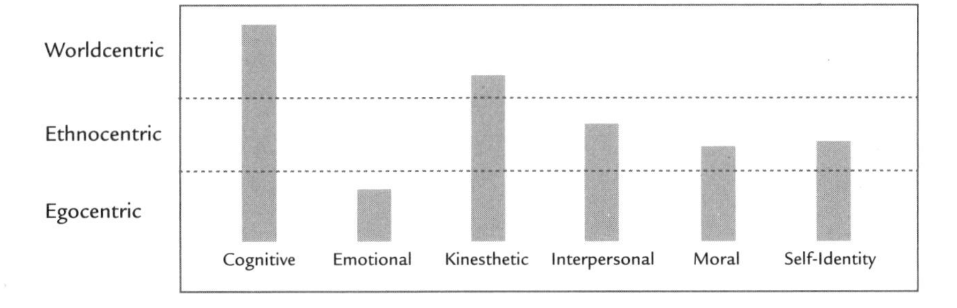

Figure 4.2. Psychograph.

Individual Centers of Gravity

Even though different lines develop at different rates, we can speak abstractly of an overall center of gravity. For instance, a 2-year-old will have a very different emotional and moral capacity than an 18-year-old or a 56-year-old. One of Wilber's schema names nine possible centers of gravity, which are divided into three groups: prepersonal, personal, and transpersonal. Prepersonal centers of gravity—using the cognitive line as a marker—begin with the undifferentiated or primary matrix (child in utero), and move to the sensoriphysical (first few months of infancy), phantasmic-emotional (first 2 years), and representational-mind (2 years and beyond, with onset of linguistic capacity).

The personal centers of gravity include concrete operational (the prepubescent child internalizes rules and roles), formal reflexive (where the adolescent discovers that rules and roles are changeable and even conventional), and centauric/vision-logic (where some adults become capable of integrating mind and body, and entertaining several competing perspectives).

Transpersonal centers of gravity include the illumined mind, intuitive mind, and overmind.[13] These levels include and transcend the limits of the personal centers of gravity. Moving from one developmental center of gravity to the next is fraught with difficulty and can take many years.[14]

In figure 4.3 individual development is depicted in colors drawn from Wilber's use of the rainbow to represent the spectrum of consciousness. It represents the altitude of levels in general.[15] These colors avoid the difficulties with using the terms of one developmental model (e.g., concrete operational cognition) to describe a level in another model (e.g., conformist self-identity). As a result, we can speak of amber cognition and amber self-identity as a particular complexity of consciousness.[16]

Another important way to categorize the centers of gravity of the developing self is first, second, and third tier, as indicated in the UL quadrant of figure 4.1 and in figure 4.3. People operating within levels of development known as first-tier identify almost exclusively with their center of gravity and thus tend to regard other centers as inadequate or misguided. For instance, someone operating at the orange (modern) center of gravity tends to regard people at the amber (mythic) center as naive, superstitious, and irrational. First-tier centers of gravity, or developmental waves, include instinctual (infrared), magic (magenta), egocentric (red), mythic self (amber), achiever self (orange), and sensitive self (green).

People operating at second tier no longer exclusively identify with any particular perspective. Instead they adopt an increasingly integrative multiperspectivalism, which recognizes the partial truth content, however limited, of more junior centers of gravity. Instead of eradicating orange rationality, as some green counterculturalists would prefer, or oppressing amber mythic belief, as some orange moderns want to do, second tier encourages healthy expression of all developmental levels without demanding that people vertically transform.[17] Second-tier centers or waves include holistic self (teal) and integral self (turquoise). See figure 4.3.

It is inaccurate to describe anyone as wholly "red" or "green." Development is differential in nature. People may be red in one line, orange in another, and green in still another. A man may occasionally act from the red, emotional center of gravity when a car cuts him off; from an amber, interpersonal center of gravity when he attends church; from an orange, cognitive center when he is competing for a professional promotion; and from green values when he supports a Sierra Club initiative to curb factory pollution. Typically individuals operate from both the level above and below their center of gravity 25% of the time in any one line. Hence, we can say that someone acted in a way consistent with green or orange, but not that someone *is* green or orange.

Figure 4.3 indicates that the sequentially developing UL and LL levels form a holarchy. Ideally, the orange level transcends and also includes the contributions of the amber level. Compared with the infrared, magenta, red, and amber centers, the orange and green centers have emerged relatively recently. For instance, the rational-achiever self became significant only about three centuries ago in connection with the scientific, political, economic, social, cultural, and technological developments that made mo-

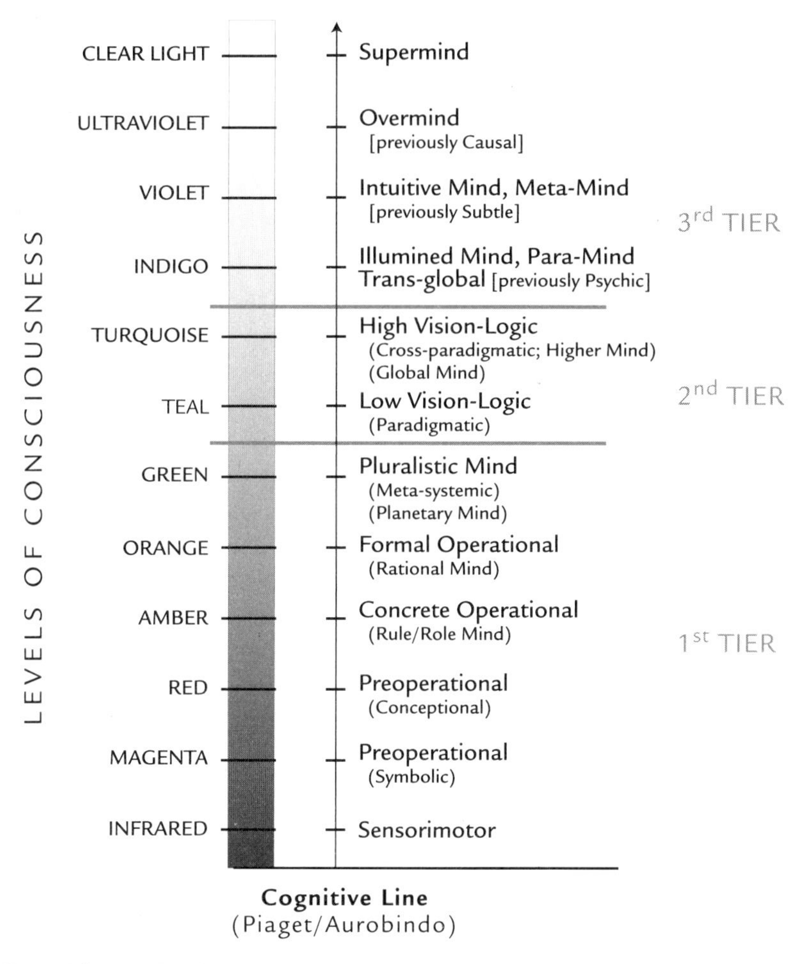

Figure 4.3. Wilber's color spectrum of consciousness.

dernity possible. As for moral development, an individual adult's center of gravity is usually consistent with the overall center of gravity of their culture. Yet some adults will operate at a center, wave, or level that either lags behind or surpasses the prevailing center of gravity of their culture. The AQAL diagram divides the culture/worldview portion of the quadrants (LL) into premodern (archaic through mythic order), modern (scientific rational to pluralistic), and postmodern (integral and beyond). In many cultures, such as the contemporary United States, no one center of gravity prevails; instead, there are three competing centers: traditional, modern, and postmodern. The first is the mythic or amber center, which includes evangelical and traditional Christians who worship a law-giving Divinity and who expect society to operate according to those laws. The second is the scientific-rational or orange center, which renounces (in principle, at least) the ethnocentrism of the mythic or amber center. Orange adopts the moral socio- and worldcentrism espoused by Enlightenment moderns, who emphasize that through rationality, individuals can attain autonomy, exhibit responsibility, and achieve personal aims. The third is the relativistic or green center, which deconstructs the meta-narratives of Western science and makes room for multiculturalism, indigenous wisdom, new age spirituality, and environmental activism. Amber and orange compose about 70% of the adult population in the U.S.A., whereas 25% of the population operates from green pluralism or multiculturalism. It now seems that 2% or 3% of the population operate at the holistic center of gravity, and much smaller percentages at integral and beyond.[18]

As we have said, each individual must pass through earlier developmental waves, prior to attaining the later ones. Likewise, *members* of cultures cannot skip entire developmental stages without serious negative consequences. (However, cultures can "skip" stages in the sense that a group of card players can have five amber players, three orange players, and two green players. The center of gravity—the dominant mode of discourse—is amber, because most of the players operate from an amber level of development. But if two amber players leave the group and are replaced by two green players, the center of gravity—the dominant mode of discourse—then "skips" to green.)

One of the most infamous attempts to force vertical transformation occurred early in the 20th century, when communist revolutionaries concluded that Russia's agrarian society, whose members operated from an amber-mythic center of gravity, could make the great leap forward into the workers' state (the very idea of which arose from an industrial orange center of gravity that was beginning to turn communist green). Marx had long maintained that such a leap was impossible, so he focused his revolutionary energies on industrial countries such as Germany, which had developed toward a modern industrial model and had a substantial class (many members) of orange capitalists, intellectuals, attorneys, managers, small businessmen, and artists.

Only later in life did Marx entertain the possibility that the members of an entire society could somehow jump from amber agrarian to green socialist without the intermediary orange capitalist-industrial-bourgeois developmental center of gravity. Lenin, and especially Stalin, committed the extraordinary blunder of incarcerating, killing, or rendering impotent much of the bourgeoisie (the only modern members!) that had developed in Russia by the end of World War I. The Soviet Union was a social, political, economic, cultural, and environmental catastrophe. We hope that this serves as a lesson: members of a society cannot skip developmental stages. Nor is it wise for a society to eradicate, imprison, or murder its individuals who have stabilized the more advanced levels of development.

Interior Individual Development: An Example

To explore the idea of development in greater detail, and to understand the ways that worldviews shape how we think, we will unpack interior individual development in this section. The evolution from egocentrism to ethnocentrism to worldcentrism is necessary for resonating with many of the noble aims of environmentalism. You must care about something more than just yourself, or more than just your family or tribe, to care for the natural environment. As we discuss in detail in chapter 7, our individual developmental scheme is largely drawn from Susanne Cook-Greuter's full-spectrum research on postautonomous ego development of the self-identity line (we also draw on William Torbert's action-inquiry research).[19] We prefer Cook-Greuter's and Torbert's models of individual psychological development because each of their models covers the first, second, and third tiers (though we will not discuss third tier until chapters 7 and 9), and because they provide the most sophisticated and extensive full-spectrum (prepersonal, personal, post-personal, and post-postpersonal) research available and are fully consistent with Wilber's synthetic framework.[20] According to Cook-Greuter, ego development refers to an integration of several developmental lines at a given altitude. The cognitive line, which drives self-understanding and the concomitant self-identity, is at the same level as a person's ego stage. Ideal ego development occurs when lines are actually integrated into a coherent self-system. When the lines aren't appropriately coordinated, mild to severe psychopathology occur.

Cook-Greuter and Tobert's levels closely coincide with those of Clare Graves's model of value systems—though Graves's research in our view focuses on one line—value development.[21] Cook-Greuter's model also closely coincides with the developmental levels described by Wilber. In the following schema, we use Wilber's colors to refer to altitude (i.e., levels of development in any line). According to Wilber, each level of altitude also represents a content-free space of consciousness that serves as a horizon of meaning within which the content of various lines of development can be manifested and/or generated. Integral Theory uses the cognitive line as the first source of content within any given altitude due to the evidence that the cognitive line is necessary but not sufficient for the other self-related lines.

Each self has a unique way of relating to itself, others, and the world.[22] In addition, each self has a healthy and an unhealthy ("mean") expression. We use these descriptions first to show *how* individuals think at each level. Then we briefly apply their worldview to how they might register, assess, and respond to environmental problems, such as endangered species. We apply this map further in chapter 7 when we present the 8 ecological selves. We begin with centers in the first tier.

First Tier

1. Infrared—Instinctual/Symbiotic. The *symbiotic self* is an individual who focuses entirely on surviving in an incomprehensible world. This level mainly pertains to infants, but perhaps also to some people who, in extreme situations, revert to this early level. People at this level are nonverbal and heavily dependent on others for care. The main task for the individual of this first stage is to construct a stable world of objects so as to separate from their surroundings. At this center of gravity, the concept of an endangered species cannot and would not even arise.

2. Magenta—Magical/Impulsive. The *impulsive self* balances dichotomous forces such as good and evil. Mostly children and some adolescents occupy this level. Relatively few contemporary adults remain at this level of development.[23] People at this level or altitude have a strong concern for creating safety and satisfying basic needs. They also have a sense of unlimited power combined with superstitious and magical notions. Their activity is often highly repetitive. Moreover, they view other people primarily as a source of self-gratification, and feel confused and made anxious by the complexity of the world. This is the beginning of the 1st-person perspective.

At this altitude, the individual's impulses and needs are most important, and those needs are deeply influenced by their communities' time-honored rituals, taboos, superstitions, folkways, and lore. Mean magenta is the dark side of magenta, when blood oaths, ancient grudges, strong ethnic identification, and an impulsive readiness for violence (although often ritualized) prevail. Projection and introjection are their common defenses. Neither constructive nor mean magenta have access to the concept of endangered species. Furthermore, few if any contemporary groups function primarily at this level.

3. Red—Egocentric/Self-Protective. The *self-protective self* is impulsive, but impulsivity is now placed in the service of supporting an incipient self-structure, and not just to satisfy immediate needs and wants. Red individuals typically identify themselves in terms of will, ideas, and wishes. They project all their feelings and cannot yet self-reflect. Hence, overgeneralization is rampant. Seeing others as competitors for space, goods, and dominance, people operating at red have little capacity for explicit insight into self and others. Because they experience the world as a dangerous place filled with perilous risk, they often cross others' boundaries in a crusade of low trust and hypervigilance.

At red, self-assertive individuals break free from constraint placed on them by group rituals and codes. Red selves vent their desires, demand respect and attention, hate being dissed, and experience no guilt or remorse for their actions. Examples include a child's temper tantrum, mercenaries, some epic heroes, warlords, the shipwrecked students depicted in *Lord of the Flies*, test pilots, and in some cases America's rugged individualistic mentality. People riding the red wave also possess the growing capacity to see the world more "realistically," that is, with fewer projected spirit forces and with greater distinction between self and other, including totem animals, and to direct their impulses into longer-term goals.

At this wave of development, individuals would support the protection of endangered animals insofar as it gave them status or leverage toward their own self-serving goals. For example, some individuals in developing nations are aware of the value that industrialized nations place on protecting endangered species. As a result, these individuals become involved in protective projects to bolster their own status within their community. This same developmental level can support blatant disregard for endangered species. For instance, a self-protective individual who enjoys trophy hunting might take a safari to East Africa to hunt big-game species before they are placed on an endangered species list, in which case hunting them would become more difficult.

Operating as mean red, power-hungry individuals care nothing about the fate of animals and plants, apart from exploiting them for enhancing the status and power of the self. Threatening species are to be either dominated or extirpated, whereas useful ones are to be cultivated and exploited. These individuals may exhibit remorse if extinction causes short-term costs to individual pleasure, power, wealth, and status, although anger, frustration, and blaming are more likely. In any event, such individuals do not yet experience guilt, because they cannot conceive that they have acted in a morally unethical way. Lacking compassion, they cannot conceive of the inherent worth of species or of the long-term negative consequences of wiping them out. Mean red displays the same attitude toward human beings—dominate one's opponents, and make friends so long as they are useful. The consciousness associated with this altitude can be atavistic and gang-oriented, as it is in some urban areas where both amber and orange authority has broken down.

4. Amber—Mythic/Conformist. The *conformist self* is traditional, rule-oriented, and concerned with group membership. People whose center of gravity is at this level define themselves through others. They have no stable and clear boundaries between the self and the group. Displacement, reaction formation, and suppression are common defenses. They suppress negative feelings and overemphasize positive ones. They have a strong need to be accepted and to reject those who do not conform to group norms. They tend to view their world through a concrete-literal lens that includes concrete beliefs in many mythic realities. Clare Graves called this level "absolutistic." Despite such limitations, amber marks the beginning of a 2nd-person perspective in that the individual's own perspective (red) becomes subordinate to the groups perspective.

Perhaps in reaction to the havoc wreaked by power-craving individuals, this level seeks to establish overarching values and truths that all must obey. At this stage people rely on sacred scripture, prophets, and similar inspired authority to discover meaning, purpose, and direction in life. Coercive social power enforces codes of conduct based on eternal, absolute principles. By controlling personal impulses and living righteously, people are guaranteed a future reward. Everyone must assume their proper place in the social hierarchy, at the top of which is an external authority whose commands must be obeyed. This authority can be God, the president, the king or queen, the czar, the leader, or the boss. Respect for the law and exercise of discipline are needed to build character and moral fiber. Examples of amber-mythic cultures include contemporary Singapore (Confucian culture in modern garb), Hasidic Judaism, the European Middle Ages, and contemporary evangelical Christian groups. As a premodern center of gravity, amber-mythic culture remains ethnocentric, even though sacred scripture often points the way toward a universalism. This universalism is misinterpreted as the universalism of their view: the one and true God.

Members of amber-mythic culture can be concerned about and can take constructive steps toward preserving other species. Many indigenous communities at an amber level protect endangered animal species, not as biological placeholders within ecosystems, but instead as associates (or even as extensions) of the tribal self (a version of this can also occur at the red altitude). Amber-mythic shows special respect for species with traits—such as courage, resourcefulness, nobility, obedience, faithfulness, even humor—that have symbolic, allegorical, and inspirational value. Amber monotheism may affirm that species deserve respect because they are creatures of God.

On the other hand, mean amber can be strongly anthropocentric, using scripture to justify human interest, especially those connected with the quest for salvation. For instance, some contemporary evangelicals maintain that instead of letting Earth lie fallow, in the form of "habitat protection," people should follow scripture by "developing" the planet, even at the cost of plant and animal species. Endangered species can also be placed at risk within secular communities at an amber center of gravity because they often don't see the value of protecting them. Individuals within those communities are often afraid to speak out, fearing that they will be ostracized or face other group consequences, or they trust the community consensus and fail to think for themselves about the issue.

5. Orange—Achiever/Conscientious. The *conscientious self* emphasizes linear causality and objective thinking in the service of a newly emerging separate self-identity, which competes for autonomy, wealth, power, and status. People at orange have great independence and confidence. Although interested in their emotional life, they emphasize rationality. They associate with others who are also drawn to achieve through concision, efficiency, and efficacy. Still, they have a genuine interest in others, independent of their own needs and values. They experience the world as predictable and measurable. At this level, people develop the capacity to hold 3rd-person perspectives. That is, they can disidentify with their 1st-person (personal) perspective and with the 2nd-person (cultural) views, thereby gaining the ability to examine both self and culture from a 3rd-person, objective, "scientific" standpoint.

At this stage of development culture moves from premodern to modern. The orange altitude advocates that people should act in their own self-interest. Liberal capitalism is the major sociopolitical expression of orange. (Socialism, another major expression of modernity, shares a great deal with liberal capitalism, including a largely anthropocentric, exploitative attitude toward nature. According to socialists, however, collective ownership of the means of production is required to achieve individual self-realization.)

Typically risk takers, optimistic, and self-reliant, orange moderns use strategy, technology, and competition to win. Despite their strong competitive streak, orange moderns are at least in principle worldcentric, as they demand that everyone be granted the same legal protections and access to markets. By positing universal human rights, orange took a giant step forward, even though centuries later not all humans have acquired such rights. Typically anthropocentric, orange moderns regard nonhuman species as objects to be dominated for sake of personal or communal advancement. Gifford Pinchot, founder of the U.S. Forest Service, promulgated an orange-based management approach that attempted to reconcile the competing goals of maximum exploitation of natural resources and long-term sustainability. Orange is concerned about endangered species mostly if some utilitarian value is at stake. For instance, if the species might be the source for new pharmaceuticals or might be an amenity for an exclusive housing development, orange could be persuaded to protect it. From an orange perspective, change—whether extinction of a species or global climate change—is inevitable and often costly. Change, however, is also frequently an excellent opportunity to define and serve new markets.

According to environmental historian Roderick Nash, as certain moderns continued to expand the circle of beings deserving of moral standing, they began demanding rights for nonhuman beings, thereby making possible John Muir's preservationist environmental ethics, which is to be contrasted with Pinchot's utilitarian environmental ethics.[24] With Muir, however, orange individualism begins to move toward green pluralism.

Some free market environmentalists use orange economic incentives and market mechanisms to achieve increasingly green aims, including preservation of species. For moderns, however, the value preference for such preservation cannot be imposed on others. Hence, according to moderns, it is not only expedient but also just to compensate landowners who are legally required to preserve habitat needed for endangered species.

Mean orange: Radical individualism combined with anthropocentrism can lead to a very shortsighted and abusive attitude not only toward other people but also toward one's own property. For instance, if my interests as a mean orange property owner are the only ones that I consider, then I will be unaffected by arguments that I have an obligation to future generations of humans, much less to present generations of animals and plants. If my decision to clear-cut my forest property incidentally contributes to the extinction of a species, I as mean orange am unconcerned, unless of course someone can demonstrate that the market value of the species is greater than the market value of my trees.[25]

6. Green—Sensitive Self/Individualistic. The *individualistic self* emphasizes connectivity between people especially by sharing experiences, acknowledging contextual aspects of relationships (e.g., gender, class, race), and systemic dynamics of reality. Paradoxically, because they are so sensitive to individual feelings they become strong advocates of community and egalitarianism. Also, because they are aware of the conditioning dynamics of culture and context, and because they take into account multiple viewpoints, those at the green level empathize with others and are willing to entertain alternative truth claims. Although they appreciate objectivity and logic, they tend to emphasize subjective and more holistic and organismic approaches to meaning-making. They value feelings and their expression.

People operating at green condemn the greed and unhealthy materialism of orange individualism that can legitimate environmental destruction and social injustice. However, their hyperfocus on orange leaves them blind to the oftentimes worse injustices committed by red and amber, and when they do register these injustices they often mistakenly attribute them to orange. Greens promote an egalitarian agenda of justice, equity, and participation by all. Affirming consensus and eschewing hierarchy of all kinds, greens explore alternative forms of spirituality and more community-oriented lifestyles. They discern a link between environmental and social exploitation, and so they insist that there is more to environmentalism than just the wilderness protection favored by deep ecologists, whom greens often regard as not only too macho but also ethnocentric. Greens would seek to protect species, although in conjunction with broader social justice initiatives. Greens see the value of all species and believe that if we lose one species, we lose an important member of our ecological community.

Green egalitarianism leads to multiculturalism, which celebrates diversity, including the unique contributions, values, and beliefs of all cultures. In the United States, multiculturalism helped to reveal ongoing prejudices that privileged Anglo culture in comparison with other cultures. Because of their worldcentrism, greens seem to complete what Habermas called the incomplete project of the Enlightenment, which defines modernity. In fact, however, greens are antimodern in some respects.

Mean green:[26] Because of green's concern for the natural environment, and because green is but one step away from second tier, green has much in common with Integral Ecology, but there are crucial differences as well. Some critics contend that Wilber does not give enough credit to green and is too critical of their value system. Wilber, however, has time and again emphasized the many vital contributions made by the green wave. His criticism seeks to awaken in green the possibility that they are blind to important possibilities.

Like all first-tier altitudes, green usually regards itself as the possessor of the truth and believes that other memes are deeply misguided. Hence, green often dislikes modern orange, thereby undercutting legitimate economic development. Green also frequently ignores or looks down upon traditional amber, thereby undermining the conventional foundations necessary for individual development and the communal solidarity important for many members of culture.

Depth is a form of hierarchy, but green regards all hierarchy with suspicion. In its suspicion, it tends toward an egalitarianism that often tries to eliminate cultural depth in a way that mimics how natural science ignores nature's interior depth. This suspicion of hierarchy can lead green to uniquely unsupportable assertions, such as that there are no developmental levels, no evolution, and no progress. According to this philosophy, there is no better or worse (yet they hold their view as better than others and are thus trapped in a performative contradiction). Rejection of hierarchy leads to problematic positions such as "biocentric egalitarianism," according to which a member of one species is no more important than any other. Integral Ecology identifies this as ground value, which as noted earlier is only one of three ethical values relevant to the natural world. As a result of this exclusive use of ground value, there is no way to decide whether to kill a gazelle or a termite, a tiger or a virus, much less a human being or a turtle, because they all can be equal in green's eyes. Green's critique of hierarchy impedes recognition of the fact that green constitutes an advanced level of consciousness, when compared with other first-tier stages.

Second Tier

7. Teal-Holistic Self/Autonomous. The *autonomous self* welcomes chaos and multiple variables in the service of self-development. At this altitude, individuals understand that the self is embedded in many contexts and dimensions. They accept many aspects of self and integrate shadow material. They tolerate others in spite of their negative traits and differences of opinions or values. They experience their world as multidimensional with overlapping contexts and systems.

Teal-holistic is the first stage of second tier. Here, people recognize the value of all previous levels as necessary for healthy human development. This stage and all others after it are still "under construction," because relatively few people have attained them. The teal-holistic center of gravity has not yet formed a strong Kosmic habit or a well-established morphogenetic field. Respecting and seeking to integrate prior centers of gravity, rather than dissociate from them, teal is prepared to listen to everyone, to forge strategic alliances whenever possible, and to speak in ways that sincerely respect others. In *Spiral Dynamics*, Don Beck and Chris Cowan use the term "spiral wizard" to describe an individual at this level, as they are capable of interacting effectively with individuals and groups operating at different centers of gravity. Teal-holistic is committed to transdisciplinarity, to organize multiple perspectives in order to characterize and propose innovative solutions for complex problems. They display unusual leadership qualities. They neither seek nor shun the limelight, and take great satisfaction in service, especially in facilitating effective outcomes to which many different kinds of people have lent support.

Individuals who operate at teal seek to reintegrate body and mind, emotion and reason, sense and soul, Descent and Ascent. Hence, teal-holistic is also called the centauric wave of development, involving a more integrated personhood. People at teal operate with vision-logic or "holistic aperspectivalism," a mode of consciousness that lets people understand, appreciate, and consider how things appear from several different, even conflicting perspectives.

At the teal level, the holarchic, ever-changing character of the Kosmos manifests itself. To navigate such a complex world, teal-holistic prioritizes flexibility, spontaneity, knowledge, functionality, and competence, while minimizing whenever possible rank, power, and status.[27] Graham Linscott, using a developmental perspective, comments here that this wave of development helps integrate previous waves of development in some African countries:

> Whereas the individual at Red will despise tribal order [Magenta], reject the rules of [Amber], and accept entrepreneurial Orange only to the extent he can exploit it (typically, gangsterism), the individual at [Teal] can appreciate the need for a reasonable sharing of wealth and a caring for the less fortunate (Green); for an entrepreneurial economy (Orange) without which there can be no wealth; for rules and regulations [Amber] without which there is anarchy; for a channeling of the raw individual energies (Red) of detribalized individuals into constructive pursuits; and for communities which desire it to be tribally organized [Magenta].[28]

Linscott's remarks shed light on how teal-holistic would address the issue of endangered species. Decisions are based upon the situation and a perspectival view. Teal would ask questions such as: What species is at stake? What factors threaten it? Who are the stakeholders, in addition to the species and life forms in the species' habitat? What resources—material, monetary, social, cultural, personal—are available? What legal and regulatory issues come into play? How can we hear conflicting perspectives, so that more authentic dialogue can occur? What kinds of solutions have already been proposed?

Resolving complex issues in a way that works for at least most parties is not at all easy, as Darcy Riddell and others discovered negotiating a resolution to the Great Bear Rainforest controversy, discussed in part four.[29] Anyone who has been engaged in resolving an environmental problem knows that one of the most daunting tasks is to create common ground. Even if common ground is achieved for a time, this ground is anything but stable. As a complex adaptive system, feedback constantly alters the conversation. This evening's consensus may become tomorrow morning's dissensus. Learning to ride such a roller coaster (and tolerate despair), while keeping one eye on the prize is essential for an Integral Ecologist.

Mean teal: Teal-holistic constantly considers the developmental spiral and entertains multiple-quadrant perspectives. Thus they find truth via the sum total of all systems and through their own authentic autonomy. Should either one of these approaches fail, teal-holistic people are prone to existential crises and paralysis. If there is no final truth, then at the end of the day how is one to assess the best course of action? Teal-holistic is multiperspectival, and this developmental level does not come with a ready-made set of values. Without a team of teal-holistic ecologists to work with, individuals are particularly subject to burnout and despair. Teal-holistic can reinstate hope by establishing goals without clinging to them. Because perspectives are continually shifting, however, there is no possibility of perfect or permanent solutions, but only solutions that are good enough given the circumstances. By surrendering attachment to permanent solution, teal-holistic can keep despair and burnout at bay. One of the difficulties in naming the shadow side of teal and turquoise is that they both are so new that their dark side isn't as apparent as that of the other worldviews. Those forging the teal-holistic path must pay attention to this issue. However, when assessing second-tier shadow we must distinguish between those who use teal and turquoise rhetoric, concepts, and ideas but whose center of gravity is below these levels. Otherwise, we will mistakenly assign first-tier shadow to second tier.

8. Turquoise-Integral Self/Integrated.[30] Evolving to the integrated self wave helps alleviate teal's existential anxiety. At turquoise, the heart opens and increases the individual's awareness of the widespread suffering around the planet. The capacity for tolerating such suffering without being overwhelmed by it is a crucial aspect of the compassion that arises with the turquoise wave. Turquoise accomplishes, with ever-greater skill, the integration of Ascent and Descent. Comprehending the extraordinary interpenetration of energy throughout all levels and domains of the Kosmos invites Ascent to nondual integralism. Here there is a marriage of wisdom and compassion, which allows turquoise-integral to apprehend that nothing needs to be done, because everything is already perfect, which in turn paradoxically serves as the basis for a profound commitment to action. Turquoise is connected to the Divinity penetrating all beings at all levels in all quadrants. Hence, people at turquoise remain centered and at ease, even while engaged in frequent and demanding acts. The despair that threatens teal-holistic is replaced by a profound, experientially based conviction that Eternity has always already interpenetrated the Finite. Recognizing that no manifestation, emanation, or creature endures forever, people at turquoise

appreciate each moment for what it is, without clinging to ideas about how things ought to be.

At turquoise-integral, all phenomena appear as elements of the whole, not just the "good" things, but also the "bad" things, including industrial pollution, atomic weapons, logging, and burning carbon fuels. Divinity is present in everything at every moment. Having experienced this firsthand, turquoise-integral can also help generate conditions in which violence does not occur, in which species do not become extinct, and in which terrestrial life forms can prosper in a way that both sustains span and encourages depth. Yet turquoise-integral remains unattached to outcome, over which no one has very much control. Being unattached to the outcomes is not indifference to the joys or sorrows that accompany the outcome. Turquoise-integral celebrates joy and shows compassion for suffering. Clearly, this developmental wave has not been fully stabilized on any wide cultural basis.[31] This book is an articulation of how a turquoise-integral approach would address endangered species as well as the other major issues facing the planet.

Third Tier

Beyond turquoise are levels that open to the third tier of development, which has been embodied and explored by very few people. Two third-tier levels are indigo (ego-aware) and violet (unitive), which we discuss in some detail in chapters 7 and 9.[32] For now the above discussion on first- and second-tier levels will serve our needs.

Development: Progressive but Not Preordained

Many postmodern theorists and environmentalists are suspicious of grand narratives of progress, especially those that claim we are justified in exploiting the natural environment to actualize our alleged potential. We believe some of these claims are justified, albeit confused. It is true that in the name of progress, people calling themselves communists and liberal capitalists committed countless crimes against humanity and the natural environment. Such crimes are often viewed as representing the darkest side of modernity. But we believe these critiques fail to see that even though these perpetrators lived in modern cultures and supposedly adhered to modern ideologies, in fact they operated from preconventional or perhaps conventional levels of moral development. Just because people can fly passenger jets into high-rise buildings does not mean that such people are worldcentric moderns.

According to postmodern critics, modern meta-narratives of progress are erroneous for many reasons, but two in particular stand out. The first reason is that they are metaphysical—they assume a hidden, preordained outcome, as in mid-19th-century America, when "Manifest Destiny" seized the imagination of a young nation on its way to claiming a continent coast to coast. Wilber now agrees that a satisfactory developmental model must be post-metaphysical. It cannot claim that a developmental path is preordained.[33]

Hence, Wilber introduced in 1999 the distinction between involutionary and evolutionary pregivens: the developmental path is not laid out in advanced by involution, but instead, as evolution occurs, Kosmic habits are created that are then inherited by future generations. Calling on the work of philosopher Charles S. Peirce, Wilber defines both natural and sociocultural developmental waves as habits.[34] Habits constitute morphogenetic fields that make it easier for subsequent generations to accomplish something that was more difficult for earlier generations. Enduring patterns of sociocultural organization persist because they are adaptive. Tribalism is an effective organizational model for certain situations. Humans participated in tribalism so often that it has become ingrained. Hence, inner-city youth have no trouble re-creating tribal patterns, attitudes, and practices (even if it is often the unhealthy manifestation of tribalism). Population growth creates different circumstances and usually forces tribes to develop different sociocultural systems, usually agricultural.

Sociocultural development is *not*, however, analogous to the developmental stages of an organism, even though Hegel and Marx—strongly influenced by Aristotle—interpreted social-cultural stages as actualization of human potential. Like Whitehead, Wilber explains the emergence of integrative novelty not merely in pragmatic terms ("It worked"), but also as having been stimulated by the erotic lure of the Kosmos that draws us toward ever-more complex wholes, forms, and relationships. Each new wave or development is an attempt to respond to the limitations and partiality of the previous wave, which itself responded to previous partiality. These habits establish well-trodden paths for those who come later. There is no certainty about the developmental levels yet to come, although hints of possible futures are discernible.

Second, as many feminists and ecofeminists have argued, meta-narratives of progress are often motivated by a desire to regain a lost, hidden, and glorious unity. According to this line of thinking, historical existence, with all its nagging uncertainties, manifest imperfections, moral demands, and messy corporeality, will be overcome when alpha meets omega, whether this be the worker's paradise or ecotopia, or ethnically cleansed nations in which we eradicate all otherness. These demented ideas of "progress" surely deserve criticism. Our notion of progressive development does not include some final moment in the future when all opposites are reconciled and all differences overcome. Instead, we expect that existence will always be incomplete, messy, uncertain, and finite.

Technological Innovation: Is It Really Responsible for Killing Nature?

Major sociocultural transformations are frequently linked to unanticipated and far-reaching technical innovations, which required humans to redefine themselves economically, socially, individually, and culturally. Hence, in a passage bound to please social theorists, Wilber writes that

> the mode of techno-economic production is the single strongest determinant for the average level of consciousness in society. Thus, if the mode of production is foraging, the average level tends toward magical [magenta]; if industrial, rational (orange); if information, pluralistic (green). Individuals in those societies can be higher or lower than the average (precisely because the social mode does not determine the consciousness), but the average itself (or the cultural center of gravity in the LL) parallels those systems in the LR: cultural center of gravity and social level of development are often isomorphic.[35]

For instance, machine technology rendered the physical advantages that evolution conferred upon men less important and made it possible for women to compete for many jobs once reserved for men. Hence, it is no surprise that the rise of feminism coincided with the industrial revolution. In that same vein, Wilber writes that James Watt and other innovators of machine technology did more to free slaves than anyone else. Such emancipation, however, would not have occurred without corresponding developments in the cultural LL domain. Members of the industrial working class suffered enormously during the transition from the agrarian to the industrial mode of production. Even Marx regarded this unfortunate situation as unavoidable, at least until capitalism maximized industrial productivity. The natural environment also experienced significant new kinds of damage because of global industrialism, but, although it is often unmentioned, industrialism has occasionally reduced other kinds of environmental damage.[36]

In view of the sometimes destructive environmental, social, and cultural consequences of industrial technology, many environmentalists are technophobic.[37] Albert Borgmann observes that many technological innovations are simultaneously beneficial and harmful. Many environmentalists are technophobic because industrial technology has been used in ways that have caused untold environmental damage. These concerns are not limited to environmentalists. Consider the bioethical issues raised by cloning, genetic engineering, human creation of new species, and the (potential) invention of intelligent machines. Throughout modern times, what generations of humans regarded as "natural" has been repeatedly transgressed. As our species has evolved, we became capable of undreamt-of technological feats. To condemn our powerful technological capabilities as unnatural creates a problematic duality between humanity and nature. If everything that is is natural, there is nothing unnatural about anything that humans do or make. Even H-bombs and chain saws are natural phenomena. The only basis upon which many feel they can condemn such artifacts is to assert that humans are profoundly different from nonhumans. This assertion, however, undermines interpretations of humans as just another animal species. After all, we don't condemn the artifacts of other species (e.g., chimpanzee tools, wasp nests). *Humans are different* because they have more freedom to choose which artifacts to create and how to use them. Humans have no choice, however, about making and using artifacts—they are a *sine qua non* of human existence. This noospheric aspect of Nature is not locatable in the biospheric level of Nature, as we explained in chapter 2.

Modern technology, then, is characterized by both dignity and disaster, beneficence and harm. The oil stove warmed the house better, but something was lost when the family hearth disappeared.[38] Around the same time, the automobile solved the serious public health problem generated by tons of horse manure deposited on city streets, but the internal combustion engine generated new forms of pollution, urban sprawl, and countless other problems. Taking this into account, Ulrich Beck has argued that techno-industrial societies have entered a phase called "reflexive modernity," in which they are increasingly preoccupied with identifying and managing the many risks that those societies themselves have manufactured.[39] Hence, some people call for adopting the "precautionary principle" to require adequate risk assessment before allowing implementation of potentially harmful technological innovations.[40]

Given that those with greater economic and political power are not always the most morally enlightened; given that executives of publicly held corporations are legally required to maximize returns to their investors, not to think of the long-term future; given that the unanticipated harmful effects of technological innovations often wreak greater harm on those with less money and power—we can understand the motivation behind the precautionary principle. Yet, we also understand the point of those who emphasize the complexities involved in risk assessment, and who warn against the drawbacks involved in establishing a government agency charged with screening new products and processes and deciding which should be allowed into market.[41]

Although highly successful in terms of innovation and productivity, the modern industrial paradigm has until recently ignored the fact that humans depend in important ways on the biosphere. How humans collectively handle the problems generated by the phase-specific modern industrial worldview, as Wilber remarks, "will determine whether a new and more adequate worldview emerges to defuse these problems, or whether we are buried in our own wastes."[42] Some people argue that corporations and even whole economies move toward "ecological modernization." These people hope that ecological modernization will create a bridge to a postmodern era guided by an environmentally sustainable worldview, ecological modernization, new technologies, market incentives, and emerging marketing strategies, and that corporations will generate far more eco-friendly production and distribution processes.[43]

Critics like the social theorist Allan Schnaiberg, however, charge that the "treadmill of production," that is, the international commitment to economic growth at almost any cost, dwarfs any alternative, including ecological modernization.[44] Schnaiberg draws on the Marxist concept that the means of production (economic base) strongly conditions class structure and the cultural formations arising from the techno-economic base.[45] Because individual and collective consciousness is profoundly conditioned by the prevailing means of production, and because industrialism has emphasized the "it" domain, the leading ideologies of late modernity were socialism and capitalism—each of which were committed to human self-actualization through economic productivity. Unfortunately, this productivity tears up the biosphere. In this context, human self-actualization was not aimed at deepening subjectivity and intersubjectivity; it was aimed at increasing human control over exteriors. Arguably, then, the disaster of modernity stems from the fact that science, technology, and industry collapsed consciousness into the "it" domain. Drawing in part on the work of Habermas, Wilber remarks:

> The it-domain was growing like a cancer—a pathological hierarchy—invading and colonizing and dominating the I and the we domains. The *moral* decisions of the culture were rapidly being handed over to science and *technical* solutions. Science would solve *everything*. . . . And thus science (theoretical and technical [including scientific socialism and neo-classical economics]) would not only solve all problems, it would decide what was a problem in the first place—it would decide what was real and what was not.[46]

Spurred in part by subjective and intersubjective development that demands greater respect both for human culture and the biosphere, environmental concerns have become integrated into mainstream political processes. As a result, many North American and European governments and corporations are making at least some move toward ecological modernization. Even if this move is considerably greater than treadmill advocates say it is, the emerging industrial giants India and China are not prepared to step off the treadmill of production until far more economic growth is achieved.

Interior Development, Modern Identity Crisis, and Environmentalism

All shifts from one level of development to the next involve a crisis of the self. The self is letting go of old ways of interpreting and seeing the world. It is a death of an old self and a birth of a new, more inclusive self. We imagine that the crisis in moving from a modern self to a postmodern self and beyond has also affected environmentalism.

As the modern industrial age gives way to the postmodern information age, many moderns are not able to cope with this upheaval without making major personal changes. In some cases, however, the evolution from a modern orange self to a postmodern green self has created a painful decentering or even fragmentation.[47] For moderns, there is only one true way of defining nature. Mind is the mirror that reflects nature. As Ingolfur Blühdorn and others have argued, however, once postmodernism undermined "nature" as an independent, objective other, then the correlate of such "nature"—that is, the modern subject—is also automatically undermined. Unified nature and unified subject are correlates within modernity's realistic, epistemological regime.

In contrast, postmodernism is a hall of mirrors, each differently reflecting nature. If Blühdorn is correct, beginning in the 1960s, moderns must have experienced an *interior challenge to a unified* self that correlated (and still correlates) with the external challenge to a unified nature. Seeking to defend a unified, objective, and to some extent pristine nature, some people have effectively fought on behalf of modernism's mode of centered subjectivity and realistic epistemology.[48] We believe that many environmentalists projected their existential crises onto nature, which was undergoing a crisis of its own—one of habitat loss, species extinction, and industrial pollution. As a result, many people perceived their own death as the Death of Gaia.

Many deep ecologists are epistemological realists who condemn postmodern environmentalists for speaking of nature as a social construct. In so doing, deep ecologists call upon the very same modern, scientific worldview that they otherwise blame for contributing to environmental destruction.[49] Moderns and Eco-Romantics alike usually adhere to a one-dimensional, flatland, industrial ontology, which equates reality with the totality of depthless, material phenomena and systems of such phenomena. Some environmentalists associate their felt interconnectedness with nature (an UL reality) with the material interconnections in ecosystems as described by scientific ecology (a LR reality). The personal *feeling* of interconnectedness, however, is by no means identical with the industrial *map/grid* of interconnectedness. This is a classic confusion of a 1st-person phenomenological experience with a 3rd-person objective analysis. Furthermore, interconnection is not just all good: interconnections are also *killing* the planet, making almost everyone complicit in its demise. The Buddhist biologist Michael Soulé points this out in a discussion with Clark Strand. Here is an excerpt:

> [Clark:] So it turns out that interdependence is a lot more complicated than we think. On the one hand, we talk about being interrelated

> and interconnected, and that gives us Buddhists a warm, glowing feeling. On the other hand, we're interconnected economically as well, which creates all kinds of intractable problems as far as production and consumption are concerned. Correct?
>
> [Michael:] Exactly. But there's another problem, which is the denial of nature's dark, violent side. The danger with such illusions is that they can lead us down the path of smug romantic certainty about the way the world works. For example, Buddhists may want to perceive balance, harmony, and nonviolence in ecosystems because such a view would seem to justify nonviolence in human communities. But murder and infanticide are common in mammals. . . . The same kind of violence occurs in invertebrates. . . . So there's this false romanticism, too, in contemporary Buddhism, this feeling that everything is mutualistic and synergistic and interconnected.[50]

In *Integral Spirituality*, Wilber explains that both phenomenological and systemic interconnection are important, and an Integral approach makes room for each. The danger lies when one is reduced to the other. For example, it seems natural to have a unitary experience of the planet and label it "Gaia." In this experience Gaia *feels like* a single organism with a single consciousness. Gripped by such a powerful experience, environmentalists then confuse this experience with interobjective facts. They think that the planet *is* a single organism; but we contend that it is *not* a single organism ("Gaia"). It is a planetary collective or global ecosystem.[51] In other words, while Gaia might feel like a single, unified organism in moments of open awareness, such a feeling does not mean that Gaia is a single organism in the interobjective world.[52] Although the contents of the interior and exterior domains have definite correlations, they must not be collapsed into one another.

Many Eco-Romantic modernists and rational-purposive modernists implicitly reject the Ascent tradition and insist that only this world—that is, the material world revealed by the senses—is truly real. As we discussed in chapter 1, modernism and Eco-Romanticism are alternative routes that each exclusively affirm the Descent tradition. Modernism began with a dualistic ontology, but in practice it has increasingly committed itself to a materialist account of the world. Although unable to explain the place of interiority in material reality, moderns ultimately identified with rationality to such an extent that they dissociated from their bodies.[53] Paradoxically, moderns deny the interiority that makes rationality possible. The two most explicitly materialist movements of the 20th century, Soviet Marxism and National Socialism, campaigned explicitly against bourgeois subjectivity and individualism. Elements of this campaign, given theoretical expression by Nietzsche and Heidegger, reemerged in different guise in French poststructuralism, which criticizes individual subjectivity and agency.

Romanticism emerged in part as a revolt against the hyperrationalism, individual alienation, and nature destruction associated with modern Ascent. Instead of dissociating from and dominating material nature, however, Romantics chose to identify with it—Descent. In *Sex, Ecology, Spirituality*, Wilber argues, in a way too complex to reproduce here, that such an embrace of material-empirical nature not only failed—as did modernism—to account for human interiority, but also involved narcissism and thus regression to earlier stages of emotional and cognitive development. Eager to deny the Ascent tradition's claim that the cosmos involves nonmaterial, interior dimensions, modernists and Romantics ended up—*mutatis mutandi*—sharing the same one-dimensional materialism of Descent.

Systems Theory: Is It Really Holistic?

We can understand why many environmentalists would react angrily to Wilber's contention that their pro-nature stance shares ontological foundations with nature-assaulting modernity. Some respond with claims that systems theory can help us overcome our divided atomistic worldview of modern science. Yet Wilber maintains that systems theory is thoroughly modernist because it ignores interiority and is therefore materialistic. Although systems theory is important, it cannot account for interiority, consciousness, or awareness—let alone the development of consciousness.[54] According to Wilber, systems theory says

> nothing about ethical standards, intersubjective values, moral dispositions, mutual understanding, truthfulness, sincerity, depth, integrity, aesthetics, interpretation, hermeneutics, beauty, art, the sublime. . . . All you will find are the objective and *exterior correlates* of all that. All you will find in systems theory are information bits scurrying through processing channels, and cybernetic feedback loops, and processes within processes of dynamic networks of monological representations, and nests within nests of endless processes, all of which have *simple location*, not in an individual, but in the social system and network of objective processes.[55]

So we cannot explain ecological consciousness with systems theory, or by any other Right-Hand quadrant methodology. Ecological consciousness can only be understood by acknowledging interiors.[56] Wilber writes:

> In order to have sustainable economies living in harmony with ecosystems, human beings must have interior levels of development that can hold ecological consciousness: there is no sustainable exterior development without correlative interior development, no exterior landscape that can survive without an interior landscape capable of holding it. It does no good to emphasize the worldcentric web of life if people are still at egocentric and ethnocentric levels of interior development—which an alarming 70% of the world population is.[57]

Systems theory adheres to the representational or mapping paradigm, criticism of which is central to postmodernism.[58] According to postmodern theory, knowledge claims are always situated within contexts or perspectives. Even data-driven cognition generates conclusions and systems whose ontological status is heavily dependent on the perspective of the generator. Wilber argues that systems theory overcomes subject-object (and thus human-nature) dualism only by "*reducing* all subjects to objects in the 'holistic system.'" The systems view may be represented as "very holistic and all-inclusive," but in fact it "guts the interiors of the entire Kosmos" by eliminating the "lifeworld of all holons."[59] Hence, systems theory is a form of "subtle reductionism." Expanding the universe from individual behavior (most often of atoms) to that of complex systems does not mean that systems theory is truly holistic (see figure 4.4). Ideally, holism requires the inclusion of interiors and exteriors, but systems theory does not accomplish this.

However, this doubling (from atoms to atoms + systems) is misleading because while it greatly expanded what are considered real phenomena in the cosmos, it doesn't include any individual or collective interiors. We applaud systems theory's initial expansion as a step away from gross reductionism but note that it still adheres to subtle reductionism—subtle because it is difficult to spot in light of the fact that it includes more exteriors than gross reductionism. As noted in figure 4.4 Integral Ecology includes interiors as well as exteriors, and as a result expands the cosmos into the Kosmos.

Even our map is a reduction of this complex and mysterious universe![60] Nonetheless, the 4 quadrants represent the four basic dimensions and perspectives that must be considered to have a more complete picture of ourselves and the world we live in. Any perspective that is given unjustifiable priority over the others is also a form of reductionism or quadrant abolutism (see figure 4.5).

Transcendence: Friend or Foe of the Environment?

In reawakening the Ascent tradition by emphasizing interiority, Wilber legitimates the idea of transcendence, which is best understood in terms of development. Thus, we humans are not transcending Nature; rather,

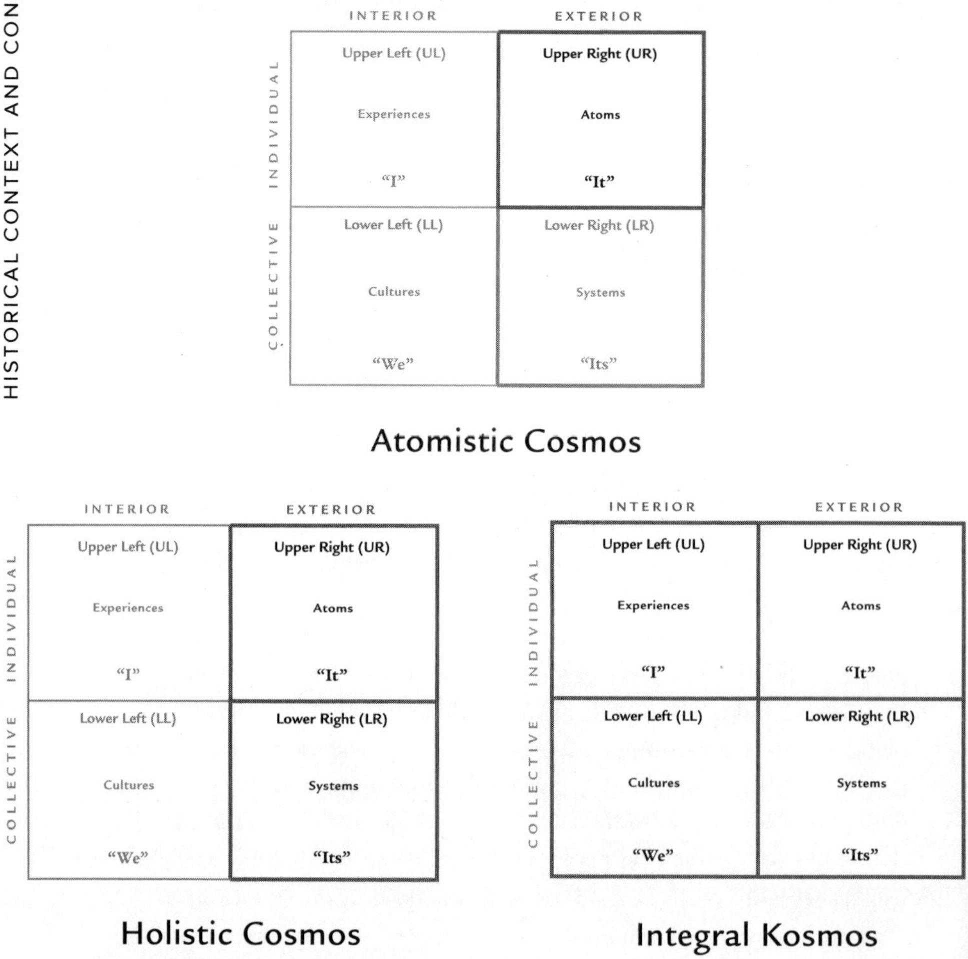

	INTERIOR	EXTERIOR
INDIVIDUAL	Upper Left (UL) Experiences "I"	**Upper Right (UR)** **Atoms** **"It"**
COLLECTIVE	Lower Left (LL) Cultures "We"	Lower Right (LR) Systems "Its"

Atomistic Cosmos

	INTERIOR	EXTERIOR
INDIVIDUAL	Upper Left (UL) Experiences "I"	**Upper Right (UR)** **Atoms** **"It"**
COLLECTIVE	Lower Left (LL) Cultures "We"	**Lower Right (LR)** **Systems** **"Its"**

Holistic Cosmos

	INTERIOR	EXTERIOR
INDIVIDUAL	**Upper Left (UL)** **Experiences** **"I"**	**Upper Right (UR)** **Atoms** **"It"**
COLLECTIVE	**Lower Left (LL)** **Cultures** **"We"**	**Lower Right (LR)** **Systems** **"Its"**

Integral Kosmos

Figure 4.4. Three views of the universe.

one level of Nature is transcending and including another level of Nature: the noosphere transcends and includes the biosphere, which in turn transcends and includes the physiosphere. Development of the healthy aspects of each individual and collective level of human interiority is crucial to transforming human treatment of the natural world. The global, techno-industrial economic system causes great ecological damage, but we think it is inefficient to try to change the systems and institutions of the LR without simultaneously creating conditions that make the healthy manifestation of each interior level of development possible.[61] Much environmental destruction results not because people have mean-spirited attitudes toward the natural environment, but rather because people either do not express healthy manifestations of their level, have not developed to at least a level of awareness that spontaneously cares about the natural world, or have not had environmental concerns connected to their level. Likewise, lack of respect for people other than one's own kind partially causes large corporations to move to the developing world and engage in socially and environmentally

	INTERIOR	EXTERIOR
INDIVIDUAL	Upper Left (UL) **Solipsism** (only individual experiences) I	Upper Right (UR) **Atomism** (only atoms) IT
COLLECTIVE	WE **Relativism** (only the social construction of reality) Lower Left (LL)	ITS **Holism** (only parts interwoven into systems) Lower Right (LR)

Figure 4.5. Quadrant absolutisms.

destructive practices that would usually not be tolerated in Europe, North America, or Japan. Changing these destructive practices will certainly require instituting different social and economic systems, but more is required. Individual attitudes and cultural mores must encourage the healthy expression of each level of development, and must encourage an interior evolution toward worldcentrism. How are we to do this if we ignore the reality of individual and collective interiors?

Caribbean-born Maureen Silos argues that without interior change, the people of the developing world will be unable to define their own path and will continue to pursue some version of the modernist program socially, culturally, and environmentally. Externally imposed systemic change will not do the job. Silos argues that long-term, significant change cannot occur without simultaneous change in interior domains.[62]

The Evolving Interiors of Environmentalists

In the mid-1990s the Green movement started losing steam. People spoke openly of the end of old-time environmentalism.[63] Unlike their counterparts from the 1960s and 1970s, the environmentalists of the 1990s are increasingly green, communitarian postmodernists, rather than Eco-Romantic, individualistic moderns. The interiors of environmentalists are evolving. Since they are more comfortable experimenting with multiple and even conflicting modes of selfhood, most no longer conceive of nature as a unified, stable "other." Instead, they regard nature as moving between stability and instability, order and chaos, change and stasis. Humans inevitably perturb natural processes, but perturbations also arise from nonhumans. Hence, many environmentalists have ceased aiming to return to a pristine, unified nature, and have become institutionalized as one voice negotiating with others (even within one individual) in regard to defining, relating to, protecting, and utilizing nature. It seems that many members of the environmental movement are looking forward, not back.

One of the major insights that has accompanied these postmodern environmentalists is the understanding that nature is never a straightforward given but arises within a context of cultural and social systems. Even ecological "problems" become such only through a sociocultural perspective. Nature is a social construct, in the sense that it is interpreted and experienced within the competing clearings of different social categories and cultural norms.

We maintain that environmental problems are also in important respects symptoms of distorted and immature subjectivity and intersubjectivity, so we applaud those environmentalists who are developing their interiors. There are laws, regulations, and institutions to protect the environment from harmful behaviors, but if people act in accordance with worldcentric morality, they would not knowingly harm other people or the environment.[64]

We must create circumstances for interior development. Were a critical mass of people to operate from worldcentric morality, they would most likely demand that social and political institutions conform to worldcentric values. It is difficult to imagine a worldcentric population demanding ethnic cleansing, religious intolerance, divine right of kings, blood-based aristocracy, second-class status for women, or environmental degradation. Constitutional democracy, with its commitment to universal human rights, provides the template for countries making the move from the mythic-ethnocentric amber to the worldcentric orange wave of individual and cultural development.

Is There Any Hope?

We face a critical time. Life as we know it is threatened by human behavior. Yet we maintain that human history exhibits development on many different fronts. Life is in many respects better than it was. More possibilities are available to people in the 21st century than there were in the 2nd century. However complex and flawed our development has been, it has led at least in some places toward more integrated, comprehensive, and just social formations and cultural norms. There is a growing and worldwide concern for animal welfare, environmental protection, and human rights, however imperfectly these concerns are expressed. True enough, the stakes are greater than ever, but there are also countless opportunities for people to rise to these challenges. More people live in poverty than at any time in human history, but the general population is larger than ever before. The percentage of people at the poverty level continues to shrink to the lowest recorded levels. Although slavery still persists in parts of the world, at least the institution of slavery has been universally condemned and no longer enjoys legitimacy.

It has been extraordinarily difficult to define and set in motion programs of sustainable development, yet increasingly more people recognize that business as usual will have to change. Human activity may sufficiently perturb our world and cause widespread misery and uncertainty. It was, however, less than a generation ago that the gravest environmental concern was nuclear winter. The prospect of making a difference appeals to many people. We recognize the seriousness of our current situation, but we are not put off by it. We may not develop in time, but we are developing. We have hope.

Here we have examined some major features of personal and cultural development as they relate to Integral Ecology. Far more remains to be said. We have not addressed the vast contributions made by environmental philosophy, environmental literary studies, eco-theology, and other LL domains. While many of these approaches will be discussed in the coming chapters, their relevance to the topics discussed here are worthy of consideration. We hope, however, that we have demonstrated that development in personal and cultural domains are of vital importance to address our ecological survival. If we can acknowledge that individual and collective interiors exist, that they develop, that there are multiple ways of conceiving of nature, we maintain that we can learn to bring forth the health of each and every level of development and communicate to each level to create healthy ways to relate to this wild, vast world.

PART TWO

The *What*, *Who*, and *How* of Ecological Phenomena

5

Defining, Honoring, and Integrating the Multiple Approaches to Ecology

Today, a unified ecology is more of a goal than a reality. However, this dream is only one way of looking at modern ecology. Another way of looking at ecology is to take the view that it is made up of a number of different subdisciplines. In this model, ecology is made of several distinct concept-based subdisciplines, each with its own set of concepts (or world view). These subdisciplines are different ways (or perspectives or views) of looking at the same thing, ecology. Because the perspectives are different, they produce different questions, require different techniques, and result in different conclusions about the relationships, distribution, and abundance of organisms, or groups of organisms, in an environment.

—STANLEY DODSON[1]

Looking Back

Since Ernst Haeckel first defined the field of ecology in 1866 with his book *Generelle Morphologie der Organismen*, the field of ecology has multiplied, divided, and morphed into numerous schools and subschools. Each approach captures a perspective or phenomena excluded by others. With the emergence of new schools of ecology, there is a tendency for the nascent approach—the "new kid on the block"—to define itself against the existing approaches; some even pair up to discredit the other "misguided approaches." In this way, scholars, practitioners, and activists build fences between approaches when we are in ever more need of bridges. The result is a fragmented field of approaches that are pitted against one another.

As we outlined in part one, Integral Ecology provides a framework capable of organizing and integrating the myriad perspectives and their multiple fields into a complex, multidimensional, postdisciplinary approach that defines and provides solutions for environmental problems.[2] The AQAL approach is the only organizing framework currently available that can honor this radical multiplicity. AQAL provides a map for understanding the relationship between *who* is perceiving nature (epistemology), *how* the perceiver uses different methods, techniques, and practices to disclose nature (methodology), and *what* is perceived as nature (ontology).[3] By including these three intertwined aspects (the Who, How, and What), Integral Ecology provides the first post-metaphysical approach to the natural world—an approach that avoids asserting perceiver-free ontological structures and thereby avoids falling prey to the myth of the given.[4] This post-metaphysical approach stipulates the position of the perceiver and the perceived. Following Wilber, we refer to these positions collectively as the "Kosmic address."

By considering the perceiver-perceived dynamics of each school of ecology, an Integral practitioner can honor and integrate the important perspectives about ecological phenomena. By understanding the prominent worldviews in which different approaches are embedded, we can understand why they emphasize certain realities to the exclusion of others. We can use our understanding to see which aspects of reality an approach underemphasizes or neglects, and take steps to minimize the impact of their omission. We can also coordinate between approaches so that we can consider more aspects of reality.

In this chapter we examine some of the major historical and contemporary perspectives on the science of ecology.[5] These differences demonstrate that definitions of ecology have not been static. There is no single definition of ecology to which all or even most ecologists subscribe. In fact, there are currently over 100 unique definitions of, approaches to, and perspectives on ecology![6] Most have their own academic journals, research institutions, and field practitioners.

The Aristotelian Roots of Scientific Ecology

We consider Aristotle (384–322 B.C.E.) the first Western "ecologist."[7] His ideas about nature and categories of natural phenomena dominated Western science for nearly 2,000 years.[8] For our purposes, four of Aristotle's postulates were particularly important. He maintained that

- species are eternal and therefore fixed;
- each species has a perfect essence (*eidos*), so that variations in species are inconsequential;
- the Kosmos is organized along a spectrum of complexity (*scala naturae*, or the Great Chain of Being), according to which humans are at the top of biological complexity and perfection;
- the natural world is in perpetual equilibrium: ecological variation is unimportant.

The first three postulates discouraged evolutionary thinking until the 17th century, and as a result of the fourth postulate, 19th-century ecologists were mainly concerned with static patterns instead of dynamic processes. A combination of factors undermined Aristotelian orthodoxy. European exploratory voyages exposed people to new biota, fossils of extinct species, and geographical evidence that the world was old. Along came the invention of the microscope and the creation of new taxonomies to account for the variety of nature. Then the roar of the industrial revolution demonstrated that the natural world could be altered in a short period of time (within one human life span). As the Aristotelian worldview of nature gave way to an expanding and exploring Europe, a new understanding of the natural world emerged.

The Birth of Ecology

As noted, the first definition of ecology ("*oekologie*") was made by Ernst Haeckel.[9] Haeckel was a German zoologist and a prolific writer who helped popularize Darwin's evolutionary theory in Germany. His writing during the 1860s and 1870s indicates that his definition of ecology was inspired by Darwin's discussion of the "economy of nature" in the *Origin of Species*.[10] In 1869 Haeckel presented his definition to the philosophical faculty of the University of Jena.[11] An excerpt from this talk became the first widely circulated English-language definition of ecology:

> By ecology we mean the body of knowledge concerning the economy of nature—the investigation of the total relations of the animal both to its inorganic and its organic environment; including, above all, its friendly

> and inimical relations with those animals and plants with which it comes directly or indirectly into contact—in a word, ecology is the study of all those complex interrelations referred to by Darwin as the conditions of the struggle of existence.[12]

This is the most common lay understanding of ecology today: the study of interrelationships between organisms and their environment.[13] Haeckel's "organism-environment" definition was Darwin inspired, but it was not situated within an evolutionary perspective. (It was not until the 1940s that ecology was placed within the context of evolution, when the Darwinian synthesis of natural selection and genetics occurred.)[14]

The term "ecology" became more popular in the 1890s, triumphing over the term "bionomics" proposed by the famous 18th-century French naturalist Georges-Louis Leclerc, Comte de Buffon. By the early 1900s, the first scientific ecological societies and journals appeared. In 1913, the British Ecology Society established and published the *Journal of Ecology*. Two years later, the Ecological Society of America was formed.

Four Influential Definitions of Ecology

At the end of the 19th century a number of influential definitions of ecology emerged. While the history of scientific ecology is filled with overlapping and divergent trajectories, all worthy of consideration, we will focus on four of the more important definitions of the last century: *organismic*, *economic*, *ecosystem*, and *chaotic*.[15]

Organismic Ecology Definition: Nature as Super-Organism

In the 1900s two main schools of ecology emerged: plant ecology and animal ecology.[16] Plant ecology developed in the early 20th century and focused on distribution, because unlike animals, plants are sedentary and easily identified and mapped. One of the most important figures of plant ecology was the American prairie ecologist Frederic Clements. He introduced the concept of succession and vegetation climax (e.g., an "old-growth" forest) in his 1916 book, *Plant Succession*, and adhered to an organismic philosophy that viewed interdependent plants as forming a super-organism: "The unit of vegetation, the climax formation, is an organic entity. As an organism, the formation arises, grows, matures, and dies. . . . The climax formation is the adult organism, the fully developed community, of which all initial and medial stages are but stages of development. Succession is the process of the reproduction of a formation, and this reproductive process can no more fail to terminate in the adult form in vegetation than it can in the case of the individual."[17]

In *Bioecology* (1939), Clements extended his concept of the biome to animals.[18] Victor Sheldon then applied Clements's concepts of super-organism and climax to animal communities, focusing on predator-prey relationships.[19] Warder Allee built upon Sheldon's organismic approach and emphasized group selection and cooperation in contrast to dominance hierarchies and competitive individualism. His approach is summarized in the 1949 coauthored text *Principles of Animal Ecology*.

Economic Ecology Definition: Nature as Economic Machine

Plant and animal ecology developed along separate lines until the 1950s. Both schools originally defined themselves in organismic terms, only to be overshadowed by the economic versions that gained prominence in the 1940s and 1950s.[20]

In reaction to Clements's super-organism perspective, Henry Allan Gleason championed an individualistic plant ecology. He asserted that plants lived where they could, and believed that regions were best described as areas of continual change, competition, and probability, rather than holistic communities.[21] The British biologist Charles Elton also emphasized competition. His 1927 textbook, *Animal Ecology*, popularized the notions of food chain, "pyramid of numbers," and niche, and promoted an economic metaphor (competition for food) to explain fluctuating populations:[22] "The present book is chiefly concerned with what may be called the sociology and economics of animals, rather than the structural and other adaptations possessed by them. . . . I have laid a good deal of emphasis on the practical bearings of many of the ideas mentioned in this book, partly because many of the best observations have been made by people working on economic problems (most of whom, it may be noted, were not trained as professional zoologists). . . ."[23]

In 1946, the British-trained ecologist George Evelyn Hutchinson promoted mathematical models for ecology. Building on the biogeochemical approach of the Russian V. Vernadsky (who coined the term "biosphere"), he used an economic metaphor instead of the super-organism metaphor to explain communities. He highlighted the importance of feedback loops, which stabilize systems in the face of environmental change. The view of ecology as a stable system maintained by feedback loops soon became associated with the next prevailing definition: cybernetic web.

Ecosystem Ecology Definition: Nature as Cybernetic Web

This definition is associated with Eugene Odum (and to some extent with his brother, Howard Odum), who founded ecosystem ecology in the 1950s.[24] Relying on the idea of emergent properties, and the metaphor of a super-organism, Odum explored the structure and function of the ecological system. His approach combined the super-organism and economy metaphors into the cybernetic one: an economic-like, self-regulating machine. Odum's widely used textbook, *Fundamentals of Ecology*, moved the concept "ecosystem" into the ecological discourse.

Haeckel had emphasized the relationship between organisms and their environment as a complex community of individuals that develops in predictable ways in a stable environment (organismic ecology); Gleason and Elton emphasized individual competition in a haphazard and accidental environment (economic ecology). Now Odum emphasized the parts of a complex system formed by the energy flows that form a homeostatic environment. According to Odum, an ecosystem is "any . . . entity or natural unit that includes living and nonliving parts interacting to produce a stable system in which the exchange of materials between the living and the nonliving parts follows circular paths is an ecological system or ecosystem. The ecosystem is the largest functional unit in ecology, since it includes both organisms (biotic communities) and abiotic environment, each influencing the properties of the other and both necessary for maintenance of life as we have it on the earth."[25] Odum's textbook, along with a few other books that appeared around the same time, represents the shift from ecology as an economic machine to a cybernetic treatment of ecosystems and organisms informed by the new fields of systems theory and computer science.[26]

Since many North American scientists were displeased with the ways organismic ecological theory was used in Nazi Germany, they increasingly drew from ecosystem studies, which went beyond an economic understanding and emphasized information theory, computers, and mathematical modeling in the context of a mechanistic model of nature.

The emerging ecosystem view promised to manage resources by mapping the structure and function of natural systems, and predicting their responses to disturbance. Ecosystem ecology flourished, in part because its mechanistic and dynamic approach mirrored the prevailing cultural zeitgeist. Initially, the major source of funding for ecology came from the U.S. Atomic Energy Commission, which wanted to know the effect of nuclear weapons on organisms and food chains. During this period, the growing importance of cybernetics (self-regulating machines) led to the establishment of systems ecology.

As the environmental movement grew in the late 1950s and early 1960s, ecosystem ecology became increasingly popular because it emphasized the balance of nature and ecosystem integrity and stability. Environmentalists were drawn to the Romantic notions of an Eden-like paradise, and ecologists appreciated the complex computer models of energy flows and feedback loops. Funding increased.

Ecosystem ecology enjoyed an unrivaled popularity during the 1960s and 1970s and is still the most common understanding of ecology among environmentalists. It was also during this time that the chaos and complexity sciences emerged, which eventually provided an entirely new understanding for ecology.

Chaotic Ecology Definition: Nature as Chaotic

Established in 1961 by Edward Lorenz's climate studies, the concept of nature as chaotic led to the formulation of the famous (though overstated) "butterfly effect." Chaos theory wasn't embraced by many ecologists until the mid-1970s. The ecological historian Donald Worster notes, "Despite the growing popularity of the new ideas, however, ecologists were among the last to become interested in them and only a few ever made a full conversion to the science of chaos."[27] One of the first was the mathematical ecologist Robert May, who used modeling to show, contrary to popular belief, that more species inhabiting an area did not automatically create a more stable ecosystem. Chaos ecology has challenged many long-held notions of orderliness, balance, harmony, predictability, and stability of natural systems.[28] Daniel Botkin's *Discordant Harmonies* is a great example of an approach to ecology that is based on chaos-order.[29]

> Wherever we seek to find constancy we discover change. . . . The old idea of a static landscape, like a single musical chord sounded forever, must be abandoned, for such a landscape never existed except in our

imagination. Nature undisturbed by human influence seems more like a symphony whose harmonies arise from variation and change over every interval of time. We see landscape that is always in flux, changing over many scales of time and space, changing with individual births and deaths, local disruptions and recoveries, larger scale responses to climate change from one glacial age to another, and to the slower alterations of soils, and yet larger variations between glacial ages.[30]

Of course, there are other important definitions of ecology, but the aforementioned four have had the most influence.[31] The first, Clements's organismic definition of plant ecology, set the stage for ecological research for the first half of the 20th century. The second, Elton's economic definition of animal ecology, took advantage of the growing desire to use, manage, and manipulate natural resources. The third view, Odum's "systems" definition, emphasized cybernetic (self-regulating) processes, and continues to guide ecological understanding. The fourth, Botkin's chaotic definition, is increasingly popular and contributing to a reevaluation of many long-held truths about ecology. See figure 5.1. The four major metaphors (super-organism, economics, cybernetics, and chaos) associated with each definition all share an industrial ontology.[32] They all view "nature" as a great interlocking order of exterior sensory data (i.e., Nature). This view denies interiority; so even those individuals and schools that champion a holistic view of ecology inadvertently deny the existence of interiors. Those who recognize interiors

TIME PERIOD	APPROACH	KEY FIGURES
1910s–1940s	Organismic Ecology **Plant Ecology** Animal Ecology	Frederic Clements Victor Sheldon
1940s–1960s	Economic "New" Ecology Plant Ecology **Animal Ecology**	Henry Gleason Charles Elton
1960s–1970s	Ecosystem Ecology	Eugene Odum
1970s–2000s	Chaotic Ecology (Non-equilibrium)	Daniel Botkin

Figure 5.1. Some historical trends of scientific ecology in the United States.

(as do many proponents of Odum's ecosystem ecology) are forced to reduce them to their interobjective correlates. Ironically, these holistic approaches are not very holistic—they deny half of the Kosmos.

The science of ecology has by and large been used to manage natural resources. Consequently, many of the concepts, terms, and metaphors that have informed the development of scientific ecology are actually at odds with the desire to protect nature. Contrary to popular stereotypes, all four definitions have been used to exploit the environment.[33] Using the science of ecology as a basis to respect nature is a recent development. It was not until the 1960s that the science of ecology became associated with environmental protection.[34] Even now ecologists differ widely in their relationship to environmental protection.[35]

Six Contemporary Schools of Ecology

While the four historical definitions outlined represent influential currents within ecology, they do not capture the variety of contemporary definitions and scientific schools of ecology. In the postwar years, from 1945 to 1960, the number of ecologists doubled, and then doubled again in the 1970s following the emergence of the environmental movement. Not surprisingly, there was a correlative expansion of distinct approaches to ecology and a diversification within the field. Multiple theoretical schools emerged, each critical of their rivals, each with a unique perspective.

Thus, the study of ecology can be approached from a multitude of angles, perspectives, scales, units of analysis, topics, and methods, depending on the scientist's goals and orientation. These variables lend themselves to unending distinctions between definitions.[36]

As mentioned in chapter 1, ecologist Stanley Dodson conceived of a way to categorize all these approaches. He identified four major ways to define ecological approaches (perspective of the scientist, organism under study, habitat under study, and the application of the research).[37] The first way defines schools by the organizing perspective or concept of the scientist (e.g., ecosystem, population, and community). The second defines ecologies by the organism we study (e.g., plant, animal, microbe, zooplankton, human, deer, tree). Third, we can define the school by the habitat we study (e.g., terrestrial, lakes and streams [limnology], marine, arctic, rainforest, benthic thermal vents, urban). Finally, we can define schools of ecology by the application the research serves (theoretical, conservation, agricultural, public policy, academic, management, restoration). We can combine these different ways of categorizing ecology to create various distinct and overlapping approaches (a landscape ecologist specializing in arctic tundra for restoration purposes).

Dodson focused on six major kinds of ecologies defined by the organizing concept or perspective held by scientists about realities associated with the Lower-Right quadrant: landscape, ecosystem, physiological, population, behavioral, and community. He uses these major perspectives to organize his excellent anthology. He chooses *perspectives* as the organizing framework because he believes they represent the foundational differences between ecological schools. Each school has its own foundation, its own perspective or conceptual foundation, its own methods, and its own domain of inquiry. No one of the six perspectives has the last word, because each highlights different structures, relationships, and dynamics, while downplaying, ignoring, or denying others.

Here are his definitions of each perspective. We also provide a representative question each approach would ask about a rural North American landscape, with an example of the tools and techniques each approach would use to answer such questions.[38]

1. *Landscape Ecology:* We can conceive of landscape as different "patches," or types of land, characterized by different organisms and environments. Landscape ecology examines the interaction between this pattern of patches and ecological process—that is, the biological causes and consequences of a patchy environment.

Typical Question: How does the two-dimensional pattern of forest, field, and farm buildings affect the ability of deer to move from one forest patch to another?

Tools and Techniques: Satellites, photos, maps, and computers are essential, especially for geographic information systems (GIS).

2. *Ecosystem Ecology:* Studies the interactions of organisms with energy and matter. The "system" aspect describes how energy or matter moves among organisms and parts of the environment. Ecosystem size and shape depend how we frame questions about energy flow or chemical cycling.

Typical Question: In this watershed, how much phosphorus is stored in the soil of the forest and fields? How much is applied to the fields each year? How much moves annually into the stream?

Tools and Techniques: Calorimeter pressure bomb, quantitative chemical analysis.

3. *Physiological Ecology:* Studies how individual organisms interact with their environment—their biochemical processes and their behaviors that lead to homeostasis and survival. Homeostasis in this definition is the maintenance of time, matter, and energy that allows for individual growth and reproduction.

Typical Question: Is the local climate optimal for the genetic strain of the corn growing in the fields?

Tools and Techniques: Respirometer, treadmill, infrared gas analyzer (IRGA), stable isotope chemistry, light sensors, thermocouples.

4. *Behavioral Ecology:* The goal of behavioral ecology is to understand how a plant or animal adapts its behavior to its environment; behavior is understood as the result of an evolutionary process.

Typical Question: How do the size, condition, and age of male redwing blackbirds affect their ability to defend breeding territories along the stream bank? How, in turn, does this impact their breeding success?

Tools and Techniques: Sampling traps, computer, greenhouse.

5. *Population Ecology:* A population is a collection of individuals from the same species that occupy a defined area. Population ecology focuses on how and why populations change in size and location over time.

Typical Question: What factors control the size of the trout population in the stream?

Tools and Techniques: Video equipment, event recorder, binoculars, radio tags, geographic position satellites, computer, DNA fingerprinting.

6. *Community Ecology:* Community ecologists examine communal patterns and intersections within and between groups and species. Distributions of species are influenced by biological interactions (such as predation and competition) and environmental factors (such as temperature, water, and nutrient availability).

Typical Question: How many species of native plants and insects live in the woodlot, and are there enough pollinators to maintain the plant diversity?

Tools and Techniques: Quadrant sampling, species identification book, enclosures.

As you read these six definitions (let alone the 200 that you could generate using our appendix!), it becomes clear that each of these approaches represents a different perspective on ecology and the environment. All of these perspectives simultaneously reveal and conceal different aspects of the environment. No one approach has the last word, because each one highlights various phenomena, relationships, and dynamics. They are the result of the concepts, perspectives, questions, tools, and techniques that ecologists bring to bear on the natural and social world. Dodson explains: "Because the perspectives are different, they produce different questions, require different techniques, and result in different conclusions about the relationships, distribution, and abundance of organisms, or groups of organisms, in an environment."[39]

Given the variety of prominent schools, many ecologists (e.g., Allen and Hoekstra) have attempted to unite the various scientific approaches to ecology.[40] One of the problems is that while ecologists generally agree that there are basic ecological principles underlying ecology, they disagree about what those principles are! This has led ecologists such as Dodson to a perspectival approach, according to which each school of ecology represents a valid perspective that should be considered on its own terms.

Although these six contemporary schools of scientific ecology represent different perspectives, they all focus on the *exterior* realities of individuals (behaviors) and/or collectives (systems). Therefore, these six schools are capable of making scientifically warranted assertions about the exterior domains. However, there are also many approaches to ecology that draw their inspiration from subjective (e.g., psychology or art) or intersubjective (e.g., ethics or religion) dimensions of reality. Previous attempts toward a unified concept of ecology only unified natural scientific approaches to ecology, which focus on objective (and interobjective) realities. Integral Ecology unites objective, interobjective, subjective, and intersubjective approaches to ecology and the environment. In so doing we include the six schools that Dodson identifies as well as other approaches that specialize in the exteriors, and we make abundant room for the myriad other schools that have important insights into ecology from interior perspectives.

Thus the classical definition of ecology (the study of the interrelationship between organisms and their environment) becomes the study of the interrelationship between organisms' experiences and behaviors, and their cultural and systems environments. In other words, *Integral Ecology is the study of the subjective and objective aspects of organisms in relationship to their intersubjective and interobjective environments at all levels of depth and complexity*. In fact, we can use the classical definition as long as we understand that by "organism" we mean an individual holon with subjectivity, and by "environment" we mean a collective holon with intersubjectivity. Thus, Integral Ecology doesn't require a new definition of ecology as much as it needs an Integral interpretation of the standard definitions of ecology, where organisms and their environments are recognized as having interiority. Integral Ecology also examines developmental stages in both nature and humankind, including how nature shows up to people operating from differing worldviews.

If Nature Knows Best, Why Does She Keep Changing Her Mind?

Our understanding of what nature is has changed over the past century and continues to do so (see figure 5.2). The science of ecology evolved from a *descriptive* ecology in the 19th century (with naturalists describing what and when), to a *functional* ecology in the early 20th century (with a stronger emphasis on experimentation in the lab and field), to a *theoretical* ecology in the mid-20th century (using calculus, models, and simulations to explain the manner in which it evolved), to a *practical* ecology in the late 20th century (with ecologists entering the political area and informing policy). The shifting sands of ecological understanding raise an important issue: if our views about ecology are ever-changing, are we a moving target? Is nature a moving target? Are they both a moving target?

Adherents to any given approach tend to make their understanding of nature prescriptive—they often look to nature for insight into how we should live and organize our societies. This sentiment is expressed in the cliché "Nature knows best!" which implies that we should model our behavior and societal organization on natural laws and dynamics. According to this line of thought, if we discover nature's hidden principals, we can live in accordance with them either for harmony's sake or for the sake of utilitarian gain, or both.

Of course, this assumption isn't a problem when you assume you have the best view and others are misguided. However, if you take an Integral approach and study the history of our understanding of nature (see figure 5.2), you must wonder, "If nature knows best, why does she keep changing her mind?" Is it nature that is changing, or is it our perception of her that is shifting? Or is it both? As figure 5.2 illustrates, different worldviews and their scientific bases disclose and conceal different aspects of nature and ecological systems. They each highlight distinct qualities based on their perspective.

What will be the next perspective in this unfolding procession of how nature shows herself to us? We propose there is no single nature or environment. There is no single ecology that studies these manifold natural domains. Ecology is not a domain in nature to study; it is a way of studying the environment (wild, rural, urban). Thus, ecosystems are not lying around "out there" waiting for ecologists. Every definition of ecology is, as we've said, the result of what part of reality ecologists look at, how ecologists look at that part of reality, and who they are as they look.

By locating each of these views holarchically, and providing an address with a road map so that others can arrive at the same location through vari-

ECOLOGY APPROACH	SCIENTIFIC FOUNDATION	PRIMARY METAPHORS & QUALITIES
Organismic Ecology 1910s - 1940s	Botany	*Super-organism: Nature as Divine Order* Community, succession, maturation, managerial, cooperation, developmental, directional, predictable, lawful, harmonious, holism, passive humans separate from but in balance with nature
Economic Ecology 1940s - 1960s	Physics	*Machine: Nature as Economy* Individualistic, accidental, chance, haphazard, parts, energy transfer, entropy, systems, controllable, productivity, food chain, utilitarian, active humans separate from but in control of nature
Ecosystem Ecology 1960s - 1970s	Systems Theory Computer Science	*Cybernetics: Nature as Web* (Super-organism + Machine) Balance, homeostatic, mature, management, directional energy transfer, humans as primary disturber
Chaotic Ecology (Non-equilibrium) 1970s - 2000s	Chaos Theory Complexity Theory	*Butterfly Wings: Nature as Chaotic* Disorder, unpredictability, nature as primary disturber, no balance, non-equilibrium, patch dynamics, complex

Figure 5.2. Primary metaphors of several schools of ecology.

ous injunctions, we can avoid postulating that ecological realities exist "out there" in some sort of naive empiricism that falls prey to the myth of the given. There is no such thing a perceiver-free natural environment.

25 Main Approaches to Ecology

One of the first steps in Integral Ecology is to identify the perspectives needed for a comprehensive approach to the environment and its ecology. As mentioned, there are over 100 distinct perspectives of and approaches to ecology (see appendix, which also lists over 100 distinct perspectives on ecological thought and environmental studies). Like Dodson, we wanted to find a means to sort and organize this maddening multiplicity of ecologies. We developed a classification system that allows us to group schools into more general approaches.[41] At this point, we have identified 25 main approaches among the 200 various perspectives on ecology, the natural world, and the environment (see figure 5.3).

Each of these approaches contains a number of distinct schools or perspectives (at least 3 and up to 20 in some cases!) that serve as examples of each approach. Figure 5.4 shows three representative schools for each of the 25 main approaches, for a total of 75 unique approaches to the natural world and its ecological processes. If you are not convinced of the need for an integral ecology yet, look at the longer list in the appendix.

The categories of the 25 main approaches are not meant to serve as exclusive distinctions. In fact, many of them overlap with several other categories. They are merely a heuristic device to highlight general themes and patterns between various schools. Many schools can be placed within several

Scientific	Social	Philosophical
Economical	Technological	Cultural
Acoustic	Evolutionary	Ethical
Medical	Ecological	Religious
Aesthetic	Psychological	Esoteric
Behavioral	Agricultural	Somatic
Representational	Geographical	Therapeutic
Historical	Complexity	Spiritual
	Linguistic	

Figure 5.3. 25 main approaches to ecology.

different approaches, depending on which author, book, or research project you analyze. In other words, depending on the Who, How, and What. The point is not to freeze any school or perspective into one approach, but to use the framework of the 25 main approaches to organize the hundreds of distinct perspectives and relate them to each other. Feel free to adjust, augment, and change the list. The 25 main approaches serve as a reminder of the variety of complex perspectives we must consider in any truly Integral approach to ecology and environmental studies.

Scientific Chemical Ecology Physiological Ecology Cognitive Ecology	**Technological** Nanoecology Industrial Ecology Ecological Architecture	**Philosophical** Ecological Ontology Ecological Hermeneutics Ecological Philosophy
Economical Environmental Economics Ecological Economics Ecological Modernism	**Evolutionary** Developmental Systems Ecology Evolutionary Ecology Evolutionary Psychology	**Ethical** Liberation Ecology Environmental Justice Environmental Ethics
Acoustic Acoustic Ecology Music Ecology Bioacoustics	**Ecological** Ecosystem Ecology Population Ecology Community Ecology	**Religious** Spiritual Ecology Ecological Theology Process Ecology
Medical Molecular Ecology Clinical Ecology Building Ecology	**Psychological** Emotional Ethnology Ecopsychology Organic Psychology	**Esoteric** Deva Gardening Archetypal Ecology Design Ecology
Aesthetic Ecopoetics Environmental Aesthetics Romantic Ecology	**Agricultural** Agricultural Ecology Biodynamic Agriculture Permaculture	**Somatic** Feminist Ecology Ecological Phenomenology Architectural Phenomenology
Behavioral Behavioral Ecology Restoration Ecology Environmental Psychology	**Geographical** Subtle Ecology Geopsychology Landscape Ecology	**Therapeutic** Horticultural Therapy Psychoanalytic Ecology Ecotherapy
Representational Mathematical Ecology Theoretical "Pure" Ecology Ecological Modeling	**Complexity** Living Systems Theory Chaotic Ecology Developmental Systems Theory	**Spiritual** Deep Ecology Nondual Ecology Transpersonal Ecology
Historical Paleo "Ancient" Ecology Historical Ecology Ecological Anthropology	**Cultural** Ethno-Ecology Cultural Ecology Information Ecology	**Linguistic** Biosemiotics Ecosemiotics Linguistic Ecology
Social Political Ecology Social Ecology Human Ecology		

Figure 5.4. Some representative schools.

In the face of such multiplicity, we need a definition of ecology that is robust enough to contain the myriad of approaches to ecology, and thrifty enough to support meaningful discourse and avoid the reductionism associated with other definitions of ecology. Thus, Integral Ecology asserts that the 4 quadrants are 4 "terrains" of ecology, which:

- are comprised of and influenced by natural and social systems;
- involve the individual behaviors of organisms at all scales (including microbes and humans), which are understood as members (not parts) of ecosystems;
- include members that have various degrees of interiority (perception, experience, intentionality, and awareness); and
- allow members to interact within and across species to create horizons of shared meaning and understanding.

These 4 terrains of an ecological occasion co-arise and mutually influence each other in complex ways; none of them are granted ontological priority. In short, as we stated above, Integral Ecology is the study of the subjective and objective aspects of organisms in relationship to their intersubjective and interobjective environments at all levels of depth and complexity.

Our definition of ecology provides the recognition that any ecological system is a complex eco-social system whose members not only exhibit behaviors but also possess interiors that comprise overlapping horizons of culture. In other words, an Integral approach to ecology is not just interested in the complex ways members of eco-social systems interconnect with each other or behave individually and in groups, but is also interested in the interiors of members of ecoystems and the complex meshworks of overlapping horizons of meaning that occur within and between members. This is the *enactive* web of life, not just a web of exterior interconnection and behaviors but also a web of interior experiences and cultures of all sentient beings![42] Integral Ecology recognizes that different approaches to ecology and the environment (wild, rural, and urban) are situated in various worldviews and focus on diverse aspects and different levels of complexity within these four dimensions, using a variety of methods and techniques.

This comprehensive definition allows us to organize the 25 main approaches to ecology within the 4 quadrants or 4 terrains of ecology. Note: We are using each approach in a way that is consistent with the terrain it is

	INTERIOR	EXTERIOR
INDIVIDUAL	**Somatic** **Psychological** **Therapeutic** **Aesthetic** **Spiritual** **I**	**Scientific** **Acoustic** **Behavioral** **Medical** **Representational** **IT**
COLLECTIVE	**WE** **Cultural** **Linguistic** **Philosophical** **Ethical** **Religious** **Esoteric**	**ITS** **Historical** **Social Economical** **Technological** **Evolutionary** **Ecological** **Agricultural** **Geographical** **Complexity**

Figure 5.5. Approaches within the 4 terrains.

located within. For example, "Spiritual" could be used to describe collective-interior phenomena, but we are using it specifically to point to the terrain of experience. Likewise, we use "Religious" only to refer to the terrain of culture. Further, we can use the 4 terrains to group the 24 approaches (see figure 5.5) and the respective schools of each approach (see figure 5.6).[43]

The Kosmic Address of Ecosystems

Integral Ecology provides a roadmap for arriving at the coordinates of an ecological perspective. This road map and the resulting vista are what Wilber refers to as the Kosmic address. We can summarize the Kosmic address with three main points.

First, given that the universe has no fixed center and is expanding in all directions (what cosmologist Brian Swimme calls "omnicentric"), and given that it lacks a foundational level (atoms give way to quarks, which give way to strings . . .), then we can specify any environmental phenomena,

	INTERIOR	EXTERIOR
INDIVIDUAL	**Terrain of Experiences** Feminist Ecology Ecological Phenomenology Architectural Phenomenology Ecopsychology Organic Psychology Emotional Ethnology Ecotherapy Horticulture Therapy Psychoanalytic Ecology Ecopoetics Romantic Ecology Environmental Aesthetics Deep Ecology Nondual Ecology Transpersonal Ecology I	**Terrain of Behaviors** Chemical Ecology Physiological Ecology Cognitive Ecology Acoustic Ecology Music Ecology Bioacoustics Behavioral Ecology Restoration Ecology Environmental Psychology Molecular Ecology Clinical Ecology Building Ecology Mathematical Ecology Theoretical "Pure" Ecology Ecological Modeling IT
COLLECTIVE	WE **Terrain of Cultures** Information Ecology Ethno-Ecology Cultural Ecology Linguistic Ecology Biosemiotics Ecosemiotics Ecological Ontology Ecological Hermeneutics Ecological Philosophy Animal Rights/Welfare Environmental Justice Environmental Ethics Spiritual Ecology Ecological Theology Process Ecology Deva Gardening Archetypal Ecology Design Ecology	ITS **Terrain of Systems** Paleo "Ancient" Ecology Historical Ecology Political Ecology Social Ecology Environmental Economics Nanoecology Industrial Ecology Developmental Systems Ecology Evolutionary Ecology Ecosystem Ecology Population Ecology Community Ecology Agricultural Ecology Permaculture Subtle Ecology Landscape Ecology Living Systems Theory Chaotic Ecology

Figure 5.6. Representative schools within the 4 terrains.

object, or process only in relationship to another phenomena, object, or process.

Second, there is no pregiven world that exists independently of perception. Perceptions are actually perspectives. Perceptions arise only within a perspective. The manifest world, the world of environmental phenomena, is an unfolding of perspectives of perspectives of perspectives. Wilber explains: "As far as we know or can know, the manifest world is made of sentient beings with perspectives, not things with properties, nor subjects with perception, nor vacuum potentials, nor *dharmas* [i.e., phenomena], nor strings, nor holograms, nor biofields, etc. Those are all perspectives relative to some sentient being."[44]

Third, we must specify the relationship of both the perceiver and the perceived in order to provide the location of any occasion. In its simplest formulation this involves providing the developmental level and the perspective (quadrant) through which, or in which (quadrivium), we encounter the occasion. To locate a phenomenon, we need at least the altitude and the perspective of both the perceiver (s = subject) and the perceived (o = object). So, following Wilber, we can represent the Kosmic address in this way: Kosmic address = $(\text{altitude} + \text{perspective})_s \times (\text{altitude} + \text{perspective})_o$.[45] Given that a subject's perspective is a quadrant (the perspective through which it encounters phenomena), and an object's perspective is a quadrivium (the perspective from which phenomena are encountered), we can also diagram these perspectives as: Kosmic address = (altitude + quadrant) x (altitude + quadrivium).[46] This is the general address of a phenomenon. It doesn't tell you who is home, what color the house is, or the furniture they have inside the house. The Kosmic address, however, is a powerful reminder that environmental phenomena are enacted.[47]

So where is a global holarchical ecosystem? As a highly complex, dynamic, temporal phenomenon, ecosystems emerge when an individual with at least a turquoise level of cognition "looks" through the Lower-Right perspective of natural systems: ecosystem = (turquoise altitude + Lower-Right quadrivium). This address only specifies the object of ecosystems. We must include the address of the subject: ecosystem = $(\text{any altitude} + \text{any quadrant})_s \times (\text{turquoise altitude} + \text{Lower-Right quadrivium})_o$. Of course, an individual at any level of consciousness can look through any quadrant to "see" the Lower-Right dimension of an occasion disclosed at a turquoise level. But it is only when the altitude of the subject matches the altitude of the object (in this case, turquoise) that the subject can actually see the object on its own terms. Anything short of that level involves a translation "down" of the object to the perceiver's level of consciousness and therefore becomes increasingly distorted.

Thus, an individual whose center or gravity is red will perceive nature through their five senses as their immediate environmental surroundings. An amber center of gravity will perceive nature through a folk ecology of local regions based on community theories and traditions. An orange center of gravity will perceive nature in a scientific way (e.g., behavioral ecology), with causal relationships and measurable behaviors and processes. This altitude is where the beginning of systems theory occurs, albeit an interlocking order that is static, with one universal trait (e.g., nature as a machine or giant organism).[48] A green center of gravity will perceive nature through a pluralistic lens (e.g., social ecology), as a multiplicity of bioregional ecosystems that are in delicate balance, with multiple traits and no unifying center.[49] A teal center of gravity perceives nature as the systemic ways that bioregional ecosystems link to larger global ecosystems. Teal uses multi-system approaches like planetary ecology to identify patterns that connect thousands of eco-social systems and their multiple traits. A turquoise center of gravity perceives nature's holonic structure at multiple scales (spatial and temporal) and uses approaches like developmental systems ecology to map the complex systemic dynamics and their relationships to individual and collective interiors. In addition to recognizing that there are thousands of systems connected in a global holarchical ecosystem, turquoise perceives the luminous dimension of the natural world.[50] Thus, Wilber explains, "a 'global holarchical ecosystem' is a signifier whose referent is a very complex multidimensional holarchy existing in a turquoise worldspace; this actual referent can be directly cognized and seen by subjects at a turquoise altitude, in 3rd-person perspective, who study ecological sciences."[51] (See figure 5.7.) Also, at the turquoise altitude the essential aspects of all previous levels are contained. Thus, humans as compound individuals contain every level of ecology. Each of these levels in an individual is in relational exchange with that level in the ecosystem. In short, humans have the potential to contain eight major levels of ecology. Thus, we are very ecological beings.

If an individual does not, in some way, specify the Kosmic address of a given phenomenon, they likely subscribe to the metaphysical assumption

Altitude Level (the Who)	A Typical Approach (the How)	View of Lower-Right (the What)
Turquoise	Developmental Systems Ecology	Global Holarchical Ecosystem
Teal	Planetary Ecology	Global Ecosystem
Green	Social Ecology	Multiple Ecosystems
Orange	Population Ecology	Ecosystem
Amber	Folk Ecology	Local ecological regions
Red	The five senses	Immediate natural surroundings

Figure 5.7. The view of the Lower-Right quadrant of nature from different altitudes.

that the phenomenon is pregiven. System holarchies and Great Web of Life maps are not representations of a world out there. They are realities that exist only within specific worldspaces. Thus, if someone says, "When I think about how the planet's ecosystems are being destroyed by greed and human arrogance, I feel a deep sadness for all beings on the planet," we could specify (assuming a turquoise subject) the Kosmic address as follows: "When I think (i.e., take a perspective: cognitive line) about how the planet's ecosystems (turquoise altitude, LR) are being destroyed by capitalistic greed (pathological orange altitude) I feel (emotional line) a deep sadness for all beings on the planet (turquoise altitude, UL and LL)."[52]

One of the implications of this approach (which is quite powerful when you contemplate it) is that we cannot say that ecosystems existed 50,000 years ago. Why? Because even if humans occupied a magenta or red level of development, they (like any other sentient being) could not conceive of ecosystems; hence, they could not perceive (i.e., enact) them. Postmodernism asserts time and again that we cannot postulate a pregiven, ahistorical world. After all "what we think of today as 'eco-systems' will probably be

Figure 5.8. No single tree!

understood, a thousand years from now, to be energy sinks of dark matter controlling access to an 11th dimensional world of hyperspace. . . ."[53] Wilber asserts that this post-metaphysical stance neither slips into subjective idealism nor prevents us from talking meaningfully about something like ecosystems in prehistory. The rejection of the myth of the given still allows us to describe "intrinsic features" of sensory experience. Thus "we can say that if ecosystems did not ex-ist or stand forth in the magenta worldspace, they nonetheless 'subsisted' in it, or were present as intrinsic features of the kosmos not cognized by magenta."[54]

The intrinsic features themselves are not pregiven; they are the result of the altitude making the claim. Thus, the intrinsic features themselves are "interpretive and con-structed." They are then retrofitted into earlier times or granted to lower levels. Thus, they are not intrinsic features of the world "back then"; they are intrinsic features of a turquoise worldspace. Consequently, what is intrinsic to the Kosmos and is said to ex-ist or sub-sist is different for each altitude and tetra-constructed. Thus global holarchical ecosystems do not exist in any worldspace below turquoise and cannot be found anywhere in their less than turquoise phenomenology. Real objects are not seen from a perspective—they are within that perspective!

Imagine a juniper in a forest with a number of sentient beings perceiving the tree: an ant, a blue jay, a mule deer, an amber lumberjack, a green backpacker, and a turquoise scientist doing fieldwork (see figure 5.8). Now, as each individual "looks" at the tree, they all enact a different tree. The ant feels the rough bark as it travels the scent trail leading to sugary sap. The blue jay experiences the tree as a source of insects to take care of its midday hunger. The mule deer sees the tree as protective shade from which to observe the nearby meadow. The lumberjack sees 1,200 board feet. The backpacker sees the intrinsic value of the tree. The scientist sees the evolutionary trajectory of the tree as it struggles with this year's drought cycle. This drought cycle has been exacerbated by climate change and a 10-year history of devastating forest fires in the region. In short, there simply is no such thing as "one tree"! Rather, there are different layers of trees enacted by each perceiver and the evolving consciousness associated with it.[55]

By adopting a post-metaphysical approach, we can honor and integrate the myriad approaches to ecology. Environmental phenomena are so complex that anything less than an Integral approach will deliver temporary solutions at best, ineffective results at worst. For truly integral, long-term solutions, we need an ecology of perspectives that combines the insights, approaches, concerns, techniques, and methods from the 200 distinct perspectives of our environment. This is accomplished by recognizing that the 200 different approaches to ecology and the environment represent the many combinations of: ecosystem = $(\text{any altitude} + \text{any quadrant})_s \times (\text{turquoise altitude} + \text{Lower-Right quadrivium})_o$. When we consider the eight major worldviews of individuals and their 4 quadrants (the Who), and combine that with the eight ways (the How) to research environmental phenomena (the What), it is no surprise there are many divergent and overlapping schools of thought. Such a meta-perspective allows one to assess, rank, and organize the various eco-perspectives in a truthful, sincere, just, and functional way that avoids being just another perspective. The three terms of Who x How x What respectively define the phenomenological *space* in which a phenomenon is enacted, a particular *mode* used to enact the phenomenon, and the specific *dimension* being registered. "The first term asserts existence, the second, the mode of existence, and the third, the dimension of existence."[56] Let us give a quick example of looking at a tree.[57] When you state, "I see a tree," you are inevitably invoking a Who (I) x How (see) x What (a tree). You have singled out a particular occasion (in this case, a tree) and begun a 1st-person knowing process using your 3rd-person objective vision of a 3rd-person object over there. All sentient beings enact a world in this manner by focusing their awareness on some occasion; they momentarily stop the flow of perspectives and a world is literally born. But the tree doesn't have to be just a 3rd person. If you recognize a level of shared vitality between you and the tree, it becomes a Thou, a 2nd person. This works both ways. If you approach the tree, and if it has any degree of sentience, it becomes the 1st person and you become the 2nd person. As Wilber explains:

> Whenever the agency or intentionality of any holon—cell to ant to ape—is directed anywhere—and it is always directed somewhere—it is directed toward or within a world of other sentient holons, and this is why, if one atom bumps into another atom, then, from the point of view of that atom, a first person just encountered a second person, who in turn responded as first person to the second person of the first; if they influence each other in any way, that is a type of communication, and that communication is not merely a dynamic web but a third person, and so on.[58]

In the following chapters we'll examine the various distinct schools of ecology through the simplified three-term Kosmic address of (Who) x (How) x (What): the 8 ecological selves' developmental and holarchical dynamics, the 8 modes of ecological research those selves utilize to investigate the 4 terrains of ecology and their 12 niches. Thus, using the Who x How x What formula, we can represent the Kosmic address of this book as turquoise x AQAL x NATURE! In other words, this book is written from our embodiment of turquoise using the 5 elements and 8 zones of Integral Theory as praxis to explore all facets of the natural world. We will now turn our attention to the 4 terrains of ecology and their 12 niches.

We have decided to present the Who x How x What framework in the following order: What (chapter 6), Who (chapter 7), and How (chapter 8). In our experience this order facilitates understanding. The official order (Who x How x What) is based on the linear process of a looker (the Who) using a tool (the How) to look at something (the What). However it should be noted that all three dimensions are conascent and arise together (i.e., they enact each other).

6

Ecological Terrains: The *What* That Is Examined

The environment is so complex and our abilities to understand it so imperfect that our appreciation of it will always be limited. The best we can do is assemble numerous partial glimpses—partial natures. Each nature appears to us through a different lens of human culture: science, art, religion, and so on. Each nature has people who see and appreciate it, and each nature defines a role for humans to play. Some natures thrive independently of humans; other natures thrive only through human understanding. Some natures define limits that humans exceed only at great risk; other natures are malleable and can be reconfigured through a respectful and deliberate partnership.

—R. BRUCE HULL[1]

A defining characteristic of Integral Ecology is that it reflects on what part of reality one focuses on (what quadrants are privileged), who is looking at that part of reality (what levels of development are represented), and how one investigates that reality (what perspectives or methods of inquiry are used). Integral Ecology examines the relationship between knower and known: in which reality (the What), the onlooker (the Who), and the method (the How) interact in complex ways (Who x How x What). Integral Ecology proposes that the What consists of at least 4 terrains and their 12 niches, the Who consists of at least 8 ecological selves, and the How consists of at least 8 ecological modes or methodologies. Integral Ecology strives to honor all niches of environmental concern, all selves of environmental identity, and all modes of environmental inquiry: all-niches, all-selves, all-modes. These ecologically explicit categories are a unique theoretical contribution to Integral Theory, operationalizing it in the context of ecological practice and discourse.

In this chapter we focus on the What. We provide an overview of the 4 terrains and three examples using the 4 terrains at different scales (nonhuman and human) from both quadrants and quadrivia.[2] This demonstrates the flexible and responsive ways we can use the 4 terrains in multiple contexts. Then we introduce three levels of complexity within each terrain, thereby creating the 12 niches. We describe each niche and the various approaches to ecology that specialize in them. The result is a tour of dozens of schools of thought. We then apply these niches to more detailed examples.

The 4 Terrains

The richness and complexity of environmental phenomena cannot be understood or described solely through objective modes of inquiry. To honor and include all aspects of wild, rural, and urban ecologies, we must recognize that ecosystems are only one of 4 terrains of any environmental occasion. For every interobjective understanding of complex interacting energy flows, balances, cycles, patterns, and networks, there are objective observations of movement, behavior, activity, and form; subjective realities of experience and perception; and intersubjective spaces of shared meaning and mutual resonance. To fully understand the (inter)objective dimensions, we must understand the subjective and intersubjective dimensions. If we only study ecosystems through objective modes of inquiry, then we only study one-fourth of environmental phenomena! This (inter)objective-only approach, the mainstay of most ecologists, is akin to studying a pond by observing only its surface (e.g., the coming and going of birds, the shifts in water color, the types of waves that form). If we do not explore the depths of the pond (e.g., the interiors of the beings who live in or interact with the pond), its true complexity is left uncharted. Likewise, we discover very little of environmental phenomena if we confine our inquiry to behaviors and systems, and exclude the sophisticated methodologies needed to explore the felt experiences and cultural horizons of those beings (human and nonhuman) who are members of any ecosystem.[3] Ecosystems are enacted—brought fourth—through tetra-mesh. This does not mean they are simply created or constructed by individuals or organisms. Ecosystems contain intrinsic features (i.e., are objective and interobjective), but those intrinsic features are disclosed through an interpretive encounter between an organism and its environment.

Integral Ecology inquires into all 4 quadrants, or 4 terrains: behavioral terrain (behaviors at all levels of organization), experience terrain (experiences at all levels of perception), systems terrain (systems at all levels of ecological and social intersection), and cultural terrain (cultures at all levels of mutual resonance and understanding), and explores the complex ways that distinct and irreducible networks of experiences, networks of cultures, networks of behaviors, and networks of systems co-arise in complex, ever-encompassing networks at all levels of ecological manifestation.

The experience terrain includes the subjective experiences (e.g., somatic, emotional, cognitive, spiritual) of humans and nonhumans. The level of experience enacts the level of the complexity of the ecosystem perceived. Examples of the experience terrain include how a person feels in his or her body while hiking up to a bluff; how a crane experiences a diminishing food supply as its wetlands are drained for a local housing project; how a red-legged frog experiences the electromagnetic discharge from power lines placed over its pond; how someone living in a large city experiences disconnection from the natural world; and how a redwood tree experiences acid rain.[4]

The cultural terrain includes the shared horizons (e.g., morals, symbol systems, meaning, affect, experience) that exist between and across humans and nonhumans. Different types of nondiscursive cultural structures or habitus (à la Bourdieu) with which an organism approaches the environment determines the kinds of practices they use to enact the environment. Examples of the cultural terrain include: how a herd of elk makes sense of a local logging project; how different aspects of the natural world come to symbolize complex realities; how a male mountain lion understands a mating call from a female in heat; how different human communities relate to the natural world during particular historical periods; how some bird species borrow song segments from other birds (or cell phones!) to create their own melody; and how a backpacker can have mutual understanding with a bear encountered on the trail.[5]

The behavioral terrain includes the physical boundaries or surfaces (e.g., skin, cell membranes, organs, and tissues), the actions and movements (e.g., growth, digestion, flight, sleep) of humans and nonhumans, and their exchanged artifacts. Examples of the behavioral terrain include the pH and chemical composition of river water that winds through an industrial agriculture area; the distance between branches on a palm tree; the time intervals between a sheep's feeding patterns if it has been injected with growth hormones; the act of recycling; the average number of eggs a hen mallard lays; a pack of wolves exchanging material signifiers; and the metabolism of slugs.

The systems terrain includes the functional interaction (e.g., food chains, mating rituals, migration patterns, competition) and influence (e.g., pollution, seasons, and weather patterns) of humans and nonhumans. Examples of the systems terrain include how economic development affects a watershed; how deforestation relates to drought cycles; how succession patterns in a temperate rain forest are altered due to road building in the area; how over-hunting deer influences the nutrition flows within various biozones; and how urban traffic patterns contribute to climate change.

The 4 terrains are represented by figure 6.1. Within each quadrant, we list the primary mode of knowing: subjective realities are most accurately revealed through felt experience (e.g., direct perception, introspection, phenomenological investigation, meditation, body scanning); intersubjective realities are most accurately revealed through mutual resonance (e.g., dialogue, energetic connection, shared depth, participant-observer techniques, interpretation and understanding); objective realities are most accurately revealed through observation (e.g., measurement, laboratory observation, field research, chemical testing, sensory perception); and interobjective realities are most accurately revealed through functional fit (e.g., study of part-whole relationships, observation of systemic dynamics, analysis of instrumental function, diagramming energy flows and feedback loops, statistical analysis, investigation of systems adaptation, and documentation of pattern maintenance). Each of the 4 terrains employs different techniques, injunctions, and methods to inquire into its respective dimension.

To illustrate the 4 terrains we provide three examples. The first discloses an oak tree from a human perspective, describing it with the language associated with each quadrant (I-Language, We-Language, It-Language, and Its-Language). The second discloses a view of reality from the perspective of a frog (we use the term "tetra-hension" to refer to the capacity of human and nonhuman organisms to perceive all 4 terrains).[6] The third example is a quadrivial description of toxic emissions. These examples show the distinct flavors, textures, and images of each terrain. For an additional example of

Figure 6.1. The 4 terrains.

the 4 terrains, consult the case study in part four by marine ecologist Brian N. Tissot, "Integral Marine Ecology: Community-Based Fishery Management in Hawai'i."

The Integral Ecology of an Oak Tree: A Quadrivia Perspective

The following descriptions each represent only one of many possible descriptions. Each description is from the perspective of humans looking at and perceiving the oak tree (a quadrivia perspective) (see figure 6.2). But we could also describe each terrain from various disciplines (a quadrivial perspective) or from the perspective of the oak tree itself (a quadrant or quadradic perspective).

Terrain of Experience: Feelings Connected to the Tree

When I looked at the oak I got goose bumps and a sense of joy came over me. I climbed into its branches to get a better view of the valley. As I grabbed the lowest branch I could feel the rough, leathery bark scrape my skin. By the time I climbed to the crotch of the trunk, I was out of breath and happy to be cradled by this living giant. I was satisfied as I sat in there. I felt close to life, and at peace with the world in the arms of this great old tree. It had been many months since I had gone hiking, and it was long overdue. I had forgotten how nourishing I found these walks in the wild.

Terrain of Culture: Shared Meaning About the Tree

We came across the large oak on the north slope of the valley. Wherever the oak grows, it has been held in reverence due to its awe-inspiring qualities. We talked about how the oak was a symbol of power and strength to the Druids. My sister said that the biblical mother-shrine Mamre at Hebron included a sacred oak and that the Old Testament scribes claimed it was the home of Abraham. Mark mentioned that in Roman times, when a soldier saved the life of a citizen, he was awarded the most coveted crown, made of oak leaves and acorns. After exhausting our oak trivia, we sat down in the shade of the oak and had our picnic. That was the last time we were all together, and now whenever I see an oak I am reminded of that wonderful day we shared.[7]

Terrain of Behavior: Measurements of the Tree

The oak is 72 feet tall, has coarse brown bark that is deeply furrowed, and is leaning a bit to the right. It is an evergreen tree with a short, stout trunk and many large, crooked branches. It has a broad, rounded crown. There are three large branches that frame the entire tree, and the leaf length ranges from ¾ to 2¼ inches. Its acorns are long and narrow. They are egg-shaped and enclosed by a deep, thin cup with many brownish, fine hairy scales outside and silky hairs inside. The botanical name is *Quercus agrifolia*. This tree is 79 years old and has survived three droughts and two brush fires. It is located on the north slope of a small valley in an open grove with canyon live oak and California black oak surrounding it. This tree is composed of plant cells, molecules, strings, and quantum vacuums.[8]

Terrain of Systems: The Functional Fit of the Tree

The oak is an essential part of the ecosystem. It prefers dry areas and often thrives on steep, rocky slopes, but it can also be found in canyon bottoms or across rolling hills. The size and shape of the tree is largely determined by its access to water during its growing years, as well as the soil conditions and space available. Oaks are sometimes found in pure stands but are usually intermixed with various pine trees. The oak's seeds (acorns) are essential food for a variety of forest animals such as rodents and birds. Many indigenous tribes also used the acorn as a source of protein, sometimes using them to make flour (after leaching out the bitter tannins). The wood of the oak is heavy and strong and has been used for a variety of products and ornaments. Indigenous groups used acorns and oak products medicinally to combat diseases such as tuberculosis.[9]

The Integral Ecology of a Frog: A Quadradic Perspective

Like with the oak tree, we can approach the 4 terrains of a frog from quadrants or from quadrivia. If we focus on the 4 quadrants of a frog, we see

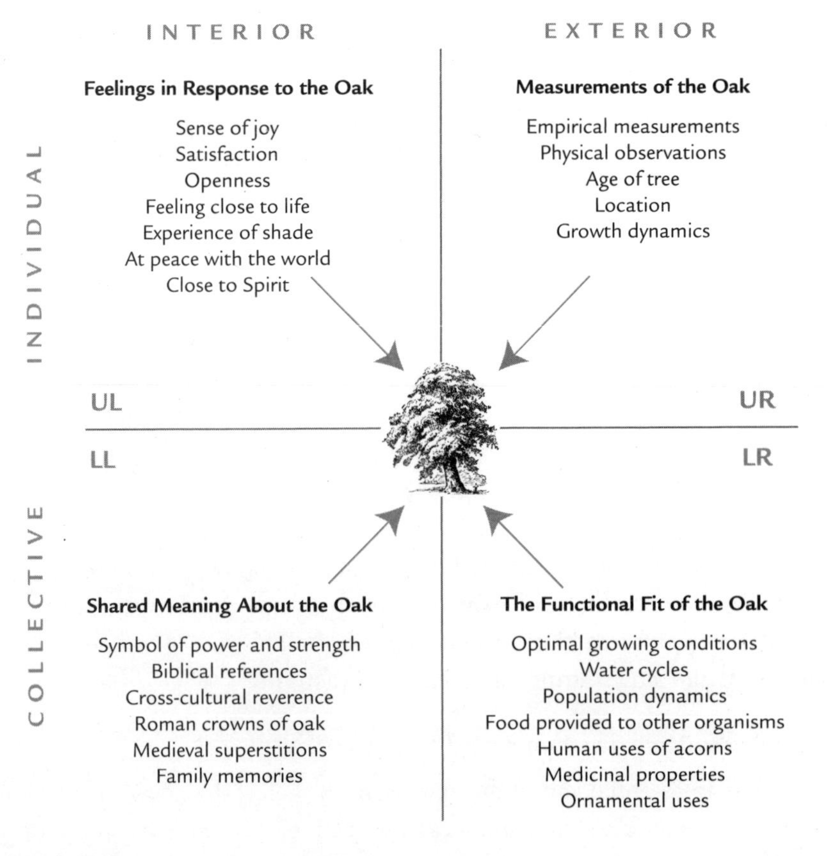

Figure 6.2. Quadrivia of an oak tree.

how a frog perceives its own world through each perspective: tetra-hension. Tetra-hension occurs at all levels within a frog and consists of four distinct modes of nonreflective perception of an individual holon. These include the subjective perception of proto-sensation, the objective perception of the five senses, the intersubjective perception of resonance, and the interobjective perception of functional fit ("functional apprehension"). Each of these modes of tetra-hension reveals a different world: a subjective world, an objective world, and so on. If we were to focus on quadrivia, we would see how humans use different disciplines (e.g., psychology [UL], biology [UR], and so on) to perceive frogs through each of the four perspectives. Since the oak tree example illustrated quadrivia, this example will illustrate the quadrants of a frog (see figure 6.3). Sentient beings have quadrants and can look through (quadrivia) at an object (either the quadrants of an individual or the quadrivia on an occasion). These three terms are none other than our Who x How x What.[10]

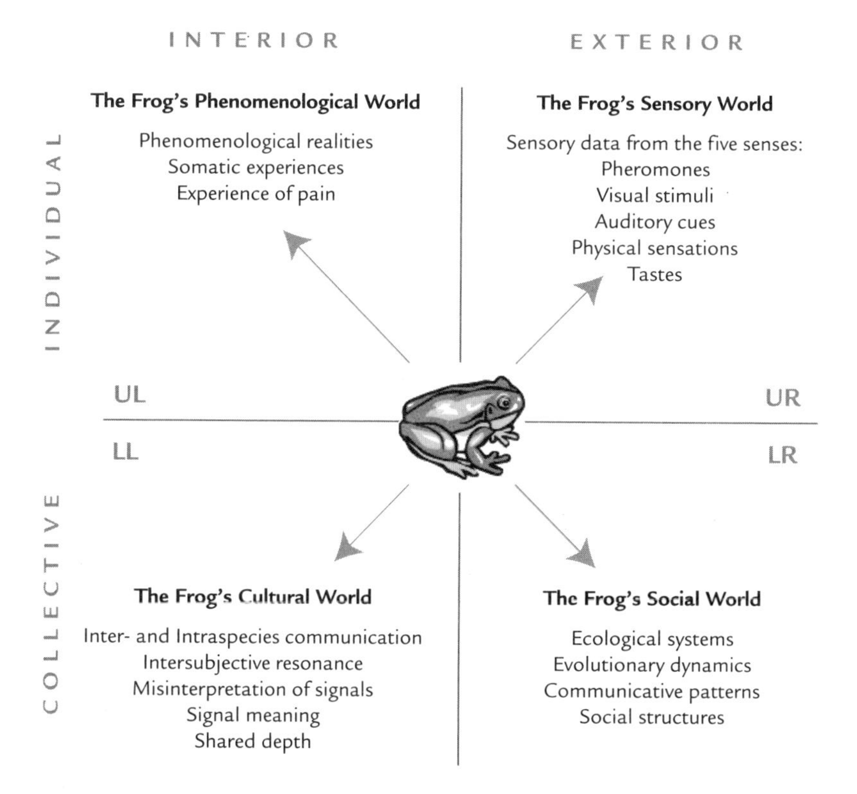

Figure 6.3. Quadrants of a frog.

Terrain of Experience: The Frog's Phenomenological World

The terrain of experience includes the frog's phenomenological experiences—its subjectivity. The German biologist Jakob von Uexküll pioneered work in the phenomenological world of animals during the early 1900s. His work serves as a foundation to biosemiotics. Jesper Hoffmeyer, a Danish leader in the field of biosemiotics, explains: "We need a theory of organisms as subjects to set alongside the principle of natural selection, and Jakob von Uexküll's umwelt theory is just such a theory."[11]

This terrain represents the frog's 1st-person awareness—its somatic experience of hot and cold water, physical pain, and impulses. The frog may not have a self-conscious relationship to these experiences, but it does have an interior that supports a variety of experiences, even if relatively simplistic.

Terrain of Behavior: The Frog's Sensory World

The terrain of behavior includes the objects of the frog's senses and capacity to perceive movement and differentiate its surroundings. Sensory ecology provides insight into the sensorial capacity of organisms and how they register pheromones, visual stimuli, auditory cues, physical sensations, and tastes.[12] Accurate perception is crucial for the frog's survival.[13] This terrain also includes how the frog cognizes its environment and structurally couples with it as a result.[14]

Terrain of Culture: The Frog's Cultural World

The terrain of culture includes the frog's communication and exchange of meaning with frogs and other animals such as snakes, birds, insects, mice, and foxes. When organisms communicate and interpret each other's signals, they create a "semiotic niche," or an intersubjective space of meaning.[15] Biosemiotics calls the sum total of all these niches the semiosphere.[16] The semiosphere is an autonomous sphere of communication and meaning that exists between all organisms.[17] For Integral Theory this semiotic network is actually a tetra-occasion: LL semantics, LR syntax, UL signifieds, and UR signifiers.[18] However, the LL intersubjective aspect is what we have in mind here when we describe the semiotic niche.[19]

Frogs, like all sentient beings, have a specific semiotic niche. That depth meshes or collides with the depth of meaning in other sentient beings. A frog that misunderstands the intentions of a roaming fox—jumping at the wrong moment—is likely to end up as dinner. Consequently, interpretation plays an important role in an organism's survival and reproductive success. Hoffmeyer understood the importance of shared interiors when he wrote that "one can never hope to understand the dynamic of the ecosystem without allowing for some form of umwelt theory."[20] In Integral Ecology we speak of the frog's "culture" as a general, intersubjective space between individual frogs, and embrace a pansemiotics like that associated with Charles Peirce.[21] Frog culture includes all the signals frogs use to communicate meaning (vocalizations, pheromones, movement, visual display, touch). It includes the ways frogs interpret inorganic features and other beings within their world. As Wilber explains, "Even electrons have to interpret their environment—not to mention bacteria, worms, and wolves. . . . The deer watching a hunter must interpret the hunter's actions, and not merely react to each of the them like, say, a falling rock. Precisely because all holons (all the way up and down) contain a moment of sentience, they will always have to interpret their environments and therefore interpret each other's interpretations. Needless to say, adequate interpretation therefore demands same-depth translation. If one holon attempts to interpret a holon of greater depth, something will definitely get lost in the translation."[22] We do not assume any degree of self-reflectivity on the part of the frogs. But frogs do share a collective interior!

Terrain of Systems: The Frog's Social World

This terrain includes the various systems of norms and rules that frogs perceive and participate in: ecological, evolutionary, social, and communicative. Frogs unconsciously participate in all kinds of syntactical elements. The sum total of the exchanges of material signifiers among frogs and between frogs and their environment constitutes an important part of the ecological niche of a frog. In addition there are various social structures and regulations that frogs adhere to that are informed by ecological pressures and evolutionary dynamics. These various systems make up the frog's social world.

In short, a frog, like other organisms, has four distinct perspectives or lived worlds, which we, drawing on biosemiotics, refer to as its umwelt. Various subschools of biosemiotics tend to use "umwelt" as a primary referent for phenomena associated with different zones (i.e., phenomenological [#1], hermeneutic [#3], cognitive [#5], and communicative [#7]). As noted in chapter 2, we expand "umwelt" to refer to all of these: the organism's capacity for quadratic perception. So not only does an organism subjectively

(1p) perceive its environment (3-p), but it also subjectively (1p) perceives others (2-p) and itself (1-p). "Umwelt" in its narrow usage typically refers to just the organism perceiving the UR (the subject's perceptual world).[23] But in addition to a perceptual or sensory world (objective), an organism has a phenomenological world (subjective), a cultural world (intersubjective), and a social world (interobjective). Since umwelt is often framed as "all the meaningful aspects of an organism's world," this would include not just 3-p realities but also 2-p and 1-p perspectives. Thus, the "subjective universe" of an organism in an Integral approach to umwelt theory becomes a fourfold perspectival world (i.e., organisms have four dimensions-perspectives as a result of their being-in-the-world). Their four irreducible dimensions allow them four distinct perspectives. By expanding umwelt theory to refer to the quadrants of an organism, we make explicit the four dimensions of an organism's "perceptual" world. Thus, in umwelt theory not only do organisms have an "outer world" (umwelt) but they also have an "inner world" (innenwelt), and an "others world" (sozialenwelt).

Each of the frog's 4 rich terrains is already studied by various scientific

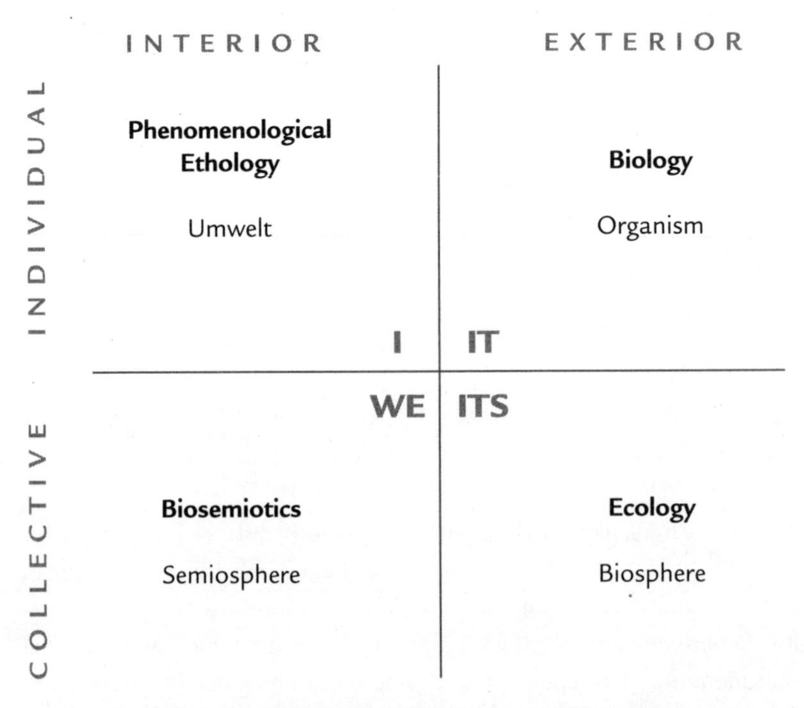

Figure 6.4. Qualitative and quantitative sciences used to study organisms.

disciplines.[24] We can use biology to study the objective organism (UR); ecology to study the interobjective biosphere (LR); phenomenological (e.g., cognitive and emotional) ethology to study the subjective perspectives of organisms (their umwelts) (UL); and biosemiotics to study the intersubjective semiosphere (LL)—see figure 6.4.[25] Clearly, each of these disciplines studies more than just these terrains, but our placements serve to reflect the methodological strength of each discipline. Integral Ecology integrates these 4 terrains and their respective disciplines so as to understand the depth and complexity of an organism's fourfold world.

The Integral Ecology of Toxic Emissions: A Quadrivia Perspective[26]

Figure 6.5 is a quadrivial example of an Integral understanding of the problem of toxic emissions. Each section examines toxic emissions from a different terrain and their associated disciplines.

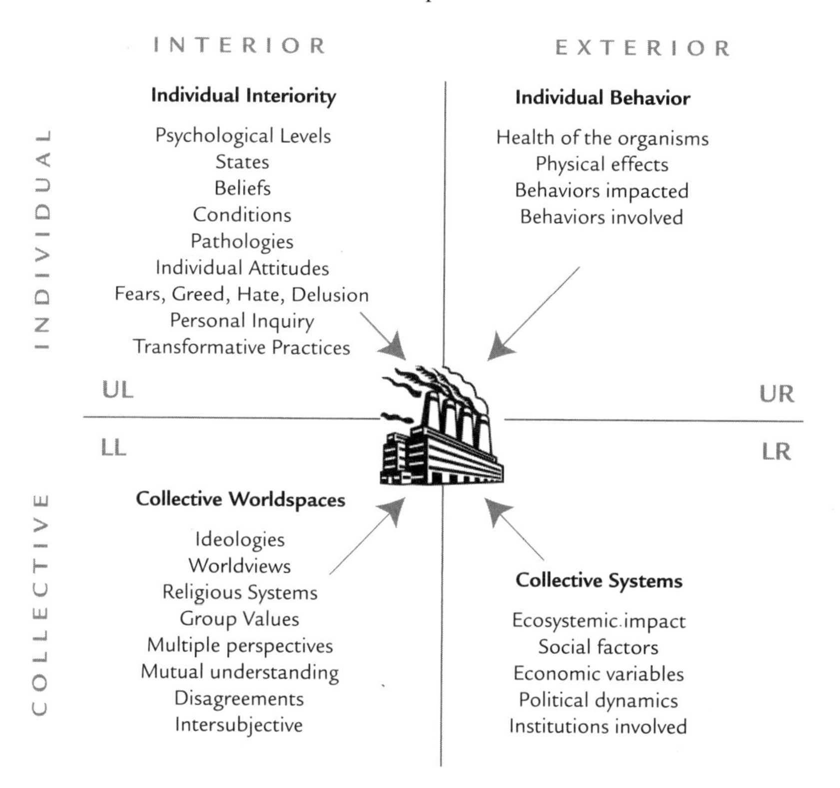

Figure 6.5. Quadrivia of toxic emissions.

Upper-Right Quadrant: Terrain of Behavior

Toxic chemicals can cause (or trigger) various deleterious effects in individual cells, organs, and organisms (plants, animals, and humans). We must study, measure, and describe these so that more comprehensive grounded recommendations can be made about limiting their release into the environment. In other words, it is important to both understand how individual behavior and health are effected by toxins at all levels of ecological organization (from cells to organs to organisms) and look closely at how human behaviors in our daily activities contribute to and sustain environmental toxicity.

Lower-Right Quadrant: Terrain of Systems

In the Lower-Right quadrant we examine how toxic chemicals affect ecosystemic processes (watersheds, food chains, nutrient cycles, etc.) and individual (cellular/organic/organismic) processes. We also study the various social, economic, and political factors involved in the production and release of toxic chemicals. We must understand how sociopolitical institutions contribute to and maintain toxic emissions at all levels (e.g., local water boards, county commissions, state agencies, and federal committees). By understanding their role, we can understand the ways these institutions can contribute to ecological recovery.

Lower-Left Quadrant: Terrain of Culture

Integral Ecology examines the ways ideologies, worldviews, religious systems, and value systems encourage, discourage, or are neutral about toxic emissions. Each level will be motivated to take corrective action by different values. Integral Ecology encourages us to understand the worldview of different levels of development. Developing mutual understanding between individuals and worldviews is critical to resolving this problem, and it is no easy task. To do so, we must explore the role of intersubjectivity in its many dimensions (proximity, emotional, linguistic, contextual) within and between the human sphere and other organisms.

Upper-Left Quadrant: Terrain of Experience

Psychological levels, states, beliefs, conditioning, and pathologies shape individual attitudes, and so these levels shape individual attitudes about toxic emissions. As Integral Ecologists we must understand these different levels, states, and types, and their motivations and beliefs about the environment. Integral Ecology holds that transformative practices such as therapy, contemplation, meditation, and community service help individuals discover the roots of their attitudes, beliefs, and emotions that give rise to care for or damage the environment. We stress the importance of individual (UR) and collective (LR) transformative practices so as to support individual development (UL), which in turn can affect collective attitudes and practices (LL), leading to new institutions (LR) that further support interior development (UL). Until we can create healthy expressions of each level of development and until we have leaders who embody worldcentric and planetcentric levels of development, we will continue to misuse nature.

The 4 terrains provide a way to explore all of the conditions that give rise to environmental issues. Each terrain highlights a very different approach, and a unique dimension of ecology that we must consider if we want a comprehensive understanding and comprehensive solutions. Yet each terrain is more complex than what we have described above through the three examples, so we have additionally identified at least three broad levels within each terrain. These levels represent increasing interior depth and exterior complexity of phenomena. The 4 terrains and their respective levels constitute the 12 niches of environmental realities.

The 12 Niches

The 12 niches can represent different aspects of reality that various environmental approaches specialize in.[27] For example, *ecopsychology* specializes in psyche: the psychological dynamics of ecological grief (a sense of despair over damage to the environment) and disconnection from the natural world; *environmental justice* specializes in institutions and action: the relationship between social systems and intentional conduct; *eco-phenomenology* specializes in soma and communion: somatic realities and intercorporeal dimensions; Lovelock's *Gaia theory* specializes in movement and intersections: physical movement and natural systems; and *natural capitalism* specializes in institutions and intersections: the integration of social systems and natural systems. The boundaries between the 12 niches, as with all ecological niches, are permeable and fluid. It is not easy or always necessary to define where one ends and another begins. We summarize the 12 niches in figure

6.6 with a short description of the domain (e.g., psychological dynamics) and a single word to refer to that domain (e.g., Psyche).

From a quadratic perspective, a niche at any level (e.g., somatic realities) does not occur without the other niches on that same level (e.g., intercorporeal dimensions, physical aspects, or natural systems). The 4 niches of every level arise simultaneously, because each niche of that level is simply a different aspect of the very same phenomenon. For example, a mule deer simultaneously has somatic realities, intercorporeal dimensions, and physical movements, and is a member of natural systems. One aspect does not occur before or after the others; these niches co-arise. A change in one creates a change in the others. These niches are interrelated phenomena, although each niche has its own irreducible traits.

More complex levels transcend and include less complex levels with each level being a concentric circle (see figure 6.7). For example, compassionate perspectives transcend and include shared horizons, which in turn include intercorporeal dimensions. Likewise, there can be a physical (gross) body without an intentional mind, but there cannot be an intentional mind without a physical body.[28] There can be an intentional mind without transpersonal spiritual experiences, but if there are spiritual experiences, then there must be an intentional mind and a physical (or subtle) body. In other words, a single niche does not occur in isolation; all the

	INTERIORS		EXTERIORS	
	EXPERIENCES	CULTURES	BEHAVIORS	SYSTEMS
3rd Level of Complexity	**Pneuma** Spiritual Realization	**Commonwealth** Compassionate Perspectives	**Skillful Means** Worldcentric Actions	**Matrices** Global Systems
2nd Level of Complexity	**Psyche** Psychological Dynamics	**Community** Shared Horizons	**Action** Intentional Conduct	**Institutions** Noetic Systems
1st Level of Complexity	**Soma** Somatic Realities	**Communion** Intercorporeal Dimensions	**Movement** Physical Movements	**Intersections** Environmental Systems

Figure 6.6. The 12 niches of environmental concern.

niches of the same level, along with all the niches below, co-arise, even if they are outside our awareness. "Higher" or more complex doesn't just equal better—each level is simultaneously of equal value, more significant than, and more fundamental than other levels of complexity.[29] One level is only more complex insofar as it includes the essential aspects of the previous level and adds something new.

A very important thing to keep in mind with regard to the examples of approaches within each level is that in this chapter we are focusing on the What of each approach and not the How or Who—though the object of focus of an approach often indicates both the methods and the worldviews involved. In particular, the level of Who largely determines the level of What can be enacted. While the levels of altitude are typically used to describe the level of complexity of the Who, they can also be used to indicate

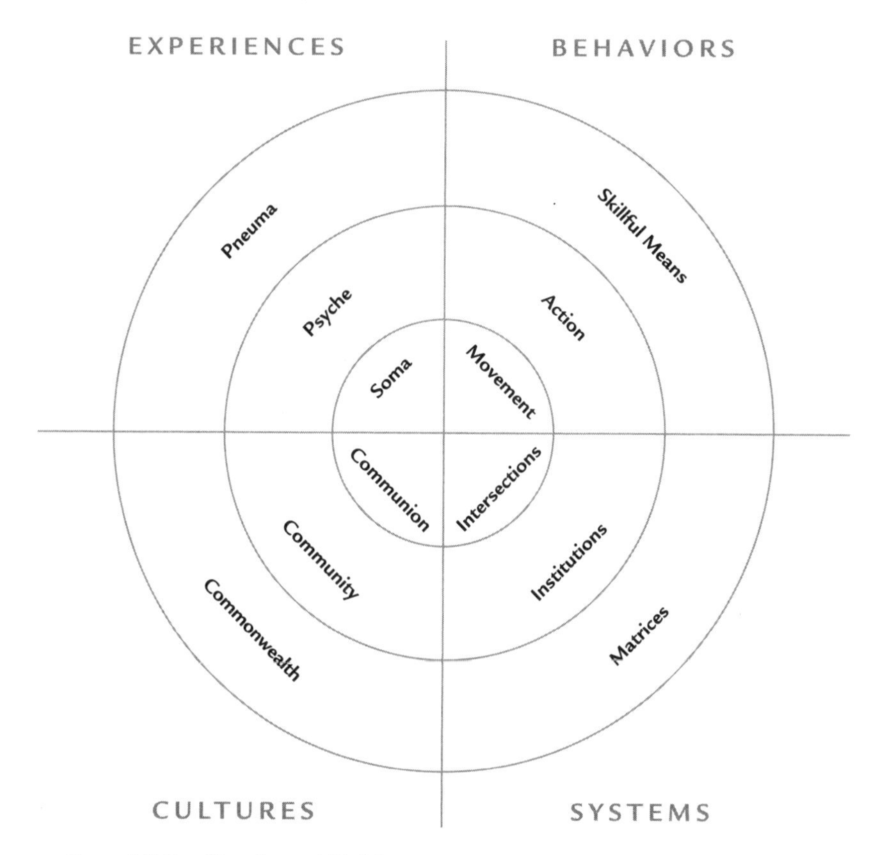

Figure 6.7. The 4 terrains and 12 niches.

the level of complexity of either the What or the How (see endnote 2). Thus, the first level of niches represents magenta to red complexity, the second level represents amber to green complexity, and the third level represents teal to indigo complexity. This is particularly important with regard to the third level, as we are focused on approaches that illustrate stablized worldcentric experiences, behaviors, cultures, and systems. Consequently, we are not including many New Age approaches that are "spiritual" in the third level of complexity, because in our view they are too often dependent on states, which are temporary. As a result, there are often only a few examples at the third level of complexity; we feel this is appropriate, as there are few authentic worldcentric expressions in each of the 4 terrains.

In order to honor the importance of the state-based approaches, we include a section, "State-Based and Subtle Energy Approaches in Each Terrain," after the 12 niches that highlights where they fall within the 4 terrains. If you prefer to place these state-based approaches within the third level of complexity, we ask that you keep in mind the important distinctions between states and stages. In fact, we invite readers to use the 12 niches in ways that best suit their own integral applications without feeling they have to adhere to our particular placements. The usefulness of the 12 niches is not dependent on our specific employment of them. However, we feel there is value added in separating out state-based approaches to the What.

The niches represent aspects of reality that are not independent of an observer, though they might be independent of any particular individual observing them.[30] In other words, the existence of ecological phenomena is not dependent on being perceived by any particular individual, but our understanding of it occurs within our observation and subsequent language about the phenomena. Integral Ecologists commit to holding all niches within their field of attention, because all 12 niches are present. Obviously, some situations require placing more attention on one or a few niches, but the other niches cannot be ignored or reduced to their correlates. Each niche is irreducible to any other and can only be fully understood on its own terms. Each niche has a tradition of experts (a community of the adequate) that have documented its contours and provided methods to access and understand its realities. The more niches that are acknowledged and included, the more sustainable any given project becomes.

For a more comprehensive picture of the niches and the different ecological approaches associated with each, we provide several representative examples for every niche. Some of these approaches are actual schools of thought, while others are labels or categories we have used to designate a distinct perspective. The following examples are intentionally brief in order to cover a lot of approaches in a respectable space and in so doing illustrate just how many legitimate perspectives on ecology there are. We encourage readers to follow up on sources provided in the notes for a fuller description of these approaches. Also, most of these schools are described in the appendix.

Approaches to Ecology That Emphasize the Terrain of Experience

Subjective approaches include all those movements or positions that recognize interior dimensions of individual holons. There are three levels inside Nature: soma, psyche, and pneuma.

Soma

The soma niche explores our felt experience of the natural world as it occurs within the lived body. There are three main categories within this domain: *eco-phenomenology*, *eco-embodiment*, and *architectural phenomenology*. Eco-phenomenology draws heavily on the philosophy of Merleau-Ponty along with a few other thinkers in the phenomenological tradition.[31] Eco-embodiment is similar, but it draws more on the field of somatics and emphasizes emotions, movement, and practices that highlight the nature-body connection.[32] Architectural phenomenology explores how individuals directly experience the built environment and is concerned with "dwelling" (à la Heidegger).[33] As a result it actually links individual phenomenology, intersubjective space, and the surrounding environment.

Psyche

The psyche niche explores the psychological dynamics that connect or disconnect us from the natural world, and the types of sentience that exist in the natural world. There are at least seven main categories within this domain: *eco-self development*, *children and nature*, *environmental psychology*, *ecological psychology*, *ecopscyhology*, *psychology and ecology*, and *animal consciousness*. Eco-self development studies the developmental relationship between individual minds and the natural world.[34] Children and nature studies the relationship between children's interior realities and their natural surroundings.[35] Environmental psychology focuses on the effect that human-made environments, such as office buildings or a housing complex, have on human psychology.[36] In contrast, ecological psychology generally studies evolutionary theories of perception—often with an emphasis on individual exteriors.[37] Ecopsychology is a loose collection of activists, psychologists, and ecologists who study the roots of our alienation from the natural world due to material consumption, technological addiction, ecological grief, and depression.[38] Psychology and ecology refers to any approach that studies the connection between aspects of psychology (e.g., emotions, identity, values) and one's natural surroundings.[39] Animal consciousness (e.g., cognitive ethology and emotional ethology) studies the interior lives and emotional/mental capacities of nonhuman animals.[40]

Pneuma

The pneuma niche represents permanent development of worldcentric and planetcentric perspectives. It explores the spectrum of spiritual realization and unitive experiences individuals at these levels have with members and aspects of the natural world. There are at least two main categories within this domain: *the ecological self* and *eco-spiritual writing.* The ecological self approaches believe that individuals possesses the capacity to expand their self-identity until they identify with the entire natural world.[41] Thus, this approach studies the transcendence of the egoic self, and the structural capacity to connect with all living beings and the planet. Eco-spiritual writing includes 1st-person poetry and prose that describe the loss of self-boundaries with the natural world.[42]

Approaches to Ecology That Emphasize the Terrain of Culture

The intersubjective approaches to the environment include all those movements or positions that recognize the interior dimensions of social holons.

Communion

The communion niche studies the patterns of cultural practices that create shared somatic experiences in relationship to the natural world. There are at least two main categories within this domain: *environmental sensitivity* and *ecological intercorporeality*. Environmental sensitivity studies the ways environmental toxicity has altered individuals' somatic realities.[43] As a result of the growing number of individuals with various environmental sensitivities, the capacity for certain kinds of intercorporeal experiences is diminished. Eco-intercorporeality, like eco-phenomenology, draws on Merleau-Ponty's and Whitehead's philosophy, as well as the somatic work of Anna Halprin's environmental dance, Emily Conrad Da'oud's continuum, and Bonnie Bainbridge's body-mind centering.[44] Eco-intercorporeality studies the ways our somatic worlds are shared, and the subtle, somatic intersubjective dimensions of human and nonhuman relationships.[45]

Community

The community niche studies the various worldviews that determine how we experience place and establish relationships with the natural world. There are at least 22 distinct categories within this domain, which makes it the largest grouping among the 12 niches. This is not too surprising, as the environmental movement has largely had a social-cultural emphasis. This niche includes *historical concepts of nature, nature and culture, sequential ecological worldviews, postmodern nature, continental philosophy and nature, critical theory and nature, interspecies relations, human geography, sense of place, process ecology, landscape studies, bioregionalism, eco-literature, eco-feminism, woman and nature, gender and science, cultural appropriation of indigenous traditions, animal cultures, landscape and language, traditional knowledge / indigenous wisdom, environmental ethics (biocentric, extension, land, ecocentric),* and *Goethean science.*

Three approaches deal with worldviews in various ways. Historical concepts of nature traces concepts of nature as they have appeared in different cultures and across time. This approach has contributed immensely to our understanding of the shifting conceptual relationships between humans and nature.[46] Nature and culture studies the relationships and intersections between cultural realities and the natural world.[47] Sequential ecological worldviews maps the developmental sequence of ecological worldviews.[48]

Three approaches highlight postmodern philosophical positions. Postmodern nature studies the social construction of nature.[49] Continental philosophy and nature studies the connection between European continental thought and ecological issues.[50] Critical theory and nature studies the value of critical theory for environmental solutions. It draws on the work of Habermas and examines the intersubjective conditions needed for authentic subject-subject relations that are free from subtle and overt forms of domination.[51]

Four approaches emphasize ways humans interact with their landscapes. Sense of place studies how people make meaning of their geographical location.[52] Human geography studies the ways in which people and communities feel and think about space—home, neighborhoods, and nations.[53] Landscape studies focuses on urban, rural, and wild landscapes and how they reflect changes in our culture.[54] And bioregionalism cultivates ecological intimacy and support of local biozones.[55]

Two approaches capture the influence of language on the natural world and vice versa. Eco-literature uses the novel or prose to promote ecological ideals including ecocriticism.[56] Landscape and language explores the relationship between the landscape and the words we use.[57]

Three approaches connect language, gender, and cultural power dynamics. Ecofeminism studies the historical, cultural, and ideological relations between the treatment of women and the treatment of nature. The ecofeminist movement often focuses on the implications of relationality (e.g., men-women, rich-poor, human-animal, north-south, spirit-matter) for our environment.[58] Woman and nature, and gender and science, both uncover the assumptions and biases within cultural dynamics and in particular within science.[59]

Three approaches highlight the cultural role indigenous peoples have played within ecology. Cultural appropriation of indigenous traditions studies the ways groups, particularly middle-class white Americans, portray and appropriate other cultures (especially Native American cultures).[60] Traditional knowledge documents ethnobotanical and ethnoecological information within traditional communities.[61] Indigenous wisdom studies indigenous cultures (e.g., Native American, Australian Aboriginal) that contain an earth-friendly cosmology.[62]

Two categories focus on animals—our relations with them and their relations with one another. Interspecies relations studies interspecies relationships, often between humans and nonhuman animals.[63] Animal cultures studies the existence of "culture" within animal species.[64]

Process ecology develops an ethical approach to the environment based on the work of Alfred North Whitehead.[65] Whitehead's "philosophy of organism" attributes interiority to all compound "individuals" within nature (including atoms). Not surprisingly, many environmentalists use Whitehead's process philosophy to argue for the intrinsic value (i.e., significance based on depth) of nature's compounded individuals.[66]

Environmental ethics studies the multiple ethical positions toward the environment.[67] There are at least four: *biocentric ethics* values the individual lives of all beings and highlights their intrinsic value as individual members of the biotic community;[68] *extension ethics* accords ethical status to beings based on their level of sentience;[69] Leopold's *land ethic* values the integrity of the land as support for the community of beings;[70] *ecocentric ethics* values the entire ecosystem above individual members.[71]

Lastly, there is the category of Goethean science, which treats objects of study as subjects and allows these subjects to speak for themselves. Renewed appreciation of Goethean science has created a new understanding of the possible intimacy between subjects. Goethean science has been described as a "phenomenology of nature" and as "conscious participation in nature."[72]

Commonwealth

The commonwealth niche studies the shared worldviews of spiritual relations with the natural world and the Kosmos. There are at least seven main categories within this domain: *ecofeminist theology*, *liberation theology*, *creation spirituality*, *the new cosmology*, *ecology and religion*, *tantric wisdom*, and *transcendentalism*.

Most of these categories re-vision current religious traditions into ecological contexts. Keep in mind that we are emphasizing the post-conventional aspects of these categories. Ecofeminist theology applies ecofeminist principles to Christian theology so as to understand God in ecologically appropriate ways.[73] Liberation theology draws on emancipatory traditions to radicalize the Catholic Church and connect the plight of the Earth with social issues. This movement has been very successful in Latin America.[74] Creation spirituality draws on Meister Eckhart and other Christian saints and sages to develop an approach to Christianity that celebrates Creation (i.e., nature).[75] The new cosmology draws on the work of Pierre Teilhard de Chardin to reenchant the cosmos and our relationship to it.[76] Ecology and religion is a broad category that includes all attempts to "ecologize" the various world religions.[77] Of particular interest in this approach is the Yale-based Religions of the World and Ecology Project, which, among other things, published a 10-volume encyclopedic collection of essays that examine the connections between our globally inherited religious traditions and ecological issues.[78] Also of interest within this category is the hermeneutics of animals in the context of religious traditions.[79] The highest teachings from various religious traditions comprise tantric wisdom, which has been used to situate our planetary situation within a nondual context.[80] New England transcendentalism, which is part of the Romantic tradition, studies nature as the abode of Spirit.[81]

Approaches to Ecology That Emphasize the Terrain of Behavior

Objective individual approaches to the environment include all those movements or positions that recognize the exterior dimensions of individual holons.

Movement

The movement niche studies physical behaviors that are connected to the natural world as well as the natural dynamics of exterior process. There are at least five main categories within this domain: *physical and life sciences*, *the new biology*, *wilderness skills*, *environmental illness*, and *pollution*.

Physical and life sciences include all the hard sciences (e.g., physics, chemistry, biology, hydrology, geology, zoology, ornithology) that we use to study individual phenomena of the natural world. The new biology expands the field of biology from its mechanistic commitments into a more holistic science, often drawing on Goethe's methodology.[82] Wilderness skills are the programs that teach survival skills and how to live off the land.[83] Environmental illness studies the relationship between environmental toxins and human and nonhuman health.[84] Pollution studies the behaviors of individuals, communities, and corporations that pollute the biosphere.[85]

Action

The action niche studies intentionality-driven behaviors. All of these approaches choose to act in ways to benefit the environment. There are at least four main approaches within this domain: *eco-friendly behaviors*, *voluntary simplicity*, *environmental justice*, and *eco-social action*.

Eco-friendly behaviors teach individuals about ecologically destructive and beneficial behaviors.[86] Voluntary simplicity encourages us to reduce the "busyness" in our lives so that we may live more simply and in accord with the innate flows and rhythms of the natural world.[87]

Environmental justice explores how racism, classism, and sexism contribute to ecologically damaging behaviors.[88] Eco-social action explores personal action that can effect global change.[89]

Skillful Means

The skillful-means niche studies individual behaviors that result from or create experiences of being "one with nature" or that foster right relationships that stem from worldcentric or higher stages and foster development to those levels. There are at least two main approaches within this domain: *outdoor trips and pilgrimages* and *planetary activism.*

Outdoor trips and pilgrimages are "journeys" into the wild or sacred sites to nourish one's relationship to the Sacred.[90] Planetary activism places eco-social activism into global and spiritual contexts.[91]

Approaches to Ecology That Emphasize the Terrain of Systems

Interobjective approaches to the environment include all those movements or positions that recognize the exterior dimensions of social holons.

Intersections

The intersections niche studies the measurable exteriors of natural systems. Due to its affinity with modernity (a focus on the functional fit of systems), many current environmental approaches are located within this niche. There are at least six main categories here: *ecological sciences, systems and complexity sciences, planetary ecology, biodiversity and species extinction, permaculture and gardening,* and *food and agriculture.*

Four categories deal with the various system sciences. Ecological sciences includes all the hard sciences used to study various aspects of ecosystems.[92] Systems and complexity sciences study the parts, the interconnections, and the holistic nexus of relationships.[93] Planetary ecology is the study of the Earth as one large ecosystem.[94] Biodiversity and species extinction documents the variety of life on Earth, where it occurs, and its threatened status.[95]

The following categories study food production. Permaculture and organic gardening studies sustainable design in small-scale ecosystems that produce food for human consumption.[96] Food industry and agriculture studies the ecosystem dynamics related to food production, the use of herbicides and pesticides, monocropping, transportation, water usage, and so on.[97]

Institutions

The institutions niche studies eco-social systems and institutions that maintain and result from levels of awareness, ranging from symbolic to rational, and their resulting worldviews. There are twenty-two main categories within this domain: *eco-nomics, sustainable development, urban planning, arts and crafts movement, social theory, Marxism and ecology, Foucault's biopower, green politics, environmental law, globalization, environmental law, environmental education, eco-design, green architecture, natural building, technology, human population studies, state of the world, environmental history, ecology and civilizations, the shadow of indigenous,* and *ancient cultures.*

Eco-nomics studies the ways that economics can serve the environment and shows that environmentally friendly business practices can serve, rather than hinder, the bottom line.[98] Sustainable development describes development that meets the needs of the present without compromising the ability of future generations to meet their own needs. As the name suggests, it encourages development that a particular place, and the planet, can sustain.[99] Urban planning is a branch of architecture that creates ecologically wise designs for communities within urban spaces.[100] Particularly noteworthy in this context is the work of Patrick Geddes, which is quite integral in a variety of ways.[101] The arts and crafts movement studies the link between labor, art, and social action.[102] Social theory addresses social inequities to redress ecological imbalances.[103] Marxism and ecology studies the relationship between class structure, production dynamics, technology, and ecology.[104] Foucault's biopower connects power dynamics to ecological truths.[105] Green politics seeks to influence politics for the benefit of the environment.[106] Environmental law writes and fights legislation at local, regional, national, and global levels.[107] Globalization studies and critiques the trend of investment funds, businesses, and governments to move beyond national borders and markets to other markets around the globe.[108] Environmental education is any curriculum, program, or school that educates students about the environment.[109] Eco-design takes cues from the natural world to create nontoxic (or less toxic) sustainable and effective designs.[110] Green architecture designs buildings suited to their natural surroundings. These buildings are less impactful or harmful to the environment, are often built of recyclable materials, and use such technology as energy-efficient appliances and compost toilets.[111] Natural building uses natural products within the home as well as natural materials to build the home.[112] Technology studies the impact of technology on either the biosphere or our lives.[113]

Human population studies calculates the carrying capacity of the Earth or regions of the planet based on regional resource dynamics, and makes predictions on how populations will affect and be affected.[114] State of the world uses statistical science to analyze the status of the Earth in numerous categories (e.g., forest cover, water access, biodiversity). Some reports suggest that things are getting better and others suggest that things are getting worse.[115]

Three categories focus on history. Environmental history studies the evidence of the history of relationships between humans and nature.[116] Ecology and civilizations studies the ways that entire civilizations have impacted and been impacted by ecosystems.[117] The shadow of indigenous and ancient cultures studies all the anthropological, cultural, and scientific research of the destructive behaviors of previous cultures and contemporaneous indigenous communities, including cannibalism and human sacrifice.[118]

Matrices

The matrices niche studies worldcentric and planetcentric systems associated with the Earth. While these consciously created post-formal systems are rare, there are at least three main approaches within this domain: *developmental systems ecology*, *spiritual aesthetic systems*, and *sacred geography*.

Developmental systems ecology describes the nested structures of natural systems, which contain more and more information.[119] Spiritual aesthetic systems, such as feng shui and wabi sabi, organize space and objects in a spiritual context.[120] Sacred geography describes and often locates "power places" and sacred sites that are usually the pilgrimage sites of various traditions.[121]

State-Based and Subtle Energy Approaches in Each Terrain

As mentioned at the beginning of our discussion of the 12 niches, there are approaches in each terrain that draw heavily on nonordinary states of consciousness. Many of these approaches put individuals into contact with the subtle-energy dimension of the planet. These subtle-realm experiences are an important part of the nature mysticism lattice discussed in chapter 9.[122]

Terrain of Experience

There are four main approaches in this terrain: *shamanism*, *psychedelics*, *experiencing nature spirits*, and *merger states*. Shamanism studies the capacity to journey into other dimensions of the natural world and emphasizes individuals from indigenous cultures who have mastered the techniques of "journeying."[123] Psychedelics as a category is often considered similar to shamanism. However, psychedelics is a more general category that has no indigenous roots and refers primarily to the use of drugs such as mushrooms or LSD to expand consciousness and connect with nature.[124] Experiencing nature spirits studies the ways individuals can cultivate the capacity to see, hear, and experience nature spirits, elementals, and devas.[125] Merger states studies the modes of consciousness that allow an individual to momentarily identify with minerals, plants, or animals.[126]

Terrain of Culture

There are two main approaches in the terrain of culture: *interspecies communication* and *nature archetypes*. Interspecies communication studies the telepathic communications between humans and animals.[127] Nature archetypes studies eco-spiritual archetypes, such as the Green Man, which serve as symbols for entering into direct relationship with the natural world.[128]

Terrain of Behavior

There are four main approaches in the terrain of behavior: *neo-pagan rituals*, *indigenous ceremonies*, *vision quests*, and *occult science*. Neo-pagan rituals draws on paganism to create rituals and spells for working with natural forces.[129] Indigenous ceremonies study the rites of passage and activities for connecting with the Earth and one's ancestors.[130] Vision quests is a Native American solo spiritual journey, in which one fasts and prays to open to guidance from the spirit world.[131] Occult science explores the reality and behavior of nature spirits and elementals.[132]

Terrain of Systems

There are four main approaches in the terrain of systems: *deva gardening*, *biodynamic agriculture*, *mystical systems*, and *natural farming*. Deva gardening is the capacity to work on subtle energetic levels with nature intelligences, often referred to as devas, elementals, and nature spirits.[133]

Biodynamic agriculture is a process to work with natural subtle-energy systems developed by Rudolf Steiner.[134] Mystical systems attempts to marry mysticism with physics (e.g., Capra's *Tao of Physics*) or living systems (e.g., Macy's *Mutual Causality in Buddhism and General Systems Theory*). Natural farming is a farming system developed by Japanese farmer Masanobu Fukuoka, whose techniques have been widely acclaimed. He attributes his success to his connection to the spiritual dimension of the land.[135]

Now let us return to the 12-niche chart (figure 6.8) and place a representative approach in each domain. This will allow us to capture the flavor of each domain, and highlight how such a framework can help organize hundreds of ecological positions and perspectives. These examples are mere representatives. Imagine that you can "double-click" on any of these complex eco-approaches and find all the schools that they represent, and their proponents in many other domains. Ecofeminism, for example, is most often associated with the second level of complexity of cultures (e.g., interconnections of cultural thought patterns), but during the course of its development over the last 30 years, ecofeminists have increasingly inhabited additional niches of environmental concern, emphasizing "interconnection" via the terms of each niche. So we can further break down the entire school and evolution of ecofeminism according to the 4 terrains (see figure 6.9). Note: we have placed a question mark in the third niche for each

	INTERIORS		EXTERIORS	
	EXPERIENCES	CULTURES	BEHAVIORS	SYSTEMS
3rd Level of Complexity	**Conservation Psychology**	**Process Ecology**	**Planetary Activism**	**Global Ecology**
2nd Level of Complexity	**Ecopsychology**	**Ecofeminism**	**Environmental Justice**	**Environmental Economics**
1st Level of Complexity	**Eco-phenomenology**	**Eco-intercorporeality**	**Environmental Health**	**Living Systems Theory**

Figure 6.8. Representative ecological approaches per niche.

<table>
<tr><th></th><th>INTERIOR</th><th>EXTERIOR</th></tr>
<tr><td>INDIVIDUAL</td><td>Terrain of Experiences
PNEUMA:
Spiritual Realization
?
PSYCHE: Psychodynamics of Human-Nature Relationships
Sacred Interconnections
(e.g., Christ, Spretnak, Macy)
Psychological Interconnections
(e.g., Keller, Sewall)
Epistemological Interconnections
(e.g., Harding, Guren, Haraway, Mills)
SOMA: Somatic Responses to Nature
Somatic Interconnections
(e.g., Olson)
I</td><td>Terrain of Behaviors
SKILLFUL MEANS:
Worldcentric Behaviors
?
ACTION: Behavior Resulting from Thinking Patterns
Ritual Interconnections
(e.g., Starhawk, Brodle)
Power/Race/Class Interconnections
(e.g., Gebara, Roddick)
MOVEMENT: Physical Behaviors That Have Ecological Impact
Empirical Interconnections
(e.g., Carson)
IT</td></tr>
<tr><td>COLLECTIVE</td><td>WE
Terrain of Cultures
COMMONWEALTH:
Compassionate Perspectives
?
COMMUNITY: Human-Human and Human-Nature Relations
Religious Interconnections
(e.g., Reuther, McFague, Gray)
Historical Interconnections
(e.g., Merchant)
Conceptual Interconnections
(e.g., Fox, Plumwood, Griffin)
Ethical Interconnections
(e.g., Cuomo, King)
COMMUNION: Shared Somatic Meaning
Intercorporeal Interconnections
(e.g., Bigwood, Holler)</td><td>ITS
Terrain of Systems
MATRICES:
Global Systems
?
INSTITUTIONS: Systems that result from mental productions
Linguistic Interconnections
(e.g., Adams, Dunayer)
Political Interconnections
(e.g., Lahr, Sturgeon, Plumwood)
Socioeconomic Interconnections
(e.g., Shiva, Mies, Mellor)
INTERSECTIONS: Exteriors Fitting Together
Systemic Interconnections
(e.g., Oyama)</td></tr>
</table>

Figure 6.9. An Integral analysis of ecofeminist schools.

terrain to indicate the current lack, in our view, of Integral ecofeminist theorists in these areas. This presentation is based in part on the work of Karen J. Warren.[136]

The 12 Niches of a Stream Restoration Project

To further illustrate the value of the 4 terrains and their respective niches, we will use another example of a stream restoration project. To provide a realistic context, we created a fictitious case study. The project is described and then followed by niche-specific questions to help design a comprehensive restoration project. Many of the questions are multivalent and could be asked from other niches.

Lacey is a small junior high school in Fisher, Oregon. Behind it runs Anderson Creek, named after Paul Anderson, who settled near the creek in the 1860s. Currently, the stream winds through a number of fields and backyards, and alongside the school's property (its soccer and baseball fields, and the playground and tennis court).

The school's biology teacher, Ms. Buonocore, became motivated to create a stream restoration project when she discovered an old picture book that contained black and white photographs of the area during the "early" days of the town's history. These pictures made her aware of how much the stream had changed during the last 150 years. Since then she spoke with everyone she knows about a restoration project. Her goal is to create something that could benefit the entire community and also serve educational purposes.

She generated enough interest and support to enable her to form a committee—A.C.T. (Anderson Creek Taskforce). It is composed of eight people (Ms. Bunocore, two parents, a city planner, two neighbors from the area, a woman who owns a restaurant near the stream, and a student). The committee has contacted an Integral Ecologist to serve as a consultant for the project. Here is a series of possible questions from each of the 12 niches that our Integral Ecologist asks to help her create the most comprehensive restoration project possible.

Subjective Dimensions to Consider

Soma: Somatic Realities How can we design walkways, and place signs, benches, etc., to support or create a positive interior experience for visitors?

How can we construct walkways, etc., so as to minimize the somatic impact on local flora and fauna?

Psyche: Psychological Dynamics What forms of consciousness exist in or around Anderson Creek? How do they affect one another? How will they be affected by the proposal?

What are the features of the landscape that are most aesthetically pleasing and are most likely to inspire people?

What will be concealed from view and revealed by any design we utilize? What do we want to reveal? Is there anything we want to conceal?

What kinds of signs or information booths will connect people to the history, landscape, and ecology of the area?

Pneuma: Spiritual Realization How does the proposed design serve the development of consciousness or spiritual state experiences?

What kinds of spiritual experiences are associated with the project area?

Intersubjective Dimensions to Consider

Communion: Intercorporeal Dimensions Which kinds of meaningful encounters can we facilitate between species, and which ones should we avoid?

What are the rights of each species, and how might we honor them?

How does the design of the area (e.g., walkway direction, sitting arrangements) create shared somatic space?

Community: Shared Horizons What are the community values and perceptions of the creek area? How can we accommodate different values in visiting the area?

What are the power dynamics (e.g., race, gender, and class) in the community that are relevant to the proposed project?

How might the community become motivated to take short- and long-term responsibility for the restored area?

What is the eco-cultural history of the area? What are the ways that history has been documented (e.g., pioneer journals, folk songs, drawings, surveys, native myths, local legends)?

Commonwealth: Compassionate Perspectives How can this place be a sanctuary for divinity?

How does this area connect to the local indigenous tribe's cosmology?

Objective Dimensions to Consider

Movement: Physical Movements Which species are native? Which ones are introduced?

What are the health conditions of various components of the ecosystem (e.g., the soil's nutrient content, the pH balance of the water)?

How can individuals be sure to not introduce toxins into the stream?

How are the nearby homes, businesses, and school activities impacting the stream and surrounding area?

What kinds of behaviors do we want to encourage and discourage around the stream?

Action: Intentional Conduct What kinds of local organizations would volunteer or contribute time and energy to this project?

How might school activities interface with the area?

Can we use recycled materials for construction?

Should we plant a garden? Should it be organic? Where should we place it?

Skillful Means: Worldcentric Action What can we do to encourage different individuals with different worldviews to use this area?

How is this project promoting actions that consider multiple perspectives?

Should a multifaith altar or shrine be established as part of the design?

Interobjective Dimensions to Consider

Intersections: Environmental Systems How is the stream connected to the larger watershed?

What are the ecological dynamics of the area, and what ecosystems does the stream travel through?

What migratory birds pass through this area?

Is there a drought cycle?

Are there any ley lines running through or near the site?

Institutions: Noetic Systems How do the school system and local politics facilitate and complicate the project?

What are the dynamics of the local economy? Where will financial maintenance come from?

What environmental laws or ordinances exist in the community?
How will the public be informed and educated about the area?

Matrices: Global Systems How will the project fit with local and global issues?

Which world- or planetcentric organizations could be enlisted as a sponsor of this project?

We find the symbolic range of questions that result from the 12 niches instructive. These are only a handful of the many potential questions that could inform such a project, but it creates a more comprehensive and perhaps more creative vision for any project.

Conclusion

The 4 terrains and their respective 12 niches represent the ontology of Integral Ecology: the What of knowing. Now we will turn toward the epistemology of Integral Ecology: the Who of knowing. Integral Ecology always takes a "participatory" approach—the known is never understood apart from the knower, the perceived always arises in the context of a perceiver. To understand the complexity of enacted or co-created spaces, we will explore each pole (ontological and epistemological) separately. After outlining the 8 ecological selves and their respective epistemologies, we will focus on the various modes of inquiry: the How of knowing. These modes connect the 8 eco-selves (epistemology) to the 12 niches (ontology). Only after exploring the What, the Who, and the How can we understand the complex tetramesh and recursive relationship that exist between these three aspects of all environmental phenomena.[137]

7

Ecological Selves: The *Who* That Is Examining

Before you start exploring the essence of the things that surround you, investigate first what you yourself as a subject carry into nature. Before judging the things you look at, examine your own perception.

—JAKOB VON UEXKÜLL[1]

My thinking is not separate from objects; that the elements of the object, the perceptions of the object, flow into my thinking and are fully permeated by it; that my perception itself is a thinking, and my thinking a perception.

—JOHANN WOLFGANG VON GOETHE[2]

The general public has been saturated with ecological information, and yet has not dramatically altered behaviors responsible for serious eco-psycho-social problems. Additional information in and of itself is clearly not enough! We cannot continue to believe that education about the natural environment will change people's behavior. And if it is difficult to change individual behavior, how will we influence social and economic change?

Our view is that integral awareness of developmental dynamics and the capacity to take multiple worldviews are crucial elements in achieving behavioral changes and altering our current treatment of the biosphere. By understanding the dynamics and structure of levels of psychological development, we can tailor our communication to each perspective while holding the larger view. If we encourage the healthy manifestations of each level of development, we can inspire and motivate individuals with any worldview, especially amber and higher, to participate in environmental solutions.[3]

Cultivating mutual understanding between perspectives is an essential component in addressing our environmental problems. Mutual understanding, as we use it here, is dependent on the cognitive, emotional, and interpersonal capacities to take many perspectives, even those contradictory to one's own. These capacities emerge most fully at specific levels of awareness, which have been described by developmental researchers: Robert Kegan's 5th order of consciousness;[4] Jean Gebser's integral-aperspectival worldview;[5] Jürgen Habermas's domination-free discourse;[6] and Ken Wilber's vision-logic (i.e., turquoise altitude).[7] As discussed previously, Integral Ecology asserts that leaders embodying anything less than a worldcentric level of development will cripple comprehensive environmental solutions. Leaders can facilitate solutions to our environmental problems by increasing their capacity to transcend and include partial perspectives and coordinate between them.[8]

Integral Ecology is therefore committed to exploring developmental psychology and its relationship to the self (subjectivity), culture (intersubjectivity), and nature in both individual organisms (objectivity) and the systems they are members of (interobjectivity). Precisely because Integral Ecology acknowledges both interior and exterior realities, it offers more comprehensive and effective responses to environmental problems, as it avoids reducing environmental phenomena and solutions to either of those two domains.

In this chapter we focus on the relationship between interior psychological development and understandings of ecological realities.[9] We review previous, albeit incomplete, attempts to account for the relationship between interior development and views of nature. We then explore the possiblility of an ecological line of development. Next we present our model of ecological self-identity (the eco-selves) using the developmental research of Susanne Cook-Greuter. Finally, we present one example that illustrates the value of the framework of the 8 eco-selves.

Psychological Development and Ecology

The relationship between psychological development and systems ecology is a very crucial one for Integral Ecology, in large part because ecological consciousness cannot be accounted for or explained by the frameworks of ecology. This makes sense methodologically when you consider the 8 modes, but this point is all too often lost on environmentalists and ecologists.[10] In fact, a successful approach to exterior ecology is dependent in important ways on the interior development of individuals toward worldcentric and planetcentric identities. Just because two people share the same exterior landscape in no way means they must inhabit the same interior cognitive or moral landscape. Given the importance of constructive-developmental structuralism for ecology, it is amazing that it has been ignored by most ecologists and ecotheorists—even those who explicitly talk about the value of expanded identities. Ironically, developmental psychology is what provides a road map of personal transformations needed to arrive at the destination of ecological awareness prized by so many environmentalists. It is not enough to memorize 3rd-person descriptions of the web of life, as individuals at any level of development can do that: *knowledge by description.* What is needed are permanent 1st-person transformations into wider identities: *knowledge by acquaintance.* Given that such transformations are arduous and take time, we can't simply dialogue ourselves into eco-awareness.

Recently, a relatively small number of researchers have begun to investigate the relationship between psychological stages of development and ecological attitudes. Two pioneers, Harold Searles and Urie Bronfenbrenner, highlight how the environment impacts development. Their work is an important step in the recognition of how interiors interface with one's surroundings. In *Nonhuman Environment in Normal Development and in Schizophrenia* (1960), Searles examines, among other things, the infant's desire to merge with his or her environment—an impulse Searles claims occurs throughout life.[11] Bronfenbrenner, writing almost two decades later in *The Ecology of Human Development: Experiments by Nature and Design* (1979), advances the idea that the developing person is affected by the surrounding environment and restructures it.[12] He refers to his model as the bioecological model.[13] (When he uses the term "ecology," he is primarily referring to the human environment—mass media, neighborhood play area, extended family, and ideologies of the culture at large).[14] Although Searles and Bronfenbrenner do not use a vertical model of development and do not focus on the human-nature relationship, this research clearly contributes toward a fuller understanding of this relationship.

The human ecologist Paul Shepard, whose work we return to in chapter 9, was arguably the first ecologist to apply vertical models of human development, including those researched by Jean Piaget, Erik Erikson, and Erich Fromm. He connected their developmental stages to different perceptions of nature. He argues that modern society is perpetuating an arrested development or "ontogenetic crippling" that prevents adults from fully maturing.[15] With this argument, it seems that Shepard promotes a return to an ecological yesteryear that actually never existed. Shepard puts a problematic spin on the data and spirit of the models he uses, in order to drive home his point that the developmental progress of individual psychology has been interrupted and has largely ceased since the advent of agriculture (this is clearly not the case, and there is a mountain of evidence to the contrary).[16] We applaud Shepard for applying developmental models, especially considering that he is a founder of both deep ecology and ecopsychology, and few prominent individuals within those schools advocate or make use of developmental models (partly because such models are inconsistent with their anti-hierarchical postmodernism).

Arne Naess, another founder of deep ecology, is also one of the few prominent individuals within that school whose notion of "Self-realization" stems from a developmental model.[17] He asserts that the self develops toward an ever-increasing identification with all life forms, an identification that moves from egocentric to ecocentric.[18] Warwick Fox, who describes himself as a transpersonal psychologist, also uses some form of a developmental model. Influenced by Naess, Fox draws loosely on a Freudian distinction of id, ego, and superego to create a tripartite conception of the self that includes a "desiring-impulsive self," a "rationalizing-deciding self," and a "normative-judgmental self."[19] He then contrasts these "narrow, atomistic, or particle-like conception[s] of self" with the "transpersonal self," which refers to "a wide, expansive, or field-like conception of self."[20] He implies that the self develops from a narrower one to a more inclusive one, yet offers no explicit vertical dimension in his model.

For three decades, the renowned social ecologist Stephen Kellert has been researching individual values about nature in the United States, Europe, Asia, and Africa. Adopting both subjective and intersubjective perspectives, Kellert explores the relationship between life-span development, moral judgment, and ecological perception. In *The Value of Life: Biological Diversity and Human Society* (1996), Kellert draws on his extensive research to present the nine universal and cross-cultural values of nature and living diversity. He acknowledges the similarities between his findings and the developmental stages of Jean Piaget and Lawrence Kohlberg.[21] Kellert writes: "Independent of [their] findings, my colleague Miriam Westervelt and I encountered an analogous developmental sequence in the emergence of children's values of nature and animals." He explains that "an important difference, though, seems to be the later development of nature-related values in children—perhaps indicative of the lesser significance of human/nature versus human/social relations in the normal developing child, at least in modern society."[22] Kellert explains that they were initially surprised by their results given the way society frequently romanticizes children's relationship with nature. However, he explains that "our results [similar to Piaget's and Kohlberg's] show that young children typically view animals and nature in highly instrumental, egocentric, and exploitive ways, largely responding to short-term needs and anxiety toward the unknown."[23] Based on his cross-cultural research, Kellert presents a three-stage model of the development of values (see figure 7.1).

Kellert supports the view that these basic values emerge differently at various ages or stages, as in other developmental models (e.g., Piaget's cognitive development), and identifies four characteristics of this developmental

STAGES & AGES	VALUES EMPHASIZED	QUALITIES OF RELATIONSHIP TO NATURE AND ANIMALS
3rd 13 - 17 years Late Childhood or Adolescence	Moralistic Naturalistic Ecologistic	Maturation of more abstract, conceptual reasoning. Cognizant of larger spatial and temporal scales (e.g., ecosystems, landscapes, evolution). Fuller capacity for ethical considerations. Can take the perspective of animals (e.g., suffering). Systematic understanding of complex relationships.
2nd 6 - 12 years Middle Childhood	Humanistic Symbolic Aesthetic Scientific	More appreciation of nature near the home. Animals recognized as having their own needs. More curious about and exploratory with nature. Feelings of responsibility toward nature emerge. Critical thinking and problem-solving emerge. Knowledge about nature becomes important.
1st 3 - 6 years Early Childhood	Utilitarian Dominionistic Negativistic	Satisfying material and physical needs. Avoiding threat and danger. Establishing feelings of control and security. Subordinate animal needs to personal desires. Indifference and anxiety toward most of nature.

Figure 7.1. Kellert's stages of children's values of nature and animals.

progression: an evolution from relatively concrete and direct perceptions or responses to more abstract forms of expression; a shift from egocentric to sociocentric; a shift in attention from one that initially can only embrace local settings (concrete and direct perception) to one that can include regional and then global understandings (abstract); and the emergence of emotional values first, followed by more abstract and logical values.[24]

While we agree with Kellert's progression of values, we don't agree with how he has correlated them with his three stages. Basically, we feel that he has developmentally compressed the values and assigned each set of them to the previous stage. For example the humanistic, symbolic, aesthetic, and scientific values are characteristic of orange altitude, but he places them at the second stage of middle childhood, which is when concrete operations are forming (i.e., amber altitude). In other words, we would not expect to see those values fully emerge until the third stage. Likewise, current developmental research indicates that the moralistic, naturalistic, and ecological values as presented by Kellert rarely occur in children between 13 and 17 years of age. Thus, while Kellert's model offers an important contribution to a developmental approach to environmental values, we situate his findings in an eight-level framework (i.e., the eco-selves) and move all his values as represented in figure 7.1 up one level.

Kellert and associates also discovered that there are gender differences in how we value nature. This is consistent with Carol Gilligan's suggestive research on moral development in women.[25] Kellert found that women stressed humanistic values and exhibited strong emotional attachment and moral inclination to protect parts of nature, especially animals. Women also had moderately high scores on negativistic values (fear of nature). Again, this is consistent with Gilligan's findings that women have strong relationships with friends and family but are not as interested in remote relationships. Men, on the other hand, were moderately low on humanistic and moral values, moderately high in utilitarian and dominance values, and low in negativisitic values (less fear of nature). Men were more inclined to use logic, emphasize mastery and pragmatic uses of nature, and understand nature in an abstract sense. At the same time, men were less inclined to connect emotionally to animals or natural spaces. Kellert asserts that while such findings remain preliminary, they can help us understand the types of values predominant in traditionally male-based professions (e.g., wildlife management) and female-based organizations (e.g., various animal welfare organizations).[26] Like Kellert, Integral Ecology recognizes the importance of including gender (and types in general) in understanding various approaches to and perspectives on ecology and the environment.[27]

Another pioneer in this area is Olin Eugene Myers.[28] He has researched human-animal interactions, the moral development of environmental care and responsibility, and the value of environmental education.[29] He is considered a founding figure of the emerging field of conservation psychology, which embraces a developmental perspective and explicitly integrates multiple schools of psychology and ecology so as to understand the relationship between human psychology and the health of the planet.[30]

The psychologist Peter Kahn, Jr., has also contributed to the exploration of the relationship between vertical development and ecology.[31] His highly acclaimed book, *The Human Relationship with Nature: Development and Culture* (1999), is the first and currently only book-length treatment of the relationship between psychological development and perceptions of nature.[32] His understanding is based on eight years of original research with children and adults in the United States (Houston and Prince William Sound), the Brazilian Amazon, and Portugal. Kahn's findings are too numerous to mention here. His more important discoveries include the existence of universal structures of moral reasoning that appear to guide individuals' views of nature across urban, rural, and wild environments; and the fact that, while children across cultures are primarily anthropocentric in their reasoning, they do demonstrate some capacity for authentic biocentric reasoning (see below). Kahn continues to make important contributions to this area.[33] In particular, he coedited a scholarly interdisciplinary volume on children and nature with Stephen Kellert.[34] Kahn and Kellert's recent research implies the existence of semi-independent lines of development such as biocentric reasoning or environmental values.

Ecological Lines of Development

In a striking discovery, Kahn found that indigenous children living in the Amazonian jungle surrounded by pristine nature, and urban children living in the United States surrounded by very little nature (i.e., the "concrete jungle"), both demonstrate the same developmental progression of moral reasoning about nature. Children from both geographical settings predominately used anthropocentric reasoning (~95% of the time). The urban children demonstrated the same basic percentage of biocentric reasoning as the Amazonian children (~5% of the time).[35] Kahn's findings point to the cross-cultural occurrence of both biocentric and anthropocentric reasoning among children at early ages regardless of their environment (e.g., wild or urban). These findings highlight that all children progress through basic stages of development, regardless of the richness of their biotic surroundings. In this sense, Kahn's research confirms what most developmentalists already know: capacities like morality and cognition, and even environmental reasoning, develop in a series of invariant cross-cultural stages of complexity. On the other hand, these findings are surprising, especially to these developmentalists, because they suggest that some form of biocentric reasoning is available at each level of development and is therefore not a capacity that emerges solely at far more complex levels in the developmental journey. According to Kahn's research, a person would not have to become one with all humans (worldcentric) before that person could become one with some or all animals or nature (ecocentric).[36]

These findings might explain how so many environmentalists can be bio- or ecocentric (concerned for the natural world) yet care very little for their human neighbors. The environmental justice activist Carl Anthony captures this well when he said to Theodore Roszak—the founder of ecopsychology: "Why is it so easy for [environmentalists] to think like mountains and not be able to think like people of color?"[37] The answer may be that those environmentalists have a higher biocentric line of development then they do an anthropocentric line. Kahn's research provides preliminary and suggestive evidence of a biocentric line of development that unfolds relatively independent of an anthropocentric line. In fact, Kahn suggests as much when he claims that there is a dialectical relationship between children's moral relationships with humans and with animals such that each form of reasoning (anthropocentric and biocentric) is informed by the other.

For Kahn, *anthropocentric reasoning* "refers to an appeal to how effects to the environment affect human beings."[38] The main types of these appeals include economic welfare (it saves money to recycle), physical welfare (people get sick when the water is polluted), human psychological welfare (pets make people happy), personal human interests (we like to swim in a clean lake), human-centered justice (public access to wilderness is important), and aesthetics (no one likes litter on the beach—it's ugly!). In constrast, *biocentric reasoning* "refers to an appeal that the natural environment has moral standing that is at least partly independent of its value as a human commodity."[39] The main types of appeal in this form of reasoning include

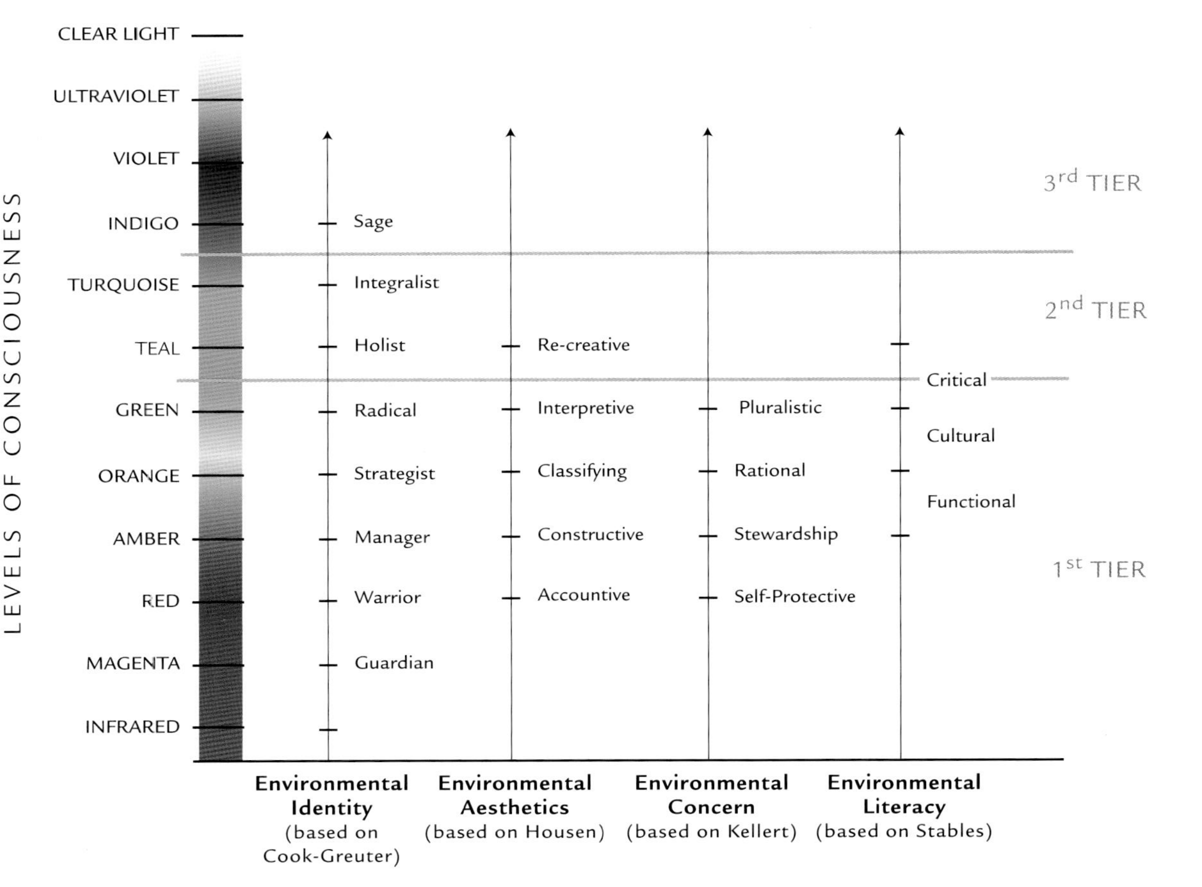

Color Plate 1. Various ecological lines.

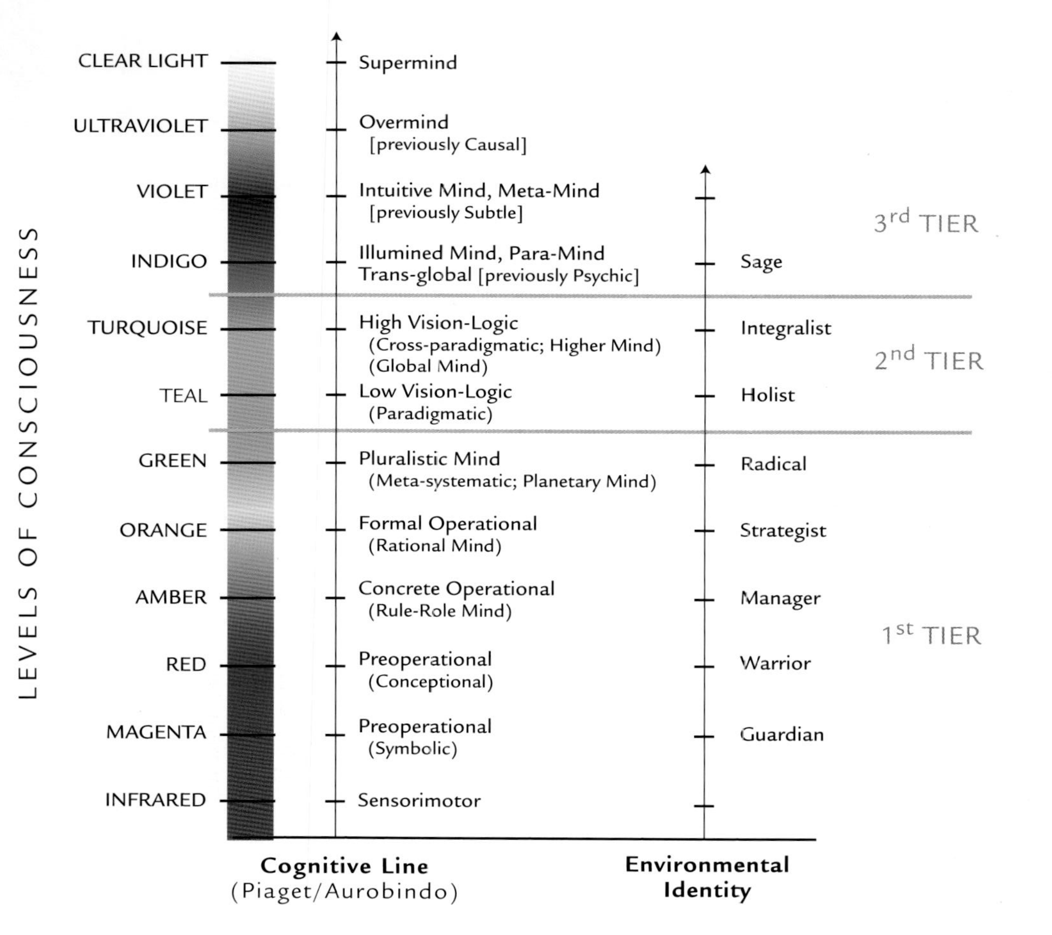

Color Plate 2. The 8 ecological selves.

"intrinsic" value (animals have their own role in nature), nature has rights (trees grow like us, so they should be protected), respect or fair treatment (animals can feel sad), and freedom (animals have a right to live their life).

Kahn then makes a number of distinctions within biocentric reasoning. He calls early biocentrism "isomorphic reasoning," which is an appeal that recognizes a correspondence between nature and humans by means of a direct or a conditional isomorphism. *Direct isomorphism* is symmetrical: animals have hearts like humans. *Conditional isomorphism* is a symmetrical relationship established through an if-then understanding: if we do not like dirty water, then fish must not like it either. He names more complex biocentrism "transmorphic reasoning." Transmorphic reasoning extends isomorphism through the use of either *compensatory transmorphism* (which emphasizes similarities and differences between humans and nature) or *hypothetical transmorphism*, which uses reason to take the perspective of nature. Kahn proposes that transmorphic reasoning hierarchically transcends and includes isomorphic reasoning. In other words, the capacity to reason biocentrically expands into a more comprehensive structure that also includes asymmetrical characteristics. Kahn explains that "biocentric reasoning emerges, if at all, more fully in older adolescents and adults."[40] When we place anthropocentric and biocentric reasoning on a sample psychograph, with our terms of general development on the left and Kahn's terms on the right, it looks like figure 7.2 for an individual with a higher capacity for anthropocentric reasoning.

This psychograph suggests that there is a corresponding form of biocentric reasoning for each form of anthropocentric reasoning. Hence, direct isomorphism mirrors egocentrism; conditional isomorphism mirrors ethnocentrism; compensatory transmorphic reasoning is like sociocentrism; and

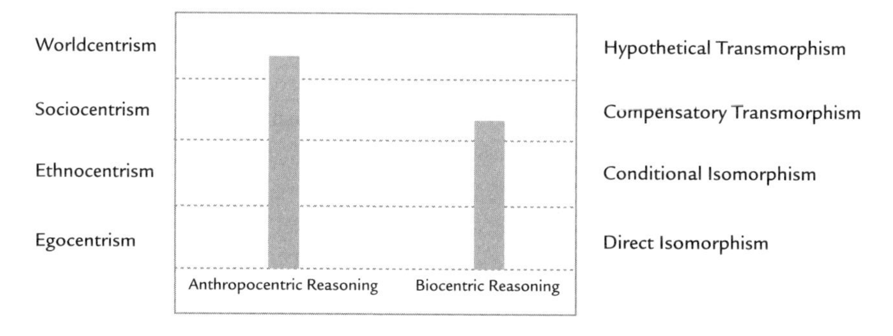

Figure 7.2. Kahn's two possible environmental lines of development.

hypothetical transmorphism is a more complex and abstract form of biocentrism, just as worldcentrism is more complex and abstract than sociocentrism. Consequently, we can think of biocentrism as a line of development and as a specific level of development within that line: hypothetical transmorphism.[41] This helps illustrate how confusing the term "biocentrism" can be if individuals do not clarify whether they are speaking of a general capacity that develops over time or of a particular level of maturation of that capacity. In addition, the anthropocentric reasoning and biocentric reasoning capacities can be seen as different aspects of the moral line.

As a result of the multiple meanings of "biocentrism," it is possible for an individual to be biocentric (conditional isomorphism) and hate their neighbor (ethnocentrism). According to our general schema an individual must inhabit a worldcentric level of development (embracing all people) before they can inhabit a planetcentric worldview (embracing all beings). Otherwise, one's biocentric conceptions don't fully and unconditionally include humans, thereby trapping one in the dualisms of culture versus nature, and humans versus animals. At a planetcentric level of development, these two streams of moral judgment may become inseparable, though generally separate lines do not merge at any point.

We need more research before we can discern whether Kahn has discovered a new line of development (biocentric reasoning), or even two lines—biocentric and anthropocentric environmental reasoning—or whether he has simply discovered how an existing line (morality) expresses itself in relationship to the environment. We think it is plausible that such research may reveal hitherto unknown developmental lines predominately concerned with nonhuman phenomena. In fact, Howard Gardner, who pioneered the notion of "multiple intelligences," has recently added "naturalist intelligence" as another form of intelligence to his original list of seven.[42] Richard Louv dedicates an entire chapter ("The 'Eighth Intelligence'") of *Last Child in the Woods* to Gardner's naturalist intelligence. Unfortunately, Louv doesn't present this intelligence as a line of development with discrete stages. In contrast, Gardner explains that the other intelligences unfold in stages. As a result, Louv does not make it clear how this intelligence develops. This might be due in part to the fact that Gardner's eighth intelligence is postulated without any solid research currently backing it.

Further research will articulate how already established lines of development manifest in regard to the environment. For instance, how does the cognitive line contextualize environmental knowledge? How does the interpersonal line understand interspecial relations? How does the aesthetic line contextualize environmental aesthetics? How does the moral line express ethics in relationship to the environment? We think that many currently established lines of development could, in principle, be translated to or situated within an ecological context (see below).[43]

For example, in "Experiencing Nature: Affective, Cognitive, and Evaluative Development in Children," Kellert provides a theoretical sketch of three ecological lines of development.[44] He explores the role of three types of nature experiences (direct, indirect, and symbolic) on three modes of learning (cognitive, affective, and evaluative) in childhood and early adolescence. For each capacity, Kellert draws on existing developmental models. He uses Benjamin Bloom's six stages of hierarchical intellectual reasoning for cognitive development; he uses Krathwohl, Bloom, and Masia's five stages of emotional maturation for affective development; and finally, he draws on his own research, and those of his doctoral students, for evaluative development.[45]

Jon Geselle of Integral Institute has begun to articulate what particular lines of development might look like in relationship to the environment. Geselle builds on Wilber's distinction that psychological lines can generally be divided into the categories of subjective, intersubjective, and objective lines—depending on the content that is focused on. He proposes three content-specific lines: *environmental aesthetics* (subjective), or the ability to see beauty in nature, which is based on the work of Abigail Housen and her stages of accountive, constructive, classifying, interpretive, and re-creative; *environmental concern* (intersubjective), or the ability to care for nature, which is indebted to the work of Kellert discussed above and recalibrated (in color plate 1) in light of recent research and summarized with our labels as protective (dominionistic and negativistic), stewardship (utilitarian), rational (humanistic, symbolic, aesthetic, and scientific), and pluralistic (moralistic, naturalistic, and ecologistic); and *environmental literacy* (objective), or the ability to know nature, which is indebted to the work of Andrew Stables and his stages of functional (i.e., scientific knowledge), cultural (i.e., the social construction of nature), and critical (i.e., holding multiple perspectives) environmental literacy.[46] While Stables's model is not informed explicitly by developmental psychology, he does recognize a developmental relationship between the three forms of environmental literacy, with each subsequent form building on the previous ones. Color plate 1 (following page 222) illustrates how Integral Ecologists might begin to draw on existing developmental research to explore various lines of development and their relationship to nature.

The 8 Eco-Selves

One of the most important lines of psychological development is the self-identity line due to its defining relationship with the other self-related lines (cognition, interpersonal, and moral). As we mentioned in chapter 4, according to Cook-Greuter, ego development refers to an integration of several developmental lines at a given altitude. The cognitive line (i.e., the capacity to take perspectives), which drives self-understanding and the concomitant self-identity, is always at the same level as a person's ego stage or higher. Ideal ego development occurs when lines are actually integrated into a coherent self-system, so they are a good indicator of an individual's center of gravity. In this section we apply Cook-Greuter's research to 8 ecological identities (see color plate 2, following page 222).[47] These eco-selves describe how an individual at specific levels of ego development identifies with aspects of the natural world.[48] In other words, we are taking a single line and exploring what it looks like when directed at the natural world (context versus content).

Cook-Greuter identifies at least cight basic levels of development that arise individually and collectively. Each level corresponds to an ecological self. These selves represent the various ecological perspectives that can exist nonexclusively within all individuals. Most people embody multiple identities, while others are more embedded in a single perspective. A growing number of integrally aware individuals are able to relate to all eight of these perspectives. Different ecological selves tend to gravitate toward different aspects of the natural world and often prefer specific methods of investigating or making contact with nature.

Each eco-self has a unique way of relating to itself, others, and the natural world.[49] In brief, the Eco-Guardian respects and fears nature; the Eco-Warrior wants to conquer nature (or in some cases culture); the Eco-Manager is dedicated to managing nature from a religious or secular position; the Eco-Strategist not only wants to manage nature but wants to use nature, and in many cases exploit it for some kind of profit (usually capital); the Eco-Radical wants to save nature for all of humanity and often for its ground value; the Eco-Holist wants to unite nature's multiple flows so the complex system can flourish; the Eco-Integralist celebrates nature as holonic and honors all ecological perspectives; the Eco-Sage is "one with" nature (and Nature and NATURE).

All eight of the eco-selves have strengths and weaknesses, dignities and disasters.[50] They all have an environmental ethos appropriate to their worldview and the capacity to be ecologically destructive. One ecological self is not necessarily more environmentally friendly than another. This is an extremely important point because most green-environmentalists are not aware of how they themselves are harming the planet, nor are they aware of how amber-fundamentalists and orange-capitalists are sometimes helping to save it. See figure 7.3.

	Environmental Ethic	**Ecological Violation**
Eco-Sage	Experiences the unity of All and identifies with the totality of manifest creation.	Is too otherworldly. Can be disembodied and removed from pragmatic action in the world.
Eco-Integralist	Honors and integrates multiple approaches to the environment. Sees value in all perspectives.	Includes too much and gets bogged down in conflicting views.
Eco-Holist	Maps the complexity of relationships within and between ecosystems.	Overrelies on exterior systems and as a result commits subtle reductionism.
Eco-Radical	Promotes eco-social justice. Exposes the disaster of modernity.	Advocates a flatland ecology through guilt and shame tactics. Ignores the dignity of modernity.
Eco-Strategist	Conserves resources for consumption over the long term.	Exploits nature as a result of greed and a focus on short-term profits.
Eco-Manager	Passes laws and establishes institutions to act as stewards over nature.	Promotes domination of humans over the natural world.
Eco-Warrior	Challenges the system through tactical and non-conventional ways.	Mistakes one's own will for nature's. Can be aggressive: striving to conquer nature.
Eco-Guardian	Performs rituals to maintain control and power. Sees nature as ensouled.	Is one with aspects of nature but not one with humanity. Approaches nature with slash-and-burn tactics.

Figure 7.3. The dignity and disaster of each ecological self.

As figure 7.3 demonstrates, the shadow of the Eco-Radical (e.g., browbeating people with guilt, shame, and apocalyptic messages) is arguably less in service of a sustainable ecology than the virtue of the Eco-Manager (e.g., passing and enforcing laws and establishing institutions to protect the natural world). Making the distinction between the dignity and disaster of each ecological self provides a nuanced framework for analyzing environmental problems. It is important to cultivate the capacity to understand as many of these perspectives as possible and the developmental dynamics that guide them.

These examples are intended to be illustrative, not exclusionary. Examples here (and throughout this book) are to serve as pointers to important distinctions and qualities, but not at the expense of the complexity held by each approach. Many approaches to the environment and their proponents can and do inhabit multiple sites of ontology, epistemology, and methodology. We are not interested in compartmentalizing or restricting the multifaceted nature of any approach. On the contrary, we want to identify which voices are most qualified to speak on behalf of which aspects of reality. In so doing, we can coordinate and build bridges between divergent but essential perspectives.

The rest of the chapter is devoted to providing a brief overview of each eco-self, building on the discussion of developmental psychology in chapter 4. Figure 7.4 highlights the relationship between environmental identity and other important lines of development. At the end of the chapter we provide an example of different eco-selves being honored within an outdoor youth corps.

The Eco-Guardian (Romantic Ethos)

The Eco-Guardian is an impulsive self who connects with the cosmos by balancing good forces against evil dynamics. This self focuses on creating safety and satisfying basic needs. It has a sense of unbridled power mixed with superstitious and magical belief. Eco-Guardians view other people in light of self-gratification. The complexity of the world makes them uneasy. Very few adults have this exclusive identity, though many approaches to the environment, especially New Age and Romantic schools, make use of the content of this structure of identity.

When applied to the natural environment, this self often focuses on returning to a lost ecological paradise. Sometimes the "Fall" from

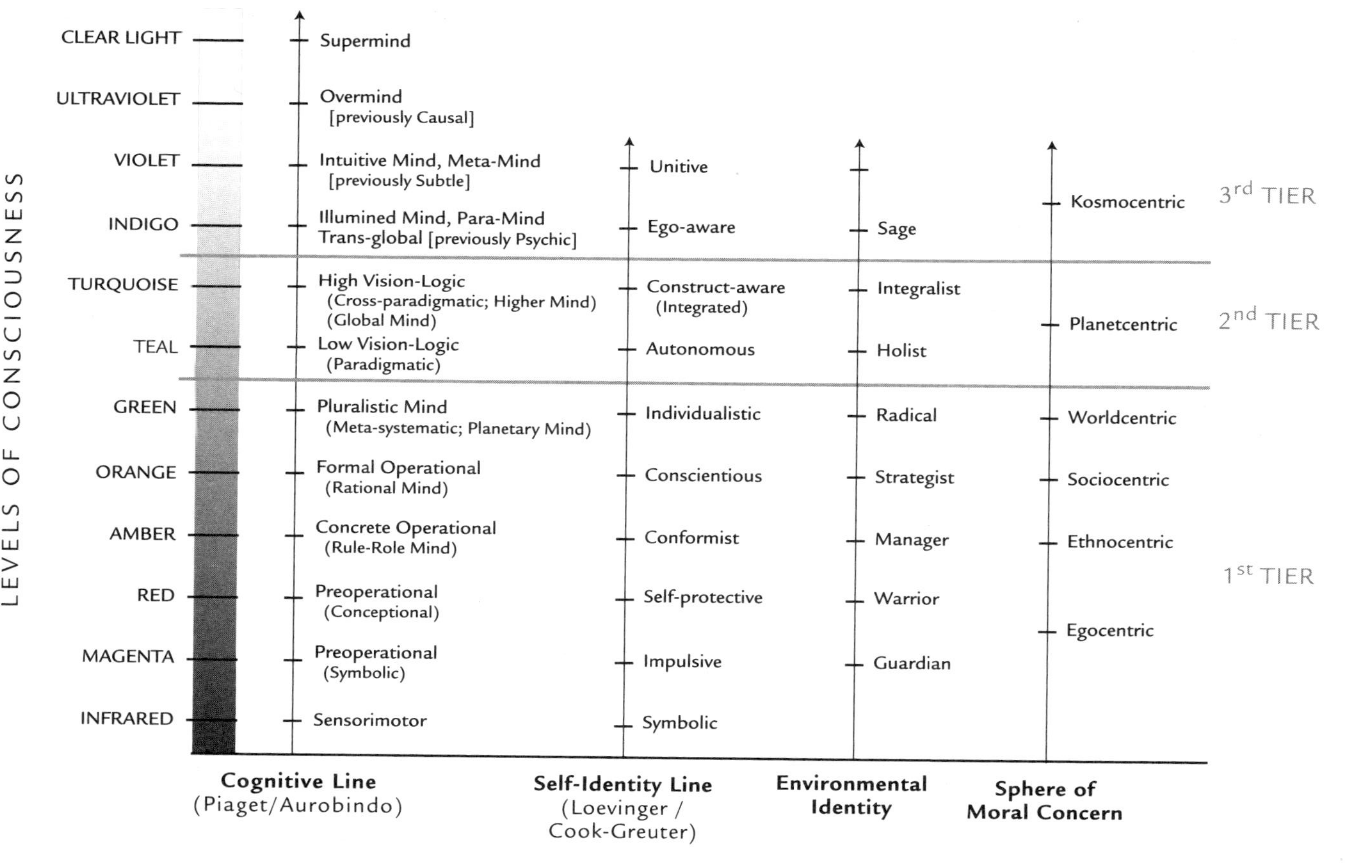

Figure 7.4. Important lines of development.

ecological grace is associated with horticulture (some deep ecologists), agriculture (some ecofeminists), or industrialization (some social ecologists). Eco-Guardians often emphasize magic or unseen forces. This approach is very "tribal" in that they place importance on ancestral ways; they hold naive animistic beliefs and maintain customs such as ceremonial rituals and rites of passage as important and a way to connect with the natural world. They appeal to the mystery of nature, especially through signs and omens. They respect councils, especially of elders, and lineage connections. Leadership is often based on age. Shamans and witches are seen as the gatekeepers of the world of Spirit.

Examples of the Eco-Guardian can include *aspects* of earth goddess groups;[51] nature worship;[52] totemism;[53] eco-rituals;[54] wicca;[55] paradise lost perspectives;[56] the cultural appropriation of indigenous practices;[57] and some forms of deep ecology[58] and ecofeminism.[59]

The Eco-Warrior (Heroic Ethos)

Eco-Warriors are self-protective and self-serving. Their impulsive nature is now placed in service of supporting an idea of the self, rather than just the self's immediate needs and wants. People operating at this altitude often perform heroic acts, which serve to magnify their own status. They identify the self in terms of effort and preferences. Self-preservation is central. Their feelings are guarded and inaccessible. Overgeneralization is very common, with many judgments and simple ideas. They see others as competitors for space, goods, and dominance, and have little capacity for insight into self and others. Due to their lack of trust of others they are hypervigilant and bullying. For them the world is a risky game that can be quite dangerous.

As such Eco-Warriors take a heroic approach to the environment. They focus on the assertion of the self over the system or nature. They are driven by impulsivity and immediate reward. Leaders establish themselves through power and strength. They often have a "to hell with others" attitude. They emphasize obtaining power and not being constrained. They desire respect and have an appreciation for the "law of the jungle" and "nature red in tooth and claw." They have a macho quality that feeds a heroic images of themselves as one person against everything. They highlight toughness and their groups are often ganglike. They value "hands on," "survival," and "street" skills. Various types of turf-wars are common for Eco-Warriors, and they experience minimal guilt.

Examples of the Eco-Warrior can include *aspects of* EarthFirst!;[60] monkey wrenching;[61] ecotage;[62] ecoterrorism;[63] the stoic mountain climber;[64] extreme sports such as mountain biking, river kayaking, rock climbing;[65] trophy and sports hunting;[66] frontier mentalities;[67] survival skills;[68] off-the-grid housing;[69] social Darwinism;[70] and Warwick Fox's "desiring-impulsive self."[71]

The Eco-Manager (Stewardship Ethos)

The Eco-Manager is a conformist self who is rule-oriented and concerned with group membership. Eco-Managers get their self-identity from others. It is hard for them to differentiate between themselves and the group. Projection and introjection are common defenses. Positive feelings are used to suppress negative ones. "Us" versus "them" drives their sense of group belonging. Their world is based on concrete-literal interpretations.

Thus Eco-Managers take a stewardship approach to the environment. They focus on maintaining order and following the law, either the divine order or the laws of the state. They believe that order must be maintained to keep harmony and stability. They manage nature now so the future will hold nature's bounty. People follow a higher authority (God, the law, a political or religious leader) and comply with rules and regulations to avoid punishment. Leaders are those who have seniority or those who are in the rightful position. Honor and obedience are prized attributes. Justice and fairness are provided to those who follow the rules.

Examples of the Eco-Manager can include *aspects of* the Earth viewed as Garden of Eden;[72] Puritan ethos;[73] Boy and Girl Scouts;[74] Environmental Protection Agency;[75] environmental legislation;[76] fish and game wardens;[77] national and state parks;[78] wildlife management;[79] Endangered Species Act;[80] Ducks Unlimited;[81] and the Audubon Society.[82]

The Eco-Strategist (Rational Ethos)

The Eco-Strategist is a conscientious self who is defined by an orientation toward scientific empiricism. This approach is placed in service of a newly emerging separate self-identity, which competes for wealth, influence, and social standing. This eco-self values independence and confidence. Eco-Strategists lead with rationality but are interested in their emotions. They emphasize efficiency and efficacy as a means for success. The world is viewed as being measurable and predictable. They value others for their own sake (e.g., supporting universal rights).

The Eco-Strategist employs a rational approach to the environment. They use technology to enhance the standard of living. They emphasize progress and seek the "good life." They value autonomy and independence. Life is a game to be played and won. They measure success by financial achievement. There is a desire to make things better and to use competition to accomplish this. They highly value science and universal rights for humans, and embrace an opportunistic vision of the future. They respect the invisible hand of the economy.

Examples of the Eco-Strategist can include *aspects of* natural capitalism;[83] conservationism;[84] resourcism;[85] the Lockean worldview;[86] the science of ecology;[87] deontological ethics;[88] urban planning;[89] utilitarian perspectives;[90] environmental pragmatism;[91] environmental psychology;[92] behavioral approaches;[93] industrial agriculture;[94] and Warwick Fox's "rationalizing-deciding self."[95]

The Eco-Radical (Equality Ethos)

The Eco-Radical is an individualistic self who highlights how we are all connected through similar experiences, shared contexts such as race or gender, and various systems (e.g., political and ecological). Eco-Radicals are sensitive to people's experiences and are willing to consider contradictory truth claims. They supplement objectivity and logic with subjective and more holistic approaches. They value personal experience and express feelings easily.

Eco-Radicals take a postmodernist approach to the environment. They focus on the liberation of humans and animals from greed and domination. They promote community, unity, and sharing resources across class, gender, and racial divisions. They make an effort to explore the interiority of other people and beings and to connect with Spirit. They prize consensus as a way of making decisions and avoiding hurt feelings. They highlight participation and teamwork. They expect social responsibility, political correctness, sensitivity, and tolerance. Often the community comes before the individual. Socially engaged activism is used to overcome oppressive hierarchies and power structures.

Examples of the Eco-Radical can include *aspects of* deep ecology;[96] ecofeminism;[97] social ecology;[98] animal rights;[99] biocentrism;[100] ecocentrism;[101] ecopsychology;[102] environmental justice;[103] green politics;[104] David Abram's eco-phenomenology;[105] the analysis of historical concepts;[106] bioregionalism;[107] various doomsayers and apocalyptic approaches;[108] and the social construction of nature.[109]

The Eco-Holist (Holistic Ethos)

The Eco-Holist is an autonomous self who is comfortable amidst complexity. Eco-Holists recognize that individuals occupy multiple contexts. They embrace the many layers of self (including shadow material) through a complex psychology. They see the importance of various, even contradictory, values and perspectives and have a high tolerance of others' "negative" traits. Their world is multidimensional and dynamic.

Eco-Holists approach the environment from a holistic-complex perspective. They focus on the dynamic systems that overlap in any given situation. They are capable of holding conflicting truths. The Eco-Holist demands a flexible, open system that allows for the full range of reality to express itself. There is an existential emphasis on being and personal responsibility. Hierarchies are replaced with holarchies. They grant leadership to those who can hold a multiplicity of perspectives. The diversities of people and perspectives are celebrated on their own terms. Eco-Holists see partial value in all perspectives. They use skillful means to meet people where they are. They understand complex systemic interactions. Chaos and complexity are valued and paradoxes are embraced. Nonlinear capacities are cultivated. Transparency becomes important.

Examples of the Eco-Holist can include *aspects of* Félix Guattari's three ecologies;[110] the new cosmology;[111] Teilhard de Chardin's noosphere;[112] the Gaia hypothesis;[113] Gregory Bateson's ecology of mind;[114] the system sciences of chaos and complexity;[115] Charlene Spretnak's ecological postmodernism;[116] Aldo Leopold's land ethic;[117] sustainable development;[118] Edgar Morin's complex thought;[119] biodynamic agriculture,[120] Duane Elgin's awakening earth;[121] Daniel Kealey's application of Jean Gebser's work to environmental ethics;[122] process ecology;[123] Leonardo Boff's liberation theology;[124] Dodson's perspectival approach to ecology;[125] and Warwick Fox's "normative-judgmental self."[126]

The Eco-Integralist (Inclusive Ethos)

The Eco-Integralist wave addresses the existential anxiety of meaning associated with the Eco-Holist. For the Eco-Integralist the heart opens and increases the individual's capacity to feel the widespread suffering around the planet. This capacity for remaining open to such suffering without being consumed by it is an important quality of the compassion that emerges with this Eco-Self. Wilber has captured this paradox with the phrase "hurts more, bothers you less." The Eco-Integralist is deeply committed to the integration of transcendence and innocence. This is a marriage of wisdom and compassion, which recognizes that nothing needs to be done, because everything is always already perfect. The Eco-Integralist perceives the luminous nature of all life forms and manifestations. Thus, individuals at this altitude have the capacity to stay open and relaxed even while involved in arduous tasks. This eco-self has ongoing access to the experiential insight that the manifest realm is Divinized. The Eco-Integralist recognizes that no ecological reality lasts forever, thus they appreciate each phenomenon, without clinging to a view of how it should be but working hard to change things for the better.

Because both the Eco-Holist and the Eco-Integralist are expressions of the two newest waves of consciousness they are often collapsed into one wave or confused. Thus, it is worthwhile to note some important differences between the Eco-Holist and the Eco-Integralist (see figure 7.5). The Eco-Integralist uses multidimensional thinking/feeling (systems of systems of systems) without viewing the realities of one system through the realities of another system, whereas the Eco-Holist tends to make use of only one or two systems, usually through an interobjective framework. An Eco-Integralist embraces a participatory, nonrepresentational (i.e., post-metaphysical) perspective, while the Eco-Holist tries to map the world more accurately through modeling. Eco-Holists look at the maps they create, whereas Eco-Integralists place themselves in the map. Eco-Integralists are acutely aware that current solutions may contribute to future problems in ways they cannot imagine or recognize, while Eco-Holists tend to think that their multiperspective approach is the best solution to the problem. An Eco-Integralist embraces the interiority (experience and culture) at all levels of sentient beings, while the Eco-Holist often restricts interiority to the "higher" animals. The Eco-Integralist makes use of or honors all eight methodological families for disclosing reality, while the Eco-Holist honors just a few. The Eco-Integralist not only acknowledges that things are getting ecologically worse (e.g., increased planetary destruction), but that they are also getting better (e.g., increased planetary protection and ecological awareness). The Eco-Holist usually emphasizes one of the poles. Eco-

Integralists seek and value paradox, while Eco-Holists try to account for it. Eco-Integralists' unwavering commitment to the biosphere is grounded in the recognition of the emptiness of all phenomena: they recognize the pneumasphere as transcending and including the noosphere and biosphere, whereas the Eco-Holist is often identified with just the physiosphere (matter), biosphere (life), and noosphere (mind).

Examples of the Eco-Integralist include Bhutan's "Middle Path" to development;[127] Brian N. Tissot's work with marine fisheries in Hawai'i;[128]

Eco-Holists	Eco-Integralists
Make use of only one or two systems, usually through an interobjective framework.	Use multidimensional thinking/feeling (systems of systems of systems) without viewing the realities of one system through the realities of another system.
Try to map the world more accurately through modeling.	Embrace a participatory non-representational (i.e., post-metaphysical) perspective.
Look at the maps they create.	Place themselves into the map.
Think and feel that their perspective is the best solution to the problems.	Are accurately aware that their perspective and approach, while providing solutions for problems, is actually contributing to future problems in ways they cannot imagine or recognize.
Restrict interiority to the "higher" animals.	Embrace the occurrence of interiority (experience and culture) at all levels of sentient beings.
Use only one or a few methodologies.	Use and/or honor all 8 methodological families for disclosing reality.
Only emphasizes that things are getting ecologically better or that they are getting worse.	Acknowledge that things are getting ecologically worse AND they are getting better.
Try to account for or explain paradox.	Seek out and value paradox.
Do not see the apparent emptiness of the biosphere.	Recognize the emptiness-luminosity of all phenomena.
Integrate the biosphere and noosphere.	Integrate the biosphere, noosphere, and pneumasphere.

Figure 7.5. Some general differences between Eco-Holists and Eco-Integralists.

Michael Zimmerman's environmental philosophy;[129] Darcy Riddell's eco-activism in Canada's Great Bear Rainforest;[130] Brian Eddy's Integral Geography;[131] Cameron Owens's analysis of waste reduction in Calgary;[132] Joel Kreisberg's environmental medicine;[133] Kevin Snorf's integral eco-design;[134] Gail Hochachka's Integral Community Development;[135] Wade Prpich's analysis of the organic standard of Canada;[136] David Johnston's market transformation in Alameda County, California;[137] Daniel Wahl's "saluto-genic and scale-linking" design;[138] Ian Wight's "placemaking";[139] and Stan Salthe's developmental systems ecology.[140]

The Eco-Sage (Unity Ethos)

We did not discuss this level of development in chapter 4, as it is so rarely stably reached (less than 1% of the U.S. population).[141] This level corresponds to both Susanne Cook-Greuter's ego-aware self, and the unitive self. The Eco-Sage is an ego-aware self who integrates multimodal and multidimensional elements across contexts in the service of humanity. Eco-Sages are aware of the subtle ways the ego filters experience. They recognize paradox and the limits of "mapping." They desire to work through their own limits and blind spots and increase their capacity to witness themselves in the moment. They understand others in developmental terms and encounter them without judgment. They have a profound understanding of others' complex and dynamic personalities. They experience the world as a place full of potential and paradox. At this stage the environmental identity becomes even more of a transparent manifestation of Being, completely spontaneous and open. They have stable access to transpersonal realities such as the capacity to witness their experience and keep their boundaries open. They view others as manifestations of Being-Spirit. They experience the world as an immanent expression of timeless Spirit.

The Eco-Sage approaches the environment from an authentic transpersonal perspective that transcends and includes the previous eco-selves. Thus, it takes more than having peak states of union with the natural world to be an Eco-Sage. Eco-Sages focus on the subtle ways of being connected with the natural and human realm. They have an increased capacity for self-identification with members of the natural world. A variety of unitive states are experienced with Gaia in its gross, subtle, and causal manifestations. In fact, any of the eco-selves can have a peak experience of gross, subtle, causal, or nondual union with Gaia. The distinction between stages of development and states of consciousness is crucial to navigate the complexity associated with the 8 eco-selves and their multiple ways of relating to the natural world (see chapter 9).

For the Eco-Sage, there can be the experience of subtle-realm beings both of the Earth plane (e.g., elementals and nature spirits) and other dimensions (e.g., the archetypal realm). The Eco-Sage has a deep commitment to all sentient beings (seen and unseen) and an increased capacity to work with the energetic systems of the manifest and subtle realms.

Examples of the Eco-Sage can include *aspects* of transcendentalism;[142] J. W. Goethe's *Urpflanze*;[143] St. Francis of Assisi's Canticle of Brother Sun;[144] Ken Wilber's Eco-Noetic Self;[145] Joanna Macy's ecological self;[146] Chris Bache's species mind;[147] some neo-pagans;[148] nondual spiritual activism;[149] McClellan's nondual ecology;[150] and Warwick Fox's "transpersonal-ecological self."[151]

There are numerous individuals or approaches that many people would be inclined to place in the list above. For example: Black Elk;[152] Matthew Fox's creation spirituality;[153] Masanobu Fukuoka's natural farming;[154] Ralph Metzner's green psychology;[155] ayahuasca visions;[156] shamanism;[157] vision quests;[158] and deva gardening;[159] just to name a few. These individuals and approaches include many extremely important "spiritual" qualities, insights, and dimensions, which are considered essential to Integral Ecology. Any of the eco-selves can experience altered states. In fact, as mentioned, each eco-self has access to gross, subtle, causal, and nondual experiences of the natural world. However, they will interpret these nonordinary states according to their "center of gravity" of psychological development. The distinction we are emphasizing is that the Eco-Sage represents a stabilized capacity (indigo altitude and above) to experience transpersonal dimensions and the distinguishing capacities of the other eco-selves. As a result there are at least 32 distinct varieties of nature mysticism—only 4 of which are associated with the Eco-Sage (see chapter 9).

The examples provided for each ecological self are not fixed. Almost any example provided for any eco-self can be held from many of the levels. For example, the rhetoric of deep ecology, which is listed as an example under the Eco-Radical, can be used to support and justify neo-pagan rituals (magenta worldview), monkey wrenching (red worldview), environmental legislation (amber worldview), natural capitalism (orange worldview), and social activism (green worldview). Deep ecology, as with the other examples, tends to embrace or express one of the 8 eco-selves more than the others. This does not deny the many "camps" or variations within any one school of thought, especially given that individuals can use the rhetoric of one altitude to support their own. In other words, the 8 ecological selves correspond to researched levels of psychological development; placement is dependent on interior psychological motivation and value systems, not just outward behavior.

As individuals evolve into more complex waves of being, they enfold previous waves and optimally have access to them all. As an individual assimilates new value systems, aspects of the previous structures remain available when their expression is appropriate. Consequently, people often contain multiple developmental structures within themselves but primarily identify with one.

To conclude this chapter and illustrate the value of the 8 eco-selves, Sean will provide an actual example from his work with young adults in a residential outdoor program.

Northwest Youth Corps (NYC)

During the spring and summer of 1999 I (Sean) spent five months working with young adults in the wildness of Oregon and Washington. I was a crew leader for Northwest Youth Corps (NYC), a residential work program for high school students. While working there I had the opportunity to see the potential for individual and collective transformation when many of the ecological selves are honored and included.

NYC offers four sessions during the spring, summer, and fall. Each session is five to six weeks long and has around 40 young adults divided into four crews. Crews travel all over Oregon and Washington doing projects in national parks or national forests and sometimes for the Bureau of Land Management. As a crew leader, I was in charge of 10 people for the entire five to six weeks. I was in multiple roles: boss, father, brother, mother, friend, coworker, doctor, teacher, and disciplinarian. The relationships with crew members that spring forth in this rich context are multidimensional.

Crews spent every hour of every day together. We worked eight-hour days Monday through Thursday, and a half day on Friday. Our project could be anything from pruning trees to placing plastic over little saplings to protect them from deer and elk, from maintaining trails to building a new trail.

Each project lasted for one work week, and then we broke down camp on Friday and headed to a weekend site, which is a larger campsite where all four crews gathered from Friday night through Sunday morning. This was a great chance for people from different crews to get to know each other, and it helped build a larger community. There were a variety of activities on the weekend that allowed people to get to know others (such as recreational trips to nearby areas, sharing circles, playing Frisbee or Hacky Sack, and eating in a big group). On Sunday, crews were informed of their new work project and headed out to their new locations. Many times there was a long drive and hike to get to the work site. We started working bright and early on Monday.

A typical day started at 5:00 A.M., when the crew leader and the chosen breakfast person began cooking the morning meal. By 5:30 everyone was awake and had 30 minutes to get dressed, eat, and be ready to work. At 6:00, we either hiked to the work site or drove, depending on the situation. Once at the work site, everyone participated in a "safety circle," where each crew member provided a stretch and a safety tip for the day. We worked for five hours with one 15-minute break, ate lunch, then worked another two hours, with another 15-minute break, and then worked the last hour. Then we went back to camp. People had a half hour to change out of work clothes, eat a snack, read or journal, before they were expected to do their chores. Chores included fetching water for the camp, making lunch for the next day, preparing dinner, collecting firewood, cleaning tools, or writing in the group journal. Dinner was served around 5:30 or 6:00 P.M. After dinner, most crews had some free time before SEED (the environmental curriculum) started. SEED was usually taught around the campfire. It was a 30-minute to one-hour lesson taught by the crew leader. The topics ranged from community issues to ecology. During the first few weeks, crew members presented their "personal histories" after SEED. Each individual spent a half hour or so telling his or her life story to the group. After each personal history, we all said something positive about the person who shared. After SEED, crew members were allowed to go to bed or hang out a little before "lights out" at 9:00 or 9:30 P.M.

Every day was exhausting, and we felt as if we had done two or three days' worth of living in each one. People were tired every day; then they got up and did it all over again. This intense schedule facilitated breaking down boundaries and building community. There was an intense level of intimacy because of the close quarters. In a six-week session we developed friendships that normally take years to create. As a result, members were drawn closer to those around them.

The Process of Thought Transformation at NYC

There were a number of features embedded in the fabric of the NYC program that allowed for the expression of many of the ecological selves. Particularly noteworthy was the emergence of Eco-Radical perspectives over the course of the six weeks. This perspective usually does not emerge in youth in this age group until after they have gone to college. I attributed this "early emergence" to the ways in which the overall program recognized each of the preceding eco-selves and provided channels for their healthy expression.

For a member's experience to become potentially "transformative," the leader had to create a supportive environment. It is this support that acts as a catalyst for people to allow their latent potential to bloom. Crew leaders spent the first week of each session laying the groundwork for this context. We discussed the meaning of community and co-created ground rules such as commitment, communication, respect, compassion, and listening, and then signed a contract. This created ownership and solicited participation in creating a communal space.

The success of initial efforts to establish a community identity and create a supportive place is largely dependent on the crew leader. The crew leader position demands a level of responsibility that is unmatched. I was the guiding force for the community and had to establish parameters of what would be expected and tolerated. It required firm authority, since adolescents push boundaries as part of their individualization process, and compassionate, unconditional support.

A good crew leader knew when to support and validate a crew member's frustration or to challenge them beyond their comfort zone. If people do not feel loved, respected, challenged, and supported, they are unlikely to allow themselves to be vulnerable and take the risks necessary to actualize unrealized parts of their being.

Having provided a general overview of the structure of this program, I will now highlight the ways in which the program included many of the ecological selves.

The Eco-Guardian: Rituals emerge among the group and are encouraged (such as offering a reading before dinner, singing a song during dish washing). Totems and mascots are often created by the crew members. Each crew is also associated with a color, which takes on special meaning over the course of the session. Many stories are told and songs sung while working. Cultural myths and legends are shared and created.

The Eco-Warrior: Aggression was channeled through competitions (who can lay more new trail in the next hour) and heroic acts (allowing crew members to return to the van miles away to fetch a forgotten item). At times crew members became angry at the constraints of the program and threw tools down. For the most part, I honored these displays as important expressions. Each week there was also a different crew task that provided a way to accomplish something (like building a bridge over a large creek). Over the course of the program many crew members overcame fears of being outdoors.

The Eco-Manager: The program was highly structured. There were clear rules and consequences for transgressions. The crew leader was the authority and held most of the power. Often the crew members turned their crew leader into a mythic or godlike figure who laid down the law and enforced the rules. We also managed resources on our site in a hands-on way. Crew members learned about various environmental regulatory bodies (e.g., national parks, the Bureau of Land Management), and a stewardship approach was discussed and adopted around the campsite.

The Eco-Strategist: The program served as summer employment for many participants. For most crew members this was either their first "real" job or the most challenging one they had ever had. Thus, many crew members cultivated self-esteem and were given many opportunities to shine and demonstrate their talents and self-sufficiency. Consequently, many of them experienced a level of achievement and independence that far exceeded any of their previous reference points. Their individual expression was valued and encouraged; and we discussed economic relationships to the environment.

The Eco-Radical: We strongly emphasized community. We respected differences, established a hate-free zone, and talked about gender and racial issues. Consensus was often used as a decision-making tool. Crew members were encouraged to tell their personal stories during campfires, which provided an avenue for their voices to be heard. Members' feelings were respected. The intrinsic value of nature was discussed and demonstrated.

At the beginning of each session, crews were predominately filled with Eco-Warrior and Eco-Manager energies. Over the course of the six weeks, I repeatedly witnessed an incorporation of Eco-Strategist and Eco-Radical perspectives. I was continually amazed at the degree of transformation that occurred. Over many sessions I realized that the beginning emergence of Eco-Strategist and Eco-Radical horizons was possible because the healthy aspects of previous selves (Eco-Guardian, Eco-Warrior, and Eco-Manager) had been honored and included within the very structure of the program. Of course, it was no surprise that the Eco-Holist, Eco-Integralist, or Eco-Sage worldview never emerged, since those horizons usually do not appear until much later in adulthood if at all. However, many crew members did report various states of connection, openness, and spirituality with the natural world. I was impressed by the simple fact that high school adolescents demonstrated and embodied healthy Eco-Strategist and emerging Eco-Radical perspectives.

These ecological selves represent the Who of Integral Ecology's enactive model. Mind is not a mirror of nature. Rather, mind contributes to one's construction and interpretation of nature. Now that we have examined the What and the Who of Integral Ecology, we will explore the How by examining the 8 eco-modes (or methodologies) of environmental research.

8

Ecological Research: *How* We Examine

Different kinds of ecology require expertise in using and understanding different technologies. These different technologies require different tools.

—STANLEY DODSON[1]

Each of the important methodologies (from empiricism to collaborative inquiry to systems theory) are actually types of practices or injunctions—in all cases, they are not just what humans think, but what humans do—and those practices therefore bring forth, enact, and illumine a particular dimension of one's own being—behavioral, intentional, cultural, or social. . . . (This is why different forms of praxis yield different theoria.)

—KEN WILBER[2]

In order for the eco-selves to examine any of the 12 niches of reality, they must employ a particular methodology. Integral Methodological Pluralism (IMP) acknowledges that there are at least 8 ecological modes we can use to study our environment. In chapters 1 and 2 we introduced IMP. In this chapter we build on that material by reviewing the current mixed methods approaches in the field. Then we introduce Integral research, which is based on Integral Methodological Pluralism, and next provide an overview of some considerations of Integral research. We then discuss how different approaches to ecology use different methods, and provide an overview of the modes of inquiry each ecological terrain uses. To illustrate IMP's ability to provide a framework for mixed methods research, we present a few examples of interdisciplinary ecological research.[3]

Integrative Research

In many respects the empirical study of ecosystems and environmental dynamics is predisposed toward an interdisciplinary or transdisciplinary approach because such complex systems are best understood from multiple perspectives and methods. While many ecologists and environmentalists recognize this, they rarely include subjective and intersubjective methods (Left-Hand methods) and techniques. For example, the Integrative Ecology Group (IEG) out of the Doñana Biological Station in Sevilla, Spain, explains on their website:

> The mission of the IEG is to understand the mechanisms by which biodiversity is organized and how biodiversity responds to human-induced perturbations. We pursue this by developing interdisciplinary research in ecology. We combine field work, mathematical models, computer simulations, genetic analysis, and statistical analysis of large data sets. Our research involves an international program of collaborators world-wide. Our research lines integrate evolutionary ecology, theoretical ecology, and population genetics.[4]

They provide examples of their research interests, including "Evolutionary ecology of plant-animal interactions (pollination and seed dispersal)," "Complex topology of food webs and plant-animal mutualistic networks," and "Population genetics of endangered vertebrate species."

Similarly, a number of graduate ecology programs are based on integrative ecology, which allows students to combine various methods, theories, disciplines, data sets, and scales. The University of California at Davis, one of the world's top-ranked graduate ecology programs, provides a degree in integrative ecology that supports students "with interests that are either focused on basic or theoretical aspects of ecology, or on integration of inquiry across various scales or disciplines."[5] But the program literature, faculty biographies, and student research interests are associated only with the terrains of behavior and systems.[6]

Lastly, consider the academic journal *Ecology and Society: A Journal of*

Integrative Science for Resilience and Sustainability. According to its website: "The journal seeks papers that are novel, integrative and written in a way that is accessible to a wide audience that includes an array of disciplines from the natural sciences, social sciences, and the humanities concerned with the relationship between society and the life-supporting ecosystems on which human wellbeing ultimately depends."[7] This sounds promising from an integral perspective, as the humanities recognize subjective and intersubjective perspectives. However, the website continues:

> In general, papers should cover topics relating to the ecological, political, and social foundations for sustainable social-ecological systems. Specifically, the journal publishes articles that present research findings on the following issues: (a) the management, stewardship and sustainable use of ecological systems, resources and biological diversity at all levels, (b) the role natural systems play in social and political systems and conversely, the effect of social, economic and political institutions on ecological systems and services, and (c) the means by which we can develop and sustain desired ecological, social and political states.[8]

What initially appears as integral science turns out to be predominately objective and interobjective (Right-Hand methods) integrative science.[9]

The need for the integration of methodologies within the field of ecology has been recognized by many theorists and practicing ecologists.[10] Attempts thus far have predominately focused on exteriors approaches (behavioral and systems-based). We applaud these integrative efforts, but they fall short of what we propose. In our opinion we need a mixed methods approach to ecological research that combines qualitative and quantitative methods across all 4 terrains.

Mixed Methods Research

Since the emergence of the Chicago School of sociology in the 1920s, there has been a growing debate between the usefulness of qualitative and quantitative approaches to research. This debate became even more sharply delineated in the 1970s as academia embraced postmodernism and expanded the application of qualitative techniques beyond the confines of anthropology and sociology into a variety of new disciplines such as women's studies, education, psychology, and medicine. For the last 30 years, it seems, this debate has only grown ever more bitter, with qualitative approaches often absolutizing subjective (experience) and intersubjective (culture) approaches, and quantitative approaches exclusively championing objective (behavior) and interobjective (systems) approaches.[11]

While the disputes between qualitative and quantitative research agendas continue, a new option emerged, which combines these methods—mixed methods research. R. Burke Johnson and Anthony Onwuegbuzie, two mixed methods researchers and theorists, provide a useful definition of mixed methods research: "The class of research where the researcher mixes or combines quantitative and qualitative research techniques, methods, approaches, concepts or language into a single study. . . . Mixed methods research also is an attempt to legitimate the use of multiple approaches in answering research questions, rather than restricting or constraining researchers' choices (i.e., it rejects dogmatism). It is an expansive and creative form of research, not a limiting form of research."[12]

Johnson and Onwuegbuzie use the label "integrative research" synonymously with "mixed research," both of which they prefer over "mixed methods research," as these terms have a more inclusive and paradigmatic nature.[13] John Creswell, a pioneer of mixed methods, provides another definition of mixed methods research:

> a mixed methods approach is one in which the researcher tends to base knowledge claims on pragmatic grounds (e.g., consequence-oriented, problem-centered, and pluralistic). It employs strategies of inquiry that involve collecting data either simultaneously or sequentially to best understand research problems. The data collection also involves gathering both numeric information (e.g., on instruments) as well as text information (e.g., on interviews) so that the final database represents both quantitative and qualitative information.[14]

Both of these definitions illustrate the inclusive nature of mixed methods research. In addition to including both quantitative and qualitative, mixed methods provides a flexible way to combine quantitative and qualitative research techniques in such a way as to minimize the limitations and maximize the strengths of each method. Johnson and Turner refer to this as the "fundamental principle of mixed research."[15]

Beginning with Brewer and Hunter's book *Multimethod Research: A Synthesis of Styles* (1989), mixed methods approaches have been receiving increasingly more attention.[16] In spite of the growing interest and development of mixed methods approaches, Johnson and Onwuegbuzie point out, "Much work remains to be undertaken in the area of mixed methods research regarding its philosophical positions, designs, data analysis, validity strategies, mixing and integration procedures, and rationales, among other things."[17]

Integral Methodological Pluralism: A Philosophical Position

In chapter 1 we explained that IMP is guided by three principles: *nonexclusion, unfoldment,* and *enactment.* These principles recognize that within each of the 4 quadrants there are two types of methodologies, those that examine the inside and outside of each quadrant (see figure 8.1).

Wilber developed an integral mathematics of perspectives where 1st-, 2nd-, and 3rd-person perspectives interact with 1st-, 2nd-, and 3rd-person perspectives interacting with 1st-, 2nd-, and 3rd-person perspectives.[18] For the purposes of this chapter we will be using the following three notations: 1st-person (1-p) or 3rd-person (3-p) x inside (1-p) or outside (3-p) x interior (1p or 1p*plural) or exterior (3p or 3p*plural). Recall that this is our Who x How x What formula. When referring to 2nd-person realities, the "1p*plural" variable is

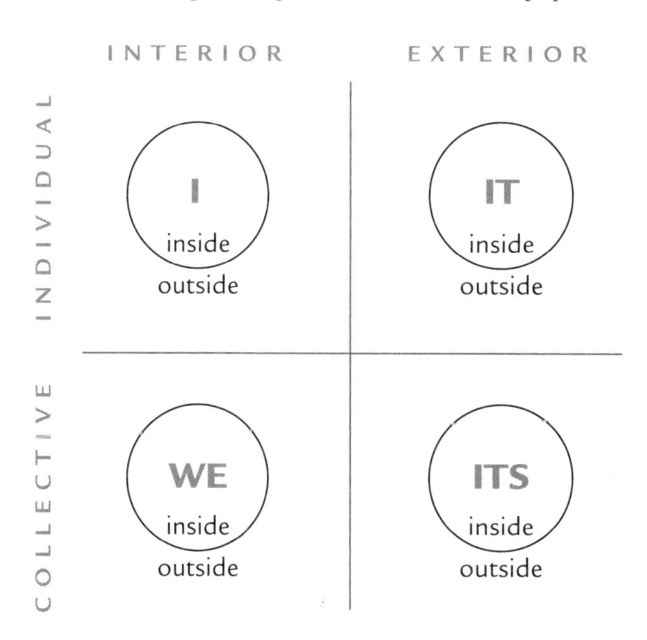

Figure 8.1. The 8 fundamental perspectives.

used as the third variable since 2nd-person is more technically understood as 1st-person plural. Likewise, "3p*plural" refers to interobjective realities. These three variables, as noted in chapter 5, also represent quadrant x quadrivium x domain (where domain can be either a quadrant or a quadrivium). For example, the perspective that emerges from meditation or psychological introspection is represented as a 1st-person generalization of a 1st-person view of the reality in a 1st person: (1-p x 1-p x 1p). Integral math produces a powerful approach to research that accounts for and integrates the different methodologies of various perspectives.

The 8 methodological families Wilber identifies, as noted in chapter 2, are as follows: [19]

Zone 1: phenomenology (1-p x 1-p x 1p): a 1st-person perspective (1-p, Who) using a 1st-person approach (1-p, How) to explore 1st person (1p, What), which explores direct experience (the insides of individual interiors).

Zone 2: structuralism (1-p x 3-p x 1p): a 1st-person perspective (1-p, Who) using a 3rd-person approach (3-p, How) to explore 1st person (1p, What), which explores reoccurring patterns of direct experience (the outsides of individual interiors).

Zone 3: hermeneutics (1-p x 1-p x 1p*plural): a 1st-person perspective (1-p, Who) using a 1st-person approach (1-p, How) to explore 1st person plural (1p*pl, What), which explores intersubjective understanding (the insides of collective interiors).

Zone 4: cultural anthropology or ethnomethodology (1-p x 3-p x 1p*pl): a 1st-person perspective (1-p, Who) using a 3rd-person approach (3-p, How) to explore 1st person plural (1p*pl, What), which explores recurring patterns of mutual understanding (the outsides of collective interiors).

Zone 5: autopoiesis theory (3-p x 1-p x 3p): a 3rd-person perspective (3-p, Who) using a 1st-person approach (1-p, How) to explore 3rd person (3p, What), which explores self-regulating behavior (the insides of individual exteriors).

Zone 6: empiricism (3-p x 3-p x 3p): a 3rd-person perspective (3-p, Who) using a 3rd-person approach (3-p, How) to explore 3rd person (3p, What), which explores observable behaviors (the outsides of individual exteriors).

Zone 7: social autopoiesis theory (3-p x 1-p x 3p*pl): a 3rd-person perspective (3-p, Who) using a 1st-person approach (1-p, How) to explore 3rd person plural (3p*pl, What), which explores self-regulating dynamics in systems (the insides of collective exteriors).

Zone 8: systems theory (3-p x 3-p x 3p*pl): a 3rd-person perspective (3-p, Who) using a 3rd-person approach (3-p, How) to explore 3rd person plural (3p*pl, What), which explores the functional fit of parts within an observable whole (the outsides of collective exteriors).

Following Wilber, we refer to each of these methodologies as a numbered zone. See Figure 8.2.

We use each of the names for these methodological families as an umbrella term that covers dozens of distinct methodologies. For example, zone 1, the phenomenological family (1-p x 1-p x 1p), includes the specific methodologies of Husserlian phenomenology, introspection, meditation, contemplative prayer, and so on. All of these methodologies provide an interior view of an inside view of an individual's subjectivity using various methods

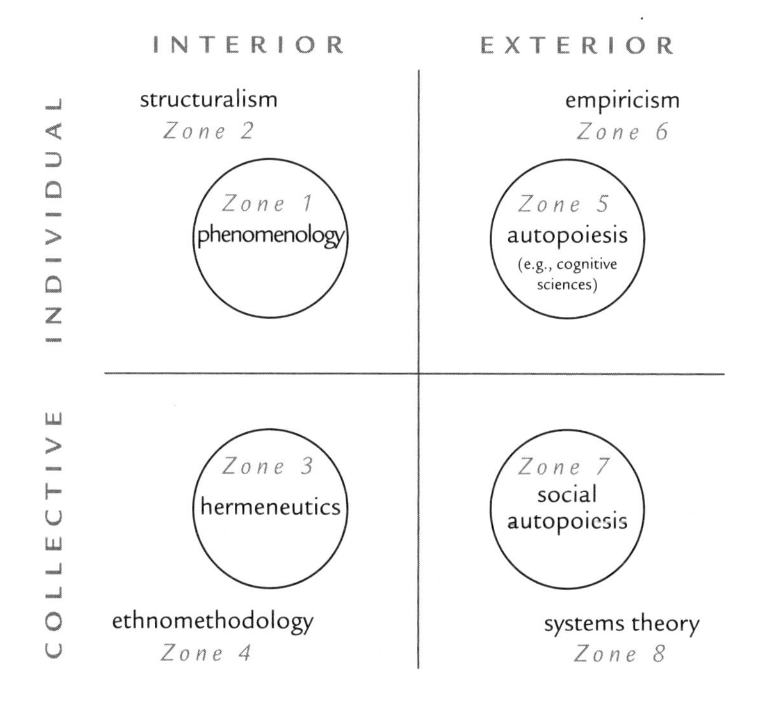

Figure 8.2. The 8 methodological zones.

(or injunctions). Likewise, zone 4, the ethnomethodological family (1-p x 3-p x 1p*pl), includes semiology, genealogy, archaeology, grammatology, cultural studies, poststructuralism, neostructuralism, and semiotics. All of these methodologies use different methods to provide an interior view of an outside view of intersubjectivity. In short, IMP provides a powerful framework that can coordinate multiple methods and integrate their findings into a coherent understanding. When we use IMP this way, the result is Integral research.

Integral Research

Integral research, a multimethod approach, draws primarily on IMP and Kosmic perspectivalism.[20] Additional influential approaches include William Tobert's developmental action inquiry, which emphasizes the integration of 1st-, 2nd-, and 3rd- person folk and formal methodologies to serve the vertical developmental transformation of self, others, and world;[21] Goethean science, a powerful mixed methods approach that includes 1st-, 2nd-, and 3rd-person investigative methods and recognizes the need for the researcher to stabilize postrational levels of psychological development;[22] and William Braud's integral inquiry, which identifies four types of research questions, for both qualitative and quantitative research that correlate with Integral Theory's 4 quadrants.[23]

One of the basic premises of Integral research is that any phenomena should be examined simultaneously from 1st-, 2nd-, and 3rd-person methodologies. The resulting data must be presented in terms and concepts associated with their respective methodology. This avoids reducing the data of one method to terms associated with another method. The Integral Model triangulates and tetra-correlates the data from different methodologies and their methods into a coherent presentation.

There are numerous ways to include 1st-, 2nd-, and 3rd-person methodologies in a research project. In fact, one of the strengths of Integral research is that it is scalable. You can triangulate the data with as few as three different methodologies (one from each major category) or as many as eight (one from each of the major methodological zones). At the very least, Integral research would involve choosing a method from the subjective realm: phenomenology or structuralism; from the intersubjective realm: hermeneutics or ethnomethodology; and from the (inter)objective realm: empiricism, autopoesis theory, systems theory, or social autopoesis theory.

For example, to investigate the barriers to environmental sustainability within a company, one could use phenomenology to interview fifteen employees (phenomenology: 1-p x 1-p x 1p), three focus groups composed of different levels of management (hermeneutics: 1-p x 1-p x 1p*pl), and an outside consultant to analyze organizational dynamics (systems theory: 3-p x 3-p x 3p*pl).[24] While these three methods are not exhaustive, and we could use many others methods within each of these zones, or methods from entirely different zones, they at least ground the research within the three major domains of reality—1st-, 2nd-, and 3rd-person dimensions. Consequently, the research is less likely to miss crucial information, while it is more likely to provide an understanding of what was not researched.

Below are examples of techniques for 1st-, 2nd-, and 3rd-person research and some of the validity considerations of each method. Note that autopoiesis and social autopoiesis are the most recent methodological families and as such have fewer methods and are less developed than the other six. The following lists are not exhaustive, but they do provide an illustrative overview of the practices, injunctions, and validity issues that can contribute to Integral research.

Z1: Phenomenological-Inquiry Techniques (1-p x 1-p x 1p)

Self-inquiry, phenomenological investigation, sensory and/or somatic awareness practices, visualization, prayer, reflection, introspection, journaling, artwork, poetry, movement, autobiographical writing, Gendlin's Focusing, perspective-taking, meditation, yoga, t'ai chi, vision quests, dream journaling, shadow work, and mindfulness practices.

Validity: Positionality, honesty, authentic expression, thick descriptions, sincerity, integrity, vulnerability, epoche, identify assumptions, acknowledge bias, and transformative.

Z2: Structural-Assessment Techniques (1-p x 3-p x 1p)

Various kinds of development tests (i.e., psychometric measures); personality type tests; interviewing friends, family, and colleagues about oneself; journaling to reveal patterns of awareness; subject-object interviews; sentence completion tests; obtaining evaluations from mentors and teachers; feedback from others; psychograph analysis; watching videotapes of oneself; listening to audiotapes of oneself; noticing speech and thought patterns.

Validity: Established psychological tests, use of established developmental models, critical subjectivity, long-term inquiry, thick descriptions with analysis, triangulation, self-observations, and feedback from others.

*Z3: Hermeneutical-Interpretative Techniques (1-p x 1-p x 1p*pl)*

Interviews, role-playing, dialogue and debate, small-group work (e.g., dyad and triad), group rituals and activities, discussion groups, storytelling, attending performances such as plays, interpretive analysis, textual analysis, collective reflection and visioning exercises, collective introspection, various participatory methodologies, focus groups, trust-building exercises, group facilitation, nonviolent communication, interspecies encounter, making love, mediation and conflict resolution, improvisational acting, dancing, jazz, and martial arts.

Validity: Giving voice to others, serving community, reciprocity, honoring others, trust, member checks, emancipatory, mutual understanding, resonance, meaningful, and symbolic.

*Z4: Ethnomethodology Techniques (1-p x 3-p x 1p*pl)*

Participant-observer techniques, participatory evaluation, forms of cultural analysis, appreciative inquiry, cultural anthropological techniques, forms of structural analysis, feedback about one's role on a team, family or couples therapy, actively observant parenting, polling, coaching and mentoring, analysis of semiotic codes, and use of mock scenarios.

Validity: Cross-cultural, observation of group dynamics, symbolic coherence, well-documented observations, prolonged engagement, member checks, and acceptance by group.

Z5: Autopoiesis Techniques (3-p x 1-p x 3p)

Imaginatively and scientifically projecting oneself into a biological organism's perspective, diagramming cognitive inputs and outputs, modeling vision and perceptual systems, identifying pattern recognition capacities of organisms, and mapping structural couplings between organisms/environments.

Validity: Pragmatic, advances the field, adequate account of the scientific observer, matches the biological phenomena, empirical, logical, predictive, and explanatory.

Z6: Empirical-Observation Techniques (3-p x 3-p x 3p)

Surveys; documentation; exams; fieldwork; observation; 3rd-person descriptions and reporting; using charts, graphs, and statistics to present information; exercise and diet; open-ended interviews; participant observer techniques in activities and sessions; review documentation; writing and dissemination of case studies; gap analysis; diagnostic testing; appraisal; skill-building; technical/social capacity development; anonymous reviews of writing, and opinion polls.

Validity: Repeatable, controlled conditions, empirical, logical, measurable, persistent observation, use of multiple senses, high response and return rate, clear questions/form, and representative samples.

Z7: *Social Autopoiesis Techniques (3-p x 1-p x 3p*pl)*

Analysis of senders and receivers of data, accounting of the perspective of the observer and the multiple senders and receivers, holonical mapping, diagramming networks and channels of communication, and identifying the binary language structures that allow systems to register and respond to different phenomena.

Validity: Provides an adequate description of the functional fit of the phenomenology of systems, includes multiple perspectives of a system's members, includes a detailed understanding of the observer of the system(s), and explains why different systems can and cannot "see" certain realities.

Z8: *Systems Analysis Techniques (3-p x 3-p x 3p*pl)*

Statistical analysis, mapping, scientific studies, library research of previous studies, monitoring and evaluation, data flow modeling, spray and tree diagrams, normalization, mathematics, and computational analysis.

Validity: Functional fit, repeatable, controlled conditions, empirical, logical, reputable sources, multiple sources, and direct experience with system.

Now let us consider Integral research in the context of ecology and the environment.[25] Clearly some methodologies and their techniques reveal certain aspects of the environment better than others. In the following section we discuss the kinds of ecological research currently occurring within each terrain of Integral Ecology.

Modes of Inquiry in the Terrain of Experience

In the terrain of experience, phenomenology (Z1) and structuralism (Z2) are two important modes of inquiry, and both have important applications for ecology.

Phenomenology examines the inside of experience and reports felt qualities as they arise in immediate awareness (a subjective perspective using an inside approach to investigate individual interiors).[26] David Abram's work connects Merleau-Ponty's work to ecology and is a well-known example.[27] Andy Fisher also draws heavily on the tradition of phenomenology.[28] Much of ecopsychology and deep ecology are grounded in the mode of felt experience, and recently eco-phenomenology has emerged as a distinct field.[29] There is also some interesting research on animal and plant senses.[30]

Structuralism examines the outside of experience and maps the developmental stages of interior realities (a subjective perspective using an outside approach to investigate individual interiors). The environmental work of the psychologist Peter Kahn is a great example of structuralism,[31] as is Marc Bekoff and Dale Jamieson's research in the field of animal consciousness.[32] Others are applying the findings of the world-renowned psychologist Robert Kegan[33] to environmental issues.[34]

Modes of Inquiry in the Terrain of Culture

In the terrain of culture, hermeneutics (Z3) and ethnomethodology (Z4) are two important modes of inquiry, both of which have important applications for ecology.

Hermeneutics studies the inside of cultures and the shared meaning that occurs between humans or animals (a subjective perspective using an inside approach to investigate collective interiors). In general, many explorations of the concept of nature or wilderness use hermeneutics.[35] Edward Casey's important work on "place" uses this mode.[36] H. Peter Steeves's edited volume on *Animal Others* employs hermeneutics in the context of interspecial connection.[37] David Keller's exploration of technology and the "lifeworld" draws on Hans-Georg Gadamer's theory of science to propose an "ecological hermeneutics."[38] John Van Buren applies philosophical hermeneutics to environmental ethics and proposes a "critical environmental hermeneutics."[39] Robert Mugerauer and Keith Basso explore the hermeneutic relationship between landscape and language.[40] Other approaches that use hermeneutics include the Harvard project on religion and ecology;[41] the study of traditional knowledge systems;[42] environmental ethics;[43] and Goethean science.[44]

Ethnomethodology examines the outside of cultures and maps the developmental or structural aspects of collective interior realities (a subjective perspective using an outside approach to investigate collective interiors). Duane Elgin's presentation of environmental worldviews is a good example of this approach.[45] Much of the work in historical ecology and environmental anthropology also employs these methods.[46] Some ecofeminists use these methods to compare concepts such as nature, body, and woman.[47]

Modes of Inquiry in the Terrain of Behavior

In the terrain of behavior, autopoiesis theory (Z5) and empiricism (Z6) are two important modes of inquiry—both have important applications for ecology.

Autopoiesis theory examines the inside of behavior (an objective perspective using an inside approach to investigate individual exteriors) and is predominantly associated with the work of the Chilean scientist Francisco Varela and his work in cognitive science and biophenomenology.[48] This mode is relatively new and is currently being developed by many researchers inspired by Varela's pioneering insights.[49]

Empiricism examines the outside of behavior and relies on the senses, especially sight, and their extensions (microscopes and telescopes) to record data (an objective perspective using an outside approach to investigate individual exteriors). Empiricism consists of the "hard" sciences: biology, zoology, botany, and chemistry (of course, there are others). Additional approaches that use this mode include eco-tourism, natural building, energy efficiency, and the analysis of environmental behaviors such as recycling and buying "green" products.

Modes of Inquiry in the Terrain of Systems

In the terrain of systems, social autopoiesis theory (Z7) and systems theory (Z8) are two important modes of inquiry—both have important applications for ecology.

Social autopoiesis theory examines the inside of systems (an objective perspective using an inside approach to investigate collective exteriors). Niklas Luhmann primarily developed this methodology.[50] Felix Geyer and J. van der Zouwen applied his work to the study of social systems in their work on sociocybernetics.[51]

Systems theory examines the outside of systems, and how parts fit in a complex dynamic whole (an objective perspective using an outside approach to investigate collective exteriors). Most approaches to ecology (population ecology, community ecology, conservation ecology, ecosystem ecology, and landscape ecology) employ techniques associated with this methodology.[52] Systems sciences, such as Ludwig Von Bartalanffy's general systems theory,[53] and Susan Oyama's developmental systems theory,[54] also employ these methods.

Each methodology and their methods reveals unique aspects of reality. One cannot, for instance, discover the behavioral realities of a red-legged frog using phenomenology—only empirical methods will inform you about that frog's territorial range. Similarly, empirical observation (Z6) will not reveal phenomenological realties (Z1), but that does not mean that the frog does not have some interior experience of perception and a basic capacity for awareness. There are some researchers who understand this (that the methods of one domain cannot reveal the truths of another). Marc Bekoff has dedicated his life to finding scientific methods to investigate the interiors of animals (cognition, emotions, self-referencing) and integrating those findings with behavioral and evolutionary approaches to ecology.

For the rest of this chapter we will unpack two examples, neither of which uses the AQAL model, although both are multimethodological, include interiors and exteriors, and are aware of the importance of individuals and collectives. We think these pioneering approaches could be even more effective if they used an explicit Integral orientation. The first example is Marc Bekoff's "deep ethology," which honors animals by examining their phenomenological, cultural, behavioral, and ecological aspects in a developmental context. This example provides an important illustration of how ecological science can begin to incorporate the interiors of animals in a scientifically rigorous way. One of the most important contributions of Integral Ecology is providing a framework for honoring the subjective and intersubjective dimensions of organisms, and IMP is at the heart of this framework. The second example will focus on the "integrative framework" developed by the phenomenologist Ingrid Leman Stefanovic for "eco-research" and environmental sustainability. This example is less radical than the previous one but is just as valuable because it helps situate IMP within a human context and illustrates how human interiors can be more fully included in environmental projects.

Marc Bekoff's Interdisciplinary "Deep Ethology"

Bekoff began his career as a behavioral ecologist and ethologist in the mid-1970s at the University of Colorado in Boulder. Initially he focused on the ecology and social behavior of coyotes. Soon his focus included domestic dogs and wolves, and before long he was interested in the interiors of animals. In fact, he explains that since he was a boy he always "minded animals"[55]: "I always wanted to know what [animals] were thinking and feeling—What is it like to be a dog or a cat or a mouse or an ant? and What do they feel? And to this day, learning about the behavior of animals—all animals—has been my passion. When I study coyotes, I am coyote, and when I study Steller's jays, I am jay. When I study dogs, I am dog."[56]

Bekoff's commitment to understand the interiors of the animals sets him apart from most ecologists. He has been brave enough to ask questions like "Do animals forgive each other?" "Do whales experience love?" "If they do, what kind of love?" and "Which animals have the capacity for self-consciousness?" He is adamant about the existence of animal interiors: "That many animals have subjective and intersubjective communal lives—other animals are in their thoughts and feelings—and a personal point of view on the world that they share with other individuals, seems beyond question."[57]

Arguably, his research has done more to integrate animal interiority into ecology than the work of any other ecologist.[58] Over the last three decades Bekoff has emerged as a leader in the field of cognitive ethology, which he defines as "the comparative, evolutionary, and ecological study of animal minds, including thought process, beliefs, rationality, information processing, and consciousness."[59] Following Darwin, Bekoff views animal interiors—their cognitive, emotional, and moral capacities—along an evolutionary continuum that emphasizes differences of degree and not kind.[60] For instance, he points out that "we now know that individuals of many species use tools, have culture, are conscious and have a sense of self, can reason, can draw, can self-medicate, and show very complex patterns of communication that rival what we call 'language.'"[61] As a result, many of the characteristics that we thought distinguished humans from nonhuman animals are merely lines in the sand that are continually washed away with ever more effective techniques for inquiry.[62] Thus, we seriously consider how we are animals (i.e., have much in common physiologically, psychologically, and culturally with animals) and how animals are humans (i.e., have capacities usually attributed solely to humans).

Inspired by deep ecology and ecopsychology, Bekoff uses the term "deep ethology" to describe his concerted effort to inhabit the worldspace of the subjects he studies: "I, as the see-er, try to become the seen. I become coyote. I become penguin. I try to step into animals' sensory and locomotor worlds to discover what it might be like to be a given individual, how they sense their surroundings, and how they behave and move about in certain situations."[63] We call this practice "perceiving perceiving" (see chapter 10). Bekoff is very aware that most of the scientific community is resistant to such an approach. Therefore, throughout his work, he provides sophisticated scientific and philosophical responses to naive and legitimate critiques.[64] For our purpose, we focus on Bekoff's methodological commitments.

Bekoff identifies himself as an interdisciplinary researcher who uses a comparative approach to animal behavior that draws on ethology, behavioral ecology, evolutionary theory, neurobiology, endocrinology, psychology, philosophy, biochemistry, genetics, anatomy, theology, philosophy, religion, cognitive science, ethics, and consciousness studies. He is clear that "animal minds can be studied rigorously using methods of natural science and will not ultimately have to be reduced or eliminated."[65] He also champions the inclusion of fieldwork and laboratory research. Moreover, he works at multiple levels of analysis: micro (e.g., frame-by-frame film analysis of animal behaviors and statistical analyses) and macro (e.g., long-term field observations in many locations). He uses the terms "pluralistic" and "holistic" to describe his overarching approach. In fact, he even calls for "methodological pluralism," explaining that such "a pluralistic approach should result in the best understanding of the nonhumans with whom we share the planet."[66]

We resonate with Bekoff's methodological pluralism. His work integrates 1st-, 2nd-, and 3rd-person perspectives and acknowledges the quadrants of animals (i.e., their own capacity to take 1st-, 2nd-, and 3rd-person perspectives). Bekoff recognizes animals' capacity for tetra-hension and explores their world from all four of their quadrants: their subjectivity and experience (self-cognizance, emotions, beliefs, and desires); their intersubjective relationships with each other (love and forgiveness, play, moral considerations, "wild justice," and cooperation); their objective perceptions of their world (visual, auditory, and olfactory capacities); and their interobjective communications (vocalizations, gestures, signaling, social manners, and scent-marking). He also explores the animal world from all four of his quadrivia. He can study their interior mental and emotional states; their exterior behaviors and physiology; their shared interior feelings and experiences; and their exterior social dynamics. Bekoff draws on all four terrains of ecology and makes use of most of the 8 zones of Integral research. We propose that Bekoff's impressive interdisciplinary approach and methodological pluralism could be even more robust and clear when situated in the Integral Ecology framework. Our framework accomplishes this by providing an explicit way to ensure that all 8 major zones of research are included, and it highlights how the data from each zone are to be understood in relationship to one another.

Consider the following: cognitive ethologists often claim to use the same method, but they apply that method to very different aspects of the phenomenon under investigation. An Integral framework helps clarify this confusion. For example, the pioneering ethologist Donald Griffin used the term "cognitive ethology" to refer to the study of the phenomenology of animal consciousness and awareness (Z1).[67] As cognitive psychology has developed over the last 30 years, researchers increasingly use the term "cognitive ethology" to refer to the study of animal's cognitive process (Z5). As a result, some cognitive ethologists focus on animal consciousness (Z1), while others focus on animal cognition (Z5). In view of this situation, we refer to the former as phenomenological ethology or conscious ethology and the latter as cognitive ethology.[68]

Bekoff also addresses this split by distinguishing between what he calls "weak cognitive ethology" and "strong cognitive ethology." Those who employ the methodologies associated with weak cognitive ethology study animals' interior states by using information processing and computer models (Z5) associated with cognitive science (they do not study the animals' interior experience). This form of ethology is often associated with behavioristic functionalism (Z6) because both fields are concerned with individual exteriors. The main difference is that those who employ the methods of behavioristic functionalism approach animal cognition from the outside (Z6) and cognitive ethology from the inside (Z5). Similarly, the fields of classical ethology, behavioral ecology, neuroethology, and zoology all rely heavily on zone 6 methodologies.

Researchers who employ strong cognitive ethology do so through multiple zones. They often employ zone 6 methods to document behaviors and their neurological underpinnings; zone 5 methods to understand cognitive processes and informational dynamics; and zone 1 methods to explore capacities for experience and self-awareness. Additional zone 1 methods include what we call emotional ethology (the study of animal experiences of their emotions) and zoophenomenology (the study of the full spectrum of animal interiority from simple sensations to complex experiences).[69] Also, biosemiotics' umwelt theory is an important approach that integrates methods from zones 1, 5, and 6. In fact, biosemioticians use methods associated with all the inside zones (1, 3, 5, 7).[70]

If we were to study animals from a zone 2 perspective, we would use the structural aspects of comparative psychology, which compares the mental capacities of various species in an effort to identify the various structures (vertical and horizontal) involved in animal psychology.[71] We would also use developmental ethology, which explores the various interspecial (between species) and intraspecial (within species) developmental stages that occur in cognitive, emotional, and moral capacities.[72] So far we have focused on those methodologies that focus on the individual animal. Researchers who want to understand animal culture and society use methods that are associated with the four collective zones (3, 4, 7, 8). At least three important disciplines investigate zone 3. Zoohermeneutics explores how animals interpret their environment and other animals.[73] Anthrozoology studies animal-human relationships.[74] And zoosemiotics studies animal communication and emphasizes the interpretive dimension of signaling.[75]

Zone 4 examines the exciting and controversial area of animal culture. Currently, researchers are conducting important studies on social animals such as whales, dolphins, elephants, hyenas, and various birds.[76] For the emerging field of animal culture, including capacities that are socially learned and transgenerational, we propose the name *zooethnography*, which is informed by the fields of ethnography and cultural anthropology. The animal psychology and trauma recovery research of Gay Bradshaw is a great example of a multizone approach that explores animal culture (especially in social vertebrates such as elephants and parrots). In particular she brings together disciplines such as ecopsychology (Z1), psychiatry (Z1 and Z6), ethics (Z3), attachment theory (Z4), ethology (Z4 and Z6), neurobiology (Z6), and landscape ecology (Z8).[77]

Zone 7 focuses on the social autopoietic dimensions of reality. This is a complex and obscure methodological zone, and disciplines that focus on animals from this zone have only recently emerged. Over the last decade, the Dutch biosemiotician Søren Brier has developed (bio)cybersemiotics, which integrates Niklas Luhmann's communication theory, Charles Peirce's semiotics, and Jakob von Uexküll's umwelt research.[78] This is a very new field, yet Brier's theory and research are an important contribution to an Integral understanding of animals and their relationship to their environment.

Zone 8 (the outside of exteriors) methodologies are more familiar to the average person. They include the various systems sciences, such as ecology and economics. In regard to animal research, there are at least three disciplines worth mentioning. Socioecology studies the social relations between and within species.[79] Population ecology studies the population dynamics of a single species. Community ecology studies the various social interactions between species within any particular ecological community.

This completes our brief overview of how the 8 ecological modes can support an Integral understanding of animals. In figures 8.3 and 8.4 we use the Who x How x What framework to summarize how the 8 eco-modes can be used to provide an quadrivial analysis of animals.[80] The questions presented in figures 8.3 and 8.4 are the kinds of questions that an existent

	INTERIOR	EXTERIOR
INDIVIDUAL	WHO: Integral Researcher (1-p) HOW: Comparative Psychology (3-p) In what ways is the animal's awareness structured? WHAT: Animal as a psychologically structured organism (1p) STRUCTURALISM (1-p x 3-p x 1p) I	WHO: Integral Researcher (3-p) HOW: Direct Observation (3-p) In what ways does the animal react to its world? WHAT: Animal as a physical and behavioral organism (3p) EMPIRICISM (3-p x 3-p x 3p) IT
COLLECTIVE	WE WHO: Integral Researcher (1-p) HOW: Long Term Field Studies (3-p) In what ways does the animal share cultural membership with its own species? WHAT: Animal as a culturally regulated organism (1p*pl) ETHNOGRAPHY (1-p x 3-p x 1p*pl)	ITS WHO: Integral Researcher (3-p) HOW: Functional-Fit Analysis (3-p) In what ways does the animal fit into the many systems it is part of? WHAT: Animal as a systems-bound organism (3p*pl) SYSTEMS THEORY (3-p x 3-p x 3p*pl)

Figure 8.3. The outside quadrivia of an animal.

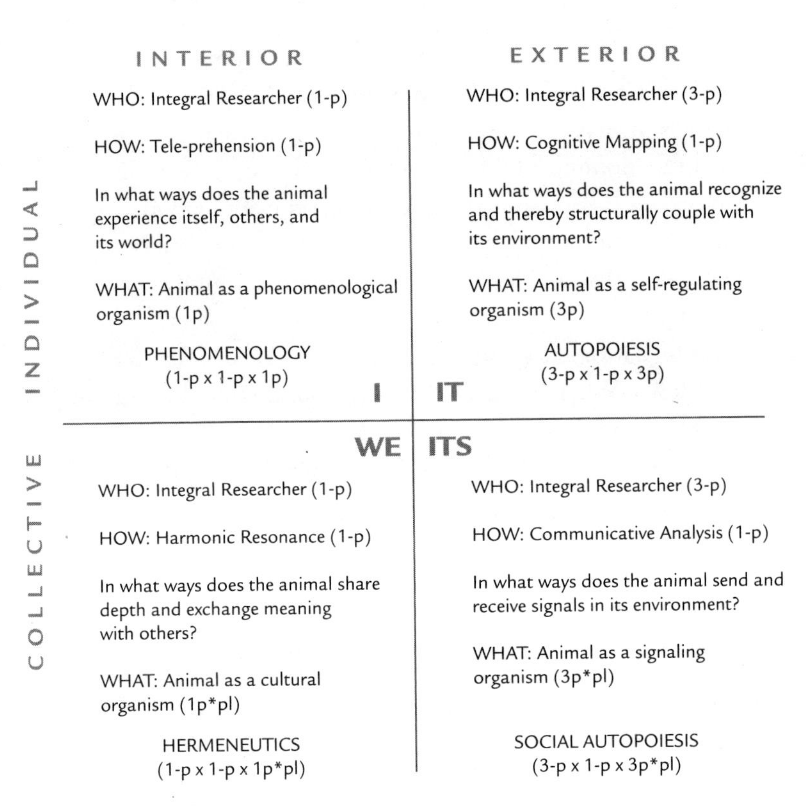

Figure 8.4. The inside quadrivia of an animal.

approach associated with each zone actually asks from within the context of its methodological expertise. In figure 8.5 we list the kinds of disciplines within each zone that are asking and capable of answering these sorts of questions. Many disciplines listed in figure 8.5 occupy multiple methodological zones (e.g., comparative psychology, listed in zone 2, is also associated with both zones 5 and 6). So we have placed disciplines according to the zone we want to emphasize.

The Problem of Animal Minds

An advantage to the methodological framework outlined above is that it helps overcome the problem of other minds.[81] It does this in a number of ways. Arguably the most persuasive way is by triangulating ("quad-" or even "octangulating") data from each of the methodological zones. Thus, even if information from zone 1 is vague and unclear, we could still marshal evidence from the seven other zones to provide a fairly comprehensive view.

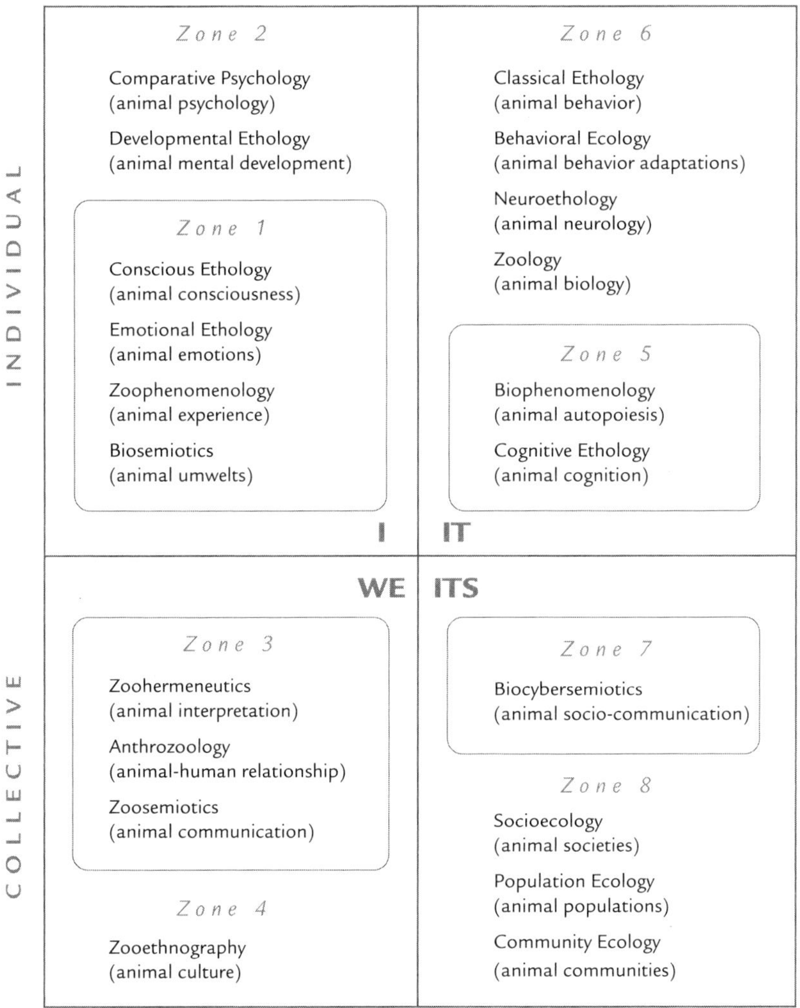

Figure 8.5. An IMP approach to animals.

This approach becomes even more powerful when we can identify correlations between zones. While we might not be able to describe conclusively the texture of animal interiorities, we can come to some meaningful and sophisticated conclusions. Each zone provides an important piece of the puzzle. For instance, zone 6 can provide information about the behaviors and neurological networks shared by humans and various mammals. Likewise, researcher employing methods from zone 2 can identify various structures of consciousness that appear across or within species. Zone 4 can provide anecdotes of unique animal relationships that appear to defy behavioral or environmental explanation. As Bekoff points out, "the plural of anecdotes is data."[82] Methodologies from zone 3 can reveal information about animal communication and interpretation, and the occurrence of mutual resonance or intersubjectivity between humans and animals.[83] We refer to this intersubjectivity as shared depth.

Because intersubjectivity is the joining of two subjects, we do not feel that zone 1 is a "black box." In part, we feel the "problem of other minds" as it is often construed is based on a metaphysics of isolated atomistic awareness, which renders even human subjectivity inaccessible. In fact, Wilber advances a radical post-Kantian position that overcomes the epistemological chasm of how a subject can know an object.[84] Wilber points out that in many cases the observer contains internally similar kinds of holons to those it is trying to know externally. The example he gives is this: a scientist looking at a single-cell organism through a microscope actually contains many such organisms throughout his own body. The compounded evolutionary nature of the scientist actually contains that which he is seeking to know. In other words, he has shared depth with the organism. "The gap between subject and object (including object as thing-in-itself) is fundamentally bridged theoretically: they share, at that level of the AQAL matrix, a cellular intersubjectivity or cultural solidarity that allows knowing and understanding to occur."[85] The Kantian divide is crossed because both the observer and the observed are individuals whose interiors and exteriors co-arise within themselves and together. You don't have simply a subject viewing an object. You have a subject with exteriors encountering an object with interiors (technically each is a 4-quadrant occasion). Just as the 1st- and 3rd-person dimensions of the scientist arise simultaneously within his or her own awareness, so do those dimensions within and between the single-celled organism. There is no fundamental divide between observer and observed.

So while we acknowledge that organisms, including humans, have a dominant monad—or an individualized location of awareness—we do not think that this is an insurmountable barrier to direct contact with those interiors. It is difficult to understand the interiors of another, but we do not believe it is impossible. One of the most effective bridges between an animal's interiority and ours is through shared depth or harmonic resonance.[86] Wilber points out that "the agency of each holon establishes an opening or clearing in which similar-depthed holons can manifest to each other, for each other: agency-in-community (all the way down)."[87] This shared space of unmediated experience (no exchange of language) is what Wilber calls a worldspace. In *A Brief History of Everything*, Wilber gives this example:

> When you interact with your dog, you are not interested in just its exterior behavior. Since humans and dogs share a similar limbic system, we also share a common emotional worldspace ("typhonic"). You can sense when your dog is sad, or fearful, or happy, or hungry. And most people interact with those interiors. They want to share those interiors. When their dog is happy, it's easy to share that happiness. But that requires a sensitive interpretation of what your dog is feeling. Of course, this is not verbal or linguistic communication; but it is an empathic resonance with your dog's interior, with its depths, with its degree of consciousness, which might not be as great as yours, but that doesn't mean it's zero.
>
> So you empathetically interpret. And the dog does the same with you—you each resonate with the other's interior.[88]

It is possible to have an unmediated experience of an animal's emotional state.[89] However, you do not share a conceptual worldspace with your dog, so when you think about GATT and the environment, your dog cannot have a direct, unmediated experience of that, even if you explain it in detail. For two subjects to share an unmediated, direct experience, they have to share a common worldspace (physical, energetic, or emotional). Because humans share different kinds of worldspaces with many beings, they can have direct, unmediated experiences with a wide range of subjects, including animals. Wilber provides an example of a wolf pack to illustrate this:

> Wolves, for example, share an emotional worldspace. They possess a limbic system, the interior correlate of which is certain basic emotions. And thus a wolf orients itself and its fellow wolves to the world through the use of these basic emotional cognitions—not just reptilian and sensorimotor, but affective. They can hunt and coordinate in packs through a very sophisticated emotional signal system. They share this emotional worldspace.
>
> Yet anything outside that worldspace is not registered. I mean, you can read Hamlet to them, but no luck. What you are, with that book, is basically dinner plus a few things that will have to be spat out.

> The point is that a holon responds, and can respond, only to those stimuli that fall within its worldspace, its worldview. Everything else is nonexistent.[90]

Unmediated, direct experience occurs only within a shared worldspace/worldview, but it still occurs in an interpretive context regardless of whether it is articulated in language. In this shared depth we can understand an animal's interiors. Of course, we do not share complex verbal language with animals—though a few animals, such as chimps, dogs, and African grey parrots, have mastered between 200 and 700+ human words.[91] A limited vocabulary does not mean that animals cannot self-disclose their interior experiences—if anything, it means we must be more attentive to their signals if we want to understand them. In fact, many pet owners and ethologists are already good at this.[92] Wilber explains that this intersubjective circle of self-disclosure is not only true for humans, "but also for all sentient beings as such. If you want to understand your dog—is he happy, or perhaps hungry, or wanting to go for a walk?—you will have to interpret the signals he is giving you. And your dog, to the extent that he can, does the same with you."[93]

Self-disclosure is typically interpreted through analogy—we infer an animal's interiority based on its behavior. While analogies are helpful starting points and should be included in an Integral approach, we believe we can learn more about animal interiority by providing "thick descriptions" of animal signals and behavior. We could use zoophenomenology to document the range of signals animals use to self-disclose their interior states.

To accurately understand shared depth, researchers must train in phenomenological discrimination. The more researchers can distinguish between slight variations of bodily sensations, emotional and mental states, and even nonordinary states, the more those researchers will be able to interpret the shared depth.[94] Phenomenological training also lends itself to cultivating what Gordon Burghardt calls "critical anthropocentrism."[95] Bekoff expands on Burghardt's notion and asserts that using anthropocentrism to understand the relationship between animal interiors and their behavior does not necessarily involve "losing our perspective of who animals are in their own world."[96] In other words, anthropomorphism does not have to involve a total eclipse of the animals' points of view. He states that being anthropocentric is actually unavoidable simply because we rely on human language and our own experiences to relate to and describe animal behavior and animal feelings. He echoes our view of shared depth when he claims that "anthropomorphism can help make accessible to us the behavior and thoughts and feelings of the animals with whom we are sharing a particular experience."[97]

We think that Integral Methodological Pluralism provides a number of keys for opening the so-called "black box" of animal subjectivity and discovering what lies on the inside. To do so we must "octangulate" our data, or use data from all 8 zones to research animal subjectivity: social dynamics (Z8), animal communication (Z7), flexible behavior (Z6), cognitive structures (Z5), anecdotes (Z4), shared depth (Z3), psychological structures (Z2), and animal self-disclosure (Z1). We must also train researchers in phenomenological inquiry to provide them with the power to discriminate within their own interiors, and cultivate nonordinary states so that they can register and language the shared depth they experience with animals and demonstrate a critical anthropocentrism. Creating a legitimate path to animal interiors is important for providing an approach to ecology that includes the interiors of the nonhuman members of ecosystems.

Just as Bekoff recognizes the need to include multiple methods and disciplines to understand animal interiors, Ingrid Leman Stefanovic highlights the importance of integrative frameworks for understanding the intersection of human interiors with ecosystems.

Ingrid Leman Stefanovic's Integrative Eco-Research

Ingrid Leman Stefanovic is Director of the Centre for Environment and professor of philosophy at the University of Toronto. Her 30-year research and teaching career has emphasized interdisciplinary approaches to ecosystems and the environment. In particular she has focused on environmental ethics, including individual values and perceptions in environmental decision-making, sustainable development, environmental phenomenology, and urban planning. Her book *Safeguarding Our Common Future: Rethinking Sustainable Development* (2000) uses phenomenological and hermeneutical methods (Z1 and Z3) in the area of sustainable development, which is typically only approached from behavioral and systemic perspectives (Z6 and Z8).[98] In this book she criticizes the positivist approaches to sustainable development and then proposes methodological guidelines for developing phenomenologically informed environmental indicators of sustainability. Her goal is to deepen our understanding of sustainability beyond the definitions of the Brundtland Report in 1987. In so doing, she provides a number of examples of how interiors have been successfully included in several projects. One of those projects is known as the Ecowise Study of Sustainablity of the Hamilton Harbour.

In 1991 the Ecowise study received $2.1 million in federal funding from the Canadian government for a three-year (1992–1995) large-scale interdisciplinary project to study the sustainability of the Hamilton Harbour ecosystem in western Lake Ontario. This ecosystem had been identified as polluted, in part due to industrial waste from the local steel companies. The Ecowise team consisted of over 30 university researchers from five Ontario universities, organized into four working teams and representative disciplines: Human Values and Perceptions (sociology, social psychology, and philosophy); Contaminants (genetics, geomorphology, hydrology, biochemistry); Biotic Recovery (biology and medicine); and Policy Analysis and Economics (economics and political science).[99] A fifth "was responsible for communication, education, and dissemination of research findings, as well as for 'integration.'"[100] It was Stefanovic's task to develop an integrative framework for the entire project: "It was clear to most researchers that the project was to consist of more than a multidisciplinary collection of discrete viewpoints. The integrative framework for the project, therefore, was to seek to do more than catalog individual researchers' work. Instead, our aim was to elicit a holistic overview of the research study, as well as of the ecosystem itself, as it was perceived by researchers."[101]

Stefanovic and her assistants developed a phenomenological interview process to discover how researchers perceived Hamilton Harbour (in terms of both their sense of place and how they intended their research to benefit the ecosystem). They used a six-step process to identify key concepts and distill themes, and rechecked their findings through additional interviews. Over 80 descriptors of the ecosystem were plotted on a computerized matrix so as to track and identify the relationship between ecosystem components and the various research projects (over 5,000 interactions were generated by the matrix), indicate areas of further study, identify areas in need of more or better communication between projects, and help generate new data through new questions.[102] One of the significant findings that emerged through this process was that in spite of the fact that more of the researchers came from the social sciences and humanities (disciplines associated with the Left-Hand zones), the bulk of the key concepts were related to the physical environment (realities associated with the Right-Hand zones). Stefanovic concluded that the "importance of the human settlement side of the equation was underrated in favor of the harbor itself."[103] Overall, many interviewees appeared to dislike the city or avoided it. Stefanovic noted that in these interviews many important subjects, such as urban planning, community, social services, and religion, were rarely mentioned. In contrast, such topics as habitat preservation, water quality, and wildlife were more frequently mentioned.[104] Consequently, a result of the phenomenological interviews was that even though researchers recognized the importance of including urban environments, they often prioritized natural science perspectives involved with the natural ecosystem. Another benefit of the interviews is that they linked the community to the project, making it more responsive to the local context and not simply a one-size-fits-all, top-down approach.

Stefanovic's work is pioneering in many respects, not the least of which is connecting phenomenology and hermeneutics to the natural sciences in an ecological context. Drawing on the hermeneutic theory of Hans-Georg Gadamer, she asserts that only when we distinguish the horizons of interpretation associated with different disciplines can we genuinely speak of an overlap or fusion of horizons.[105] This is exactly the impulse guiding the presentation of the 8 eco-modes of Integral Methodological Pluralism. If we use the 8 zones we can more clearly see the success of her efforts with the Ecowise project, and see areas that went unaddressed by her interdisciplinary approach. If we place each of the four focus groups and their activities into the zone that they seem most related to, an interesting picture emerges (see figure 8.6).[106]

The Ecowise project integrated the insides of the interior zones and the outsides of the exterior zones. These are the extremes of both sides: the very interior on the left and the very exterior on the right of each quadrant. Even so, the primary emphasis of the project still lies in the Right-Hand quadrants. As Stefanovic herself points out, not only do most of the researchers favor the ecosystem (even over LR social factors), but so do the phenomenological interviews, which identified 80 descriptors of the Harbour (exterior dimensions) rather than interior ones like aesthetics, ethics, sense of place, psychological well-being, somatic experiences, or religion. Even though the

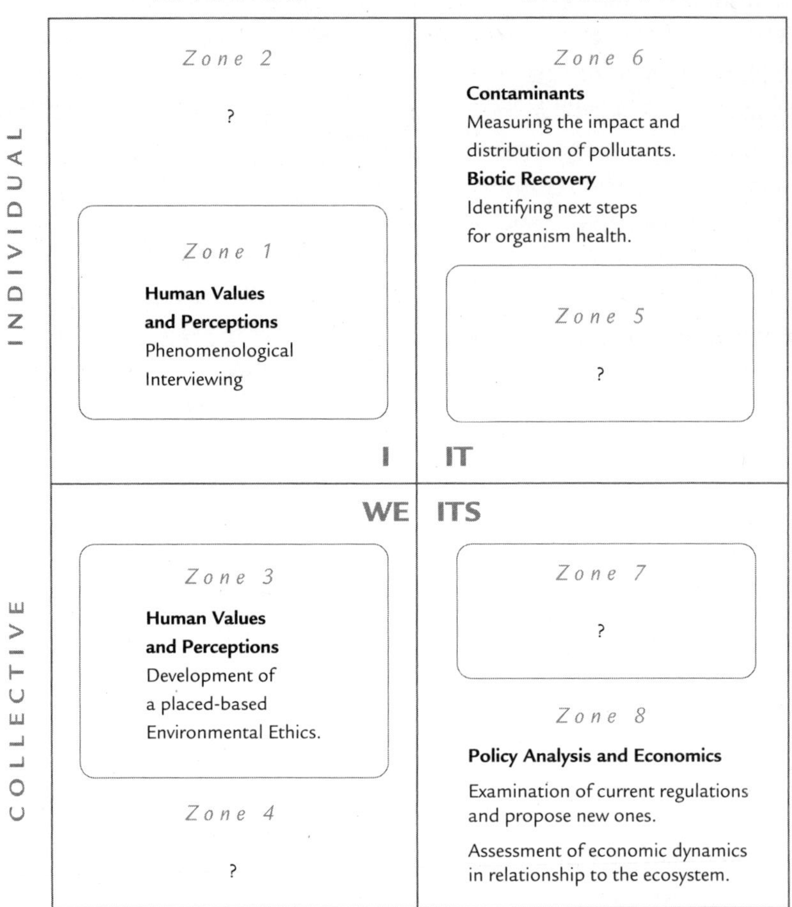

Figure 8.6. An IMP analysis of Ecowise.

project included people's interiors, their interiors were focused on exteriors. Thus, the power of including the subjective and intersubjective elements of the researchers was short-circuited by their own apparent lack of interest in interiors. So while phenomenological interviews played an important and even pioneering role in the project, we think that additional methods should be used, especially ones that explore and document interiors on their own terms.

More zone 1 and 3 realities must be considered. Researchers could structurally analyze (Z2) the prominent worldviews (traditional, modern, and postmodern) of the community members and researchers. It may be important to understand the cultural practices (Z4) in and around the Harbour. Cognitive psychology (Z5) could identify ways to work with people's unconscious behaviors and heuristics to increase sustainable behaviors and reduce destructive ones.[107] And a socio-communication analysis (Z7) could highlight how the social systems connected to the Harbour are (or are not) communicating about its sustainability. We provide an example of such an analysis in the next chapter when looking at multiple constructions of the environmental crisis. In short, while the Ecowise project breaks new ground and successfully includes interiors, we think that more work is needed to fully integrate methodologies associated with the terrains of experience, culture, behavior, and systems. We think the 8 ecological modes provide an effective means to do just that.

The examples of both Bekoff and Stefanovic highlight the fact that many individuals recognize the need for more integral understandings of organisms and ecosystems. Such researchers are building new bridges between methodologies. We think that one way to facilitate such integrative efforts is to explicitly use the AQAL framework to research organisms and ecosystems and integrate the findings of multiple disciplines.

IMP and the 8 modes of ecological research looks promising as a way of integrating across and within so many disciplines because it is postdisciplinary, as we discussed in chapter 1. Thus it avoids the pitfalls of inter- or transdisciplinary approaches. Another reason the IMP framework is valuable is that it articulates the relationship between the various methodological zones. This creates a coherent integration of multiple understanding.

These examples illustrate that Integral Ecological research is not impossible. In fact, it is much easier than we might think. Having the Integral Model makes it all the more feasible, because it coordinates various methods. The more of reality we acknowledge, the more sustainable our solutions will become, precisely because the project will be more responsive to the complexity of that reality. We cannot exclude major dimensions of reality and expect comprehensive, sustainable results. Eventually those realities that have been excluded will demand recognition and incorporation as the design falters and is abandoned for more nuanced and comprehensive strategies. Hence the need for an Integral approach.

Much work remains to further develop Integral research, yet even in its incipient stages, Integral research promises to unite more forms of established research than any other approach thus far.

Review and Looking Ahead

In part one we explained Integral Theory. In part two we applied Integral Theory to ecology. We demonstrated that Integral Ecology consists of the Who, How, and What, and that each of these highlights an essential aspect of an environmental phenomena. The 4 terrains of ecology and their respective 12 niches of environmental concern represent the ontological dimension of environmental phenomena. The 8 ecological selves represent the epistemological dimension. The 8 ecological modes represent the methodological dimension. The 12 niches of ecological concern, the 8 eco-selves, and the 8 modes of environmental knowledge comprise the three legs of the enactive crucible, where epistemology connects to ontology through methodology. In other words, the various eco-selves make use of the different modes of inquiry to disclose, through participation, the various environmental phenomena associated with the 12 niches. The result is a comprehensive, post-metaphysical approach to environmental phenomena. Thus, Integral Ecology is uniquely situated to honor and include all the various insights of over 200 distinct perspectives on ecology and the eco-social environment.

Having discussed the triad of observer, observation, and observed, we now turn to part three to explore some of the implications and applications of the Integral Ecology framework for self, community, and world. We begin part three with an Integral understanding of ecological harmony and environmental crisis in a post-natural world. We use the term "post-natural world" to refer to the fact that we live in a time in human history when nature and culture are so intermixed that we cannot rightly speak of the natural world without implicating ourselves in it. As opposed to reifying nature "out there" or "over here," we champion a post-naturalism in which nature is intertwined with culture, culture is shot through with nature.[108] We explore the many ways we can be in (and out of) harmony with Gaia, unpack multiple perspectives on the environmental crisis, and speculate on how can we transform our own awareness and embodiment (e.g., mind-body integration) for the betterment of consciousness, culture, and nature.

Thus, in the subsequent chapter we provide a variety of 1st-person practices that help cultivate the Integral awareness needed for today's complex world. We then turn to some examples of the ways that Integral Ecology is currently being applied throughout the world.

PART THREE

The *Who*, *How*, and *What* Framework Applied

9

Ecological Harmony and Environmental Crisis in a Post-Natural World

Our challenge is to stop thinking . . . according to a set of bipolar moral scales in which the human and the non-human, the unnatural and the natural, the fallen and the unfallen, serve as our conceptual map for understanding and valuing the world. Instead, we need to embrace the full continuum of a natural landscape that is also cultural, in which the city, the suburb, the pastoral, and the wild each has its proper place, which we permit ourselves to celebrate without needlessly denigrating the others. . . . To think ourselves capable of causing "the end of nature" is an act of great hubris, for it means forgetting the wildness that dwells everywhere within and around us.

—WILLIAM CRONON[1]

In this chapter we are interested in the questions "How do we align ourselves in beneficial ways with nature?" and "How do we collectively avoid environmental catastrophe?" To answer these we take the Who x How x What framework presented in part two and apply it to the occurrence of nature mysticism, or the many ways that individuals can be "one with nature." Our views on this topic are quite different from other approaches to ecology and spirituality, which generally lack or avoid a developmental understanding. Then we use the same framework to show how various waves of development define the causes and qualities of our environmental crisis. Here we emphasize the perspectival aspect of the crisis, which demonstrates that there is not a single ecological crisis, but many different perceptions and conceptions of ecological crises.

Thus, this chapter explores important aspects of humanity's relationship with nature in terms of both individual experience (nature mysticism) and our collective activities (the environmental crisis). At the heart of these issues is the age-old question of what is the proper relationship between human culture and wild nature. Do we have to leave the city to have an experience of nature mysticism? Does human culture need to be downsized in order for us to protect the planet? Our answer to both of these questions is "no." From an Integral Ecology perspective, nonhuman nature is culturally inflected, whereas human culture is profoundly influenced by and affects nature. In other words, we avoid a nature-versus-culture stance by recognizing that every occasion has a cultural dimension (LL) and a natural dimension (UR and LR). Thus, we can say culture is the collective interior of aspects of nature, and nature is the exterior expression of aspects of culture. To pit nonhuman nature against human culture is to make a fundamental mistake with regard to humanity's relationship with the so-called "natural" world—not to mention that such a binary denies the existence of animal culture.

By emphasizing the ways nature and culture arise together and are tetra-enacted, Integral Ecology avoids the pitfalls of viewing nature strictly in terms associated with either naive realism or social construction.[2] Integral Ecology is a post-natural approach: nature is not simply given, independent of our observation, nor are we the sole generators of nature. All environmental phenomena are the result of the Who x How x What; the result of 1st-, 2nd-, and 3rd-person perspectives interfacing with each other. Integral Ecology also avoids splitting nature or culture into a simplistic good/bad opposition. Consequently, Integral Ecology offers a robust framework with which to understand humanity's relationship with nature. In particular, Integral Ecology provides a complex way in which to make sense of the myriad ways humans can experience and contribute to ecological harmony and the ways we can prevent the ecological crises that we are helping to propagate.

To begin with, we explore the various ways an individual can have a unitive experience with the natural world. Then we use the Wilber-Combs Lattice, which presents the relationship between vertical levels and horizontal states, to present the varieties of nature mysticism that an Integral approach includes.[3] This leads us to discuss the occurrence of nature mysticism in children and prehistoric peoples and the idea that humans have fallen from ecological grace at some point in the historical record. Building on the discussion of the Eden myth of environmentalists, we turn our attention to the environmental crisis and the ways it is perceived by different worldviews. In so doing, we suggest that there is no single crisis but rather multiple crises and in some respects no crisis at all. Finally, we present an important slogan of Integral Ecology that requires holding a threefold paradox.

One with Which Nature?

What does "being one with nature" mean? Is it understanding the web of life, expanding your self-identity, believing in the power of the Goddess, going hiking on weekends, seeing nature spirits, eating only vegetables (and a little chocolate), experiencing the luminosity of the natural world, realizing that there is nothing to save, buying 10 acres of rainforest every year? Well, it depends at least on three things: what you mean by "one with," what developmental level you operate from, and what you mean by "nature." In other words, mystical union with the natural world is the result of the many combinations of Who x How x What.

"Being one with nature" can refer to a particular kind of experience, a specific set of behaviors, a certain type of relationship with other beings, or a particular role within eco-social systems. In fact, an individual can be one with nature in all 4 terrains: behavior, experience, culture, and systems. It is possible to *feel* one with nature but not *act* as one with nature. It is possible to be part of systems that are one with nature but a member of a culture that is not, and so on. We need a "tetra-mesh" understanding of "being one with nature." It is not enough to have only a phenomenological experience of unity with the natural world.

Peter Kahn's research on how children and adults report being in harmony with nature supports a quadratic understanding of unity with nature. In several studies his team asked children to describe what it meant to live in harmony with nature.[4] Five categories emerged from the data: physical (doing something to, for, or with nature); sensorial (apprehending the natural world with one's senses); experiential (having a particular state experience or feeling); relational (having a particular type of relationship with nature); and compositional (being in balance with nature as a system).[5] Each of these types of harmony describes a different terrain or quadrant (see figure 9.1).

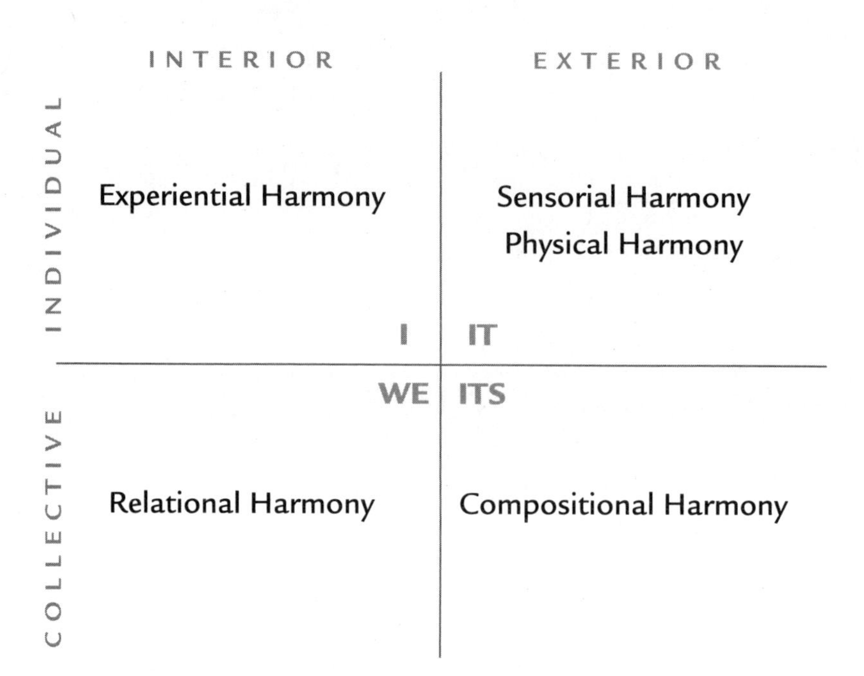

Figure 9.1. Five types of harmony with nature.

Different worldviews also understand "being one with nature" differently and can emphasize different terrains of harmony. For example, not only would an Eco-Warrior and an Eco-Radical have different phenomenological experiences of oneness with the natural world, but a Warrior might gravitate toward physical harmony, whereas a Radical might emphasize compositional harmony.

Nature mysticism as a stable structure-stage realization of unity with the planet and its myriad forms emerges with the indigo (third-tier) altitude of psychospiritual development, which very few individuals attain. However, each previous level has its own form of nature mysticism, which is to say that at any level of development one can have a unitive experience with some aspect of the natural world. Nature mysticism, often associated with shamanism or 18th- and 19th-century romantics, has many other rich sources: Native American spirituality, Australian aboriginal dreamtime, European neo-pagans and hermeticism, New England transcendentalists, and African spirituality.[6] Most of the scholarship and research into nature mysticism suffers from not distinguishing stages of development from states of consciousness.[7] In other words, there are both structural unions with nature and state

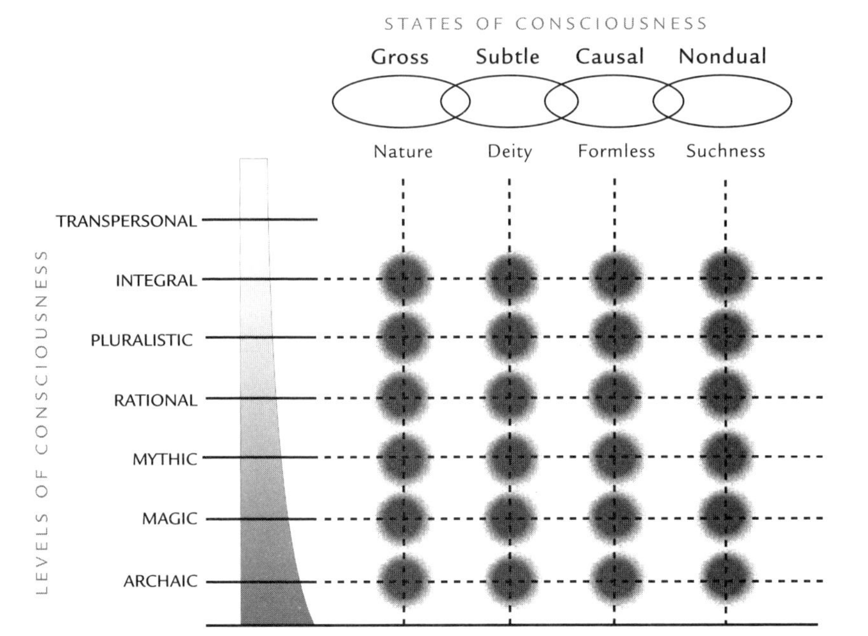

Figure 9.2. Nature mysticism lattice.

unions. This is represented in figure 9.2, with structural unions occurring along the vertical axis and state unions occurring along the horizontal axis.[8]

So an Integral approach to unification with the natural world needs to distinguish between quadratic types of harmony (i.e., how people unify with nature), levels of psychological interpretation of mystical states of union (i.e., who is unifying with nature), and the four major types of unification with the planet: gross, subtle, causal, and nondual (i.e., what part of NATURE are they unifying with and how is NATURE being reflected in Nature or nature?).

Correlated with these four types of unification (gross, subtle, causal, nondual) there are four major realms of NATURE.[9] There is NATURE as the gross realm (NATURE as what is revealed by the senses; Life, Nature, and Senses), as the subtle realm (NATURE as a powerful blessing of luminosity; Light, Love, and God/dess), as the causal realm (NATURE as nonexistent or empty of inherent existence; Dark, Empty, and Nonexistent), and the unification of all three realms as nondual expression (NATURE as Always Already; union of emptiness and form; Spaciousness, Flow, and Openness).[10] See figure 9.3.

"Being one with nature" can thus occur within any of the terrains (as experiential, behavioral, cultural, and systems harmonies), at any level of development, and with any of the four major realms of NATURE (gross, subtle, causal, and nondual). So the next time you are backpacking or having a cocktail and someone says, "I recently had a powerful experience where I was one with nature!" be sure to ask that person, "What terrain of harmony did you emphasize?" "What structure are you interpreting your experience from?" and "Was that with gross, subtle, causal, or nondual NATURE?" When you receive a funny look, just smile!

Varieties of Nature Mysticism

In discussing nature mysticism, it is important to distinguish it from *earth religion* (generally used to describe neo-paganism), *natural religion* (generally used to describe religion based on reason and ordinary experience), and *nature religion* (an umbrella term for religious groups and practices that emphasize the sacredness of the natural world).[11] Clearly, practitioners of either earth religion or nature religion are likely to have state experiences (and in some cases even stabilized stages) of nature mysticism.[12] However, the experience of nature mysticism states (gross, subtle, causal, or nondual) or the development of permanent access to indigo awareness is not dependent on worshiping nature. In fact, for many individuals, nature mysticism (states or stages) occurs outside of an intentional or religious context.

Now, if we take representative individuals at the five most common levels of psychological development and assume that they can have a unitive experience in any of the four realms of NATURE (figure 9.3), we arrive at 20 possible distinct "one with nature" experiences.[13] Figure 9.3 is a table that provides short, suggestive descriptions of each of these varieties of nature mysticism.[14] This table is designed to provide a quick overview of how an Integral approach to nature mysticism might begin.

In order to give some illustrative examples of these various states of nature mysticism, let us start with a description of how an individual at each of these levels would relate to nature based on his or her structure of consciousness. After all, it is the structure of one's awareness that interprets a given state of unity with nature. So in order to present the various states of nature mysticism, we need to highlight the structures that filter and enact those states. After we discuss structure, we will discuss the various states.

	Gross union with NATURE LIFE / GAIA NATURE as senses.	**Subtle union with NATURE** LIGHT / GODS / LOVE NATURE as powerful beings.	**Causal union with NATURE** DARK / EMPTINESS NATURE as non-existent.	**Nondual union with NATURE** SPACIOUSNESS / FLOW NATURE as Always Already.
Kosmocentric Trans-conventional Nature viewed in light of a Divine Mystery.	Experience of merging with natural phenomena (plants & animals) and animal communication (telepathy).	Experience of archetypal deities (e.g., the Earth Goddess or Pan), and of planetary etheric patterns and dynamics.	Planet as Perfect! Planet as the Creative Void. One of many planets. More planets will come...	Tao: Marriage of Heaven and Earth. Experience of Kosmic Flow.
Planetcentric Post-post-conventional Nature viewed in light of all beings.	Experience of the clarity and simplicity of NATURE and of the complexity of global systems. Quarks discovered.	Experience of the luminosity of the entire planet (all beings as love-light) and the power-force of nature.	Experience of the infinite depth of NATURE. Nature as shimmering image of NATURE but empty.	Experience of planet Flow.
Worldcentric Postconventional Nature viewed in light of the Global Commons.	Experience of nature as valuable on its own terms. Rights extended to natural phenomena. Atoms discovered.	Experience of the luminosity of particular areas/regions of Earth and power spots of nature.	Nature as Illusory: Gaia equals samsara. Society as Illusory: Gaia equals nirvana.	Experience of world Flow.
Ethnocentric Conventional Nature viewed in light of one's group.	Experience of superstitions, omens, evil eye, etc. The use of rituals (e.g., blood sacrifice, fertility rites) to connect with nature.	Experience of Divine order as expressed through the blessings of the Earth Mother and nature spirits.	Experience of demonic forces directed at one's group, community, tribe.	Experience of group Flow.
Egocentric Preconventional Nature viewed in light of one's self.	Experience of self is embedded with and nondifferentiated from nature. The use of magic (e.g., spells, charms, wishes) to connect with nature.	Experience of nature as bountiful blessings for oneself.	Experience of demonic forces directed at oneself.	Experience of self Flow.

Figure 9.3. Varieties of nature mysticism.

Reviewing these common structures in this context highlights how feelings of connection to and a sense of unity with nature can show up at both the prepersonal and personal stages of development. Just because an individual has strong feelings of connection to the environment or certain aspects of it does not necessarily mean they are having a state experience of unity with the natural world or that they are stabilized at the indigo altitude. In other words, when discussing nature mysticism we must distinguish between feeling close to nature based on one's current level of development, on the one hand, and due to a transitory state experience, on the other.

As a way to compare and contrast each of these levels and their respective states, we provide example statements regarding whale protection from different levels of development (egocentric through Kosmocentric, and their representative eco-selves).[15]

Egocentric (Eco-Guardian or Eco-Warrior)

"I love whales and don't want them to all die." The egocentric level involves projecting one's own needs onto Gaia and defending them as if they were Gaia's own desires. Here Gaia is an extension of one's self. Gaia is basically valuable only insofar as she provides what you want (e.g., provides good crops, favorable weather, food and shelter, challenging bike trails).[16]

An individual at this level who has a gross union with nature might, for example, interpret a close visual encounter with an orca whale while fishing as a personal sign that they should take some particular action they have been deliberating on. Also, they might relate to the orca as their totem (or power animal) and wear a whale necklace as a way of feeling connected to this prestigious animal. A subtle union—where the individual not only sees the orca up close but has an intense feeling of cosmic love wash over them—could lead the person to interpret that feeling as a powerful personal blessing for the remainder of the fishing season. In contrast, a causal union, where the encounter of the whale is accompanied by a sense of vast space, would likely result in the person taking the visit by the whale as personally threatening and terrifying. Lastly, a nondual union during this encounter would provide a wonderful sense of how life is perfect and would be an affirmation of their personal flow with all things.

Ethnocentric (Eco-Manager)

"Whales are important to our community's livelihood and must be protected." The ethnocentric level also involves projecting one's group or community values onto Gaia, while often disguising the projection as "nature knows best." Experienced primarily as an extension of one's community, Gaia here is really only valued insofar as she contributes to and confirms one's own value system (e.g., "nature is symbiotic," or "No! Nature is competitive.").[17]

A gross realm union at this level would likely take the same orca encounter and view it as either a good or bad omen for the entire community ("This orca is a sign that we will not have good fishing this season, and many families will suffer as a result"). In response, the person might be inclined to perform some simple ritual such as avoiding fishing near the area where they saw the whale or praying to God daily that their community be spared a bad fishing harvest. With a subtle union, the individual would most surely have a positive experience and could relate to the sighting as a blessing for the community by one of God's beautiful creatures ("I saw an orca today when I was fishing for salmon and I had this overwhelming feeling that God is going to watch over this community and make sure we don't have another difficult fishing season"). However, with the experience of causal union (i.e., an accompanying experience of supreme vastness or emptiness) at this level, one is likely to draw the opposite conclusion. The reason for this, as with the previous level, is that such an experience of the Void at this level of psychological maturity is so disorienting to the self-sense that it is typically interpreted in negative terms. In contrast, an experience of the light and love that is usually part of a subtle union is almost always related to in positive terms. And a nondual union would bestow a sense of community flow for the person. ("When I saw the whale that close it struck me like lightning that even though we have a lot of problems in this town, everything is simply perfect.")

Worldcentric (Eco-Strategist or Eco-Radical)

"Whales are valued by people for many reasons: aesthetic, emotional, spiritual, economic, and evolutionary. So we should work together to save them." This level involves a recognition that an individual's values of Gaia are only one of many ways of relating to her, and that Gaia's complexity can support multiple, even contradictory understandings of the Earth. Here, individuals also recognize that to protect Gaia is to protect the multiple perceptions of her in their healthy forms (e.g., the 8 worldviews of the eco-selves). In addition to an emerging understanding that Gaia is not a pregiven ontological reality, one begins to sense an increasing spaciousness around one's own understanding of the Earth.

An individual at this level would be inclined to interpret a gross union in terms of the whale itself without references to self or community. For example, the person would see the whale as a conscious individual with its own rights to hunt salmon, even if that means they or their community's fishing season is impacted. With a subtle realm union, such a close encounter could result in an experience of the whole planet being filled with love. ("As I watched this orca surface several times near my boat I began crying for no reason, just because she was so beautiful, and as I was crying—for about two minutes—I became aware of how much love there is on this planet!")

Causal union at this level of psychological development marks the boundary between those at the lower levels, who tend to interpret it in negative terms, and those at the higher levels (i.e., planetcentric and Kosmoscentric), who tend to interpret it in positive terms, as we will see. Consequently, an individual at this level can have either a wonderful or an awful causal nature mysticism experience. The main reason for this is that the causal experience—through its infinite emptiness—essentially undermines the taken-for-granted ontology of everyday reality. For some people at this level, such deconstruction is a good thing (e.g., society is illusory and therefore nature is paradise) and for others it is a bad thing (e.g., nature is illusory and therefore nature is hell). So an individual at this level who had a causal union experience might say: "Watching that killer whale was really intense—this vast spaciousness opened up, I felt free, and all my worldly concerns evaporated. It even took me a moment to land back in my body. That whale took my breath away!" Alternatively, they might say: "As I sat on the boat, this huge killer whale came up behind us—really close. I was so surprised by this. I've never seen a whale that close. Then I went totally blank. It was as if I was gone, the whale was gone, the boat was gone. But I could still see the whale and the boat. But still, everything was gone? I couldn't help but feel, once it was over, that nothing exists—NOTHING! It freaked me out." In contrast to the dramatic effect of the causal union, nondual union is interpreted at this level as a spaciousness that infuses the entire world: "I've seen many whales, even around here, but something very interesting happened last week when Roy and I came upon a whale in the bay. At first I was just excited to see him so close, but as I watched him swim, dive, and breech, I was stuck by his beauty of movement and for the next ten minutes everywhere I looked I felt a sense of everything flowing around me: the other boats, the pelicans, the clouds, the people on the beach. It was so cool."

Planetcentric (Eco-Holist and Eco-Integralist)

"Whales have rich lives and have a right to carry on, though we must also consider how to honor that in relationship to human needs." At the planet-centric level there is a recognition that in addition to the many worldviews of humans, each being on the planet actually discloses the natural world in a different way. Therefore, Gaia is understood to be the ever-evolving result of many perspectives. An experience of gross union with NATURE at this stage tends to emphasize two distinct qualities of nature: its simplicity and its complexity. One reason for this is that this stage is more capable of holding paradox. Thus, when encountering a whale so close, an individual having this kind of mystical experience might say: "I couldn't help but be aware of how simple that whale was gliding through the water, AND at the same time I kept having these images of how much complexity there is in their ability to swim like that. Hell, our aircraft can't even maneuver like that."

Whereas the gross realm union usually focuses on physical and observable aspects of the encounter, the subtle realm union, as we have seen, introduces the experience of luminosity. At this level, the luminosity of NATURE often extends to the entire planet (Nature). The power-force of nature—the creative potential of the physical-biological dimension—is often very prominent in experience at this level: "For at least five minutes after we snuck up on that whale, I couldn't help but have this powerful sense of the force of evolution. And what is strange is it had this quality of love."

Causal union at this level provides an individual with an experience of the infinite depth of NATURE. Nature is perceived to be an expression of the Divine/NATURE. Often nature experienced here has a shimmering quality to it. "I often feel close to the Source of All Things when I watch wildlife in their native habitat—but when we came that close to the whale this morning, I had this utterly profound sense of Presence and Pureness." A nondual union at this level simply extends one's experience of spacious flow to an even wider context: the planet.

Kosmoscentric (Eco-Sage)

"Whales are an amazing expression of Eros. I hope they are around for a long time." With this level comes the direct perception of every organism and nature in general being a manifestation of the Divine. Not only is there a profound compassion that arises out of this awareness but there is also a paradoxical letting go, such that if whales were to become extinct, this otherwise tragic loss would be understood as yet another expression of Godhead (i.e., NATURE). This stance is anything but a passive acceptance of biodiversity loss. Rather, it actually lends itself to a more fierce commitment to preserve "God's creatures." The difference is that one is not attached in a limiting way to any particular outcome. If whales were to become extinct, an individual at this level would simultaneous grieve deeply for our collective loss and celebrate the mysterious unfolding of the natural world.

The four unions at this level represent what is most often associated with nature mysticism. This is the case because this level represents the transpersonal altitudes (indigo and higher). Gross realm union at this level involves features such as the experience of merging with natural phenomena, including plants and animals, as well as moments of telepathy between oneself and animals: "When the whale and I made eye contact, I felt as though I had been sucked into her, as if I were now the whale. It was so bizarre! I swear I could read her thoughts for a few moments—but now I couldn't tell you what they were."

With a subtle realm union at this level, the sense of luminosity that was described earlier often takes the form of archetypal beings (e.g., some experiences of the Earth Goddess or Pan). It can also manifest as the perception of etheric patterns and dynamics associated with various aspects of nature. Thus, an individual encountering a whale at the indigo level with this subtle state could report, "As the whale rolled to her side and held her flipper out of the water, I had the strongest sense that she was the Earth Goddess manifest. There was something about her size, her power, her gentleness, that triggered this hypnotic state of witnessing Mother Earth herself."

A causal realm union at this level has the telltale sign of experiencing nature (and Nature) as the Great Perfection. Here, nature is totally perfect and includes such things as H-bombs, toxic pollution, species loss, and sprawling cities. There is an experience of the Creative Void within which the natural world arises moment by moment. This union gives rise to an intense love and compassion for the planet and all its aspects. Simultaneously, one becomes wholly unattached to the fate of any natural (and social) phenomena. There is a recognition of the Kosmic dance: many planets are born and many die. Here is revealed the sacred union of divine compassion with divine wisdom. "I could have watched that whale all day! I have never felt so at peace. On the one hand, as I sat there, I was so painfully aware of how we are poisoning our oceans, overfishing them, and in the process killing beautiful creatures like this. On the other hand, I knew that somehow everything was all right, even if all the oceans were to become just dead bodies of water in another fifty years."

Nondual union at this level is the crème de la crème! The highest level (in this schema) with the most profound state. Here is the fullest marriage of Heaven and Earth, God and Goddess, Sun and Moon. And like the Tao that has no name, there is only Flow.

Do We Have to Regress to Childhood to Be "One with Nature"?

One of the benefits of the above illustrated lattice of states and stages is that it demonstrates the perspectival aspect of nature mysticism and clarifies the types of nature mysticism that children and young adults have access to. Children's experience of nature has produced a large body of literature and research and remains one of the more active areas of environmental scholarship. Of course, there are a variety of conflicting opinions, most of which may be divided into the Romantic camp and Developmental camp. We honor the insights of both positions and are committed to avoiding the limitations of each. We affirm the unique relationship children have with the natural world, while exploring the ways in which their state experiences are mediated by their psychological structures (levels, lines, and types). In this way we can avoid the pre/post fallacy in the context of children and nature mysticism.

Early childhood encounters with nature and nonhuman beings can play an important role in individual development in many different lines, including moral, cognitive, kinesthetic, and aesthetic.[18] Such encounters do not require growing up in the country or in a near-wilderness area.[19] Playing in an urban park that has vegetation and small animals, exploring the flora and fauna near a railroad track, or raising and caring for pets provides opportunities for young people to experience the beauty, complexity, and depth-dimension of nonhuman life forms. Children find many different ways in which to explore their environment: built, natural, and everything in between.[20]

Following a tradition that goes back at least to Jean-Jacques Rousseau, contemporary ecopsychologists argue that modern individuals are often developmentally warped and stunted, partly because urban-industrial civilization deprives them of appropriate relationships with nature as children. There is considerable debate, however, about whether childhood experience with nature is a precondition for healthy human development in general, and for developing an ecological sensibility in particular.[21] Even defining the debate is difficult because of changing conceptions of childhood, psychological health, and the relation between supposedly innate and culturally induced attitudes. As Theodore Roszak maintains, ruptures in the humanity-nature relationship undermine the well-being of person and planet. Ecopsychologists argue that because humans lack satisfactory corporeal, sensory, and emotional connections with nature, they seek satisfaction where it cannot be found, in mindless consumerism, otherworldly yearnings, lust for power over others, or rampant exploitation of nature. Roszak writes: "The core of the mind is the ecological unconscious. For ecopsychology, repression of the ecological unconscious is the deepest root of collusive madness in industrial society. Open access to the ecological unconscious is the path to sanity."[22] Roszak adds that "the crucial stage of development is the life of the child. The ecological unconscious is regenerated, as if it were a gift, in the newborn's enchanted sense of the world. Ecopsychology seeks to recover the child's innately animistic quality of experience in functionally 'sane' adults."[23] Yet developmentalists have argued for years that a child's level of interior development is actually quite undifferentiated and self-absorbed, and they have a great deal of evidence to support this claim. Even Kahn's evidence supports the fact that children are largely egocentric. Clearly a return to egocentrism and self-absorption would do nothing for our environment and less for our own personal development.

Do We Have to Go Back in Time to Become "One with Nature"?

Roszak, like many critics of modernity, insists that the European Enlightenment led not to greater rationality but instead to irrationality, because moderns—and modern males in particular—have dissociated themselves from their organic, emotional, feminine, and planetary roots. Although this claim has some truth, it is overstated in that it totally condemns modernity, or offers a problematic conception of human consciousness, or both, as in Roszak's case.[24] Another leading ecopsychologist, Paul Shepard, argues that two factors turned Western civilization into a breeding ground for pathologically immature human beings: first, the onset of agriculture about 10,000 years ago, and second, Judeo-Christian antipathy toward the material world. As we have mentioned, Shepard is one of the few ecopsychologists and deep ecologists to employ an explicit developmental model: Erik Erikson's model of psychosocial development. While we commend Shepard for recognizing the importance of such development, we disagree with his interpretation of it and his polarized writing style, which often relies on splitting nuanced territory into simplistic binaries.

According to Shepard, in traditional tribal cultures the primary fantasies, impulses, fears, and anxieties that children and adolescents experience are resolved and integrated by way of frequent childhood contact with complex, mysterious, beautiful, and sometimes terrifying natural phenomena, and by means of initiation rites carried out by wise and caring elders who are themselves mature people. Healthy human ontogenesis, according to Shepard, the kind of development that leads to mature human beings, requires good primary mothering, and good secondary mothering in the form of immersion in nature. Internalizing the landscape's flora and fauna, byways and paths, provides an analogy for harmonious integration of powerful emotions and fantasies.

Like many other critics of modernity, including deep ecologists such as George Sessions and ecofeminists such as Jim Cheney, Shepard claims that the rise of agriculture undermined authentic human relationships with nature, by extirpating the habitat along with the hunter-gatherer tribes dependent on it. Shepard claims that Americans have played out this "wrenching alienation" by destroying the rich continent to which they migrated from Europe.

Possessed of "the world's flimsiest identity structure," today's adults are childish "by Paleolithic standards."[25] Arrested development leads to "massive therapy, escapism, intoxicants, narcotics, fits of destruction and rage, enormous grief, subordination to hierarchies that exhibit this callow ineptitude at every level, and, perhaps worst of all, a readiness to strike back at a natural world that we dimly perceive as having failed us."[26] Having amputated and botched child development for thousands of years, Western civilization amounts to "a tapestry of chronic madness," where "history, masquerading as myth, authorizes men of action to alter the world to match their regressive moods of omnipotence and insecurity."[27] So thoroughgoing is this dementia, we are told, that even attempts to break out of it are merely madness in a different guise. Consider the environmentalist who wants to protect owls and their habitat from oil exploration and drilling. According to Shepard:

> In our society those who would choose the owl are not more mature. Growing out of Erik Erikson's concept of trust versus nontrust as an early epigenetic concern and William and Claire Russell's observation that the child perceives poor nurturing as hostility—a perception that is either denied and repressed (as among idealists) or transferred in its source so as to be seen as coming from the natural world instead of from the parents (as among cynics)—there arises an opposition that is itself an extension of infantile duality. Fear and hatred of the organic on one hand, the desire to merge with it on the other; the impulse to control and subordinate on one hand, to worship the nonhuman on the other; overdifferentiation on one hand, fears of separation on the other—all are two sides of a coin. In the shape given to a civilization by totemically inspired, technologically sophisticated, small-group, epigenetically fulfilled adults, the necessity to choose would never arise.[28]

Having concluded that several thousands years of human history, particularly Western history, have been an enormous error, the consequences of which afflict us so completely that efforts to change are merely further symptoms of universal pathology, Shepard then surprisingly concludes that all is not lost. Beneath the veneer of deranged modernity we still contain within us the "genuine" and "authentic" capacity for mature development. We need only to change child-rearing practices everywhere in ways consistent with our evolutionary heritage, as exemplified in Paleolithic tribes. Shepard does not explain how such changes could possibly be initiated in the face of such universal pathology. Even stranger is his contention that the trappings of modern civilization, including advanced technology and social organization, could somehow be retained, even while children are raised in conditions that seem so antithetical to modernity. (Elsewhere, Shepard called for the U.S.A. to shift to high-tech food production, which would completely eliminate the need for land-based agriculture.[29] In this way, the U.S. population could move to the coastlines, while farmland would revert to wild habitat of the kind needed for thousands of groups of young people to bond with and hunt resurgent wildlife.)

Shepard does not describe how he himself was able to recover from culturally induced madness to diagnose this pathology with such assurance.

Still, Shepard has certainly put his finger on something important, namely, that contemporary individual pathologies are often profoundly related to encompassing cultural ills. Moreover, there is much to be said for his notion that the urge to merge with nature, on the one hand, and the desire to flee from nature, on the other, are two sides of same developmental coin. Finally, Shepard rightly emphasizes the importance of UL and LL factors in influencing how people have defined, used, and related with the nonhuman world for millennia.

In *The Dream of the Earth*, Thomas Berry diagnoses the history of civilization in terms similar to those laid out by Shepard. Humankind has fallen from a condition in which tribal peoples lived in a natural "wonderworld," into a condition, which he calls "wasteworld," in which alienated people are destroying their environment and themselves. Understood in these terms, all subsequent human history is not development but decline and even degeneration. Berry's and Shepard's vision accords in important—not to say disturbing—ways with the "decline and degeneration" discourse that was so widespread toward the end of the 19th century and well into the 20th century. Perhaps as an indication of interior personal and collective development, neither author couches his ideas in terms of the racist vocabulary so intrinsic to the late 19th-century European books, journals, and social movements, which called upon northern Europeans to rescue themselves from the crippling, emasculating, and enervating effects of urban-industrial life, so that the white race could compete successfully against the vigorous, if spiritually inferior, "colored" races of the world.

Nevertheless, the contemptuous attitude toward modernity, the depiction of civilization as a decline and fall from pristine and noble origins, and the failure to consider the dark side of tribal and premodern sociocultural conditions are all present to some extent in the work of many authors including Shepard and Berry. The rise of modernity is interpreted as the final stage in the process by which an increasingly deranged humankind desperately but futilely attempts to avoid dealing forthrightly with personal and cultural problems generated by declining from authentic origins.

Our Fall from Ecological Grace? Did We Ever Stop Being "One with Nature"?

Strangely, most of the Western ecological crisis narratives recapitulate the Christian mythos of the Fall, the Apocalypse, and the Redemption.[30] It is ironic that environmentalists use this Western, Christian narrative, even while abhorring amber-traditionalists. These narratives usually identify a historical moment when humans were expelled from the ecological garden of a pristine Earth; we are told about the forthcoming Apocalypse that is in large part due to our ecological sins, and are asked to repent and mend our ways before it is too late; further, it is assumed by many that only a full-blown crisis will arouse us to get our ecological act together and redeem ourselves. Nevermind that social science and history repeatedly have demonstrated that crisis makes people more conservative and change resistant.[31]

There are many such accounts of the decline of the West. Almost every major historical moment has been blamed for the current ecological crisis. Just for a sample, the following moments in history have been identified by environmentalists as the point of no return or the primary contributor to our current crisis:[32]

- Emergence of the human species (William Allman and Edward O. Wilson)
- The first use of tools and the emergence of language (Murray Bookchin)
- Early agriculture: the domestication of plants and animals (Paul Shepard)
- The shift from horticulture to plow-based agriculture (many eco-feminists)
- Large-scale agriculture (Daniel Quinn)
- Greek rationalism and the alphabet (David Abram and Leonard Shlain)
- The emergence of the Judeo-Christian tradition (Lynn White, Jr.)
- Dualistic language (Robert Greenway)
- Literacy (Michael Cohen)
- The scientific revolution (Carolyn Merchant)
- The rise of rationalism in Europe in the 18th century (Deborah Winter)
- The industrial revolution (David Suzuki)
- Industrial mind (David Orr)
- Cartesian-Newtonian paradigm (Fritjof Capra)
- Modern science (Ralph Metzner)
- Disenchantment of the cosmos resulting from science (Morris Berman)
- Modern technology (Jerry Mander)
- Modern modes of perception (Laura Sewall)
- Mismanagement of the global commons (Garett Hardin)
- Growth economies (Donnella Meadows)

- Capitalism (Tibor Scitovsky)
- Consumerism (Alan Durning and Duane Elgin)
- Population explosion (Paul Ehrlich)
- Globalization (Vandana Shiva)
- Transnational corporations (Helen Caldicott)

Some environmentalists do not point to a specific historical moment but rather cite a more general etiology:

- Consciousness (Peter Russell)
- Discontinuity between humans and the nonhuman (Thomas Berry)
- Spirituality (Al Gore)
- A lack of character (Wendell Berry)
- The repression of the ecological unconscious (Theodore Roszak)
- Personal dysfunction resulting from civilization (Chellis Glendinning)
- A limited concept of self (James Hillman)

Given this variety, it seems that individuals pick their favorite moment in history to blame, based on their own particular values and perception. It is interesting to note that many of the different root causes are more strongly associated with each of the four terrains than others. For example: consciousness (UL), consumerism (UR), capitalism (LR), and the Cartesian-Newtonian worldview (LL). When we plot the six most commonly cited reasons for our Fall (figure 9.4), an interesting picture emerges: either the

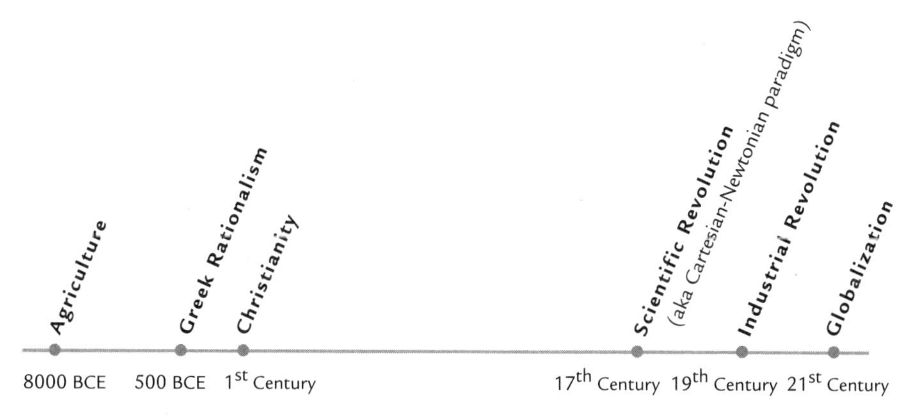

Figure 9.4. Commonly cited historical causes of the eco-crisis.

Fall occurred a long time ago, between 10,000 B.C.E. and the 1st century, or it occurred relatively recently, between the 17th and 21st centuries. By far the most common of all six is the scientific revolution, which involved such luminaries as Rene Descartes, Francis Bacon, Galileo Galilei, Nicolaus Copernicus, and of course Isaac Newton.[33]

In general these views ascribe our separation from nature as either inherent in being human (sociobiology), the result of having lost something we once had (romanticism), or the result of healthy ego development, which leads to a more mature unification (developmentalism).[34] Clearly some of the causes listed are more responsible than others. However, they all provide insight into why the relationship between humans and nature is strained. While all of the root cases listed are worth discussing, we will explore only a few of the more popular ones.

David Abram argues that humans were separated from nature largely due to the invention of the Greek alphabet, when people moved from picture language to abstract words that bear no resemblance to what they represent.[35] Likewise, Heidegger maintains that since ancient Greece, Western history has been sliding toward techno-industrial nihilism. This degeneration was abetted by wedding Christian theology to Western metaphysics, which was itself degenerating in part because crucial Greek terms were translated into Latin.[36] Max Horkheimer and Theodor Adorno point to instrumental rationalism as the culprit of civilizations' self-destructive tendencies. Some ecofeminists have agreed with Shepard that agriculture, not the vicissitudes of Heidegger's "history of being" nor Horkheimer and Adorno's "dialectic of enlightenment," is responsible for the rise of patriarchal, nature-despising civilization. These ecofeminists assert that we need to recover not the hunter-gatherers celebrated by certain eco-masculinists, but rather the nonhierarchical horticultural societies that ostensibly lived harmoniously with nature.[37]

Eco-Romantic moderns often view ancient tribal cultures and other premodern cultures through rose-tinted glasses. Those glasses reveal positive aspects of the culture—including close intersubjective relationships, common cultural purpose, inspiring rituals, an enchanted cosmos, and easy access to wild or semi-wild nature—but conceal the negative—including slavery, second-class status for women (in many cases), destructive superstitions, ethnocentrism, shorter life spans, tribal warfare, and human sacrifice, to name a few. Premodern cultures, including tribal ones, are not invariably successful, either intersubjectively or in their capacity to deal with the natural environment.[38] There is a strong possibility that Pleistocene tribes hunted several species of North American megafauna to extinction. If true, this reality should shake the foundations of theorists who believe tribal cultures are automatically or prone to be in harmony with their habitats.[39] Ecological "respect," such as that demonstrated by some indigenous peoples, does not automatically provide ecological knowledge and vice versa.[40] While prehistoric tribal members might have been "one with" their local environs, we should be careful not to confuse this with complex systems cognition or postconventional compassion toward all sentient beings (including all humans). Postmodern ecotheorists often compare the dignity of the premodern world with the disaster of the modern world, and then call for some form of revival of the premodern. Integral Ecology avoids this kind of splitting into "all good" and "all bad," and recognizes the dignity and disaster of both premodern and modern times (see figure 9.5).[41]

Integral Ecology compares the good and bad news of many different time periods. Clearly, modernity had spoiled the natural environment in certain ways—out of the same ignorance of earlier times. The big difference was one of means. Indigenous tribes could inflict only so much damage with slash-and-burn tactics, whereas modern culture could cause much damage with industrialization. Clearly, premodernity also had some problems, including environmental ones. But regressing to a childish state, or regressing in time, is both impossible and undesirable. Our solutions lie in honoring and integrating the dignities of the premodern, modern, and postmodern, not in demonizing any of them.

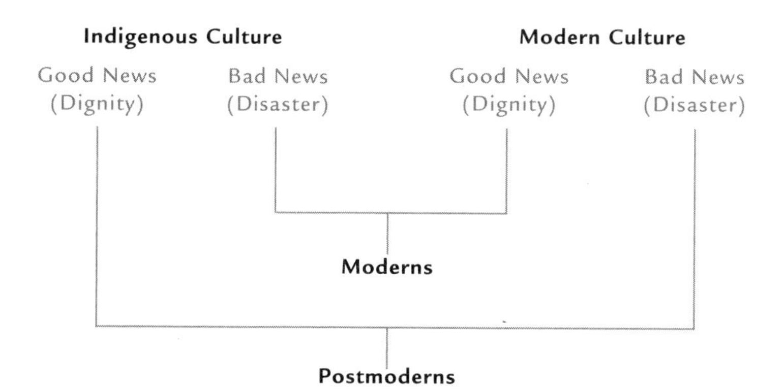

Figure 9.5. Cultural splitting of modernity and postmodernity.

An Integral Approach to Indigenous Peoples

In light of the tendency of environmentalists to mismatch the good news and bad news of modernity and premodern peoples, Integral Ecology offers a number of considerations for avoiding romanticizing indigenous and premodern peoples.

1. Avoid committing the dignity/disaster fallacy of comparing the dignity of one culture with the disaster of another. It is important to always consider the dignity and disaster of each culture when comparing two or more cultures.

2. Avoid using the general labels "indigenous" and "Western," as these umbrella terms prevent a more nuanced conversation. Instead, refer to actual indigenous tribes or specific Western cultures. There are as many differences between Apache, Navajo, Samburu, Masai, Murri, Wangkai, and Maori tribes as there are between French, Australian, English, Canadian, Japanese, and United States cultures.

3. Be able to name at least five different tribes and describe important aspects of their culture. For example, their cosmology, main hunting/agricultural practices, gender relations, and war dynamics. All too often, those who talk a lot about "indigenous peoples" lack specific ethnomethodological knowledge of actual tribes.

4. Clarify whether you are talking about premodern or contemporary indigenous peoples. While there are important similarities between them, there are arguably more differences. All too often, environmentalists have a static view of indigenous peoples.

5. Distinguish between the average mode of consciousness and the highest mode within a tribe in both structure-stages and state-stages. It is important to recognize that while many premodern tribes had shamans, it is unlikely that many of these shamans were at postrational waves of development, which means that they were appropriately ego- and ethnocentric. Even contemporary shamans are rarely postrational (this applies equally to spiritual leaders in any tradition or linage). To the extent that there were or are postrational shamans, it still is inappropriate to assume that the majority of their tribes were similarly developed. In other words, it is problematic to

make general statements about a culture or group of people based on the few highly developed individuals from that culture. Imagine if we romanticized the United States as a great nation filled with loving people because of Martin Luther King, Jr., Walt Whitman, or Jane Goodall.

6. Be aware of the occurrence of cultural appropriation. For example, the oft-quoted famous speech made by Chief Seattle ("All things are connected . . .") was exposed in 1987 as coming from a 1970s movie script by Ted Perry.[42] Nevertheless, many environmentalists continue to use this quote approvingly, either as though it were Chief Seattle's own words or as a symbolic expression of environmental awareness. In either case, such usage propagates cultural appropriation of indigenous cultures.[43]

7. Be aware of how indigenous people have internalized romantic images of themselves for cultural status. This is a well-documented unconscious phenomenon with marginalized people. The most obvious example is how Native Americans have played into the image of having ecological wisdom because the counterculture movement has elevated them from the cultural sidelines to center stage.

8. Become familiar with historical evidence in the fields of historical ecology and environmental anthropology that challenges romantic views of indigenous peoples. Over the last decade an immense amount of scholarship and research has accumulated that provides a much-needed correction to outdated Eden-like views of indigenous cultures. In fact, some indigenous activists have pointed out that environmentalists are engaged in a form of racism when they paint a picture of the nature-loving "Indian."

9. Recognize that ecological consciousness as it is understood today did not exist even 100 years ago, let alone 10,000 years ago. This is not to say that premodern peoples did not care for their local environment. Rather it is a reminder that we have to be careful not to project our current ways of understanding onto yesteryear. There is a big difference between being one with one's local bioregion and being one with the entire planet and all humans.

10. Different cultures can be more developed in various social lines, such as kinesthetic and polyphasic.[44] The capacity for this kind of differentiation is only achieved by individuals with postconventional or vision-logic

interior capacities. This is the case simply because only at the teal altitude can an individual hold enough complexity in their mind's eye to compare and contrast the good and bad parts of two systems such as social lines. Thus, an individual needs to be able to take a meta-perspective on both cultures: premodern and modern. This is a very important developmental achievement.

11. Keep in mind that a lack of technological means to inflict ecological damage on a large scale is not the same as having a mature environmental ethic. Many tribes have had a negative ecological impact on their surroundings, but since they lived in small areas, for example, they were able to move to new areas and left behind their ecological footprint, which could then recover over the next 50–200 years. In many cases, tribes appear to have a mature environmental ethic due to a nature-loving cosmology, when in fact the average mode of consciousness for the tribe was actually not what we would consider a planetcentric worldview. Thus, an environmental ethic is assumed as a result of a lack of documentation of historical ecological impacts, the lack of means to inflict more ecological damage, and a nature-based cosmology.

12. Similarly, having an impressive knowledge about nature, as many indigenous peoples do, is not the same as having worldcentric respect for nature. It is reasonable to assume that many indigenous peoples' respect for nature was motivated by fear and dependence as much as by spacious love and admiration.

One Crisis, Many Crises, No Crises?

Integral Ecology avoids pointing to a single cause as the proverbial wrong turn in humanity's unfolding. In fact, Integral Ecology agrees that there are many different root causes for our current situation. After all, reality is an "all quadrant, all level" affair. There is also no decisive historical moment at which anthropogenic ecological problems became critical. In fact, as the historical ecologist Charles Redman points out, the ecological crisis has been around for a long time:

> From the earliest years of their profession, archaeologists have been concerned with how humans were able to adapt successfully to their surroundings, but it is only in recent years that they have begun to focus on

> how these adaptations may have degraded the environment. The results of these studies, taken from many parts of the world, have provided us with surprisingly clear examples of serious human impacts on the environment reaching back thousands of years. A careful review of this newly available literature finds that the environmental crisis is not a new problem, but its basic human-environmental relationships have been with us for millennia. It is only the technology with which we operate, the size of the population, and the extent of the impact that have changed since prehistory.[45]

Redman's perspective highlights that today's environmental crises do not have a specific origin. Rather, such problems have arisen inevitably for millennia as humans make a living from local resources. What has changed is the scope, degree, and severity of the crisis. Yet Redman's analysis does not assert that environmental crises have only been gaining momentum from the earliest times. Historical ecology and environmental anthropology demonstrate that there are cycles of environmental crisis and balance throughout history in many different cultures: indigenous, modern, Western, Eastern, North, and South. Not only have there been multiple environmental crises throughout history, but the current crises are so complex and multifaceted that we cannot conceive of them in the singular. Frederick Buell writes:

> Though much has been (and is daily being) written about what constitutes environmental crisis, most of it is devoted to urgent present assessments and warnings. Little has been written that surveys how and why these assessments and warnings have changed over time. The result is that people tend to speak of the environmental crisis—as if "it" were a clear, stable, and ahistorical concept. To do so, however, is unfortunate, because it suppresses the complexity, diversity, and dynamism of accumulating environmental problems.[46]

We feel that the AQAL framework provides a powerful analysis of the many currents involved in creating, perpetuating, and responding to our historical and contemporaneous environmental crises. Integral Ecology highlights that any anthropogenic ecological crisis is the result of a complex tetra-mesh of the 4 terrains and their various levels of complexities. Thus, any crisis is the result of a complex and unique mixture of fractured consciousness, unsustainable behaviors, dysfunctional cultures, and broken systems. To identify only one or a couple of these contributing factors and hold them up as the main culprit will not help anyone to effectively address these crises. Figure 9.6 provides a representative illustration of contributors to any ecological crisis, based on the 4 terrains.

Thus, an ecological crisis is a 4-quadrant crisis. It is also an all-level crisis. Each level of depth or complexity within the quadrants can contribute to a crisis and be a leverage point to help solve a crisis. While it is crucial to consider the levels in all quadrants, here we focus on the 8 eco-selves and their conception of the crises. Because each eco-self has a different relationship to and perspective of the natural world, they each perceive a different crisis. What follows is a brief but illustrative overview of how the eco-crisis is viewed from each eco-self: Eco-Guardians experience a crisis of balance between good and bad forces; Eco-Warriors encounter a crisis of power and fear of losing power over their domain; Eco-Managers perceive a crisis of management—they wish they could manage nature better; Eco-Strategists perceive a crisis of resources—they worry about what will happen to the economy if we run out of oil; Eco-Radicals are alarmed by the loss of bio-

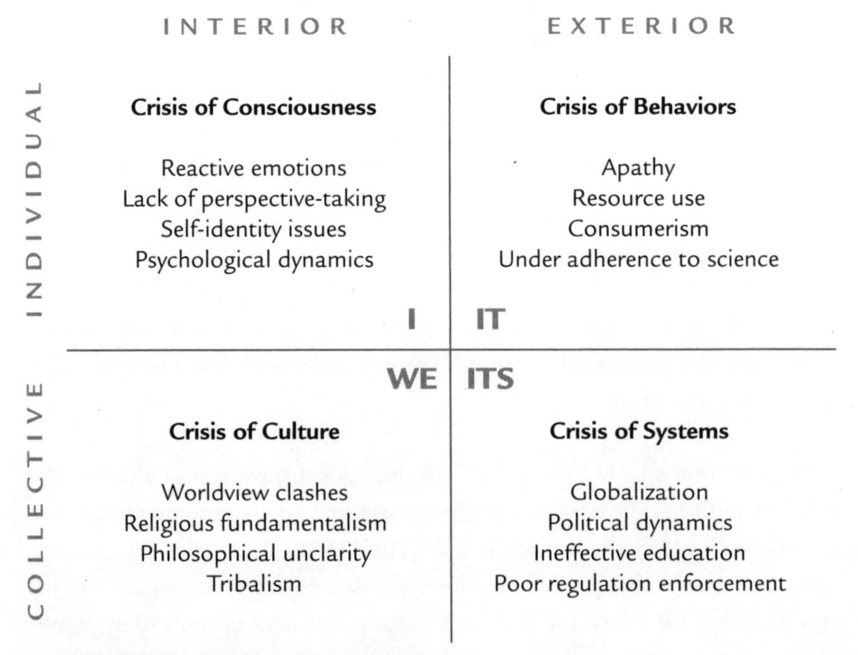

Figure 9.6. The 4 terrains of an eco-crisis.

diversity—species loss is a crime against nature; Eco-Holists perceive a crisis of global systems—local regions and global networks are being impacted by multiple factors; Eco-Integralists recognize the truth of all these perspectives because they see a crisis of perspectives; and Eco-Sages, although deeply committed to preserving the greatest depth and the greatest span, recognize that in an absolute sense there is no crisis—things are perfect as they are; planets will die, planets will be born. To further illustrate these various relationships to the crisis, we provide below a representative quotation about the current environmental crisis from the perspective of each eco-self (these are summarized in figure 9.7). Keep in mind that our placement of quotes does not necessarily reflect the authors' level of development; rather their quote is simply representative of that level.

Eco-Guardian: Crisis of Harmony

> Since the imagination arises from the child's contact with nature, each child is a born ecologist. Thus save the children to save the imagination to save the planet.
>
> —JAMES HILLMAN[47]

Eco-Warrior: Crisis of Power

> Are you tired of namby-pamby environmental groups? Are you tired of overpaid corporate environmentalists who suck up to bureaucrats and industry? Have you become disempowered by the reductionist approach of environmental professionals and scientists? . . . Our front-line, direct action approach to protecting wilderness gets results.
>
> —EARTH FIRST![48]

Eco-Manager: Crisis of Management

> Faced with the widespread destruction of the environment, people everywhere are coming to understand that we cannot continue to use the goods of the Earth as we have in the past. . . . [A] new ecological awareness is beginning to emerge. . . . The ecological crisis is a moral issue.
>
> —POPE JOHN PAUL II[49]

Eco-Strategist: Crisis of Resources

> Besides exhausting the unrenewable and impairing the renewable resources, we have left unused vast resources which are capable of adding enormously to the wealth of the country. . . . It has not occurred to

us that a stream is valuable not merely for one, but for a considerable number of uses; that these uses are not mutually exclusive, and that to obtain the full benefit of what the stream can do for us we should plan to develop all uses together.

—GIFFORD PINCHOT[50]

Eco-Radical: Crisis of Biodiversity

The richness and complexity of the natural world is declining at an ever-accelerating rate, as the earth's burgeoning human population strives for a steadily rising technological standard of living. Natural diversity is being brutally simplified to make way for a dizzying blend of artificial landscapes—villages, housing developments, parking lots, roads, factories, mines, shopping malls, schools, parks, gardens, golf courses, plantations, and croplands. The biggest threats to the diversity of life on the earth are habitat loss, introduction of alien species into communities, and fragmentation of natural areas caused by bulldozing, paving, plowing, draining, dredging, trawling, dynamiting, and damming.

—YVONNE BASKIN[51]

Eco-Holist: Crisis of Global Systems

From a wider perspective, the crisis of the [noos]phere and that of the biosphere are mutually implicative, as are the crises of the past, present, and future.

Many of these crises can themselves be looked on as polycrisical of interwoven and overlapping crises. . . . Thus one is at a loss to single out a number one problem to which all others would be subordinated. There is no single vital problem, but many vital problems, and it is this complex intersolidarity of problems, antagonisms, crises, uncontrolled processes and the general crisis of the planet that constitutes the number one vital problem.

—EDGAR MORIN[52]

Eco-Integralist: Crisis of Perspectives

Gaia's main problems are *not* industrialization, ozone depletion, overpopulation, or resource depletion. Gaia's main problem is *the lack of mutual understanding and mutual agreement in the noosphere* about how to proceed with those problems. We cannot reign in industry if we cannot reach mutual understanding and mutual agreement based on a worldcentric moral perspective concerning the global commons. And we reach that worldcentric moral perspective through a difficult and laborious process of interior growth and transcendence.

—KEN WILBER[53]

Eco-Sage: No Crisis

We are losing a nice local version of reality we've been basking in for several million years, the lovely landscapes, the fauna and flora of the late Cenozoic, the Age of Flowering Plants and of Mammals. These have been sweet indeed, and it is sad to see them go. Difficult goodbyes must be said. But we won't miss them for long—there's plenty more where they came from. The unbridled, fecund wildness that lies at the heart of co-arising emptiness-luminosity will not disappoint us. A really deep

Eco-Sage	No Crisis
Eco-Integralist	Crisis of Perspectives
Eco-Holist	Crisis of Global Systems
Eco-Radical	Crisis of Biodiversity
Eco-Strategist	Crisis of Resources
Eco-Manager	Crisis of Management
Eco-Warrior	Crisis of Power
Eco-Guardian	Crisis of Harmony

Figure 9.7. The various eco-crises of the 8 eco-selves.

ecologist has understanding of this, and faith in it. This fertile, dangerous, healthy and real wildness is where we should be resting our hopes and our hearts and our minds. We have nothing to lose.

—JOHN MCCLELLAN[54]

These sketches of the various perspectives indicate that it is misleading to speak of *the* ecological crisis, when in fact there are multiple understandings of what *the* crisis is. Additionally, the notion of a single crisis reinforces a naive empirical understanding that the crisis is simply "out there," as opposed to recognizing that it is as much "in here," inside of us.[55] To further highlight the importance of recognizing the multiple constructions of the environmental crisis, we turn to a sociocybernetic analysis of the environmental crisis.[56]

A Sociocybernetic View of the Eco-Crisis

Drawing on Niklas Luhmann's theory of social autopoiesis, David Connell discusses the multiple constructions of the environmental crisis within the context of sustainable development.[57] He opens his article by explaining that

> simplicity, linearity, and predictability were once the norm in scientific studies of the environment in the West. Recent developments in complex systems thinking have challenged this approach. Theories of emergence, self-organisation, and autopoiesis, to name a few, take paradox, unpredictability, non-linearity, and complexity as a basis for scientific study. These theoretical developments have contributed to multiple constructions of the environmental crisis.[58]

Connell explains that it is nearly impossible to reconcile these various perspectives of the crisis within "normal science" (i.e., objective, 3rd-person perspective, or a science driven by the orange altitude) because it is a "problem of reference." This level and perspective of science fail to register multiple constructions of the environmental crisis, because reality is understood as objects viewed by subjects. In other words, a science driven by the values of an orange level of development cannot account for the observer. The challenge, according to Connell, is to avoid its opposite—postmodern conceptions of the subject as center of the world in an effort to accommodate multiple constructions and points of view. He then turns to Luhmann's work to explain how and why multiple constructions of the environmental crisis are possible.

Connell explains that the process of societal differentiation has created multiple observing systems (e.g., law, education, religion, economy, science). These functional systems are organized around binary codes, which provide constraints and possibilities to our construction of reality. For example, the legal system can construct the environmental crisis only in terms of legal/illegal, as in the question of who is legally obligated to clean up a toxic spill. Similarly, the economy uses the cost/no cost binary for how much environmental damage costs. Thus, it is the "existence and persistence of organizationally closed binary codes" that allow for multiple constructions.[59]

An ecological crisis is too complex for any single individual to comprehend, so people select the information to include in their construction of that crisis. The construction of a crisis enables a society (or a group of individuals) to cope with the uncertainty that the crisis represents. It allows individuals to communicate about the crisis and to simplify the problem by selecting what is possible. "Constructions, as generalized symbolic media of communication, make it possible to share meaning with different people in different situations, which in turn allows people to come to the same or similar conclusions."[60] Thus, multiple constructions of a crisis serve as connections between the complexity of the situation and the socially regulated dynamics of selecting particular elements about the situation. Multiple constructions occur because there are multiple ways to observe the environmental crisis. We discussed this in the historical shifts of understanding ecology in chapter 5.

Unlike the typical modernist subject-object approach or subject-centered postmodern approach, a sociocybernetic approach to the environmental crisis includes the observer in the realm of science. This movement is known as a second-order observation: "Including the observer marks a difference between first-order observations of constructions that describe the environmental crisis and second-order observation of social constructions recognizing that a describer (i.e., the observing system) is implied in the construction. One is no longer seeking to understand the environmental crisis as an object, but seeking to understand the observing system that constructs the crisis. . . ."[61]

Connell explains that "at a global scale, the environmental crisis can be constructed as a single problem with a single perspective."[62] But he highlights that events like the United Nations World Summit on Sustainable Development make us more aware of the possibility that there is no single construction of "the" environmental crisis, but many crises. After all, he explains, if there were just a single environmental crisis, then society would know how to respond to it. There would be no need for conferences that bring together world leaders in order to work together on *the* environmental crisis. In fact, because there is no single agreed-upon understanding of the world, there are inevitably multiple perspectives on what constitutes a crisis. This is why leaders and heads of state cannot speak for all of society—society consists of multiple systems, and a politician, for example, can only speak for the political system. In other words, in a differentiated society such as ours, there is no single body that represents all the other systems. This is why the definition of the problem, and the distinctions we make about it, determine the ways the problem can be addressed, and why the different perspectives of the crisis can only see their proposed solutions as viable. Connell writes: "In the absence of a single binding representation of society, not only the constructions of the environmental crisis but also the responsibilities for addressing the crisis become fragmented along functional lines. We may have a common crisis, but we also have different ways of interpreting the problem and different ways of assuming responsibility for the problem."[63]

So the question arises: how might one begin to integrate these multiple perspectives? Connell proposes that "to reconcile multiple constructions of the environmental crisis one must determine what distinctions guide the observations of environmental crisis."[64] Admittedly this is a daunting if not impossible task. However, the Integral Model provides a very sophisticated framework that can identify the major distinctions that individuals and schools of thought make. Therein lies the power of Integral Ecology to reconcile the many perspectives on ecology that are often at odds with each other.

In sum, an Integral approach to an ecological crisis recognizes that different perspectives understand the crisis or lack of crisis from their own worldview. It is also important to keep in mind that, in terms of subjectivity, many people make their crisis the planet's crisis; in terms of intersubjectivity, everyone generally assumes they all know what one another means by "ecological crisis"; in terms of objectivity, scientific data can be marshaled to support almost any position for or against a particular view of a crisis. It is for reasons like these that Integral Ecology takes the position that *things are getting worse, things are getting better, things are Always Already perfect.* In other words, there is bad news, good news, and no news.

The Integral Ecology Slogan: Things Are Getting Worse, Better, and Are Perfect

In general it does not take much to convince the average individual that the conditions of the environment are getting worse.[65] The many ways that humans negatively impact the Earth are almost taken for granted these days. It will take more effort if you want to build a case that our ecological systems are actually getting better. In fact, such a position will likely make you an ecological heretic with almost any self-identified environmentalist.[66] Even more difficult will be to point out how the our global commons is getting neither worse nor better but is perfect—things just do not get better than this. Now imagine that all three positions—things are getting worse, are getting better, and are perfect—are all equally true! While this might seem preposterous and is at least paradoxical, this is Integral Ecology's view on environmental crises.

One of the defining characteristics of Integral Ecologists is the capacity to hold multiple and even contradictory perspectives in their hearts and mind simultaneously. Typically, people fall prey to one perspective or another. When individuals can see only that the environment is being destroyed faster than we can save it, they tend to become doomsayers and apocalyptic. When individuals can see only environmental solutions, they are hopelessly idealistic and become blinded by their own optimism.[67] Many of our best environmental advocates and role models recognize the importance of balancing these perspectives on the environment. For example, in his well-received book *Red Sky at Morning: America and the Crisis of the Global Environment*, the renowned environmental leader James Gustave Speth explains: "I have tried to be both realistic and hopeful. It is not always a happy marriage. Some may look at the difficulty of reversing current trends and despair; they are stuck in the abyss. Others may blithely assume that things will work out ('they always do'); they are being wishful. The right answer, I believe, lies at neither extreme."[68]

Similarly, Amory Lovins, co-CEO of the Rocky Mountain Institute, reflects on the relationship between things getting worse and getting better: "I am more hopeful now than I was a few years ago. I think the speed and importance of things getting better outweighs the speed and importance of things getting worse. One of the most hopeful developments is the co-operation between the North and the South in the global civil society. We have much richer expertise available now than we had before."[69]

As these quotes highlight, there is room for the view that our planet needs serious protection, and there is room for the view that we are developing the motivation and the solutions to do so. There is also a profound truth in the claim that the environment is perfect and that nothing needs to be done, but if individuals adhere only to this truth, they become passive and naive at best, and dangerously irresponsible and in denial at worst.

There is an important distinction between the first two positions (worse and better) and the third position (perfect). Both "things are getting worse" and "things are getting better" are relative truths. They are directed at the time-bound manifest plane. In contrast, "things are Always Already perfect" is an ultimate and timeless view. Consequently, its truth is not judged by the pragmatic and scientific standards associated with environmental assessments, professional opinion, NGO statistics, and state-of-the-world reports. The recognition of the planet as the Great Perfection is the result of wisdom and compassion cojoined in embodied awareness. The wisdom provides the recognition that there is nothing to protect, and the compassion provides the commitment that everything must be protected! Clearly the rational mind cannot grasp this conflict of ecological interests: protect nothing, protect everything. That is why grasping and understanding the truth of the planet as totally perfect can only occur in *transrational* modes of development. In his article "Nondual Ecology," which appeared in the Buddhist magazine *Tricycle*, John McClellan captures this provocatively: "There is only One Thing happening, not some things that are good and others that are bad. This includes fragrant ecosystems, fresh and unsullied in wilderness areas on spring mornings, and it includes urban industrial megagrid, ghettos and famine zones, materialist mind greed, the extinction of wild species and the slavery and torture of 'domesticated' ones. Life and death. Even television."[70]

One of the ironic outcomes of authentically recognizing that there are no old-growth forests or Bengal tigers to protect or save is that you are liberated from your own egoic clinging. As a result, you are more capable of skillful action in the world, because as long as your environmental action is driven by such understandable human tendencies as personal preferences, likes and dislikes, fear, projection, anger, and ambition, you will react, not respond. The Pulitzer Prize–winning environmental poet and practicing Buddhist Gary Snyder states this succinctly: "Knowing that nothing need be done, is where we begin to move from."[71] Nonduality frees you from dualistic thinking, from hope and fear. It does not mean things are getting better or that things are getting worse. It means you are free from thinking like that: you are free from the contraction that clings to either side of polarity. Also it allows you to get out of your own way because it loosens up your exclusive identity with any given perspective so you can actually see all the perspectives more clearly. This is why you can see the relative truth of how the planet is in danger and at the same time the planet is recovering—even flourishing!

With these perspectives in mind, Integral Ecology emphasizes that any act to save, protect, or promote nature might ultimately have the opposite effect. Integral Ecologists are all too aware that perspectives are often limited by their context (historically, personally, culturally, developmentally). As a result, we can act with good intentions that ultimately are detrimental to our cause. Thus, we must continually cultivate humble action and say to ourselves, "Based on my understanding in this moment I am going to do ——(fill in your favorite environmental cause or activity), but I realize that I might actually be contributing to the problem in ways I cannot see or understand."

Many ecologists and activists unreflectively think that because they are committed to making a difference, their activity is inevitably beneficial. While immediate benefits might be observable in the short run, over the long run such activity may well prove to be problematic. In fact, it is helpful to assume that whatever you are doing to protect the planet is in some way also harming it. Very few actions (and even nonaction!) are simply good or bad. There are consequences to what we do and don't do. By realizing that we, in spite of our Integral Ecological commitments, are actually harming the planet in various ways that we are not even aware of, we cultivate a much-needed humility that actually serves to build bridges between ourselves and others whom we might otherwise judge as being destroyers of the earth.[72]

The capacity for being able to hold all three positions of the Integral Ecology slogan is a developmental achievement. Interestingly, orange individuals tend to advertise that things are getting better, whereas green individuals are more likely to cry out that things are getting worse. At the teal level of altitude, individuals can see the truth of both perspectives—largely because they can see the context out of which each position is issued. It is only with the emergence of turquoise awareness that individuals can access all three perspectives continuously. Not only do they comprehend the myriad ways the planet is both in peril and flourishing, but also at the very core of their being they simultaneously surrender to all the destruction and beauty arising throughout the Earth and vow to be in nonattached service to all organisms and ecological processes as well as humanity and civilization.

The cultivation of this kind of Integral Ecological awareness is essential. For we will only be effective as change agents and community leaders to the extent that we have developed worldcentric and planetcentric capacities: perspective-taking, critical thinking, emotional discrimination, shadow awareness, interpersonal skills, embodiment, and moral motivation. Consequently, we have devoted the entire next chapter to providing a series of practices that facilitate the kind of ecological awareness this planet so desperately needs and will be fine *without*.

10

Practices for Cultivating Integral Ecological Awareness

The conventional notion of the self with which we have been raised and to which we have been conditioned by mainstream culture is being undermined. What Alan Watts called "the skin-encapsulated ego" and Gregory Bateson referred to as "the epistemological error of Occidental civilization" is being unhinged, peeled off. It is being replaced by wider constructs of identity and self-interest—by what you might call the ecological self or the eco-self, co-extensive with other beings and the life of our planet.

—JOANNA MACY[1]

There are many spiritual paths available in today's postmodern world, all of which teach various practices for cultivating discrimination and compassion. As discussed in chapter 9, one of those paths is nature mysticism. Its roots reach back to the Paleolithic past. Contemporary expressions of nature mysticism are alive in Native American spirituality, Australian Aboriginals, European neo-pagans, and African spirituality. These indigenous-based approaches to Spirit take nature as their path toward higher and wider modes of being and knowing, and prescribe a host of practices that cultivate aspects of ecological awareness.[2]

Additionally, some major religions have created transformative practices that are designed to help individuals realize their connection with nature. For example, Taoist qigong and t'ai chi are often based on animal movements, as are many Hindu yoga poses. Mahayana Buddhists vow to save all sentient beings, seen and unseen—this is a powerful commitment to serve the natural world. Christianity's "Canticle of the Creatures" prayer is another example of devotion to the natural world.[3] In short, there are many spiritual paths that offer practices and perspectives for connecting with nature for the transformation of self, other, and world.[4]

Integral Ecology recognizes that authentic ecological awareness (a worldcentric or planetcentric self-sense) is an amazing developmental achievement that few adults realize, let alone stabilize. As a result, it is not enough to provide people with an integral map or an integral nature mysticism; we must also articulate a comprehensive path and practices of transformation. This path must meet practitioners wherever they are, help them develop healthy attributes at their level of development, and, should they desire it, carry them to the farther reaches of transpersonal development. The practices we describe here are designed to support the development of:

- an Integral Ecological Identity: a sense of self that includes all other humans as well as the natural world;
- Integral Ecological Action: maintaining as small an ecological footprint as possible;
- Integral Communion with All Beings: a capacity to resonate with the diverse perspectives of humans and nonhuman life; and
- membership in an Integral Web of Life: contributing to positive systems change at multiple scales within numerous contexts (e.g., educational, political, economic).

The development of these four capacities represents an actualization of each of the 4 terrains of Integral Ecology (see figure 10.1).[5] The cultivation of Integral ecological awareness is a 4-terrain affair. Not only is there one's self-sense and the various kinds of thoughts and emotions that accompany such a developmental achievement, but there are also particular kinds of behaviors, relationships, and roles within systems that manifest and are an expression of such awareness.

These 4 terrains of ecological awareness serve as a foundation for a unique approach to nature mysticism, which we call Nature as Transformative Path. Informed by the Integral Model and the various spiritual tra-

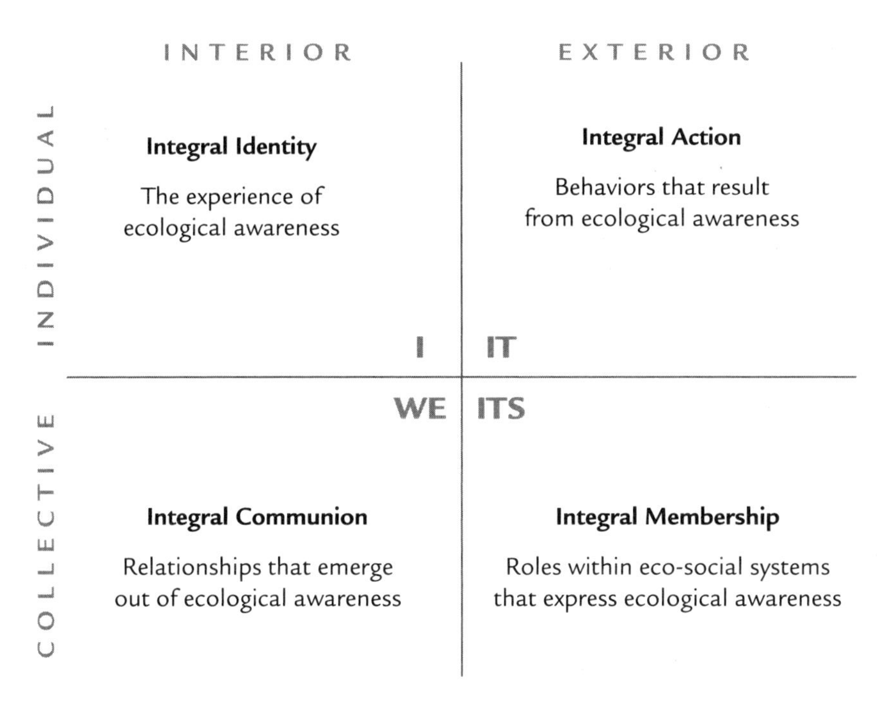

Figure 10.1. The 4 terrains of ecological awareness.

ditions, such as those mentioned above, it is primarily a path of practice, supporting individuals to cultivate and stabilize profound experiences of Integral ecological awareness. These practices are designed to use our interaction with and inhabitation of the natural world as an authentic way of transformation.

Practice, Practice, Practice . . .

There are two main categories of practice: *perspective practices* and *depth practices*. Perspective practices strengthen our capacity for taking multiple perspectives simultaneously. Many of these practices increase our ability to perceive more of each of the 4 terrains or to see the natural world from various worldviews. Depth practices cultivate interior qualities of experience (e.g., embodiment, transcendence, clarity, fullness). Many practices exercise various aspects of these categories simultaneously. These categories are not exclusive; rather they serve as a general framework to identify the strengths and limitations of various practices so as to help us cover all the bases of a truly Integral practice.

You can choose to incorporate a few of these practices into your life in an unstructured way, or you could choose to take a more systematic approach and create something that we call an Ecological Integral Life Practice (Eco-ILP), modeled after a process developed by Ken Wilber and Integral Institute.[6] Below, we will first introduce several general practices that we have developed and that we recommend. Then we will introduce practices modeled after ones that Wilber has created, and discuss how to incorporate them into an Eco-ILP.

In figure 10.2, we outline the practices detailed in this chapter. The general practices we have developed can be done on their own or as part of a more formal Eco-ILP. Our Eco-ILP practices are based on the 4 core modules (Body, Mind, Spirit, and Shadow) associated with Integral Life Practice.

Although all of these practices can support an Integral ecological awareness, one practice is particularly important—ROPE Weaving and Climbing—because it exercises many aspects in a comprehensive way. As we will illustrate, this rich practice contains the principles used in most of the other practices. In fact, many of the other practices are variations of this one.

The rest of the chapter is devoted to providing brief descriptions of the powerful practices listed in figure 10.2. The more you explore them and engage in them, the more you will cultivate new and dynamic ways of perceiving and experiencing the natural world. We encourage you to adopt

General Practices	Eco-ILP Practices
ROPE Weaving & Climbing	Integral Kata
Walking in Four Places at Once	3-Body Hiking
Integral Nature Observation	Integral Ecology Framework
Perceiving Perceiving	3-2-1 Eco-Shadow Process
Integral Breathing	Perfect Planet
Who, How, What Reflection	Sky Gazing
AQAL Scanning	1-2-3 of Gaia
Eco-Self Tour	
Integral Empiricism	

Figure 10.2. Examples of Integral Ecology practices.

these practices in ways that best support your own development of Integral ecological awareness. Enjoy!

ROPE Weaving and Climbing

Integral ecological awareness weaves at least four distinct but related strands of awareness (see figure 10.3). The acronym ROPE serves as a reminder of the various modes of awareness, but the modes can be explored in any order:

Resonating: using mutual understanding to directly resonate with self, other, and world (terrain of culture);
Observing: using the five senses to observe self, other, and world (terrain of behavior);
Patterning: using systemic analysis to recognize patterns of self, other, and world (terrain of systems); and
Experiencing: using introspection to directly experience self, other, and world (terrain of experience).

Any one of these strands of awareness is a powerful mode of experiencing the natural world. However, when woven together into a single "rope" of awareness, they are exponentially insightful and reveal an unmatched fullness of reality. In its most basic form, this practice involves shifting your awareness from one terrain of ecology to another and noticing the distinct qualities of each and how they show up in your awareness. Tap into each strand of awareness and shift among them until you are able to hold all four in your awareness simultaneously (a process that usually takes many years to develop and stabilize).

As we have mentioned in previous chapters, within the 4 terrains of ecology (behaviors, experiences, cultures, systems) are 12 niches, three per terrain, each representing another level of complexity. Thus, not only can one weave a ROPE of awareness, simultaneously perceiving the 4 terrains, but one can also climb that ROPE of awareness, perceiving each terrain at different levels of complexity. For example, you can experience your physical body, your thoughts, or the Spirit out of which you arise. Just as you develop the capacity to hold the awareness of multiple terrains, you can develop the capacity to hold multiple levels of reality in your awareness too. When you understand the process, you can climb the ROPE by shifting through each domain and cycling through the levels of body, mind, and Spirit.

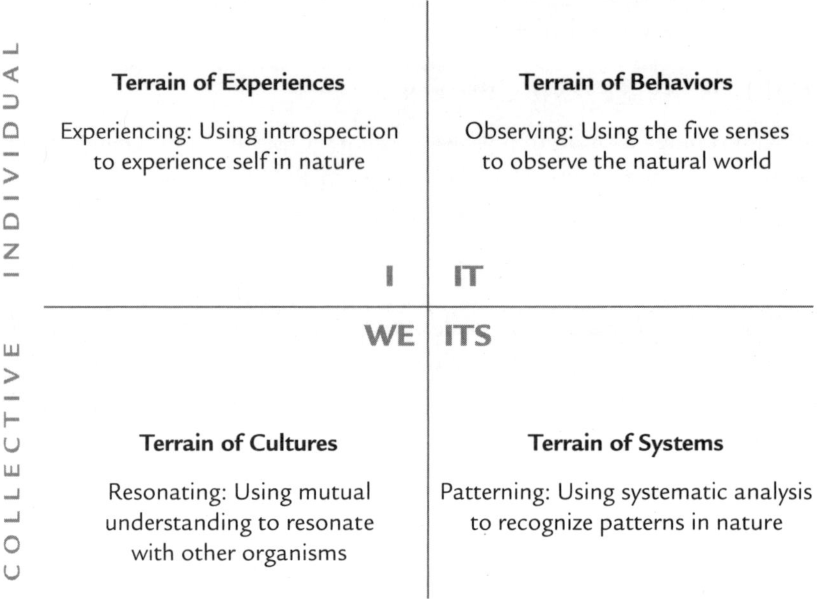

Figure 10.3. ROPE weaving and climbing.

Let's look at an example (see figure 10.4). Suppose you are outside sitting on a log or hiking and you want to practice ROPE climbing. To begin, start with the level of body. First center yourself. Then notice the ways you are physically *resonating* with the various beings around you (crickets, birds, trees). Now *observe* the physical movements of the wind, sun, clouds, plants, and animals. Next notice the basic ecological *patterns* that surround you (e.g., plant clusters, erosion patterns, mice trails around a brush pile). Lastly, become aware of your *experience* of your own body in that moment. In general you should spend a few minutes on each step (resonating, observing, patterning, experiencing). The more time you spend on each one, the better, as it will sharpen and deepen your awareness in that context.

Then begin the cycle again at the level of mind, spending a few minutes inhabiting each terrain of awareness. Start by becoming aware of how you are *resonating* with the various expressions of mental activity around you (a deer looking at you on the edge of the timberline, the bluejays waiting for you to feed them). Then become aware of and *observe* the more intentional aspects of the plants and animals around you, such as a chipmunk sounding

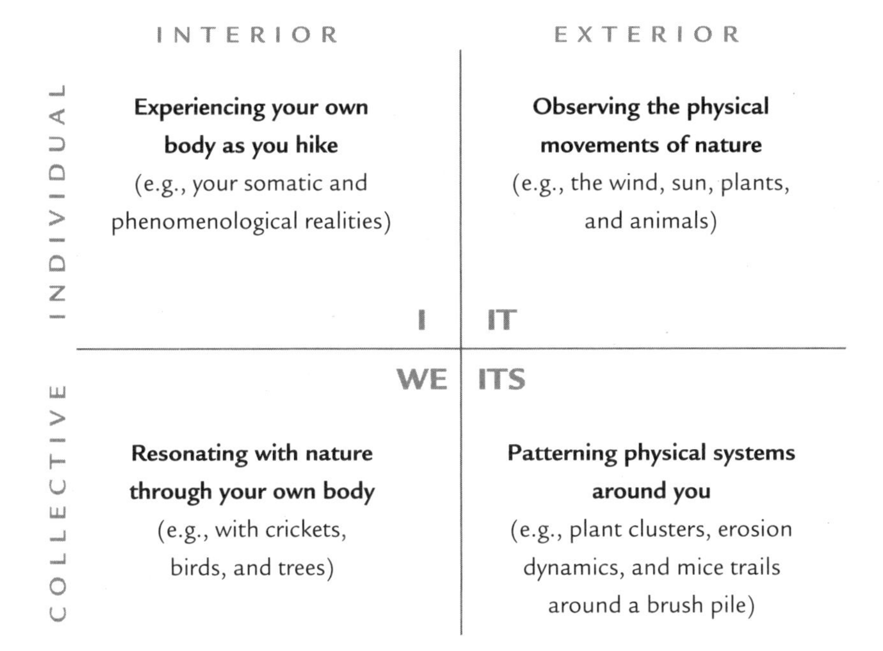

Figure 10.4. ROPE climbing: body.

her "chip chip chip chip" alarm. Next bring your attention to the intentional *patterns* in the environment such as the trail system constructed by the national park system. Finally, become aware of your *experience* of emotions and thoughts as you take all of this in. (See figure 10.5.)

Begin the cycle yet again at the level of Spirit. As before, spend a few minutes inhabiting each terrain of awareness. In total you should plan on spending *at least* 30 minutes and up a full hour doing this entire practice (approximately 10–20 minutes per level). First, *resonate* with the miracle of all the beings around you. Then *observe* the Great Perfection in all the activity of those beings and other natural processes. Next, notice the co-arising *patterns* of all ecological phenomena. Lastly, *experience* your own timeless awareness as the container for all this that you experience in all 4 terrains. You are aware of a subject, yet that awareness has no location. How long has it been present? (See figure 10.6.)

In this way, you can cycle through all 12 niches, or the 4 terrains at three levels of complexity. ROPE climbing is very rewarding and always reveals something interesting and important. It can take as little as 30 minutes or as long as you want.

	INTERIOR	EXTERIOR
INDIVIDUAL	**Experiencing your emotions and thoughts as you encounter nature** (e.g., the joys, grief, insights, and memories that emerge as you are hiking) **I**	**Observing intentional behaviors around you** (e.g., a chipmunk sounding her "chip chip chip" alarm) **IT**
COLLECTIVE	**WE** **Resonating with the mental activity around you** (e.g., a deer watching you, birds calculating your direction)	**ITS** **Patterning the intentional natural systems around you** (e.g., the trail system you are on and human use patterns in the area)

Figure 10.5. ROPE climbing: mind.

	INTERIOR	EXTERIOR
INDIVIDUAL	**Experiencing how your own timeless awareness is the creative space out of which all this is manifesting** (e.g., experiencing yourself as the entire Planet Earth) **I**	**Observing the Great Perfection in all the activity around you —even human** (e.g., loving how perfect the trash on the side of the trail is—and then picking it up) **IT**
COLLECTIVE	**WE** **Resonating with the Pure Spirit of all the beings around you** (e.g., perceiving the translucent essence of the birds singing in the nearby trees)	**ITS** **Patterning the co-arising of all ecological phenomena** (e.g., noticing the divine connection of the Great Web of Life)

Figure 10.6. ROPE climbing: Spirit.

Walking in Four Places at Once

The next practice is a variation on ROPE weaving and climbing. The main difference is that it moves back and forth from the various terrains of awareness organically (in whatever order you are drawn to), as contrasted with the more systematic flow of ROPE weaving. This practice is typically performed while hiking or walking in nature. As a result, we hope that with practice you can begin walking in four places not only in the natural world but anywhere you might be in your everyday life—on the street, in your office, at the mall.

Try this practice the next time you are walking in nature. This can be in your backyard or on a longer hike in the mountains or the woods. If you live in a more urban setting, you can still try this as you walk down the street. As you begin walking, you will shift your awareness every few minutes between each of the 4 terrains of ecological awareness. For example, move your attention from a subjective perspective (experiencing) to an objective perspective (observing) to an intersubjective perspective (resonating) to an interobjective perspective (patterning). Any order is fine. The point is to become aware of each of these foundational perspectives. After you have spent a few minutes inhabiting each terrain, you can begin to inhabit several of them at the same time. For instance, you might start by being aware of what you are *observing* and then notice what you are *experiencing*, then try to perceive both *observing* and *experiencing* at the same time. As your awareness relaxes enough to allow you to do this, try to shift your awareness back and forth between the two terrains while remaining connected to each of them. In other words, you have both terrains in your field of awareness, but you keep shifting which one is foreground and which one is background. Continue to practice holding two perspectives at the same time by bringing together different combinations of two terrains (e.g., resonating and observing, patterning and experiencing, patterning and observing). As you become more proficient at resting in two perspectives at the same time, you can begin to include a third, and eventually the fourth. For instance, as you hike across the meadow, continue noticing what you are experiencing *and* observing, and now stop trying to be aware of what you are observing and instead become aware of what kinds of organisms you are resonating with. So now you are experiencing and resonating. Keep shifting your awareness like this as you hike until you are able to maintain conscious contact with all three terrains: experiencing, observing, and resonating. It is like juggling balls of awareness: you start out with two and you keep introducing a third until you get really good at keeping all three flowing at the same time; then you begin to introduce a fourth awareness ball. As with juggling, including each additional perspective is exponentially harder. As you practice this, ask yourself: What keeps you from doing this? What is difficult about shifting and inhabiting various points of view? What part is easy for you? The benefit of this practice, as with all of these perspective practices, is that it takes our undifferentiated awareness, which contains all 4 terrains, and consciously differentiates them into irreducible perspectives that can then be more deeply integrated into a conscious whole. For a deep integration of these perspectives to occur in our own selfhood, we have to strengthen our capacity to access any one of them independently of the others. In short, you become an embodiment of multiperspectival awareness, which increases your intimacy with reality because you are in more conscious contact with it through multiple modes of being and knowing. In turn, this presence allows you to be more timely and skillful in responding to circumstances and situations.

Integral Nature Observation

Like Walking in Four Places at Once, this practice will help you develop the capacity to see natural phenomena from multiple perspectives at the same time. Go into your backyard, a nearby park, or any place where you can observe nature. Make yourself comfortable, then on blank sheets of paper, following the example shown in figure 10.7, create four headings—Terrain of Experience, Terrain of Culture, Terrain of Behavior, Terrain of Systems—with space below each to jot down notes, drawings, or poetry.

Spend at least 5 minutes within each terrain using the modes of data collection appropriate for each (experiencing, observing, resonating, patterning) to describe a natural phenomena such as a river, the sky, an anthill, or a stand of pine trees. In other words, cycle through the 4 terrains of awareness as described above, and then write down what you notice from each of these perspectives. For instance, if you are sitting in a redwood forest, you might write in the Terrain of Experience section about how you feel calm, open, sad, in awe, or curious as a result of being surrounded by these mammoth trees; in the Terrain of Culture section about your sense of communion with these ancient guardians of the coast, fragments of local stories about these trees or this ravine, cultural symbols that you can see from where you are sit-

The 4 Terrains of ______

TERRAIN of EXPERIENCE

Subjective (e.g., experiencing the natural world)

TERRAIN of CULTURE

Intersubjective (e.g., shared horizons with the natural world)

TERRAIN of BEHAVIOR

Objective (e.g., measurements of the natural world)

TERRAIN of SYSTEMS

Interobjective (e.g., functional fit of the natural world)

Figure 10.7. Sample format of Integral nature observation.

ting; in the Terrain of Behavior about all the facts you know regarding redwoods, estimations of their age, height, and width, when this area was last logged, the names of any animals or plants you can see or know live in the area; and in the Terrain of Systems about the weather patterns in this valley, the watershed dynamics, how forest fires (or their suppression) over the last 50 years have played a role in growth cycles, the impact of human visitors on the delicate root systems of these towering giants.

Among the perspective practices that have you explore and notice multiple perspectives within your own awareness, this practice is unique because it has you write down what you notice. The act of writing serves an integrative function by supporting you in making these various perspectives objects of your awareness. It also links these insights to the body's knowing through the use of your hand to write. In addition, this practice has you explore one single phenomenon from multiple perspectives, which is a very worthwhile and rewarding endeavor.

Perceiving Perceiving . . . Through Their Eyes

This practice is designed to deepen your resonance with other beings by taking the perspective of other members of the ecosystem.

The next time you are in a natural setting, as you look around, reflect upon the notion that everything you see—insects, flowers, birds, lizards—is perceiving the world in its own way. Perhaps through simple prehension, perhaps through signaling, but they all are having an interior experience of their environment. Don't just perceive them—perceive them perceiving. Then go further and perceive them perceiving one another. For example, if you notice a lizard basking in the sun, stop and first notice that "I'm looking at the lizard." Then become aware of the lizard perceiving its world. Move closer to the lizard until it becomes aware of you and then stop. Feel into the lizard perceiving you. Now place your awareness on the tension between you and the lizard where you can sense it perceiving you perceiving it. Now just hang out in this moment of uncertainty and enjoy the perception of perception that each of you are engaged in.

This exercise helps transforms your perspective from merely observing (onlooker consciousness, which is typical of ecologists) into an Integral consciousness. When you experience both the exterior and the interior of members of an ecosystem this way, something truly amazing happens. It will transform your understanding of ecology in a deep way.

Integral Breathing

Integral Breathing is a practice that can be done at any time throughout the day either alone or with other people. In fact, it is ideal to do with a group of friends during a 5-minute break on a hike or when you all gather around the firepit for a group meditation. Like other practices, it can be done at home or at work, but there is something unique about connecting to our breath when we are outdoors. The goal of this practice is to use our breath as a vehicle for Integral awareness. Many contemplative traditions use the breath in a variety of ways to cultivate awareness. Thus, it only seems appropriate to use it to cultivate Integral awareness, especially when we are in nature. One of the benefits of this practice is that it is so easy to do and only takes 1 minute (in its "4-breath" short form). Also, for such a short and easy practice, it is surprisingly powerful in its capacity to renew and transform.

You can do this practice standing up, walking, or sitting down. Basically, anytime you are breathing, you can transform that physiological process into a consciousness-raising practice of cultivating intimacy with your surroundings by being fully awake to the subjective, intersubjective, and objective aspects of that exact moment.

To begin with, why don't you do this as you are sitting there reading this passage. This current moment is as good as any for you to try this out. Later you can do it in other contexts. But since you are reading this anyway, we might as well start this practice now. There is no time like the present.

Start by taking a deep inhalation. Exhale, and on the exhale notice what arises or is already present within you. What are the somatic sensations, emotions, thoughts, or illuminations within your "I" space? Inhale again and on the next exhale notice what arises or is present between you and others in your life or with individuals who happen to be near you in the moment. What are the shared contexts, stories, experiences, history, resonance within your shared "We" space? Inhale again and on this next exhale notice what sensorial information arises or is present in your awareness with regard to your physical surroundings. What are the sounds, smells, tastes, sights, and physical feelings within this "It" space? Inhale again, and on this last exhale notice what ecological or planetary information arises or is present in your awareness with regard to the bioregion or watershed you are currently in. What are the economical, legal, industrial, educational systems and dynamics within this "Its" space?

That completes one cycle of Integral Breathing, which can be done in about 1 minute. We find that it is often worthwhile to do three cycles at a time. Notice how during the cycle of breathing your awareness travels out from within you to the people (or organisms) physically near you, to the immediate space around you, to the larger environmental context. Thus your awareness, with each breath, is expanding from your own body-mind to larger and larger physical contexts (i.e., span): I into We into It into Its. As a result, one cycle of 4 breaths can often feel like a single Kosmic breath—a breath that is breathing all 4 terrains of awareness.

Here it is in short form:

I Breath: Become aware of what is arising within you. Rest within.

We Breath: Notice what arises between you and others. Connect with others and rest in those connections.

It Breath: Notice the qualities of the room or space you are in. Open your senses to the world. Rest there.

Its Breath: Notice the larger context beyond the room or space you are in. Perceive the systemic patterns around you. Rest everywhere.

Who, How, What Reflection

A simple reflective exercise that is extremely powerful is to take a few minutes to reflect on the Who, How, and What of the moment. Often it can work best to start with what you are looking at, then notice how you are doing the looking, and then reflect on who you are as the looker. In other words, stop what you are doing and identify what part of reality, which terrain or niche, you are focused on. Next, become aware of the methodologies (folk or formal) that you are using to encounter the natural world in that moment. Lastly, notice the perspectives you have been inhabiting and resisting while focusing on this part of the natural world.

You can also connect this simple reflection with Integral Breathing, using a single breath (inhale/exhale) as a support for each of these inquiries.

AQAL Scanning

This practice provides a simple but comprehensive way to quickly scan your awareness, behaviors, connections with others, and interactions with your surroundings in an Integral way. This is another practice you can do just about anywhere—anywhere, that is, where you want to take 10 minutes to explore your own awareness and become more awake to the current moment. Being in nature is a perfect context for such reflection and rejuvenation. By spending a few minutes reflecting on and answering the questions associated with each terrain of awareness, you cultivate an embodied awareness of multiple dimensions. This embodied awareness in turn supports you as an ecological being by making you more aware of the many interconnections that constitute your own existence.

Below we provide a number of questions within each of the 4 quadrants. They serve as symbolic reminders of the types of questions you can ask. Feel free to generate additional or substitute questions for your own unique situation.

UPPER LEFT: EXPERIENCE

How am I experiencing my body, my emotions, my thoughts, and my soul?

Where have I focused my awareness, and what have I missed as a result?

UPPER RIGHT: BEHAVIOR

How am I taking care of my physical needs (e.g., stretching, drinking water)?

What behaviors have contributed to or prevented understanding and progress in this situation?

LOWER LEFT: CULTURE

How am I resonating with the bodies and the energies of the group?

How many different perspectives am I inhabiting or avoiding?

In what ways am I trying to understand others?

LOWER RIGHT: SYSTEMS

What is my awareness of the natural world or my environment (e.g., time of day, weather, temperature, lighting)?

What are the social roles and dynamics of this situation?

In what ways am I instrumental, detrimental, or irrelevant to the success of this project?

Eco-Self Tour

A powerful and effective way of learning to take multiple perspectives is, in any given moment, to inhabit any of the various ecological worldviews associated with the eight eco-selves. (See pages 226–238 for a reminder of these.) This can be done while you are out in nature: how would an Eco-Warrior relate to these campsite rules, how would an Eco-Radical feel about there being so many people fishing on the lake today, what would an Eco-Manager recommend for dealing with growing trash loads in the surrounding urban centers? Or you can focus on a single issue such as the development of a multi-use open space area near your house and explore within your own awareness, and in conversations with others, what each of the eco-selves would want to see happen with such a public space. What would be their contributions to a successful project? Where would their blind spots lie?

Integral Empiricism (aka Transforming Perceptions into Perspectives)

One of the strongest habits of perception that we have is to assume that everyone else sees and perceives the world the same way we do. Also, we are chronic naive empiricists, assuming that what we see is what is objectively lying around "out there" in the world. We mistake our perspective for a perception. In fact, it is quite challenging to be aware of your perspective while it is unfolding in the moment-to-moment interaction with the world. However, the more we can become aware of how our perception is actually our own perspective, the more capable we will be in experiencing how all phenomena are the result of perspectives colliding and aligning.

In order to become increasingly aware of how your perception is in fact a perspective, practice the following. Look at some natural phenomena, such as a lake, a field of wildflowers, trees next to your house, "weeds" coming up through the concrete sidewalk. As you are looking at this object of perception, reflect on the fact that this object does not exist "out there" in the world independent of your observation of it in this very moment. Notice that the object is in fact enacted, brought into being, by your very encounter with it. You are not creating the object but rather disclosing it through your unique psychological structures and worldview. One way to understand this is to recognize that the object is always more complex than your particular perception of it. That is why your perception is actually a perspective. You are not seeing all there is to see about this object. How could you? Thus, the phenomenon is complex enough to support multiple, even contradictory perspectives of it. Notice, too, that other people and beings would have a different perception/perspective of this phenomena.

In summary, this practice is designed to help you catch yourself in the act of being a naive empiricist so that you can transform your perceptions of the natural world into perspectives: Integral Empiricism.

Ecological Integral Life Practice (Eco-ILP)

Ken Wilber and colleagues at Integral Institute have developed a unique and Integral approach to personal development that combines a mix of practices to exercise multiple dimensions of our being. They call this approach Integral Life Practice. Their approach emphasizes four essential modules—kinesthetic, cognitive, contemplative, and psychodynamic—which they refer to as Body, Mind, Spirit, and Shadow. Additional important modules include areas of focus such as Moral, Affective, Interpersonal, and Service.

The idea is that, in order to take a rounded, Integral approach to personal transformation, one needs to attend, at a minimum, to the four core modules. One can supplement this later by addressing other modules, but at a minimum, attention to the areas of Body, Mind, Spirit, and Shadow is required. You can create an Integral Life Practice for yourself by finding one or two practices in each of these core areas and incorporating them into your life.

Here we are going to focus on how each of the four core modules can be worked with using ecological practices. Here are some of the practices we have developed that you can use to create an Eco-ILP:

Body Module: Integral Kata (done in nature), 3-Body Hiking
Mind Module: Integral Ecology Framework: Who x How x What
Spirit Module: Perfect Planet, Sky Gazing, 1-2-3 of Gaia
Shadow Module: 3-2-1 Eco-Shadow Process

Body Module

Integral Kata (aka The 3-Body Workout) As part of Integral Institute's approach they have created the Integral Kata, a series of physical movements that takes between 15 minutes in its short form and up to an hour in its long form. Either the short or the long form can serve as a daily practice for exercising your physical, subtle, and causal bodies. The short form uses simple movements, breathing, and visualizations to shift your awareness from causal to subtle, and then into gross awareness. The long form uses more elaborate practices—qigong, yoga, breathing, and meditation—to move from causal to subtle to gross, and then return to subtle and causal. Here we maintain that essential form taught by Integral Institute and simply situate the kata within a natural context such as in your backyard or at a nearby park. The kata is powerful in and of itself, but it becomes even more charged when done outside with your bare feet touching the earth. Below is an overview of the "1-Minute Module" developed by Integral Institute.[7] For a fuller overview, see the Integral Life Practice starter kit.

Begin by standing and breathing normally with your palms together in "namaste" gesture in front of your heart. Starting with the causal body, notice the suchness of this moment—the pure present is-ness of Now. Feel your whole body alive with presence. Say to yourself or out loud, "I am this suchness. I am the openness in which all things arise." Then inhale, exhale, and inhale. Place your palms crossed over your chest, and then on the last exhale open up both arms with palms flat as if you were starting a push-up and say, "I release to infinity."

Next connect with your subtle body. Still standing and breathing normally, place your hands with fingers interlaced in front of your belly as if you were holding a ball up to yourself. In this position use your hands to feel and gather energy, saying, "I breathe into the fullness of life." Then exhale and move your hands up the front of your chest and then over your head with fingers still interlaced, saying, "I breathe out and return to light." Next, inhale and bring your hands back down along the side of your body, returning them to the position in front of your belly with fingers loosely interlaced. At this point say, "Completing the circle, I am free and full." Do this cycle of arm movements and sayings 8 times: exhaling, hands move up the front to the sky; inhaling, hands circle back out and down. Doing it this many times serves to generate the subtle energy, or *ch'i*, that is associated with this body.

Lastly, connect to your physical body. Still standing and breathing normally, place one hand over the other on your belly. Inhale and exhale. Then say, "Infinite freedom and fullness appear as this precious human body." After this, squat down as you inhale and exhale, placing your palms on the earth, saying, "Touching the earth, I am connected to all beings." To complete the Integral Kata, perform the dedication by bowing in the four directions (turning right, clockwise) and saying, "May my consciousness and my behavior be of service to all beings in all worlds, liberating all into the suchness of this and every moment."

3-Body Hiking The Integral Kata provides one way of working with and accessing the three major bodies we have: physical, subtle, and causal. Another way of working with these layers of our embodiment is to do what we call 3-Body Hiking, which you can practice anytime you are hiking, or even if you are just taking a short walk. To do this, you simply shift your awareness between the three bodies as you are moving around. For instance, you begin by feeling your gross physical body: the movement of muscles, the flow of blood, your breathing and sweating, the development of thirst, aches and pains. Feel the many sensations that are alive in your corporeal self.

After doing this for at least a few minutes, begin noticing your subtle energetic body: energy moving throughout your limbs, the experience of love and light occurring as you walk. Notice the different energies associated with your chakras or meridians; feel energy being exchanged between your feet and the ground or between your body and the trees you pass.

Finally, place your attention on your timeless causal body: the pure empty space of your entire physical and subtle body, the creative potential of the presence that pervades you, the simple Now that is walking. As you continue to do this, you will discover that you always walk with all three bodies—you just haven't maintained conscious awareness of it.

Shadow Module

3-2-1 Eco-Shadow Process This practice is inspired by the ILP practice known as the 3-2-1 Shadow Process. However, it is different in some important ways in the use of the 2nd- and 1st-person perspectives. Like the ILP practice, this practice is also designed to reintegrate into our awareness the shadow material (those aspects of the self that have been cut off or disassociated) that we have in relation to the environment. This practice is very important, especially since we often project our shadow onto individuals who are doing damage to the environment or just have a different relationship to nature than we do. We also project our shadow onto the natural world: disliking certain landscapes or responding in particular ways to different kinds of animals. The more we become aware of and own our ecological shadow, the more authentic a relationship we can have with others who value the natural world for different reasons than ourselves, and with the natural world itself.

In general this practice works by bringing to your awareness some situation or person that you have had or are having a strong negative reaction to. For example, a stakeholder in a project you are involved in is continually blocking progress of the group with their stubborn position, or a local mom and pop shop is being forced out of business by Wal-Mart's recent arrival in town. You work with this "trigger" by adopting three different perspectives of it and in that way reintegrate the part of you that has been split off and projected outward onto the situation, person, or natural world.

It is often helpful to use a journal when doing this process. After you have done it many times in written form, you can do it in your mind or speaking out loud to yourself. This process can also be done in dyads, where you describe the three perspectives to a partner while they listen with openness. For our purposes we will assume you are using a journal to engage in this practice.

First become the OBSERVER of the situation. In great detail describe the situation or the material that triggers you, from a 3rd-person perspective ("For three years, this guy has been showing up to every meeting and doing everything he can to . . . ," or "Recently, a local forested area was sold to a real estate company that said it was going to . . ." Basically at this point you are simply laying out the so-called facts of the situation. This is the easiest part of the entire process but serves the function of clarifying what it is you are going to be working with.

Next you become the EXPLAINER of the situation. Now you describe the same scenario and set of circumstances in 2nd-person by taking the perspective of the person(s) or thing involved having a dialogue with you, explaining to you why they are doing what they are doing ("The reason I'm holding my ground is . . .") and then imagine how you would respond. Continue the dialogue and respond back and forth until the whole situation has been explained to you by the person whose perspective you are taking. In other words, the goal here is to take the perspective of the other person (or animal) so fully that you can then explain that perspective to the part of yourself that doesn't understand this perspective. Don't worry too much whether what you are saying on their behalf is accurate; use your imagination and perspective-taking capacities to give them the benefit of the doubt. This is a projective exercise, because that is what you do with your shadow—you project it. Becoming the explainer and taking the perspective of the person who is annoying you is difficult, but when done with integrity it makes possible the most powerful part of this practice: owning your projection.

Lastly, you become the OWNER of the situation. Now describe the scenario from your 1st-person perspective, explaining what you are doing to contribute to the problem or, better yet, cause the problem in the first place ("The reason you are sand bagging is that I keep subtly placing you on the defensive . . ." or "I'm getting frustrated with you! I don't like it when the process gets bogged down by one person because . . ."). This is the hardest part of the entire exercise: facing up to the ways in which we are creating the situation that is negatively impacting us. In effect, this exercise is helping us to move from the habitual role of victim ("This is what happened to me!") to the self-authoring (to use Robert Kegan's terminology) role of dynamic awareness ("This is how I am manifesting what is happening to me"). Of course, this is not to minimize the ways in which people do screw up and impact us, nor is it an invitation to collapse into a simplistic "It's my fault!" mentality. The trick is to follow what irks us to its source and witness the myriad ways we co-create the very things we despise, as a result of being unaware of our shadow.

In summary, you want to "See it. Take it. Own it." Inhabit a 3rd-person perspective and see the situation. Inhabit a 2nd-person perspective and take the viewpoint of the person (or event) you are triggered by. Inhabit a 1st-person perspective and own the ways you are creating the situation or are reacting negatively to it as a result of your own fears, anxieties, blindspots, and anger.

This is a very effective and powerful practice, especially in the context of environmental activism, where there is a lot of projection on the spoilers of the Earth such as big business, the Republican party, oil companies, trophy hunters, and even other environmentalists who have different values. Consider the following three brief examples as illustrations of this process.

Example 1: *Observer*—I get annoyed at my brother because he refuses to recycle. He is a conservative Christian and doesn't see a connection between such environmental behaviors and the "Good Book." *Explainer*—I don't have anything against people who recycle; I just don't think it makes a difference. Besides, if there is going to be an ecological crisis, it is out of our hands—it will be God's will. *Owner*—I see how my judgment of traditional Christianity and a long-standing competitive dynamic with my older brother are working together to trigger me with regard to my brother's behavior in this situation. While it would be nice if he did recycle, it isn't that big a deal that he doesn't.

Example 2: *Observer*—I hate how Mary Joe at work is always talking about those "crazy environmentalists" and that they should just "chill out" because technology is going to catch up with the issues and "market forces" will prevent a peak oil crisis—even if we run out of oil. *Explainer*—I'm worried about the state of the environment and our impact on it, but I also believe in the creativity of the human spirit and feel that no matter what comes our way, we will be able to respond effectively. Of course, there might be some major bumps in the road—I'm not saying it will look pretty or be easy. But I have no doubt that people will come together, industries will come together, and solutions will be found. *Owner*—I have to admit I feel so scared about an approaching environmental crisis that I find it hard to be around anyone who doesn't mirror that back to me. I too think that we are very creative and will rise to the challenge, but when I perceive people as being naively optimistic, I feel they are living in a different world than mine and I have a hard time tolerating that.

Example 3: *Observer*—I couldn't believe a story I heard the other day about a well-known environmentalist who refused to use a plastic mat at a yoga studio because it was a "petroleum-based product"! This is so typical of these radical types. *Explainer*—I realize that it might have come off as extreme and that the person who was offering me the mat was just trying to be hospitable. But you know it is an important principle to me to do everything I can to not support petroleum-based products. I don't need to shove this in people's faces, but it is a standard I live by. *Owner*—I feel triggered by her refusal to use the kind of yoga mat I use every day—it makes me feel like I'm not "eco" enough, but I care for the environment just as much as she does. There is something about this dynamic that makes me feel less than worthy, and that hits this tender core in me that I don't like to face.

Mind Module

The Integral Ecology Framework: The What, Who, and How In a way, this whole book has been a description of the Integral Ecology framework practice. In particular, the "Applying Integral Ecology" section in the next chapter provides a summary of how to exercise your mind using the insights of the Integral Model in the context of ecology. The very use of the Integral Ecology framework (the 4 terrains, the 8 eco-selves, the 8 eco-modes) is psychoactive in that it activates capacities of discrimination and compassion. We feel that there is no better way to exercise your mind around ecological content and environmental issues than to make use of the Integral principles outlined in this book—even if you come to different conclusions! In other words, to practice the mind module for your Eco-ILP, engage with the framework presented in this book, add to it, extend it, deconstruct it, recontextualize it.

Spirit Module

Perfect Planet: Feeling, Caring, Dissolving The following practice is based on *tonglen*, a Buddhist practice in which we allow pain to touch and open our hearts. In response we send back care and concern, so it is often called a practice of "taking and sending." We have developed Perfect Planet, a variation of this practice, as a way of simultaneously cultivating compassion for the suffering occurring around the globe, forgiving those who are inflicting such destruction on the Earth (including ourselves!), and recognizing that the planet is an expression of the Great Perfection. As a result, this is a particularly powerful practice because it asks us to do three very hard things: feel the immense depth of pain that is occurring all around us; cultivate compassion and care for those who are the hardest to love—the ones causing this suffering; and then recognize how absolutely perfect this drama is as a manifestation of radiant Spirit. So this is not for the faint of heart. Consequently, it is important to pace yourself and allow the practice to be nourishing and supportive of wisdom and compassion. We encourage you to spend many weeks on one phase of the process before adding the next phase to your meditation.

This practice is best done sitting down on a nice rock or log (or inside on a cushion or chair) with a meditation posture (feet on the ground, hands on your knees or in your lap, back straight, etc.). Plan on spending about 10 minutes on each of the three parts of this practice. Once you are settled, become grounded by focusing on your breath and your body. When you are ready, you can begin the practice by following the general guidelines outlined below.

Part One: *Gaia Body*—Feeling the suffering of the planet. To start, just breathe for a few minutes to become grounded. Visualize your mind and body becoming clearer and clearer. You might want to imagine your thoughts falling away, or imagine and connect with the pristine feeling of nature. Or you might imagine white light being poured into your body through your head, but particularly through or into your heart. Allow this feeling, however you come by it, to rest in your chest.

Once you feel open in the chest and heart, bring to mind an aspect of an ecological or environmental crisis that causes suffering. Maybe the crisis hurts someone you know; maybe it kills certain kinds of animals or causes them to suffer; maybe it makes you suffer to think of it. Or you could call to mind the image of someone or a group of people who are suffering. Now, as you are able, gently breathe in that pain to the bright and open feeling in your chest. Then, as you exhale, breathe out brightness and openness. Imagine that you are offering this openness, this brightness, to help those involved, even yourself.

It is easiest to start by focusing on one just person (such as a loved one who is suffering) or one particular situation (such as urban sprawl in your area and the resulting loss of local habitat). When you become more adept at this practice over time, you can broaden the practice, expanding and relaxing your awareness to include more and more suffering that is occurring across the planet. Continue to breathe in the suffering, transforming it into light in your heart center and exhaling love. Only take on what you feel comfortable and capable of doing. If at any time during this practice you feel too overwhelmed to continue, you can always reground yourself by returning your focus to your breath. Be sure to go at a pace that feels appropriate to where you are at in the moment.

Part Two: *Gaia Heart*—Cultivating compassion and care for those who cause suffering to the planet. While in the first phase of the process you focused on the those who are suffering as a result of environmental destruction and ecological crisis, in the second phase you turn your attention to those individuals who you perceive to be causing the suffering. Start by focusing on a single person, such as a friend of yours who has a large ecological footprint and often is doing things like littering and driving a "gas guzzler." Recognize that even though this person might be causing some difficulty for you or others in the world, she is still a human being who experiences suffering the same as we all do. Inhale her suffering into your heart space and exhale into her heart space love, understanding, and compassion. As you feel comfortable, begin including more and more people and circumstances that are ignorantly or intentionally harming the planet. Again, go at a pace that is manageable for you and allows you to really feel deeply into the process. Quality of awareness and strength of intention are what matter here.

Part Three: *Gaia Spirit*—Dissolving the suffering of the planet. In this final phase, you will take the profound ecological suffering you have connected to through focusing on those experiencing it and those creating it, and dissolve it into timeless awareness. After you have completed the first two parts of this practice and cultivated authentic compassion toward the multitudes of people who are, in various ways, destroying our planet, bring back into your awareness all those who are suffering as a result of this destruction. Hold both ends of the suffering spectrum in your loving attention. Breath deeply into the fullness of this suffering for many minutes. Continue to become more and more present, anchored in the Now, as you breathe into this fullness. As your awareness becomes pure Presence, extend this presence out in all directions, enveloping all those plants, animals, and people who suffer and all those who perpetuate this suffering.

As your awareness encapsulates the suffering, allow it to transform, dissolve it, into pure timeless Presence: a Presence that is free from all suffering, a Pureness that knows neither victim nor destroyer, friend nor enemy, "tree hugger" nor logger, environmentalist nor multinational corporation. As this Presence washes over the Earth, rest in a place where you recognize that the planet and all her inhabitants are the Great Perfection. When you are ready, slowly bring your awareness back to your breath and open your eyes. Bow to the Perfect Planet.

This practice teaches us many things. For one, it shows us that running from or reacting to pain is not always the most helpful response. Further, it awakens us to open our heart when we want to shut it down, or ignore particular people, or when we feel downright hatred toward someone. As you can imagine, it is a very useful practice for those tough moments at the negotiation table. Even using a few breaths to perform this practice can help reground yourself and make you more capable of taking other perspectives compassionately.

Sky Gazing Sky Gazing (*'namkha arted*) is a Buddhist practice from the Dzogchen tradition and adapts well to an ecological context. To begin, sit comfortably in the outdoors—on the ground, on a blanket, or in a chair—where you have a good view of the sky. Place your hands on your knees with your palms turned upward. Keep your eyes wide open and look straight ahead with a gentle, unfocused gaze. Relax your jaw, leaving your mouth slightly open. The idea is to have all of your senses physically open to Reality. Just sit, breathe, and bear witness to your surroundings and notice what happens. Do not focus on birds, trees, or clouds; gaze into the space of the sky. The goal is to be relaxed using this external sky-space to mirror back to us and connect us with our internal sky-space so that we can become present to the Suchness of reality *as it is*. This internal-external sky-space is luminous clarity; rest in that freedom and fullness. This practice can be a wonderful way of absorbing the beauty of the natural world in a nonconceptual way.

The 1-2-3 of Gaia 1-2-3 of Gaia is a practice of Integral Ecology that honors the complexity and depth of the planet.[8] The basic idea is that there are at least three distinct perspectives one can take when relating to the sacred dimension of the planet. One can take a 1st-person perspective, where your own self-identity expands to become the world soul or the entire planet: I AM. Following Wilber, we call this the Eco-Noetic Self.[9] This is also what most deep ecologists are referring to when they talk about "self-realization" or the "ecological self."

One can also take a 2nd-person perspective, where you are in a profound devotional and surrendering I-Thou relationship with the planet: YOU ARE. In this perspective the planet is personified as the Great Earth Mother or Goddess. She is both giving of life and wrathful. This is the sacred perspective that many ecofeminists and neo-pagans emphasize.

Lastly, one can take a 3rd-person perspective, where you are so awestruck by the physical complexity of the planet and its systems that you perceive it as a divine manifestation or what can be called its Great Perfection: IT IS. From this perspective the process of evolution, chaotic systems, dynamic ecosystems, all appear as expressions of a mysterious and divine process. Many systems theorists get close to this perspective, but they tend to stay on the secular side of the street. The cosmologist Brian Swimme and Thomas Berry capture the sacred version of this perspective wonderfully in their book, *The Universe Story*.

All three of these divine expressions of our relationship with the planet are equally important and valid. While different approaches often emphasize one of these over the others, Integral Ecology puts equal emphasis on all three of them. It is important for us to cultivate all three of these sacred perspectives in relationship to this amazing planet we live on. Only by inhabiting the intersection of all three perspectives can we come to know the full divinity of the earth. These three perspectives can be summarized in the following mantra: *I AM, YOU ARE, IT IS*.

By way of concluding this chapter, we will provide a guided meditation based on a transcription of a 1-2-3 of Gaia meditation that Sean led at an Integral Ecology seminar on the University of Colorado campus in Boulder.[10] Spaces between sections represent silence. As you read it, allow yourself to feel into these three profound ways of relating to Gaia.

Get into a comfortable position. Place your feet on the ground with your back straight. Use your breath to ground yourself and become present to your body. Allow your awareness to move with your breath. Allow your awareness to include people in this room, then the people outside this room; include all the animals and plants and beings on this property, throughout the state, throughout the country.

Use your breath to relax into that part of your awareness that is the planet, that is already identified with all beings, that is none other than all beings. Embrace and include all manifestations, all glorious expressions of your own body, your own soul, your own spirit. You have so many expressions: you are elk; you are beaver; you are trees; you are African; you are Asian; you are Latino.

There is no being on this planet who is not you. And, in fact, there are no nonbeings that are not you.

For a few moments, just rest in this place of deep embrace.

Now take a breath and shift your awareness: feel your unique self, your ordinary identity.

Feel into your relationship with the planet Earth, or Earth as Goddess, or God as nature. Feel your love of this planet, your deep commitment to justice and protection, your desire to fight and preserve it.

Embrace this connection, where you humbly surrender and prostrate before this great planet, and say, "Thank you, thank you, thank you for All." For a few moments, rest in this place of divine surrender to the great Other, to that which we call Gaia, Earth, the World Soul, the beautiful Goddess.

Now take another deep breath and shift your awareness. Become aware of the planet and all of its complexities, all of its expressions: the wild forests, the noisy cities, the clean rivers, the dirty rivers, the weeping children, the frightened animals, the dancing fox, the singing birds.

Allow your awareness to encounter all of this and more. And rest in a place of complete acceptance. The Great Perfection of this planet: how beautiful it is in its vulnerability; how dynamic and complex it is in its systems; how balanced, chaotic.

Allow your awareness to circle the globe, bearing witness to all people, plants, animals, and angels. How beautiful it all is.

And as you bear witness to the Great Perfection of the planet, if you are willing, say these words in your mind: "It is so. It is simply so." There is destruction and there is renewal. There is fear and there is openness. It is so. It is simply, simply so.

Rest in this place of acceptance and appreciation for all that is, throughout our global community, around our globe, in our homes and in our forests, with the wind whispering, "It is so."

11

Integral Ecology in Action

There is nothing so practical as a good theory.

—KURT LEWIN[1]

The ultimate most holy form of theory is action.

—NIKOS KAZANTZAKIS[2]

Integral Ecology provides a robust theoretical framework and is very practical. In fact, Integral Ecology places an equal emphasis on theory and praxis: each informs the other in an integral dialectic. The Integral Ecology framework is so successful partly because it is applied across many contexts and scales. In general, there is no single solution to any given problem, but rather a variety of Integral solutions to every problem. What might be a solid Integral solution in one context might not be a possibility or even a good idea in another.

For example, two communities facing critical drops in their native salmon stream runs might successfully respond by including different niches (from the 12 discussed in chapter 6) or the same niches in different ways, working with similar worldview conflicts in unique ways, drawing on different traditions of research to analyze and make sense of the needed course of action.

In this chapter we present some guidelines that will assist you in applying the Integral Ecology framework. After introducing these guidelines, we apply them to global climate change to illustrate the Integral considerations of such a complex problem. Next, we examine the act of recycling from the Integral Ecology framework as a way of demonstrating the comprehensive analysis it can provide. Both of these examples are designed to provide a general overview of how to apply Integral Ecology and its value. The rest of the chapter is devoted to presenting 20 examples of Integral Ecology being used by practitioners across the globe. It is our hope that these inspiring examples will support you in your own Integral ecological action.

Principles of Applying the Integral Ecology Framework

The following guidelines are presented according to their association with the 8 ecological selves (the Who), the 8 ecological modes of research (the How), and the 4 terrains and 12 niches (the What):

THE WHO

Identify the main perspectives involved and those excluded (and why).
Acknowledge your own perspective.
Take as many perspectives as you can, especially those that seem most foreign to you.
Communicate in a way that speaks to multiple eco-selves (especially the Eco-Manager, Eco-Strategist, and Eco-Radical).
Find the common ground among perspectives, and build alliances between them.
Be aware of the dignity and disaster of each perspective, including your own.

THE HOW

Identify which methodological zones and specific techniques have been employed.
Involve experts from different zones.
Make use of 1st-, 2nd-, and 3rd-person practices and techniques.
Be explicit about how you are collecting, validating, recording, and communicating "data."
Be aware of the research design—its limits and strengths.
Explore the best way, in your situation, to combine multiple methods to investigate the phenomenon.

THE WHAT

Identify the terrains represented in the problem or the proposed solution.

Identify which terrains are being privileged or neglected in the process.
Explore how more terrains or niches can be included in addressing the issue.
Clarify confusions between quadrants and quadrivia.
Consider the terrains of both humans and nonhuman organisms.

While we could include additional guidelines, we have intentionally kept their number limited so as to concentrate your efforts. These guidelines represent the minimum considerations to apply Integral Ecology. We will now illustrate each of these guidelines with a single example.

An Integral Approach to Global Climate Change

Arguably the most pressing issue facing humanity is the predicted occurrence of global climate change. There are few topics as complex and controversial as climate change, sometimes referred to as global warming. In fact, one of the reasons it is so controversial is exactly because it involves so many complex issues, variables, scales, and perspectives.

Within U.S. popular culture the two sides of the debate have each obtained a well-known representative. On the skeptical side is the American author and film producer Michael Crichton, who challenges many orthodox positions on climate change through his fictional thriller *State of Fear*.[3] On the convinced side is former vice president Al Gore, who captured the attention of the American public and the world with his successful 2006 documentary *An Inconvenient Truth*.[4]

In between and within the general positions represented by Crichton and Gore lies a lot of variation.[5] On the one hand, skeptics of climate change are not a unified front. They cast doubt on the occurrence of climate change for many different reasons, including the limits of consensus science, the unhelpful sensationalism of environmentalists, the weakness of current climate models and computer programs, the lack of long-term data on natural cycles, the unclear role of human activity, and the difficulty of predicting climate change. On the other hand, even supporters of the global warming theory, such as mainstream scientific institutions—e.g., the United Nation's Intergovernmental Panel on Climate Change (IPCC), the Union of Concerned Scientists, and the United States National Academy of Sciences—have individual scientists who don't agree on the details or focus on divergent areas of interest. In addition, even those who do agree on the details often disagree over what steps need or ought to be taken to respond to the perceived crisis.

Due to the global scope of the alleged problem and its perceived impending impacts on many aspects of human society, controversy surrounds almost every aspect of the climate change conversation. One way to illustrate the complexity of global climate change is to consider many important questions.[6] The following list of two dozen representative questions, arising from the 4 terrains, illustrates the bare minimum of an Integral analysis.

TERRAIN OF BEHAVIOR

In what ways have humans impacted climate, and how are they continuing to do so?

What are the roles of volcanism and solar activity on temperature cycles?

To what extent are current temperature changes the result of natural cycles and/or human activity?

In what ways are correlations (e.g., between rising temperatures since the industrial revolution) being confused with causation?

What are the most reliable sources of temperature measurement: weather balloons, land stations, ocean sensors, or satellites?

What is the long-term scientific relationship between climate change and carbon dioxide in the atmosphere?

TERRAIN OF EXPERIENCES

To what extent do scientists or qualified individuals feel fear about speaking out on either side of the issue?

In what ways are proponents overstating the evidence and opponents understating it?

To what extent are opponents of climate change suspiciously connected to the fossil fuels industry? And likewise, to what extent are proponents of climate change unfairly influenced by political, professional, economic, and social pressures?

How do people in different geographical locations directly experience climate change?

What psychological and emotional dynamics are preventing nuanced discussions of and collaborations on climate change?

How are the self-identities of community and global leaders playing a negative and positive role in making progress in addressing climate change?

TERRAIN OF SYSTEMS

What are the short-term and long-term policies needed to deal with climate change?

Which policies can be justified in the face of so much uncertainty of predictions about climate change? What are the economic consequences of various policies?

How are politics and business preventing justifiable action from being taken?

What is the range of small-scale and large-scale systems effects, both positive and negative?

What kinds of economic and political impacts could the Kyoto Protocol have on the global community?

How much insight do paleoclimatic studies shed on our current situation?

Will a cutback in emissions create a setback in gross domestic product?

In what ways might climate change impact some people negatively but have beneficial consequences for other people?

Could climate change be beneficial for Gaia as a planet in both the short and long term?

TERRAIN OF CULTURE

Why is climate change a partisan issue in the U.S.A. and abroad, with conservatives often hesitating to take action while liberals generally want to take action immediately?

How is climate change perceived, responded to, and argued about differently by various countries and cultures?

How much scientific consensus is there about today's apparent climate change?

How are we to interpret the petitions that suggest more scientists are refuting global warming?

How are various concepts and perspectives on climate change the result of specialists using different time frames and amplitude scales?

Clearly, this list of key questions highlights that the scope and size of the issues surrounding global climate change are so large that to try to provide an Integral assessment is beyond the scope of this chapter.[7] In fact, doing so would be an important book. So, instead of attempting to comprehensively analyze climate change using the Integral Ecology framework (Who, How, and What), we will apply the above-mentioned Integral Ecology guidelines to the issue. By doing this we can sketch out the beginnings of an Integral response to climate change.

The Weatherperson Says . . .

One of the crucial insights of an Integral approach to global climate change is that any statement of "fact" made about the situation has a context and is made by a person with a perspective. There is no neutral weatherperson. In other words, all claims issued about global warming are issued from a perspective, one that may be true within one context but partial within a larger context. Consequently, the framework of the 8 ecological selves provides an important way to contextualize various statements made about climate change. This allows us to identify the main perspectives that are involved and the ones that are excluded. Examining the power dynamics of inclusion/exclusion can reveal a lot about what is and isn't being stated in the conversation. Not only is it important to try to understand the worldviews behind various claims about the situation, but you need also to be aware of and acknowledge your own perspective.

By doing this, by being up front and continually self-inquiring into your own motives, shadow projection, defenses, insights, and preferences, you can break down the tendency to erect an "us-versus-them" barrier that characterizes so much of the debate about climate change. Becoming aware of and owning your own perspective also place you in a more effective position to inhabit the various perspectives involved in the conversation.

It is important to develop ways to communicate about these complex and emotionally charged topics that addresses multiple worldviews, especially those associated with traditional, modern, and postmodern values.[8] The reasons a conservative Christian, an up and coming corporate executive, and a gender studies professor will support a local climate change initiative will most likely be very different from one another. Yet all of these people can support the initiative if you take the time to help them see the value of it on their terms, and if you manage to build alliances between them. These different worldviews will perceive and misperceive issues around climate change very differently. Everyone's view of the issues—including your own—will contain valid insights and distorted understandings. By leading with your own limits and wearing them on your sleeve, you help create space for a deeper conversation. By being actively aware of the perspectival nature of global climate claims, you avoid uncritically accepting what any expert or organization presents as being "just the facts."

Forecasting the Weather

In many respects the climate change discussion is a conversation about scientific data: "What does the research say?" Within an Integral approach, scientific data are not confined to the Right-Hand quadrants. Consequently, the 8 ecological zones of research provide a useful framework for broadening the research base for climate change endeavors.

We believe this is true for at least two reasons. First, empirical science (behavioral and systemic) can be used to support each side of the debate in sophisticated and persuasive ways. Second, systems analysis continues to be incapable of providing accurate predictions of large-scale, multidimensional systems such as those involved with climate change.[9]

Given that the major forms of data collection used in climate change analysis face these limitations, research into climate change would benefit from using more methodological zones. Not only would this provide more reliable data points, but it would also likely lead to new insights that would strengthen our understanding and response. It will be important to include specific techniques of inquiry and research from each of the zones and to consult experts associated with them: climatologists, systems analysts, cognitive psychologists, cultural anthropologists, phenomenological and hermeneutic researchers, and developmental psychologists. At the very least, Integral efforts should include 1st-, 2nd-, and 3rd-person practices for understanding climate change.

For example, at the community level you can combine personal stories of individuals experiencing climate changes in the area (1st-person), establish local focus groups to explore policy changes that can be presented to the city council (2nd-person), and include statistical analysis of temperature trends for the bioregion for the last century (3rd-person). In doing so, we have to be aware of what are the best and most practical research designs for specific needs and goals. A small community project will differ in many ways from a federally funded research grant.

In either case we need to become aware of the best research designs and their limits. Only by drawing on all 8 ecological zones of research can our understanding of the complexity of climate change begin to match the intricacy of the phenomenon. We are kidding ourselves if we think we can predict and respond to climate change dynamics by examining the data from only a few zones.

What Is the Weather Today?

The 4 terrains and their 12 niches provide a very useful lens from which to understand global climate change. Often the problem of climate change is couched in the language of the systems terrain, while solutions are often presented solely in terms of the behavior terrain. Thus, the Right-Hand exterior terrains dominate the conversation, thereby occluding the roles that experience and culture play both in generating climate change and in finding solutions to it. Thus, it is important to use the 12 niches to include more perspectives. For example, analysts could ask if and how psychological dynamics propagate destructive behaviors, and how social systems reinforce those dynamics. Or, how we can use cultural perceptions to obtain more support for proposed policy changes.

The causes and solutions of global climate change are a 4-terrain affair. Hence, it is important both to examine the phenomenon of climate change from the 4 terrains (a quadrivium) and to study how climate change impacts the 4 terrains (the quadrants) of individuals, both humans and nonhuman organisms. Only by considering all 4 terrains in these ways can we really tell what the weather is today. Using the 4 terrains and their 12 niches provides a solid foundation for a nuanced understanding of this phenomenon.

The Who, How, and What of Recycling: A Simplified Example

As a way of further illustrating the Integral Ecology framework, we take up another example: recycling. In this case, we will limit ourselves to showing how a single action could be understood from each niche, eco-self, and zone. By exploring a single example through each of these distinctions, we hope to give you a better sense of how taking into account the Who, How, and What can provide a comprehensive view.

Eight Good Reasons to Recycle

Different levels of ecological identity can support the same behavior. Therefore, it is difficult to ascertain an individual's motivations based on exterior behaviors. Below we explore an example of the hypothetical motivation to recycle for each of the 8 ecological selves. Using the framework of the 8 ecological selves, we can demonstrate a variety of motivating value sets and dispositions to support a single action such as recycling. Brief statements symbolically elucidate each eco-self, which can serve as an illustration of the kinds of motivations that can support a single behavior.

Eco-Guardians are motivated to recycle because it is seen as a ritual to keep the balance of nature satisfied. If they do not recycle, they risk creating disorder in the mysterious balance of things.

Eco-Warriors either refuse to recycle, because they view recycling as a form of control over their own will, or they recycle as an act of heroism, fighting for the Earth's salvation. Whether or not they recycle, they perceive themselves as engaged in the battle against the industrial juggernaut.

Eco-Managers are motivated to recycle because they feel it is their duty, either to God, their community, or to the nation-state. If they do not recycle, they will be subject to punitive measures by authorities or feel guilty for putting the order of the system at risk.

Eco-Strategists are motivated to recycle because it makes good long-term sense. It conserves resources for consumption at a later date and therefore supports the bottom line of profit. If they do not recycle, they are passing up an opportunity to save or make money.

Eco-Radicals are motivated to recycle because it is good for the community and the planet. It saves resources, which can help address social and economic imbalances. If they do not recycle, they are faced with apocalyptic consequences that amount to ecocide.

Eco-Holists are motivated to recycle as a way to keep the energies of the Earth in dynamic flow. Recycling for them makes systemic sense and is viewed as part of complex feedback mechanisms. If they do not recycle, it could produce disastrous and unpredictable systemic results.

Eco-Integralists are motivated to recycle because it is important for the Earth, for humanity, for the nation-state, for members of the community, and for themselves. Here, recycling is an act of dynamic synergy that is performed for multiple, even contradictory reasons. If they do not recycle, it could disrupt natural, cultural, social, and personal harmony.

Eco-Sages are motivated to recycle because for them it is a beautiful and rightful act that simply flows from their being. If they do not recycle, they feel pain as if they had transgressed their own body.

These simplistic examples illustrate that various worldviews understand and respond to environmental issues for completely different reasons, so we must consider individual and cultural interiors when searching for viable solutions to complex environmental issues. It is ineffective to impose the values of one worldview upon another: one cannot convince an Eco-Strategist to recycle for the reasons that an Eco-Radical does. Instead, we must translate the meaning of one value set to another, so that the terms of one perspective can be assimilated into another. Translation begins to occur only at an Integral level of psychological development, because it is the first worldview that can hold multiple worldviews. As a result, the Integral Ecologist is uniquely positioned to serve environmental problem-solving, especially when multi-stakeholders (with different levels of development) are involved.

The Eight Methods to Understand Recycling

To further illustrate the application of the 8 eco-modes, we explain how each methodological family would serve a comprehensive understanding of recycling. Of course, more could be stated regarding each zone. However, this quick overview should help explicate the various methods.

Phenomenological methods could be used to enact, bring forth, and illumine subjective aspects of recycling, such as the felt experience of somatic realties, psychological dynamics, and spiritual experiences for individuals involved with recycling.

Structuralist methods could be used to enact, bring forth, and illumine subjective aspects of the experience of recycling, such as the patterns and sequential stages of experience for different individuals as they recycle.

Hermeneutical methods could be used to enact, bring forth, and illumine intersubjective aspects of recycling, such as the various meaning systems (cultural, religious, philosophical) that manifest through and guide recycling.

Cultural anthropological methods could be used to enact, bring forth, and illumine intersubjective aspects of recycling, such as the patterns and reasons various cultures historically and currently reuse products.

Empirical methods could be used to enact, bring forth, and illumine objective aspects of recycling behaviors, such as the activities involved with recycling by individuals at all stages of the process.

Autopoietic methods could be used to enact, bring forth, and illumine objective aspects of recycling, such as the autonomous recycling behaviors and heuristics of individuals interacting with their environment.

Systems methods could be used to enact, bring forth, and illumine interobjective aspects of recycling, such as its ecological benefits and how it conserves energy flows as well as the political, economic, and legal dimensions of recycling.

Social autopoietic methods could be used to enact, bring forth, and illumine interobjective aspects of recycling, such as institutional forces and forms of communications within multiple systems that self-regulate to prevent or promote a society to recycle materials.

The 12 Niches of Recycling

The 12 niches of environmental concern can serve as an ecological checklist for any environmental project, issue, or situation. If we analyze recycling from an Integral perspective, we explore the ways recycling manifests in each niche. At the first level of complexity there exist: soma (the way the body feels during the act of recycling); communion (the shared cultural practices associated with recycling, such as washing out milk cartons and crushing aluminum cans); movement (the physical acts and behaviors associated with recycling); and intersections (the ecological impact of recycling). At the second level of complexity are psyche (the different psychological states and dynamics connected with recycling, such as pride, guilt, and fear); community (the conceptual and ideological dynamics supporting or preventing recycling); action (the power, race, and class realities involved in recycling); and institutions (the laws, politics, education, and economics supporting or preventing recycling). At the third level of complexity are pneuma (the various worldcentric and planetcentric experiences encountered through recycling); commonwealth (how recycling resonates with various worldcentric or religious-symbolic systems and values); skillful means (the act of recycling as modeling sustainable behavior); and matrices (the subtle energetic benefits for the Earth due to recycling).

Integral Ecology Is Already Happening

For the past decade a growing number of people have been exploring Integral Ecology by drawing on Integral Theory and applying it to the environment (wild, rural, and urban). Integral Ecology weaves together the myriad approaches to the natural world in an effort to respond as effectively and timely as possible to the complex ecological problems that face us, our communities, and our world in an evolving universe. In effect, Integral Ecology unites consciousness, culture, and nature in the service of sustainability. People who are utilizing the Integral Ecology framework recognize that it is not enough to integrate ecosystems and social systems (e.g., economies, laws, schools) since both of these realities are interobjective systems. Instead, what is needed is an integration of subjective (e.g., psychology, art, phenomenology), intersubjective (e.g., religion, ethics, philosophy), and interobjective (e.g., behavior, science, systems analysis) realities.

For the rest of the chapter we will provide a synopsis of various projects that are occurring around the planet that are informed by the Integral Model in general or are actively using specific aspects of the Integral Ecology framework. These inspiring examples of Integral Ecology in action are arranged by Integral research and Integral organizations.

Integral Research

Sustainable Consumption in Calgary, Canada (Cameron Owens)

Cameron Owens presents three years of research, in the city of Calgary, on the various barriers to sustainable consumption.[10] "The focus of my particular investigation was to discern the barriers inhibiting solid waste reduction in Calgary and to utilize the Integral Model as a framework for ordering and understanding the many diverse factors perpetuating waste."[11] By using an Integral approach, Owens was able to identify three distinct realms of barriers to waste reduction: subjective ("where 'I' am the barrier"), intersubjective ("where 'we' are the barriers"), and objective ("where 'it/s' are the barriers"). In addition, Owens uses the 12 niches to explore three levels of complexity within each realm, providing a comprehensive assessment of what is preventing sustainable behaviors, which in turn lends itself to discovering ways of removing these barriers. Figure 11.1 illustrates in brief, as Owens's case study does in more detail, how each of the 12 niches can be used to more fully understand the factors involved with creating or preventing sustainable consumption and waste reduction. In his conclusion, Owens summarizes the value of including all 4 terrains:

> An Integral understanding exposes the problem of focusing our attention and energy on solely one dimension. For example, merely changing regulations or providing facilities does not address some of the deep-rooted personal barriers, such as cognitive dissonance. Providing facilities does not automatically imply that the facilities will be used. As well, without establishing cultural dialogue about sustainability, there may be fervent resistance to government attempts to regulate. In turn, inspiring individual and group transformation may be precluded by not adequately developing fair political/economic institutions or implementing appropriate regulations and infrastructure.[12]

He goes on to explain that "a successful sustainability strategy" must honor the different levels of reality and developmental dynamics. Thus, "sustainability, to the Integral Ecologist, involves encouraging personal, cultural, and structural transformation that allows a wider and deeper embrace of the Kosmos."[13]

INTERIOR

EXTERIOR

INDIVIDUAL

(UL) Personal
In what ways am "I" inhibiting or enabling sustainability?

Somatic Realities
(e.g., feelings, emotions, sensations)

Psychological Dynamics
(e.g., beliefs, attitudes, awareness)

Spiritual Experiences
(e.g., spiritual realizations, wisdom, empathy)

I

(UR) Behavioral
Individual behaviors that inhibit or enable sustainable consumption and waste reduction.

Influence(s) behavior(s)
(e.g., recycling, buying "green" products, shopping locally, supporting regional products, reusing products, using public transportation, composting food scraps, energy conservation)

IT

COLLECTIVE

WE

(LL) Cultural
In what ways are "we" (group/culture) inhibiting or enabling sustainability?

Intercorporeal Dimensions
(e.g., group feelings, shared somatic realities)

Worldviews
(e.g., group beliefs, stigmas, cultural norms)

Compassionate Perspectives
(e.g., ethics, religious community)

ITS

(UR) Structural
In what ways is/are "It/s" inhibiting or enabling sustainability?

Physical & Natural Systems
(e.g., ecosystems, watersheds, built environment, roads)

Social Systems
(e.g., economics, institutions, laws)

Subtle Systems
(e.g., subtle energy, architecture, sacred gardens)

Figure 11.1. Categorizing barriers to sustainable consumption.

An Integral Analysis of Canada's Organic Agriculture Standards (Wade Prpich)

In his case study, Wade Prpich reveals that Canada's national policy concerning organic agriculture has sustainable and ethical aims that it fails to achieve.[14] This case study is a summary of the research Prpich did to receive his master's degree in environmental design, specializing in environmental science. Using Integral Theory, Prpich explains that the failure occurs because of an overemphasis on objective dimensions of livestock treatment (behaviors) and agriculture (systems). The policy and its institutionalization neglect the developmental process that produces the interiors that support ethical behaviors. In addition, Prpich develops and presents a variety of Integral policy recommendations that would more effectively support the aims of the national standard by incorporating subjective and intersubjective understandings of ethical concern and behavior.

First, he proposes that an "All-Quadrant Step" would involve recognizing that "the farmer is an individual subjective I who is immersed in an intersubjective cultural We. Therefore, the National Standard must address the I and We within its policy to complement and support the It/s strategies."[15] Prpich provides a number of examples of how the national standard could incorporate the quadrants into the policy (e.g., the usage of 1st-person and 2nd-person language): including personal accounts of farmers implementing various guidelines, including emotional or spiritual stories of people using the national standard, and explicit discussion of community identity, aims, and goals.

Next, Prpich outlines an "All-Level Step" that would "promote the value of the levels within the [national standard] policy and communicate the importance of cultivating individual and collective interior growth to develop the necessary radius of concern to achieve its stated goals of environmental quality and sustainability and the ethical treatment of livestock."[16] This step, Prpich suggests, could be accomplished, for example, by having Integral workshops and training be part of organic certification or renewal. Such training could facilitate personal and cultural reflection, examine the role of interior capacities in accomplishing the policy goals, and examine the development of an expanding moral radius of concern. Prpich concludes by stating that Integral policy must incorporate both quadrants and levels:

> If the NSCOA [National Standard of Canada for Organic Agriculture] is to be an indisputably sustainable and ethical policy, all the quadrants (i.e., experience, behavior, culture, and systems) and levels (e.g., body, mind, soul, and spirit) must be acknowledged and made accessible as a potential to all individuals who are informed by its guidelines, expressed intent, and philosophical underpinnings.[17]

Case studies such as Prpich's are invaluable in showing how we might begin to reconceptualize and institutionalize Integral policies that better serve our environment.

Integral Research on the "Invasive" Pepperweed in Sebastopol, California (Kevin Feinstein)

Kevin Feinstein recently completed a year-long case study on the issue of the pepperweed in the Sebastopol Laguna.[18] Sebastopol is a small northern California community with a strong commitment to environmental awareness and action. In fact, it is home to one of the only city councils in the United States with a Green party majority. Thus, a public outcry arose when the Sebastopol city council, convinced by the reports of the ecological danger posed to its laguna by an invading non-native plant (*Lepidium latifolium*, or perennial pepperweed), took action to lift the five-year ban imposed on spraying herbicides on city property in order to eradicate or at least control the pepperweed. As a result, the city council agreed to uphold the ban under the condition that the community organize itself and remove the pepperweed by hand.

This situation provided an ideal case study. Feinstein's research focused on the environmental hermeneutics (e.g., the various perspectives involved) but also looked at the pepperweed plant and provided a critique of the field of invasion biology. Combining an analysis of scientific data with an exploration of the worldviews of key players involved on both sides of the issue, Feinstein used Integral Methodological Pluralism to explore the multidimensionality of the pepperweed issue:

> I have used the eight mode methodological pluralism for the primary reason of reaching a more detailed comprehensive view of this issue, attempting to expand beyond the typical flatland (only one or two views) found in most ecological studies. Although, I do not claim my study

to be the final word or all inclusive, I do think the use of the integral framework provides for a larger perspective than any other studies of the issue. The eight modes within themselves are also not set in stone, they merely are an attempt to be more expansive, borrowing from the western philosophical traditions in which Ken Wilber and other integral theory purveyors are so well versed.[19]

Of the many themes that emerged in this multimethod approach, "Paradigms/Worldviews/Perceptions" was one of the most salient. Feinstein explains:

> In many ways, the entire study is about how we hold the natural world, the Laguna, the ecology, and the pepperweed in our individual and collective consciousness. This theme is largely what attracted me to this case, for it was inherent in the discussion. It kept coming up as different viewpoints were argued amongst environmentalists, with and often without the awareness of the different values/memes/paradigms/frameworks in which the arguments were embedded. Not only does the difference in paradigms underlie the entire issue, but a significant amount of attention was given to this theme by the participants. They spoke in these terms often and freely, either specifically naming worldviews or demonstrating them purposefully by their descriptions.[20]

One of Feinstein's primary conclusions was that sound management practice requires the inclusion of various worldviews and perceptions surrounding the pepperweed issue. As Feinstein's research shows, the Integral Ecology framework holds much promise in responding to complex community issues such as those faced by the community of Sebastopol.

Integral Ecology and Environmental Policy Development (Brad Arkell)

Brad Arkell is a doctoral student with a background in zoology, ecology, and environmental studies.[21] He is working under Peter Hay, author of the comprehensive *Main Currents in Western Environmental Thought*, to delve into the world of environmental ethics and values.[22] Also, he currently works full time as a government officer, investigating and assessing the natural values of public land and assisting with the development of protected area policy in Tasmania and Australia.

His dissertation, "Integral Ecology and Environmental Policy Development: Understanding Complex Natural, Social and Political Systems in the 21st Century," investigates the potential for integrating or unifying a wide range of ecological, economic, social, and political (policy) "perspectives" using Integral Theory and the Integral Ecology framework. In particular, he uses Integral Methodological Pluralism to demonstrate how an Integral approach can dramatically increase our understanding of ecological, economic, social, and political systems. His work aims to demonstrate that even having an awareness of the Integral approach can be incredibly useful in the wide range of disciplines, professions, and fields that contribute to policy development.

To demonstrate the efficacy of the Integral analysis, he uses an AQAL approach to analyze a number of theoretical and real-world case studies. These scenarios elucidate the main argument of his thesis: that until we are using (or have) an Integral view or definition of biodiversity, sustainability, ecology, and, ultimately, politics and policy, we will not be able to move first toward a green state and then to an integral one. Arkell explores three main themes to produce strong arguments for this assertion. The first theme presents a general Integral approach, focusing on Integral Ecology in particular and drawing on and synthesizing all of the up-to-date work on both Integral Ecology and the latest evolution of Ken Wilber's work: Integral Methodological Pluralism.

The second theme examines the political and policy environment under which a green and then integral state could arise. This includes an Integral approach to participatory democracy, and a discussion and analysis of green politics and policy in Australia from an Integral view. The second theme draws heavily upon Robyn Eckersley's monograph *The Green State*, which examines how the current liberal democratic state could evolve into the "ecological" or green state.[23]

The third theme explores a number of approaches to policy development using an Integral Model. This includes suggesting how natural resource management could evolve into something more integral, analyzing perspectives of biodiversity to facilitate effective communication with stakeholders, and indicating an Integral approach to environmental policy development. The thesis also provides valuable short appendixes outlining practical tools for use by natural resource managers, extension officers, policy advisers, and environmental educators.

Integral Geography (Brian Eddy)

For over a decade, the geographer Brian Eddy has been exploring cutting-edge approaches to the mapping and representation of ecological, geographical, and social/cultural realties. He has research interests in cybercartography, information theory, GIS science, Integral Theory, multimethodological frameworks, and developing the concept of Integral geography. In his case study, "Integral Geography: Space, Place, and Perspective," Eddy draws on Integral Theory to situate the various components of geography into a more comprehensive geographical approach.[24] In turn, he makes use of the discipline of geography to improve Integral Theory (by making a spatial dimension more explicit): "Integral Theory and Integral Ecology benefit from the use of multiple forms of expression, representation, and understanding of the world. . . . Augmenting Integral narratives with geographical maps may help communicate similarities and differences in perspective, and will be helpful in examining contemporary issues such as globalization, development, sustainability, biodiversity, and geopolitics."[25] Toward these ends, Eddy proposes an Eco-AQAL model (see figure 11.2) that shows the holarchical relations between six primary global ecosystem components. He explains that this variant of the AQAL model is used to connect Integral Theory and Integral Ecology more explicitly with the spectrum of sciences used among geographers into a more inclusive and comprehensive framework: "One variant of an AQAL ontology for addressing human-environment relations is presented here as an Eco-AQAL model, where the ontological realms are represented by primary global ecosystem spheres differentiated against the AQAL matrix [figure 11.2]. In addition to ensuring the differentiation between external and internal dimensions of each level (and their respective modes of inquiry), this holarchical ordering of ecosystem spheres also highlights the intrinsic dependencies of one level upon another, and corresponds to their relative evolutionary emergence on Earth."[26]

Eddy suggests that there are several advantages to this model. First, it improves on many conventional ecosystem models by including the astrosphere (universe) and lithosphere as well as the "anthroposphere" (the human realm). As such, Eddy claims that "this extension of depth in both directions presents the Eco-AQAL model as a Kosmocentric model, in contrast to biocentric or anthropocentric approaches."[27] This model also makes the holarchical relationships between these spheres explicit. Thus,

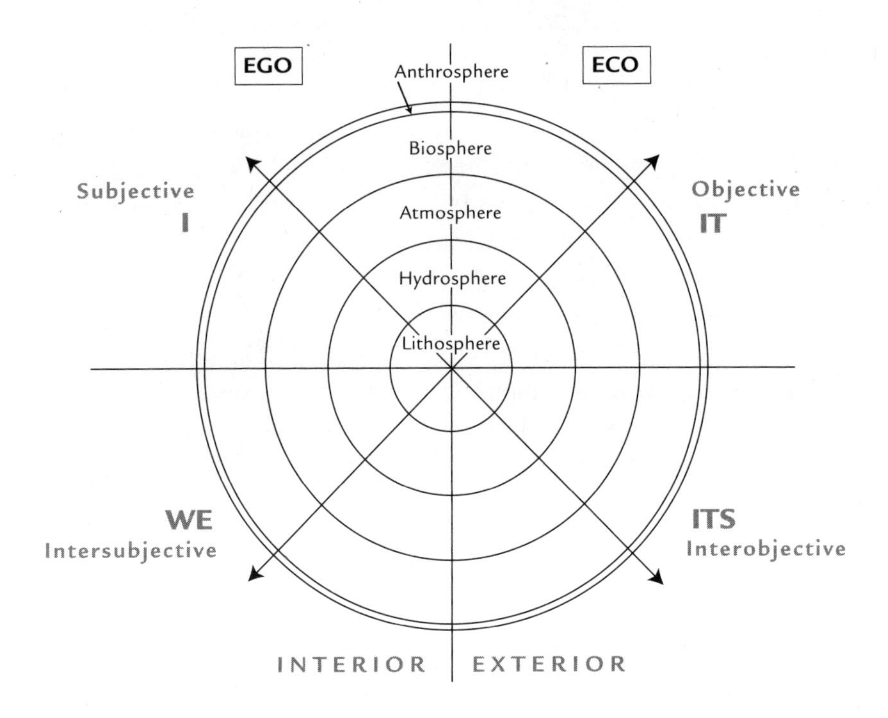

Figure 11.2. Eddy's Eco-AQAL model.

the biosphere can be destroyed leaving the atmosphere, hydrosphere, and lithosphere.

Another advantage lies in the use of anthroposphere to avoid some of the confusions that result from the more general term "noosphere," which can apply to mental capacities in various nonhuman species.[28] For Eddy, the anthroposphere "constitutes the place of human activity, dwelling, and infrastructure, including the many non-sentient holons (artifacts and heaps) humans have constructed."[29] In other words, it is the geographical domain that is primarily the result of human influence, consciousness, and activity.

For Eddy, the Eco-AQAL model also supports Integral Methodological Pluralism by situating various earth, life, and human sciences in holarchical relationship to one another, with more fundamental sciences at the bottom (e.g., earth science) and more significant ones at the top (e.g., human sciences). As a result, scientific knowledge resulting from the "hard" sciences must be situated "within the moral, ethical, and aesthetic realms of decision making concerned with human and ecosystem well-being."[30] For Eddy, Integral geography is concerned with studying various moments that occur across "a variety of geographical scales of interaction."[31] He explores ways to situate the spatial aspects of involutionary and evolutionary currents within various geographical moments that include individuals and their communities in integral places. As a result, an Integral mapping system is detailed by Eddy that combines cosmospheric, biospheric, and anthropospheric dimensions to create integral places.

Eddy's research further demonstrates that one of the key contributions Integral geography can make to collaborative Integral inquiry is to highlight the importance of "scale" in human knowledge and in our observations of, and interaction with, ecological phenomena. He frames this argument in a concept of "geo-ontological contingency" by arguing that the appropriateness of human knowledge about the world is contingent upon both our internal representations (wherein scale pertains to levels of detail in human knowledge) and external representations (wherein scale pertains to a distance factor between the observer and the observed). Extending the philosophy of Gilles Deleuze, his research illustrates how in dealing with issues of sustainability and well-being, it is impossible to avoid both objectivity in discerning "what there is" and our subjectivity in discerning "how it might be." He then suggests that Wilber's AQAL map is not only useful but necessary for navigating geo-ontological contingency. Eddy's research is currently looking into the implementation of the Eco-AQAL model in a Google Earth–type "infrastructure" and "infostructure" for Integral practitioners to map, analyze, and share information on projects around the world.

The Integral Sustainability Center (Barrett C. Brown, David Johnston, and Cynthia McEwen)

Since 2002 Barrett Brown, David Johnston, and Cynthia McEwen have worked with a global network of researchers and practitioners to begin fleshing out an Integral approach to sustainability. As co-directors of the Integral Sustainability Center at Integral Institute, they have documented numerous projects and organizations in the areas of sustainable development, international development, corporate social responsibility, and market transformation that use the Integral Model to foster social, environmental, and economic sustainability. For example, Brown has documented how the HIV/AIDS Group at the United Nations Development Programme (UNDP) is using an Integral approach as part of their Leadership for Results Programme.[32] This program is a global response to the HIV/AIDS epidemic being carried out in over two dozen countries.[33] The following excerpt comes from the group's handbook:

> The four-quadrant framework, adapted from the work of Ken Wilber, is an analytical tool that can be used to explore the relationship between intentions and values, on the one hand, and actions on the other hand. It does so at both the individual and collective levels. It is possible to gain a deeper, more profound understanding of the epidemic by identifying, analyzing and reviewing the causes and origins of actions. By placing current responses to HIV/AIDS in their respective quadrants, we can reflect on how holistic our response has been to date.[34]

This program uses an Integral framework for leadership and capacity development, development planning and implementation, and advocacy and communication.[35] In addition to this initiative, there are a variety of other programs within the United Nations Children Fund (UNICEF) and UNDP that have senior staff using an Integral framework to inform and advance their projects.[36]

One of the major results of the Integral Sustainability Center's networking and promotion of the Integral Model for sustainability is a growing collection of published and unpublished papers, case studies, and articles by various sustainability practitioners. In 2006 they made this massive collection of resources available when they launched—under the auspices of Integral Institute—a free online library that holds the work of 60+ integrally informed sustainability scholar-practitioners, totaling thousands of pages.[37]

Research at the Integral Sustainability Center has focused on translating the AQAL Integral vision into the nuts and bolts of sustainability practice. Brown, Johnston, and McEwen are driven to make the Integral framework as practical as possible to serve positive social and environmental change. Their research is guided by a number of key questions:

> Our work on Integral Sustainability has focused on answering a number of important questions, including: What are the key dynamics in each quadrant in the context of sustainability? What are the quadrant-specific tools and practices which can be used to influence those dynamics? What are the essential steps to performing an Integral analysis and an Integral intervention in a sustainability initiative? How does one effectively

communicate about sustainability to people with different worldviews? How can we organize today's sustainability knowledge, practices, and tools using the Integral framework? What does Integral leadership for sustainability look like in practice? How do we help sustainability leaders develop their internal capacities and external skills so they can better respond to complex global issues? What are the best practices concerning Integral Sustainability in business, government, and civil society? How can we apply Integral Methodological Pluralism to make sustainability research more effective?[38]

Through public workshops, private consulting to individuals and organizations, leadership briefings, and writings, the co-directors at the Integral Sustainability Center are exploring these questions. In the process, they are empowering people with the insights of Integral sustainability.[39]

Hester Brook Retreat (Will Varey)

The Integral Model is providing the operating platform for an Integral refuge and restoration project in the Southwest of Australia. Hester Brook Retreat was established by the Forsyth Trust in 2004 as a 50-year Integral Ecology project.[40] The land consists of a single block of 150 acres (60 hectares) of rejuvenating Jarrah-Karri forest within an agricultural district. The location is on a valley system and has a semipermanent watercourse running through it, providing a diversity of micro-ecosystems. Located on a corridor between other conservation zones, the block is a refuge for many endangered and rare species.

This area has had a rich variety of uses since the 19th century. The area was partially cleared in the 1890s, was heavily logged of old-growth trees in the 1920's, was used for cattle grazing in the 1960s, was neglected and suffered heavy weed infestations in the 1980s, and was burnt out in the 1990s. As Will Varey, the main Integral practitioner behind Hester Brook Retreat, explains, "When [the area was] acquired, things were both getting better in terms of natural reforestation and worse in terms of non-indigenous intrusions and infestations."[41]

Varey has been instrumental in using AQAL to inform a diversity of projects of inquiry and restoration using different ways of knowing, including "the empirical scientific, the ecological systemic, the ethnographic cultural, and phenomenological experiential dimensions of the land, its inhabitants and its visitors."[42] These projects, as Varey explains, "are informed by AQAL/ Integral Theory as the epistemological basis for understanding the continuously emergent properties of the location in terms of its whole/parts and its parts as a whole."[43] As a result of its Integral foundations, Hester Brook Retreat provides a unique opportunity for Integral Ecological research and inquiry.

Integral Design (Mark DeKay)

Mark DeKay is an associate professor in the College of Architecture and Design at the University of Tennessee in Knoxville.[44] He has been a leading figure in the development of approaches to Integral design at Integral Institute.[45] DeKay has been applying Integral principles to a variety of design contexts for many years.

For example, he recently coauthored a paper with fellow architect Mary Guzowski, "A Model for Integral Sustainable Design Explored Through Daylighting," for the 2006 American Solar Energy Society Conference (ASES). After presenting an overview of Integral Theory, this article discusses Integral Ecology's 4 terrains and applies them to sustainable design by asking, "From an integral sustainability perspective, how shall we shape form?"[46]

From the perspective of behaviors (UR), the design question is, How shall we shape form to maximize performance? In this terrain, good form minimizes resource consumption and pollution while maximizing preservation and recycling.

From the perspective of systems (LR), the design question is, How shall we shape form to guide flow? In this terrain, good form supports ecological patterns by creating structures in the built environment that best accommodates ecological process through mimicry of and fitness to the context of natural ecosystems.

From the perspective of cultures (LL), the design question is, How shall we shape form to manifest meaning? In this terrain, good form reveals and expresses "the patterns that connect" in ways that celebrate the beauty of natural order, place inhabitants into relationship with living systems (or the idea of nature), and situate human habitation in bioregional place.

From the perspective of experiences (UL), the design question is, How shall we shape form to engender experience? In this terrain, good form orchestrates rich human experiences and creates centering places conducive to self-aware transformation, in which we can become most authentically who we are. Summarizing this in figure 11.3, they explain that "this expanded multi-perspectival view can enable designers to more comprehensively address the complexity of today's ecological challenges by including the individual, cultural, and social dimensions that contribute to the creation of a sustainable world."[47]

Building on the the 4 Terrains of Sustainable Design

DeKay and Guzowski present the 12 niches of daylighting to illustrate how they "could integrally-inform sustainable design thinking, allowing a designer to move beyond objective physical performance to also consider rich experiential, ethical, moral, and social implications of light."[48] After presenting an overview of the 12 niches of daylighting (see figure 11.4), they apply them to daylighting at the United Theological Seminary's Bigelow Chapel in New Brighton, Minnesota.

In addition, DeKay organized the 2006 retreat (July 15–19) for the Society of Building Science Educators, which represents the group of faculty in architecture schools who deal most directly with sustainability issues. The theme was "Integral Sustainable Design: Re-integrating What Modernism

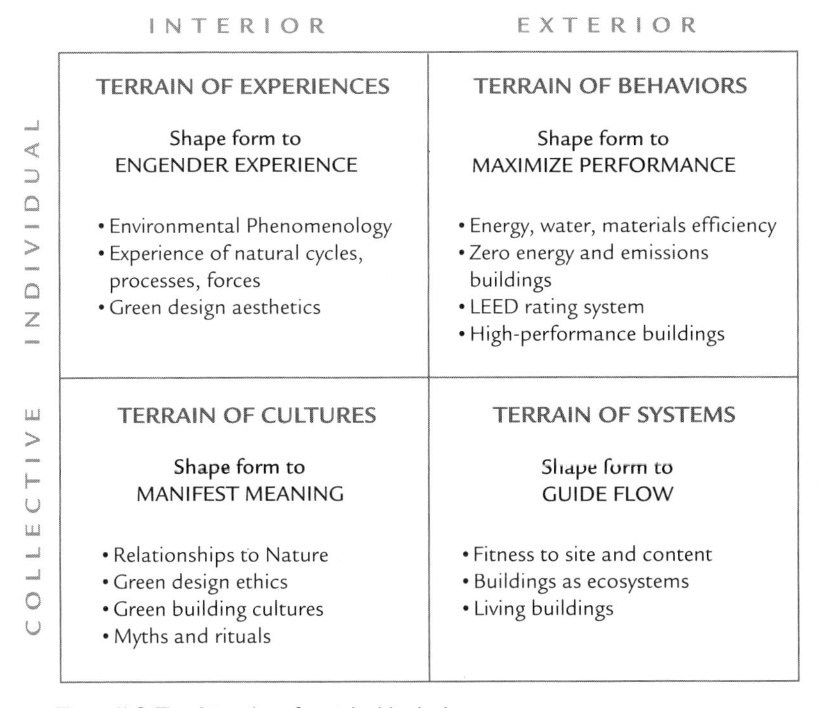

Figure 11.3. The 4 terrains of sustainable design.

INTERIOR | EXTERIOR

INDIVIDUAL

TERRAIN OF EXPERIENCES

[L3] Pneuma: *Spiritual Realization*
- Experience of subtle or spiritual aspects of light
- Mystical light and enlightenment
- Experiencing Inner Light

[L2] Psyche: *Psychological Dynamics*
- Environmental psychology of light
- Phenomenology of light
- Aesthetics, beauty, poetics of light

[L1] Soma: *Body Feelings*
- Physical sensations of sunlight and daylight
- Archetypal responses (diurnal, seasonal)
- Experiences of delight, comfort, health, and well-being

Upper Left (UL) I

TERRAIN OF BEHAVIORS

[L3] Skillful means: *Effective Actions*
- Models for sustainable action
- Designing healing environments
- Building transfiguration (alchemy of light)

[L2] Action: *Intentional Conduct*
- Dynamics modeling and simulation (quantitative and qualitative)
- Technological systems, constructions, codes
- Architectonic systems and spatial order with light

[L1] Movement: *Physical Behaviors*
- Physics of light in buildings
- Biology of vision
- Physiological response to light and visual comfort

IT Upper Right (UR)

COLLECTIVE

Lower Left (LL) WE

TERRAIN OF CULTURES

[L3] Commonwealth: *Compassionate Perspectives*
- Perspectives on nature
- Cosmology and light
- Spiritual / religious symbolism and interpretation

[L2] Community: *Shared Horizons*
- Values for daylight and conservation
- Ethics of light (right-to-light, and healthy places)
- Building cultures and pattern languages of daylight

[L1] Communion: *Shared Cultural Practices*
- Daylight savings time
- Rituals of light (celebrations, seasonal holidays, lighting candles, sunsets, bonfires)
- Meaning of characteristics of light and dark (positive and negative)

ITS Lower Right (LR)

TERRAIN OF SYSTEMS

[L3] Matrices: *Subtle Systems*
- Effect of light on subtle energy patterns
- Color properties of Inner Light
- Alexander's Field of Centers

[L2] Institutions: *Social Systems*
- Urban planning for light access, politics
- Education, building and energy codes
- Social activity patterns (work schedules, programming, etc.)

[L1] Interaction: *Natural Systems*
- Ecological impacts and fitness of architectural daylight
- Bioregional and contextual response
- Complex building systems integration

Figure 11.4. The 4 terrains and 12 niches of daylighting.

Differentiated and Post-modernism Dissociated."[49] One of the guiding questions for the retreat was: "Can we discover a meta-theory of sustainable technology? How do we make sense of the multiple understandings of technology—for instance, can we reconcile Chris Benton's love of measuring all things measurable, Marietta Millet's fascination with the experiential quality of light, Lances LaVine's cry in the wilderness (a song of rich symbolic meaning), and the complex building-as-a-whole-systems logic of John Reynolds?"[50]

DeKay describes the Integral theme for the retreat as follows: "For the 2006 SBSE Retreat we will explore how a developmental approach that accounts for stages, levels, or waves of evolution among the many different lines of human capacity—including our highest potentials—might help us open new ways to teach and learn sustainable design."[51] The retreat featured a variety of presentations and discussions that explored Integral themes and applications. For most at the retreat this was their first exposure to the Integral Model, and while there were many questions, there were also fruitful conversations that will serve the ongoing explorations of Integral design.

DeKay is also using Integral design principles to guide the Beaver Creek Watershed Green Infrastructure Plan.[52] This project is connected to the Green Vision Studio at the University of Tennessee's College of Architecture and Design. DeKay and his colleague Tracy Moir-McClean are the principal investigators. The project is a visionary exercise with the goal of creating a document that can be used by various individuals and institutions to support their decision-making process around issues of preservation, conservation, and land use development. According to DeKay, "This is probably the most integrally informed project we have. Our use of Integral Theory and Integral Ecology shows up in various places in the language of the report and in the issues we chose to address and the methods we used to address them. While the Integral Model is not overt it nevertheless is the guiding framework."[53] While the details of the project extend beyond the scope of this section, figure 11.5 provides good overview of how Mark and his colleagues are using the Integral Model in the context of this project.

Integral Placemaking (Ian Wight)

Since the mid-1990s, the urban designer Ian Wight has been exploring the place and purpose of an Integral approach to regional and city planning.[54] In fact, Wight was one of the few individuals at the first Integral Ecology

	INTERIOR	EXTERIOR
INDIVIDUAL	**UL: EXPERIENCE** **Individual-Interior: Self and Consciousness** The invisible subjective interior reality of an individual **Intention:** Rich, full-sensory experiences. Facilitate individual development to higher levels of ecological self. Provide opportunities to relate to nature at each level. **Areas Addressed:** Human experience of landscapes, ecological education, ecological aesthetics, landscape and urban aesthetics, environmental psychology, personal identity, 1st-person phenomenology. **Tools:** Integral aesthetic awareness of the designer, role-taking, design for nearby nature, access to natural recreation, visual impact assessment, view-shed analysis, environmental education, and educational landscapes. **I**	**UR: BEHAVIOR** **Individual-Exterior: Organism and Part** The visible, objective, external reality of an individual **Intention:** Settle the land with minimum impact and maximum health. Conserve resources and reduce flows to environmental sinks. **Areas Addressed:** Resource conservation and efficiency, reducing water and air pollution, habitat conservation, farmland, wetland, and forest preservation. **Tools:** Best Management Practices (BMPs), ecological monitoring, performance standards such as TMDL, conventional engineering, flood zones, stream buffers, slope protection rules, conservation easements, NRCS techniques, urban forestry. **IT**
COLLECTIVE	**WE** **LL: CULTURE** **Collective-Interior: Meaning and Worldviews** The invisible, intersubjective, internal realities of groups **Intention:** Come to agreement on how to design and live with nature. Manifest rich symbolic human-ecological relationships. **Areas Addressed:** Civic dialogue, fitness to cultural context, cultural regionalism, sense of place, Genius Loci, historical and cultural landscapes, collective values, meaning of form languages, ideas of and relationships to nature(s). **Tools:** Charettes, community visioning, public input processes. Pattern Languages of the building culture, multiple intelligence communication, visual preference surveys, symbolic design languages, civic design, urban design with nature for community identity, shared knowledge bases.	**ITS** **LR: SYSTEMS** **Collective-Exterior: Social Systems and Environment** The visible, interobjective, external realities of groups **Intention:** Fit settlement pattern to natural pattern using principles of ecological order. Making cities work like nature. **Areas Addressed:** Bio-regionalism, spatial pattern of habitat, ecological restoration, greenway and park networks, ground-surface water system, pedestrian-bicycle networks, building-transit-open space patterns. **Tools:** GIS landscape analysis, urban design, zoning, including special historic, cultural and environmental overlays, comprehensive planning, site review processes, design guidelines, development and building codes. Pattern Languages, land trusts, parks foundations, Low Impact Development (LID), Smart Growth, ecological engineering, science of landscape ecology, contextual thinking, network thinking.

Figure 11.5. Green infrastructure in the 4 quadrants.

gathering held by Integral Institute in 2000. Wight is an associate professor and head of the Department of City Planning at the University of Manitoba, Canada. In his case study "Placemaking as Applied Integral Ecology: Evolving an Ecologically-Wise Planning Ethic" he describes the work he has been doing:[55]

> Planning as placemaking is therefore hypothesized as a serviceable bridge between past (basically spiritually-bereft, exterior-situated) planning agendas and the projected necessary spirit-affirming and spirit-embracing integrative postmodern planning. Such ecologically-wise planning, featuring the integration of the "It," "We" and "I" domains, entails what Wilber calls "Integral practice." It is contended that place and placemaking can be usefully conceived as the manifestation of Integral practice in a planning context—especially a planning that sets its fundamental purpose in terms of being an active part of "getting subjects to agree on how to live in accord with nature."[56]

In providing a critique of typical notions of growth and sustainability, Wight conceives of Placemaking as a postmodern Integral practice:

> Placemaking—it is suggested—comes much closer to capturing the full dimensions of the challenge, and encompasses the spiritual fuel to take us even deeper and wider. Such a "placing" of planning practice also cautions against too easy a focus on just "its," as in cities or regions of any kind. These are Flatland formulations—which can easily marginalize the "I" and "We" perspectives if we are not consciously "Integral" in our planning practice. "Integral practice" would respect all perspectives by situating them in the AQAL framework. Here is the new challenging terrain for a constructive postmodern planning theory, fully mindful of the integration of mind, culture, and nature.[57]

He goes on to outline how this practice would be one that brings together subjective and intersubjective aspects of planning with the more conventional objective approaches and provides a place for Spirit to be included. Wight does a particularly good job of showing how many theorists and practitioners connected to the study of place provide important parts of such an Integral Placemaking practice.

Integral Organizations

The Institute for Sustainable Futures (Chris Riedy)

Chris Riedy is a research principal at the Institute for Sustainable Futures, University of Technology, Sydney, Australia. Reidy has energy sector and greenhouse policy expertise, experience with futures tools and methods, and knowledge of participatory approaches to policy development.

At the Institute for Sustainable Futures, Dr. Riedy and colleagues are using an Integral framework to guide a transdisciplinary approach to environmental policy and sustainability.[58] In his PhD research, completed in 2005, Riedy used Integral Theory to explore behavioral, psychological, systemic, cultural, and developmental perspectives on Australian energy and greenhouse policy.[59] As part of this research, he has developed strong familiarity with stakeholder perspectives in the Australian energy sector and with methods for capturing multiple perspectives in policy development. Following from this work, the Institute for Sustainable Futures is using Integral Theory to develop and apply innovative participatory approaches that seek to include multiple individual and cultural perspectives in environmental decision-making. This approach was trialed in two projects during 2006, one focused on catchment management and the other on regional climate change response.

Drishti—Centre for Integral Action (Gail Hochachka)

For over a decade Gail Hochachka has worked in ecological research, environmental advocacy, public engagement, community development, and local resource management. She is a research associate of the POLIS Project on Ecological Governance at the University of Victoria, working on the methodology and practice of Integral community development. In 2003 Hochachka founded the Drishti—Centre for Integral Action "with a vision for transformational eco-social change that blends current approaches to environmental and social issues with practices of contemplative action."[60] This nongovernmental organization (NGO) is based in British Columbia, Canada, and is composed of board members with experience in business, environmental groups, government, and community development. The Drishti homepage explains:

> Our approach to eco-social issues is guided by integral theory. . . . This approach brings together a myriad of disciplines and strategies into a dy-

> namic whole, such that they inform and complement each other. This approach also integrates the "exterior," practical aspects of life (such as ecology, economics and social systems) with the "interior," subtle aspects of humanity (like psychology, culture and spirituality). By uniting disciplines and by acknowledging the role of interiority in society, the integral approach includes more of reality in its embrace. Thus, it offers a more comprehensive framework for understanding eco-social issues and more appropriate methods of working with such issues.[61]

Projects currently being carried out by members include "environmental activism, youth empowerment and capacity building, workshops and conferences on integral applications, and community development."[62] As an organization, Drishti focuses on the following four project areas: research and writing, public engagement, Integral community development, and art and Integral Ecology.[63]

Drishti serves to support global well-being in two key ways. First, Drishti is a learning community for dialoguing and deepening understanding about Integral praxis. Activities in this area include research and writing, public engagement, and being a node for Integral dialogue. Second, Drishti is also a platform for working with an Integral approach to global well-being. Its work in this latter area consists primarily of research, writing, workshops, and capacity-building initiatives that profile or promote an Integral approach to community development, sustainability, international development, ecology, and leadership. Drishti also uses the Integral Model to support applications in leadership and management, organizational development, social marketing and messaging, strategic planning and project implementation, and group process work.[64] In addition, Drishti provides a number of important resources for Integral practitioners.[65] The most noteworthy is an electronic version of Hochachka's recently published and pioneering work, *Developing Sustainability, Developing the Self: An Integral Approach to International and Community Development.*[66]

While Drishti's global activities and interests extend beyond just Integral Ecology (to include such areas as Integral community development and Integral international development), much of the work Drishti is involved in is directly connected to or has important implications for Integral Ecology. Some noteworthy examples include Drishti's recent project entitled "Case Studies on an Integral Approach to International Development," which has some important findings for any attempt to include human interiority within ecological and environmental efforts; Integral Ecology projects in Nigeria and Ethiopia; and ecological aspects of Integral community development in El Salvador. Also, see Hochachka's case study in part four, which presents her research and community work in El Salvador.

Case Studies on an Integral Approach to International Development

Drishti's recent project—"Case Studies on an Integral Approach to International Development"—explored how organizations and practitioners are integrating interiority in international development work, looking at both integrally informed organizations (those explicitly using an Integral approach) and folk-integral organizations (those implicitly using Integral principles and concepts).[67] This research project sought to learn more about how organizations and practitioners are engaging interiority, in a culturally sensitive or culturally specific way, and as an integrated response to a given community development challenge. There were seven participating organizations from around the world.[68] The four key findings of this research as presented by Hochachka included:

1. All participating organizations recognized that human interiors have enormous influence on the process and outcomes of sustainable development.

2. Organizations and practitioners use diverse methodologies to work with human interiority.

3. Organizations' and practitioners' approaches to working with interiority often moved through stages in accordance with beneficiaries' changing needs, which may correspond with the various stages in the spectrum of psychological development.

4. The integration of interior and exterior methodologies was an important, if not necessary, component to the success of development projects for these participating organizations.[69]

These findings highlight that ecological and environmental projects will be more successful if they include human interiors in their efforts.

Integral Ecology in Nigeria

Over the last few years Drishti has become familiar with and supported the ways that Integral Ecology is being used in Africa. In 2004, Drishti carried out an organizational assessment of One Sky, the Canadian Institute for Sustainable Living, which works in human rights and the environment with partner organizations in Sierra Leone and Nigeria (and formerly also El Salvador).[70] This work involved a site visit to Nigeria and a follow-up visit in 2006. Over that time period, Drishti witnessed One Sky increasingly using aspects of Integral Ecology to articulate why they work in coalition with other partner organizations and how this type of trans-individual approach gives rise to a larger vision and moral underpinning for sustainability. Even though not all members of the organization and its partner organizations fully embrace or use the Integral approach, leadership does use the Integral framework to be better able to make sense of the complexity involved in the process and activities of their Nigeria project. Within this leadership there is a growing awareness of the importance of including an understanding of interior stages and the need for capacities and methods to work with them.

Integral Ecology in Ethiopia

In Ethiopia, Hochachka works with Negash Shiferaw, a consultant for an integrally informed leadership development program (spearheaded by UNDP HIV/AIDS Group). Shiferaw applied aspects of an Integral framework in his master's degree program in sustainable development. In his fieldwork, he trained six facilitators in a methodology called Community Conversations, which includes the quadrants of Integral Theory.[71] These facilitators are using this methodology with communities in the Zeway Watershed of the Rift Valley in Ethiopia. The facilitators explained to Hochachka that they find the quadrants very useful for understanding more clearly the reasons and motives for environmental degradation. One facilitator, Hussen Urgesa, a diplomat in natural resources management and now a development agent in the Abine Germama Kebele (district) near Zeway town in the Rift Valley, explained that, by using quadrants, he can see that many reasons may motivate the behavior of a man who cuts down a tree. Some of these reasons include "concern for the family, fears around economic security, social norms in the village, etc., in other words, the reasons for the behavior show up in all quadrants. This has given [the facilitator] a way to identify the deeper causes of ecological degradation, and also enabled him work more effectively to address those deeper causes."[72] Urgesa states that the Integral approach helps him to identify the many factors contributing to development issues like deforestation and sustainability.

Integral Ecology in El Salvador

While working in El Salvador over the last decade, Hochachka had the opportunity to work with Centro Bartolomé de las Casas, a popular education organization that works toward community well-being in a postwar context (see part four). Integrally informed and appreciative of the AQAL model, Centro Bartolomé members use an Integral approach to support their efforts at raising environmental awareness and community development. They include Integral themes in workshops that help people connect in new ways to nature. Larry José Madrigal, the project coordinator, explained to Hochachka the various Integral applications regarding the environment, including their use of multiple methodologies, such as reflection, art and symbolism, subjectivity exercises, and embodied practices. These activities are guided by the Centro's insight that "the change in behaviors that impact the environment takes place through a change in one's sense of self in relation to the environment."[73] Centro Bartolomé de las Casas has also used integrally informed reflection exercises with El Salvador's main environmental organization. For example, they had participants spend intimate and reflective time with a tree in the forest, to touch, smell, and hold the perspective of that tree, relating differently to it, and to themselves, in the process. This is a very subjective and intersubjective methodology, to "raise awareness" on the environment, compared with the more common approach of objective information sharing.

One unique methodology used by the team of Centro Bartolomé is called "subjective scripture," in which they facilitate a process whereby participants read passages from the Bible in an embodied manner. Centro Bartolomé explains that ecological degradation is sometimes encouraged by Upper-Left and Lower-Left factors, such as a belief that "the Bible tells us to dominate nature." Subjective scripture (or embodied readings) enable individuals to revisit those sections of the Bible to experience a different sense of their meaning. Through this process, participants discover that the Bible (e.g., Genesis 1:26) can be interpreted as saying that humans should "have dominion over" other life forms, but it can also be interpreted to mean "administrate" and "care for" Creation. In such processes, Centro Bartolomé members engage the interior dimensions of individuals in a way that works to deepen and transform their faith.

The Permaforest Trust (Tim Winton)

The Permaforest Trust was founded by Tim Winton and operates as a not-for-profit sustainability education center located 200 kilometers south of Brisbane on the subtropical east coast of Australia.[74] The center is located on a 100-acre property of regenerating rainforest, organic gardens, and eco-forestry and permaculture systems in a bioregion defined by the lush volcanic landscapes of the Mount Warning shield volcano. The Permaforest Trust hosts an annual 40-week residential training program in Accredited Permaculture Training and related sustainability and ecology disciplines.

The Permaforest Trust uses Integral Theory through the application of the AQAL framework as both their underpinning philosophy and the basis of their entire operational practice. Students, management, and supporting members of the Trust all undertake core training in the AQAL framework at the beginning of each semester. The program is an intensive community and learning environment where individuals seek to model strategies for shrinking their environmental footprint and lowering their resource use. The focus at the center is on learning to live within natural limits while maintaining high levels of personal, communal, and ecological well-being.

In joining the program at the center, students voluntarily enter a world where electricity, fuel, water, personal space, transport, and many other material resources as well as personal freedom are all limited by natural resource flows and community commitments. This poses distinct challenges to people accustomed to living in a society with unprecedented amounts of ever-growing resources. Using the AQAL framework has allowed the participants to analyze the nature of these challenges and to develop strategies to deal with them effectively. From the perspective of Winton, one of the most important contributions of an Integral approach is the recognition that ecology and sustainability have important personal (UL) and community (LL) dimensions. By shifting some of the focus to these domains, individuals at the center have been able to limit community conflict and enhance group dynamics. Moreover, they have introduced a critical awareness of self-care. Both of these initiatives have freed up energy to increase ecologically positive outcomes in land restoration practices (LR). In addition, as Winton explains, the various elements of the Integral Model have served as powerful distinctions for the goals of the Permaforest Trust:

> Developmental stages and lines have helped the centre's leadership understand patterns in the challenges faced by students and to create solutions; quadrants have been used to develop a whole new language of "I Space," "We Space," and "Eco Space," which has helped raise awareness of the scope our work; and masculine and feminine typologies have allowed us to monitor and balance energies and approaches to community life over time. In general, Integral Theory has translated into effective ecological practice in numerous and often surprising ways. At the Permaforest Trust it has proven over and over again to help us create solutions, innovations, and strategies for challenges that were intractable using other approaches.[75]

Centre for Human Ecology and the Falkland Centre for Stewardship (Nick Wilding)

Nick Wilding, an organizational consultant, action researcher, and transformational educator, uses Integral Theory and the Integral Ecology framework to inform the design, content, and process of his work in a variety of contexts and with several environmental organizations in Scotland.[76] In particular, there are four major areas where Wilding applies Integral principles.

1. Wilding runs a variety of scheduled programs in human ecology and action research through the Centre for Human Ecology, which is based in the department of geography and sociology at the University of Strathclyde, Glasgow. He also teaches courses within the Master's of Science in Human Ecology and supervises master's theses. Core and option components of this master's program in human ecology are explicitly based on Integral principles, and he introduces Integral Ecology concepts in a four-day workshop that is part of the program. Through the center he has recently developed a professional certificate in action research, which explicitly works with 1st-, 2nd-, and 3rd-person practices and methods of inquiry.

2. Alongside the MSc Human Ecology and the Action Research professional certificate, Wilding directs a new program for emerging leaders living and working in rural communities. This Rural Leadership Programme began in autumn 2006 and is based at the Centre for Stewardship in Falk-

land in Fife. Throughout this program Wilding applies an Integral lens to the concept of stewardship that is central to its curriculum. The purpose is to "offer action-inquiry learning programmes for rural people who, building on their passion for their places, are leading the way to resilient and sustainable rural community futures."[77] Wilding emphasizes a "head, heart, hands" approach grounded in Integral action research.

3. Wilding has incorporated a developmental levels approach within an 18-month learning journey called "Get Your Voice Heard." This program is designed for emerging black/minority ethnic activists engaging with local democracy in North Edinburgh.

4. In all of his consultancy work with Scottish Natural Heritage, local authorities, and NGO networks, Wilding is employing Integral Theory, Integral Ecology, and Integral action research.

Integral City (Marilyn Hamilton)

Marilyn Hamilton uses the Integral framework and Spiral Dynamics to design ecologically informed and operationally sound change in cities.[78] She is the founder of www.integralcity.com and TDG Global Learning Connections. Her focus is on integrating personal and cultural value systems, social structures, and infrastructures in the city, within the context of the climatic and geographic life conditions of eco-regions.[79] Hamilton has developed Integral City Systems (ICS), a decision-making framework for ecological interfaces of the built environment and quality of life that embrace aesthetic, ethical, and scientific realities. ICS is designed to address leaders, teams and groups, organizations, communities, cities, and metro-regions, and includes assessing, mapping, capacity building, facilitating processes, and planning and tracking. Figure 11.6 presents her approach to integral cities.[80]

She has also written a number of articles promoting Integral meta-mapping as a common language for urban change using the 4 quadrants and eight levels of complexity within them to reveal the emergence of capacities at multiple scales, and supporting the need to balance interior abilities with exterior aptitudes.[81] She has used these arguments to address issues of bio-security (especially avian flu), urban risk and resilience, and vital signs monitors. In addition, Hamilton teaches, facilitates, and researches integrally based leadership, organization change, and systems thinking with

	INTERIOR	EXTERIOR
INDIVIDUAL	**Quality-of-Life Citizen** A Quality-of-Life Citizen is the "I" and "mind" of the city. They seek more quality, less conflict, more creativity, less inefficiency, and more inclusivity. They hope for a bigger picture, a better way. Thus they: • Motivate intentions • Discover the complexity of community • Demonstrate leadership • Work for quality of life	**City Managers** City Managers act like the "brains" of the city. They need meshworks, to make hierarchies and networks effective, because so much is dependent on them. Thus they: • Develop leadership capacities • Negotiate change and diversity • Meshwork effectively with council, staff, agencies, and others • Plan strategically for the global village
COLLECTIVE	**Foundations** City Foundations represent the "we" and "heart" of a healthy city. They focus on quality of life, asset mapping, community development, and social capital to create new agendas. Thus they: • Develop board capacity • Map visions and cultural assets • Fund effective projects • Meshwork strategically for a healthy city	**Developers** City Asset Developers build the "body" of the city. They integrate sustainability. ecology, and spirituality in a way that achieves elegance, order, flexibility, and flow. Thus they: • Mesh design, creativity, and knowledge • Build compassionate bridges to the city of the future • Develop and manage technology systems • Structure, pattern, and process for the future

Figure 11.6. Hamilton's Integral City framework.

such institutions as Royal Roads University, California Institute of Integral Studies, the Banff Centre, and Adizes Graduate School.

SalmonBerry Designs (Kevin Snorf)

SalmonBerry Designs is an ecological design and consulting company based in Santa Cruz, California. The majority of current clients are suburban homeowners. In addition, SalmonBerry Designs does larger-scale projects in the rural areas of Santa Cruz County and in the greater Bay Area. Design and consulting services include business consulting and organizational design, land design, garden design, interior and home design consulting, and housing-land interfaces. The company was founded by Kevin Snorf to develop a method of introducing sustainable design techniques to populations that would not normally take interest in environmental values.[82]

In its fledgling stages, SalmonBerry Designs struggled to take flight, just like any business. With the recognition that he was constantly pushing his own bias for sustainability and ecology and that clients were not responding, he began to draw on Integral Theory and tried a different method. Instead of trying to sell the postmodern concepts of "ecology," "organic landscaping," or "sustainability," he tried asking clients—many of whom had more traditional or modern values—what they truly desired in their home and yard. He found a pattern of answers that was surprisingly consistent. Clients repeatedly asked for the same things. They wanted homes and yards that were affordable, easy to care for, and aesthetically appealing—"like *Sunset* magazine" (an amazingly common phrase).

Snorf considered how to make his methods and techniques in ecological design, permaculture, and organic landscaping work for homeowners in ways that satisfied their stated needs. As a prerequisite, projects were designed to be affordable, maintenance-free, and aesthetically appropriate. As he started consciously taking his clients' perspectives and inhabiting their worldviews and values, he began marketing the business accordingly and SalmonBerry Designs came into demand.

During client interviews, other values and preferences showed up that were easily integrated with Snorf's new methodology. He considered types of plants, aesthetics, colors, textures, and formation of space. As Snorf came to know his clients better, they started to discuss their lives, interests, and hobbies. Deeper values and meanings in the design process surfaced. While working in the landscaping industry as an undergraduate, Snorf saw the creation of million-dollar organic landscapes that were never used. He realized that if people could not find a way to make such landscapes part of their everyday lives, they were far less likely to use such spaces. Sustainability had more to do with the practical value people associated with a space than with the materials used or ecological value derived. Snorf started his business because he liked watching people come alive while relating to nature, not because he liked landscaping or home design. He wanted to create space that people used and enjoyed, while connecting them with the outdoors. As a result of using Integral Theory and the Integral Ecology framework, Snorf has come to realize that "the job of an integral designer is to dig for what people are really striving for and create space that meets those needs, wants and values. People understand, perceive, and relate to nature and ecology in very different ways. Thus, effective designs have to reflect diverse perspectives instead of mirroring the designer's own perceptions and preferences. People's values, beliefs, needs, and hopes need to become part of the design so that their interiors interface with the exteriors of the nature around them. Integral ecological design considers both how to design for people's interiors and how designs will interface with their interiors."[83]

In order to get a sense of his clients' interiors and their relationships to nature and the aesthetic they liked, Snorf asks clients to name their favorite places to spend time in nature. This is helpful in making plant choices and designing the "feel" the client hopes to get from the space. Snorf also explains ecological technologies or methods, often without talking about the environmental benefits of those technologies. For example, he often mentions that a drip system saves tons of time and money on watering in California's Mediterranean climate without also highlighting that such a system also saves water. Most clients care about having to do less maintenance and keeping costs down, and aren't all that interested in the fact that drip irrigation saves water. Consequently, it is often only at the end of a design project that he will casually mention the "green" dimensions to the design installed or created, pointing out to clients that the project was environmentally friendly and that they were making a major impact on resource savings; that they were going to have more wildlife (butterflies, birds, bees, hummingbirds, etc.) than they would know what to do with; and that many of the new plants in their yard were medicinal, edible, or aromatic.

Designing environmental and ecological benefits into a project while tending to the client's explicit design needs validates the need for designers to hold multiple perspectives. An integrally informed ecological design process allows the designer to translate ecological benefits into the perspectives of the various eco-selves. When designers take the perspectives of all potential clients, they increase effectiveness and create more dynamic designs. Integral design weaves the interior terrains into the design process, asking designers to be skillful about how they create cultural and psychological space. Snorf explains:

> The most technologically sustainable building in the world is not sustainable if it looks horrendous and people do not enjoy it. For better or worse, every person has some relationship to nature. The confusing part is that people care about nature, ecology, and sustainability for different

reasons. As designers, we have to engage those concerns. This is why SalmonBerry Designs shifted its focus from designing environmentally only in the systems terrain, to creating habitat in all four terrains.

Spaces need to be created that are sustainable in the objective right hand terrains of ecology and organisms, socially acceptable and enticing, culturally palatable and meaningful, and psychologically fulfilling and engaging. To be truly effective in any Integral design endeavor, all of these perspectives and considerations must be encountered.[84]

Snorf is currently taking time to write and develop the theory and methods of Integral design. In addition, he is expanding the scope of the business to engage large-scale commerce and industry, and also putting together a team of designers and professionals to look at intractable environmental problems in corporate business and industry. Using eco-strategist incentives to engage those interests, Snorf is moving Integral design beyond the residential scale. He is also planning an Integral Design Center as a retreat space for designers to problem-solve and learn more about themselves and their designs. In addition, Snorf teaches yoga and martial arts, applying Integral Theory and Integral coaching in those disciplines as well.

Teleosis Institute (Joel Kreisberg)

In 1999 Joel Kreisberg founded Teleosis, a not-for-profit organization dedicated to health-care professionals serving the global environment.[85] Kreisberg is the executive director of Teleosis and a practicing chiropractor, homeopath, and environmental educator. He also teaches courses in holistic health and education at John F. Kennedy University.[86] Kreisberg has drawn on the Integral Model for many years to inform his work and life. One of the results is that he has been a pioneer in establishing the field of ecologically sustainable medicine (ESM), which is committed to identifying and promoting health-care practices that do not harm people or the environment. In 2004 Kreisberg launched *Symbiosis: The Journal of Ecologically Sustainable Medicine*.[87]

Not only has Kreisberg drawn on the Integral Model in general, he has forged new ground in Integral Ecology. In 2005 he completed a master of arts in Integral Ecology from Prescott College.[88] Throughout his studies he applied the Who, How, and What framework to explore issues related to plants and humans. For example, he has used the 12 niches to generate a series of questions about medicinal plants: "To create an integral ecological understanding of medicinal plants requires . . . using an integral methodology. Using the [12 niches], the following discussion begins this inquiry. The first step is generating a series of questions. Using the four [terrains] and their three levels of complexity [i.e., the 12 niches] allows us to develop a map of what is required to initiate a discussion of the ecology of plant medicines in an integral manner."[89]

In another article he used Spiral Dynamics to explore different worldviews of redwood forests: "The purpose of an integral ecology is the integration of a variety of perspectives. The goal is to ensure a balance of viewpoints with an appropriate venue for each. . . . The subject of this [article] is redwood forests, with its dominant species *Sequoia sempervirens*. Many perspectives of *Sequoia sempervirens* ecology are examined using the organizing principle of Spiral Dynamics. One could examine any ecosystem or biome with this method. . . ."[90]

For his thesis he used Integral Methodological Pluralism to provide an impressive study of the ecology of sudden oak death, which is ravishing the native oaks in northern California:

> The integral ecology of Sudden Oak Death is an exercise in using eight approaches to our understanding of the ecological phenomenon Sudden Oak Death. The approaches are: natural history; structuralism, the structures of self-identity in relationship to nature; ethnobiology, how traditional peoples of California lived with oak trees; autopoiesis, how oak trees are self-regulating; complexity theory, how oak trees are part of a self-regulating system called Gaia; environmental hermeneutics, our ethical valuation of oak woodlands; and finally phenomenology, a personal understanding of the oak woodland.
>
> By focusing on oak woodlands of California and Sudden Oak Death, this paper offers the reader the opportunity to apply integral methodology to solving ecological and social problems. Integral ecology is a 21st century tool for academic and social research, offering powerful perspectives that will translate into novel solutions to the ecological problems we face as we continue to learn how to care for our first home, the planet Earth."[91]

Kreisberg concludes his Integral research by observing that "an integral ecology provides a powerful tool from which we can and will foster our understanding of nature and ourselves, encouraging openness to the potential of the 'self-disclosing and self-enacting Kosmos.'"[92] The insights Kreisberg has gained using the Integral Ecology framework have helped make Teleosis the unique and successful organization that it is.

Teleosis has three major goals: to educate health professionals about the principles and practices of Ecologically Sustainable Medicine, to build a community-based network of Green Health Care professionals, and to provide affordable high-quality, sustainable medical services benefiting underserved populations and the environment in which we live.[93] To accomplish these goals Teleosis has a variety of programs, including the publication of *Symbiosis: The Journal of Ecologically Sustainable Medicine*, and the Green Health Care Program, an environmental health educational program that supports medical professionals in establishing green health clinics, becoming environmental health advocates, and practicing medicine more sustainably.[94]

Throughout its many activities, programs, and organizational structure, Teleosis draws on the Integral Model in various ways to support its mission and goals. The influence of the Integral Model can readily be seen in the Green Health Care Program. For example, the following overview appears in the Executive Summary of the Green Health Care Program:

> The Green Health Care Program offers an integral approach to health care and the environment. An emphasis is placed on the continuous transformation of the medical professional by attending to multiple perspectives of physical well-being, mental and emotional health, meaningful relationships, and environmental sustainability. Throughout the three components of the program [greening one's workplace, becoming an environmental health advocate, and providing sustainable medical care], medical practitioners highlight the links between themselves, their patients, their communities, and the environment. As a result, practitioners benefit from integral action that reconnects personal health with vibrant communities and a flourishing environment.[95]

With programs like this, Teleosis provides the service of demonstrating how to translate Integral principles into action.

Open Sky Wilderness Therapy (John Dupuy)

Open Sky Wilderness Therapy is a licensed outdoor treatment program based in Durango, Colorado.[96] It uses a holistic and Integral wilderness treatment program that combines traditional clinical modalities, naturopathy, and wilderness therapy. Employees work with teenagers between the ages of 13 and 17 and young adults from 18 to 28 on a variety of issues, including drug and alcohol abuse, ADD, low self-esteem, depression, anxiety, and anger management. Programs ranging from 5 to 12 weeks in length are based on a "multidimensional foundation" of body, mind, heart, and soul and recognize student growth stages that move out from self to clan (program participants and staff) to family back home, and finally to the larger community.

John Dupuy, the recovery services director for Open Sky, is a pioneer in the field of wilderness therapy and was the primary designer of Passages to Recovery, the first adult wilderness recovery program. Since the spring of 2006 he has been using the Integral Model to support the work at Open Sky:

> We have used the Integral map at Open Sky to approach the problem of chemical dependency. The impetus for the Integral approach was that standard treatment was usually not very successful, and even when programs were trying a more holistic approach, they often lacked a unifying map to guide the treatment process. This is one of the great gifts of what the AQAL map provides.
>
> The first step is to take a look at the client, using the 4 quadrants. Once that has been done, we look at the work that needs to be done, once again using the quadrants. The client is then explained and taught the AQAL map to provide coherence and understanding about the recovery process. This usually takes place over three or four days in a wilderness setting.[97]

Once clients have an understanding of the Integral Model, the program has them begin with the challenges that are present in each of the 4 quadrants and the work that must be done in each. The next phase is to talk about lines of intelligences and an "Integral Recovery Practice." There is a clear delineation within the program between altered states and stage development. Dupuy explains:

> This is not a hard sell, as our drug-taking students know all about altered states. We explain how in the past wilderness treatment was often an altered-state experience. Like the Esalen workshops in the 1970s, after

> X amount of wilderness time the students would leave feeling radiantly healthy and renewed, but upon return home the experience would fade, the old behaviors would emerge, and relapse would happen. That is the challenge. The wilderness can wonderfully facilitate an awakening, but how do you keep from going back to sleep when one leaves the wilderness? Integral practice![98]

Dupuy explains to the participants that they need to begin a practice that they will want to continue for the rest of their life—one that will convert their altered state experiences into stage growth. Dupuy feels this is extremely important because the problem must be "up-leveled": "In chemical dependency and addiction this means that the student must go from egocentric to at least ethnocentric, from the locus of control going from the reptilian brain to the neocortex. From 'I want what I want' to 'my recovery is not just about me.' And how we achieve this is an ongoing four-line (body, mind, heart, and soul) Integral Recovery Practice.[99]

The Integral perspective becomes the focus of day-to-day treatment that includes Body (nutrition, resistance training, yoga/stretching, cardio work); Mind (learning the AQAL map, discussions, reading great books); Heart (group, individual and family therapy, including trauma work); and Soul (an ongoing contemplative practice). Many Open Sky students, who use binaural beat entrainment technology for their meditation practice, are making remarkably rapid spiritual and emotional progress. The contemplative practice is adapted to meet them where they are. One student, for example, prays and reads her Bible, while another meditates. Therapists also teach participants the three major states of consciousness: gross, subtle, and causal, or waking, dreaming, and deep sleep, and their corresponding brain waves (beta/alpha, theta, and delta) and how these figure into meditation and recovery.

With regard to developmental stages they begin with a basic explanation of egocentric, ethnocentric, worldcentric, and Kosmocentric. Later, for more advanced students, they include an overview of the Spiral Dynamics value memes. The vertical dimension is very important to the programs at Open Sky. Dupuy explains, "In treatment we move from very primitive addictive 'I've gotta have drugs' to taking responsibility for the addiction. The addictive 'subject' becomes 'my addiction.' The addiction is transcended, but owned."[100]

Discussion of types is also included in the programs at Open Sky. Both the basic masculine/feminine types with their attendant strengths and pathologies, and the enneagram personality system have proven very useful. Thus, the Integral Model with its five elements has provided a powerful framework for the important wilderness therapy work provided by Dupuy and his colleagues at Open Sky.[101]

Green Light Trust (Chris and Ilsa Preist)

Green Light Trust is a small-scale environmental organization working in the Eastern Region of the United Kingdom.[102] Its main role is to encourage communities to create and maintain new woodlands using traditional wildlife-friendly management techniques. Its ethos is very much one of empowerment; it does not aim to create these woodlands, but rather seeks to train and mentor groups of local people in hands-on woodland management, biodiversity, community engagement, and self-organization. This results in the creation of long-term sustainable woodlands that can be managed by the local people. In this way, Green Light Trust has catalyzed over 30 new community-owned woodlands in East Anglia, one of the most intensively farmed areas of the UK.

The organization has been very successful on the ground but has been struggling to attract support from governmental and other large funding bodies. For this reason, it requested support from Chris and Ilsa Preist, two Integral practitioners, to identify the issues and transform the organization. To do this, they used a combination of Integral Ecology, Integral organizational development, and Torbert's action inquiry.

The Integral Ecology framework was used to identify the different benefits of a new community-owned woodland, both to the natural world and to humankind, according to the AQAL model. For example, the Preists used the 12 niches to identify some of the benefits of a community-owned woodland:

> In applying Integral Ecology to small-scale community-managed woodlands we have used the 12 niches as a helpful framework to understand the multifaceted nature of our work. For example, carrying out physical work outdoors leads to a feeling of physical well-being (soma) and improved physical health (movement). Being involved in a team learning together how to plant and manage trees (communion) results in monoculture agricultural land becoming a new woodland with high biodiver-

sity (intersections). A feeling of psychological balance (psyche) comes from practical hands-on outdoor work (action) and associated improvement in mental health. A shared community project, involving diverse members and age groups in the community, results in a strengthening of communal feeling, making the community feel more safe and secure, and a sense of connecting community with nature (community). A local group with long-term existence results in drastically improved long-term volunteer rates beyond the involvement of Green Light Trust, and therefore a very cost-efficient way of generating new natural green space, contributing to regional and national governmental sustainability strategies (institutions). In some cases, by being involved in a long-term project with regular contact with nature, people make a connection with being part of something larger (pneuma). Their actions spread beyond the project, resulting in private land in the area becoming managed in a more sustainable and biodiverse way, and other sustainability projects emerging (skillful means). In some cases, communities have reached out to other cultures involved in protecting or creating woodlands, such as villagers in Papua New Guinea, and formed a "'global kinship'" link of friendship between the communities (commonwealth) contributing to shared understanding and appreciation across cultures, and practical mutual assistance and support in caring for the planet's energetic systems (matrices).[103]

In addition, the Preists have been using the Integral Model to identify how best to communicate about Green Light Trust's projects not only to Eco-Radicals within the community, who naturally resonated with the organizations goals, but also with the more traditional Eco-Manager sensibilities of community members and the Eco-Strategist perspectives of funding bodies. This process has also allowed the Preists to support the organization's own self-assessment. Consequently, they have focused their efforts on boosting the Eco-Strategist, orange rational aspect of the organization. This involved catalyzing a broadening of organizational culture, to ensure that the directors and staff of the organization could appreciate the Eco-Strategist perspective, even if it wasn't what primarily motivated them. It also involved identifying Right-Hand evidence-based correlates of the Left-Hand benefits the organization had traditionally emphasized, presenting this to funders depending on their different priorities, and the gathering of evidence using a combination of empirical, systems, phenomenological, and hermeneutical approaches.

For example, one funder was particularly concerned about community but wasn't convinced by the initial phenomenological accounts of the benefits given by Green Light Trust. By collecting data and anecdotes from existing communities, the Trust was able to give additional hermeneutical (e.g., improved social structures as evidenced by a stronger sense of community that was formerly fragmented), empirical (e.g., reduced vandalism and antisocial behavior in the area), and systemic (e.g., the formation of a lasting and effective local community organization) evidence, resulting in a significant three-year funding arrangement. Clearly, the work of Chris and Ilsa Preist demonstrates that Integral Ecology can be a highly effective way of improving the way an environmental organization tailors its message to its different stakeholders, as well as ensuring that different values and ways of thinking are understood and appreciated by members of the organization. This can result in significant improvements in organizational performance.

What's Working (David Johnston)

In 1990 David Johnston was named Builder of the Year by the Washington, D.C., chapter of the National Association of the Remodeling Industry. That same year he was also inducted into the *Remodeling* Magazine Hall of Fame as one of the Top 50 builders in the country. Two years later Johnston founded What's Working, an Integral green building consulting company.[104] This company specializes in such things as cost/benefit analysis for environmental features of building, sustainable building materials, and energy and environmental policy development, communicating to various value systems. Johnston is a well-known expert in the field of sustainable construction.[105] For years his work has been explicitly informed by the Integral framework: "By using integral methodology, Johnston has shown both builders and homeowners alike the importance of integrating exterior social, economic, and political systems with the interior motivations and value systems of each set of stakeholders. Once people and organizations realize that they can integrate their deep values (i.e., the desire of many re-modelers to protect old growth forests) with their own businesses, Johnston's integral approach to green building is embraced with unbridled enthusiasm (so much so that 120 groups have been certified in the course of just nine months)."[106]

Johnston has used the Integral approach to guide a number of other projects he is involved in. For example, he has developed an Integral plan for market transformation for a New England green building residential program. He has also been working on transforming the entire U.S. construction industry based on Integral principles. In 1995 he played a major role in starting the City of Boulder's Green Points Program, which requires that anyone seeking a building permit must meet at least 65 environmental standards (out of a checklist of 280). He has worked with the Denver Metro Home Builders Association, helping them develop their Built Green program for new homes, which has certified over 13,000 homes. Johnston has also worked with the Colorado Office of Energy Conservation, developing their statewide green building program, and for the City of Los Angeles, developing their Sustainable Buildings program.

One of the most interesting examples of Johnston's application of Integral principles occurred in his involvement with the Alameda County Waste Management Authority in northern California. He used an Integral methodology to develop a residential green building program for the cities of Oakland and Berkeley. In the process he developed an innovative approach to market transformation, using an "Integral catalyst" model. His efforts resulted in the creation of the Bay Area Build It Green program (www.builditgreen.org), which is sponsored by key players (manufacturers, builders, remodelers, architects, and various public agencies) from nine counties that are part of the Bay Area:

> The organization is now an example of a horizontally and vertically integrated nonprofit, connecting as it does the vertical axis of manufacturers, distributors, and the retailers of green building products with the horizontal axis of homeowners/buyers, builders, re-modelers, architects, and realtors (all of whom have been trained in the integral model by Johnston himself). This has allowed businesses to take the burden away from public agency intervention, transforming the market in the process, and it is no surprise that this is one of the fastest growing "green built" programs in the country.[107]

The success of this program, largely due to its Integral methodology, has led the State of California Building Industry Association to adopt aspects of it with the Energy Star program. The result is an integrally based, statewide green building program sponsored by homebuilders. Without a doubt, Johnston has been a pioneer in using Integral principles to serve the environment.

Next Step Integral (Stephan Martineau)

For fifteen years Stephan Martineau has worked in the areas of community development, ecosystem-based planning, and watershed management. In 2003 he founded the international nonprofit Next Step Integral (NSi), which is "dedicated to the advancement of human consciousness and to the integral embodiment of our human potential."[108] The main emphases of NSi are on applications of Integral Theory and the AQAL model in the context of Integral education, Integral parenting, Integral community, and Integral Ecology.[109] For our purposes we are going to focus on their work in Integral Ecology in general and Integral forestry in particular.

In January 2004 Martineau joined forces with Lisa Farr, the director of a local watershed association, to begin the arduous task of establishing an Integral approach to a community forest project in the Slocan Valley of British Columbia, Canada. This goal was particularly daunting given the historical tensions over a 35-year period between various worldviews within and outside of the community (e.g., loggers, miners, farmers, environmentalists, First Nation individuals, artists, practitioners of several religious faiths, government workers, and a multinational corporation)—not to mention the nine failed attempts by the BC government to establish a workable solution to the divisions within the community between stakeholders connected to the forest. The guiding principles of their initiative included recognizing and honoring the diverse perspectives about the forest among Slocan Valley residents; recognizing that these perspectives are informed by lenses associated with each of the quadrants (e.g., cultural, psychological, historical); and recognizing that any viable and long-term solutions would have to integrate the many conflicting views within the community. In addition, Martineau has identified a number of "main capacities" explicitly grounded in the Integral Model but used implicitly to support their initiative: holding and inhabiting multiple perspectives; an awareness of and an ability to work with the multiple lines of individuals; a commitment to personal growth and shadow work; creating shared motivations; balancing empathy, engagement, and impartiality; and cultivating qualities, attitudes, and capacities that supported mutual understanding.[110]

On January 14, 2007, NSi submitted an application for a Community Forest Agreement to allow the local community to manage 35,000 acres of contested forest.[111] In July 2007 their proposal was approved! Thus three years of negotiations and grass roots work guided by the Integral Model resulted in the creation of the "world's first large-scale integral forestry co-operative."[112] This community forest project has the support of an impressive 95% of the inhabitants in the valley. Aptly named the Slocan Integral Forestry Cooperative (SIFCo), this project is a true testimony to the power of the Integral Model—even as an implicit guide—in working with diverse perspectives to achieve a common goal that other approaches failed to manifest. For Martineau, "the most exciting part of the story has been to facilitate and witness the coming together of multiple worldviews and value systems to embrace an Integral vision."[113] In summarizing his involvement with this Integral initiative, Martineau explains, "It has been my experience that it is only by using an integral approach that a diverse and divergent collective can come to a resolution, a vision, and an action plan forward."[114] Now that SIFCo has been granted tenure over the land, the coming years will be an important testing ground and source of clarification of the tenets of Integral Ecology. We are excited to watch the Integral efforts of NSi and SIFCo as they continue to unfold.

Conclusion

This collection of examples is wonderfully far-ranging in its topics of exploration as well as its illustrations of applied Integral Ecology. This variety of examples of Integral Ecology in action speaks directly to the flexibility and coherence of its theoretical foundation: Integral Theory. By acknowledging the multifaceted nature of ecological systems and problems, Integral Ecology creates a space for multiple perspectives to contribute to the sustainability conversation. In our ever-evolving universe, each of us should strive toward inhabiting multiple perspectives—especially those that stand in contrast to our own habits of thinking and feeling. Only through developing such a capacity for worldcentric perspective-taking can we adequately achieve the mutual understanding so desperately needed on a planet fragmented by conflicting worldviews and approaches.

Practitioners of Integral Ecology are committed to honoring and including the multidimensionality of reality as well as cultivating their own capacity for worldcentrism. It is this dual commitment of comprehensive embrace and perspective-taking that allows the individuals in the above examples to be so successful in their Integral endeavors. Each one of them is serving as an exemplar of what a more Integral world looks and feels like.

Having provided a brief overview of nearly two dozen examples of Integral Ecology in action, we close this book by providing more in-depth case studies. In part four we have chosen three case studies that provide a great overview of the Integral Ecology framework in action.[115] The first case study is by Gail Hochachka, whose organization, Drishti, was discussed above. In it she explores an Integral approach to sustainable development, drawing on years of researching and living in a coastal community in El Salvador. She illustrates how current approaches to community development can be more effective through an Integral methodology that acknowledges and works with the interiors of individuals and communities. In particular, Hochachka explains the methodology that she employed with two focus groups (a women's group and a fisherfolk cooperative) to foster psychological and community development from egocentric modes toward worldcentric perspectives.

The second case study is provided by the marine ecologist Brian N. Tissot, who uses the Integral Ecology framework to analyze the complex eco-social dynamics surrounding the lucrative aquarium industry connected to Hawai'i's beautiful coral reefs. In applying Integral Ecology to fishery management, Tissot has been able to analyze and work more effectively with the multiple-use conflicts that have occurred between tourists/snorkelers, native Hawai'ians, marine ecologists, and fish collectors.

The last case study is by Darcy Riddell, an environmental activist and leadership trainer, who utilizes the framework of Integral Ecology to explore effective conservation strategies. In her case study, she demonstrates the importance of the Integral capacity to understand and work effectively with multiple layers of systems (natural and social) and values (personal and collective) from an evolutionary and developmental perspective. This Integral capacity leads Riddell to situate environmental activism within the 4 terrains as a way of expanding the sphere of activity that activists work within for effective change.

PART FOUR

Applications of Integral Ecology in Self, Other, and World

CASE STUDY I

Integrating Interiority in Sustainable Community Development:

A Case Study with San Juan del Gozo Community, El Salvador*

GAIL HOCHACHKA
Drishti Centre for Integral Action, Vancouver, Canada

This case study explores the global need for, and the process toward, an Integral community development; an approach that integrates exterior needs (such as economic growth, resource management, and decision-making structures) and interior needs (such as cultural, spiritual, and psychological well-being). Including "interiority" in development is unique to conventional and alternative development practices, and this analysis suggests it is necessary for sustainability. In other words, while the planet is surely in need of sustainable ways of living, sustainability is not arrived at merely by systemic interventions; rather it also requires a deeper understanding of, and engagement with, interpersonal and personal dimensions such as worldviews, values, and motivations. Integral community develop-

*This chapter is an expanded and revised version of "Integrating Interiority in Community Development," which originally appeared in *World Futures: The Journal of General Evolution* 61 (2005), nos. 1–2: 110–26. Address correspondence to Gail Hochachka, Drishti—Centre for Integral Action, 4211 Doncaster Way, Vancouver, BC V6S 1W1, Canada. E-mail: gail@drishti.ca

ment works in three domains of Practical (action/application), Interpersonal (dialogue/process), and Personal (self-growth/reflection), and engages the developmental nature of worldviews. Using this approach in a case study in El Salvador, research outcomes showed increased collaboration and self-reflection, where economic objectives merged with equity and environmental concerns.

KEYWORDS: community development; community economic development; integral development; healthy communities; expanding worldviews; values; interiority; developmental psychology; sustainable development; Jiquilisco Bay, El Salvador.

Expanding and Deepening "Development"

Our canoe silently moved across the lagoon as we embarked on the nightly fishing trip—lanterns from other canoes glowing in the distance, mangrove trees on either side, and the melodious *slap* of the *ataraya* fishing nets softly reaching our ears as we paddled. My awe of this experience was echoed in the fisherfolk around me—nothing in the capital city of San Salvador compares to this, they explained, smiling, even if the city is more "developed." One wonders exactly what kind of "development" has taken place in the urban centers anyway. According to UNDP Human Development Report for El Salvador (Rodriguez, 2001), although the country has had a growth rate of 4%–5% a year since 1992, inequality in income distribution remains one of the most pronounced in the world, where 20% of the most wealthy receive 18 times that of the poorest 20% (Rodriguez, 2001, 2, 10–11). Minimum wage is lower than what it was in 1996, and civil violence is still rampant, making the country one of the most violent and insecure in the world (Rodriguez, 2001, 1–3). From international economists to these local fisherfolk, analysts agree that increases in economic growth in El Salvador have resulted in a *certain kind* of development, but not in a comprehensive or sustainable development of the country (Rodriguez, 2001, 10–11; Torres-Rivas et al., 1994).

This partiality in the focus on development on primarily economic and material indicators results in an asymmetrical development. Attention and investment from development interventions goes toward building hard capacity, technology, infrastructure, and the techno-economic base, yet less focus and resources are directed toward developing the soft capacity of the society, cultural well-being, social capital, and the psycho-cultural base for the country. This asymmetry in development's focus may be one of the key barriers for sustainability as well, as we will soon explore. This elucidates the central questions of this case study, namely *what is development, and how can development expand its focus to integrate more of reality in its definition and process, particularly to include communities, ecosystems, and "interiority"?*

For my master's thesis, I explored this question within the community of San Juan del Gozo, El Salvador, for nine months from 2000 to 2002 (see figure I.1). This research was later presented in a book, entitled *Developing Sustainability, Developing the Self: An Integral Approach to International and Community Development* (2005). In 2006, our nonprofit organization, Drishti—Centre for Integral Action received a grant from Canada's International Development Research Centre (IDRC) in Ottawa for further research into the role of interiority in international development (this project will be referred to as the Drishti Case Studies). This case study focuses primarily on the study in El Salvador, although at certain appropriate points, I draw also on the Drishti Case Studies research project (www.drishti.ca/resources.htm, retrieved January 3, 2008).

Through this Salvadoran case study and in the subsequent Drishti case studies with six organizations in Africa, Asia, and Latin America, it is becoming increasingly clear that a comprehensive engagement in "development" includes material, exterior dimensions such as economic growth,

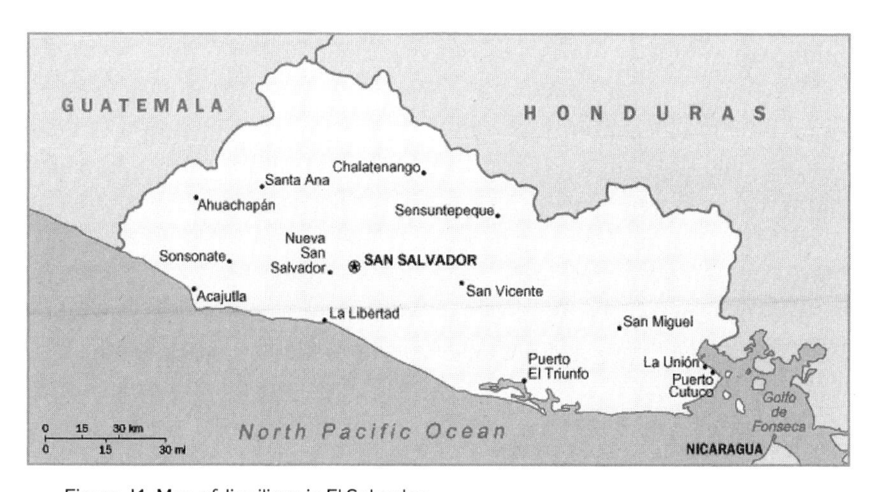

Figure I.1. Map of Jiquilisco in El Salvador.

institutions, and built infrastructure, and also the *interior* aspects of a particular community or nation. Often these interior aspects of development, such as social relationships, cultural understanding, and psychological well-being, are not only as important as the material needs, but also supportive of stabilizing these exterior interventions over time. For example, long-term success in forest conservation cannot exist only in exterior initiatives such as policies or codes of conduct, but is supported synergistically by the emergent understanding, values, and/or interior motivations of individuals and communities that carry out that policy or code of conduct. Far too often, it is believed in development that quick fixes consist of new material "things," be it a new policy, new infrastructure, new institutions, or new funding sources. Repeatedly, across the planet, we see the vestigial remains of these isolated interventions: a latrine to support access to clean water is misunderstood and stands unused; a fancy building built as offices for highly trained workers remains empty; a policy envisioned by one politician is overridden by the subsequent politician and the deforestation continues. These interventions are not inherently wrong, but they are partial, and that partiality does not give rise to long-term sustainable results. My research has found that when the thinking behind development shifts to integrate exterior and interior dimensions of social change, the methodologies and approaches used also shift, including more of reality in the development process, and often the results actually begin to stick. Thus, an Integral community development—in which communities, ecosystems, and interiority are included in development practice—holds more potential for environmental sustainability. We will get into further examples momentarily.

This research also found that when practitioners themselves are engaged in personal growth (i.e., with conscious practices that support their own personal development), their ability to work toward an Integral community development increases. It is through their own personal growth that practitioners are more able to consistently access expansive, less self-contracted awareness (i.e., less anthropocentric and egocentric) and are able to let go of personal and/or professional agendas. This seems to first manifest as humility, a willingness to not know all the answers, and to meaningfully work with the community in addressing local concerns. This reflects one of the core assumptions of Integral Ecology, which describes how a worldcentric capacity to see through and beyond our ideological, cultural, racial, and gender differences is needed to find and sustain viable solutions to environmental degradation (Esbjörn-Hargens, 2005). We shall need new technologies and systems to address environmental deterioration, as well as the interpersonal and moral capacities to relate skillfully and compassionately to one another; the latter of these is truly an interior aspect of the self's own development.

While in this first case study in El Salvador and throughout the subsequent Drishti Case Studies it became evident that this approach to development practice is rare, I am also hopeful that the field of international development is unusually oriented toward its own refinement and development. From the community level to the international sphere of influence, development practitioners increasingly seem to recognize this need for integration in reflection of the global complexity today, and some also recognize the deep role that personal growth plays in this endeavor.

In the following subsection, I explore further the interior and exterior dimensions of change, looking at why partiality has been the norm for so long and how we may come to greater freedom and fullness, attuning to and moving with the gestalt of community unfolding.

Interior and Exterior Dimensions of Change

Since the inception of the formal field of international development post–Bretton Woods, what we will call conventional development has focused on exterior aspects of human society that relate to material needs, and has tended to exclude the more qualitative interior components of human life, such as ethical, cultural, psychological, and spiritual needs. The epistemology of modern economic theory that underpins conventional development purports to offer objective "truth," and yet at best provides a partial truth and understanding of a multifaceted whole. Not surprisingly, the field of development has been only partially effective, and oftentimes outright ineffective or harmful, over the past five decades.[1] Here, I will briefly look more closely at why this unilateral focus on exteriority arose as it did.

In many respects, this focus on measurable, material, and tangible aspects of development is understandable—it was truly the dignity of modernity, which differentiated religion from science, that allowed both to more fully provide their own unique offerings to society (Wilber, 2000). Surely the innovations of science, such as modern medicine, technology, and communications, would not have been possible without this differentiation. However, this differentiation was prone toward a disassociation in certain cases, and as a result, one of the legacies of modernity is a tendency to preference objectivity over subjectivity. In some ways, this too was understandable. Without the right ways to engage in subjectivity, it can be a simply bewildering domain of human life—how does one deal with beliefs, faith, values conflicts, and psychopathologies, especially if one's training has been in a single scientific field? Some scientists working in community development have explained to me their *understanding* of the need to work with personal and interpersonal issues, and yet they hesitate to do so since subjectivity is so "messy" compared with the "cleanliness" of science. Understandably, yet unfortunately, the field of conventional development tends to just leave this subjective dimension out, perhaps even merely for lack of means to work with it. The interior domain of subjectivity does not go away, however, and continues to factor into every moment; whether or not we decide to include interiority into the development process does not take away the fact that *the domain of "I," or of the self, exists.*

Examples of this run a vast spectrum. For example, when interiority is ignored, instances occur where individuals suffering from personal trauma are implicitly inhibited from fully participating in community development processes. Or, in another more extreme example, when interiority is not included, situations can arise in which religious fundamentalist groups colonize the domain of subjectivity, giving a place for community people to express their beliefs and faith in ways they were not able to within a development project; and yet this does indeed impact on that development project's effectiveness. A final example of this can be seen with forest conservation efforts that focus primarily on the interlocking social, economic, political, and ecological systems of the issues. This approach may initially seem quite successful. However, without the "interior buy-in" (such as alignment with values, motivation, cultural resonance) from community people, from the larger society, and from the consumers, often the gains made for systems change is quickly rolled back.

This modern, conventional approach (what would be called "orange altitude" in Integral Theory terms) emphasizes the "expert" with his or her interventions and engaged in a one-way relationship that treats local people more as objects than subjects in their own futures. The interventions are primarily embedded in scientism and technologically focused, which is not necessarily wrong, but rather, the interventions are not also embedded in the psychological and cultural context. Without engaging and affirming the local interpretations of meaning, these interventions of money, resources, and infrastructure often go unused and misunderstood, and are ultimately ineffectual. These interventions lack connection to the context, culture, and consciousness of the people living there.

Paul van Schaik, an international development consultant for organizations such as UNICEF, reiterates how development "activities have largely operated in the upper and lower right-hand quadrants (objective and exterior) . . . and have to a large extent ignored the interior and cultural quadrants."[2] The International Development Research Centre author William F. Ryan, S.J., also explains:

> Security, sustainability, and stability often depend on a system of values that has taken centuries to develop within a specific society. Current development strategies, however, tend to ignore, often underestimate, and sometimes undermine cultural values or the cultural environment, which are essential to healthy human development. The question, then, becomes: How can human values and belief systems be properly integrated into the modern economic development paradigm? (Ryan, 1995)

Seeing these limitations, many civil society organizations attempted to fill in the many gaps left by conventional practices, reframing "development" with more postmodern, postcolonial, and participatory approaches (Another Development, EcoDevelopment, Community Economic Development).[3] The importance of what we will call the alternative development approach cannot be overstated—the alternatives make room for communities and ecosystems in the primarily economic objectives of conventional development. And yet these alternatives also tend to privilege their own part of the greater whole.

This postmodern alternative approach (what would be called "green altitude" in Integral Theory) brought validity to knowledge on traditional, place-based knowledge, often differing in both in method and outcome from the conventional development approach. This emphasis on a dialogical relationship between external practitioner and local people helped to embed interventions and projects in local realities, infused with cultural meaning. These horizontal relationships between subject and subject in a co-creative process brought their own important and notable advances. The horizontality that is important for such a relationship was applied across the board in an attempt to address and rectify the persistent power inequities in society. Addressing power issues is definitely achingly important, yet by reducing all hierarchies for fear of *power* hierarchies, the interior natural holarchies (nested whole-parts) of self-development were left out of consideration.

Today, while such terms like "personal growth" and "self-empowerment" are often used in a postmodern discourse (especially in the fields of sustainability and international development), surprisingly few practitioners actually ask the question "How do people grow?" Interestingly, that question cannot be answered without acknowledging a holarchy of being and becoming—which is something that the alternative approach is hesitant to accept for fear of opening the door to dominating hierarchies.[4] And so, almost via an implicit association, the question "How do people grow?" is not even *asked* in the first place.[5] In securing a more dialogical relationship for connecting projects to the local context and culture, what this alternative approach leaves out is the importance of consciousness and the dynamics of its growth.

Today, both modern-orange and postmodern-green approaches still are evidenced throughout the field of international development and sustainability. The global situation today that differs from the mid-1900s is that the world is becoming more complex at a dizzying rate. Most practitioners remain at orange/green altitudes and use the tools and approaches found at those stages. However, for others, this complexity calls practitioners to find more integrative and holistic ways to engage in their projects. A key aspect of integration is the interior and exterior dimensions of change. Often in this field the terms "soft" and "hard" capacities are used to refer to the Left-Hand quadrants and Right-Hand quadrants, respectively. Goldman Award winner Ricardo Navarro, a Salvadoran environmental advocate, describes these as the software and hardware of social change, underlining that both are important (Navarro, personal communication, 2004).

In many cases, these "soft" capacities such as values, motivations, beliefs, and qualitative well-being are identified as being important, but *how* to go about integrating them is less apparent. Neilson and Lusthaus (2007, 5) explain how these softer capacities are critical for any other investments in international development work, particularly those that aim to build the capacity of individuals working in the field. However, they go on to discuss how "softer capacities are *more difficult to capture* since they are not usually as visible or as easily discernable for monitoring or evaluation purposes as the harder capacities" (ibid.; italics added).

In the current field, with donors insistent on funding interventions that are measurable (i.e., able to be captured), this perceived lack of measurability of interior "soft" dimension of change may alone be a barrier to focusing and funding these aspects of the work. The preference for the Right-Hand quadrants, then, is not necessarily only due to the perception of their critical importance, but also due to the familiarity with the objective methods of working with those dimensions. In these cases, the barrier is in *how* to go about working with the more subjective and intersubjective dimensions, not *whether* they are important. For many—even those practitioners who agree the domain of subjectivity is critical for this work—the rationale for *not* engaging interiority seems to be the lack of familiar, socially acceptable, or trusted ways to do so.

This hesitancy in alternative approaches to include and work with subjectivity in a meaningful way could be also due to one of the positive legacies of postmodernity. Postmodern studies call for a solid understanding of how intersubjective networks co-create the individual such that no individual is ever essentially "free" of the cultural contexts in which they are embedded. One would never encounter these intersubjective networks purely via subjective reflection or phenomenology (Wilber, 2006, 275–301). Perhaps this important demand for injunctions that could give rise to awareness and understanding of intersubjectivity was why postmodern alternative approaches emphasized dialogue and interpretation, often downplaying or ignoring subjectivity. Perhaps the way forward is to integrate subjectivity while also continuing to include and engage understanding of intersubjectivity and objectivity, which is precisely one of the key insights of an Integral approach.

The Integral Ecology approach expands and deepens the practice of sustainable community development, providing a rigorous framework for including both the *dimensions* of context, culture, and consciousness as well as the *methodologies* we will need to employ to do so. It balances the emphasis on transferring technology or boosting economic growth, to also include the dialogical connection to culture and context, and to also acknowledge and work with the nuanced holarchy of personal and cultural transformation. Without these latter arenas of practice, community development efforts are bound to partiality. And partial solutions simply cannot address very complex challenges.

Integrating these approaches and their methodologies so as to enact an Integral approach in coherence with the complexity of any development issue today is the challenge before the field of sustainable development. This challenge involves building on existing conventional and participatory approaches that address certain components of development (such as economic security, decision-making and governance, social and technical capacity, sustainable natural resource management) to integrate "interior" psycho-cultural components (such as community well-being, moral and emotional capacity, awareness, and worldviews) in an Integral approach to sustainable community development).

Development as Evolution of Consciousness

An Integral approach to development is new, so it has yet to become mainstream or widespread. However, there are some examples, both past and present, that have been effective in balancing both qualitative and quantitative dimensions of community development.[6] Although these examples come from different parts of the world and differ in many ways, there are common elements. In particular, in all examples emphasis is placed on individual and collective shifts in worldviews and value systems, which have profound impacts on how the community or society operates as a whole.

Liberation theologists in Latin America call this a "dynamic action of awakening."[7] Developmental psychologists and cultural theorists describe this "awakening" with empirical research on how our worldviews, values, epistemologies, and "orders of consciousness" unfold in nested stages moving toward greater complexity.[8] Other theorists and practitioners suggest that empowered, sustainable, and equitable societies emerge as this movement toward more complexity, or this "awakening," occurs throughout society (Silos, 2000; Wilber, 1996). Ken Wilber (2000, 42–44) describes and summarizes the various bodies of research on this process of unfolding, explaining how it occurs both in individuals (reflected in psychological structures) and in the collective (reflected in morals and perspectives).[9]

I have used a simplified and practical model of expanding worldviews (based on Wilber, 1999, 111) for working in community development. The essence of this pragmatic theory is that moving toward a sustainable development, or developing sustainability, involves shifts in our worldviews from an egocentric focus to include greater and greater spheres of awareness, care, and action. As one's sphere of concern begins to transcend and include more than the *egocentrism* of immediate self-needs, it transforms into *sociocentrism*, where we also care about our group, our community, and our society.[10] As sociocentrism expands to *worldcentrism*, consideration extends even further to include not just one's self and one's people, but all peoples and all beings (Wilber, 1996, 183–87). In this way, worldcentrism situates the healthy aspects of egocentrism and sociocentrism in a larger context of concern.[11] At the farther reaches of worldcentric awareness, new perceptions of self-in-the-world emerge where "one becomes conscious of the interconnectedness of all humanity at all stages of development and starts integrating all humanity within oneself" (Kimura, 1998, 117). With further expansion into Kosmoscentric awareness, one experiences a release of the attachments of the gross realm and a radical recognition of evolutionary processes (referred to by some as Spirit's unfolding), such that while one is even more compassionately called to action, there is a simultaneous letting go of the gravity of outcomes. This Kosmoscentric view reflects another core component of Integral Ecology: while the state of the environment is likely getting worse, aspects of this change are getting better, and simultaneously it is all simply perfect.

Being able to take the perspective of "other," which is a phenomenon of rationality at an early worldcentric perspective, whether a neighbor, a member of another family, other nations, or other species, enables compassionate action. Further integration and coordination of these "other" perspectives, through the emergence of trans-rationality, is needed to meet complex needs and address interconnected problems (Wilber, 1998, 229–31). At the same time, rough estimates suggest a large perentage of the population has a worldview at sociocentric or lower. Considering even this estimate, since most of the environmental and/or development messages used in communities today actually come from a worldcentric perspective of the problems at hand, these messages are not necessarily connecting with where the majority of people are coming from.[12] In other words, the worldcentric communication is often over the heads of community people, which is not only disrespectful but ultimately ineffective.

Thus, in this case study, I am suggesting two key points regarding the developmental nature of worldviews for sustainable development. First, sustainable development lies not only in new, ecologically sound institutions, management, and laws, but also in our collective and individual growth toward and beyond worldcentrism. At the same time, practitioners need to be able to work better with where individuals and communities are actually coming from: using skillful means in communicating to different worldviews so that community people can hear and resonate, and/or creating conditions for healthy translations at the stages of consciousness that are present. Integral community development is a map for such a process, offering an expansive and inclusive approach that is embedded in the infinite process of consciousness unfolding.

Toward an Integral Model

An Integral approach to sustainable community development includes all quadrants and all levels, as per Wilber's AQAL framework. While more of the AQAL model can be brought in by the practitioner (such as lines, types, states, and zones), at the very least it includes quadrants (or the Big Three) as interconnected domains of Practical (*action/application*; "It/Its"; UR and LR), Interpersonal (*dialogue/process*; "We"; LL), and Personal (*self-growth/reflection*; "I"; UL) (figure I.2), and levels in terms of translation and transformation.

The Big Three: Practical, Interpersonal, and Personal

The Practical domain of *action/application* includes fulfilling ecological, economic, social, and political needs through various types of 3rd-person perspective interventions, such as infrastructures, management plans, institutional designs, and technical capacities. There are many progressive practical and interdisciplinary tools that are low-impact and site-specific, and employ appropriate technologies in addressing complex issues. This domain of "It/Its" is necessary in finding solutions, and most effective when used with the other domains of the Integral Model.

The Interpersonal domain of *dialogue/process* involves the 2nd-person perspective of shared meaning, which is important for negotiating values and ethics, arriving at a common vision, and deciding upon appropriate actions. This can be done using various communicative processes, participatory frameworks, and social capacity–building activities. By including dialogue in development, local people become active subjects in, rather than passive objects of, the development process. This "We" space in community development not only fosters political empowerment but also creates space to explore concerns, ideas, and goals with each other, and to really hear each other's situations, values, and stories.

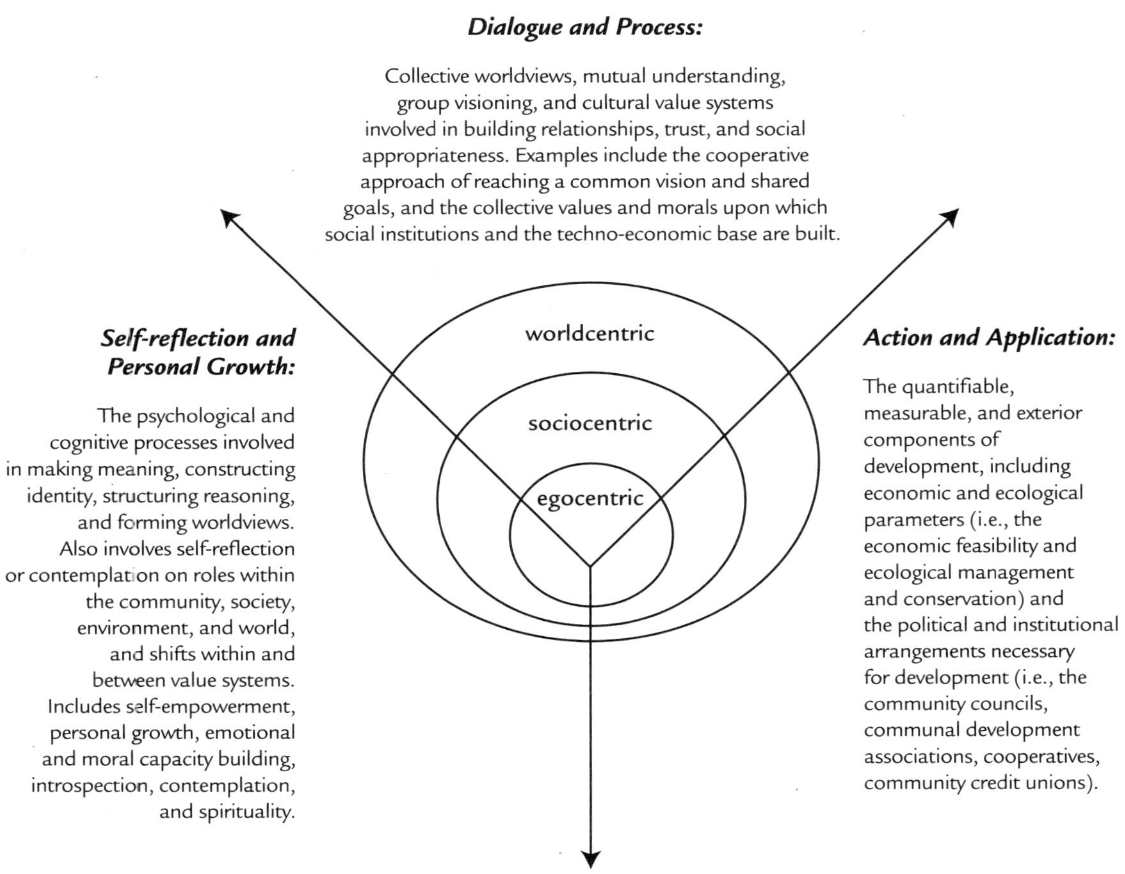

Figure I.2. An Integral approach to community development.

The Personal domain of *self/experience*, includes 1st-person perspectives that contribute to individuals' personal engagement in a community development process. Processes that invite this 1st-person perspective can include reflection, introspection, visioning, embodiment, art, contemplation, and other practices. This is also the domain of self-development that enables the process of moving from a "small" (embedded or identified) sense of self to a more encompassing sense of self (de-embedded or expanded).[13] Including the domain of "I" with the other domains of the Integral Model enables individuals to better understand their current and potential role in effecting positive change in their community, as well as their individual impact on each other and the environment. When these three components are used in community development, as my research in El Salvador attests, a more sustainable result is achieved.

Levels: Translation and Transformation

A key aspect of an Integral community development approach is how to engage with the developmental processes of transformation and translation. Processes of human development include the horizontal development at each stage (i.e., *translating* that stage healthfully), as well as vertical development of transcending and including the essential aspects of the previous stage to stabilize at the subsequent stage (i.e., *transformation*).

In other writings I have discussed these dynamics, their difference, and considerations for engaging them (see Hochachka, 2006, 21–22; Hochachka, 2005, 55–59). Below I discuss some ways of working with transformation and translation; this section is not comprehensive and rather serves to point out some general ways to begin understanding and working with the developmental unfolding of worldviews. Much of what I draw from here comes from Wilber's synthesis of developmental psychology, as well as insights from popular education, liberation theology, and sustainable development theory.

I use the term "worldview" to refer to the "center of gravity" of an individual's self stage, keeping in mind that an individual's center of gravity consists of many other lines of development, including their cognitive, affective (emotional), moral, interpersonal, and spiritual lines, to name a few. Wilber explains that these worldviews are better understood as *probabilities* in people: if someone responds with sociocentric awareness 50% of the time, they are likely to respond with egocentric and worldcentric awareness 25% of the time, respectively. Worldviews are astoundingly complex, and they should not be used as definitive categories for labeling or for quick judgments that can quickly fall into inaccuracy and become perjorative.

With that caution in mind, there are two general approaches for working with worldviews. One involves creating conditions for the transformation between one stage and the next stage, a literal shift in worldview. The other involves fostering healthy translation of the existing stage of development an individual is currently at. This latter approach also includes translating key communications (about sustainability, for example) into the local worldviews of community people.

Transformative Processes

Shifts in worldviews are part of a transformative process in which one's frame of reference shifts profoundly, and what was once *subject* ("part of me") becomes perceived as *object* ("something I can observe, reflect, and act on") (Kegan, 1994; Wilber, 2000). Thus, our understanding of and relationship to our selves and our surroundings change with each developmental stage.

Transformation surely occurs throughout a life span, but developmental psychologists report that it makes up very little of the self-development process. Theorists from the disciplines of social science explain how changes in worldviews often begin when individuals are deeply moved by something and have thus begun acting differently (Kotter and Cohen, 2002, 1–12). Personal transformation can also be propelled by a "disorienting dilemma" (which can be a traumatic and dramatic experience) (Taylor, 2000, 298–301), or by an "integrating circumstance" (which can be conscious or unconscious, and usually is a culmination of a process of searching for a missing element in one's life) (Schugurensky, 2002, 70). Development practitioners in Sri Lanka say that through meditative practice, individuals are motivated and mobilized to act for the development of the village, with a more a connected and compassionate perspective; "only through inner transformation can the outside world change."[14] Finally, developmental psychology research has found that transformation tends to occur in some of the following circumstances: (1) when life conditions (in all quadrants) enable growth to a subsequent stage, (2) when the self experiences irreconcilable cognitive dissonance, (3) by simply living life itself, and (4) through conscious practice (therapeutic as much as spiritual).

The length of this case study does not permit extensive discussion on each of these. Suffice to say here that most developmental psychologists seem to concur that a person cannot "grow" another person. While this shows up as a well-intentioned impulse at orange and green altitudes (for different reasons and motivations), it has not been confirmed or validated by data on psychological development to date. Rather, for the most part, self-development runs its own course through life, following a "transcend and include" format, where the previous stage is integrated into existence at the subsequent stage.[15]

Considering this, what can a practitioner of Integral community development actually do to support transformation? A practitioner can create conditions that may be more conducive to transformation, and the more a practitioner understands about the stages of self-development, the more skillful this creation of emergent conditions for change could be.

There are myriad ways to work toward creating conditions for transformations in worldviews, some of which are highly intuitive, others that are "proven" empirically and well documented. Popular education techniques involving *conscientization* (Freire, 1972) have been found to enable shifts in individuals' sense of self (both oppressors and the oppressed) such that they become sensitized to their current and potential role in effecting positive change in their community, society, and environment. Popular education techniques seek to radicalize local people through this type of shift in perspectives, where what was once accepted and not questioned by local people becomes something that can be analyzed, critiqued, and acted upon.

Wilber describes other practices that help to foster shifts in self-stages and worldviews as "Integral Life Practice" (ILP). An ILP includes engaging all quadrants in one's practice, working with psychological shadow material and states of consciousness, as well as community service (Wilber, 1999, 121). The psychological premise behind this is that the balancing quadrants supports one's transformation to the next stage.

In recent work in an Amazon rainforest conservation context in Peru, the quadrants were used with practitioners to develop their own "personal ecology practices." Individuals used quadrants to identify gaps in their personal ecology practice and to devise a more integrated approach to their own self-development process, again with the premise that balanced quadrants create conditions for transformation (Simpson, personal communication, October 2007).

Translation Processes

However, the fact remains that it is challenging to work toward transformation; shifts in worldviews and ways of thinking take time and specific life conditions, and this is often an extremely personal process that just cannot be spurred along by a third party (Kegan, 1994; Beck and Cowan, 1996). Moreover, self-transformation is really hard work and involves a dying to the existing stage identity as the ego re-identifies with the next stage identity. This is not an easy process to undertake at the best of times, let alone when combined with the ongoing demands of community life in impoverished regions or when engaged in the pressing work to abate environmental deterioration. Although a widespread cultural transformation to worldcentric awareness is desirable, it is also unlikely in the near future.

This points to another way of working with worldviews. Rather than foster transformation in worldviews, one can take the approach of coming to know and honor how people see the world, to foster full health at that stage of development. The vast majority of self-development involves this translation of one's existing stage of development. Researchers have found that until that existing stage is fully lived out, there is little developmental pull to the next stage. Assisting this translation process could involve helping people balance their quadrants, it could involve the patience of allowing perspectives to run their course dry, or it could involve encouraging perspective-taking exercises to help fill out (or fully translate) that existing stage.

A key aspect of engaging in this translation process involves communicating more connectively to their worldview. Integral sustainability practitioner Barrett Brown (2006) explains this further:

> If our economic, environmental and social challenges require values that will drive new behavior—values which are fundamentally different than most people have—then we might not be successful in our efforts, as those values may not change fast enough on a large enough scale. Yet if we can learn to work with the values that people hold and translate what needs to be done so that it resonates with those core values, then we may go much farther, much faster toward sustainability. This is fundamentally a process of truly honoring people for who they are—not trying to force a change in values upon them—yet simultaneously explaining shared goals (like sustainability) in ways that are meaningful to them. (430)

This ability to appropriately translate an intervention (e.g., the meaning of sustainability) into terms that resonate with local worldviews makes a crucial difference in the ultimate effectiveness of a project. One could design the perfect sustainable development intervention, but if no one in the local region understands or values it, then it will most likely fail. Beck explains that the question is not "how do you motivate people," but "how do you relate what you are doing to their natural motivational flows?"[16] Understanding the local psychology and the meanings behind behaviors can assist practitioners in communicating complex concepts (such as democracy, equity, and sustainability) into the local worldviews of the community.

To do this translation into local worldviews, one must first be able to identify the worldviews of others. Wilber (2005, personal communication) refers to this as "coming to know the bandwidth or the altitude at which people are flying." Because worldviews are so complex, this identification can be challenging. However, it is important to distinguish between deep structures and surface structures of worldviews; the former being universal aspects of these stages of development and the latter shining quadratically through different contexts, cultures, languages, types, states, circumstances, etc. In this process of identifying the deep structure stages, some questions to consider are: Where are people coming from? How are they making sense of the world? What does this person care about? What motivates him to take action? Why is she doing what she is doing? How is he making sense of his surroundings? Why does she believe what she believes? From those questions, one can get a general understanding about a person's worldview, to which communications can then be tailored.

As a practitioner stabilizes her or his own Integral stages of development (i.e., teal and turquoise), the intuitive ability to see or feel what stage a person is at seems to become more heightened (van Schaik, personal communication, 2007). Harvard research reports statistical evidence for people's intuitive capacity to gauge the developmental stages of others (described in Stein and Heikkienen, 2008, 17). However, more accurate tools for identifying these deep structures exist, including Kegan's Subject-Object Test (orders of consciousness), Susanne Cook-Greuter's Sentence Completion Test (ego development), and the Lectical™ Assessment Survey (LAS). Other approaches are less empirical but nevertheless useful, such as Beck and Cowan's Spiral Dynamics (values development), Esbjörn-Hargens's (2005) work on Eco-Selves (ego development in environmental contexts), and Brown's (2006) work on values development and "hot/cold buttons" in sustainability work.

Putting Theory into Practice: San Juan del Gozo, El Salvador

Given this background in the current development context and this brief outline of Integral community development, I now shift toward how this approach lands in practice.

For nine months from 2000 to 2002, I worked in a coastal community in Jiquilisco Bay, El Salvador, a region of mangrove forests and small resource-dependent communities. My research explored an Integral approach to community development. I worked with a research assistant in collaboration with CESTA Amigos de la Tierra in the community of San Juan del Gozo for six months in 2000–2001.[17] The project then included a follow-up visit one year later for three months (November 2001 to January 2002) to further evaluate the project outcomes. When I refer to "us" or "we" in the research process, I am referring to my research assistant, Concepción Yesenia Juarez, and myself.

We used participatory action research (PAR) methodology in an Integral community development framework that proceeded in three project phases.[18] The *initial phase* of the research consisted of house-to-house interviews that gave us an opportunity to learn about community values and dynamics, and to build trusting relationships with community members. These interviews were quite intimate, lasting 1–2 hours, in which conversations expanded beyond the research questions to discuss many small but important aspects of community life. In the *second phase*, we worked with two focus groups using dialogue, group visioning, appreciative inquiry, and community mapping, to discuss common needs and visions, as well as to collaborate in responding to the community's pertinent concerns. Our discussion flowed into action in the *third phase*, including training workshops, meetings, fund-raising, cross-community exchanges, and soliciting assistance for specific initiatives. Throughout all phases, the Integral framework helped us to pay attention to exterior and interior dimensions of the research (see figure I.3).

The Integral approach was less of a specific "methodology" than an implicit guide to ensure that as much of reality as possible (i.e., subjective, intersubjective, and objective dimensions) was honored and included in the process of using the (more specific and explicit) PAR. However, this was more due to the preliminary use of Integral Theory, and in subsequent work, such as the Drishti Case Studies in 2006, I have directly drawn upon

Phase 1 **Interviews and Meetings**		***Phase 2*** **Focus Groups**		***Phase 3*** **Exchanges and Actions**	
EXTERIOR	INTERIOR	EXTERIOR	INTERIOR	EXTERIOR	INTERIOR
Open-ended questionnaire and meetings with leaders and local councils to understand the socioeconomic, ecological, political, and historical context of the community.	Using a conversation-style interview process, enabling interviewees to share personal stories, beliefs, and feelings. Learning about the "interior" context of the community (local beliefs, worldviews, and values). Building trust between our research team and the community. Sharing conversations on spirituality and (with some community members) prayers.	Dialoguing on issues and concerns and collective problem-solving. Building capacity for group dialogue and collaboration. Using focus groups as a venue for social and technical capacity building (workshops on organization, fund-raising and cooperative training).	Creating a "safe" and trusting space for exploring "self-in-relation." Bringing new ideas into dialogue, fostering an atmosphere of exploration, activating the "what if" mind. Building moral and emotional capacity (self-esteem and confidence to engage in the focus groups, and facilitating connection with others).	*Exchanges:* Sharing experiences and resources, learning other groups' challenges and successes. *Action:* Enabling research to be useful to meeting community's material needs.	*Exchanges:* Connecting different groups; fostering appreciation of "other" ways of being and perspectives; and providing opportunities to truly "see" these other perspectives. *Action:* Making action only one component of the project to point out the interior components.

Figure I.3. Exterior and interior dimensions of methodologies used in each phase of the research.

the Integral approach for methodological design and for guiding the process of the project. For the Salvadoran case study, rather than employing empirical psychological tests[19] to track changes in worldviews, we used an approach akin to Jordan's (1998) methodology of identifying key phrases that point to inner understandings of community processes.[20] I combined this approach with participant-observer methodology and triangulated it with my research assistant and other community members.

Community concerns (expressed in phases 1 and 2) included deforestation of the surrounding mangrove forest for fuel wood and construction materials, depletion of fish stocks, and limited economic alternatives for community members, particularly women. This was compounded by inadequate organization or capacity to address these problems. We decided to work on these interrelated concerns in two focus groups, with the fisherfolk and the women's council.[21] Below, I first describe our work with the focus groups in turn, and then I discuss the Integral aspects of the process and outcomes.

Fisherfolk Focus Group

The community lagoon has a central role in the local economy and links to ecosystem and community health. Our focus group was made up of six members of the fishery council, Brisas del Mar—the leaders who oversee the management and use of the lagoon. In the focus group, the fisherfolk discussed the lagoon's history and current state, and explored their key concerns and possible solutions.

Their main concern was low resource production of the lagoon. Through discussions and community mapping activities, the focus group decided that the two important and interrelated issues related to low production were that (1) the lagoon border had been eroded during Hurricane Mitch in 1998, which resulted in the escape of shrimp and fish back into the surrounding mangrove, and (2) users of the lagoon were possibly overfishing. However, the leader of the fishery council, Oscar William Durán Martínez, also explained how these exterior concerns required a stronger social organization—that is, more attention to the interior dimensions of group dynamics. He explained how "we [the fisherfolk] need to organize better to be able to deal with any other [material] problem." Thus, actions proposed by the group included (1) collaboration in fixing the border of the lagoon, (2) limiting the use of the smallest net size and/or implementing fishing restrictions, and (3) boosting participation and motivation in problem-solving by strengthening internal organization.

Working together, and guided by the Integral framework, we began implementing these solutions. We sought financial support to fix the lagoon border and simultaneously worked to build capacity for fund-raising and networking. Through dialogue, the fisherfolk agreed to limit the use of the *smallest* net size, to aim for a fishing restriction in the future, and to share the most productive part of the lagoon equally among all fisherfolk. These strategies provided a foundation for long-term sustainable and equitable management of the fishery. Our weekly focus group dialogues were important in improving internal organization, and in addition to this, we also arranged a workshop series on leadership and organizational management presented by a Salvadoran NGO who specializes in this area. The focus group was a catalyst for more specific scientific research on low lagoon production (as part of CESTA's regional project).

Women's Focus Group

The focus group of six women—individuals who also made up the community's first-ever Women's Council—was primarily concerned with the lack of stable work opportunities for women in the community, which limited their abilities to provide for their families.

We began exploring ideas to improve the local economy for women using dialogue and appreciative inquiry, and by creating a supportive space for self-reflection. The group discussed the following possible projects: (1) individually developing community economic development (CED) initiatives to benefit two to three families, (2) buying and raising livestock with bank loans or other credit mechanisms, (3) creating an artisanal fishing lagoon next to the mangrove forest of Jiquilisco Bay, and (4) developing an eco-tourism project in the community and the surrounding area. After months of deliberation they decided to combine the latter two ideas: since fish and shrimp would enter the lagoon naturally with the tides, the women could operate a small restaurant for fresh seafood dishes, and with reforesting of the area surrounding the lagoon, they could offer bird-watching canoe trips into the mangroves.

Using appreciative inquiry, the focus group discussed what skills and resources they already had and what they needed to carry out this initiative. Their actions included forming a cooperative of 20 women, soliciting land for the initiative, improving group collaboration and organization, seeking financial support, and contracting technical studies for the construction, environmental impact assessment, and management of the lagoon. We collaborated with the women on this process. The cooperative began holding meetings with the 20 potential members in 2001, and the Salvadoran Institute for Cooperative Promotion (INSAFOCOOP) trained the group in cooperative management and organization. With our support, the women held various meetings with the Salvadoran Institute for Land Transfer (ISTA) and inquired into the necessary technical studies and environmental impact assessments. We began training the cooperative on fund-raising for the initiative and, with key leaders, built capacity for a community-based form of eco-tourism.

Integral Community Development in Action

One year after the project, we did quantitative and participatory evaluations (personal interviews and group discussions) to gauge the exterior and interior outcomes of the research. The exterior results were immense considering the short length of the project. The fisherfolk focus group had (1) mitigated disputes between fishers by organizing more equitably, (2) effectively discussed and began implementing (short- and long-term) sustainable management strategies for the lagoon, (3) sought financial and technical support to identify and address the reasons for low production, and (4) boosted collaboration amongst fisherfolk. The secretary of Brisas del Mar explained that this improved collaboration throughout and beyond the project was particularly significant. Such internal organization is crucial for collaborative resource management. The women's focus group (1) had begun the formation of a cooperative, (2) were attending capacity-building workshops, and (3) were working on raising funds and capacity to develop their lagoon and eco-tourism project. They had gone from being excluded in decision-making (and thus also from issues regarding family security) to having an unprecedented formal place in municipal governance and a nationally recognized legal presence as a cooperative. In December 1999 the Women's Council met for the first time with municipal government representatives to discuss the community's development needs, and today continue to work with other development NGOs in the region.

The "interior" outcomes of the research were evident throughout the project as well. People had come together to discuss their differences and seek solidarity in their similarities, to address seriously their need for collaboration and organization, and to recognize each other's needs and value systems even if different from their own. The fisherfolk focus group recognized that family health was the common value that transcended their differences; thus, decisions regarding resource management took into consideration the poorer fisherfolk. In five months, the women had gone from discussing economic initiatives for two to three families to a larger, more inclusive project of 25 women that recognized the needs of the poorer families in the community.

The interior dimension of our work also related to the kinds of "capacity" that was developed through our process of working together. Developing or building capacity is a complex process that includes, not just technical and social capacity, but also "emotional capacity (the ability to relate healthily at an emotional level to the world), the moral capacity to treat other beings with respect, the capacity for humour, the capacity to love, and the capacity to engage in a spiritual life" (Rooke, 2004, 1). This "soft" capacity often involves self-esteem and confidence in this type of disenfranchised community (coming from largely an amber altitude in Integral terms), and becomes the cornerstone for community leadership.

In our project, building technical and social capacity was facilitated through workshops, training, and (informal) mentoring; moral capacity was fostered through collective visioning and dialogue; and emotional capacity occurred through self-reflection and sharing (see figure I.3). This "multifaceted" capacity-building was important in enabling the community to move beyond the dependency model of development to a more self-empowered and sustainable process. Some individuals who had never been formal "community leaders," such as Digna de Jesús Andrade (president of the Women's Council) and Rosa Telma Flores de Zetino (vice-president of the Women's Council), at first expressed nervousness about their leadership capacities but then gradually took on such roles with self-esteem, confidence, and vigor. By my last month in the community, the Women's Council was holding meetings with the other cooperative members, conducting meetings with land-tenure representatives from ISTA, and interacting confidently with trainers and officials from other government institutions (such as INSAFOCOOP and the Ministry of the Environment and Natural Resources) and other nongovernmental organizations (such as the Salvadoran Women's Movement).

During the participatory evaluations and through our work together, participants shared their own recognition of the complementary "interior" and "exterior" dimensions of development. The fisherman Samual Rivas explained that, while economic security is important, "you do not come with money, but instead with thoughts and ideas, with knowledge. Sometimes money is not worth anything, and what is more valuable is the knowledge of how to improve community development." The women's first two choices for a name for the cooperative were Fuente de Jacob (Jacob's Well) and La Visión (The Vision), both of which hint at an ecclesiastical root and reflect the inner dimension of our work together.

Possibly the most profound interior outcome of the research was the subtle but meaningful shifts in worldviews. Throughout the project, we observed shifts in self-identity, morals, and perspectives, toward a more inclusive worldview that demonstrated more consideration for the less fortunate members of the community, for women in decision-making, for the surrounding ecosystems, and for neighboring communities (see figure I.4). Also, the increase in self-esteem and confidence signifies changes in self-identity, such that individuals could meaningfully participate and connect with others. These shifts toward worldcentrism reflect the process in which stabilizing one developmental stage allows for the next stage to emerge (Beck and Cowen, 1996; Kegan, 1994).

A Note on Methodology

To truly engage with inhabitants in community-directed work, the development practitioner must be able to "meet people where they are," in terms of both their value systems and their ways of "making meaning," building a bridge between existing worldviews and the emerging ones (as described by Kegan, 1994). Combining support and challenge, in each of the three phases of the projects methodology, was significant in creating the conditions for health at each stage of personal and collective development, and for growth through those various stages. In all phases of the methodology, honoring the interior dimensions of the process enabled us to create connections between people and between groups that tetra-arose with the exterior aspects of arriving at outcomes and carrying out solutions.

In this regard, the PAR methodology only went so far. The Integral approach, even as an implicit methodology, was unique compared with other (previous and current) development activities in the community. Rather

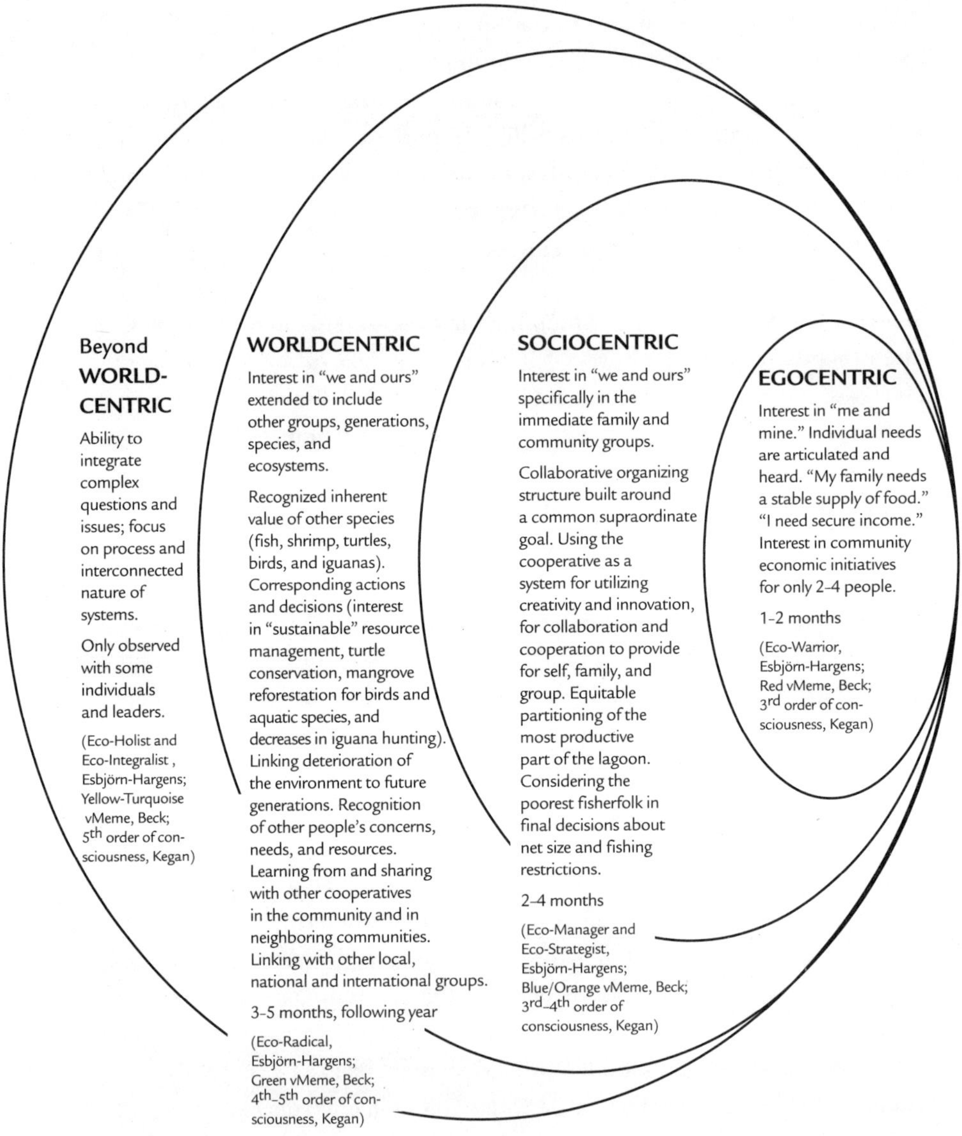

Beyond WORLD-CENTRIC

Ability to integrate complex questions and issues; focus on process and interconnected nature of systems.

Only observed with some individuals and leaders.

(Eco-Holist and Eco-Integralist, Esbjörn-Hargens; Yellow-Turquoise vMeme, Beck; 5th order of consciousness, Kegan)

WORLDCENTRIC

Interest in "we and ours" extended to include other groups, generations, species, and ecosystems.

Recognized inherent value of other species (fish, shrimp, turtles, birds, and iguanas). Corresponding actions and decisions (interest in "sustainable" resource management, turtle conservation, mangrove reforestation for birds and aquatic species, and decreases in iguana hunting). Linking deterioration of the environment to future generations. Recognition of other people's concerns, needs, and resources. Learning from and sharing with other cooperatives in the community and in neighboring communities. Linking with other local, national and international groups.

3–5 months, following year

(Eco-Radical, Esbjörn-Hargens; Green vMeme, Beck; 4th–5th order of consciousness, Kegan)

SOCIOCENTRIC

Interest in "we and ours" specifically in the immediate family and community groups.

Collaborative organizing structure built around a common supraordinate goal. Using the cooperative as a system for utilizing creativity and innovation, for collaboration and cooperation to provide for self, family, and group. Equitable partitioning of the most productive part of the lagoon. Considering the poorest fisherfolk in final decisions about net size and fishing restrictions.

2–4 months

(Eco-Manager and Eco-Strategist, Esbjörn-Hargens; Blue/Orange vMeme, Beck; 3rd–4th order of consciousness, Kegan)

EGOCENTRIC

Interest in "me and mine." Individual needs are articulated and heard. "My family needs a stable supply of food." "I need secure income." Interest in community economic initiatives for only 2–4 people.

1–2 months

(Eco-Warrior, Esbjörn-Hargens; Red vMeme, Beck; 3rd order of consciousness, Kegan)

Figure I.4. Trends observed in worldviews of participants.

than fostering dependency and the expectation that external entities (NGOs, government, etc.) would solve the community's problems, which was particularly the case in Jiquilisco Bay, the Integral framework made room for "self" (the interiors of individuals) in the process of community-directed development. We opened up space for 1st-person perspectives to be shared throughout the project, by engaging all quadrants throughout the phases of the project (figure I.3). Interiors were included during the house-to-house visits, where community inhabitants discussed their lives, families, daily activities, thoughts, and perspectives, and our research team cultivated openness to truly hear what was said. People's experience was also honored in the focus groups, where we created a trusting and expansive space in which participants shared inner reflections, fears, worries, and hopes. Perhaps in doing so, we created conditions for translation at an existing stage and, for some individuals, transformation to the next stage.

As the research progressed, participants shared with me their interior self-reflections (e.g., about their individual roles and sense of themselves within focus groups, as leaders in the community as a whole and regarding the surrounding environment). Their reflections often emphasized shifts in their worldviews and perspectives (figure I.4). These outcomes suggest that this Integral approach fostered in participants a renewed sense of their own potentials in effecting change in their community, moving from a more passive, dependent stage (sociocentric amber) toward either a healthier embodiment of that same stage or emergence into a more active, independent yet relational stage of development (worldcentric orange). While more empirical psychometric techniques and tools are definitely useful, our outcomes suggest that an embodied manner of engaging the domain of "I" (phenomenological, intuitive, and qualitative) nevertheless had a significant role in all aspects of this project.

A final point on methodology relates to the immense responsibility of development practitioners to work on their own developmental unfolding and expansion of worldviews, to dis-center from an exclusive egocentric, ethnocentric, sociocentric, and even worldcentric perspective. The more a practitioner can authentically stabilize an Integral stage of consciousness (teal and turquoise), the more fully present, appropriately responsive one can be. This is particularly important when working in situations of great suffering; how we relate with suffering, holding while also releasing our attachment to it, also relates with our quality and depth of service to those who suffer. The development researcher Majid Rahnema (1997, 8–9) explains how:

> The most significant quality [of working in development] is to be open and always attentive to the world and to all other humans. . . . *Attentive* implies the art of listening, in the broadest sense of the word, being sensitive to *what is*, observing things as they are, free from any preconceived judgment, and not as one would like them to be, and believing that every person's experience or insight is a potential source of learning. . . . Intervention should therefore be envisaged only in the context of a constant exercise of self-awareness. . . . [italics in original].

My own self-development practices (yoga and meditation) during this project helped to foster the expansion of my own awareness, to be clear on my intention, to be receptive to intuition regarding the project, to surface my assumptions and locate my biases, and to be open to the differences in perspective that I encountered. I would return every evening after a full day's work with consideration for the community and thoughts about our work together streaming through my awareness. It was a quality of ongoing reflection, of ongoing regard, that I have come to realize is central to applying an Integral approach. This constancy of presence with the process—of gazing fully into *what is* and not turning away, of giving myself over fully to what is arising while still cultivating precision where needed, and of wholly embracing those whom I interacted with—this is simply part of what is means to practice integrally. Of course, I would often long to bring material goods to the community, from pragmatic things like money and food to less pragmatic things like a bouquet of flowers. Yet, I resisted bringing material gifts of that nature, not wanting to replicate the dependency pattern, in which charity creates someone dependent and in need of charity. Instead I drew from my very being that flowering of care and concern into each and every moment.

This depth of personal practice shows its impacts. Those I worked with often remarked that there was an unnamed qualitative uniqueness in *how* I engaged with them, even when in some circumstances *what* we did was similar to others. Unlike other development practitioners who I would later hear had been robbed or treated badly, the community people were very close to me; on the infamous corner of the road where such robberies occurred, I would naturally stop and chat with those I met there. Years later, on several occasions I would return unannounced and would be warmly welcomed into their homes, with some remarking, "I was just thinking of you the other day, wondering when you'd be here to visit us." Upon hearing this, I would smile inside, thinking about the astounding nature of *embrace*—as one's worldview and very fabric of being expands and deepens to embrace others, is it not true that one truly contains multitudes? And I have come to realize how I experience embrace so fully with this community, these people, that in truth I have never actually left them nor have they left me.

Conclusion: An Integral Sustainable Development

Integral community development is premised on an understanding that community well-being in a cultural, behavioral, and systemic sense tetra-arises with the health of human consciousness as it moves through developmental stages. "Development," therefore, means much more than accumulated wealth, built infrastructure, or economic growth, although it can include these, in one form or another. It is about aligning and attuning to the evolutionary unfolding of self, community, and environment, and recognizing where people are at and where their deepest vision can carry them, creating spaces to explore self-in-relation, and consciously enabling self-growth and new discourse to flow through the collective into more compassionate action.

As I paddled with the fisherfolk on their community lagoon, I realized how an Integral approach offers the meta-mosaic for bringing together community development practices. The practical outcomes of this research suggest that this approach could be useful, beyond this case study, in moving toward sustainability in other communities and regions. Our evaluation one year after the project showed how socioeconomic and ecological objectives were merging, cooperative institutions began blending exterior (technical, social) and interior (moral and emotional) capacity building, and there was recognition of others' perspectives with a more expanded and inclusive awareness.

However, twice upon return I found different circumstances. In February–March 2004 and then again in November 2005, I returned to El Salvador to work briefly with the same community. These two trips allowed me an opportunity to do an informal assessment of our prior work from 2000 to 2002.

What I witnessed when I arrived in 2004 brought to light how the global political economic dynamics of the LR quadrant hold a disproportionate amount of power in social systems. Thus, however successful our work was in the LL quadrant (shared visioning and group organization) and in the UL (changed attitudes, individual capacity-building, and motivation), recent political and economic trends have undermined some of the successes associated with the other quadrants. The Jiquilisco Bay region had begun to be developed as "the next Acapulco" for conventional tourism, which was then (and still is) a grave concern for local people whose main income comes from their natural habitat. This has caused the fishing cooperative to be wary of whom they work with, causing what appeared to be a shift back toward sociocentric awareness, rather than a continued move toward worldcentric. The women's group has had an ongoing struggle with governmental institutions to legalize their cooperative, and the frustratingly slow and bureaucratic process has provoked frustration, lack of confidence as well as distrust with the NGO CESTA. This experience validates their concerns that "rural communities are forgotten" in a political system that does not (or cannot) adequately support their efforts to foster community economic development initiatives.

However, returning again in 2005, I found that the women's cooperative discontinued their work with CESTA, demonstrating an empowered stance that I commend them for, and yet continued collaboration with another NGO, the Salvadoran Women's Movement. The cooperative is now fully legalized and has secured access to land near their farms and adjacent to the mangrove, and is moving ahead with one of its original plans to raise livestock for dairy products, engage in the national economy, and support their families and the community as a whole. Without full regard for the environmental implications of this initiative, perhaps this is not yet a demonstration of a fully worldcentric approach to their community development needs; however, it is certainly an empowered and immense step for a group of women who formerly had no access to jobs, income, and thus no ability to affect family security. The women have entered into a form of community action that they are able to sustain, stepping daringly into a more empowered awareness of themselves, seeing their place in the future of the community more expansively over time, and working collaboratively with a large group of others in this endeavor. I suggest that this could be an expression of a healthy sociocentrism, if not the emergence of an early worldcentrism, although confirmation for this would require more time with the community.

From these two opportunities to assess the initial work together, it seemed that the community focus groups had really stretched under the conditions of our initial work together in 2000–2002, perhaps dipping into a potential not yet fully realizable, then reverting into a less healthy expression of sociocentrism in 2004 due to the political, economic, and institutional pressures, before then settling in 2005 to a healthy sociocentrism or early worldcentrism that they could actually stabilize. It is likely that the cognitive line of development is leading certain individuals into worldcentric, supported by the surrounding NGO discourse on worldcentric ideas, topics, and technologies. In time and when ready, the group will stabilize worldcentric awareness. If we consider the entire pathway, namely from the former contraction of egocentrism, to then arrive at a healthy sociocentrism or to peak into early worldcentrism is surely an immense developmental leap. If anything, the Integral approach has disclosed to me, as a practitioner, the profound need for patience in this evolutionary process and deep regard for any progress made. In that light, I am astounded and grateful by all I have witnessed in this community of San Juan del Gozo.

The other interesting aspect to this informal assessment was my lack of foresight when I left El Salvador in 2002 for sharing the approach I was using with other practitioners who could continue follow-up support with the community. When I left, the buoyancy of our initial work together was not maintained, the discourse of socio- to worldcentric dissipated in the community, and thus the very supports that would have provided emergent conditions for stabilization at worldcentric were not present. However, in 2005, CESTA showed interest in the Integral approach to community development, and our NGO Drishti—Centre for Integral Action returned to hold a two-day capacity-building retreat with CESTA. From this two-day retreat, one of the project coordinators is now holding a discussion group with students at the university about an Integral approach, applying it in his own master's fieldwork in a neighboring community in Jiquilisco Bay, and exploring how it discloses important aspects of environmental and community development work. Again, I am reminded of the astounding pace and flow of evolution, as we each are brought forth along the currents of change.

Through the lens of Integral Ecology, it is clear that the world's complex problems achingly call for an Integral approach. The complexity and depth we find in these global issues need to be met with an equal complexity and depth for perceiving, understanding, and working with these problems. An Integral approach provides for us a framework and a "psychoactive" pathway for engaging fully in this challenge before us. We are called to surrender our categorizing minds that seek partiality to allow in an integrated view, a deep recognition of the multiple perspectives arising in every moment, and an equally spacious action that makes room for these multiple ways of viewing and being in the world. Answers to community needs worldwide are not just found in scientism, nor solely in participation and process, nor merely in spiritualism or psychotherapy, but rather via coherence among them. This tetra-arising of perspectives and dimensions is the mandala that we engage today, as we engage in community development anywhere.

In San Juan del Gozo, our outcomes suggest that by integrating interiority with the more exterior community development practices, individuals became empowered in ways that contribute to community action and to averting ecological and social crises, and the community became more able to address and move beyond these crises. Our experiences and outcomes in this case study suggest that sustainable community development is neither a far-off pipe dream nor a theoretical study, but an unfolding reality that is most successful when the interior and exterior aspects of individuals and communities are included and honored.

NOTES

1. From the maximization of growth and the trickle-down effect of the 1950s and 1960s, to the widespread targeted-aid strategies of the 1970s, to the idea of sustainable development of the 1980s up until today.
2. www.vanschaik.demon.co.uk/ischaikp.html, April 6, 2004.
3. Such as: Dag Hammarskjöld Foundation, 1977; Riddell, 1981; Ghai and Vivan (eds.), 1992; Sirolli, 1999; Sachs, 1999; Ekins, 1986; Moffat, 1995; and Torres-Rivas and Gonzales-Suarez, 1994.
4. The use of the phrase "being and becoming" relates to Tantric philosophy that describes being as the essence of Shiva and becoming the dynamism of Shakti, or in other philosophies it is that which is Always Already present and that which is continually evolving forth. Bringing them together, we include both the aspect of being that is Always Already present with the evolutionary impulse to complexify, grow, and transform. In a more practical sense, "being and becoming" is used by Dr. Tam Lundy of BC Healthy Communities Initiative,

in which she has been developing an Integral Capacity Building framework with a team of facilitators for working in communities in British Columbia, Canada.

5. Ironically, in so many cases, it is the very growth of consciousness that leads to liberation, along with the change in systems and institutions. As Stephen Biko, the great South African martyr, famously put it, acknowledging the untold power of the mind and consciousness, "The greatest weapon in the hand of the oppressor is the mind of the oppressed."
6. These examples include liberation theology in social change work in Latin America (Gutierrez, 1973), the work of Sri Aurobindo and the Mother in community engagement in Pondicherry and Auroville, India (Satprem, 1970); the village-based social action of Gandhi (Khoshoo, 1995); the Sarvodaya Shramadana movement in Sri Lanka, http://sarvodaya.org/index.html (April 6, 2004); http://sarvodaya.org/Library/Essays/sarvmacy.html (April 6, 2004).
7. Liberation theology unites popular movements with Christian theology in the struggle for social and political liberation, with the ultimate aim of complete and integral liberation (Gutierrez, 1973, 81–100, 113).
8. Wilber writes: "It should be remembered that virtually all of these stage conceptions—from Abraham Maslow to Jane Loevinger to Robert Kegan to Clare Graves—are based on extensive amounts of research and data. They are not simply conceptual ideas and pet theories, but are grounded at every point in a considerable amount of carefully checked evidence. Many of the stage theorists . . . (Piaget, Loevinger, Maslow, and Graves) have had their models checked in First, Second and Third World countries" (Wilber, 2000, 40–41). See Kegan, 1994; Loevinger and Wessler, 1978; Piaget, 1966; Torbert, 1991; Beck and Cowen, 1996; Torbert, http://www2.bc.edu/~torbert (April 6, 2004); Fisher, Rooke, and Torbert, 2000; and Gebser, 1949/1953/1985.
9. See Wilber, 2000, 42–44, charts 5a–c, 206–8, for a compilation of empirical research on psychological development.
10. Wilber uses *sociocentric* and *ethnocentric* often to refer to the same, or a very similar, worldview. I have chose to use the term "sociocentric" in this article, since it seems to me to more accurately reflect the worldview I am describing, namely, an identification of the self with the surrounding social group. This can expand in degrees from family, to community, to ethnic group, to nation, to transnational social group such as the Catholic religion or a corporate business, etc. The term "ethnocentrism" seems to offer a more specific definition, that is, the worldview built on a self-identity in relation to ethnic group.
11. Worldcentrism has been described by other theorists as postconventional (Kohlberg), universal care/hierarchical integrative (Gilligan), existential/ironist (Torbert), and 4th-order-self-authoring/5th-order-integral (Kegan). See Wilber, 2000, charts 5a–c, 206–8.
12. See examples of this in Barrett Brown's work on "hot/cold buttons" in relation to sustainability.
13. Kegan (1994, 55) describes this as a move toward more complex orders of consciousness as "a continuous stepping outside of one's own view, such that one's view is perceived as *object* rather than *subject*."

14. See www.foundation.novartis.com/sarvodaya_movement.htm.
15. Pathologies occur when there is either repression and denial, or attachment and fixation, to a previous stage. For more information on this, see Wilber's *Integral Psychology* (2000).
16. Spiral Dynamics Mini-course, 6; see www.spiraldynamics.org/pdf_resources/SDMC.pdf.
17. The "EcoMarino team" in CESTA was crucial for this research and included Concepción Yesenia Juarez, Rafael Vela Nuila, Sofía Baires, Rosibel Acosta Cantón, Hamish Millar, and Ricardo Navarro.
18. Financial support came from the International Development Research Centre's John G. Bene Fellowship in Community Forestry
19. Such as Kegan's Subject-Object Test, Susanne Cook-Greuter's Sentence Completion Test, or other tools for identifying meaning-making. In this regard, our methodology could be greatly improved.
20. Jordan (1998) outlines a methodology based on cognitive-development theory and constructive-developmental psychology.
21. Fisherfolk: Luis Alonso Martínez, Juan-Roberto Rodriguez, José-Roberto Hernandez, Samuel Rivas, José Guadalupe Garcia Flores, José-David Esquiel. Women: Digna de Jesús Andrade (president of Women's Council), Rosa Telma Flores de Zetino (vice-president of Women's Council), Delmi del Carmen Villarta de Palacios (secretary of Women's Council), Edith Andusol Plineda (treasurer of Women's Council, fish merchant), Graciela del Carmen Rivas (vocal of Women's Council).

REFERENCES

Beck, D., and C. Cowen. 1996. *Spiral Dynamics: Mastering Values, Leadership and Change*. Malden, MA: Blackwell.

Brown, B. 2006. "Integral Sustainable Development Part 2." *AQAL: Journal of Integral Theory and Practice* 1(2): 422–40.

Dag Hammarskjöld Foundation. 1977. *What Now: Another Development*. Uppsala, Sweden: Dag Hammarskjöld Foundation.

Ekins, P. 1986. *The Living Economy: A New Economics in the Making*. London and New York: Routledge.

Esbjörn-Hargens, S. 2005. "Integral Ecolgy: The What, Who, and How of Environemental Phenomena." *World Futures: The Journal of General Evolution* 61 (1–2): 5–49.

Fisher, D., D. Rooke, and B. Torbert. 2000. *Personal and Organizational Transformations through Action Inquiry*. Rev. ed. Boston: Edge Work Press.

Freire, P. 1972. *Pedagogy for the Oppressed*. Harmondsworth, UK: Penguin.

Gebser, J. 1949/1953/1985. *The Ever-Present Origin*. Translated by N. Barnstad and A. Mickunas. Athens: Ohio University Press.

Ghai, D., and J. M. Vivan, eds. 1992. *Grassroots Environmental Action: People's Participation in Sustainable Development*. London: Routledge.

Gutierrez, G. 1973. *A Theology of Liberation: History, Politics and Salvation*. Maryknoll, NY: Orbis Books.

Hargens, S. 2002. "Integral Development: Taking the Middle Path towards Gross National Happiness." *Journal of Bhutan Studies* 6 (Summer): 24–87.

Harper, S. M. P, ed. 2000. *The Lab, the Temple, and the Market: Reflections at the Intersection of Science, Religion and Development*. Ottawa: IDRC and Kumarian Press

Hochachka, G. 2005a. *Developing Sustainability, Developing the Self: An Integral Approach to Community and International Development*. POLIS Project on Ecological Governance. University of Victoria, BC.

———. 2005b. "Integrating Interiority in Community Development." *World Futures: The Journal of General Evolution* 61 (1–2): 110–26.

———. 2006. *Overview and Synthesis: Case Studies on an Integral Approach to International Development*. Drishti—Centre for Integral Action. www.drishti.ca/resources.htm (accessed April 10, 2007).

Jordon, T. 1998. *Constructions of "Development" in Local Third World Communities: Outline of a Research Strategy*. Occasional Papers 1998: 6. Kulturgeografiska Institutionen, Göteborgs Universitet, Sweden.

Kegan, R. 1994. *In Over Our Heads: The Mental Demands of Modern Life*. Cambridge: Harvard University Press.

Khoshoo, T. N. 1995. *Mahatma Gandhi. An Apostle of Applied Human Ecology*. New Delhi: Tat Energy Research Institute.

Kimura, Y. G. 2000. *Think Kosmically, Act Globally: An Anthology of Essays on Ethics, Spirituality and Metascience*. Waynesboro, VA: University of Science and Philosophy.

Kotter, J. P., and D. S. Cohen. 2002. *The Heart of Change: Real Life Stories of How People Change Their Organizations*. Boston: Harvard Business School Publishing.

Loevinger, J., and E. Wessler. 1978. *Measuring Ego Development*. Vols. 1 and 2. San Francisco: Jossey-Bass.

Moffat, I. 1995. *Sustainable Development: Principles, Analysis and Policies*. New York: Parthenon Publishing Group.

Neilson, S., and C. Lusthaus. 2007. *IDRC-Supported Capacity Building: Developing a Framework for Capturing Capacity Changes*. Ottawa: Universalia Management Group.

Piaget, J. 1966. *The Psychology of the Child*. New York: Harper Torchbooks.

Rahnema, M. 1997. "Signposts for Post-Development." *Revision: A Journal of Consciousness and Transformation* 19 (Spring): 8–9.

Riddell, R. 1981. *Ecodevelopment: Economics, Ecology and Development: An Alternative to Growth Imperative Models.* Farnborough, UK: Gower.

Rodriguez, W. A. P., ed. 2001. *Informe Sobre Desarrollo Humano: El Salvador 2001.* Programa de las Naciones Unidas para el Desarrollo. San Salvador: PNUD, 2, 10–11.

Rooke, D. 2004. *Organisational Transformation Requires the Presence of Leaders Who Are Strategists and Magicians.* www2.bc.edu/~torbert/21_rooke.htm (accessed April 6).

Ryan, S. J. W. F. 1995. *Culture, Spirituality, and Economic Development: Opening a Dialogue*. Ottawa: IDRC.

Sachs, W. 1999. *Planet Dialectics: Explorations in Environment and Development.* Halifax, NS: Fernwood Publishing.

Satprem. 1970. *Sri Aurobindo or The Adventure of Consciousness.* Delhi: The Mother's Institute of Research.

Schugurensky, D. 2002. "Transformative Learning and Transformative Politics. The Pedagogical Dimension fo Participatory Democracy and Social Action." In *Expanding the Boundaries of Tranformative Learning: Essays on Theory and Praxis,* edited by E. V. O'Sullivan, A. Morrell, and M. A. O'Connor. New York: Palgrave.

Silos, M. 2000. "The Politics of Consciousness." Unpublished manuscript.

Sirolli, E. 1999. *Ripples from the Zambezi: Passion, Entrepreneurship and the Rebirth of Local Economies.* Gabriola Island, BC: New Society Publishers.

Stein, Z., and K. Heikkinen. 2008. "On Operationalizing Altitude: An Introduction to the Lectical Assessment System for Integral Researchers." *Journal of Integral Theory and Practice* 3 (1) 105–138.

Tamas, A. 1996. *Spirituality and Development: Concepts and Categories; Dialogue on Spirituality in Sustainable Development.* Ottawa: Canadian International Development Agency. June 19–20.

Taylor, E. W. 2000. "Analyzing Research on Transformative Learning Theory." In *Learning as Transformation: Critical Perspectives on a Theory in Progress* by J. Mezirow et al., San Francisco: Jossey-Bass, chap. 11, pp. 295–328.

Torbert, W. R. 1991. *The Power of Balance: Transforming Self, Society and Scientific Inquiry.* Newbury Park, CA: SAGE.

———. 2004. Homepage. www2.bc.edu/~torbert/ (accessed April 6).

Torres-Rivas, E., and M. Gonzales-Suarez. 1994. *Obstacles and Hopes: Perspectives for Democratic Development in El Salvador.* Democratic Development

Studies, International Center for Human Rights and Democratic Development, San José.

van Schaik, P. 2004. "The Process of Integral Development." www.vanschaik.demon.co.uk/ischaikp.html (accessed April 6).

Wilber, K. 1995. *Sex, Ecology, Spirituality: The Spirit of Evolution*. Boston: Shambhala Publications.

———. 1996. *A Brief History of Everything*. Boston: Shambhala Publications.

———. 1998. *The Eye of Spirit: An Integral Vision for a World Gone Slightly Mad*. Boston: Shambhala Publications.

———. 1999. "An Approach to Integral Psychology." *Journal of Transpersonal Psychology* 31 (2), 109–136.

———. 2000. *Integral Psychology: Consciousness, Spirit, Psychology, Therapy*. Boston: Shambhala Publications.

CASE STUDY II

Integral Marine Ecology: Community-Based Fishery Management in Hawai'i*

BRIAN N. TISSOT
Washington State University, Vancouver, Washington

Successful fishery management requires that a dynamic balance of disciplines provide a fully integrated approach. I use Integral Ecology to analyze multiple-use conflicts with an ornamental reef-fish fishery in Hawai'i that is community-managed via the implementation of a series of marine protected areas and the creation of an advisory council. This approach illustrates how the joyful experiences of snorkelers resulted in negative interactions with fish collectors and thereafter produced social movements, political will, and ecological change. Although conflicts were reduced and sustainability was promoted, the lack of acknowledgment of differing worldviews, including persistent native Hawaiian cultural beliefs, contributed to continued conflicts.

*This case study originally appeared in *World Futures: The Journal of General Evolution* 61 (2005), nos. 1–2: 79–95. Address correspondence to Brian N. Tissot, Washington State University, Vancouver, WA 98686. E-mail: tissot@wsu.edu.

I thank my colleagues in West Hawai'i: Bill Walsh, Leon Hallacher, Sara Peck, Brent Carmen, and Bob Nishimoto. Ideas in the article were enriched by discussions with Claudia Capitini, Noelani Puniwai, Bill Walsh, Ken Wilber, and the Integral Ecology group. The manuscript was improved by comments from Jennifer Bright, Sean Esbjörn-Hargens, and Samantha Whitcraft.

KEYWORDS: aquarium fish, fishery management, Hawaiian culture, Integral Ecology.

Introduction

Marine ecosystems are renowned for their abundant and seemingly endless resources. However, despite the long-term importance of these ecosystems in protecting shorelines, controlling climate, and providing food and inspiration, the world's oceans are currently in crisis (POC, 2003). This situation is particularly clear with respect to fisheries, which are declining globally (NRC, 1999). Commercially important species are under increasing fishing pressure (FAO, 2000), and ecosystem structure and function are compromised (Jennings and Kaiser, 1998). Efforts to manage fisheries have largely met with failure. These management failures are primarily due to our limited understanding of marine ecosystems, uncertainties between fishing intensity and stock depletion, underestimation of the complex interactions with social systems, and lack of political will (Botsford et al., 1997). Catches are primarily driven by economic forces that eventually overwhelm slowly replenishing stocks. In some cases, specific stocks have been so severely overexploited that they are now listed as endangered species (e.g., abalone: Hobday and Tegner, 2000).

This case study illustrates a new approach to fishery management using Integral Theory (Wilber, 1995) to examine community-based management of marine protected areas (MPAs). A major goal of MPAs is to establish a network of areas closed to fishing (marine reserves) that promote sustainable fisheries outside their boundaries (Murray et al., 1999). MPAs are currently of wide national and international interest (Allison et al., 1998; Bohnsack, 1998; NRC, 2000), as they have been shown to benefit fishery populations, support fishery management, enhance nonextractive human activities such as tourism, protect ecosystems, and increase scientific understanding of marine communities (Hastings and Botsford, 1999; Murray et al., 1999).

Community-based management is a process that empowers local communities to manage their resources by letting individuals in the community contribute to decisions that affect local resources. One of the major benefits of community-based management is the development of strategies compatible with the unique environment, with the specific resources, and with the cultural and historical context of the local area (White et al., 1994). Community-based management also aids in resolving conflicts over limited fishery resources among multiple stakeholders (Capitini et al., 2004).

Environmental conflicts are notorious for their complexities stemming from a combination of biological uncertainty, multiple stakeholders and issues, multiple and unique values and worldviews, and clashes between scientific and traditional knowledge (Daniels and Walker, 2001). Effective conservation and management require the dynamic incorporation of multiple disciplines including biology, ecology, political economy, and sociology to create an integrated management approach (Honing, 1978; Michaelidou et al., 2002).

In 1995 Ken Wilber published *Sex, Ecology, Spirituality*, in which he presented an Integral Model that describes evolution as co-occurring in four dimensions or quadrants: the exterior-individual (behavioral) quadrant, the exterior-collective (systems) quadrant, the interior-collective (cultural) quadrant, and the interior-individual (experience) quadrant. Within each quadrant lies an unfolding holarchy of components that embrace and transcend each other in complexity. Integral Ecology (IE) is one application of Wilber's Integral analysis applied to ecological issues (Wilber, 2000). IE can also provide effective tools for addressing ecological issues through increased explanatory power by integrating divergent domains and by connecting with Wilber's comprehensive research (Hargens, 2002). The approach used here is to explore and acknowledge each of the quadrants with all of their attendant complexity, thereby applying the AQAL (all-quadrants, all-levels) model (Wilber, 1995). The IE AQAL approach to ecological issues takes into account all perspectives and their respective knowledge claims, thus examining all interests, and providing recommendations for solutions that honor each perspective while maximizing the sustainability of the system as a whole (Esbjörn-Hargens and Zimmerman, *Integral Ecology*, 2006).

IE is particularly well suited to examine the complex interactions associated with the management of coastal fisheries. Most studies of fishery management acknowledge the roles of biology, ecology, sociology, economics, and politics while paying little attention to important cultural dimensions (Dyer and McGoodwin, 1994; Friedlander et al., 2003). Furthermore, no one, to my knowledge, has integrated experiential or spiritual dimensions into the discussion. In this article I use IE to analyze a coral reef fishery in Hawai'i that uses community-based management of MPAs as a process to resolve conflicts and develop sustainable resources. The example presented is unusual in scope and complexity in that it involves the harvesting of live reef fish for the aquarium trade in areas where viewing reef fish is part of local recreation and a high-volume tourist business. Thus, in addition to the normal complex interactions associated with fishery management, there exists an additional multiple-use conflict over the extraction of these resources that involves differing worldviews regarding the appropriate use of the coral reef fishes. Intermeshed with these issues are the sociological, cultural, and spiritual dimensions of native Hawaiian culture, which persists in many of the more rural communities of Hawai'i.

Aquarium Fishery in West Hawai'i

Global trade in marine ornamental fishes is a major international industry involving an annual catch of 14–30 million fish (Wood, 2001). Almost all tropical marine ornamentals are collected live from coral reefs, and many originating from the United States are captured in Hawai'i, which is known for its high-quality fishes and rare, high-value endemics (ibid.). In the 1970s aquarium collectors along the west coast of the island of Hawai'i (hereafter, West Hawai'i) first developed conflicts with the rapidly growing dive-tour industry selling views of fishes on the reef. The conflict developed around the perception by the dive-tour industry that colorful reef fishes were dwindling due to collecting activities, thus diminishing the aesthetic value of the reef—a classic clash of conservationists' (i.e., sustainable yield) versus preservationists' (i.e., aesthetic beauty) worldviews (Capitini et al., 2004). One rallying point of the conflict for preservationists in voicing their concerns was the abundant, colorful yellow tangs (*Zebrasoma flavescens*) that form large schools at natural high densities and swarm over the reefs. Yellow tangs account for over 72% of the aquarium trade harvest in West Hawai'i and thus numerically dominate the collector's take (Miyasaka, 1997). Significantly, the dominant and aesthetic presence of these bright yellow schools of fish along the coastal reef is one reason why West Hawai'i is often referred to as the "Gold Coast."

Although the conflict was recognized in the 1970s, the issue was not fully addressed for two decades (Tissot and Hallacher, 2003). However, by 1997 the situation had grown into a serious multiple-use conflict bordering on violence (Dybas, 2002). Because the state agency charged with managing fishery resources, the Department of Land and Natural Resources' Division of Aquatic Resources (DAR), repeatedly failed to resolve the conflict, pressure by local citizens' groups resulted in several bills submitted to the Hawai'i state legislature to ban collecting or to establish MPAs. In 1998 one of these bills passed to become Act 306, creating a fishery management area along the entire 120-kilometer coastline of West Hawai'i and mandating substantive involvement of the community to help manage reef resources. One of the specific mandates required that a minimum of 30% of the West Hawai'i coastal reef be designated Fishery Replenishment Areas (FRAs)—MPAs that specifically prohibit aquarium fish collecting. The law also required a scientific evaluation of the effectiveness of the FRAs after five years (Tissot, 1999).

To create the FRAs, DAR and the University of Hawai'i Sea Grant Extension established the West Hawai'i Fisheries Council (WHFC), a community-based group composed of the diverse stakeholders associated with reef resources in West Hawai'i (Capitini et al., 2004). The WHFC included aquarium collectors, the owners of aquarium retail stores, commercial dive-tour operators, a hotelier, commercial and recreational fishermen, shoreline gatherers, recreational divers, and several general community representatives—some of whom were members of the Lost Fish Coalition, a grass roots organization that aimed to shut down the aquarium industry (Capitini et al., 2004). Two council members had fishery degrees, and 40% of the council identified themselves as native Hawaiians. In addition to stakeholder representation, the WHFC also attempted to balance membership among the diverse geographic areas in West Hawai'i (Walsh, 1999), based in part on traditional Hawaiian land divisions, or *ahupua'a*, thus building one aspect of the council on the community- and *'ohana* (extended family)-based traditions of Hawaiian culture. *Ahupua'a* are ecological, sociological, and political land divisions created by native Hawaiians. These divisions are generally delineated by natural watersheds running from the mountains to the sea and out onto the reef (Handy and Pukui, 1958; Kirch, 1984; Callicott, 1994).

One of the major goals of the WHFC was to establish the location of the FRAs using a consensus-based approach (Capitini et al., 2004). Thus, the council provided not only a process to generate the location of the FRAs, but also a means to resolve conflicts among the diverse group of stakeholders. After considerable, often contentious debate, the council proposed to DAR that a series of nine FRAs be spread out along the 120-kilometer coastline of West Hawai'i (figure II.1). The West Hawai'i community rallied around the proposal, providing a 93% approval at a public meeting. The proposal was then approved by the Governor's office and, as a result, the series of FRAs were officially closed to collectors in January 2000 (Walsh, 1999).

After the FRAs designation but before its closure, I helped organize and coordinate a group of researchers focused on the design and development of a coral reef fish monitoring program in and around the designated FRA sites. The resulting West Hawai'i Aquarium Project (WHAP) was created to evaluate the effectiveness of the FRAs in recovering depleted aquarium

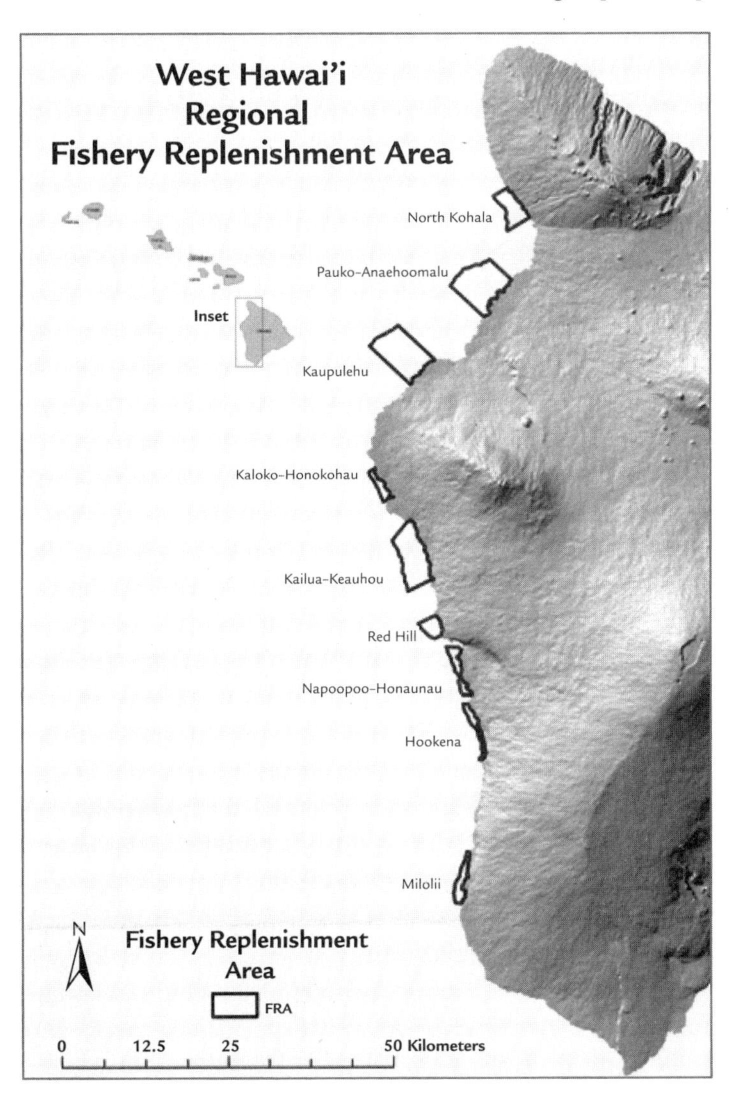

Figure II.1. Map of West Hawai'i illustrating the location of the nine Fishery Replenishment Areas (FRAs).

fish stocks. Results of our surveys are summarized quarterly and provided to DAR, the WHFC, and the public in an attempt to provide near real-time awareness of biological changes in the FRAs and engage the community in management issues (see http://coralreefnetwork.com/kona).

After five years of closure, the overall abundance of aquarium fishes significantly increased in the FRAs, relative to areas closed for over 10 years and those still open to collecting (Tissot et al., 2004; Walsh et al., 2004). Although only two of the ten most commonly collected species have significantly increased since 1999, one of these is the yellow tang, the hallmark of the aquarium industry in West Hawai'i (Tissot et al., 2004). Equally important, the catch and catch-per-unit-effort of aquarium fishes in West Hawai'i has not significantly declined after FRA establishment, indicating that a productive aquarium fishery can coexist with a large network of FRAs (Walsh et al., 2003, 2004).

In order to illustrate the IE components of the West Hawai'i aquarium fishery, I will provide an Integral analysis of the issues surrounding the aquarium fishery. I will use the 4 quadrants as a framework for highlighting the multiple dimensions and their respective issues/levels involved: behaviors, systems, cultures, and experiences (figure II.2). I include in the analysis components of native Hawaiian culture that played a role in the complex process. After presenting key elements associated with each quadrant, I provide a discussion of fishery management in Hawai'i from an Integral Ecological perspective.

Behavioral Dimension (Exterior-Individual)

The behavioral dimension explores the behaviors of individuals within the system. These may include the actions and movements of fish and humans. For example, the tendency of yellow tangs to form colorful schools of "golden" fish at high densities is an indicator of natural high abundances. When the schools are reduced by collecting pressures, their absence impacts the aesthetic quality of the reef overall, eliciting protests by people who view the fish on a regular basis. The response of different stakeholders to abundant tangs and other reef fishes also invoked differing behaviors. Tourists and locals enjoy observing the fish in their natural environment—a preservationist view. In contrast, aquarium collectors react to an abundant resource by harvesting—a conservationist view. Furthermore, individual collectors take different approaches to capturing live fish. Some collectors are yellow tang specialists and use barrier nets to capture high numbers of

	INTERIOR	EXTERIOR
INDIVIDUAL	**Psychological** • Joy in viewing colorful fish on the reef • Anger when missing local fish • Horror of connecting with fish's journey to aquarium • Anger and frustration of collectors singled out by community **Spiritual** • Connection between people and marine life • Connection with ancestral spirits or *na 'aumakwa* **I**	**Behavioral** • Schooling behavior of yellow tangs at high density • Novice snorkelers who observe schools of colorful fish • Professional scuba divers who target rare fishes and behaviors • Aquarium collectors who target rare and endemic fishes of high value **IT**
COLLECTIVE	**WE** **Cultural** • Aquarium collectors' concept of sustainability clashing with preservationist attitude of fish viewers • Interest- vs. identity-based values in conflict resolution process **Religious** • *Hawaiian Kumulipo* and connection with nature • *Aloha 'āina* reinforcement of *pono* • *Mālama 'āina* concept of taking only what is needed with respect	**ITS** **Ecological** • Response of ecosystems to decline in herbivorous aquarium fish abundance **Economic** • Importance of marine tourism to economy **Political** • Failure of DAR to manage the aquarium fishery • Pressure by citizens' groups to create laws to develop FRAs **Educational** • Research on collector harvesting • Monitoring of the effectiveness of FRAs to replenish aquarium fishes **Sociological** • Representation of individuals in traditional *ahupua'a* • Establishment of FRAs in *ahupua'a* • Availability of information on marine resources to WHFC representatives and local community • Recognition and integration of *pono, ahupua'a, konohiki*, to reinforce management approach

Figure II.2. 4-quadrant analysis of West Hawai'i community-based management of the aquarium fishery.

fish for the bulk wholesale market. Conversely, other collectors target uncommon or rare fishes of higher individual value, such as Tinker's butterfly fish (*Chaetodon tinkeri*).

Concomitantly, varying collecting strategies impact specific sectors of the dive tour industry in different ways. A major component of West Hawai'i's dive tour industry consists of large vessels transporting hundreds of novice snorkelers out to the reefs to experience the reef and its organisms as a whole. In contrast, other dive operations specialize in repeat, quasi-professional scuba divers who target specific organisms for new species sightings to add to their dive logs or for marine photography enthusiasts. In the latter case, value is placed on dive operations that have detailed, long-term ecological knowledge of the locations and habits of rare marine species. Therefore, when collectors target and remove rare, resident fishes from the reef, the more specialized dive tours are directly affected. Conversely, the bulk removal of whole schools of yellow tangs and other colorful reef fishes has a more pronounced effect on the larger-scale diving operations and ultimately on the snorkelers' aesthetic experience on the reefs. Thus, the behaviors of individual fish species and human usergroups have direct consequences on each other and play a major role in the kinds of interactions between stakeholders in West Hawai'i.

Systems (Exterior-Collective)

Another dimension included in an Integral analysis is the overlapping natural, social, and political systems, and their interactions. This dimension is well recognized and analyzed in traditional fishery management.

At the ecosystem level, yellow tangs are herbivorous fish that may play an important regulatory role in controlling the abundance of algae in coral reef communities. However, although several groups used this interplay to oppose aquarium collectors, there are no observed increases in the abundance of algae in areas subjected to aquarium fish harvesting (Tissot and Hallacher, 2003). On the human level, local aquarium collectors are a small component of a large international trade network involving wholesalers, retail store owners, and worldwide hobbyists. Fish collected on reefs in Hawai'i are sold to local wholesalers for several dollars each and then shipped cross-country; they can, for example, end up in a store in Kansas retailing for more than $60 each. In West Hawai'i there are approximately 60 active collecting permits and, based on voluntary reporting, the annual harvest in 2001 was 708,000 fish at a total value of $1.06 million (Walsh et al., 2003). However, it is likely that the catches and values are underrepresented and the actual value of the fishery may be considerably larger than reported (ibid.).

Coral reef aquariums, ranging from small private displays to massive public exhibits, are sophisticated operations requiring detailed knowledge of environmental conditions, nutrient dynamics, and habitats of coral reef fish. Subsequently, many hobbyists own large aquaria stocked with a variety of reef fishes and living corals from around the world in complex, closed aquaria costing thousands of dollars to stock and maintain (SPFS, 1999). These displays can have high educational value and may provide the opportunity for people far removed from Hawai'i to enjoy the wonder and beauty of coral reef fishes in their own homes or at their local public aquarium. As hobbyists promote sustainable practices in the aquarium industry, educational opportunities provided by the aquariums they create can lead to increased awareness of and appreciation for the oceans' resources, and thereby help promote worldwide conservation efforts to protect coral reefs (ibid.).

At odds with the aquarium industry is the much larger tourism industry, which is the second largest generator of revenues for the State of Hawai'i. A major component of tourism in Hawai'i centers on marine dive tours, which account for a large component of the $3 billion a year ocean industries revenue in Hawai'i (Cesar and van Beukering, 2004). Dive tour operations are closely linked to a wide variety of other industries that capitalize on the beauty of local reefs, including hotels and restaurants, apparel, jewelry and art, and eco-tourism that combines land and sea tours. Over the last 30 years both the dive tour and aquarium industries have experienced dramatic growth in Hawai'i (Cesar and van Beukering, 2004; Walsh, et al., 2003; Miyasaka, 1997). Overall, however, revenues from dive tourism dwarf those of the aquarium industry, which may well explain why aquarium collectors were unable to develop the political support to oppose the establishment of FRAs in West Hawai'i. In contrast, as discussed in the addendum (page 448), the more traditional recreational and artisanal fisherman have tremendous political support and have been very effective in preventing MPA establishment for consumption-based fishing practices.

A wide variety of institutions and organizations also played important roles in the development of the fishery management plan for West Hawai'i. Although DAR is responsible for state fisheries management, dissension within DAR over management issues and a consistent laissez-faire attitude about management in general ultimately led to a legislative resolution of the conflict. The legislature became involved because the West Hawai'i community strongly protested the lack of effective management of the aquarium fishery by DAR. In 1997, the Lost Fish Coalition presented a 4,000-signature petition to state legislators and requested a total ban on collecting. Thus, during the 1997–1998 legislative session there were two competing bills moving forward: one creating a total ban on collecting in West Hawai'i and another mandating a minimum of 30% of closed areas managed with community participation. Understandably, the collectors endorsed the latter, joined the WHFC, and helped influence the location of the FRAs.

Working with several university-affiliated marine ecologists and DAR, I also was involved in an important role by conducting a study to document the extent of harvesting by aquarium collectors in West Hawai'i and provide objective data to the legislature and the public. The results of the study documented significant 38%–75% declines in 7 of the 10 aquarium fishes studied (Tissot and Hallacher, 2003). The results of these studies were presented to DAR, the legislature, and the West Hawai'i community, and summaries were provided during committee hearings and at public meetings. Moreover, once the FRAs were demarcated, studies were established by WHAP to monitor the effectiveness of the management plan to replenish aquarium fishes and provide information to all stakeholders (Tissot et al., 2004).

Significant contributions to the conflict resolution process were also made by acknowledging and building on aspects of traditional native Hawaiian fishery management. The concept of resources management is implicit in native Hawaiian culture and was traditionally embedded in the overarching sociopolitical and spiritual construct of *pono*, or "balance" between the community and the ecosystem. *Pono* is recognized as the dynamic balance between the *Ali'i nui* (high chiefs), the common people, the gods, and the sacred *'āina* (land and sea), from which all food and water and thus all life is provided and maintained by the just rule of the *Ali'i* through strict laws and rituals, and ancient sacred traditions. Thus *pono* is consistent with the purposeful management of natural resources to promote sustainability; a concept clearly at odds with modern fishery management in Hawai'i today (Friedlander et al., 2003).

Recognizing the importance and value of these traditional concepts, the WHFC membership and FRAs are associated with traditional Hawaiian *ahupua'a* and with native Hawaiian populations that supported the aquarium ban and provided community-based support for enforcement, which is severely lacking in Hawai'i (ibid.). Moreover, periodic closure of fishing grounds was a common fisheries management technique in ancient Hawai'i, where *kapu* (strict laws) were used in resources management to enforce no-take areas, to restrict hunting and fishing seasons, and to establish sacred or forbidden species, often with severe penalties for violating such laws—including banishment and death. These laws were passed after the *konohiki* (a resources manager/steward appointed by the *Ali'i)* consulted with the *po'o lawai'a* (master fishermen), who had generations of intimate knowledge of the status of marine resources in their *ahupua'a.* Two functions of the WHFC mirrored this arrangement by allowing a two-way flow of information from individuals in the community to the council, scientists, and DAR, and from these groups back to the community. The attempt to provide near real-time data from WHAP to the WHFC with representation of individuals from multiple *ahupua'a* recognizes the value and builds on the flow of information in this traditional model.

Cultural Dimension (Interior-Collective)

The cultural dimension includes collective attitudes and beliefs that shape the behaviors and action of groups within the social systems. Although this dimension is acknowledged in the arena of environment conflict resolution, it is a complex and often neglected dimension of fishery management that deserves significantly more attention.

The origins of the conflicts over aquarium collecting and the lack of complete resolution were intricately intertwined with the different ethics/values, or belief systems, of the various stakeholders. Although few people in Hawai'i take issue with catching reef fish for consumption, collecting live fish for exportation is viewed as a wholly different matter. At issue are several divergent ideas. Onc is the local acknowledgment that supplementing your family's food by living off of the land is a common and accepted lifestyle in Hawai'i, and subsistence catch still provides a vital component to the household food budget for many *'ohana* today. However, the practice largely involves eating or sharing with your neighbor what you catch, and never taking more than you need. Clearly, selling live fish for solely economic gain stretches the traditional Hawaiian concept of subsistence catch. In addition, some communities more heavily dependent on subsistence catches wanted collectors banned due to a perceived competition for food fish (Walsh, personal communication) and because collectors may have been perceived as "greedy" in a Hawaiian cultural context, when compared with the local subsistence fishermen.

Another issue often debated in public and presented in the newspapers was the potential negative effects of fish harvesting on the reefs. Collectors held that their industry was sustainable and not causing harm to the reefs, a contention partially supported by scientific study (Tissot and Hallacher, 2003). However, collectors were frequently attacked with accusations that their collecting activities caused long-term reef damage and that their operations promoted unsustainable fisheries. Thus, the lack of clearly demarcated opinions, worldviews, ethics/values, and cultural perspectives, combined with a community-wide debate over the issues, confounded and prolonged the conflict.

Ethical conflicts also emerged in the consensus-based approach used to develop the FRAs. One of the major goals of the WHFC was to achieve consensus among stakeholders on the location of the FRAs and simultaneously resolve the multiple-use conflict using an alternative process of environmental dispute resolution, or EDR. Alternative EDR is a growing field wherein the psychology and behavior of conflicting interest groups is recognized and developed into self-generated conservation tactics that acknowledge and preserve personal goals (Daniels and Walker, 2001). When conflicts are complex, as was the case in West Hawai'i, they can occur at multiple levels, each of which needs to be acknowledged, understood, and honored. Conflicts often revolve around different levels of perception of the issues: so-called interest-versus value-based conflicts (Capitini et al., 2004). From an IE approach these are conflicts in an intersubjective context (i.e., traditional versus modern worldviews). In West Hawai'i the interest-based component of the conflict, a rational and scientifically based dispute over the allotment of fish resources between collectors and tour boats, was clearly recognized and understood. Although framed as a sustainability issue by many, this conflict was resolved by designating the FRAs, which prohibited collectors from operating in areas containing popular dive sites, while leaving the majority of the coastline open for harvesting. In contrast, value-based components are often more complex and derived from long-standing differences and concerns, and stem from psychology, culture, and threatened beliefs (Rothman, 1997). Known also as identity-based conflicts, these disputes are characterized by an unclear determination of their parameters and boundaries, as they stem from deeper personal values that are not clearly demarcated or understood (ibid.). In West Hawai'i, differing beliefs on the appropriate use of fish on the reef were a continued source of discord and dissension on the council. For example, aquarium collectors interpreted Act 306's language stating that "a minimum of 30% of the coastline be designated as FRAs" as meaning exactly 30% of the coast. When the total FRA coverage ended up at 35.2%, many collectors felt they were unfairly singled out and walked off the council (Walsh, 1999). Although it is unclear if their behavior was primarily related to the designation of specific FRAs, it seems likely that collectors felt that their values and rights as fishermen were not acknowledged in the process and thus viewed the conflict from a traditional "membership" position.

Aquarium collectors were dismayed by what they viewed as a small, harmless industry unfairly singled out and criticized for utilizing what they perceived to be a largely untapped, abundant, and seemingly limitless resource (Capitini et al., 2004). Some collectors boasted of harvesting 1,000 fish per day (Walsh, personal communication). Here again, different values and beliefs about the ocean contributed to conflict in West Hawai'i. Although many view the ocean as boundless and unlimited in its productivity and potential uses, others hold the ocean as sacred for sustaining life (the postmodern "sensitive self"). This dichotomy represents a classic conflict between the Judeo-Christian worldview that resources are for our use (i.e., the Garden of Eden) and the Hawaiian philosophy of *mālama 'āina* (caring for the land) (White, 1967; Callicott, 1994).

Thus, embedded in this controversy were threads of native Hawaiian cultural practices and traditions that still exist in the West Hawai'i community. Although present, these beliefs are often fragmented, misunderstood, and occasionally misrepresented or misapplied in modern Hawaiian communities. One such belief is *aloha 'āina*, or love of the land, a traditional Hawaiian value that connects the people to the place where they live and work (Pukui et al., 1972). The foundation of this connection is beautifully told in the Hawaiian creation chant, or *Kumulipo*, which describes the evolutionary descent and familial relationship of all living creatures, eventually resulting in the Hawaiian people (Beckwith, 1951). *Aloha 'āina* serves to reinforce the concept of *pono*, the balance between people and the *'āina*, but also gives rise to the concept of *mālama 'āina*, or caring for the land.

To the Hawaiians land was not owned as in Western cultures, but was for temporary use and considered sacred. The *'āina* was seen as a place for restoring spiritual, cultural, natural, and individual balance, a concept known as *lōkkahi*. George Helm, one of the founders of the Hawaiian cultural renaissance and respected *kumu* (teacher), expressed his thoughts about this concept in relation to the restoration of the sacred island of Kaho'olawe as a place for restoring *lōkkahi* (Whitcraft and Levin, 2003):

> The truth is, there is man and there is the environment. One does not supersede the other. The breath of man is the breath of Papa (the earth). Man is merely the caretaker of the land that maintains his life and nourishes his soul. Therefore, *'āina* is sacred. The church of life is not a building; it is the open sky, the surrounding ocean, the beautiful soil. My duty is to protect Mother Earth, who gives me life. And to give thanks with humility as well as to ask forgiveness for the arrogance and insensitivity of man.

In this context, then, *mālama 'āina is* a cultural extension of *pono* and serves to reinforce sociological and political systems. *Mālama 'āina* involves asking permission prior to fishing, taking only what you need, sharing your catch with your extended *'ohana*, or community, and having respect for the sacredness of the process. Thus, even though most of ancient Hawaiian culture was dominated by magical and warrior values, the ethics of resource management was guided by the spiritual beliefs of their priests (*kahuna*) elders (*kupuna*) and *kumu*, resulting in the potential for an integral view of their relationship with nature.

Clearly, harvesting live fish for economic gain and shipping them in a bag for a long, convoluted odyssey, potentially resulting in mortality and waste, violated the very core of these traditional values. Thus, the Hawaiian cultural worldview, expressed in various forms and at various levels in the communities of West Hawai'i, contributed to a conflict with ornamental reef fisheries. In the conflict resolution process, a lack of acknowledgment and understanding of these multiple cultural values likely presented a fundamental conflict in developing a consensus-based approach to fishery management on the WHFC.

Experiential Dimension (Interior-Individual)

The experiential dimension includes the subjective realities of all beings at all levels of awareness. These include the emotions and motivations of humans and their spiritual experiences, which are generally not considered in fishery management. It also involves the forms of experience and perception available to the fishes on the reef. For many individuals, snorkeling over a coral reef in Hawai'i is a rich, joyful, and occasionally spiritual, transformative experience. Floating freely with hundreds of brightly colored fish swarming over an intricate web of coral has inspired many people to learn to value the reef and its organisms. These experiences may also involve the development of personal connections between people and individual organisms on the reef. Some reef organisms are known to inhabit the same areas of the reef for extended periods of time. Butterfly fish, for example, form permanent mated pairs that establish territories on the reef for life (Reese, 1993). Thus, it is not uncommon for people to form long-term connections or relationships with individual fish, in specific areas. For example, in Wai O'pae in East Hawai'i the establishment of a "no take" marine reserve, generally a difficult and contentious process in Hawai'i, was greatly facilitated due to the mayor of Hawai'i island having a long-standing personal relationship with a fish in the proposed reserve. These connections, and the associated sharing of feelings with the plight of fish captured on the reef, played strongly into the animosity against aquarium collectors that developed in West Hawai'i.

People gasped at the thought of beautiful fish being captured and transported off their native reef, never to return. The capture and transportation of live fish is described as a litany of horrors, with often less than 10% successfully surviving the journey to their intended destination, and even then, often dying alone in aquariums due to the inexperience of owners (Baquero, 1995). This emotional connection with the plight of the reef fish was the principal driving force behind the Lost Fish Coalition, whose hallmark was a dead yellow tang in a haggle. Armed with their preservationist convictions and the powerful image of the dead fish, the Lost Fish Coalition was largely responsible for spearheading the passage of Act 306 through the legislature (www.lostfishcoalition.org). Thus, people's personal joy on the reef, transformed into horror, sadness, and eventual anger as their favorite fish disappeared, became the principal motivation to limit or eliminate the actions of aquarium collectors.

Discussion

The IE approach provides a far broader and deeper view of the biology, conflict, resolution, and management process in West Hawai'i than has previously been presented (figure II.2). The principal strength of the AQAL approach is that it provides a means to acknowledge and understand the levels of complexity and the depth of issues and beliefs in each quadrant and to appreciate the interconnections and feedback mechanisms among quadrants and between levels. This approach clearly captures and illustrates how individual experiences on the reef were transformed into social movements, political will, and eventual ecological change. Thus, interiors, largely unrecognized by scientists in fishery management, were ultimately driving the process that led to ecological change. Moreover, the example serves to illustrate the role of scientific research and education in providing informed decision-making at all levels. However, some of the complexity in the system, which perhaps resulted in the failure of the WHFC to reach complete consensus on a management solution, may have been lost by not acknowledging, understanding, or honoring the differing worldviews represented by the various stakeholders. In this respect, there were fundamental values-based or ethical conflicts in the community, and subsequently among members of the council, that were not sufficiently explored nor discussed. Moreover, the IE approach helps clarify and strengthen the overall sustainability of the system and provide a model of fishery management for the rest of Hawai'i.

In the case of the WHFC, building on Hawaiian social and cultural traditions helped promote solutions to management issues that previously may not have been possible. Native Hawaiians developed a management system based on social and cultural controls, and more specifically on a code of conduct and set of laws for fishing that was strictly enforced and provided for sustainable harvests of natural resources. One of the major components of this system was/is the belief that marine resources are limited and, as a result, there was/is a strong social obligation to exercise self-restraint through the process of nurturing and respect (Titcomb, 1972; Friedlander et al., 2003). These traditions and beliefs, although badly fragmented, are still present in modern Hawai'i and are a source of strength for the fishery management approach in West Hawai'i. The AQAL model explicitly honors modern ecological and social approaches—incorporating monitoring, conflict resolution, and fishery councils—along with the existing Hawaiian models by incorporating the political and cultural concepts of *ahupuaʻa, konohiki, kapu, pono,* and *mālama ʻāina.*

In smaller communities, such as on the Ho'olehu Hawaiian Homesteads on the rural island of Moloka'i, effective integration of modern and traditional cultures has allowed the community to adapt its management strategies to the specific environmental conditions of the area and develop scientific assessments alongside a traditional Hawaiian moon calendar to help govern harvesting of reef fish (Friedlander et al., 2003). Based on the success of the Ho'olehu community, one recommendation for the WHFC would be to decentralize into smaller community governing boards associated with individual *ahupuaʻa* or *ʻili* (groups of *ahupuaʻa*), thereby potentially facilitating an easier flow of information to and from the community, providing for more specific knowledge of reef resources, and allowing local communities a more direct role in the management of their resources.

Another challenge is to more thoroughly acknowledge and incorporate traditional knowledge from communities to complement scientific assessments. Efforts should be made to recruit and support *poʻo lawaiʻa* (master fishermen) and provide an avenue for the recognition and application of their knowledge by the regional boards. For example, one of the major questions of MPA design is where to locate protected areas in order to maximize the productivity of the ecosystem (Allison et al., 1998). This task is particularly difficult due to high levels of fishery exploitation in Hawai'i and thus a lack of understanding of natural, pre-exploited population levels. One idea in addressing this question is to design MPAs using information from traditional Hawaiian fishing areas. Hawaiians marked their productive fishing grounds, often based on centuries of observation, with fishing sacred shrines, or *koʻa* (Titcomb, 1972), some of which are still present in Hawai'i (Whitcraft and Levin, 2003). Archaeological and historical research on the location of *koʻa,* combined with modern scientific assessments and the extensive knowledge of living *poʻo lawaiʻa* and *kupuna,* could provide an effective means of designing networks of MPAs.

Finally, in the conflict resolution process, greater attention should be given to exploring and acknowledging the identity- and value-based conflicts present in West Hawai'i and on the WHFC. The use of a professional facilitator trained to tease out and identify these complex issues might go a long way toward resolving the conflicts still present (Capitini et al., 2004).

One possibility is to use or build on the Hawaiian method of conflict resolution, or *hō'ō pōnō pōnō* (to make things right through healing). *Hō'ō pōnō pōnō* is a process that occurs within an *'ohana* or community to resolve long-standing disputes (Pukui et al., 1972). The process is traditionally led by a *kupuna* and involves long periods of *pule* (prayer), *hala* (airing of grievances or transgressions), *hihia* (recognition of mutual negative feelings), and eventual resolution through *mihi* (apology) and *kala* (forgiveness) (Shook, 1986). Adaptation of this process to local councils might allow disputes among individuals to be more easily acknowledged and resolved, and furthermore may also allow a more complete and appropriate incorporation of spiritual and cultural dimensions into the process. In conclusion, the complexity illustrated using the IE approach clearly opens up a variety of new possibilities for resolving conflicts and promoting sustainability in resources management. IE provides an analytic framework that not only is efficient in identifying the multidimensionality of complex eco-social issues, but also helps identify key leverage points for effective change. IE is capable of recognizing these leverage points by taking an Integral meta-view of any situation and analyzing individual experiences and behaviors, as well as collective complexity. In a world as complex and multifaceted as ours it is hard to imagine using any approach that is not Integral in nature.

Addendum

Since 2004 several changes and events have occurred that further illustrate that multiple-use resource conflicts in West Hawai'i are a lack of mutual agreement and understanding within the subjective context (worldview and value-driven) rather than the objective context (data-driven). Lack of agreement occurs both within and among identified stakeholder groups, validating a full spectrum rather than a traditional stakeholder-driven EDR process. On the reefs, yellow tangs have continued their strong recovery in the FRAs: increasing 49% overall in West Hawai'i since FRA closure, and the aquarium fishery has the highest catch, economic value, and catch-per-unit-effort of its 30+-year recorded history (Walsh, personal communication). For many members of the WHFC, including some aquarium collectors, the yellow tang recovery data have been persuasive, resulting in broader support for MPAs and community-based management. Thus, just as individual and collective interiors initiated first political, then ecological change, Right-Hand data on biology and economics is feeding back, informing, and altering interiors. Ultimately, it is hoped that this AQAL feedback process will provide momentum for further community-based MPA models in Hawai'i, and indeed this is happening at various levels throughout the state of Hawai'i (Tissot et al., in review). However, as the West Hawai'i model became more well known within the state, user conflicts at larger-spatial scales quickly became apparent.

In an effort to extend the West Hawai'i model to other islands, a legislative bill (HB 2056) was submitted to the Hawai'i state legislature in 2004. Although the bill had a fair amount of support within the legislature and statewide, fishing groups (primarily recreational and artisanal fishers, a much larger group than aquarium collectors) rallied together and handsomely defeated the bill. The following year, perhaps in response to HB 2056, two pro-fishing and anti-MPA "Freedom to Fish" bills quickly followed. Although defeated, they made significant headway in committees, including gaining the support of legislators in West Hawai'i. "Freedom to Fish" bills, which are a nationwide phenomenon, use scientific language to purportedly protect fish stocks but in reality place such a high burden of proof on cash-strapped management agencies that they effectively strip away most fishery regulations, including existing MPAs. These bills largely present a traditional worldview membership argument (fishers have a right to fish that preempts other uses or values), using modern rational language (fish need extraordinary protection given their value, and current management is underfunded and sorely inadequate) and have been effective in rallying support to defeat MPA efforts. These conflicts, now manifested at the state level, mirror conflicts in the WHFC EDR process and are systematic of marine conservation in general, where scientific zeal and reductionism (in the form of data and ecological theory) has largely ignored the interior (or Left-Hand) roots of user conflicts under so-called "consensus-based EDR." MPAs have been used as a panacea for ocean ill health, an ideological scientific tool to promote marine conservation. However, in the long run, this narrow exterior and data-driven approach is likely to damage the long-term effectiveness of resolving marine conservation efforts by not resolving resource conflict issues with diverse stakeholders that encompass multiple perspectives, values, and worldviews (Agardy et al., 2003).

Interestingly, within the last two years a similar schism has developed within the scientific community, primarily in response to data on declining global fisheries. One group, represented primarily by academic marine ecologists who advocate MPAs as the central tool of a new approach to rebuilding the marine ecosystems of the world, is in conflict with scientists primarily working in fisheries agencies, who see many failed fisheries but also numerous successes. While marine ecologists advocate for a paradigm shift in fishery management, the fishery scientists argue that we need to apply lessons learned from successful fisheries and rebuild threatened fisheries. This disagreement within what many see as a unified "scientific community" represents yet another conflict as modern and postmodern ecologists press for no-take areas that preserve fish, habitat, and biodiversity. In contrast, fishery scientists are driven by traditional and modern worldviews and argue for traditional systems approaches valuing policy, management, and economics. Clearly, both sides have enormous merit, but as with the disconnect between fishers and other stakeholders, it is unlikely that mutual agreement will occur anytime soon without a process that incorporates an Integral Ecology approach.

REFERENCES

Agardy, T., and P. Bridgewater. 2003. "Dangerous Targets? Unresolved Issues and Ideological Clashes around Marine Protected Areas." *Aquatic Conservation: Marine and Freshwater Ecosystems* 13: 353–67.

Allison, G. W., J. Lubchenco, and M. H. Carr. 2003. "Marine Reserves Are Necessary but Not Sufficient for Marine Conservation." *Ecological Applications* 8 (1): S79–S92.

Baquero, J. 1995. "The Stressful Journey of an Ornamental Marine Fish." *Sea Wind* 9 (1): 19–21.

Beckwith, M. W. 1951. *The Kumulipo, a Hawaiian Creation Chant.* Chicago: University of Chicago Press.

Bohnsack, J. A. 1998. "Application of Marine Reserves to Reef Fisheries Management." *Australian Journal of Ecology* 23: 298–304.

Botsford, L. W., J. C. Castilla, and C. H. Peterson. 1997. "The Management of Fisheries and Marine Ecosystems." *Science* 277: 509–14.

Callicott, J. B. 1994. *Earth's Insights*. Berkeley: University of California Press.

Capitini, C. A., B. N. Tissot, M. S. Carroll, W. J. Walsh, and S. Peck. 2004. "Competing Perspectives in Resource Protection: The Case of Marine Protected Areas in West Hawai'i." *Society and Natural Resources* 17: 763–78.

Cesar, H., and P. van Beukering. 2004. "Economic Valuation of the Coral Reefs of Hawai'i. *Pacific Science* 58(2): 231–42.

Daniels, S., and G. Walker. 2001. *Working Through Environmental Conflict: The Collaborative Learning Approach*. Westport, CT: Praeger.

Dybas, C. L. 2002. "In Hawaii, the Age of Aquariums Raises Concerns: Collectors Clash with Islanders over Tropical Fish." *Washington Post*, June 3, A7.

Dyer, C., and J. R. McGoodwin. 1994. *Folk Management in the World's Fisheries*. Niwot: University Press of Colorado.

Esbjörn-Hargens, S., and M. Zimmerman. 2006. *Integral Ecology: Consciousness, Culture, and Nature*. Manuscript in preparation.

FAO (Food and Agriculture Organization). 2000. *The State of World Fisheries and Aquaculture*. Rome: FAO.

Friedlander, A., K. Poepoe, K. Popeoe, K. Helm, P. Bertram, J. Maragos, and I. Abbott. 2003. *Applications of Hawaiian Traditions to Community-Based Fishery Management*. Proceedings of the 9th International Coral Reef Symposium 2 (pp. 813–18), Bali, Indonesia.

Handy, E. S., and M. K. Pukui. 1958. *Polynesian Family System in Ka'u Hawai'*. Rutland, VT: Charles Tuttle Co.

Hargens, S. B. F. 2002. "Integral Development in Bhutan: Taking the 'Middle Path' towards Gross National Happiness." *Journal of Bhutan Studies* 6: 24–87.

Hastings, A., and L. W. Botsford. 1999. "Equivalence in Yield from Marine Reserves and Traditional Fisheries Management." *Science* 284: 1537–38.

Hilborn, R. 2007. "Moving to Sustainability by Learning from Successful Fisheries." *Ambio* 36 (4): 296–303.

Hobday, A. J., and M. J. Tegner. 2000. *Status Review of White Abalone (Haliotis sorenseni) throughout Its Range in California and Mexico*. U S. Dept. of Commerce, NOAA technical memorandum NOAA-TM-NMFS-SWR-035. 101 pp.

Holling, C. 1978. *Adaptive Environmental Assessment and Management*. London: Wiley.

Jennings, S., and M. J. Kaiser. 1998. "The Effects of Fishing on Marine Ecosystems." *Advances in Marine Biology* 34: 201–352.

Johannes, R. E. 1978. "Traditional Marine Conservation Methods in Oceania and Their Demise." *Annual Review of Ecology and Systematics* 9: 346–64.

———. 1981. *Words of the Lagoon: Fishing and Marine Lore in the Palau District of Micronesia*. Berkeley: University of California Press.

Kirch, P. V. 1984. *The Evolution of the Polynesian Chiefdoms*. Cambridge: Cambridge University Press.

Michaelidou, M., D. Decker, and J. Lassoie. 2002. "The Interdependence of Ecosystem and Community Viability: A Theoretical Framework to Guide Research and Application." *Society and Natural Resources* 15: 599–616.

Miyasaka, A. 1997. *Status Report, Aquarium Fish Collections, Fiscal Year 1994–95.* Division of Aquatic Resources, Department of Land and Natural Resources, Honolulu.

Murray, S., R. F. Ambrose, J. A. Bohnsack, L. W. Botsford, M. H. Carr, G. E. Davis, P. K. Dayton, D. Gotshall, D. R. Gunderson, M. A. Hixon, J. Lubchenco, M. Mangel, A. MacCall, D. A. McArdle, J. C. Ogden, J. Roughgarden, R. M. Starr, M. J. Tegner, and M. M. Yoklavich. 1999. "No-Take Reserve Networks: Sustaining Fishery Populations and Marine Ecosystems." *Fisheries* 24 (11): 11–25.

NRC (National Research Council). 1999. *Sustaining Marine Fisheries.* Washington, DC: National Academy Press.

———. 2000. *Marine Protected Areas: Tools for Sustaining Ocean Ecosystems.* Washington, DC: National Academy Press.

POC (Pew Oceans Commission). 2003. *America's Living Oceans: Charting a Course for Sea Change.* Arlington, VA: Pew Charitable Trust Report.

Pukui, M. K., E. W. Haertig, and C. A. Lee. 1972. *Nānā I Ke Kumu* (Look to the Source). Honolulu: Queen Lili'oukalani Children's Center.

Reese, E. S. 1993. "Reef Fishes as Indicators of Conditions on Coral Reefs." In *Proceedings of the Colloquium on Global Aspects of Coral Reefs: Health, Hazards, and History.* Miami: University of Miami, pp. 59–65.

Rothman, J. 1997. *Resolving Identity-based Conflict in Nations, Organizations, and Communities.* San Francisco: Jossey-Bass.

Shook, E. V. 1986. "Hō'ō pōnō pōnō: Contemporary Uses of a Hawaiian Problem Solving Process." Honolulu: University of Hawai'i Press.

SPFS (South Pacific Forum Secretariat). 1999. *Marine Ornamentals Trade: Quality and Sustainability for the Pacific Region.* Suva, Fiji: SPFS, Trade & Investment Division.

Tissot, B. 1999. "Adaptive Management of Aquarium Collectors on Reef Fishes in Kona, Hawai'i." *Conservation Biology* 17 (6): 1759–68.

Tissot, B. N., W. J. Walsh, and L. E. Hallacher. 2004. Evaluating the Effectiveness of a Marine Protected Area Network in West Hawai'i to Increase the Productivity of an Aquarium Fishery. *Pacific Science* 58 (2): 175–88.

Tissot, B. N., W. J. Walsh, M. A. Hixon, and K. Lowry. In review. "Hawaiian Islands Marine Ecosystem Case Study: EBM and Community-Based Management in West Hawaii." *Coastal Management.* 27 pp.

Walsh, W. J., S. P. Cotton, J. Dierking, and I. D. Williams. 2003. "The Commercial Marine Aquarium Fishery in Hawai'i 1976–2003." In *Status of Hawai'i's Coastal Fisheries in the New Millenium,* edited by A. M. Friedlander. Proceedings of a symposium sponsored by the American Fisheries Society, Hawai'i chapter, pp. 132–259.

Walsh, W. J., B. N. Tissot, and L. E. Hallacher. 2004. "A Report on the Findings and Recommendations of Effectiveness of the West Hawai'i Regional Fishery Management Area." Report to the 23rd Hawaii Legislature. 38 pp.

Whitcraft, S., and P. Levin. 2003. *Ocean/Cultural Orientation*. Kaho'olawe Island Reserve Commission. State of Hawai'i, Dept. of Land and Natural Resources. Wailuku, Maui.

White, A., L. Hale, Y. Renard, and L. Cortesi. 1994. *Collaborative and Community-Based Management of Coral Reefs*. Bloomfield, CT: Kumarian Press.

White, L. Jr. 1967. "The Historical Roots of Our Ecological Crises." *Science* 155 (3767): 1206.

Wilber, Ken. 1995. *Sex, Ecology, Spirituality*. Boston: Shambhala Publications.

———. 2000. *A Theory of Everything*. Boston: Shambhala Publications.

Wood, E. M. 2001. *Collection of Coral Reef Fish for Aquaria: Global Trade, Conservation Issues and Management Strategies*. Herefordshire, UK: Marine Conservation Society.

Worm, B., E. Barbier, N. Beaumont, E. Duffy, C. Folke, B. Halpern, J. Jackson, H. K. Lotze, F. Micheli, S. R. Palumbi, E. Sala, K. A. Selkoe, J. Stachowicz, and R. Watson. 2006. "Impacts of Biodiversity Loss on Ocean Ecosystem Services." *Science* 314: 787–90.

CASE STUDY III

Evolving Approaches to Conservation:*

Integral Ecology and Canada's Great Bear Rainforest

DARCY RIDDELL
Hollyhock Leadership Institute, Vancouver, British Columbia, Canada

This case study applies Integral Ecology to analyze the broad range of strategies environmentalists have undertaken to create protected areas and change forest practices in the Great Bear Rainforest, British Columbia, Canada. Rainforest conservation efforts in the region promoted holistic, transdisciplinary solutions and fostered agreement among diverse stakeholders, modeling an Integral Ecology approach. Environmentalists worked locally and globally, engaging with economic, cultural, political, and scientific systems to create change. The campaign involved transformations at personal and cultural levels that have enabled negotiated solutions involving over 20 million acres of rainforest on British Columbia's coast.

*This case study was published as "Evolving Approaches to Conservation: Integral Ecology and British Columbia's Great Bear Rainforest" in *World Futures: The Journal of General Evolution* 61, nos. 1–2 (2005): 63–78. Address correspondence to Darcy Riddell, Hollyhock Leadership Institute, 680-220 Cambie Street, Vancouver, BC V6B 2M9, Canada. E-mail: darcy@hollyhockleadership.org.

KEYWORDS: British Columbia forests, environmentalism, Great Bear Rainforest, Integral Ecology, Integral Theory, rainforest conservation, spiritual activism.

Introduction

> Ecology is pulling up a dandelion and finding that everything else is attached.
>
> —JOHN MUIR

For humanity to flourish in the 21st century and beyond, the diverse cultures, communities, and peoples of the Earth must learn to live within ecological limits. Although there is no quick fix for the myriad of ecological and social problems facing humanity, creative, integrative frameworks for solutions are emerging. Effective responses to managing the relationship of humans to nature will take as many forms as there are cultures and ecosystems. In this case study I explore innovations developed initially by environmental activist leaders, and later involving many parties, that model an Integral Ecological approach to the conservation of endangered temperate rainforests in the Great Bear Rainforest region of British Columbia (BC), Canada.

Solutions to global environmental and social crises will require an extraordinary and unprecedented scope of cooperation between people with very different knowledge, perspectives, cultures, and values. Effective solutions will involve the integration and restructuring of existing social, economic, political, and scientific institutions and the creation of new institutions that operate at local, national, and international levels. Solutions must also promote cooperation between people holding conflicting perspectives, and foster the transformation of human consciousness, values, and behavior. Integral Ecology is a comprehensive framework to guide the development of such solutions.

Integral Ecology is a transdisciplinary approach for solving complex, interlocking environmental problems and fostering long-term sustainability. This approach can be used to implement solutions that maximize whole-systems health for human and natural systems across many scales. Integral Ecology is used as a framework to understand, coordinate, and evaluate competing truth claims or perspectives, and to provide holistic approaches to transformational change. Integral Ecologists consider and integrate the multiple dimensions of ecological problems (e.g., social, economic, scientific, spiritual, psychological, behavioral, cultural, political, institutional) into inclusive solutions that are win-win. Integral Ecologists also emphasize that more comprehensive human value systems and worldviews are required to solve environmental problems, and these are fostered through psychological and spiritual development.

In this case study I assume a working knowledge of Integral Theory and its 4-quadrant, multiple-level approach.[1] Integral Ecology is based on this framework and is a meta-methodology for understanding and fostering ecologically beneficial change. It includes the four domains of subjective experience (Upper-Left quadrant: UL), behavior (Upper-Right quadrant: UR), culture (Lower-Left quadrant: LL), and complex systems including social, economic, and ecological systems (Lower-Right quadrant: LR), which make up Integral Theory's 4 quadrants, or the 4 terrains (Esbjörn-Hargens, 2005). The Integral Ecology framework illustrates that every event and object (or holon) has experiential, cultural, behavioral, and systemic dimensions, and that these dimensions are in complex, mutually evolving relationships or holarchies.[2] Subject and object are understood as interdependent, and Integral Ecologists value the multiple methodologies and ways of knowing that disclose qualitative and quantitative knowledge (Integral Methodological Pluralism).

Building on the insights of dozens of developmental psychologists, Integral Ecologists view human subjectivity or consciousness as developmental.[3] The developmental perspective of Integral Ecology acts as a "normative or analogical theory" that can inform timely and transformational action (Torbert and Reason, 2001, 10). Individuals are believed to co-evolve in relational exchange with other individuals, with cultures, and with the environment, at successively complex levels of organization or development. As new levels of development emerge in consciousness, culture, and living systems, they simplify and integrate system function and structure, while initiating new complexified and differentiated processes.

Integral Theory describes stages in the development of individual consciousness from birth (body) to pre-operational (symbol, image, feeling) to conventional-operational (rules, concepts) to formal-operational (reason) to vision-logic (multiple perspectives, systemic thinking), to transpersonal levels of consciousness. Worldviews are believed to develop in part due to increasingly complex life conditions along many lines of development. This trajectory of development moves from egocentric/pre-conventional (corresponding to magenta and red altitude) to ethnocentric/conventional (amber altitude) to worldcentric/postconventional (orange and green altitude) to Integral/vision-logic/second-tier (teal and turquoise altitude) (Wilber, 2001, 17–27).[4] As worldviews develop they become less partial and narcissistic, as the locus of care and concern grows to successively encompass self, humanity, and nature. According to Ken Wilber, Integral Theory is "critical of the present state in light of a more encompassing and desirable state, both in the individual and the culture at large. . . . The Integral paradigm will inherently be critical of those approaches that are, by comparison, partial, narrow, shallow, less encompassing, less integrative" (Wilber, 2001, 2). When individuals are able to authentically inhabit multiple perspectives, discern which are more or less partial, and act in a comprehensive manner, they have reached Integral/second-tier levels of development. Integral Ecologists hypothesize that to foster global sustainability, solutions are more effective if they come from perspectives at this Integral/second-tier level of consciousness.

Case Study: The Great Bear Rainforest

The story of the Great Bear Rainforest and the environmentalists working for its protection provides a glimpse of a unique model for resolving complex environmental conflicts, which has many hallmarks of an Integral Ecology approach. On the remote Pacific coast of British Columbia, Canada, just south of Alaska, is a temperate rainforest with wilderness valleys containing trees that are among the Earth's oldest living organisms. Steep, misty valleys hug rugged shorelines, and grizzlies and wolves thrive with little contact with humans. The region has a human population of fewer than 35,000. Small communities, many home to the region's aboriginal or First Nation people, dot the coast, accessible only by boat or air. Known as the Great Bear Rainforest, this area the size of Ireland was the largest unprotected coastal temperate rainforest in the world (see figure III.1).[5] In the late 1990s, as industrial logging proceeded in the region, the fight to save this rainforest grew to international proportions. It is now the site of an ambitious effort to preserve the ecological integrity of the rainforest, sustain aboriginal cultures, and develop new economic opportunities for coastal communities in British Columbia.

I began working on this campaign in 1997, first with the Sierra Club of Canada, BC Chapter, and later with ForestEthics (2004). Although most of the environmentalists who were part of the Great Bear Rainforest campaign are not explicitly using an Integral Ecology framework, I have

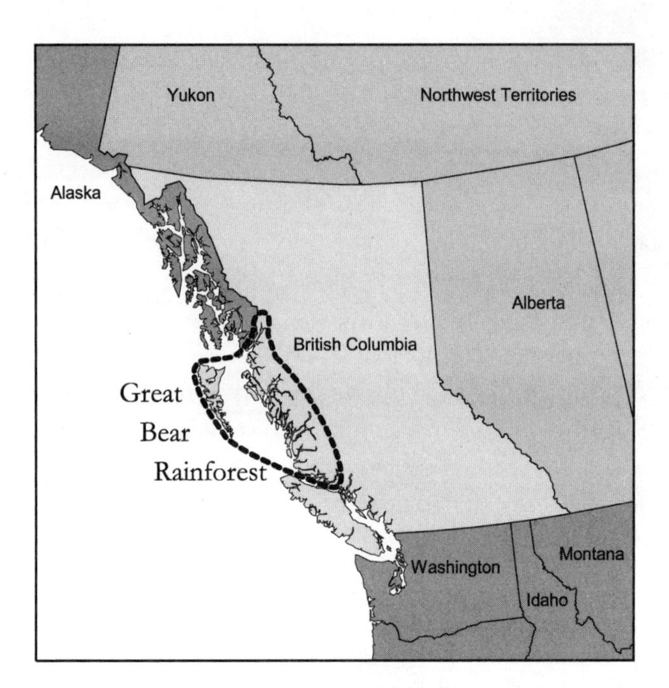

Figure III.1. Great Bear Rainforest.

observed that their efforts modeled key elements of an Integral approach. Such coordinated activism embodies what Beck and Cowan (1995) call a "brain syndicate," where people with diverse skills and worldviews come together to build holistic, integrative, and innovative solutions. The decade-long campaign to protect the Great Bear Rainforest illustrates that successful conservation efforts must engage effectively with complex economic realities, diverse cultural values, evolving scientific knowledge, shifting policy and political contexts, and the personal and interpersonal dimensions of values conflict, negotiation, and transformation. Conservation innovations in the Great Bear Rainforest fostered and integrated transformations in the 4 terrains of experience, culture, behavior, and systems, and depended on leaders' abilities to adopt and coordinate multiple perspectives—a characteristic of Integral consciousness.

Historical barriers to sustainability in BC's coastal communities have no single answer and will require decades and generations to address. However, the initiatives currently under way stand to bring the region closer to sustainability, with a scope that is unprecedented in British Columbia and unique in the world. Given the complexity of the Integral Ecology framework, and of the elements of this campaign, the following analysis represents a simplification of both and is a snapshot of my own evolving and incomplete attempts to understand events through the lens of Integral Ecology.

British Columbia's Coastal Rainforests in a Global Context

> Short of a miraculous transformation in the attitude of people and governments, the Earth's remaining closed canopy forests and their associated biodiversity are destined to disappear in the coming decades.
>
> —KLAUS TOEPFER, Executive Director, UNEP

The ongoing depletion of the Earth's closed-canopy native forests is a profound ecological loss. Forest ecosystems are a vital contributor to global climate stability, through carbon storage and climate regulation, and are vast storehouses of biodiversity. Many forests are also home to aboriginal peoples, linking native forest conservation inextricably with cultural survival. Original or "frontier" forests are endangered globally, with fewer than 22% remaining worldwide. The remaining 78% have been converted, fragmented, or replaced with plantations. Russia, Canada, and Brazil contain 70% of what remains (Bryant, 1997). Coastal temperate rainforests are more rare and endangered than tropical forests, surviving only in fragments in Chile, Tasmania, New Zealand, Norway, Alaska, and British Columbia (ibid.). These rainforests once extended down the Pacific coast of North America from Alaska through BC to northern California. By the end of the 20th century, not one intact valley remained in Washington, Oregon, or California. Only Alaska and BC contain large intact areas of this forest type, and over one quarter of the total is found in British Columbia (Kellogg, 1992).

In the mid-1990s environmentalists created the name "the Great Bear Rainforest" in reference to roughly 8.5 million hectares (21 million acres) of temperate rainforest on the BC coast previously known by government and the forest industry as the Mid and North Coast Timber Supply Areas. The region is home to a documented 230 bird species and 68 different mammals, including a genetically distinct grey wolf, dense grizzly populations, and the white Kermode or Spirit bear, which numbers fewer than 400 and is unique to the region. Over 20% of the world's wild salmon population spawns in the

Great Bear Rainforest. Most significantly, the region contains over 100 large intact valleys and many smaller ones, offering a rare opportunity for large-scale conservation and long-term protection of ecosystem integrity.

British Columbia's economy is highly dependent on the export of wood and pulp products (primarily to the U.S., Japan, and Europe), and 94% percent of commercial forest lands are publicly owned by the Province of British Columbia. The government collects stumpage fees from logging, and forest policies have historically been close in step with the desires of the major forest companies, a handful of which control the majority of the logging rights. During the 1990s the government in power introduced a Forest Practices Code and initiated multi-stakeholder Land and Resource Management Planning tables in most regions in BC.

Many parties, including environmentalists and First Nation people, felt constrained at these planning tables. Environmentalists chafed against a 12% limit on protected areas, government's "talk and log" approach, and loopholes in existing legislation that prevented ecologically viable protection of fish streams, biodiversity, endangered species, and old-growth forests. First Nation leaders resisted being classified as "stakeholders." In 1997, the Supreme Court of Canada ruled that Aboriginal Rights and Title had never been extinguished and that First Nations have unresolved claim to vast areas of land and resources in BC. These Rights and Title are the subject of ongoing treaty negotiations and litigation.

During the 1990s the rainforests of British Columbia were the scene of great controversy. Increased mechanization and years of overcutting created high levels of unemployment, and the export-driven nature of the forest industry exacerbated "boom and bust" regional economies. Government projections confirmed that regional rates of logging far exceeded long-term sustainable levels, and significant declines in the yearly amount harvested were inevitable throughout the region. Environmentalists campaigned for an end to the clear-cutting of old-growth rainforests and for protection of the coast's remaining intact wilderness valleys. Conservation campaigns fanned anger in resource-dependent communities, and environmentalists became scapegoats for the complex problems facing coastal communities.[6]

The Great Bear Rainforest campaign demonstrates a model that environmental leaders are using to create new forms of leverage and provide opportunities for large-scale, integral conservation solutions. Wilderness protection efforts in BC had previously been characterized by individual valley-by-valley conflicts, until a more coordinated and high-profile campaign to protect Vancouver Island's Clayoquot Sound emerged in the early 1990s. During this campaign, environmentalists built international boycott campaigns against BC wood products and organized blockades that were the largest act of civil disobedience in Canadian history. After much conflict, forest companies and the BC government announced a ban on clear-cutting in the rainforests of Clayoquot and initiated a local planning process that incorporated independent science and First Nations' traditional ecological knowledge. Many of the innovative tools, capacities, and institutions that originated in Clayoquot Sound were further developed during the Great Bear Rainforest campaign. These innovations increasingly embody an Integral approach. These include international markets campaigns, personal practices for transformation, expanded abilities to negotiate and collaborate, new policies and institutions, new community economic development approaches, improved mapping technologies, and holistic ecosystem-based management models. Most significantly, these innovations were eventually embraced and advanced by all of the stakeholders and governments involved in the controversy—transforming an entrenched values conflict into a global model for sustainability.

Seen through an Integral Ecology framework, these approaches imply an evolution in environmental campaigns, which is confirmed by their effectiveness in securing protection of millions of acres of rainforest via consensus from multiple stakeholders. Environmental leaders in the Great Bear Rainforest campaign have engaged in the 4 terrains or "ecologies" of experience (UL), culture (LL), behavior (UR) and systems (LR) to develop sustainable conservation solutions (see figure III.2). Furthermore, environmental leaders (and some other parties involved in the initiative) have displayed Integral capacities in arriving at negotiated solutions, derived from mutual understanding, that are win-win. International markets campaigns against products originating from endangered forests such as the Great Bear Rainforest have capitalized on changing cultural values, as people and markets begin to include more worldcentric/postconventional concerns. This evolution in the cultural quadrant (LL) has been used to leverage change in the economic system (LR) surrounding the forest industry, where environmentalists formerly wielded little or no power.

Faced with little economic power to counter the economic rationale of ongoing industrial forestry, environmentalists developed international

	INTERIOR	EXTERIOR
INDIVIDUAL	**(UL) Experiences** Individual-Interior • Identification of personal and ecological values • Vision and commitment of campaigners • Shift in self-perception to solution-builder • Transformative states (e.g., meditation, prayer, therapy) • Emotional-awareness-building (moving from anger to love, working with burnout) • Cultivation of non-attachment • Integral strategic ideas and capacities • Spiritual activism I	**(UR) Behaviors** Individual-Exterior • Consumer boycotts • Protests and other direct action tactics • Ritual and prayer • Government lobbying, political support from key individuals • Education and outreach • Behavioral change in key individuals • Dialogue and negotiation • Sustainable forestry practices IT
COLLECTIVE	WE **(LL) Cultures** Collective-Interior • Creation of a "Brain Syndicate" • Use of media to communicate values • Activating environmental values of consumers • Shift from conflict to mutual understanding in negotiations • Cultural change in environmental organization • Cross-cultural understanding • Group awareness practices, visualization • Recognition and protection of cultural values • Justice for First Nations	ITS **(LR) Systems** Collective-Exterior • Expansion of focus to include other systems beyond ecological • International markets campaigns • Effective engagement with political cycles • Investment in community development • Scientific study and mapping of ecosystem • Holistic ecosystem-based planning • Implementation of sustainable forest management policies • Legal and institutional support for conservation and First Nations Rights

Figure III.2. Examples of rainforest activism in the 4 quadrants.

markets campaigns targeting the customers of BC wood products. Markets campaigns outcontextualized the regional economic rationale for logging in endangered forests, by creating a larger economic imperative for conservation, that is, the threat of contract cancellations. Environmentalists further engaged with economic systems by creating financial incentives for conservation and by attracting investment capital for sustainable community development (LR). Such out-of-the-box thinking has been possible because of internal transformations in environmental leaders that changed their experience (UL) and their behavior (UR), cultivating their capacities to value and integrate multiple perspectives (LL) and to come to negotiated solutions around legal protection and new forest practices (LR). The emergence of new technologies and capacities to map and model ecological (LR) and cultural (LL) values expanded the terms of debate beyond narrow industrial perspectives on forestry. These examples are elaborated on in the following sections.

Markets Campaigns and Economic Incentives for Conservation

The development of international markets campaigns gave environmental leaders powerful leverage to promote conservation of the Great Bear Rainforest. Environmentalists often privilege ecological systems analysis and do not adequately take into account the economic, political, psychological, and cultural systems they are trying to influence. This privileging of one dimension can lead to a lack of credibility and power. For example, environmentalists may discount or avoid discussing the real short-term economic benefits flowing to communities or corporations from environmentally destructive forestry activities, or they may privilege environmental values and insist that an ecological justification is all that is needed for conservation. People working from an Integral Ecology perspective foster solutions and strategies for sustainability from across disciplines, rather than privileging one discipline, set of values, or methodology. In the Great Bear Rainforest, environmentalists acknowledged diverse values, using various strategies to motivate different constituents, forging solutions that had economic, political, psychological, cultural, social, spiritual, and ecological dimensions.

From the mid-1990s until 2001 there was intense conflict over the fate of the Great Bear Rainforest between environmental groups and those in support of the predominant industrial forestry model, including forest unions, various levels of government, and forest company executives. Between 1997 and 1998, roads were built into thirteen wilderness valleys in the Great Bear Rainforest. By the year 2020, almost every wilderness valley was scheduled to be clear-cut (Sierra Club of British Columbia, 1999). This information fueled blockades by Greenpeace, Forest Action Network, and several sympathetic First Nations.

A coalition of environmental organizations waged a battle of public relations at home and abroad, spreading images of vast ugly clear-cuts and labeling Canada "the Brazil of the North." They targeted international customers of BC wood products in Europe, the U.S., and Japan. In response to these campaigns, the premier of British Columbia branded environmentalists "enemies of BC" in the media, fueling anti-environmentalist anger. Despite local backlash, the campaign met with great international success, securing commitments from dozens of companies to phase out forest products originating from endangered forests such as the Great Bear Rainforest. Companies included the world's two largest wood retailers, Home Depot and Lowe's; the world's largest home furnishings company, IKEA; and fortune 500 companies such as Nike, Dell, and IBM.[7] Companies canceled contracts worth over $300 million with BC forest companies operating in the Great Bear Rainforest. Financial pressure achieved what blockades, media, and public education could not—securing a temporary moratorium on logging in the key ecological areas in the Great Bear Rainforest, while science-based planning could take place.

Between 1999 and 2001, amid ongoing controversy, negotiations took place between forest companies and environmentalists over the protection of the rainforest and the future of markets campaigns. On April 4, 2001, an agreement involving forest companies, environmental groups, First Nations, and the Province protected 20 valleys (600,000 hectares), deferred logging in 88 more wilderness valleys (900,000 hectares), and advanced ecosystem-based management principles throughout the region pending further planning. Environmental groups agreed to cease targeting companies operating in the Great Bear Rainforest through international markets campaigns. Signatories committed to mitigation and transition funding for workers and communities, and established a multidisciplinary team of experts to conduct region-wide biophysical and socioeconomic research to inform final decisions at regional land use planning tables. Furthermore, in an agreement that also included the Haida First Nation (of the Queen Charlotte Islands), eight First Nation governments supported ecosystem-based management in their traditional territories and signed a protocol with the Province giving them the right to complete their own land use plans.

Environmentalists overcame their tendency to privilege ecological systems and avoid economic systems, successfully employing market campaign strategies that resulted in unprecedented levels of protection and ongoing, innovative planning in the region. They engaged in multiple systems within the Lower-Right quadrant and wielded new forms of power by taking on the perspective of their adversaries in developing campaign strategies. Because of the export-driven nature of the BC economy, international markets campaigns created new forms of leverage on regional forest policies. By asking high-profile businesses and their environmentally conscious customers to stop buying wood from endangered forests worldwide, environmentalists operated at a higher level of the economic system that transcended yet affected the local economic rationale, forcing changes in policy. By leveraging global systems to protect specific regions, environmentalists innovated a new, more Integral approach.

Corporate commitments named not only the Great Bear Rainforest but all endangered forests, which fostered global demand for wood products that were ecologically certified by organizations such as the Forest Stewardship Council, and created a global, nonlocalized arena for leveraging protection of forests worldwide.[8] One quality of Integral consciousness is that it is not limited to localized effects, and it can implement strategies that are both local and global/nonlocal. The success of markets campaigns was made possible by the cultural readiness (LL) of large portions of the population in Europe and North America to demand ecologically friendly products and reject those coming from endangered forests. In order to create and successfully implement a marketplace strategy, environmentalists understood what would affect their adversaries, activating latent environmental values held by the public through mutual understanding (LL) and translating these values into consumer leverage upon the economic systems driving forest destruction (LR). As more local decisions are driven by the demands of the global economy, and corporations are decreasingly beholden to national laws, this evolution in social change efforts represents a critical strategic leap, by acting on multiple systems at local and global levels simultaneously.

If international markets campaigns were the economic stick, then conservation financing and investment is the carrot. First Nation communities in the Central and North Coast regions of BC suffer from unemployment rates between 70% and 90%. As part of solution-building efforts since 2001, environmentalists have used the international profile of this region to attract financial support for conservation and appropriate community economic development. Two initiatives have been spearheaded by activists to mitigate the loss of forestry income to the region—conservation financing for First Nations and socially responsible investment initiatives for resource-dependent communities. These initiatives represent risk-taking and new strategies for environmentalists in BC, further indicating their commitment to finding not just ecological but economic, societal, and culturally appropriate solutions to enable conservation. Barriers to sustainable economic development in coastal communities are significant, but prior to environmental campaigns, there were no substantial financial solutions on the horizon for coastal communities, and the development of conservation financing and incentives is a positive step in a long process. These efforts stemmed from relationships built between environmentalists and First Nation leaders, as well as the Province of BC, and are the result of environmentalists recognizing the socioeconomic needs (LR) and cultural values (LL) of other stakeholders who are affected by conservation campaigns. This approach also shows respect for the rich cultural traditions (LL) of First Nations, and respect of their legal Rights and Title (LR). Fund-raising for the Coast Opportunities Fund was successful, and in 2007 official announcements confirmed that (initiated by environmental leaders) $60 million had been donated from private U.S. and Canadian foundations and individual donors, and the government of British Columbia and the government of Canada contributed $30 million each to finance conservation management projects and business ventures in First Nation territories in accordance with ecological values. This $120 million fund is the first of its kind in Canada and is being heralded as a global model for linking cultural, economic, political, community, and conservation interests (Turning Point Initiative, 2007).

Moving Beyond the Outsider: Former Enemies in Dialogue

> Saving the biosphere depends first and foremost on human beings reaching mutual understanding and unforced agreement as to common ends.
>
> —KEN WILBER

Concern for justice, ecological health, and cultural diversity increases as individual and cultural worldviews extend from egocentric/pre-conventional to ethnocentric/conventional to worldcentric/postconventional to Integral/second-tier perspectives. When Integral capacities emerge, complex issues and diverse perspectives can be more readily integrated into holistic, long-term solutions. Leaders acting from Integral capacities serve as cultural empathizers and transformers who operate dynamically across multiple worldviews, motivating people with diverse interests toward common ecological, economic, cultural, political, and social goals. Leaders with Integral perspectives can foster healthy ecological worldviews, enabling mutual understanding and fueling individual and cultural transformations of increasing scope and depth.

In the past, fear and polarization predominated on both sides of environmental conflicts. The shadow side of North American environmentalism was well displayed in the 1990s by criticisms that were leveled at efforts of environmentalists. Small-town British Columbia mayors made reference to "cappuccino-sucking urban environmentalists" who lacked understanding of the rural way of life and did not care about community prosperity. Some First Nations resented "eco-colonial" attempts to lock up their traditional territory in parks, placing conservation ahead of hunting and other traditional uses, and ahead of the just resolution of Rights and Title. Forest industry executives accused environmentalists of being unrealistic and economically naive. Governments felt environmentalists were manipulating their political processes, and were neither responsible to a constituency nor invested in a balanced solution. Although some of these accusations were public relations ploys intended to undermine environmental protection efforts, many of these perceptions were likely justified, illuminating some of the potential blind spots and areas of environmentalism needing a more Integral approach.

As their strategies for gaining negotiating power in the Great Bear Rainforest campaign met with success, environmental leaders were challenged to shed their polarized identities and behaviors (UL and UR), and shift from being outside agitators to being solution-builders. In so doing, they became less identified solely with their ecological perspective and began to understand and respond to the differing needs and values of the First Nation leaders, community representatives, governments, and forest companies with whom they were negotiating. Company executives and environmental leaders created the Joint Solutions Project, a working group for negotiations to occur that included the six major forest companies operating on the coast, and representatives from ForestEthics, Greenpeace, Sierra Club of Canada–BC Chapter, and Rainforest Action Network.[9] Given the vitriolic history between conservation groups and forest companies, the creation of the Joint Solutions Project signaled that a remarkable shift toward dialogue and solution-building had occurred for both parties.

Negotiations provided a venue where environmentalists embraced the 4 terrains, by developing their analysis of the internal and external challenges faced by industry, First Nations, and local communities, and by addressing the full spectrum of issues critical for sustainability. Negotiations also enabled opposing sides to engage one another with humanity and mutual respect, fostering Integral capacities of mutual understanding. For environmentalists, reactive campaigning was no longer required to protect the remaining valleys in the short term, and effort shifted to building long-term solutions to address sustainable community development, First Nations Rights and Title, policy needs, corporate bottom lines, and ecological integrity. Environmentalists involved in negotiations also had to address accusations from their own constituents that they were selling out by entering into dialogue and signing agreements. One of the ways they refuted this claim was by cultivating inner integrity and new moral frameworks to guide their efforts.

Personal Dimensions of Transformation

According to the theory of Integral Ecology, ecological (world- and planet-centric) values are a product of psychological and spiritual development. Despite the common notion in environmental philosophy that the roots of the ecological crisis lie within consciousness and cultures, environmentalists most commonly focus on the external problems of economic and social systems, or on unsustainable behaviors (LR and UR), often ignoring their subjective and cultural dimensions (UL and LL), and rarely taking responsibility for personal transformation.

Integral Ecology's focus on internal development (UL) emphasizes the power of transformative practices of body, mind, and spirit in fostering the development of self, society, and culture (Wilber, 2000, 138). A commitment to spiritual growth has been central to many leaders of social change, yet it is not highly ingrained in many cultures of activism. Environmentalists often experience disempowerment due to overwork, despair at the state of the world, and the feeling of being "outsiders" to anthropocentric and modern values. Transformative practices can be a powerful antidote to the subtle superiority, alienation, and despair that can accompany activism. Integral Ecology challenges the claim that self-transformation can be put off until external injustices are fixed. Both this claim and its inverse—that one cannot act effectively or compassionately until one is "enlightened"—are fallacies. The Integral approach recognizes that internal transformations, behavioral change, cultural evolution, and social development occur interdependently and simultaneously. As activists cultivate Integral levels of development within themselves, they will increasingly facilitate Integral transformation in others.

Those of us involved in the Great Bear Rainforest campaign added meaning and depth to our work by framing the challenges we faced as part of our paths of personal and spiritual development. Interestingly, the majority of environmental leaders in the Great Bear Rainforest campaign were women. Facing anger, threats of violence, and backlash in the media and rural communities, and pushing ahead on complex new terrain, many campaigners leading conservation efforts faced burnout. During the years of intense conflict, we invested time to grow our capacities for personal leadership and authenticity, developing and sharing in spiritual and therapeutic practices.[10] This commitment to personal development was grounded in the belief that our own state of being and consciousness was affecting the perspectives of our adversaries and the outcomes of our campaign efforts. We explored spiritual traditions, developed personal transformative practices, found new ways to dialogue with our adversaries, and worked with practices to shift our motivations or responses from anger and fear toward love and nonattachment. Several of the women campaigners who were in protracted negotiations began to practice loving-kindness meditations before entering into negotiations and to visualize agreement in areas of difficulty. Many of us involved believe our commitment to personal development and the integration of spiritual practices into the campaign during its most heated and pivotal times have greatly contributed to building relationships and achieving successes.[11]

New Tools for Conservation: The Emergence of Integral Sciences

Like the "out-contextualizing" effect of the markets campaign, new scientific approaches and technologies were used to expand and change the scientific and technical terms of debate. Satellite imagery and geographic information systems (GIS) were used by Sierra Club of BC in 1997 to map the remaining coastal temperate rainforest and the extent of industrial clear-cutting on BC's coast. Prior to the application of this technology, neither the degree of destruction nor the globally significant conservation opportunity on the coast were understood as clearly.

As part of the campaign to protect the Great Bear Rainforest, several environmental groups commissioned a Conservation Areas Design for the central coast region of BC.[12] Although the phenomenon of dueling scientists is not new in environmental conflicts, their intent was to contextualize the debate in terms of the extinction and biodiversity crisis facing the planet. The big-picture approach of conservation biology provided a larger scientific perspective than the geographically narrow and technically focused forestry science that government and industry relied on to support ongoing industrial clear-cutting. Conservation biology incorporated historical data and cultural values into planning, which are important elements of an Integral science.

Due to the political and economic leverage environmental groups had generated from markets campaigns, this approach was integrated into official government planning processes. As an outcome of the 2001 agreements, an independent team of 17 scientists called the Coast Information Team (CIT) was tasked with identifying priority areas for conservation. Established by the BC government, First Nations, forest companies, and environmental groups, the CIT was to provide independent information on the region using the best available scientific, technical, traditional, and local knowledge to assist in ecosystem-based management planning.

The injection of new mapping technologies and more Integral sciences such as conservation biology and ecosystem-based management into the debate over forest management in BC undermined the narrow and technical science of sustained yield management, replacing it with a historical, multiscaled, and more integrative conception of ecosystems. These tools will increasingly enable people to make Integral decisions (UL) in planning by synthesizing complex cultural (LL), ecological, social, and economic data (LR) into decision-making models and scenarios that support sustainable practices (UR).

The CIT conducted an Ecosystem Spatial Analysis (ESA) to identify priority areas for biodiversity conservation in order to protect representative ecosystems and native species; sustain ecological and evolutionary processes; and build a conservation network resilient enough to withstand environmental change (Coast Information Team, 2003). The CIT stated in their final report that 44%–70% of the Great Bear Rainforest should be protected in order to meet conservation goals. In May 2004 a consensus recommendation was made to government that over 3.5 million acres, or 33%, of the central coast portion of the Great Bear Rainforest be placed under some form of protection, and that new forms of ecosystem-based forestry occur elsewhere. Although this fell short of the scientific recommendations, it was the product of grueling multi-stakeholder negotiation for which consensus was reached, and it is a compromise solution that integrates multiple values. On February 7, 2006, the comprehensive protection package was announced for the Great Bear Rainforest (including the central coast, north coast, and Haida Gwaii), and included four key elements: rainforest protection, improved logging practices, First Nations' involvement in decision-making, and conservation financing to enable economic diversification. This final protection resulted in 33% of the Great Bear Rainforest being off limits to logging.

Elements of the Great Bear Rainforest Agreement:

- Permanent protection—5 million acres
- New parks—3.3 million acres
- Previous parks—1 million acres
- New no-logging zones—736,000 acres
- Ecosystem-based management—21 million acres
- $120 million for conservation economy
- First Nations approve all plans
- International marketplace shift away from endangered forest products
- Model used in conservation efforts in Chile, Boreal Forest, and U.S.A.

The package includes commitments to region-wide implementation of an ecosystem-based management approach by 2009. Each First Nation approved the package, which also granted them greater decision-making power over resource development in their traditional territories. The $120 million fund will support a Community Economic Development fund, capacity-building, the establishment of a permanent conservation trust, research, and First Nation economic development projects—all guided by conservation principles, and a multi-stakeholder board.

The Great Bear Rainforest decision has been repeatedly heralded in media worldwide as a new blueprint for the resolution of complex environmental conflicts. However, with a visionary regulatory framework in place, and funding available, community and human development are even more central to successful implementation. The implementation of agreements at all levels is now the focus of work for governments and stakeholders, and each new process is still a platform for renegotiation, resurging value conflicts, and power plays. Despite this, such a comprehensive solution has better chances than many to move from vision to reality.

Conclusion

In responding to the multiple dimensions of the conservation challenge in the Great Bear Rainforest, environmental leaders have brought forth solutions that model an Integral Ecology approach. Given the region's size, its enormous cultural and ecological values, and the scope of efforts thus far, the Great Bear Rainforest is one of the most significant conservation opportunities in North American history. The decade-long work to protect it offers a comprehensive model for the resolution of environmental conflicts and the establishment of large-scale ecosystem-based management, and brings much-needed investment into local First Nation communities. Environmental leaders generated power to leverage conservation by developing campaign strategies with innovative and large-scale systems thinking that engaged the global marketplace, set the stage for conservation-based community economic development, and applied the highest scientific knowledge to conservation planning. New approaches to negotiation embodied Integral modes of consciousness and behavior. Environmental leaders cultivated mutual understanding and successfully motivated people from a variety of value perspectives to support sustainable solutions. This required that campaign goals be extended from a focus solely on rainforest conservation to include goals of conservation-based community economic development, relationship-building, social justice for First Nations, a transformation in the global trade of endangered forest products, and the personal transformation of those involved. Finally, new institutions are being built, capable of ongoing adaptive management in complex, multifaceted environments.

Given the complexity of the context in the Great Bear Rainforest, Integral approaches must be sustained and amplified in order to advance long-term solutions. Further attention is needed to ensure that new policies and institutions, funding and community development initiatives, protection, and ecosystem-based management will be comprehensive enough to ensure sustainability. This requires robust solutions that authentically address the needs of many people over time. The ecological integrity of the Great

Bear Rainforest also faces continued threats from offshore oil and gas exploration, pipeline development, large carnivore hunting, overfishing, mining, and fish-farm expansion. These threats also require an Integral approach in order to be defused. Integral community economic development will be crucial to all future conservation success, and to date these efforts do not fully address the interior aspects of development. Sustainability in the region will need ongoing attention to multidimensional solutions and increased capacities among leaders to bring forth Integral possibilities.

Integral environmental campaigns of the future must address the myriad of social, political, economic, cultural, psychological, spiritual, and ecological issues. As the ecological situation in virtually every region on Earth becomes more complex and troubled, the emergence of Integral awareness can transfigure global ecological problems from disasters into catalytic awakeners. May this awareness help all beings to flourish.

NOTES

1. See Esbjörn-Hargens, 2005 and Wilber, 1995, 1996, 2000, 2001.
2. Wilber describes this as the "tetra-evolution" of the 4 quadrants (2000, 183).
3. For a systematization of over 100 models of human development that includes psychological, spiritual, cultural, and social evolution, see *Integral Psychology* (Wilber, 2000, 197–217).
4. The color system refers to Wilber's (2006) use of the color spectrum to indicate developmental altitude (i.e., a neutral marker of increased complexity).
5. For more on the natural history of the region, see Kellogg, 1992; McAllister et al., 1998; and MacKinnon and Vold, 1998.
6. For more history on markets campaigns and forestry conflict in BC, see Austerm et al., 2002; Cashore et al., 2001; Stanbury, 2000; and Wilson, 1998.
7. For a complete listing of company commitments, see http://forestethics.org/article.php?id=597&cat=13 (May 2008).
8. In 2003, after being targeted with markets campaigns, the two largest forestry companies in Chile agreed to stop the conversion of native forests into plantations and enter into a Joint Solutions Project with environmental groups in an area covering one million acres of native forests. This success was modeled on the Great Bear Rainforest campaign. Similar efforts are now under way in the Canadian boreal forests, the largest intact forest remaining on Earth. See www.forestethics.org for more details (May 2004).
9. For more information, see Rainforest Solutions Project, www.savethegreatbear.org, and the Coast Forest Conservation Initiative, www.coastforestconservationinitiative.com.
10. Much of this occurred through the Hollyhock Leadership Institute in British Columbia.

11. As documented in the film *From New Age to New Edge* (2003), Across Borders Media.
12. For the full Conservation Areas Design report, see Jeo et al., 1999.

REFERENCES

Austerm, D., B. Painter, and Y. Posey. 2002. *Saving British Columbia's Coastal Rainforests: The Emergence of Eco-Alliances between Adversarial Interests.* Discussion paper presented to the Annual Conference of the International Association for Business and Society, Victoria, BC, June.

Beck, D., and C. Cowan. 1995. *Spiral Dynamics: Mastering Values, Leadership and Change.* Cambridge, MA: Blackwell.

Bryant, D. 1997. *The Last Frontier Forests: Ecosystems and Economies on the Edge.* Washington, DC: World Resources Institute.

Cashore, B., G. Hoberg, M. Howlett, J. Rayner, and J. Wilson. 2001. *In Search of Sustainability: British Columbia's Forest Policy in the 1990s.* Vancouver: UBC Press.

Coast Forest Conservation Initiative. 2004. www.coastforestconservationinitiative.com.

Coast Information Team. 2003. *An Ecosystem Spatial Analysis for Haida Gwaii, Central Coast and North Coast, British Columbia.* www.citbe.org.

Esbjörn-Hargens, S. 2005. "The *What, Who,* and *How* of Environmental Phenomena." *World Futures: The Journal of General Evolution* 61 (1–2): 5–49.

ForestEthics. 2004. www.ForestEthics.org.

Jeo, R., M. A. Sanjayan, and D. Sizemore. 1999. A *Conservation Area Design for the Central Coast Region of British Columbia, Canada.* Salt Lake City: Round River Conservation Studies.

Kellogg, E., ed. 1992. "Coastal Temperate Rainforests: Ecological Characteristics, Status and Distribution Worldwide." Occasional Paper Series (1). Ecotrust and Conservation International. www.ecotrust.org/publications/ctrf.html.

MacKinnon, A., and T. Vold. 1998. Old-Growth Forests Inventory for British Columbia, Canada. *Natural Areas Journal* 18 (4): 309–18.

McAllister, I., K. McAllister, and C. Young. 1998. *The Great Bear Rainforest: Canada's Forgotten Coast.* San Francisco: Sierra Club Books.

Rainforest Solutions Project. 2004. www.savethegreatbear.org.

Sierra Club of British Columbia. 1999. *Canada's Ancient Rainforest: Home of the Great Bears and Wild Salmon.* Victoria, BC: Sierra Club of British Columbia.

Stanbury, W. T. 2000. *Environmental Groups and the International Conflict over the Forests of British Colombia, 1990 to 2000.* Vancouver: SFU-UBC Centre for the Study of Government and Business.

Torbert, W., and P. Reason. 2001. "The Action Turn." http://people.bath.ac.uk/mnspwr/Papers/TransformationalSocialScience.htm.

Turning Point Initiative. 2007. *Conservation Investment Package.* www.coastalfirstnations.ca/files/PDF/Newsletters/Conservation_Investment_Package.pdf.

UNEP. 2001. *An Assessment of the Status of the World's Remaining Closed Forests.* http://na.unep.net/publications/closedforest.pdf.

Wilber, K. 1995. *Sex, Ecology, Spirituality: The Spirit of Evolution.* Boston: Shambhala Publications.

———. 1996. *A Brief History of Everything.* Boston: Shambhala Publications.

———. 2000. *Integral Psychology.* Boston: Shambhala Publications.

———. 2001. *A Theory of Everything.* Boston: Shambhala Publications.

———. 2006. *Integral Spirituality.* Boston: Shambhala Publications.

Wilson, J. 1998. *Talk and Log: Wilderness Politics in British Columbia*, 1965–96. Vancouver: University of British Columbia Press.

Conclusion: The Integral Ecology Advantage

> It may seem strange to conceive of those who espouse various ecological perspectives and who seek to overcome ecological crisis as "opponents" of one another. . . . For a number of years a "war of ecologies" raged. It would of course be naive to deny that advocates of various ecophilosophies do in fact hold opposing views on important issues, and the issues at stake should not be glossed over or obscured in any way. Yet these differences can coexist with a respect for one's opponents, an openness to the views of others, and a commitment to cooperation in the pursuit of mutually held goals.
>
> —JOHN CLARK[1]

We have described the growing field of Integral Ecology. In the first part of this book we provided an overview of the multiperspectival Integral Model that serves as the foundation for Integral Ecology. In the second part we described the three main conceptual frameworks of Integral Ecology. The 4 terrains of ecology and their respective 12 niches of environmental concern represent the ontological dimension of the environment and its members. The 8 ecological selves represent the epistemological dimension. The 8 ecological modes represent the methodological dimension. By integrating these three components, we achieved a comprehensive post-metaphysical approach that is uniquely situated to honor and include the insights of over 200 distinct perspectives on ecology and the eco-social environment. In the third part we focused on the Integral Ecology framework in action: how it complexifies our understanding of nature mysticism and the many perceptions of the current ecological crisis; how it has been used to create over a dozen personal practices for cultivating Integral ecological awareness within oneself; and how it is already being used around the planet by dozens of Integral practitioners. In the fourth part we presented the work of three Integral Ecologists to provide examples of Integral Ecology applied. As every day passes, more people are discovering the merits of this inclusive and multiperspectival approach.

The Integral Ecology Advantage

Integral Ecology provides a number of distinct advantages over other approaches, and we want to summarize some of those advantages here.

Integral Ecology Recognizes Interiority

Many environmentalists today rely exclusively on objective, scientific (3rd-person) accounts of nature. Such reliance is not wholly inappropriate, but is extremely limited, a limitation that we believe is crippling solutions to some of our toughest problems. Objective approaches, when relied on to the exclusion of other perspectives, provide no account for the interiority of humans, animals, and organisms. In contrast, Integral Ecology recognizes the reality of pan-interiority. The capacity for experience, however meager, goes all the way down.

We believe that one key to our future as a species lies in reaffirming and exploring interiority. Yet we do not believe that interiors, in and of themselves, are enough. It is not one or the other; exchanging a limited view of reality for yet another limited view of reality. We need interior-individual, interior-collective, exterior-individual, and exterior-collective perspectives and methodologies. We need them all. The Kosmos is multifaceted. We cannot expect to understand the Kosmos through partial perspectives and singular methodologies; we must have a multifaceted, kaleidoscopic view. Thus, Integral Ecology is defined as the study of the subjective and objective aspects of organisms in relationship to their intersubjective and interobjective environments at all levels of depth and complexity.

Integral Ecology Works from All-Quadrants, All-Levels

To gain that kaleidoscopic view we must cultivate an "all-quadrant, all-level," approach. Most approaches to environmental issues fail to recognize these four essential and irreducible perspectives, or the 4 terrains: the interior and exterior correlates of the individual and collective domains. Thus, they are not "all-quadrant." Integral Ecology also recognizes that each of the 4 terrains evolves in a distinct and irreducible way.

By emphasizing that all individual holons have interior and exterior, individual and collective, perspectives, and by demonstrating that any phenomenon can be analyzed or experienced in terms of these perspectives, Integral Ecology forecloses reductive, one-dimensional approaches to the Kosmos.

Integral Ecology Avoids Reductionism

Because Integral Ecology recognizes interiors and exteriors, both individual and collective, it avoids both subtle and gross reductionism.

Few environmental approaches fall prey to gross reductionism (reducing all phenomena to an objective, atomistic, 3rd-person-singular perspective). However, many approaches do collapse the natural world and its members into an interobjective (3rd-person-plural) perspective. This is subtle reductionism, as it absolutizes the methods of the Lower-Right quadrant (e.g., systems theory, scientific ecology, and economics). In more simple language, such reductionism takes human experience, animal experience, and organismic experience and reduces it to materiality. We do not believe that this vast, beautiful world is simply its constituent material parts. Were that so, neither you nor we could appreciate the beauty of an open field.

These approaches are all exterior models of relation (functional fit): for example, rain falls, grass grows, deer eats grass, wolf eats deer; or toxins in the environment affect the fertility rate of animals; or storms in South America affect weather patterns in Africa. These approaches lack comprehensive interior models of relation (e.g., development, mutual understanding, and the differences between differentiation and disassociation), and they point to exterior interconnections in hopes that people will then somehow feel interiorly interconnected. Unfortunately, these approaches fail to differentiate between exterior interrelatedness and interior connectedness, and fail to spell out the relation between the two. We acknowledge the power of Lower-Right approaches but avoid absolutizing them by emphasizing that all organisms have 1st-, 2nd-, and 3rd-person perspective-dimensions.

Integral Ecology Distinguishes between Exterior Interconnection and Interior Intersubjectivity

Environmentalists should not confuse Lower-Right interconnectedness with Lower-Left intersubjectivity (or with an Upper-Right phenomenological experience of oneness). A global community does not necessarily equal a worldcentric community. In addition to discussing how the world hangs together in its many complex and interrelated systems, we need also to understand how different worldviews, ethics, values, and intersubjectivity (each at various developmental levels) affect our evaluation and treatment of the natural world.

Integral Ecology Insists That Interior Development Is Crucial for a Deeper Understanding of Nature

Many current approaches to the relationship between humanity and nature have a shallow appreciation for the complexity of interior human development. Since many ignore or do not understand interiors, they do not understand how they develop. As a result, they do not discuss in any detail the process of interior transformation, nor do they provide practices or injunctions for interior transformation. Adherents to such approaches suggest that exterior transformation is enough. Integral Ecology asserts that our interiors must also transform.

For humans to arrive at a genuine environmental ethic, we must first stabilize a worldcentric structure of mutual understanding. In some important ways, an individual has to become worldcentric (one with all of humanity) before they become planetcentric (one with all of humanity and the natural world); for what good is the capacity to be one with nature if we are fighting with those of a different class, gender, race, or perspective? Additionally, the development of worldcentric mutual understanding allows us to identify with the natural world without falling prey to Eco-Romantic and even misanthropic forms of antimodernism.

As a result we place a premium on solutions grounded in mutual understanding between divergent viewpoints and understandings. By cultivating the capacity to inhabit other perspectives, we will be able to respond to the complex problems that face our bioregions more adequately than with our current, less comprehensive approaches.

Integral Ecology Distinguishes between Members and Parts

There is an important difference between being a part of something (and thus subject to its will and domination) and being a member of it (retaining autonomy while connecting and contributing). An organ is a part of an organism, but an organism is a constitutive member, a partner—not a part—of its ecosystem.

Integral Ecology Cautions Environmentalists Against Ecofacism

Because Integral Ecology recognizes humans as members, not parts, of an ecosystem, we avoid many forms of thinly disguised ecofascism. The ever-popular phrase "We humans are just one strand in the 'web of life'" is an attention-getter but does not hold up to careful scrutiny. Analysis of the logic of wholes and parts demonstrates that humans' noospheric dimension is not a "part" of the web of life. If humans were just parts, then individuals would be willing to sacrifice their ambitions, property, and lives for the good of the web of life. All too often environmentalists assign intrinsic value solely to the web of life, thereby concluding that the parts of the web (individual life forms) either lack value of their own or at best have equal value. Such an approach provides no criteria for making difficult moral decisions. Moreover, this approach indicates that if individual humans or classes of humans are harming the web of life, then other humans (who shall they be?) should prevent such behavior at whatever cost. This is ecofascism.

Integral Ecology Emphasizes That Spirit Is Both Immanent and Transcendent

Integral Ecology sees Spirit as both immanent and transcendent, a union of Ascent and Descent. This is the insight of the nondual traditions. Spirit is both of the world and beyond it. Many eco-approaches equate nature with Spirit and fail to see that nature is another stunning expression of Spirit (NATURE). We do not encourage such biocentrism, or an exclusive identification with nature. We do encourage nature mysticism (e.g., gross-realm union where the differentiated self extends to include the entire gross realm). In view of such considerations, Integral Ecology distinguishes between nature, Nature, and NATURE. Equivocating on the word "nature" not only leads to many confusions, but also allows people to make illegitimate claims.

Integral Ecology Distinguishes between Differentiation and Dissociation

Those who fail to distinguish between differentiation and dissociation tend to romanticize the past and condemn the present. They depict modernity as the Fall from grace, by which they mean allegedly nonduality, but premodernity was actually an undifferentiated state (e.g., children and some ancient peoples). Such an approach commits the pre/trans fallacy (aka, pre/post), which in this case means that a positive future can take the form only of a reprise of premodern sociocultural formation. Integral Ecology compares the dignity and disaster of premodernity with the dignity and disaster of modernity and postmodernity.

Integral Ecology Insists That There Is No Fall from Ecological Grace

Because we avoid the pre/trans fallacy, and because we distinguish between differentiation and dissociation, there is no single moment in which humankind has fallen from some ecological grace. Indeed, we have simultaneously fallen at every turn and grown closer to nature at the same time. Ironically, the very idea of a Fall from grace echoes a crucial theme of the Christian tradition, of which environmentalists are very often highly critical.

Integral Ecology Avoids Transpersonal Reductionism

Many theorists collapse many or all transpersonal states (psychic, subtle, causal, and nondual) into one mystical state. We argue, in contrast, that there are actually many "one with nature" states and that some are more one with nature than others. For example, some commentators see gross-realm union (e.g., many forms of shamanism) as the end point of spiritual development, when in fact there are greater, more profound and integral waves of existence. Often, people confuse gross-realm union (nature mysticism) with ultimate union (Suchness). Hence, environmentalists must do more than avoid the pre/trans fallacy; they must also avoid the "trans/trans" fallacy, in which they mistake lower transpersonal unitive stages or gross-realm states for higher transpersonal stages or nondual states.

Integral Ecology Clarifies the Meaning of "Transcend and Include"

Current approaches to the humanity-nature relation often have a problematic understanding of the concept of "transcend and include." For instance, they confuse exterior inclusiveness (the noosphere is in the biosphere) with interior inclusiveness (the biosphere is in the noosphere). As we have argued, however, the noosphere is not "in" the biosphere, but instead in important ways transcends and includes it.

Integral Ecology Maintains That the Kosmos Is Holarchical

Current environmental theories are often anti-hierarchical because they fear that acknowledgment of humankind's noospheric capacity will justify aggressive treatment of nonhuman nature. Anti-hierarchalism not only denies the very evolutionary process that allows for the pursuit of an environmental ethic, but also promotes a multicultural relativism that undermines appropriate action. All too often people fail to distinguish between a holarchy (a healthy hierarchy) and a dominator hierarchy. Integral Ecology is committed to the continual analysis and evaluation of the dynamics of holarchical integration in order to provide newer and more complete understandings of this complex process.

Integral Ecology Adheres to a Multidimensional Value System

If we fail to distinguish among ground, extrinsic, and intrinsic value, we tend to emphasize only one of these value hierarchies. In such a case, we may end by promoting radical ecocentric egalitarianism, according to which everything has the same ground value. Such a view ignores the formidable differences in complexity among the kinds of beings that have evolved on Earth, and elsewhere. Only a multitiered valuational scheme can provide us with the distinctions needed for dealing with difficult cases.

Integral Ecology Avoids "One Size Fits All" Solutions

Integral Ecology avoids taking generalized normative stances on ecological issues. We find that the natural world is too complex and multidimensional for simple answers. Without understanding the differing worldviews to which members of communities adhere, environmentalists often end up proposing "one size fits all" solutions to environmental issues. By taking into account how an issue appears in terms of the different developmental centers of gravity, environmentalists can devise motivations and justifications for solutions that are appropriate for members of each worldview. This process should be grounded in dialogue, consensus, and reflectivity between stakeholders or community members. Integral Ecologists are interested in hearing all pertinent perspectives, even if adherents to those perspectives are not wholly satisfied with the recommendations for action.

Integral Ecology Maintains That Things Are Getting Better and Getting Worse

As opposed to only thinking that things are getting better (environmental optimism) or only thinking that things are getting worse (environmental apocalypticism), Integral Ecology believes that things are simultaneously getting better and worse. We can find evidence in favor of the claim that we are making progress. At the same time, we do not have to go far to see that things are getting worse. Integral Ecologists aspire to live with this tension without being debilitated by it. They also recognize that from an absolute perspective the planet is the Great Perfection, neither getting better nor getting worse. Always already perfect.

Integral Ecology Provides a Yoga

Unlike many approaches to ecology and the environment that change the map but don't change the mapmaker, Integral Ecology not only provides a new and improved map of the natural world, it also provides a yoga, a series of practices for personal transformation that actually transforms awareness so that one can read this new map more effectively. At this point, we have provided 18 practices and meditations that can support vertical and horizontal psychological development. These practices have arisen out of our own work with the Integral Ecology framework and serve to link this framework to your direct experience. Also, these practices serve as a foundation for the development of an Integral approach to nature mysticism where our relationship with the natural world is itself a transformative path.

Integral Ecology Is Theoretically Robust and Pragmatically Grounded

Not only does Integral Ecology offer a multifaceted framework (Who, How, and What) that serves to integrate over 200 distinct schools of thought on the natural world; it also provides a scalable set of principles that allow it to be applied in a multitude of contexts. Consequently, Integral Ecology is a flexible and dynamic approach, providing concrete ways to work with assorted variables associated with any project. In fact, the framework outlined in this book, as our many real-world examples and case studies illustrate, has already provided much success, support, and guidance to ecologists and environmentalists on the ground facing complex problems and looking for Integral solutions.

Integral Ecology Is Post-Metaphysical

Perspectives on the natural world are always perspectives from somewhere by someone. There is no view from nowhere—there is always a Kosmic address. There are no pregiven environmental phenomena—only tetra-enacted occasions. As a result, nature is not something "out there" but is rather disclosed through at least eight major methodological approaches. By using Integral Methodological Pluralism, Integral Ecology avoids both the myth of the given, the naive realism that the natural world exists independently of an observer; and the myth of the framework, the naive belief that the natural world is a social creation generated solely by an observer.

The Integral Ecology Platform

To summarize the advantages and commitments of Integral Ecology, we developed a 12-point platform, inspired in part by the noble intentions of Arne Naess and George Sessions in their eight-plank Deep Ecology Platform. Our platform seeks to create a common ground for a rich variety of integral ecologies that can serve our journey forward.

Integral Ecologists recognize that human attitudes, behaviors, institutions, and practices generate complex environmental problems across the globe at multiple scales. In light of this situation, we embrace the following platform as a foundation for the most comprehensive contemplation of and response to our ecological situation.

1. Integral Ecologists use the conceptual tools and distinctions of Integral Theory's "all-quadrant, all-level" approach to analyze, characterize, and develop comprehensive solutions to environmental problems.

2. Integral Ecologists recognize that there are many ways to honor and include quadrants, levels, lines, states, and types.

3. Integral Ecologists examine the enacted nature of phenomena: Who is doing the looking? How is the looking being done? What part of reality is being looked at?

4. Integral Ecologists recognize the world-disclosing capacities of all perspectives and the methodologies they use to investigate various domains of reality.

5. Integral Ecologists are familiar with at least some version of the 8 ecological selves, the 8 methods of knowing, and the 12 niches of reality.

6. Integral Ecologists situate any domains, methods, or perspectives in which they specialize within the variety of other pertinent domains, methods, and perspectives.

7. Integral Ecologists are committed to coordinating and building bridges between various domains, methods, and perspectives, especially in the context of specific environmental problems.

8. Integral Ecologists are committed to increasing their capacity to take and hold additional perspectives in order to dismantle self-other dynamics that arise in the course of addressing most environmental issues.

9. Integral Ecologists are engaged in long-term personal transformational practices, which develop their somatic, emotional, psychological, interpersonal, moral, and spiritual dimensions.

10. Integral Ecologists recognize that all life forms experience and perceive, and create shared horizons of meaning, both within and across species, but that not all life forms have an equal capacity to do so.

11. Integral Ecologists affirm a multidimensional value ethic that suggests that an individual (human or nonhuman) or a process can simultaneously be of equal value, greater value, and lesser value than another individual or process, depending on the context and criteria used.

12. Integral Ecologists affirm the ultimate mystery of all phenomena as a way of preventing attachment to conceptualizations of reality.

We intend to revise and to improve this 12-point platform in view of criticism and feedback offered by practitioners and other interested parties.

The Road Ahead

As we look forward, we are not completely clear how Integral Ecology will develop, though we do have a comprehensive framework to guide us. The mere thought of the process excites us.

We have demonstrated that there are numerous approaches to ecology and the environment—philosophical, spiritual, religious, social, political, cultural, behavioral, scientific, and psychological. Each highlights an essential component while ignoring other dimensions. To overcome this fragmentation, Integral Ecology provides a way to weave all approaches into an environmental mandala, an ecology of ecologies that not only honors the physical ecology of systems and behaviors, but includes the cultural and intentional aspects as well—at all levels of organization. Integral Ecology takes into account the multiple worldviews within individuals, communities, and cultures, and their accompanying environmental perspectives.

Integral Ecology takes an enactive approach to the environment by recognizing that ecological phenomena are the result of an interaction between the knower, how it is known, and what is known. It allows for a comprehensive understanding of how the many ecological approaches available to us can be brought together to inform and complement each other in a complex and coherent way. This Integral framework honors the multiplicity of ecological approaches. It allows individuals to identify how various methods focus on specific ecological phenomena, and from which perspective the phenomena are being explored. By acknowledging and honoring the multivalent nature of ourselves, our communities, and our environment, we can, as global citizens, embedded in local eco-social systems, work effectively toward sustainable solutions. A premium is placed on solutions grounded in mutual understanding between divergent viewpoints and understandings. By cultivating the capacity to inhabit other perspectives and hold multiplicity, we will be able to respond more adequately than contemporary, less comprehensive approaches to the complex problems that currently face our planet.

Only by becoming increasingly aware of the Who, How, and What of environmental phenomena can we truly integrate the multiple voices calling for a more just and ecologically friendly world. Only in such a world is there the capacity to generate sustainable solutions to complex multidimensional problems. Only in such a world are all the notes of NATURE's song sung.

And yet much work remains to be done. . . .

While we have offered a nuanced framework that can advance Integral Ecology, we intend to revise this framework in light of new insights, connections, criticisms, and feedback from others. Only through reflective exploration with other Integral practitioners and their critics can Integral Ecology mature and continue to serve our planet.

While Integral Ecology provides a massive synthesis of over 200 distinct perspectives on ecology and nature, we have omitted many details. We must work together to deepen and broaden this synthesis so that it represents each approach in ways that those approaches themselves recognize. We will continue searching for better ways of defining and communicating the Integral Ecology approach. We will revise and rearrange in response to feedback from representatives of those approaches. Eventually, as other distinct and legitimate perspectives emerge, we will include them within our approach, filling out the details of this mandala with ever more care and concern.

We must further explore many issues of inter-, trans-, and postdisciplinarity to develop the most effective ways to unite disparate disciplines and to develop criteria for including some approaches and not others, depending on the circumstances of specific problems.

Although Integral Ecology presents a comprehensive understanding of ecological awareness as it relates to psychological development from birth to more comprehensive waves of development, we must further explore the ways that psychological development contributes to and hinders ecological awareness.

We have introduced an Integral approach to nature mysticism, a way to define that mysterious and beautifully overwhelming experience of being "one with nature." The approach distinguishes at least 20 distinct phenomenological and structural unitive experiences with the natural world. What a profound and joyful task it will be to carry out more scholarly and experiential research in order to substantiate these distinctions and provide methods for cultivating and stabilizing all forms of being "one with nature."

Here we have provided a platform that can serve multiple Integral approaches guided by different ultimate premises. Our commitment to becoming increasingly reflective about who is looking, how we are looking, and what we are looking at is the thread woven between all of these potentially different Integral approaches. By becoming deeply reflective individuals, and increasingly letting go of the notion that "our way is the only right way," we may be able to reach across the many divides that seemingly separate us all, and understand one another in the service of environment. May this commitment to reflective awareness serve the liberation and inclusion of all perspectives for the betterment of the natural world and its many inhabitants.

More than anything this book and the Integral Ecology framework presented herein is an invitation to think-feel differently about ecology and the environment. We have made many claims, interpretations, and distinctions which may or may not resonate with you. However, it is our hope that the details of our presentation won't overshadow the larger aim of this book: to foster an integral approach to the natural world that unites multiple perspectives in service of a better tomorrow for all beings on this amazing blue-green planet.

APPENDIX

200+ Perspectives on NATURE

Major, Minor, and Emerging Schools of Ecology, Environmental Studies, and Ecological Thought

Integral Ecology recognizes that there are four major perspectives one can take on NATURE—experience, culture, behavior, and systems—and these four resulting terrains contain levels of interior depth and exterior complexity. In addition, these terrains and their respective levels can be approached and interpreted from various worldviews using different methodologies. Consequently, the natural world is a multidimensional reality, which is revealed and concealed by the co-nascent triadic structure of the viewer-viewing-the-viewed. In other words, worldviews x methodologies x terrains = NATURE. So it should be no surprise that there are over 200 distinct perspectives on NATURE.

What follows is an encyclopedic overview of these perspectives. Each of the following major, minor, and emerging schools represent an established approach that often has its own journals, research programs, conferences, and textbooks. Together these perspectives represent a transdisciplinary intention and capacity to study the complexity and depth of the physical, cultural, and psychological relationships of human and nonhuman organisms within wild, rural, and urban environments.

The information provided for each entry is not meant to be comprehensive but rather to provide a brief description enabling readers to appreciate each distinct perspective and to suggest additional resources connected to it. In general, each entry defines the approach and tells when it began, recommends important publications that provide more information (theoretical, historical, and applied), and lists the most common methodological zones that have been used to explore the aspects of reality associated with this perspective. In some cases we have also listed the zones that are acknowledged or drawn on indirectly. Typically, each approach specializes in or emphasizes one to three zones, so when more zones are listed, keep in mind that most of these will be secondary zones and might not be easy to find represented in the associated literature of that approach. We have tried to be methodologically generous when listing the zones associated with each perspective.

The perspectives listed below represent the ones we feel are most essential for an Integral Ecologist to be aware of as well as the most important for addressing the environmental urgency of our times. In particular, we included as many substantial approaches as we could find that link interiors and exteriors. We believe such integrative approaches suggest how Integral Ecology can move forward in weaving together the natural sciences, the social sciences, and the humanities.

Entries represent a variety of categories, including schools, fields, subfields, disciplines, philosophies, social movements, labels, hypotheses, and theories. In a few cases the titles of entries are labels we have provided to refer to the intersection of two or more approaches (e.g., "critical ecology" as an umbrella term for ecological approaches informed by critical theory). Some entries use ecology as a metaphor (e.g., organizational ecology) while others use ecological theory in non-ecological contexts (e.g., business ecology). Some entries represent important historical perspectives that are not currently active (e.g., dynamic ecology). When we encountered different approaches that shared the same name, we chose the approach that was more distinct, established, and/or well known. By no means is this list exhaustive; there are many other important perspectives on NATURE, but in most cases the ones listed below will lead you to those.

We have not included schools of ecology defined by organisms (e.g., redwood ecology), biomes (desert ecology), or geographic areas (arctic ecology). Nor have we included natural sciences such as zoology or geology. In general we did not include environmental sciences such as bioclimatology, ecostratigraphy, geosociology, hydrology, petrology, phenology, phytogeography, plant sociology, zoogeography, or zoopharmacognosy. However, such fields of research are valuable to an Integral approach to ecology.

It is our sense that only by becoming familiar with the variety of schools that exist today and their perspectives on NATURE can we seriously begin the task of creating an Integral Ecology: an ecology that is an ecology of ecologies. As you read over the entries in this appendix we invite you to meditate on this question: "What kind of NATURE is it that gives rise to all these valuable perspectives?"

Lastly, this is a living document. Thus, if you see corrections or additions that need to be made to existing entries or missing perspectives that should be represented, please e-mail Sean Esbjörn-Hargens at sean@integralecology.org. Revised versions of this document will be available at www.integralecology.org. Please help us expand and improve this appendix so that it can most accurately represent the myriad perspectives on NATURE.

ACOUSTIC ECOLOGY was established in late 1960s with R. Murray Schafer's use of "soundwalks"—walking meditations devoted to a high level of sonic awareness. It then formalized in the 1970s through his field study of the soundscape of Vancouver, Canada. Acoustic ecology is dedicated to analyzing and documenting how organisms interpret and are affected by natural and artificial sounds. See Barry Truax's *Acoustic Communication* and his edited volume *Handbook for Acoustic Ecology*. Zones: 3, 5, 6, 8.

AGRICULTURAL ECOLOGY (AKA AGROECOLOGY) became a distinct field in the 1960s but can be traced to K. H. W. Klages, who explored the relationship between agriculture and ecology in the 1930s and 1940s. Agricultural ecology studies agricultural ecosystems and their components within the various natural and social systems they are embedded in. It aims to make agricultural ecosystems more sustainable through practices such as integrated pest management and intercropping. See Miquel Altieri's *Agroecology: The Science of Sustainable Agriculture*. Zones: 6, 8.

ANARCHO-PRIMITIVISM (AKA PRIMITIVISM OR GREEN ANARCHISM) has roots that extend back at least to antiquity. Anarcho-primitivism matured into a social movement during the 1960s, drawing on Marxism, anthropology, and ecology. It is based on an elaborate critique of civilization, often citing agriculture as the source of dis-ease in modern society (e.g., patriarchy, division of labor, alienation, technology, industrialism, and domestication). Anarcho-primitivists often promote "rewilding," or becoming uncivilized through the development of wilderness skills, living off the grid, and non-hierarchical community living. See Fredy Perlman's *Against His-Story, Against Leviathan*, John Zerzan's edited volume *Against Civilization: Readings and Reflections*, and Derrick Jensen's recent two-volume *End-Game*. Zones: 4, 6, 8.

ANIMAL RIGHTS (AKA ANIMAL LIBERATION) is a movement that extends back as far as 1781 when Jeremy Bentham argued for a utilitarian extension of rights to animals in his *Introduction to the Principles of Morals and Legislation*. A hundred years later, Henry Salt, an English social reformer, devoted an entire book to the idea: *Animals' Rights: Considered in Relation to Social Progress* (1892). Animal rights aims to protect animals from being considered property by extending to them the rights afforded persons. See Tom Regan's *The Case for Animal Rights* and Cass Sunstein and Martha Nussbaum's *Animal Rights: Current Debates and New Directions*. Zones: 1, 3, 4, 6, 8.

ANIMAL WELFARE SCIENCE emerged in the 1960s as people began to become more critical of animal treatment in husbandry and laboratory practices. Its roots go back at least to the establishment of the Society for the Prevention of Cruelty to Animals in 1824. Animal welfare science is the scientific study of animal welfare through experimentation, husbandry, and care within laboratories, farms, homes, and the wild. See Richard Ryder's *Victims of Science: The Use of Animals in Research* and Neville Gregory's *Physiology and Behavior of Animal Suffering*. Zones: 1, 3, 4, 6, 8.

ANIMISM is associated with the earliest forms of religion and spirituality in general and with indigenous peoples in particular. The term was coined and first used in the late 1800s by anthropologists. Animism is the belief that all objects have a soul or that all matter and nonhuman entities are animated by life. See Graham Harvey's *Animism: Respecting the Living World*. Zones: 1, 3, 4.

ANTHROZOOLOGY emerged in the 1950s and 1960s as psychologists began to document the benefits of pets for families and therapy. Anthrozoology is the scientific study of the human-animal relationship (e.g., pets, factory farming, zoos, and wild animals). See J. A. Serpell's *In the Company of Animals: A Study of Human-Animal Relationships* and Anthony Podberscek, Elizabeth S. Paul, and James A. Serpeil's *Companion Animals and Us: Exploring the Relationship between People and Pets*. Zones: 1, 3, 4, 6.

APPLIED ECOLOGY became established in the 1960s as more emphasis was placed on management issues and the practical application of ecological principles. Applied ecology uses ecological science to solve practical problems of resource management. See E. I. Newman's *Applied Ecology and Environmental Management*. Zones: 6, 8.

ARCHETYPAL ECOLOGY can be traced back to James Hillman's notion of "psychic ecology," which emerged from his 1960s research on animals in dreams and his later concept of "psychological urbanism." Archetypal ecology explores ecological relationships from a depth psychology perspective. See Daniel Noel's two-part "Soul and Earth: Traveling with Jung toward an Archetypal Ecology" and James Hillman's *Animal Presences*. Zones: 1, 3, 8.

ARCHITECTURAL PHENOMENOLOGY emerged in the 1970s as architects began drawing on the work of phenomenologists such as Martin Heidegger. Architectural phenomenology focuses on the sensory and lived experience of architectural design and landscape, often examining the role of the body and feelings in encountering natural and constructed spaces. See Christian Norberg-Schulz's *Genius Loci: Towards a Phenomenology of Architecture* and Juhani Pallasmaa's *The Eyes of the Skin: Architecture and the Senses*. Zones: 1, 3, 4, 6, 8.

ASTROECOLOGY has grown out of the nascent field of astrobiology in recent years. Astroecology studies the relationships between organisms and extraterrestrial environments through the study of comets, planets, interstellar clouds, and asteroids. See Michael Mautner's *Seeding the Universe with Life*. Zones: 6, 8.

AUTOPOIESIS THEORY (AKA BIOPHENOMENOLOGY) was introduced in 1973 by the Chilean biologists Francisco Varela and Humberto Maturana. Autopoiesis theory describes how organisms, which are self-organizing, cognize and react to their environment. See Maturana and

Varela's *Autopoiesis and Cognition: The Realization of the Living* and John Mingers's *Self-Producing Systems: Implications and Applications of Autopoiesis*. Zones: 5, 6.

BEHAVIORAL ECOLOGY began in the 1960s, building on the work of the ethologist Niko Tinbergen, which outlined the proximate and ultimate mechanisms of individual behavior. Behavioral ecology studies the evolutionary and ecological basis for animal behavior using economic models and adaptive analysis, and identifying optimization strategies. See J. R. Krebs and N. B. Davis's *Behavioural Ecology: An Evolutionary Approach*. Zones: 6, 8.

BIOACOUSTICS has been gaining momentum since the 1930s. Bioacoustics is the study of animal sounds. It explores sound production, auditory anatomy, acoustic communication, biological sonar and echolocation, and the impact of human-made noise on animals. See Brian Lewis's *Bioacoustics: A Comparative Approach* and Andrea Simmons, Arthur N. Popper, and Richard R. Fay's *Acoustic Communication*. Zones: 6, 8.

BIOCULTURAL ECOLOGY has emerged out of the field of biocultural anthropology since the 1970s. Biocultural ecology is the study of the relationships between human biology, culture, and ecology. See Alan Goodman and Thomas Leatherman's *Building a New Biocultural Synthesis* and Andrea Wiley's *An Ecology of High-Altitude Infancy: A Biocultural Perspective*. Zones: 3, 4, 6, 8.

BIODYNAMIC AGRICULTURE was developed by Rudolf Steiner in the 1920s through a series of lectures on agriculture. Biodynamic agriculture is an approach to organic farming that views farms as organisms and uses a variety of fertilization techniques to aid a holistic balance within the system. Some of the techniques work explicitly with the subtle energies of the agricultural process. See Willy Schilthuis's *Biodynamic Agriculture* and Rudolf Steiner's *What Is Biodynamics? A Way to Heal and Revitalize the Earth*. Zones: 6, 8.

BIOECOLOGY was the result of the collaboration between Frederic Clements and Victor Shelford in the 1920s to create a theoretical framework that could unify plant and animal ecology through concepts like the "biome."

Bioecology is the study of the relationships between plants and animals within a particular biome. See Frederic Clements and Victor Shelford's *Bio-Ecology*. Zones: 6, 8.

BIOGEOCHEMISTRY was founded by the Russian scientist Vladimir Vernadsky in 1926 through his book *Biosphere* and expanded by the American zoologist and geochemist G. Evelyn Hutchinson through his work as a limnologist. Biogeochemistry is a systems science that studies the relationship between biological life and the geochemistry of a region to understand complex dynamics such as climate change and soil remediation. See W. H. Schlesinger's *Biogeochemistry: An Analysis of Global Change*. Zones: 6, 8.

BIOGEOGRAPHY dates back to the 19th century and the research of evolutionists such as Alfred Russel Wallace. Biogeography is the study and analysis of how organisms are geographically distributed. It focuses on the types and historical causes of distributions. See Mark V. Lomolino, Dov F. Sax, and James H. Brown's *Foundations of Biogeography: Classic Papers with Commentaries* and Mark V. Lomolino, Brett R. Riddle, and James H. Brown's *Biogeography*. Zones: 6, 8.

BIOMIMICRY (AKA BIONICS OR BIOMIMETICS) emerged in the 1950s as engineers increasingly turned to the natural world for inspiration for innovative and functional designs. Biomimicry is the application of mimicking biological structures and systems to address technological and engineering problems. See Janine Benyus's *Biomimicry: Innovation Inspired by Nature* and Kevin Passino's *Biomimicry for Optimization, Control, and Automation*. Zones: 6, 8.

BIONOMICS was established in the 1990s as an economic theory based on biology instead of the physics that informs neoclassical economics. Bionomics views economies as self-organizing living organisms where learning is prized over capital. See Michael Rothschild's *Bionomics: Economy as Ecosystem*. Zones: 6, 8.

BIOPHILIA means "love of life" and was first formulated as a hypothesis by the psychologist Erich Fromm in the 1960s to describe a psychological attraction to anything that is alive. Twenty years later it was further developed by Edward O. Wilson to explain people's affinities for particular landscapes, baby animals, and common perceptions of nature. See Stephen R. Kellert and Edward O. Wilson's *The Biophilia Hypothesis* and Stephen R. Kellert's *Kinship to Mastery: Biophilia in Human Evolution and Development.* Zones: 1, 2, 3, 4, 6.

BIOREGIONALISM gained popularity in the 1980s as more and more people focused on living locally in the context of naturally defined areas such as watersheds. Bioregionalism is an approach to environmental and social issues based on living in an informed and sustainable way in relationship to your immediate surroundings. See Kirkpatrick Sale's *Dwellers in the Land: The Bioregional Vision* and Robert Thayer's *LifePlace: Bioregional Thought and Practice.* Zones: 1, 3, 4, 6, 8.

BIOSEMIOTICS (AKA SEMIOTIC BIOLOGY) emerged out of the work of the German biologist Jakob von Uexküll (1864–1944), who studied the phenomenal ("umwelt") world of animals. The term was first used in 1962 by the German doctor F. S. Rothschild, but it was Thomas Sebeok's increasing reference to and use of Uexküll's umwelt theory in the 1960s and 1970s that contributed the most to the development of biosemiotics. Biosemiotics is the scientific study of the way organisms interpret, communicate, and exchange information through signs. See Jakob von Uexküll's "Theory of Meaning," Jesper Hoffmeyer's *Signs of Meaning in the Universe,* and Marcello Barbieri's *Introduction to Biosemiotics.* Zones: 1, 3, 5, 6, 7, 8.

BRIGHT GREEN ENVIRONMENTALISM (AKA VIRIDIAN DESIGN MOVEMENT) was founded by the science fiction writer Bruce Sterling in the late 1990s through a series of speeches. Bright green environmentalism focuses on cutting-edge technology, aesthetic designs, and optimistic consumerism as a means of global sustainability. It is often contrasted with "dark" green approaches, which are pessimistic, anti-technology, and anti-consumerism. See William McDonough and Michael Braungart's *Cradle to Cradle: Remaking the Way We Make Things* and Alex Steffen's *Worldchanging: A User's Guide for the 21st Century.* Zones: 6, 8.

BUILDING ECOLOGY (AKA BUILDING BIOLOGY OR BAU-BIOLOGIE) was developed in Germany over 30 years ago by a doctor who noticed that many illnesses are the result of exposure to toxins connected to buildings. Building ecology examines the interrelationships between buildings (their structure and materials) and human health, often focusing on indoor air quality, electromagnetic fields, microorganisms, radiation, and toxins. See Thomas Schmitz-Guenther's *Living Spaces: Ecological Building and Design* and Athena Thompson's *Homes That Heal (and Those That Don't): How Your Home Could Be Harming Your Family's Health.* Zones: 6, 8.

BUSINESS ECOLOGY emerged in the late 1990s as organizational theorists began to use "ecology" as both a metaphor and a source of ecological principles to understand business dynamics. Business ecology is the study of the relationships between business, organisms, and their environments. See Joseph Abe, Patricia Dempsey, and David Bassett's *Business Ecology* and Amy Townsend's *Green Business.* Zones: 6, 8.

CATASTROPHE THEORY was developed by the mathematician René Thom in the 1960s as a branch of bifurcation theory to study dynamic systems. Catastrophe theory explores phenomena that shift quickly due to small changes in circumstances. See Vladimir Arnol'd's *Catastrophe Theory* and Robert Gilmore's *Catastrophe Theory for Scientists and Engineers.* Zones: 6, 8.

CHAOS THEORY was pioneered in large part by Edward Lorenz's work in the 1960s on weather prediction. Lorenz is also the one who introduced the idea of the "butterfly effect" through a paper in 1972 on predictability. Chaos theory studies behavior in nonlinear dynamic systems that are sensitive to initial conditions, are topologically mixing, and contain dense periodic orbits. See Edward Lorenz's *The Essence of Chaos* and Edward Ott's *Chaos in Dynamical Systems.* Zones: 6, 8.

CHAOTIC ECOLOGY (AKA DISEQUILIBRIUM ECOLOGY) emerged in the 1970s as the result of ecologists incorporating insights from the complexity sciences, especially chaos theory. Chaotic ecology argues against homeostasis within ecosystems (i.e., a "balance of nature" view) and studies the nonlinear dynamics of ecological phenomena. See William Drury and John Anderson's *Chance and Change: Ecology for Conservationists* and J. M. Cusing, Robert Costantino, Brian Dennis, Robert Desharnais, and

Shandelle Henson's *Chaos in Ecology: Experimental Nonlinear Dynamics.* Zones: 6, 8.

CHEMICAL ECOLOGY became established in the 1960s and 1970s with the discovery of insect pheromones and the exploration of the role of toxins in plant-insect interactions. Chemical ecology is the study of natural chemicals within and between organisms, often with a focus on chemicals used for communication and defense. See Marcel Dicke and Willem Takken's *Chemical Ecology: From Gene to Ecosystem.* Zones: 6, 8.

CLINICAL ECOLOGY (AKA ENVIRONMENTAL MEDICINE) arose during the early 1900s as doctors began to document allergies and various immune diseases caused by industrialization. Clinical ecology is the study of the impact on physical health of environmental exposure to synthetic chemicals and often focuses on allergies and chemical sensitivities. However, in recent years the field has expanded to include such issues as the impacts of UV exposure due to ozone depletion, mercury poisoning, and exposure to nuclear radiation. See Theron G. Randolph's *An Alternative Approach to Allergies: The New Field of Clinical Ecology Unravels the Environmental Causes of Mental and Physical Ills.* Zones: 1, 6, 8.

COGNITIVE ECOLOGY was introduced in 1993 through an article by Les Real, "Toward a Cognitive Ecology." Cognitive ecology focuses on integrating research on psychological and neural mechanisms to explain animal behavior. See Reuven Dukas's *Cognitive Ecology: The Evolutionary Ecology of Information Processing and Decision Making* and Thomas Grubb, Jr.'s *The Mind of the Trout: A Cognitive Ecology for Biologists and Anglers.* Zones: 5, 6, 8.

COGNITIVE ETHOLOGY can be traced to the 1976 publication of Donald Griffin's *The Question of Animal Awareness.* Cognitive ethology studies the mental states and cognitive processes of animals, often through behavioral observation. See Donald Griffin's *Animal Minds: Beyond Cognition to Consciousness* and Carolyn Ristau's *Cognitive Ethology: The Minds of Other Animals.* Zones: 1, 5, 6.

COMMUNITY ECOLOGY (AKA SYNECOLOGY) originated with plant ecology in the early 1900s. Community ecology is the study of the interrelationships between plants and animals in an ecological area, with a focus on issues such as food webs, predator-prey dynamics, competition, and succession. See Peter J. Morin's *Community Ecology*. Zones: 6, 8.

COMPARATIVE PSYCHOLOGY (AKA ANIMAL PSYCHOLOGY) began in the late 1800s when George J. Romanes applied Darwinian thought to the study of animal intelligence, which has remained one of the central foci of the field. Comparative psychology is the comparison of behaviors between species as a means of gaining insight into their structures of psychology, cognitive processes, and learning capacity. See Margaret Washburn's *The Animal Mind* and Gary Greenberg's *Comparative Psychology: A Handbook*. Zones: 2, 5, 6.

CONSERVATION ECOLOGY emerged in the 1970s following the passing of the National Environmental Policy Act (1970) and the Endangered Species Act (1973), which drew attention to species extinction and dwindling habitat. Conservation ecology focuses on the management of biodiversity and the ecological realities of species that are threatened or endangered. See George Cox's *Conservation Ecology: Biosphere and Biosurvival*. Zones: 6, 8.

CONSERVATION MEDICINE (AKA ECOLOGICAL MEDICINE OR MEDICAL GEOLOGY) was first proposed as a distinct field in a 1996 article on wildlife health in Africa. Conservation medicine studies the links between human, wildlife, and ecosystem health, often focusing on zoonotic (cross-species) diseases that result from activities such as deforestation and international travel. See A. Alonso Aquirre, Richard Ostfeld, Gary Tabor, Carol House, and Mary Pearl's *Conservation Medicine: Ecological Health in Practice* and Kenny Ausubel's *Ecological Medicine: Healing the Earth, Healing Ourselves*. Zones: 6, 8.

CONSERVATION PSYCHOLOGY (AKA PLANETARY PSYCHOLOGY) was established in 2000 by Carol Saunders and Gene Myers through a series of meetings and publications. In 2002 the Brookfield Zoo sponsored a conference on conservation psychology. Conservation psychology is dedicated to the study of the psychological dimensions of relationships between humans and nature in order to support the conservation of the natural world. See the special issue of *Human Ecology Review* 10 (2) dedicated to conservation psychology in 2003. Zones: 1, 2, 3, 4, 6, 8.

CONSTRUCTION ECOLOGY is a subset of industrial ecology that emerged in the mid-1990s. Construction ecology applies principles of natural systems to the built environment with a focus on using materials that are ecologically sound, can be replaced easily and recycled, and promote human health. See Charles J. Kibert, Jan Sendzimir, and G. Bradley Guy's *Construction Ecology: Nature as a Basis for Green Buildings*. Zones: 6, 8.

CREATION SPIRITUALITY was established in the 1970s through the writings of the theologian and priest Matthew Fox. Creation spirituality is a Christian panentheist approach to the universe that views creation as "original blessing" and promotes social activism. See Matthew Fox's *Creation Spirituality: Liberating Gifts for the Peoples of the Earth*. Zones: 1, 3, 4.

CRITICAL ECOLOGY (AKA ECOLOGICAL CRITICAL THEORY) appeared in the 1990s when proponents of Jürgen Habermas's theory of communicative action began exploring its contributions toward a critical theory of nature. Ecological critical theory applies the insights and principles of critical theory to ecological issues (e.g., public participation in environmental policies and how we communicate about nature). See Steven Vogel's *Against Nature: The Concept of Nature in Critical Theory* and Robert Brulle's *Agency, Democracy, and Nature: The U.S. Environmental Movement from a Critical Theory Perspective*. Zones: 3, 4, 6, 8.

CULTURAL ECOLOGY developed in the late 1950s as the result of anthropologists being influenced by Julian Steward's *Theory of Culture Change* (1955). Cultural ecology draws ideas from evolutionary biology (e.g., "adaptation" and "niche") to study the ways that the natural environment contributes to cultural and social realities in tribal and rural contexts. See Robert Netting's *Cultural Ecology* and Mark O. Sutton and E. N. Anderson's *An Introduction to Cultural Ecology*. Zones: 3, 4, 8.

CULTURAL LANDSCAPE STUDIES was pioneered by the landscape historian John Brinckerhoff Jackson beginning in the 1950s, largely through publications in his magazine *Landscape*. Cultural landscape studies is the exploration of meaning generated through human interactions with natural environments. There is often a focus on ordinary settings or "the vernacular landscape" such as malls, trailer parks, vacant lots, and garages. See John

Brinckerhoff Jackson's *A Sense of Place, a Sense of Time* and Paul Groth and Todd Bressi's *Understanding Ordinary Landscapes*. Zones: 1, 3, 4, 6, 8.

CULTURAL THEORY emerged in the 1980s out of the research by the social anthropologist Mary Douglas. Cultural theory proposes an understanding of risk analysis based on a typology of five biases or worldviews, wherein each worldview serves to maintain specific social structures. See Michael Thompson, Richard Ellis, and Aaron Wildawsky's *Cultural Theory*. Zones: 3, 4, 8.

CYBERNETICS began in the 1940s as an interdisciplinary effort drawing on multiple fields (e.g., biology, engineering, computer science) and became established as a distinct endeavor in the 1950s. Cybernetics is the study of communication, information exchange, and feedback loops within organisms, machines, and social systems. See Norbert Wiener's *Cybernetics: Or the Control and Communication in the Animal and the Machine* and Gregory Bateson's *Steps to an Ecology of Mind*. Zones: 7, 8.

CYBERSEMIOTICS has been developed by Søren Brier in the 1990s as an integration of phenomenology, biosemiotics, social autopoiesis, and information science. Cybersemiotics is a transdisciplinary nonreductionist approach to cognition and communication that studies the exchange of information and meaning in organisms. See Søren Brier's *Cybersemiotics: Why Information Is Not Enough*. Zones: 1, 3, 5, 7.

DEEP ECOLOGY was established in 1973 with Arne Naess's publication of the essay "Shallow and Deep Ecology." Deep ecology emphasizes the importance of expanding one's sense of self to include the natural world. Importance is placed on deep questioning and an honoring of our spiritual connection to the planet. See Arne Naess's *Ecology, Community, Lifestyle* and George Sessions's *Deep Ecology in the Twenty-first Century*. Zones: 1, 3, 4, 6, 8.

DESIGN ECOLOGY (AKA ENVIRONMENTAL METAPHYSICS) is an umbrella term proposed by Michael Riversong in the 1990s to refer to the various approaches to environmental design and architecture that are informed by working with the *prana*, life force, or *ch'i* of a built environment. Design ecology is the study of the relationship between design, subtle energies, and human use through such approaches as feng shui, wabi sabi, *vastu-shastra*, *kaso*, and aspects of Bau-biologie and Buckminster Fuller's synergetics. See Michael Riversong's *Design Ecology*, Eva Wong's *A Master Course in Feng Shui*, and Talavane Krishna's *The Vaastu Workbook: Using the Subtle Energies of the Indian Art of Placement*. Zones: 1, 3, 6, 8.

DEVA GARDENING (AKA CO-CREATIVE SCIENCE) dates back many centuries but is more recently associated with two gardens that have gained worldwide attention since the 1970s: Perelandra in Virginia and Findhorn in Scotland. Deva gardening is an approach to small-scale agriculture that works with the subtle "nature intelligences" of the surrounding environment. See Machaelle Small Wright's *Perelandra Garden Workbook: A Complete Guide to Gardening with Nature Intelligences* and Findhorn Community's *The Findhorn Garden*. Zones: 1, 3, 6, 8.

DEVELOPMENTAL ETHOLOGY is an emerging field that can be traced back to Margaret Washburn's *The Animal Mind* (1908). Developmental ethology studies the structures of awareness that develop sequentially and complexify over time. See Michael Commons's "Measuring an Approximate g in Animals and People." Zones: 2, 4, 5, 6.

DEVELOPMENTAL SYSTEMS ECOLOGY (AKA INFODYNAMICS) has been developed by biologist and natural philosopher Stan Salthe since the 1990s as an integration of thermodynamics, evolutionary theory, biosemiotics, hierarchy theory, internalism, and information theory. Developmental systems ecology is the study of how ecological systems store increasing amounts of information as they develop. See Stan Salthe's "The Natural Philosophy of Ecology: Developmental Systems Ecology" and *Development and Evolution: Complexity and Change in Biology*. Zones: 1, 2, 3, 4, 5, 6, 7, 8.

DEVELOPMENTAL SYSTEMS THEORY is a post-Darwinian approach to evolution and development associated with Susan Oyama's postmodern research since the 1990s. Developmental systems theory studies the multiple irreducible causes of heredity and enactive dynamics of organisms and their environment. See Susan Oyama's *Evolution's Eye: A Systems View of the Biology-Culture Divide* and Donald Ford and Richard Lerner's *Developmental Systems Theory: An Integrative Approach*. Zones: 1, 2, 3, 4, 5, 6, 8.

DYNAMIC ECOLOGY was pioneered in the 1900s by plant ecologist Frederic Clements. Dynamic ecology is the study of the dynamics of plant communities through such concepts as "ecological succession" and "climax state," and views such communities as a super-organism. See B. W. Allred and E. S. Clements's *The Dynamics of Vegetation: Selections from the Writings of F. E. Clements.* Zones: 6, 8.

ECOCOMPOSITION can be dated back to an article by Richard Coe in 1975 wherein he draws on the notion of ecological wholeness as an approach to writing. Ecocomposition is the study of the relationship between various mediums of discourse and environments (including imagined places). See Mary Cooper's "The Ecology of Writing" and Sidney Dobrin and Christian Weisser's *Natural Discourses: Toward Ecocomposition.* Zones: 1, 3, 4, 8.

ECOCRITICISM (AKA ENVIRONMENTAL [LITERARY] CRITICISM) started in the 1970s with a number of seminal articles and books, but it wasn't until the early 1990s that it formed into a distinct field. Ecocriticism studies the relationship between the natural world and literature, including works explicitly about nature and those implicitly informed by it. See Cheryll Glotfely and Harold Fromm's *The Ecocriticism Reader: Landmarks in Literary Ecology* and Lawrence Buell's *The Future of Environmental Criticism: Environmental Crisis and Literary Imagination.* Zones: 1, 3, 4, 6, 8.

ECOFEMINISM (AKA FEMINIST ECOLOGY) was first proposed in 1974 by the French feminist Françoise d'Eaubonne and emerged from women engaged in political action aimed at environmental issues. Ecofeminism studies the links between social inequality and the domination of nature (e.g., the oppression of nature through its feminization) and often emphasizes intersubjective connections and the role of women in protecting the environment. See Irene Diamond and Gloria Orenstein's *Reweaving the World: The Emergence of Ecofeminism* and Karren Warren's *Ecofeminist Philosophy.* Zones: 1, 3, 4, 6, 8.

ECOLINGUISTICS (AKA LINGUISTIC ECOLOGY) was forged by Einar Haugen in 1972 with the publication of *The Ecology of Language* and became established in the 1990s as linguistic studies increasingly included the ecological dimensions of language. Ecolinguistics studies the relationships between language and ecology, including viewing languages as ecosystems (i.e., linguistic ecology), the analysis of ecologically damaging and beneficial discourses (i.e., ecocritical discourse analysis), and the relationships between linguistic and biological systems (i.e., biocultural diversity). See Alwin Fill and Peter Muhlhausler's *The Ecolinguistics Reader: Language, Ecology and Environment* and Peter Muhlhausler's *Language of Environment, Environment of Languages: A Course in Ecolinguistics.* Zones: 3, 4, 7, 8.

ECOLOGICAL ANTHROPOLOGY (AKA ENVIRONMENTAL ANTHROPOLOGY) developed in the 1960s through the ethnographic research of Roy Rappaport in Papua New Guinea on the ways ritual links culture and nature. Ecological anthropology studies human populations as parts of ecosystems, with a focus on adaptability (e.g., the many cultural responses to ecological issues). See Roy Rappaport's *Pigs for the Ancestors: Ritual in the Ecology of a New Guinea People* and Emilio Moran and Rhonda Gillett-Netting's *Human Adaptability: An Introduction to Ecological Anthropology.* Zones: 3, 4, 6, 8.

ECOLOGICAL ARCHAEOLOGY (AKA ENVIRONMENTAL ARCHAEOLOGY) emerged in the 1990s as more archaeologists began to search for insights into our current environmental crisis. Ecological archaeology studies human-environment interactions from the past, including how cultures have responded to ecological change. See Eric Kaldahl's *Environmental Archaeology: Principles and Practice* and Garth Bawden and Richard Reycraft's *Environmental Disaster and the Archaeology of Human Response.* Zones: 4, 6, 8.

ECOLOGICAL ARCHITECTURE (AKA SUSTAINABLE ARCHITECTURE OR GREEN ARCHITECTURE) emerged in the 1970s largely through the "pattern language" work of Christopher Alexander. Ecological architecture is the use of ecological principles to design aesthetically pleasing, environmentally sound, and sustainable buildings. See James Steele's *Ecological Architecture* and Christopher Alexander's *The Order of Nature.* Zones: 1, 3, 4, 6, 8.

ECOLOGICAL COMMUNALISM (AKA ECO-COMMUNALISM) has its roots in 19th-century utopias and was first articulated at a gathering of the Global Scenario Group in 1995 as one of the possible future scenarios for sustainability. Ecological communalism is an approach to sustainable living that emphasizes decentralized government, green economics, and agriculture. See Paul Raskin et al.'s *Great Transition* and Gilberto Gallopin and Paul Raskin's *Global Sustainability*. Zones: 4, 6, 8.

ECOLOGICAL ECONOMICS originated in the late 1970s with the work of Herman Daly, then senior economist for the World Bank's Environmental Department. Ecological economics studies the economy as a subfield of ecology by drawing on living systems theory and thermodynamics in order to establish sustainable economies grounded in physical reality. See Herman Daly's *Steady-State Economics* and Michael Common and Sigrid Stagl's *Ecological Economics: An Introduction*. Zones: 6, 8.

ECOLOGICAL GENETICS (AKA MOLECULAR ECOLOGY) was founded by the English geneticist E. B. Ford in the 1950s and 1960s. Ecological genetics studies the ecological significance of genes in natural populations. See E. B. Ford's *Ecological Genetics* and Paul Ashton, Stephen Harris, and Andrew Lowe's *Ecological Genetics: Design, Analysis, and Application*. Zones: 6, 8.

ECOLOGICAL HERMENEUTICS is an emerging approach that takes its name from Robert Mugrauer's 1985 article "Language and the Emergence of Environment." Ecological hermeneutics explores the role of language and interpretation in the disclosure and concealment of the landscape and ecological phenomena. It also traces how concepts like "nature" and "wilderness" have been used historically or understood by different worldviews. See John van Buren's "Critical Environmental Hermeneutics" and Robert Mugerauer's *Interpreting Environments: Tradition, Deconstruction, Hermeneutics*. Zones: 1, 3.

ECOLOGICAL HUMANITIES was first articulated by the Australian political scientist Robyn Eckersley in a short 1998 article. Ecological humanities is an approach to political justice that integrates the natural and social sciences with the humanities through a "connective ontology." See Deborah Bird Rose and Libby Robin's "The Ecological Humanities in Action." Zones: 3, 4, 6, 8.

ECOLOGICAL INTERCORPOREALITY is an emerging field grounded in somatics and environmental dance. Eco-intercorporeality studies the ways individual embodiment interfaces with others and the natural world. See Andrea Olsen's *Body and Earth* and Edward Casey's *The Fate of Place*. Zones: 1, 3, 6, 8.

ECOLOGICAL MODELING began in the 1920s with modeling predator-prey relationships and oxygen balance in streams. Since that time the models have grown increasingly complex and multivariable. Ecological modeling is the representation, often through mathematical models and computer programs, of ecological realities and ecosystem dynamics. See S. E. Jørgensen and G. Bendoricchio's *Fundamentals of Ecological Modeling* and S. E. Jørgensen, B. Halling-Sorensen, and S. N. Nielsen's *Handbook of Environmental and Ecological Modeling*. Zones: 6, 8.

ECOLOGICAL MODERNIZATION is a neoliberal form of green economics that has gained attention since the 1990s. Ecological modernization promotes the idea that solutions to the environmental crisis are to be found in the process of modernization by using capitalism to promote ecological efficiency through technological advances and industrial development. See Stephen Young's *The Emergence of Ecological Modernisation: Integrating the Environment and the Economy?* Zones: 6, 8.

ECOLOGICAL ONTOLOGY (AKA GEOSPATIAL ONTOLOGY) has emerged over the last decade as a result of applied philosophical research of ontologist Barry Smith. Environmental ontology is the study of ontological (e.g., spatial) relations between organisms and their environments. In particular it is interested in token environments (opposed to types of environments) and draws on mereology, topology, and formal ontology to explore part-whole relationships and their boundaries in the context of biological and geospatial concepts such as "niche" and "mountains." See Barry Smith's "Objects and Their Environments: From Aristotle to Ecological Ontology," Barry Smith and Achille Varzi's "Surrounding Space: The Ontology of Organism-Environment Relations," and Barry Smith and David Mark's "Do Mountains Exist? Towards an Ontology of Landforms." Zones: 4, 6, 8.

ECOLOGICAL PHENOMENOLOGY (AKA ENVIRONMENTAL PHENOMENOLOGY) emerged in the mid-1980s as environmentalists began to turn toward the contributions of philosophers like Martin Heidegger and Maurice Merleau-Ponty for fresh understandings of the natural world. Ecological phenomenology utilizes the methodologies of phenomenological inquiry to "re-see" ecological and natural phenomena freed from our habitual ways of conceiving and experiencing, thereby opening up new horizons of perception and action. See David Seamon and Robert Mugerauer's *Dwelling, Place, and Environment: Towards a Phenomenology of Person and World* and Charles S. Brown and Ted Toadvine's *Eco-Phenomenology: Back to the Earth Itself*. Zones: 1, 3.

ECOLOGICAL PHILOSOPHY has been gaining momentum as a unique field since the 1990s, especially with the work of Kevin de Laplante. Ecological philosophy is the exploration of the metaphysical and epistemological issues within ecological science. See Kevin de Laplante's "Environmental Alchemy: How to Turn Ecological Science into Ecological Philosophy" and David Keller and Frank Gollev's *The Philosophy of Ecology: From Science to Synthesis*. Zones: 3, 6, 8.

ECOLOGICAL PSYCHOLOGY (AKA GIBSONIAN PSYCHOLOGY) is a post-cognitivist approach to vision developed by the American psychologist James Gibson beginning in the 1950s. Ecological psychology studies the importance of an organism's direct perception of its environment in how that organism relates to and acts in that environment. See James Gibson's *The Ecological Approach to Visual Perception* and Harry Heft's *Ecological Psychology in Context: James Gibson, Roger Barker, and the Legacy of William James*. Zones: 5, 6, 8.

ECOLOGICAL RATIONALITY is a field of thought that emerged out of Gerd Gigerenzer's research on heuristics in the late 1990s. Ecological rationality is the study of the evolutionary and ecological pressures that select heuristics used by humans to make quick decisions successfully in specific environments. See Gerd Gigerenzer and Reinhard Selten's *Bounded Rationality: The Adaptive Toolbox*. Zones: 3, 4, 5, 6, 8.

ECOLOGICAL THEOLOGY (AKA ECOTHEOLOGY) began to emerge after and as a result of the publication of Lynn White, Jr.'s famous essay, "The Historical Roots of Our Ecological Crisis," which linked Christian theology to environmental issues. Ecological theology studies the relationship between religious worldviews and environmental values. See Thomas Berry's *The Dream of the Earth* and Rosemary Radford Ruether's *Gaia and God: An Ecofeminist Theology of Earth Healing*. Zones: 1, 3, 4, 6, 8.

ECOPOETICS emerged out of ecocriticism and was formally established through Jonathan Bate's *The Song of the Earth* (2000) and the journal *ecopoetics* in 2001. Ecopoetics is the study of nature poetry and the use of poetry to raise ecological awareness. See Scott Bryson's *Ecopoetry: Critical Introduction* and Jimmie Killingsworth's *Walt Whitman and the Earth: A Study of Ecopoetics*. Zones: 1, 3, 4, 7, 8.

ECOPSYCHOLOGY developed as a result of Theodore Roszak's 1992 book *The Voice of the Earth* but has roots in early uses of the wilderness for psychological transformation in the 1960s. Ecopsychology explores the psychological and cultural aspects of how factors such as technology, consumerism, and alienated modern culture contribute to an ecological crisis. See Theodore Roszak, Mary Gomes, and Alan Kanner's *Ecopsychology: Restoring the Earth, Healing the Mind* and Andy Fisher's *Radical Ecopsychology: Psychology in the Service of Life*. Zones: 1, 3, 4, 8.

ECOSEMIOTICS (AKA SEMIOTIC ECOLOGY) began to emerge in Russia and Estonia through a network of theoretical biologists during the 1980s. In 1996 Winfried Nöth coined the term and defined it in an article published in *Zeitschrift für Semiotik*. Ecosemiotics is the study of the semiotic (sign-mediated) relationships between culture and nature, including the ways humans communicate with and about the natural world, the meaning of nature, the place and role of humans within nature, and the semiosis occurring between a human and its ecosystem. See Winfried Nöth's "Ecosemiotics" and Kalevi Kull and Winfried Nöth's *Semiotics of Nature*. Zones: 3, 4, 7, 8.

ECOSYSTEM ECOLOGY has its roots in a holistic view of nature and is often viewed as having become a distinct field in 1942 when the limnologist Raymond Lindeman published a chapter from his dissertation in *Ecology*. The Odum brothers did much to advance this systems approach with their highly influential 1953 textbook, *Fundamentals of Ecology*. Ecosystem ecology is the study of organisms within the context of the functional processes and mechanisms (e.g., the flows of energy and matter) of an ecosystem. See

Eugene Odum and Gary Barrett's *Fundamentals of Ecology* and F. Stuart Chapin III, Pamela Matson, and Harold Mooney's *Principles of Terrestrial Ecosystem Ecology*. Zones: 6, 8.

ECOTAGE (AKA MONKEY WRENCHING, ECOTERRORISM, ECO-DEFENSE) became an organized movement in the 1970s and was popularized by Edward Abbey's 1975 novel *The Monkey Wrench Gang*. Ecotage is the use of extreme measures, often violent and illegal, to achieve environmental protection. It is most often associated with radical environmental groups such as Earth First! and Earth Liberation Front (ELF). See Dave Foreman and Bill Haywood's *Ecodefense: A Field Guide to Monkey Wrenching* and Donald Liddick's *Eco-Terrorism: Radical Environmental and Animal Liberation Movements*. Zones: 3, 4, 6, 8.

ECOTOXICOLOGY was the result of the emerging public concern in the 1950s and 1960s around the ecological effects of chemicals and pesticides. The publication of Rachel Carson's *Silent Spring* in 1962 did much to establish ecotoxicology as an important field. Ecotoxicology is the study of natural and anthrogenetic pollutants in the environment and their effects on organisms. See Michael Newman and William Clements's *Ecotoxicology: A Comprehensive Treatment*. Zones: 6, 8.

EMOTIONAL ETHOLOGY has grown out of the field of cognitive ethology over the last decade, especially through the research of the ecologist Marc Bekoff. However, it is worth noting that Darwin recognized and wrote about the emotional aptitude of organisms. Emotional ethology is the study of the subjective dimension and emotional capacities of organisms. See Jeffrey Moussaieff Masson and Susan McCarthy's *When Elephants Weep: The Emotional Lives of Animals* and Marc Bekoff's *The Emotional Lives of Animals*. Zones: 1, 2, 3, 4, 5, 6, 8.

ENVIRONMENTAL AESTHETICS (AKA ECOLOGICAL AESTHETICS) began to emerge in 1984 with Allen Carlson's paper "Nature and Positive Aesthetics." Environmental aesthetics focuses on philosophical and applied issues of aesthetic appreciation of both built and natural environments. See Arnold Berleant's *The Aesthetics of Environment* and Allen Carlson and Arnold Berleant's *The Aesthetics of Natural Environments*. Zones: 1, 3, 4, 6, 8.

ENVIRONMENTAL COMMUNICATION has its roots in nature writing and has emerged as a distinct discipline in the 1990s with the establishment of the Conference on Communication and Environment. Environmental communication is the field of study that explores the communication processes involved with environmental affairs such as media coverage, public discourse, interpersonal rhetoric, and social debate. See M. J. Killingsworth and Jackie Palmer's *Ecospeak: Rhetoric and Environmental Politics in America* and Robert Cox's *Environmental Communication and the Public Sphere*. Zones: 1, 3, 4, 6, 8.

ENVIRONMENTAL ECONOMICS (AKA NATURAL CAPITALISM) became fully established in the 1970s as more and more economists were confronted with environmental realities. Environmental economics draws heavily on neoclassical economics and studies the externality (i.e., the hidden cost) of activities and policies that impact natural resources. See Charles Kolstad's *Environmental Economics* and Daniel Bromley's *Handbook of Environmental Economics*. Zones: 6, 8.

ENVIRONMENTAL EDUCATION (AKA ECOLOGICAL LITERACY) grew out of the study of natural history and outdoor education and emerged as a distinct field in the late 1960s. Environmental education is the promotion of environmental awareness and ecological knowledge within the context of educational settings, particularly K–12. See Joy Palmer's *Environmental Education in the 21st Century: Theory, Practice, Progress and Promise* and Michael Stone and Zenobia Barlow's *Ecological Literacy: Educating Our Children for a Sustainable World*. Zones: 1, 3, 4, 6, 8.

ENVIRONMENTAL ETHICS was established formally with the inaugural issue of the journal *Environmental Ethics* in 1979. Environmental ethics is the philosophical study of the ethical relationships among humans, nonhuman organisms, and the environment. See Patrick Derr's *Case Studies in Environmental Ethics* and Louis Pojman and Paul Pojman's *Environmental Ethics: Readings in Theory and Application*. Zones: 1, 3, 4, 6, 8.

ENVIRONMENTAL HEALTH is an aspect of public health that can be traced back to the 1800s. Environmental health is the study of the relationship between human health and the environment. See Howard Frumkin's *Environmental Health: From Global to Local*. Zones: 4, 6, 8.

ENVIRONMENTAL HISTORY originated in the 1960s and 1970s as global awareness emerged and historians began to look for the sources of current environmental problems. Rodrick Nash first used the phrase "environmental history" in an article in 1972. Environmental history focuses on the interaction between past cultures and their environments. See Richard White's *Land Use, Environment, and Social Change: A History of Island County, Washington* and Shepard Krech, J. R. McNeill, and Carolyn Merchant's *Encyclopedia of World Environmental History*. Zones: 4, 6, 8.

ENVIRONMENTAL JUSTICE formed primarily in the United States during the 1980s through a variety of grassroots efforts aimed at civil rights. Environmental justice is a social movement that exposes injustices toward minorities and underprivileged peoples in the context of environmental issues. The best-known example of such injustices is "environmental racism," which highlights that people of color and the poor are the most likely people to be living in heavily polluted areas due to various forms of institutional discrimination. See Joni Adamson, Mei Mei Evans, and Rachel Stein's *The Environmental Justice Reader: Politics, Poetics, and Pedagogy*. Zones: 1, 3, 4, 6, 8.

ENVIRONMENTAL LAW developed as a distinct field in the 1960s as environmentalists sought to institutionalize protection of the environment. Environmental law refers to the collection of laws, ordinances, statutes, policies, regulations, and treaties that serve to protect the environment in various forms (e.g., securing open space, regulating the use of pesticides, taxing businesses). See Bridget Hunter's *A Reader in Environmental Law* and Nancy Kusbasek and Gary Silverman's *Environmental Law*. Zones: 6, 8.

ENVIRONMENTAL MONITORING has developed at different points in various contexts over the last 25–50 years. Environmental monitoring is the study of changes in the environment through the sampling of water, soil, atmosphere, and organisms for effluents and contaminants. See G. Bruce Wiersma's *Environmental Monitoring*. Zones: 6, 8.

ENVIRONMENTAL OPTIMISM (AKA ENVIRONMENTAL SKEPTICISM) began in 1987 with the Brundtland report, *Our Common Future*, which went against the more common dire tone of environmental reports and suggested an optimistic outlook on environmental and sustainability issues.

Environmental optimism challenges the "doomsday" perspective associated with environmentalists by appealing to the scientific method to demonstrate that things are either not so bad as has been reported (e.g., in the media) or actually getting better. See Gregg Easterbrook's *A Moment on the Earth: The Coming Age of Environmental Optimism* and Bjørn Lomborg's *The Skeptical Environmentalist: Measuring the Real State of the World*. Zones: 3, 4, 6, 8.

ENVIRONMENTAL PHILOSOPHY was the result of American academic philosophers turning their attention to the environmental issues raised by the counterculture movements in the 1960s and 1970s. Environmental philosophy is the theoretical study of cultural, political, aesthetic, ethical, metaphysical, and epistemological issues related to the natural environment. See Dale Jamieson's *A Companion to Environmental Philosophy* and Frederik Kaufman's *Foundations of Environmental Philosophy: A Text with Readings*. Zones: 1, 3, 4, 6, 8.

ENVIRONMENTAL PLURALISM is an emerging approach to nature that recognizes multiple and even contradictory views of the natural world. Environmental pluralism is an engaged and self-reflective approach to working with multiple perspectives on environmental issues to arrive at common ground and support democratic action. See Bruce Hull's *Infinite Nature*. Zones: 3, 4, 6, 8.

ENVIRONMENTAL PRAGMATISM is an outgrowth of American pragmatism that began to form in the 1990s. Environmental pragmatism is the study of concrete environmental issues with a focus on practical applications and methodological pluralism. See Andrew Light and Eric Katz's *Environmental Pragmatism*. Zones: 3, 4, 6, 8.

ENVIRONMENTAL PSYCHOLOGY (AKA ARCHITECTURAL PSYCHOLOGY) emerged in the 1950s as a result of Roger Barker's research on how environments influence behavior. Environmental psychology studies the effect of the built environment (e.g., size of halls, lighting, ventilation) on human psychology and behavior. See Paul Bell, Thomas Greene, Jeffrey Fisher, and Andrew Baum's *Environmental Psychology* and Robert Bechtel and Arza Churchman's *Handbook of Environmental Psychology*. Zones: 1, 3, 4, 6, 8.

ENVIRONMENTAL SOCIOLOGY has largely been built on two shifts within sociology that occurred in the 1970s: in 1975 Allan Schnaiberg proposed a "societal-environmental dialectic," which highlights the conflictual nature between economic growth and environmental protection; and in 1978 William Catton, Jr., and Riley Dunlap introduced the "new ecological paradigm," which exposed the anthropocentrism of traditional sociology and included environmental variables. Environmental sociology is the study of the relationship between social dynamics and institutions and the environment. See Riley Dunlap and William Michelson's *Handbook of Environmental Sociology* and Deborah McCarthy and Leslie King's *Environmental Sociology: From Analysis to Action*. Zones: 4, 6, 7, 8.

ETHNOECOLOGY (AKA TRADITIONAL KNOWLEDGE OR FOLK ECOLOGY) was first outlined by Harold Conklin in the 1950s through his work on Hanunuo culture and became more established in the 1960s as anthropologists became more concerned with ecological issues. Ethnoecology is the study of how different cultures and subcultures perceive, relate, and respond to their environment through categories of ecological meaning and folk classifications. See Virginia Nazarea's *Ethnoecology: Situated Knowledge/Located Lives* and Roy Ellen, Peter Parkes, and Alan Bicker's *Indigenous Environmental Knowledge and Its Transformations: Critical Anthropological Perspectives*. Zones: 2, 3, 4, 5, 6, 7, 8.

ETHOLOGY is often thought of as beginning with Darwin and his book *The Expression of the Emotions in Animals and Men*. However, it is most commonly associated with the research of Konrad Lorenz and Niko Tinbergen beginning in the 1930s. Ethology is the study of animal behavior and its evolution in naturalistic settings, with an emphasis on physiology and anatomy. See Richard Burkhardt's *Patterns of Behavior: Konrad Lorenz, Niko Tinbergen, and the Founding of Ethology* and James Gould's *Ethology: The Mechanisms and Evolution of Behavior*. Zones: 6, 8.

EVOLUTIONARY ECOLOGY began with ecologists in the late 1950s who began to consider the role of evolution in ecosystems. In particular, V. C. Wynne-Edwards, a British zoologist, contributed to this new approach by arguing for natural selection at the level of the group in his 1962 book *Animal Dispersion in Relation to Social Behavior*. Evolutionary ecology is the study of the impact of evolution on organisms' relationship to their environment. See Eric Pianka's *Evolutionary Ecology* and Charles Fox, Derek Roff, and Daphne Fairbairn's *Evolutionary Ecology: Concepts and Case Studies*. Zones: 6, 8.

EVOLUTIONARY PSYCHOLOGY defined itself in the 1980s in part through its critiques of sociobiology, and has roots in the classical ethology of Lorenz. Evolutionary psychology is the study of how evolution has selected for adaptive cognitive processes that may or may not function well in today's environments. See Jerome Barkow, Leda Cosmides, and John Tooby's *The Adapted Mind: Evolutionary Psychology and the Generation of Culture* and David Buss's *Evolutionary Psychology: The New Science of the Mind*. Zones: 4, 5, 6, 7, 8.

GAIA THEORY (AKA GEOPHYSIOLOGY) was first proposed in an article in 1972 by the NASA scientist James Lovelock. Gaia theory claims that the Earth has a thin film of organisms (single-celled and multicellular), which serves to self-regulate atmospheric oceanic conditions in such a way as to protect and promote the conditions of life. Gaia theory is often used to promote the idea that the planet is a single super-organism. See James Lovelock's *Gaia: A New Look at Life on Earth* and Lynn Margulis's *Symbiotic Planet: A New Look at Evolution*. Zones: 6, 8.

GENERAL SYSTEMS THEORY was developed in the late 1920s by the Hungarian biologist Ludwig von Bertalanffy. In the 1950s he began to include biological systems in his theorizing. General systems theory studies the principles and dynamics common to all scientific systems (e.g., physiological, biological, social). See Ludwig von Bertalanffy's *General Systems Theory* and Lars Skyttner's *General Systems Theory: Perspectives, Problems, Practice*. Zones: 6, 8.

GEOPSYCHOLOGY (AKA PSYCHOGEOGRAPHY) was introduced by the American agricultural geographer Howard Gregor in 1963 in his book *Environmental and Economic Life*. Geopsychology is the study of the influence of geographical realities on human psychology, culture, and behavior. See Robert Sack's *A Geographical Guide to the Real and the Good* and M. A. Persinger's "Geopsychology and Geopsychopathology: Mental

Processes and Disorders Associated with Geochemical and Geophysical Factors." Zones: 1, 3, 4, 6, 8.

GLOBAL ECOLOGY (AKA PLANETARY ECOLOGY) emerged in the late 1970s as more focus was placed on the planet as an entire ecosystem. Global ecology is the study of ecological phenomena and issues on a planetary scale. See Douglas Caldwell's *Planetary Ecology* and Charles Southwick's *Global Ecology in Human Perspective*. Zones: 4, 6, 8.

GOETHEAN SCIENCE (AKA DELICATE EMPIRICISM) is based on the research of the German polymath Johann Wolfgang von Goethe (1749–1832). Goethean science recognizes that the scientist is in a participatory relationship with the object of investigation, and emphasizes the use of imagination as a way of experientially making contact with the wholeness of phenomena. See Henri Bortoft's *The Wholeness of Nature: Goethe's Way towards a Science of Conscious Participation in Nature* and David Seamon and Arthur Zajonc's *Goethe's Way of Science: A Phenomenology of Nature*. Zones: 1, 3, 6.

GREEN ECONOMICS is often associated with E. F. Schumacher's 1973 classic, *Small Is Beautiful: Economics as if People Mattered*. Green economics is a non-neoclassical approach that emphasizes social equity, bioregions as service-producing systems, the creativity of individuals, and local measurements of economic activity. Also, it often highlights the spiritual dimension of all life. See Molly Scott Cato and Miriam Kennet's *Green Economics: Beyond Supply and Demand to Meeting People's Needs* and Brian Milani's *Designing the Green Economy: The Postindustrial Alternative to Corporate Globalization*. Zones: 1, 3, 4, 6, 8.

HIERARCHY THEORY originated in 1962 when the economist Herbert Simon presented his paper "The Architecture of Complexity" at a conference. Hierarchy theory studies the holonic levels of complexity within complex systems, with particular focus on the emergent phenomena at various scales and the role of the observer at different levels. For the latter this approach draws on the work of developmental psychologists such as Jean Piaget. See Howard Pattee's *Hierarchy Theory: The Challenge of Complex Systems* and Valerie Ahl and T. F. H. Allen's *Hierarchy Theory: A Vision, Vocabulary, and Epistemology*. Zones: 1, 2, 3, 4, 5, 6, 8.

HISTORICAL ECOLOGY (AKA DIALECTICAL ECOLOGY) began to emerge in the 1970s when the anthropologist Edward S. Deevey directed the Historical Ecology Project in connection with University of Florida. During the 1980s numerous anthropologists and historians began to explore the ecological dimensions of their disciplines. Historical ecology uses an interdisciplinary approach and multiple methods to explore and document the ongoing dialectical relationship within the landscape between humans and nature. See Lester Bilsky's *Historical Ecology: Essays on Environment and Social Change* and William Balée's *Advances in Historical Ecology*. Zones: 3, 4, 6, 8.

HORTICULTURAL THERAPY (AKA THERAPEUTIC HORTICULTURE) goes back as far as the 1700s, but it wasn't until the 1950s that it became formalized through working with at-risk youth, war veterans, and the mentally handicapped. Horticultural therapy is the practice of using interaction with plants (e.g., gardening) to promote people's physical, psychological, and spiritual well-being. See Sharon Simson and Martha Straus's *Horticulture as Therapy: Principles and Practice* and Rebecca Haller and Christine Kramer's *Horticulture Therapy Methods: Making Connections in Health Care, Human Service, and Community Programs*. Zones: 1, 3, 6, 8.

HUMAN ECOLOGY began in the 1920s in association with the Chicago School of Sociology and became a fully established in the 1950s and 1960s as ecological models were increasingly used to understand human societies. Human ecology is the study of human adaptation to various environments and views humans as a part of an ecosystem. See Amos Hawley's *Human Ecology: A Theoretical Essay* and Frederick Steiner's *Human Ecology: Following Nature's Lead*. Zones: 4, 6, 8.

HUMANISTIC GEOGRAPHY was launched by the geographer Yi-Fu Tuan in 1976 with the publication of an article entitled "Humanistic Geography." Two years later a book of essays with the same title was published by David Ley and Marwyn Samuels. Humanistic geography draws on existentialism and phenomenology to study people's sense and experience of place. See Yi-Fu Tuan's *Space and Place: The Perspective of Experience* and Paul Adams, Steven Hoelscher, and Karen Till's *Textures of Place: Exploring Humanist Geographies*. Zones: 1, 2, 3, 4, 6, 8.

INDUSTRIAL ECOLOGY was introduced by Harry Zvi Evan in 1973 at a seminar in Poland and then popularized in a 1989 article in *Scientific American*. Industrial ecology is the study of industrial systems as closed loop systems wherein their waste is used as inputs for other systems. See Thomas Graedel and Braden Allenby's *Industrial Ecology* and Dominique Bourg and Suren Erkman's *Perspectives on Industrial Ecology*. Zones: 6, 8.

INFORMATION ECOLOGY has developed independently in the 1980s and 1990s out of both the information sciences and ecological anthropology. Information ecology is the study of information processes in human systems and social networks. Information scientists focus on how the informational spaces can be viewed as an ecosystem, whereas ecological anthropologists focus on how the environment plays a role in how information is cognized and communicated. See David Casagrande's "Information as Verb: Reconceptualizing Information for Cognitive and Ecological Models" and Thomas Davenport and Laurance Prusak's *Information Ecology: Mastering the Information and Knowledge Environment*. Zones: 3, 4, 5, 6, 7, 8.

INTEGRAL ECOLOGY emerged in the late 1990s as Sean Esbjörn-Hargens and Michael E. Zimmerman began applying Ken Wilber's Integral Model to scientific ecology and environmental philosophy. Integral Ecology is the study of the subjective and objective aspects of organisms in relationship to their intersubjective and interobjective environments. In addition to this volume, see Ken Wilber's *Sex, Ecology, Spirituality* and the special double issue of *World Futures* (edited by Sean Esbjörn-Hargens) devoted to Integral Ecology in 2005. Zones: 1, 2, 3, 4, 5, 6, 7, 8.

INTEGRATED ECOLOGY (AKA ORGANIC PSYCHOLOGY) is based on the life-long work of the outdoor educator Michael Cohen and was formally introduced in a 1993 article. Integrated ecology is a form of ecopsychology that uses multisensory activities to connect with nature. See Michael Cohen's "Integrated Ecology: The Process of Counseling with Nature" and *Reconnecting with Nature: Finding Wellness through Restoring Your Bond with the Earth*. Zones: 1, 3, 6, 8.

INTEGRATIVE ECOLOGY (AKA INTERDISCIPLINARY ECOLOGY) was coined by the Canadian ecologist C. S. Hollings in a 1998 editorial of the journal *Conservation Ecology*. Currently, "integrative ecology" is a label for a variety of interdisciplinary ecology research groups that have emerged since 2000. Integrative ecology is synthesizing in nature, drawing on multiple ecological schools, using a variety of scientific methodologies, and working at different environmental scales. See Lance Gunderson and C. S. Hollings's *Panarchy*. Zones: 4, 6, 8.

INVASION ECOLOGY was established by the ecologist Charles Elton through a series of radio lectures in the late 1950s, which grew into the classic book *The Ecology of Invasion by Animals and Plants* in 1958. This field took off in the mid-1980s as "invasive" species received more attention in the scientific literature. Invasion ecology is the study of the ecological impact of translocated species from one area to another, often as a result of human activity. See Charles Elton's *The Ecology of Invasion by Animals and Plants* and Julie Lockwood, Martha Hoopes, and Michael Marchetti's *Invasion Ecology*. Zones: 4, 6, 8.

LANDSCAPE ARCHITECTURE has roots in ancient times and became formalized in the 1800s, growing out of the practice of designing gardens. Landscape architecture is the practice of designing and maintaining outdoor spaces that incorporate human constructs (e.g., parks, housing developments, urban centers, wilderness areas). See Geoffrey Jellicoe's *The Landscape of Man: Shaping the Environment from Prehistory to the Present Day* and John Simonds and Barry Starke's *Landscape Architecture*. Zones: 3, 4, 6, 8.

LANDSCAPE ECOLOGY (AKA GEOECOLOGY) was named in 1939 by the German geographer Carl Troll and developed initially in Europe. In the 1980s it became an established approach through the formation of an international association and several key publications. Landscape ecology is the study of various spatial configurations across landscapes at different scales and over time. Landscape ecology often uses remote sensing, computer models, and GIS technology. See John Wiens, Michael Moss, Monica Turner, and David Mladenoff's *Foundation Papers in Landscape Ecology* and Jianguo Wu and Richard Hobbs's *Key Topics in Landscape Ecology*. Zones: 6, 8.

LIBERATION ECOLOGY began to emerge in the mid-1990s through the environmental activism of the theologian Leonardo Boff and the collaborative research of the geographer Richard Peet and the political ecologist Michael Watts. Liberation ecology focuses on developing nations and communities and draws from poststructuralism, discourse theory, and liberation theology to explore emancipatory approaches to political-ecological realities. See Leonardo Boff's *Ecology and Liberation: A New Paradigm* and Richard Peet and Michael Watts's *Liberation Ecologies*. Zones: 3, 4, 6, 8.

LIVING SYSTEMS THEORY was developed in 1978 by James Grier Miller. Living systems theory is a general theory of all living systems, which identifies 20 subsystems and 8 hierarchical levels that comprise "life." See James Grier Miller's 1,000-page book, *Living Systems*. Zones: 6, 8.

MACROECOLOGY was developed by the American ecologist James Brown in a paper he coauthored with Brian Maurer in 1989. Macroecology is the study of the large spatial dynamics and global effects on organisms and their environments. See Kevin Gaston and Tim Blackburn's *Pattern and Process in Macroecology* and James Brown's *Macroecology*. Zones: 6, 8.

MATHEMATICAL ECOLOGY grew out of population ecology in the 1920s and a 1939 work by the Russian biochemist Vladimir Kostitzin, *Mathematical Biology*. In 1969, E. C. Pielou published the first book for the field. Mathematical ecology is the study of the populations and distribution of organisms using mathematical models of individual behavior, population, and community dynamics. See Mark Kot's *Elements of Mathematical Ecology* and Simon Levin and Thomas Hallam's *Applied Mathematical Ecology*. Zones: 6, 8.

MEDIA ECOLOGY was pioneered in the early 1960s through the research and books of the communications theorist Marshall McLuhan. In 1971 Neil Postman created a program in media ecology at New York University. Media ecology is the study of media environments and their influence on people and society. See Marshall McLuhan's *Understanding Media: The Extensions of Man* and Matthew Fuller's *Media Ecologies: Materialist Energies in Art and Technoculture*. Zones: 1, 3, 4, 6, 8.

MUSIC ECOLOGY (AKA ECO-MUSICOLOGY) was introduced by the musicologist Maria Anna Harley in a 1996 article for the *Journal of Acoustic Ecology*. Music ecology is the study of the relationship between music and environment, exploring, e.g., the cognitive, symbolic, and structural relationships between natural landscapes and music. See Maria Anna Harley's "Notes on Music Ecology" and Wilfrid Mellers's *Singing in the Wilderness: Music and Ecology in the Twentieth Century*. Zones: 1, 3, 4, 5, 6, 8.

NANOECOLOGY was pioneered by the molecular engineer Eric Drexler in 1986 through the founding of the Foresight Institute and the publication of his first book, *Engines of Creation*. Nanoecology is the study of the relationship between molecular technologies and the environment. See Eric Drexler's *Unbounding the Future: The Nanotechnology Revolution* and Douglas Mulhall's *Our Molecular Future: How Nanotechnology, Biotechnology, Robotics, Genetics and Artificial Intelligence Will Transform Our World*. Zones: 6, 8.

NEO-DARWINISM (AKA THE MODERN EVOLUTIONARY SYNTHESIS) was first used at the end of the 19th century by the English naturalist George Romanes to signify evolution solely via natural selection. Neo-Darwinism has come to be associated with the modern evolutionary synthesis, which explains natural selection through genetic inheritance. See Michael Ruse's *Darwinism and Its Discontents* and Richard Dawkins's *The Selfish Gene*. Zones: 6, 8.

NEO-PAGANISM can be traced back to the Romantic movement in the early 18th century as Europeans began to explore occultism and ancient non-Christian religions. The term "neo-paganism" became popular in the 1960s with the counterculture movement. Neo-paganism consists of a variety of nature-based traditions and practices such as wicca and Slavianism, which take their inspiration from pre-Christian religions. See Graham Harvey's *Contemporary Paganism* and Sarah Pike's *Earthly Bodies, Magical Selves*. Zones: 1, 3, 4, 6.

NEOSHAMANISM has emerged since the 1960s counterculture and New Age movement have revitalized or augmented forms of indigenous religion. "Neoshamanism" is an umbrella term that refers to a variety of (post)modernist expressions of spiritual practice, which emphasize an animistic world filled with spirits, and often involves going into a trance or taking entheogens. See Michael Harner's *The Way of the Shaman* and Sandra Ingerman's *Soul Retrieval*. Zone: 1, 3, 4.

NEUROETHOLOGY became a distinct field in the early 1970s as the result of advances in technologies that allowed the identification of individual neurons. Neuroethology is the study of the neurological aspects of natural animal behavior. See Jörg-Peter Ewert's *Neuroethology* and Gunther Zupanc's *Behavioral Neurobiology*. Zone: 6.

NONDUAL ECOLOGY was proposed in 1993 by John McClellan as a critique of deep ecology. Nondual ecology is an approach that recognizes the perfected state of all things, whether natural or human-made. See John McClellan's "Nondual Ecology." Zones: 1, 3, 8.

NON-EQUILIBRIUM THERMODYNAMICS branched out of thermodynamics in the 1940s in part through the work of the Russian physicist Ilya Prigogine. Non-equilibrium thermodynamics is the study of open thermodynamic systems that are time-dependent and contain irreversible transformations. See S. R. de Groot and P. Mazur's *Non-equilibrium Thermodynamics* and Eric Schneider and Dorion Sagan's *Into the Cool*. Zones: 6, 8.

NUTRITIONAL ECOLOGY (AKA ECONUTRITION) began as a course taught in the fall of 1976 by Dr. Joan Dye Gussow at Columbia University. Nutritional ecology is the study of the relationship between nutrition and ecology. See Joan Dye Gussow's *The Feeding Web: Issues in Nutritional Ecology* and William Leonard's *Human Nutritional Ecology and Evolution*. Zones: 4, 6, 8.

ORGANIZATIONAL ECOLOGY (AKA ORGANIZATIONAL DEMOGRAPHY) developed out of the work of the Stanford sociologist Michael Hannan on organizational studies in the late 1980s. Organizational ecology is the study of how organizations emerge, grow, and die. See Michael Hannan and John Freeman's *Organizational Ecology* and Glenn Carroll and Michael Hannan's *The Demography of Corporations and Industries*. Zones: 6, 8.

PALEOECOLOGY (AKA ANCIENT ECOLOGY) was first introduced in the mid-1800s by the British naturalist Edward Forbes as a result of his consideration of the fossil record in light of his biological observations of marine life. In 1916 Fredrick Clements used the term to refer to the study of past vegetation. Paleoecology is the study of fossils in order to reconstruct ancient ecosystems. See James Dodd and Robert Stanton's *Paleoecology: Concepts and Applications*. Zones: 6, 8.

PANEXPERIENTIALISM is a metaphysical view developed by the process philosopher Alfred North Whitehead in the 1920s. Panexperientialism attributes the capacity to experience to all components of matter (e.g., atoms and molecules) and life (e.g., cells). See Alfred North Whitehead's *Process and Reality* and Charles Birch and John Cobb, Jr.'s *The Liberation of Life: From the Cell to the Community*. Zones: 1, 2, 3, 4.

PANPSYCHISM is a philosophical perspective going back to the ancient Greeks. Panpsychism argues that the constituent parts of matter and therefore all nonliving and living things have mind or consciousness. See David Skrbina's *Panpsychism in the West* and Freya Mathews's *For the Love of Matter: A Contemporary Panpsychism*. Zones: 1, 2, 3, 4.

PANSEMIOTICS is associated with the semiotic work of Charles Peirce during the late 1800s. Pansemiotics is the view that the entire universe, not just organic life, is perfused with semiosis. See James Hoopes's *Peirce on Signs* and Claus Emmeche's "The Biosemiotics of Emergent Properties in a Pluralist Ontology." Zones: 1, 3, 4, 5, 8.

PERMACULTURE was established by the Australians Bill Mollison and David Holmgren through their 1978 publication *Permaculture One*. Permaculture is an agro-ecological design theory that emphasizes an ethic of sustainability through mimicking natural patterns. See David Holmgren's *Permaculture* and Toby Hemenway's *Gaia's Garden*. Zones: 4, 6, 8.

PHYSIOLOGICAL ECOLOGY (AKA ECOPHYSIOLOGY) developed in the 1930s and 1940s through the research of individuals such as George Bartholomew. Physiological ecology is the study of an organism's physiological adaptations to its environment. See William Karasov and Carlos Martinez del Rio's *Physiological Ecology*. Zones: 6, 8.

PLACE STUDIES is indebted to the French philosopher Gaston Bachelard's *The Poetics of Space* (1958), which explored the phenomenology of intimate spaces. Place studies is a philosophical field that draws on phenomenology and hermeneutics to understand what Heidegger referred to as "dwelling." See Edward Casey's *The Fate of Place: A Philosophical History* and Tim Malpas's *Place and Experience: A Philosophical Topography*. Zones: 1, 3, 4, 6.

PLANT NEUROBIOLOGY began to form in 2000 in large part through the work of the plant biochemist Anthony Trewavas on plant intelligence. Plant neurobiology is the study of plant signaling and communication at all levels of biological complexity. See Frantisek Baluska, Stefano Mancuso, and Dieter Volkmann's *Communication in Plants* and Anthony Trewavas's "Aspects of Plant Intelligence." Zones: 5, 6, 7, 8.

POLITICAL ECOLOGY developed out of the fields of political science, anthropology, and geography in the 1970s. Political ecology is the study of the relationship between political realities and environmental issues. See Paul Robbins's *Political Ecology: A Critical Introduction* and Karl Zimmerer and Thomas Bassett's *Political Ecology: An Integrative Approach to Geography and Environment-Development Studies*. Zones: 4, 6, 8.

POPULATION ECOLOGY (AKA AUTECOLOGY) is one of the oldest schools of ecology, going back to the late 1800s and then becoming formalized in the 1920s through the development of mathematical models for population dynamics. Population ecology is the study of how populations of single species interact with the environment. See Sharon Kingsland's *Modeling Nature: Episodes in the History of Population Ecology* and Larry Rockwood and Giuseppe Bertola's *Introduction to Population Ecology*. Zones: 6, 8.

PROCESS ECOLOGY emerged in the 1970s through the application of Alfred North Whitehead's metaphysics to a Christian approach to ecology by the theologian John Cobb, Jr. Process ecology is an approach to ecological phenomena that emphasizes creative emergence, pan-experientialism, and holonic integration of the past into the present. See Charles Birch and John Cobb, Jr.'s *The Liberation of Life* and Nancy Howell's *A Feminist Cosmology: Ecology, Solidarity, and Metaphysics*. Zones: 1, 3, 6, 8.

PSYCHOANALYTIC ECOLOGY was first proposed by Rod Giblett in his 1996 book *Postmodern Wetlands*. Psychoanalytic ecology is the application of Freudian and neo-Freudian concepts to the study of "psychogeopathology" (i.e., the psychological underpinnings of destructive ecological behavior). See Rod Giblett's *Living with the Earth: Mastery to Mutuality*. Zones: 1, 2, 3, 4, 6, 8.

PUBLIC ECOLOGY has been pioneered since 2000 through the publication of numerous journal articles and the establishment of the nonprofit Public Ecology Project at Virginia Tech. Public ecology is an integrative and participatory approach to ecological science and policy that recognizes competing environmental beliefs and values of various stakeholders and takes a pragmatic approach to establishing mutual understanding between citizens, scientists, and policy-makers. See Paul Gobster and R. Bruce Hull's *Restoring Nature: Perspectives from Humanities and Social Sciences* and David Robertson's *Public Ecology: Linking People, Science, and the Environment*. Zones: 1, 3, 4, 6, 8.

RECONCILIATION ECOLOGY (AKA WIN-WIN ECOLOGY) was first proposed by the ecologist Michael Rosenweig in 2003 as a strategy for conservation biology that integrated wildlife into human-occupied landscapes. Reconciliation ecology is the study of creating and maintaining species habitats in areas inhabited by humans in order to conserve biodiversity. See Michael Rosenweig's *Win-Win Ecology: How the Earth's Species Can Survive in the Midst of Human Enterprise*. Zones: 4, 6, 8.

RESTORATION ECOLOGY has been practiced in various forms for over hundreds of years, becoming a distinct discipline, emerging out of conservation biology, in the 1980s through the work of John Aber and William Jordan. Restoration ecology is the study of using intentional activity for reestablishing an ecosystem that has been destroyed or degraded. See Donald Falk, Margaret Palmer, and Joy Zedler's *Foundations of Restoration Ecology* and Jelte van Andel and James Aronson's *Restoration Ecology*. Zones: 4, 6, 8.

REVERENTIAL ECOLOGY began in the early 2000s through the work of Satish Kumar, a former Jain monk. Reverential ecology is an approach to environmental sustainability that promotes nonviolence and reverence for all life (including human). See John Einarsen's "Satish Kumar Takes Deep Ecology a Step Further." Zones: 1, 3.

ROMANTIC ECOLOGY was coined by the British professor Jonathan Bate in 1991 with the publication of his book of the same name. Romantic ecology is the study of the ecological awareness in the Romantic poets. See Jonathan Bate's *Romantic Ecology: Wordsworth and the Environmental Tradition* and James McKusick's *Green Writing: Romanticism and Ecology*. Zones: 1, 3, 4, 6, 8.

SACRED ECOLOGY was formalized by the Canadian ecologist Fikret Berkes in the late 1990s through his work with the resource management of First Nation people in Canada. Sacred ecology is the study of the spiritual worldview and traditional environmental knowledge of native North Americans. See Fikret Berkes's *Sacred Ecology: Traditional Ecological Knowledge and Resource Management* and Howard Harrod's *The Animals Came Dancing: Native American Sacred Ecology and Animal Kinship*. Zones: 1, 3, 4, 6, 8.

SACRED GEOGRAPHY dates back to prehistory and has various modern-day expressions in different religious and occult traditions. Sacred geography is the study of the spiritual importance and subtle energies of various landscapes and geographical spaces. See Belden Lane's *Landscapes of the Sacred* and Marko Pogačnik's *Sacred Geography*. Zones: 1, 3, 4, 6, 8.

SENSORY ECOLOGY was established in 1979 by the publication of John Lythgoe's now classic book, *Ecology of Vision*. Sensory ecology is dedicated to the study of the adaptation of an organism's sensory systems (visual, auditorial, olfactory) to its lifestyle, habitat, and social interactions. See David Dusenbery's *Sensory Ecology: How Organisms Acquire and Respond to Information* and Howard Hughes's *Sensory Exotica: A World Beyond Human Experience*. Zones: 5, 6, 8.

SOCIAL ECOLOGY was established by the American libertarian socialist Murray Bookchin in the 1950s and 1960s through his writings and speeches. Social ecology is the study of the relationship between humans and their environment, and typically carries with it a critique of capitalist society. See Murray Bookchin's *The Philosophy of Social Ecology* and David Watson's *Beyond Bookchin*. Zones: 4, 6, 8.

SOCIALIST ECOLOGY (AKA ECO-SOCIALISM) can be traced back to the late 1800s and the work of the British socialist William Morris, who wrote the classic eco-socialist text *News from Nowhere* (1890). In the 1970s, individuals such as the American biologist Barry Commoner and the Polish political activist Rudolph Bahro did much to articulate the principles of this approach. Socialist ecology is an anticapitalist political approach to the environment that draws on Marxism and Green politics, and emphasizes communal ownership. See James O'Conner's *Natural Causes: Essays in Ecological Marxism* and Joel Kovel's *The Enemy of Nature: The End of Capitalism or the End of the World?* Zones: 4, 6, 8.

SOCIOECOLOGY developed as a distinct field in the 1970s when zoologists studied the effect ecologies had on animal mating systems. Socioecology is the study of how environment influences and determines the social structures and interactions of organisms. It tends to focus on social animals, especially primates. See P. C. Lee's *Comparative Primate Socioecology*. Zones: 6, 8.

SOMATIC ECOLOGY is associated with the microscopic research of the German zoologist Günther Enderlein conducted in the early 20th century. Somatic ecology is the study the relationships and semiotic networks between microorganisms and cells within a single individual. See Leo Buss's *The Evolution of Individuality*. Zones: 5, 6.

SPATIAL ECOLOGY stems from the 1980s when ecologists began including more spatial analysis in their understanding of ecological realities. Spatial ecology is the study of the spatial dimension of organisms and their environment. See David Tilman and Peter Kareiva's *Spatial Ecology: The Role of Space in Population Dynamics and Interspecific Interactions* and Marie-Josée Fortin and Mark Dale's *Spatial Analysis: A Guide for Ecologists*. Zones: 6, 8.

SPIRITUAL ECOLOGY emerged in the 1990s as a number of academics such as Bron Taylor and Roger Gottlieb began to explore in depth the intersection between religion and ecology. Spiritual ecology is the study of the relationships among religious traditions, individuals' spirituality, and the environment (e.g., how religions and spiritual experiences in nature can foster environmental action). See David Kinsley's *Ecology and Religion: Ecological Spirituality in Cross-cultural Perspective* and Mary Evelyn

Tucker and John Grim's 10-volume series on World Religions and Ecology. Zones: 1, 3, 4, 6, 8.

SUBTLE ECOLOGY is a label for approaches to ecology that emerged in the 1920s, exploring the subtle energy systems of the environment. Subtle ecology is the study of the relationships between subtle energies and the landscape, such as power spots, geomantic systems, radiaesthetic phenomena, and ley lines, through techniques such as lithopuncture, dowsing, geomancy, and the subtle energy research of Viktor Schauberger. See Alick Bartholomew's *Hidden Nature*, Marko Pogačnik's *Healing the Hearth of the Earth: Restoring the Subtle Levels of Life*, and David Cowan and Chris Arnold's *Ley Lines and Earth Energies*. Zones: 1, 3, 6, 8.

SYSTEMS ECOLOGY emerged in the 1940s as systems models began to be applied to ecological realities. Systems ecology is the study of the relationships between biological, ecological, and economic systems. See Roger Kitching's *Systems Ecology* and Howard Odum's *Ecological and General Systems: An Introduction to Systems Ecology*. Zones: 6, 8.

TERRAPSYCHOLOGY is the result of research during the early 2000s by Craig Chalquist into various kinds of eco-social traumas sustained by local regions and their inhabitants. Terrapsychology is a form of ecopsychology grounded in Jungian thought that studies the soul of place and how current residents in a particular place express and are impacted by that soul through symbolic resonance. See Craig Chalquist's *Terrapsychology: Reengaging the Soul of Place*. Zones: 1, 3, 4, 6, 8.

THEORETICAL ("PURE") ECOLOGY emerged in the late 19th century and became fully established in the 1920s through the work of the American mathematician Alfred Lotka, the British zoologist G. Evelyn Hutchinson, and the German biologist Jakob von Uexküll. Theoretical ecology is the study of the foundational forces and dynamics underlying ecological systems, often drawing on mathematical models of populations and competition. See Robert May's *Theoretical Ecology: Principles and Applications* and Robert Ulanowicz's *Ecology: The Ascendant Perspective*. Zones: 5, 6, 8.

TRANSCENDENTALISM developed in America's New England during the early 1830s as a Kantian critique of the formal religious doctrines of the time. Transcendentalism emphasizes an individual's intuition, especially in the context of nature, as the source of connection with the Divine. See Catherine Albanese's *The Spirituality of the American Transcendentalists* and Arthur Versluis's *American Transcendentalism and Asian Religions*. Zones: 1, 3, 4, 6, 8.

TRANSPERSONAL ECOLOGY began with the 1990 publication of *Toward a Transpersonal Ecology* by the Australian eco-philosopher Warwick Fox. Transpersonal ecology is the study of widening of self-identification to include the collective and natural world, often through the experience of transpersonal states of consciousness. See Warwick Fox's *Toward a Transpersonal Ecology: Developing New Foundations for Environmentalism* and Chris Bache's *Dark Night, Early Dawn*. Zones: 1, 2, 3, 4, 6, 8.

URBAN ECOLOGY grew out of sociology in the 1920s as a result of the sociological research conducted by Robert Park, Ernest Burgess, and Robert McKenzie. Urban ecology is the study of the ecological dynamics within urban environments as well as the study of urban dynamics using ecological principles. See Alan Berkowitz, Charles Nilon, and Karen Hollweq's *Understanding Urban Ecosystems* and Dianne Smith's *Urban Ecology*. Zones: 4, 6, 8.

WISE-USE MOVEMENT was established by anti-environmentalist Ron Arnold in 1988 through a conference held in Reno, Nevada. The wise-use movement is a network of industry-funded (and some grassroots) conservative groups that oppose most environmental values and argue for resource consumption. See Ron Arnold's *Ecology Wars: Environmentalism as if People Mattered* and David Helvarg's *The War Against Greens: The "Wise-Use" Movement, the New Right, and the Browning of America*. Zones: 4, 6, 8.

YOGA ECOLOGY (AKA GREEN YOGA AND ECO-YOGA) was introduced as a concept by the environmental philosopher Henryk Skolimowski in a book in 1991. Yoga ecology is the cultivation of ecological awareness and action through the practice of yoga. See Henryk Skolimowski's *EcoYoga: Practice and Meditations for Walking in Beauty on the Earth* and Georg Feuerstein and Brenda Feuerstein's *Green Yoga*. Zones: 1, 6, 8.

ZOOETHNOGRAPHY (ANIMAL CULTURE) began in the 1950s when the Japanese scientist Kinji Imanishi developed a theory to account for potato washing that emerged among monkeys in Koshima. Zooethnography is the study of cultural behaviors and dynamics in animals. See Guy Bradshaw's *Elephant Breakdown* and Frans de Waal and Petert Tyack's *Animal Social Complexity: Intelligence, Culture, and Individualized Societies.* Zones: 1, 3, 4, 6, 8.

ZOOHERMENEUTICS (ANIMAL INTERPRETATION) is an outgrowth of the umwelt theory of Jakob von Uexküll and Heidegger's work on animality in the 1930s. Zoohermeneutics is the study of animal being through animals' capacity to interpret their lifeworld. See Hans Jonas's *The Phenomena of Life* and Peter Scheers's "Hermeneutics and Animal Being." Zones: 1, 3, 4, 5, 6, 7, 8.

ZOOPHENOMENOLOGY (ANIMAL EXPERIENCE) can be traced back to F. J. J. Buytendijk's 1935 book *The Mind of a Dog.* Zoophenomenology is the study of the interiors of nonhuman organisms. See Kenneth Shapiro's "A Phenomenological Approach to the Study of the Nonhuman Animals" and Corinne Painter and Christian Lotz's *Phenomenology and the Non-human Animal.* Zones: 1, 2, 3, 4, 5, 6, 7, 8.

ZOOSEMIOTICS (AKA ANIMAL COMMUNICATION) was established by Thomas Sebeok beginning in 1963 with an article reviewing several books. Zoosemiotics is the study of intra- and interspecies animal communication and signaling from a semiotic perspective (i.e., acknowledging the interpretive dimension of animals). See Thomas Sebeok's *Perspectives in Zoosemiotics* and his *Essays in Zoosemiotics.* Zones: 1, 3, 4, 5, 6, 7, 8.

ADDITIONAL SCHOOLS OF (1) ECOLOGY, (2) ENVIRONMENTAL STUDIES, AND (3) ECOLOGICAL THOUGHT INCLUDE: (1) chronoecology, existential ecology, functional ecology, green ecology, home ecology, interface ecology, microecology, network ecology, participatory ecology, positive ecology, postmodern ecology, production ecology, psychological ecology, quantum ecology, radical ecology, radioecology, rational ecology, virtual ecology, and zooecology; (2) environmental accounting, environmental assessment, environmental biology, environmental consumerism, environmental dance, environmental engineering, environmental epistemology, environmental forecasting, environmental interpretation, environmental journalism, environmental rhetoric, and environmental science; (3) ecological design, ecological engineering, ecological evolution, ecological hermeticism, ecological physics, ecological postmodernism, ecological sanitation, ecological tourism, and ecotherapy.

FURTHER READING
The Integral Ecology Bookshelf

The following is a list of a *few* books within each of the 8 zones that we consider to be essential reading for an Integral Ecologist. This list represents those books that we have found most helpful in understanding the multidimensionality of nature and is by no means exhaustive. Most of the listed books could be placed in several zones, but we have listed them in the zone we found them to illuminate the most. (We also highly recommend Ken Wilber's *Sex, Ecology, Spirituality: The Spirit of Evolution* [Boston: Shambhala Publications, 1995], which illuminates all these methodological domains.) Top picks within each zone are indicated by an asterisk (*). For a more extensive list of important books, consult Robert Merideth's *The Environmentalist's Bookshelf: A Guide to the Best Books* (New York: G. K. Hall, 1993), which presents an annotated list of the 500 most important books, in order, on ecology and environmental studies as voted by over 200 leading ecologists and environmentalists from all over the world. The drawback is that the book is a bit dated and doesn't contain any books that have been published in the last 15 years.

Phenomenology

Bortoft, Henri. *The Wholeness of Nature: Goethe's Way towards a Science of Conscious Participation in Nature.* Hudson, NY: Lindisfarne Press, 1996.

*Brown, Charles S., and Ted Toadvine, eds. *Eco-Phenomenology: Back to the Earth Itself.* Albany, NY: SUNY Press, 2003.

Carlson, Allen, and Arnold Berleant, eds. *The Aesthetics of Natural Environments.* Orchard Park, NY: Broadview Press, 2004.

Cooper, J. W. *Panentheism: The Other God of the Philosophers—from Plato to the Present.* Grand Rapids, MI: Baker Academic, 2006.

Hurley, Susan, and Matthew Nudds. *Rational Animals.* New York: Oxford University Press, 2006.

Kohak, Erazim. *The Embers and the Stars.* Chicago: University of Chicago Press, 1987.

*Lotz, Christian, and Corinne Painter, eds. *Phenomenology and the Non-human Animal*. Dordrecht, Netherlands: Kluwer/Springer Academic Publishers, 2006.

Marshall, Paul. *Mystical Encounters with the Natural World: Experiences and Explanations*. Oxford: Oxford University Press, 2005.

Oliver, Mary. *New and Selected Poems*. Boston: Beacon Press, 1993.

*Skrbina, David. *Panpsychism in the West*. Cambridge, MA: MIT Press, 2005.

Snyder, Gary. *No Nature: New and Selected Poems*. New York: Pantheon Books, 1992.

Walsh, Roger. *The World of Shamanism: New Views of an Ancient Tradition*. Woodbury, MN: Llewellyn, 2007.

Winter, Deborah DuNann. *Ecological Psychology: Healing the Split between Planet and Self*. San Francisco: HarperCollins, 1995.

Structuralism

*Clayton, Susan, and Susan Opotow. *Identity and the Natural Environment: The Psychological Significance of Nature*. Cambridge, MA: MIT Press, 2003.

Damier, Eric. *Discourses of the Environment*. New York: Blackwell, 1999.

*Kahn, Peter H. Jr. *The Human Relationship with Nature: Development and Culture*. Cambridge, MA: MIT Press, 1999.

Kellert, Stephen R. *Kinship to Mastery: Biophilia in Human Evolution and Development*. Washington, DC: Island Press, 1997.

Parker, Sue T., Robert W. Mitchell, and Maria L. Boccia. *Self-Awareness in Animals and Humans: Developmental Perspectives*. Cambridge: Cambridge University Press, 1994.

Hermeneutics

Basso, Keith. *Wisdom Sits in Places: Landscape and Language among the Western Apache*. Albuquerque: University of New Mexico Press, 1997.

Birch, Charles, and John B. Cobb, Jr. *The Liberation of Life: From the Cell to the Community*. Denton, TX: Environmental Ethics Books, 1988.

Callicott, J. Baird. *Earth's Insights: A Multicultural Survey of Ecological Ethics from the Mediterranean Basin to the Australian Outback*. Berkeley: University of California Press, 1997.

Casey, Edward. *The Fate of Place: A Philosophical History*. Berkeley: University of California Press, 1997.

*Chase, Alston. *In a Dark Wood: The Flight over Forests and the Myths of Nature*. New ed. New Brunswick, NJ: Transaction Publishers, 2001.

*Cronon, William, ed. *Uncommon Ground: Toward Reinventing Nature*. New York: W. W. Norton, 1996.

Golley, Frank B. *A History of the Ecosystem Concept in Ecology: More Than the Sum of the Parts*. New Haven, CT: Yale University Press, 1996.

*Hoffmeyer, Jesper. *Signs of Meaning in the Universe*. Translated by B. J. Haveland. Bloomington: Indiana University Press, 1996.

Hull, R. Bruce. *Infinite Nature*. Chicago: University of Chicago Press, 2006.

Rolson, Holmes III. *Environmental Ethics*. Philadelphia: Temple University Press, 1989.

Spirn, Anne W. *The Language of Landscape*. New Haven, CT: Yale University Press, 1998.

Tucker, Mary Evelyn, and John Grim, eds. *Religions of the World and Ecology*. 10 vols. Cambridge, MA: Harvard University Press, 1994–2002.

Vogel, Steven. *Against Nature*. Albany, NY: SUNY Press, 1996.

Warren, Karen J. *Ecofeminist Philosophy*. Lanham, MD: Rowman & Littlefield, 2000.

*Worster, Donald. *Nature's Economy: A History of Ecological Ideas*. New York: Cambridge University Press, 1994.

Zimmerman, Michael E., J. Baird Callicott, John Clark, Karen J. Warren, and Irene Klaver. *Environmental Philosophy: From Animal Rights to Radical Ecology*. 4th ed. Upper Saddle River, NJ: Prentice Hall, 2004.

ETHNOMETHODOLOGY

Atran, Scott, and Douglas Medin. *The Native Mind and the Cultural Construction of Nature*. Cambridge, MA: MIT Press, 2008.

*de Waal, Frans B. M., and Peter L.Tyack. *Animal Social Complexity: Intelligence, Culture, and Individualized Societies*. Cambridge, MA: Harvard University Press, 2003.

Elgin, Duane. *Awakening Earth: Exploring the Evolution of Human Consciousness*. New York: William Morrow and Company, 1993.

Ellen, Roy F., Peter Parkes, and Alan Bicker, eds. *Indigenous Environmental Knowledge and Its Transformations: Critical Anthropological Perspectives*. New York: Routledge, 2000.

*Kempton, Willet, James S. Boster, and Jennifer A. Hartley. *Environmental Values in American Culture*. Cambridge, MA: MIT Press, 1995.

Krech, Shepard III. *The Ecological Indian: Myth and History*. New York: W. W. Norton, 1999.

Maffi, Luisa, ed. *On Biocultural Diversity: Linking Language, Knowledge, and the Environment*. Washington, DC: Smithsonian Institution Press, 2001.

Nash, Roderick. *Wilderness and the American Mind*. New Haven, CT: Yale University Press, 1982 (1967).

*Redman, Charles L. *Human Impact on Ancient Environments*. Tucson: University of Arizona Press, 1999.

Autopoiesis

Dukas, Reuven. *Cognitive Ecology: The Evolutionary Ecology of Information Processing and Decision Making*. Chicago: University of Chicago Press, 1998.

*Maturana, Humberto R., and Francesco J. Varela. *Autopoiesis and Cognition: The Realization of the Living*. Boston: Reidel, 1991.

Thompson, Evan. *Mind in Life: Biology, Phenomenology, and the Science of Mind*. Cambridge, MA: Belknap Press, 2007.

Empiricism

*Bekoff, Marc. *Animal Passions and Beastly Virtues: Reflections on Redecorating Nature*. Philadelphia: Temple University Press, 2006.

Brower, Michael, and Warren Leon, eds. *The Consumer's Guide to Effective Environmental Choices: Practical Advice from the Union of Concerned Scientists*. New York: Three Rivers Press, 1999.

Brown, Tom Jr., with B. Morgan. *Tom Brown's Guide to Nature Observation and Tracking*. New York: Berkley Books, 1983.

Buchanan, Brett. *Onto-Ethologies: The Animal Environments of Uexküll, Heidegger, Merleau-Ponty, and Deleuze*. Albany, NY: SUNY, 2008.

Dusenbery, David. *Sensory Ecology: How Organisms Acquire and Respond to Information*. New York: Freeman, 1992.

*McCully, Michael, ed. *Life Support: The Environment and Human Health*. Cambridge, MA: MIT Press, 2002.

Merkel, Jim. *Radical Simplicity: Small Footprints on a Finite Earth*. Gabriola Island, BC: New Society Publishers, 2003.

Social Autopoeisis

Brier, Søren. *Cybersemiotics*. Toronto: University of Toronto Press, 2008.

*Luhmann, Niklas. *Ecological Communication*. Translated by J. Bednarz. Chicago: University of Chicago Press, 1989.

Systems Theory

*Ahl, Valerie, and T. F. H. Allen. *Hierarchy Theory: A Vision, Vocabulary and Epistemology*. New York: Columbia University Press, 1996.

Alexander, Charles. *The Nature of Order: An Essay on the Art of Building and the Nature of the Universe.* Vols. 1–4. Berkeley: Center for Environmental Structure, 2002.

Bartholomew, Alice. *Hidden Nature: The Startling Insights of Victor Schauberger.* Edinburgh: Floris Books, 2003.

Bateson, Gregory. *Steps to an Ecology of Mind.* Chicago: University of Chicago Press, 1972.

*Botkin, Daniel. *Discordant Harmonies: A New Ecology for the Twenty-first Century.* New York: Oxford University Press, 2000.

Cowan, David, and Chris Arnold. *Ley Lines and Earth Energies: A Groundbreaking Exploration of the Earth's Natural Energy and How It Affects Health*. Kempton, IL: Adventures Unlimited Press, 2003.

*Dodson, Stanley. *Ecology*. New York: Oxford University Press, 1998.

Gunderson, Lance H., and C. S. Hollings, eds. *Panarchy: Understanding Transformations in Systems of Humans and Nature.* Washington, DC: Island Press, 2001.

Hawken, Paul, Amory Lovins, and L. Hunter Lovins. *Natural Capitalism*. New York: Brown and Company, 1999.

Keller, David R., and Frank B. Gollev, eds. *The Philosophy of Ecology: From Science to Synthesis*. Athens: University of Georgia Press, 2000.

*Lomborg, Bjørn. *The Skeptical Environmentalist: Measuring the Real State of the World*. Cambridge: Cambridge University Press, 2001 (1998).

Oyama, Susan. *Evolution's Eye: A Systems View of the Biology-Culture Divide*. Durham, NC: Duke University Press, 2000.

Pogačnik, Marko. *Healing the Heart of the Earth: Restoring the Subtle Levels of Life*. Findhorn, Scotland: Findhorn Press, 1998.

*Real, Leslie A., and James H. Brown. *Foundations of Ecology: Classic Papers with Commentaries*. Chicago: University of Chicago Press, 1991.

*Salthe, Stanley. *Evolving Hierarchical Systems: Their Structure and Representation*. New York: Columbia University Press, 1985.

Schneider, Stephen H., James R. Miller, Eileen Crist, and Penelope J. Boston, eds. *Scientists Debate Gaia: The Next Century*. Cambridge, MA: MIT Press, 2004.

Sheldrake, Rupert. *The Presence of the Past: Morphic Resonance and the Habits of Nature*. New York: Viking, 1989.

Weisman, Alan. *The World Without Us*. New York: St. Martin's Press, 2007.

NOTES

Preface

1. Individuals in attendance during this first Integral Ecology meeting included Sean Esbjörn-Hargens; Michael Zimmerman; Ian Wright, an urban planner and "placemaker"; Chris Desser, an environmental lawyer; and Gus diZerega, a well-known neo-pagan and political scientist. Sean and Michael are currently codirectors of this center. See www.integralecology.org.

2. Wilber, *A Theory of Everything*, 97–99. We will be capitalizing "Integral Ecology" to indicate an explicit use of Integral Theory and to distinguish it from "integral ecology" (i.e., other integrative approaches to ecology that do not use Integral Theory or the AQAL model as their orienting framework).

3. In the spring of 2001, the first course in Integral Ecology was offered. The course took place at Tulane University in New Orleans and was team-taught by Michael Zimmerman and John McLachlan, director of the Tulane/Xavier Center for Bioenvironmental Research. As their class project, students in the course developed a website that presents an Integral analysis of two invasive species: nutria and Formosan termites.

During the fall of 2001 Sean traveled to Bhutan for six months to study how the Bhutanese people and their government were attempting to integrate ecological sustainability, cultural preservation, and spiritual development. As a result of his research he published an article in the *Journal of Bhutan Studies* entitled "Integral Development: Taking the 'Middle Path' Towards Gross National Happiness" that provided three case studies of Integral Ecology principles being used in Bhutan. This article contains the first published definition and explanation of Integral Ecology.

In the fall of 2003 Sean taught an online course in Integral Ecology through the California Institute of Integral Studies. Recently (summer 2006), Brian N. Tissot, whose case study appears in part 4, taught a course in Integral Ecology through Oregon State University.

4. Those in attendance included Sean Esbjörn-Hargens; Michael Zimmerman; Chris Desser; Barrett Brown, the director of the Sustainable Village; and Alex

Blais, an undergraduate environmental student from the University of Victoria, Canada.

5. Between 2002 and 2004 Sean completed his dissertation, "Integral Ecology: A Post-Metaphysical Approach to Environmental Phenomena," wherein he developed the Integral Ecology framework (i.e., the Who, How, and What: see part 2). During this same period Sean edited a special double issue of Ervin Laszlo's journal *World Futures*, which was entirely devoted to Integral Ecology. This January/March 2005 issue represents the first book-length treatment of Integral Ecology. The issue is approximately 170 pages in length and contains an overview of Integral Ecology and seven case studies of applied Integral Ecology. The contributors consisted of an environmental activist, a city planner, a researcher, a geographer, a marine ecologist, a community developer, and a philosopher and represent some of the pioneers that are exploring, articulating, and applying Integral Ecology. These case studies provide the first concrete examples of the applicability of Integral Theory to ecology and environmental issues and have informed this book in important ways. Three of these case studies are reprinted in this volume. See part 4.

Introduction

1. Leopold, *A Sand County Almanac*, 189.

2. For a helpful overview of the Great Bear Rainforest debate, see Koberstein, "Journey to the Heart of the Great Bear Rainforest."

3. As of May 2006, the package still required $30 million in federal funding. For an overview of this historic achievement, see Rainforest Solutions Project, "British Columbia's Great Bair Rainforest Review."

4. To our knowledge, the phrase "integral ecology" has been used twice independently of an association with Integral Theory. At the same time as Ken Wilber's *Sex, Ecology, Spirituality* became available in America, Leonardo Boff, a liberation theologist living in Brazil, was editing (along with Virgil Elizondo) a special issue of *Concilium: The International Journal for Theology*. In the editorial that opens the issue, Boff and Elizondo called for an "integral ecology." After highlighting a variety of approaches to "ecological reflection," including conservationism, preservationism, environmentalism, human ecology, social ecology, and mental or deep ecology, they explain:

> The quest today is increasingly for an *integral ecology* that can articulate all these aspects with a view to founding a new alliance between societies and nature, which will result in the conservation of the patrimony of the earth, socio-cosmic well-being, and the maintenance of conditions that will allow evolution to continue on the course it has now been following for some fifteen thousand years.
>
> For an integral ecology, society and culture also belong to the ecological complex. Ecology is, then, the relationship that all bodies, animate and inanimate, natural and cultural, establish and maintain among themselves and with their surroundings. In this holistic perspective, economic, political, social, military, educational, urban, agricultural and other questions are all subject to

ecological consideration. The basic question in ecology is this: to what extent do this or that science, technology, intuitional or personal activity, ideology or religion help either to support or to fracture the dynamic equilibrium that exists in the overall system? (Boff and Elizondo, *Ecology and Poverty*, ix–x)

This appears to be the first printed usage of the phrase "integral ecology." As noted in the preface, even Wilber does not explicitly use "integral ecology" in published material until five years later with the formation of Integral Institute and its Integral Ecology Center and the publication of *A Theory of Everything* (see pp. 97–99), though some of his online postings did contain this phrase a year before this book was published. Inspired by *Sex, Ecology, Spirituality*, Sean Esbjörn-Hargens had been using the phrase "integral ecology" since 1997 to describe his own research and the development of the framework presented in this book.

Not surprisingly, there is a resonance between the vision that Boff and Elizondo offered and the one associated with Integral Theory. In fact, Boff's most systematic explication of his integral ecology occurs in his *Ecologia*, published a few months before the editorial appeared in *Concilium* and appearing in English in 1997 as *Cry of the Earth, Cry of the Poor*. For additional information on Boff's ecological approach to Liberation Theology, see his *Ecology and Liberation*. In spite of Boff's initial usage, the phrase "integral ecology" does not seem to occur in any other of his publications, though his work is clearly guided by an integral sensibility. This is most readily seen in his exposé of the limits of many contemporary approaches, his discussion of inner and outer ecologies, and his appreciation for Felix Guattari's *The Three Ecologies*. In spite of the compatibility that exists between Boff's vision and Integral Theory, there are some important differences. For example, Boff's approach has no model that accounts for personal, cultural, and social development. Also, Boff has a tendency to romanticize indigenous peoples, which is honorable given that he represents Catholicism, which has a long history of brutalization toward indigenous cultures. Nevertheless, Integral Theory suggests caution lest we fall into naive presentations and forms of cultural appropriation. See Berkofer, *The White Man's Indian*; Buege, "The Ecological Noble Savage Revisited," 71–88; P. Deloria, *Playing Indian*; V. Deloria, *Red Earth, White Lies*; Dilworth, *Imagining Indians in the Southwest*; Ellingson, *The Myth of the Noble Savage*; Grande, "Beyond the Ecological Noble Savage," 307–20; Huhndorf, *Going Native*; Martin, "The American Indian as Miscast Ecologist," 137–48; Redford, "The Ecological Noble Savage," 46–48; Taylor, "Earthen Spirituality or Cultural Genocide?"183–215.

It was also around 1995 that Thomas Berry referred to his cosmological vision as "integral cosmology or integral ecology" (Drew Dellinger, personal communication, July 11, 2003). The outdoor educator Michael Cohen has also used "integrated ecology" to describe his approach to wilderness therapy or nature counseling. See Cohen, "Integrated Ecology." So it seems that 1995 marks the birth of Integral Ecology. It was the year that *Sex, Ecology, Spirituality*—a foundational text for the Integral Approach to the environment—was published. In that same year the first published usage of the phrase "integral ecology" occurred. And during this period Thomas Berry, an important integral ecological writer and thinker, referred to his own work as "integral ecology." It is perhaps an indication that an idea's time has come when it appears in different contexts, each independent of the other. Perhaps this also serves as a reminder that such an approach need not be contained within any single framework. While this book presents one approach to integral ecology, we understand that multiple—even contradictory—approaches will emerge.

5. Riddell, "Evolving Approaches to Conservation," 73.

6. Ibid.

7. Ibid., 71.

8. Wilber, "Foreword to *Integral Medicine*," xii–xiii

9. The Integral Model has been applied to art (Rentschler, "Introducing Integral Art" and "Understanding Integral Art"); criminology (Gibbs, Giever, and Pober, "Criminology and the Eye of the Spirit"); education (Astin, "Conceptualizing Service-Learning Research," and Esbjörn-Hargens, "Integral Education by Design" and "Integral Teacher, Integral Students, Integral Classroom"); environmental philosophy (Zimmerman, "Ken Wilber's Critique of Ecological Spirituality"); future studies (Slaughter, "Knowledge Creation, Futures Methodologies and the Integral Agenda"); intersubjectivity (Hargens, "Intersubjective Musings"); medicine (Astin and Astin, "An Integral Approach to Medicine"); music therapy (Bonde, "Steps toward a Meta-Theory of Music Therapy"); politics (Wilpert, "Integral Politics"); psychology (Gordon and Esbjörn-Hargens, "Are We Having Fun Yet?"); psychopharmacology (Ingersol and Rak, *Psychopharmacology for Helping Professionals*); psychotherapy and counseling (Marquis and Warren, "Integral Counseling"; Mahoney and Marquis, "Integral Constructivism and Dynamic Systems in Psychoanalytic Processes"; and Marquis and Wilber, "Unification beyond Eclecticism"); research (Esbjörn-Hargens, "Integral Research"; Esbjörn-Hargens and Wilber, "Towards a Comprehensive Integration of Science and Religion"); and sustainable development (Hochachka, *Developing Sustainability, Developing the Self*). As evidenced by these examples, Integral Theory has a wide range of applicability across divergent fields of inquiry. For additional examples of the Integral Model applied, consult *Journal of Integral Theory and Practice* and Integral Institute (www.integralinstitute.org), where over 50 areas of discourse (e.g., Integral art, Integral medicine, Integral science, and Integral religious studies) are devoted to exploring Integral approaches in their respective disciplines.

10. According to Stan Salthe, a theoretical biologist with strong interests in developmental phenomena and thermodynamics, the quadrants each may be understood as embodying a different tense: interobjective involves the global present tense; subjective involves the present progressive tense; intersubjective the future progressive tense; and the objective involves past tense. An important element of this temporal scheme is that because of the preeminence of natural science, the present tense dominates almost all descriptive discourse. In their subjective aspect, however, humans are focused rarely on the present, but about what's coming next. The dominance of present-tense (interobjective) discourse eclipses other tenses, thereby concealing or putting out of play domains included in the other quadrants. See Salthe, *Evolving Hierarchical Systems*.

11. Technically speaking, 2nd-person is 1st-person plural. Thus, we are referring to 1st-person plural whenever we use 2nd-person.

12. Edwards, "Through AQAL Eyes Part 2."

13. The ecophilosopher J. Baird Callicott has promoted Leopold's insights and the value of his work for contemporary environmental studies, especially Leopold's notion of the Land Ethic. See Callicott, *Beyond the Land Ethic* and *In Defense of the Land Ethic*. He has also helped make many of Leopold's writings available. See Callicott and Freyfogle, *For the Health of the Land*; Flader and Callicott, *The River of the Mother of God*; and Callicott, *Companion to a Sand County Almanac*. Two worthwhile biographies of Leopold include Meine, *Aldo Leopold: His Life and Work*, and Lorbiecki, *Aldo Leopold: A Fierce Green Fire*.

14. In addition to Leopold, there have been a number of other important figures who have attempted more inclusive approaches to ecology. In the 1970s, anthropologist and cyberneticist Gregory Bateson presented his integrative approach to ecology in two now-classic books: *Steps to an Ecology of Mind* and *Mind and Nature*. His "ecology of mind" was one of the first ecological attempts to bring together the three value spheres of subjectivity (e.g., psychology, aesthetics, and epistemology), intersubjectivity (e.g., culture, religion, and anthropology), and (inter)objectivity (e.g., ecology, biology, and cybernetics). He made important steps toward an integral ecology. However, in his overreliance on network patterns, systems analysis, and communication dynamics, he often commits subtle reductionism (reducing interior realities to complex exterior correlates), although he was an adamant opponent of gross reductionism (reducing everything to matter).

In 1988 Thomas Berry published his book *The Dream of the Earth*, which clearly articulates an integrative approach to ecology. While he does not explicitly use the phrase "integral ecology," he uses the term "integral" in a variety of ways throughout the book. In this book, Berry discusses the evolution of consciousness, the development of cultures, and the three value spheres of subjectivity, objectivity, and intersubjectivity. It is stunning the degree to which Berry foreshadows what we present here as Integral Ecology. We think the main difference between Berry's vision and ours is that Integral Theory provides a much more sophisticated framework for operationalizing his vision (which is no small task).

There are other differences. Berry, like Boff, tends to highlight the importance of indigenous peoples without contextualizing their contributions. We believe that many people have projected onto and appropriated indigenous people to a great extent for their own ecological goals and have often romanticized their relationship to nature. There is currently a good deal of evidence that highlights how indigenous people throughout history have continually damaged and impacted their natural surroundings. See Boyd, *Indians, Fire, and the Land in the Pacific Northwest*; Broswimmer, *Ecocide*; Butlin and Roberts, *Ecological Relations in Historical Times*; Butzer, "The Americas Before and After 1492," 345–68; Cronon, *Changes in the Land*; Crumley, *Historical Ecology*; Denevan, "The Pristine Myth," 369–85; Doolittle, *Cultivated Landscapes of Native North America*; Flannery, *The Future Eaters* and *The Eternal Frontier*; Hughes, *Ecology in Ancient Civilizations* and

Pan's Travail; Isenberg, *The Destruction of the Bison*; Jackson and Jackson, *Environmental Science*; Jacobsen and Firor, *Human Impact on the Environment*; Kay and Simmons, *Wilderness and Political Ecology*; Stewart, *Forgotten Fires*; Krech, *The Ecological Indian*; Lentz, *Imperfect Balance*; Mann, *1491*; Martin and Klein, *Quaternary Extinctions*; Ponting, *A Green History of the World*; Redman, *Human Impact on Ancient* Environments; Vale, *Fire, Native Peoples, and the Natural Landscape*; Vecsey and Venables, *American Indian Environments*; White, *Land Use, Environment, and Social Change*.

In 1990, David Kealey published his 1987 dissertation from State University of New York at Stony Brook as *Revisioning Environmental Ethics*, which draws on Sri Aurobindo's Integral Yoga and Jean Gebser's integral-aperspectival structure of consciousness to propose an "integral environmental ethic." Using Jean Gebser's framework of the evolution of consciousness, he analyzes utilitarian and deontological ethics, Deep Ecology, and the eco-philosophy of Henryk Skolimowski. While Kealey never used the term "integral ecology," his approach clearly contains some aspects of the Integral Ecology we set forth here. We particularly admire his contribution in that he is one of the few voices that connects a vertical or developmental perspective with environmental approaches.

In 1993, the French thinker Edgar Morin published, with Anne Brigitte Kern, *Terre-patrie*, which was published in English as *Homeland Earth* in 1998. Morin has promoted "transdisciplinary thought" and "complex thinking," which unite biological/ecological, cultural/social, and psychological understandings. This complex approach not only recognizes and honors, in its own ways, the psychological, cultural, social, and biological dimensions but also recognizes the role of levels of complexity. Morin advances important critiques of systems theory, which allows him to avoid the subtle reductionism that assails most systems approaches. Through his notion of recursivity he discusses the mutually creating domains of psychology, culture, behavior, and systems. As is common in French thought, Morin emphasizes the social-political aspects of ecology, but not at the expense of other facets.

In alignment with Integral Ecology, Morin also emphasizes worldcentric developmental capacities as necessary for resolving contemporary problems. He calls for a level of epistemological reflection that examines the "dialectic of physical influences, cultural practices, and subjective interpretations" (Whiteside, *Divided Natures*, 108); and Morin avoids tendencies associated with system theory to prescribe web-of-life solutions (e.g., "Live like nature and it will be okay"). Morin acknowledges that our very concepts of nature are situated in a perspective. Morin does not eulogize primitive cultures, even though he holds an important place for their contributions. He is a sophisticated critic of both scientific reductionism and holism and has been since the 1980s. It is quite unfortunate that due to a lack of English translations (e.g., of his *La méthode II: La vie de la vie*, his study of the philosophy of ecology and biology—although a single chapter from this important volume has been translated; see Morin, "RE: From Prefix to Paradigm"), Morin's potential contribution to an integral ecology has been less than it could have been in the English-speaking world. Recently, Sean Kelly has written an article, "Integral Ecology and the Paradigm of Complexity," that makes an explicit link between Morin and an integral ecology. However, from our perspective, Kelly does not satisfactorily explain how the earth is an "individual-subject." Nevertheless, this remains an important article for the emerging field. In *Integral Spirituality*, Wilber acknowledges Morin's contribution, while maintaining that it remains too tied to the 3rd-person perspective at work in systems theory due to his lack of including 1st- and 2nd-person methodologies on their own terms in his writings. In other words, Wilber faults Morin for not using injunctive language in his discussion of 1st- and 2nd-person phenomena. While Morin's complex thinking obviously does honor subjectivity and intersubjectivity as foundational to biological life, he does appear to lead with the complexity sciences as his primary interpretive framework.

15. Leopold, *A Sand County Almanac and Sketches Here and There*, 129–30.

16. Confusing an "is" with an "ought" is sometimes known as the naturalistic fallacy. However, the naturalistic fallacy as formulated by G. E. Moore (often referred to as "the open question argument") in his *Principia Ethica* is importantly different from the is-ought problem associated with David Hume's *Treatise of Human Nature*. In general, both formulations are highlighting the difficulties with deriving ethical positions from natural facts. However, it is worth noting that claiming that "just because something is the case doesn't mean that it ought to be the case" is actually considered a misrepresentation of Hume's position, which was meant to challenge the derivation of *any* prescriptive position from any descriptive account. Much of natural science and philosophy has called for a complete severing of the "is" from the "ought"; this move is affectionately referred to as "Hume's guillotine." While we affirm the importance of both Moore's open question argument and Hume's is-ought problem in helping to avoid unjustified and simplistic linkages between the natural and moral spheres, we also recognize that these two spheres co-arise and are interrelated in numerous complex ways. For instance, the fact-value distinction of modern empirical science has been demonstrated to be deeply problematic by postmodern constructivist science: there are no value-free facts. Nevertheless, we are aware of the long history of ecologists and environmentalists appealing, in problematic ways, to natural facts for their moral and political positions. See in particular Daston and Vidal, *The Moral Authority of Nature*, and Nolt, "The Move from *Good* to *Ought* in Environmental Ethics." For an innovative exploration of the relationships between "must/is," "should/is," and "ought/is," see Weissman, *The Cage*.

17. Some readers may wonder why this book is not called *Integral Environmentalism*. That title has much to recommend it, because we hope the book will have a broad and lasting effect on how environmentalists conceive of their aims and of themselves. Nevertheless, we decided on *Integral Ecology*, for several reasons. First, we seek a wider audience than the community of environmental activists, although clearly we hope that our book attracts many readers from that community. Second, the term "environment" is often taken to refer to the (primarily material) phenomena that surround us. Integral Ecology, however, affirms that humans—as animal organisms—are in important respects members of ecosystems, not separate from them. That is, nature is not something "out there" that surrounds us. Third, Integral Ecology takes into account not only ecosystems and their constituent organisms,

but also 1st-person experience of and 2nd-person normative attitudes toward such phenomena, as well as social and economic approaches to them. Fourth, Integral Ecology refers to a scientific research program that aims to understand the relationship between material environment and organisms. (In this book, we *never* use "ecology" to refer to the phenomena being investigated, for example, the "ecology" of termites. This is akin to saying that "biology" *is* flora and fauna, rather than the scientific methods used by people to study flora and fauna.) Such inquiry is also a feature of Integral Ecology, which frequently challenges taken-for-granted attitudes toward nature, whether those attitudes are held by environmentalists or industrialists, modernists or postmodernists.

As we shall see, in fact, many environmentalists adhere to the same one-dimensional understanding of terrestrial nature that corporate executives do. For both, nature is all appearances, all externality, and all interlocking systems, which can be observed as objects. Nowhere is there room for or inclusion of interiority, depth, or significance. Without acknowledging such depth in nonhuman beings, we make it all too easy to regard them as lacking in any worth of their own, and hence as nothing but raw material for our purposes.

18. See chap. 8 for a discussion of the value and importance of anthropomorphism.

19. Anthropomorphism can be viewed as a form of anthropocentrism in that one is ascribing or projecting human capacities onto animals. Thus, animals are not understood in their radical otherness but are made to be similar to our human selves or interpreted exclusively on human terms.

20. For a discussion of these three types of moral value, see Wilber, *Sex, Ecology, Spirituality*, 2nd ed., 545–47.

21. Leopold, *A Sand County Almanac and Sketches Here and There*, 110. Thanks to Gus DiZerega for reminding us of this passage.

22. Panpsychism is the name given to the idea that the capacity for experience, however meager, is a basic feature of the universe. In recent times, Alfred North Whitehead was one of the most important exponents of this concept. See Birch and Cobb, *The Liberation of Life*. For a sophisticated defense of a version of panpsychism, see Chalmers, *The Conscious Mind*. Recently, Galen Strawson has argued that "physicalism entails panpsychism" (Freeman, *Consciousness and Its Place in Nature*). For a comprehensive survey and a rich philosophical treatment of panpsychism, we highly recommend Skrbina, *Panpsychism in the West*. Also see Mathews, *For Love of Matter*. Mathews's work has much to recommend it; however, she tends to polarize panpsychist and materialist positions, and her emphasis on dialogical encounter comes at the expense of an adequate understanding of the vertical dimension of depth. In fact, she misinterprets Wilber on this point in a footnote when she relies on his *The Holographic Paradigm* to claim that Wilber's nonduality is "an ethos of transcendence rather than encounter" (Mathews, *For Love of Matter*, 201). In response to the position she assigns Wilber, she explains that "the panpsychist refuses to give primacy to identification with the One, and, while indeed seeking solace in unitivity, seeks life via encounter. The erotic path, as here characterized,

is exemplified less by the great meditational traditions than by indigenous cultures, which Wilber assigns to a lower rung in the evolutionary hierarchies posited by him" (ibid.). Not only does Mathews badly misinterpret Wilber's characterization of indigenous culture, but she appears at many places in her text to be favoring Descent traditions of encounter over Ascent ones of transcendence, as opposed to, as Wilber does, including both. It is worth quoting at length Wilber's own position on panpsychism:

> This part of the [mind-body] solution (every exterior has an interior) would appear to involve some sort of panpsychism, except that, as explained in *Sex, Ecology, Spirituality*, 2nd ed. (notes 13 and 25 for chap. 4), every major form of panpsychism equates "interiors" with a *particular type of interior* (such as feelings, awareness, soul, etc.) and then attempts to push *that* type all the way down to the fundamental units of the universe (quarks, atoms, strings, or some such), which I believe is unworkable. For me, consciousness in the broad sense is ultimately unqualifiable (Emptiness), and thus, although interiors go all the way down, no *type* of interior does. I am a pan-interiorist, not a pan-experientialist, pan-mentalist, pan-feelingist, or pan-soulist. The *forms* of the interior show developmental unfolding: from fuzzy something-or-other (see below) to prehension to sensation to perception to impulse to image to concept to rules to rationality and so forth, but none of those go all the way down in one specific form. Most schools of panpsychism take *one* of those interiors—such as feeling or soul—and maintain that all entities possess it (atoms have feelings, cells have a soul), and this I categorically reject. Cells have an interior, whose form is protoplasmic irritability (fig. 5), and electrons, according to quantum mechanics, possess a "propensity to existence," but none of those are "minds" or "feelings" or "souls," but rather are merely some very early forms of interiors. (Wilber, *Integral Psychology*, 276–77)

We return to the important and controversial issue of animal and plant subjectivity in more detail in chaps. 6 and 8.

23. See Vogel, *Against Nature*, and "Nature as Origin and Difference," 169–81.

24. On the issue of outstripping environmental resources, see Catton, *Overshoot*. See the website http://dieoff.org for an extensive discussion and links to other discussions about short- and long-term problems associated with limits to the carrying capacity of the planet.

25. See Diamond, *Collapse*.

Chapter 1. The Return of Interiority

1. Wilber, "A Brief History of Everything," in *The Collected Works of Ken Wilber*, 7:97.

2. Hill, "Complexity 'Humbles' Environmental Chiefs."

3. Wilson, *Consilience*.

4. Wilber, *A Brief History of Everything*, 258.

5. Ibid.

6. Ibid., 260.

7. Ibid., 123. Many authors and theorists naturally and intuitively recognize these three irreducible domains (also see note 3, chap. 6). For example, the environmental Buddhist poet Gary Snyder divides his book *A Place in Space* into three main sections: Ethics (We), Aesthetics (I), and Watersheds (It/s). Likewise Mark Sagoff, in his essay "Has Nature a Good of Its Own?" presents three major kinds of value that humans find in the environment: aesthetic (I), moral (We), and instrumental (It/s). Also, in *Out of the Labyrinth* Carl Frankel identifies and builds his work around three domains in his integral approach to sustainability: the objective domain (It/s), the social domain (We), and the depth domain (I). A cautionary note: while Frankel's book acknowledges and is influenced by Wilber's Integral Model, he takes many liberties and as a result often ends up confusing aspects of quadrants and levels in his "triad."

8. Ibid., 275. "Kosmos" is a Pythagorean term that refers to the entire Great Chain of Being (matter, body, mind, and Spirit). Integral Theory uses it to mean the subtotal of all dimensions and perspectives of the universe.

9. Ibid., 128.

10. Ibid., 129. Many contemporary modern environmentalists speak of nature as nothing more than a totality of material phenomena interacting in complex systems. Hence, when environmentalists claim that humanity must live within the great "web of life," they repeat an 18th-century idea that is crucial to the industrial ontology of which they are otherwise so critical.

11. Greene, *Debating Darwin*, 10.

12. Habermas, *The Philosophical Discourse of Modernity*, 245. According to Habermas, drawing on Frankfurt School thinkers such as Adorno and Horkheimer, scientific rationality had as its aim the laudable goal of emancipation and production of plenty for all, but because such rationality disguised its own power drive, Enlightenment rationality became what Wilber calls "a great monological systems net" that "descended on citizens for their own 'benefit and welfare'" (Wilber, *Sex, Ecology, Spirituality*, 464). The *dark side* of Enlightenment modernity is constituted not by the mere study of the objective features of human beings, but instead by the *reduction* of human beings to those features. Rationality itself became captured by objectifying monological, positivistic modes of reasoning, thereby excluding dialogical and intersubjective modes. The hyperrational ego alienated itself from others, from its own emotions, and from the natural environment, ending up in a dangerous kind of dissociation that lent itself to attempts to "control" nature on such a gigantic scale that they appear from hindsight to be almost literally mad. Habermas summarizes Horkheimer and Adorno's assessment: "The permanent sign of enlightenment is domination over an *objectified* external nature [our Nature] and a *repressed* internal nature [our Left-Hand quadrants of nature]. Reason itself destroys the humanity it first made possible" (Habermas, *The Philosophical Discourse of Modernity*, 110).

13. Habermas, "The Critique of Reason as an Unmasking of the Human Sciences: Michel Foucault," 55.

14. For an examination of Heidegger's approach to this issue, see Zimmerman, *Heidegger's Confrontation with Modernity*. Horkheimer and Adorno's approach is at work in their *Dialectic of Enlightenment*.

15. Wilber, *A Brief History of Everything*, 284.

16. For a thorough discussion of these three kinds of "nature," see chaps. 12 and 13 in Wilber, *Sex, Ecology, Spirituality*, 2nd ed. Note that in other places, such as in *A Brief History of Everything* (pp. 286–89), Wilber only uses two terms: nature and Nature, with the latter representing NATURE. Obviously this can be confusing.

17. For an extended discussion of the relationship between nature and culture, see chap. 9.

18. This is not to suggest there are not rational or postrational environmentalists.

19. Soper also identifies three main different uses of "nature." For her there is an *environmental nature*: "an order opposed to that of humanity" (*What Is Nature?* 9) (i.e., exterior sensory nature), *a cosmological nature*: "a totality which comprises both non-human and human orders" (ibid.) (i.e., NATURE), and a *human nature*: the essence of humanity (i.e., aspects of interior feeling nature), which can be understood as either sameness with (e.g., instincts) or difference from (e.g., rationality) animality. She highlights how environmentalists and ecologists often confuse these uses when they are discussing issues (e.g., ibid., p. 156). Similarly, in chap. 8 of *Evolution's Eye*, Oyama distinguishes between internal "small-*n* nature" associated with human nature and external "big-*N* Nature" associated with biological nature. Also see Hull, *Infinite Nature*, for 12 different uses of "nature," which he identifies as a foundation for environmental pluralism. C. S. Lewis provides a fascinating 50-page etymological exploration of "nature" in *Studies in Words* (chap. 2, 24–74).

20. Wilber, *Sex, Ecology, Spirituality*, 494.

21. Ibid., 445.

22. Three important texts on human interiority and the environment include: Clayton and Opotow, *Identity and the Natural Environment*; Winter and Koger, *The Psychology of Environmental Problems*; and Milton, *Loving Nature*. For a sophisticated and heartfelt approach to animal interiority we highly recommend the work of Marc Bekoff (e.g., *Minding Animals*). Bekoff is a pioneer in both cognitive ethology and emotional ethology. His work is discussed at length in chap. 8.

23. Wilber, *A Brief History of Everything*, 273–74.

24. Ibid., 275.

25. Ibid., 287.

26. Ibid. Italics in original.

27. Ibid.

28. For an overview of the intersections between Romanticism and ecology, see McKusick, "Ecology," and with environmentalism, see Morton, "Environmentalism." Both of these chapters come from Roe, *Romanticism*, which is a current examination of the Romantics and their influence.

29. Wilber, *Sex, Ecology, Spirituality*, 470.

30. Ibid., 495.

31. Ibid., 489. This distinction is also the difference between pantheism (all *is* God/Goddess) and panentheism (all is *in* God/Goddess). For a comprehensive overview of panentheism see John Cooper's *Panentheism: The Other God of the Philosophers.* For a contemporary anthology of panentheistic thought see Phillip Clayton and Arthur Peacocke's *In Whom We Live and Move and Have Our Being.*

32. Quoted ibid., 490.

33. Wilber, *A Brief History of Everything*, 288.

34. Ibid.

35. In the last decade the fields of historical ecology, environmental history, and environmental anthropology have done much to dismantle naive understandings of prehistoric and indigenous people's relationships to their environment. In a chapter entitled "Lessons from a Prehistoric 'Eden,'" Charles Redman summarizes the findings of archeology: "The archaeological record encodes *hundreds* of situations in which societies were able to develop long-term sustainable relationships with their environments, and *thousands* of situations in which the relationships were short-lived and mutually destructive. The archaeological record is 'strewn with the wrecks' of communities that obviously had not learned to cope with their environment in a sustainable manner or had found a sustainable path, but veered from it only to face self-destruction" (Redman, *Human Impact on Ancient Environments*, 4–5, italics added). Also see Carolyn Merchant's insightful *Reinventing Eden*, which traces the Western origins of the desire for a pristine past and argues that the desire for an Edenic past has actually done more harm than good to the environment. Likewise, Evan Eisenberg's *The Ecology of Eden* traces our human obsession with a myth of paradise. For a scholarly overview of the ways native cultures destroyed their environments prior to European contact, see David Lentz's *Imperfect Balance.* Shepard Krech's *The Ecological Indian* and Cynthia Eller's *The Myth of Matriarchal Prehistory* both provide a scholarly challenge to many of the sacred cows of Romantic ecology (also see note 14 in the introduction).

36. In *Masculinities*, 134–37, R. W. Connell describes an instance of apparent psychological regression in the case of Bill Lindeman, a young man heavily involved in Australian counterculturalism and radical environmentalism. Adopting a philosophy emphasizing "undifferentiated wholeness," Lindeman came to experience a passive-receptive attitude toward nature and a "wonderfully clear, pure feeling" of communion with it. When speaking of his efforts, however, Lindeman's language became unstructured "with ideas, events and commentary tumbling out together." Connell comments that "if one follows Julia Kristeva's arguments that separation from the mother and the advent of Oedipal castration awareness are connected with

a particular phase in language, where subject and object are separated and propositions or judgments arise (the 'thetic' phase), Bill's shift in speech would make sense as the sign of an attempt to undo Oedipal masculinity" (p. 135). Bill's effort to reconstruct his masculinity by developing "an open, non-assertive self risks having no self at all; it courts annihilation" (p. 136). Wilber would argue that Bill Lindeman's effort to dissociate from masculine personhood and to embrace an unmediated union with nature achieved not a transpersonal mode of consciousness but a prepersonal one. A similar example can be viewed in Werner Herzog's documentary *Grizzly Man* (2005). This documentary tracks the last five of thirteen summers that a young man, Timothy Treadwell (a self-proclaimed "eco-warrior"), spent in Alaska living near and among grizzly bears. Ultimately, he and his girlfriend were killed by a bear in their campsite. Throughout much of the film one can see clearly the complex and at times bizarre mixture of narcissism and romantic "spirituality" that pervaded Timothy's life, worldview, and desire to "go back" to nature.

37. Wilber, *A Brief History of Everything*, 311.

38. Virtually all spiritual traditions acknowledge realms that transcend the material plane, even if those realms are somehow "intra-cosmic." It is difficult to name either a tribal or a world religious tradition that does *not* make reference to an unseen, nonlocatable domain or to an invisible generative matrix. Hence, environmentalists who explicitly or inadvertently propose a totally physiobiological, web-of-life ontology as the basis for their spiritual path cannot say that their path is somehow aligned with traditional paths, and certainly not with perennial wisdom. Even a number of adherents of eco-paganism acknowledge that there are ontological realms, including those in which the Goddess dwells, that transcend the material plane, even though such realms are somehow related to the material plane. Catherine L. Albanese maintains that today's Neo-Pagan Goddess expresses the transcendentalists' ambiguity: "Pushed one way, she celebrates the reality, the concreteness of matter. . . . Pushed another way, though, she tells us that matter is only a form of spirit, that it can be shifted and changed by spirit" (Albanese, *Nature Religion in America*, 178). Albanese also quotes the noted wiccan priestess Starhawk as saying: "The flesh, the material world, are not sundered from the Goddess, they are the manifestation of the divine. Union with the Goddess comes through embracing the material world" (p. 181).

39. A number of individuals have drawn connections between Nietzsche and ecology. See, for example, Acampora, "Using and Abusing Nietzsche for Environmental Ethics"; Hallman, "Nietzsche's Environmental Ethics"; Parkes, *Composing the Soul*, "Staying Loyal to the Earth," and "Nietzsche's Environmental Philosophy"; Zimmerman, "Nietzsche and Ecology."

For an excellent collection of essays relating Foucault's thought to environmental issues, see Darier, *Discourses of the Environment*. Also Jozef Keulartz develops Foucault's notion of biopower in the context of ecology in *The Struggle for Nature*.

Although ecological scientists seek to make their discipline objective (dispassionate and value-free), the study of ecological phenomena are inevitably influenced by unacknowledged personal agendas and cultural perspectives. In "Ecological Fragmentation in the Fifties," which appears in Cronon, *Uncommon Ground*, 233–55, Michael G. Barbour, professor of plant biology at UC Davis, argues that a paradigm shift occurred in ecological science in 1947–1959. At that time many ecologists moved from Frederic Clements's ecological holism or communitarianism to versions of Henry Gleason's reductionist individualism. Even the Odum brothers' nascent ecosystem ecology replaced communitarian concepts with functional ones. This shift was underdetermined by new evidence, because few papers on the topic were published during this time. According to Barbour, the worldview shift occurred because ecological scientists were influenced by cultural trends in postwar America, which were abandoning collectivist ideals in favor of individualistic ones. Ecological historian Donald Worster claims that the shift to the individualistic paradigm occurred about a decade later, but Worster—like Barbour—attributes the conceptual shift to larger cultural patterns, which involved moving away from socialism toward individualism. See Worster, "The Ecology of Order and Chaos," 39–48. Although most scientists who lived through the worldview claimed that their views were influenced solely by empirical evidence and scientific theory, Barbour regards such claims as naive, as does the noted ecologist Stan Rowe, who accuses his colleagues of "plain ignorance" if they believe "that science is detached, objective, factual, unmythic, and withal goal-setting" (in Cronon, *Uncommon Ground*, 253). According to Barbour, "American ecologists of the 1950s were as randomly individualistic and nonholistic as they claimed their vegetation to be" (ibid., 254). At the unconscious level, everyone—from ecologists to hod carriers—is "both holistic and individualistic, injecting broad cultural themes into their research and their lives" (ibid.). Personal and cultural factors may be particularly pronounced in ecological scientists, because they are often personally interested in conserving the living phenomena that they study, and because philosophical, religious, and mythic concepts (often tacitly) influence their understanding of the place of the human species in terrestrial nature. (Also see chap. 5 for a discussion of four major cultural shifts in defining ecology.)

40. Some of the most egregious misuses of ecological science occurred under the influence of extreme right- and left-wing ideologies. Such was the case with the influential (and controversial) German scientist and evolutionary theorist Ernst Haeckel, who coined the term "ecology" in 1866, seven years after Darwin published *On the Origin of Species by Means of Natural Selection, or The Preservation of Favoured Races in the Struggle for Life*. Haeckel promulgated a version of Darwin's evolutionary theory in the context of a more encompassing philosophical-religious framework, monistic pantheism, according to which all life not only had a common (material) origin, but is characterized by a measure of interiority from the very beginning. In addition to adhering to monistic pantheism, however, Haeckel also promoted a version of social Darwinism, according to which the white race must reinvigorate itself, in part through selective breeding (eugenics), to compete successfully with nonwhite races. (For a provocative presentation of the widespread embrace of eugenics within the U.S. prior to World War II, see Michael Crichton's appendix 1 in his *State of Fear*. Crichton uses this example to drive home his point that politicized science [i.e., environmental science] is dangerous. While we do not fully share Crichton's controversial stance on global climate change, we do value many of the important points he raises in such a crucial and contested area.) Like-

wise, David Theodoropoulos, a conservation biologist, has documented in *Invasive Biology* how the rhetoric and policies surrounding nativism and invasive species are driven by psychologies of racism and xenophobia. In fact, he makes the case that the key concepts and language used with invasive biology "are identical in all respects with their counterparts in National Socialist ideology" (p. 122). Mark Sagoff has also made connections between approaches to invasive species and xenophobia. For example, see Sagoff, "What's Wrong with Exotic Species?"

Haeckel's racist ideas were embodied in his Monist League and were later taken up by National Socialism. Sometimes, otherwise insightful ecological, religious, and evolutionary themes have been intertwined with other very dark political purposes. In the mid-1930s, for example, German National Socialism passed the most far-reaching environmental regulations until that time. Unfortunately, this concern for nature arose in an ideological context summed up by the very popular slogan "*Blut und Boden*," that is, [racially pure, unmixed] blood and [pure, German-populated] soil. The politics of racial and natural purity were major features of 20th-century fascism. National Socialism perverted Darwin's idea of the struggle for survival into a "natural" cosmic order, which preordained a global race war between the noble and blessed Aryan peoples and *all* the nonwhites. National Socialism condemned Judaism and Christianity as unnatural because of their proclamation of a transcendent Creator, and then developed a wholly this-world, blood-based idea of the sacred. Rejecting the otherworldly Ascent tradition so important to premodern European history, National Socialism replaced it with a demonic expression of the Descent tradition. For the Nazis, this transformation involved the triumph of the Aryan race, which was allegedly endowed with the purest blood, the conduit for nature's sacred power. Because Nazi Germany lacked modernity's commitment to worldcentrism, however, critics are wrong in claiming that National Socialism was an inevitable outcome of modernity. Instead, National Socialism was a premodern, ethnocentric, authoritarian social formation that had access to state-of-the-art technology. As the events of 9/11 demonstrated, antimodernist groups are quite capable of wielding advanced technology against modern nations that are, however imperfectly, committed to worldcentric moral and political views.

Left-wing totalitarianism, in the form of Soviet Marxism, also manifested traits of Ascent-despising, Descent-affirming modernity, with the same premodern underbelly. Militantly materialistic and programmatically atheistic, scientific socialism depicted human history from a 3rd-person perspective that included little or no place for interiority and that depicted society as developing according to the dialectical laws of history, as discovered by Marx and Engels. Every effort was made to stamp out bourgeois subjectivity and to replace it with class-consciousness, which was itself a superstructural feature of the economic base of society. This ethnocentric morality pegged it, too, as not fully modern. For a fuller discussion of the relationship between the Third Reich and ecology, see Bramwell's *Ecology in the 20th Century*, Biehl and Staudenmaier's *Ecofascism*, and the recent book by Bruggemeier, Cioc, and Zeller, *How Green Were the Nazis?* For an examination of contemporary racial dimensions of environmental ideals, see Gardell's *Gods of the Blood*.

41. Dodson, *Ecology*, 7. Italics added.

42. For some examples of ecologists who recognize the constructed nature of their activity, see the hierarchy theorists mentioned in note 46.

43. There are a number of worthwhile explorations of the intersection of postmodernism, ecology, and environmentalism. See Zimmerman, *Contesting Earth's Future;* Gare, *Postmodernism and the Environmental Crisis;* Soulé and Lease, *Reinventing Nature?;* Oelschlaeger, *Postmodern Environmental Ethics*. Two classic texts in this area are Simmons, *Interpreting Nature,* and Evernden and Leslie, *The Social Creation of Nature*. For a critique of deep ecology adhering to a modernist epistemology, see Van Wyck, *Primitives in the Wilderness*. Also noteworthy is Giblett's fascinating study of swamps, marshes, and wetlands from a postmodern perspective, *Postmodern Wetlands*.

44. For an engaging exploration of the constructivist currents in ecological discourse, we recommend Phillips, *The Truth of Ecology*. Also of interest are Golinski, *Making Natural Knowledge,* and Jamison, *The Making of Green Knowledge*.

45. Eco-philosophers Karen J. Warren and Jim Cheney have called on hierarchy theory to remind us that privileging a given perspective is often an excuse for unwarranted domination of those who do not share that perspective. See their essay "Ecosystem Ecology and Metaphysical Ecology."

We are aware that Integral Theory is one way of making sense of a very wide variety of phenomena as they arise and/or are generated and enacted within various perspectives. Let us be clear: Integral Theory does not present itself as the only possible integral theory. It is, however, one of the few contenders at a genuinely meta-paradigmatic level. We are aware that at least some of our truth claims will eventually be proven false, but we have done our best to investigate them. In addition, our perspectives as authors are influenced by what Joanne Hunt and Laura Divine of Integral Coaching Canada call "AQAL constellations" (e.g., how individual awareness and embodiment is structured by the five elements of AQAL). Although we point to certain limitations of Integral Theory, especially as it relates to ecology, this book is not a critical appraisal of Integral Theory. Such an enterprise is legitimate, but we leave it to others to undertake. We encourage you to make critical rejoinders to our Integral Ecology, in the hopes that your intention is to spread illumination and relationship, not confusion and rancor. We have made an effort to engage other authors and schools of thought with a hermeneutics of generosity as opposed to one of suspicion. Such an approach does not forgo a critique of others' views. We hope we are afforded the same generosity.

46. Much of the work toward a "unified" ecology has been done in the context of hierarchy theory, which is increasingly epistemological in regards to perspectives. See Allen and Starr, *Hierarchy;* Salthe, *Evolving Hierarchical Systems;* O'Neill et al., *A Hierarchical Concept of Ecosystems;* Allen and Hoekstra, *Toward a Unified Ecology;* Ahl and Allen, *Hierarchy Theory*. Despite its many merits, hierarchy theory tends to focus primarily on exteriors (e.g., behavioral and systemic phenomena) and to confuse individual and social holons. Hierarchy theory identifies various scales of complexity within ecosystems, scales that different perspectives take as primary. Once

these scales are acknowledged, their respective approaches can be unified into a single ecological framework. See Allen and Hoekstra, *Toward a Unified Ecology*. It is important to note, however, that scientific understanding decreases as the scale increases. For an interesting "inside-out" approach to working with multiple kinds of perspectives within ecosocial systems, see Waltner-Toews, Kay, Neudoerffer, and Gitau, "Perspective Changes Everything."

A number of graduate programs in ecology that are working toward their version of a unified ecology (e.g., University of California at Davis; Helsinki University; and University of Southampton, England) are based on integrative ecology, which allows students to combine various methods, theories, data sets, scales, and disciplines (see notes 4–6 in chap. 8). The field of ecological modeling has also pioneered efforts to identify patterns across ecological approaches in order to provide a comprehensive ecosystem theory; consult Jørgensen and Müller, *Handbook of Ecosystem Theories and Management*, and Jørgensen, *Integration of Ecosystem Theories*.

Also see Norton, *Toward Unity among Environmentalists*. While Norton's approach is more pragmatic than methodological, it does offer some insights into the consensual process. However, since its publication, its overly optimistic tone is now tempered by almost two decades of disagreement between environmentalists.

47. Dodson, *Ecology*, 4. Stanley Dodson has overseen the development of an ecology text that provides a very helpful overview of the various perspectives in ecological science; consult Dodson, *Ecology*, and its companion volume of primary source readings, Dodson, *Readings in Ecology*.

48. Ecological Society of America's Ecological Visions Committee, "Ecological Science and Sustainability for a Crowded Planet," 29.

49. Ibid., 15.

50. Tainter, Allen, and Hoekstra, "Energy Transformations and Post-Normal Science," 45. Emphases ours.

51. Gunderson and Holling, *Panarchy*.

52. Westley et al., "Why Systems of People and Nature Are Not Just Ecological Systems," in *Panarchy*, 119.

53. Ibid.

54. For Integral Ecology, this symbolic dimension constitutes the 2nd-person cultural or intersubjective domain. Symbolism makes certain activities available only to humans: "the creation of a hierarchy of abstraction"; reflexivity in meaning; envisioning alternative futures; and externalizing "symbolic constructions in technology..." (ibid., 105).

55. Other studies also come to the same conclusions, namely, that North American attitudes toward nature can be partly understood in terms of the threefold developmental levels, roughly traditional, modern, and postmodern. See Kempton, Boster, and Hartley, *Environmental Values in American Culture*, and Ray and Anderson, *The Cultural Creatives*.

56. See Kempton, Boster, and Hartley, *Environmental Values in American Culture*. Indeed, these days we hear much about "the end of environmentalism." Thomas L. Friedman argues that environmentalism can be shaken from its current malaise by political leaders who present solutions to major eco-problems in terms consistent with the three major U.S. cultural worldviews: religious (conservative), neocons, and greens (corresponding to our categories of traditional, modern, and postmodern). Friedman urges President George W. Bush to adopt major eco-friendly energy initiatives, because it is

> smart geopolitics. It's smart fiscal policy. It is smart climate policy. Most of all—it's smart politics. Even evangelicals are speaking out about our need to protect God's green earth. "The Republican Party is much greener than George Bush or Dick Cheney," remarked [Peter Schwartz, chairman of Global Business Network]. . . . Imagine if George Bush declared that he was getting rid of his limousine for an armor-plated Ford Escape hybrid, adopting a geo-green strategy and building *an alliance of neocons, evangelicals and greens to sustain it.* His popularity at home—and abroad—would soar. The country is dying to be led on this. (Friedman, "Geo-Greening by Example," our italics)

We could also put a positive spin on the "end" of environmentalism, namely, that liberal democracies have integrated environmental concerns into the political process, in a way unimaginable only 30 years ago.

57. In this respect, Integral Ecology has something in common with environmental pragmatism. See Light and Katz, *Environmental Pragmatism*.

58. Comparisons between human and nonhuman brains, and comparisons of human and nonhuman behaviors connected with respective brain states, offer grounds for ascribing some measure of interiority to virtually all members of the animal kingdom, and perhaps to many plants. Biosemiotics, which builds on the fact that signaling is a virtually universal feature of life, suggests that there are interior domains—however meager—corresponding to the signals that most scientists study strictly from the 3rd-person standpoint. See Hoffmeyer, *Signs of Meaning*; Kull, "Biosemiotics in the Twentieth Century"; Kull, "An Introduction to Phytosemiotics"; Sebeok and Umiker-Sebeok, *Biosemiotics: The Semiotic Web 1991*; Backster, *Primary Perception*; Ford, "Plants Have Senses." Also noteworthy is the emerging field of plant neurobiology—see, e.g., Baluska, Mancuso, and Volkmann, *Communication in Plants*, and Trewava, "Aspects of Plant Intelligence." For a general overview of plant intelligence and plant senses as well as the ways plants exhibit intentional behavior, learning, memory, and communication, see chap. 7 in Narby, *Intelligence in Nature*. For an integrative view of semiotics, see Emmeche, "The Biosemiotics of Emergent Properties" (see note 63 for more on biosemiotics).

Leopold's *A Sand County Almanac and Sketches Here and There* was so influential in part because it depicted wild animals not merely as objects to be hunted and studied but also as subjects having lives of their own, and an interiority that is both very different from, and in some ways similar to, our own.

59. See Chalmers, *The Conscious Mind*.

60. Information is a quadratic affair, which is why Wilber's integral semiotics (forthcoming in volume 2 of the Kosmos Trilogy) is so important. In general there are two major approaches to information in the universe and biosphere: there are those that emphasize syntax and those that emphasize semantics. Søren Brier is developing an integral approach to semiotics that combines N. Luhmann's communication theory, Charles S. Peirce's semiotics, Maturana and Varela's autopoiesis, and Husserl's phenomenology in a way that honors all 4 quadrants on their own terms. In fact, in his article "The Cybersemiotic Model of Communication," Brier provides a figure that presents "four main areas of knowledge" that develop (i.e., have levels of complexity: consciousness [UL]; life [UR]; energy [LR]; and meaning [LL]) (p. 78). See also Brier, "Luhmann Semioticized" and *Cybersemiotics*.

61. Biosemiotics views DNA as a semiotic process and refers to signification at the cellular and biochemical level as endosemiotics. See Pollack, *Signs of Life*, and Barbieri, *The Organic Codes*.

62. Saussure, *Course in General Linguistics*, 16.

63. Biosemiotics emerged out of the work of the German biologist Jakob von Uexküll (1864–1944), who studied the phenomenal, cognitive, and interpretive ("umwelt") world of animals (see chap. 2, notes 42–44). The term "biosemiotics" was first used in 1962 by the German doctor F. S. Rothschild, but it was Thomas Sebeok's increasing reference to and use of Uexküll's umwelt theory in the 1960s and 1970s that contributed the most to the development of biosemiotics. Biosemiotics is the scientific study of the way organisms interpret, communicate, and exchange information through signs. Key figures include Jakob von Uexküll, Thure von Uexküll, Jesper Hoffmeyer, Claus Emmeche, and Thomas Sebeok. Foundational publications include J.U. Uexküll, "Theory of Meaning"; Thomas Sebeok and Jean Umiker-Sebeok (eds.), *Biosemiotics: The Semiotic Web 1991*; and Jesper Hoffmeyer, *Signs of Meaning in the Universe*. Current publications include Anton Markoš, *Readers of the Book of Life;* Claus Emmeche, Kalevi Kull, and Frederik Stjernfelt, *Reading Hoffmeyer, Rethinking Biology;* and Marcello Barbieri, *The Organic Codes: An Introduction to Semantic Biology*. For a good overview of biosemiotics, see Barbieri's recent *Introduction to Biosemiotics;* Sebeok, Hoffmeyer, and Emmeche, *Biosemiotica;* and Emmeche, Hoffmeyer, and Kull, *Biosemiotics*. For an overview of 22 basic hypotheses that inform biosemiotics, see Frederik Stjernfelt, "*Tractatus Hoffmeyerensis*." For a similar list based on Jesper Hoffmeyer's writings, see Emmeche, Kull, and Stjernfelt, "A Biosemiotic Building."

In many ways, biosemiotics is an integrative endeavor, which is illustrated in its aims to overcome a number of dualisms, including subject-object, knowledge-information, culture-nature, mind-body, and the split between the humanities and the natural sciences (Kull, "On Semiosis, Umwelt, and Semiosphere," 307). For example, Wendy Wheeler uses biosemiotics to integrate human culture and nature (Wheeler, *The Whole Creature*). For a discussion of the far-reaching applications of semiotics in general and biosemiotics in particular, see Sebeok, *Global Semiotics*. In many ways, as will become more obvious as we proceed, this entire book can be seen as an Integral Approach to biosemiotics. Biosemiotics is explored in more detail in chaps. 6 and 8.

64. For articles on pansemiotics, see Emmeche, "The Biosemiotics of Emergent Properties in a Pluralist Ontology," and Luure, "Lessons from Uexkull's Antireductionism and Reductionism."

65. Emmeche, "The Biosemiotics of Emergent Properties," 91. In the same article Emmeche explains Peirce's position:

> For Peirce, there was no doubt that a "gob of protoplasm," say an amoeba or a slime-mould, *feels*, and that feeling has a substantial spatial extension which is subjective. Today very few philosophers (and virtually no biologists) dare to speak about feelings in single-cell organisms, and for higher organisms with specialized sense organs, the inner side of feelings [is] discussed only within philosophy of mind under the technical term *qualia* (for which Peirce's term *qualisign* may be more precise). Nevertheless, it is important to remember the phenomenological or "inner side" of matter and ask if its qualities are accounted for when someone claims to have succeeded in bringing physics and psychology on one footing. (Ibid., 90)

This same article provides a helpful overview of pansemiotics in relationship to a number of other positions attempting to explain the relationship between semiosis and evolution.

66. "The pansemiotic thesis may be read as a version of panpsychism: the idea that matter is effete mind, or that qualities of experience, sensation, pain or feeling come in degrees, and that even inorganic systems may have, eventually to very small degrees, such qualities" (ibid., 91).

67. For a revealing analysis of speciesism in both Christianity and Buddhism, see Waldau's extensive study, *The Specter of Speciesism.*

68. Jonas, *The Imperative of Responsibility.*

69. Thanks to Stan Salthe for pointing out the need for meta-discourses. Personal communication.

70. Wilber, Foreword to *Integral Medicine.*

Chapter 2. It's All About Perspectives

1. Johnson, *Body, Spirit and Democracy*, 203.

2. As we see later in connection with Integral Theory's notion of the Kosmic address, to specify the act of knowing we must identify the methodology used to disclose the phenomenon in question, and the developmental level from which the investigator is operating.

3. In using "paradigm," we follow Kuhn's original meaning and are referring to principles and procedures of inquiry or social practices that reveal and/or generate aspects of phenomena. "Paradigm" is roughly synonymous with methodology, exemplar, and injunction.

4. Klein, *Interdisciplinarity*, 66. For additional information on interdisciplinary and transdisciplinary approaches, see Klein, *Crossing Boundaries*; Moran, *Interdisciplinarity*; Nicolescu, *Manifesto of Transdisciplinarity*.

5. See Stein, "Modeling the Demands of Interdisciplinarity," for a integrally informed discussion of the differences between multi-, cross-, inter-, and transdisciplinarity and how they map onto increasing capacities of cognitive development and group competencies. Stein uses the prefix "meta" as opposed to "post" to describe the disciplinary nature of Integral Theory and Integral Methodological Pluralism. We view his use of "meta" as synonymous with our use of "post."

6. We do not say "from biomes on down" because we take the position that perspective-taking requires a kind of centered subjectivity or interiority that is (arguably, at least) lacking in social holons that have distributed subjectivity.

7. See note 21 of the introduction, on panpsychism. To understand atoms, as A. N. Whitehead pointed out, requires more than conceiving of them as nothing but exteriors that interact with other exteriors, like the billiard ball model of atomic interaction. Instead, atoms are characterized by interiority, however primitive, which allows them to "prehend" or consider their own 1st-person perspective—however basic—and other atoms as 2nd persons.

8. Wilber, "Excerpt D," part 1.

9. A human can encounter another organism both "from" and "in" these three perspectives, whereas, most organisms encounter each other only "in" these perspectives. For example, a dog has 3rd-person awareness (viewing objective phenomena via the five senses) but it cannot take a 3rd-person perspective (viewing your perspective as wrong). Thus, there is an important difference between *1st*, *2nd*, and *3rd person* perspectives (without hyphens) and *1st-*, *2nd-*, and *3rd-person* perspectives (with hyphens). The former refers to pronouns (*I, we, you, he, him, they, it*) and represents the basic ways that individuals can take subjective, intersubjective, and objective perspectives. The latter refers to developmental capacities to take a perspective on all the previous perspectives. In fact, the lack of or use of a hyphen between *1st*, *2nd*, or *3rd*, and *person* can be understood in at least three main ways: *grammatically* (e.g., has a perspective; though, as mentioned above, this is often represented by a lack of a hyphen), as *resonance* (e.g., has a capacity for resonating with other organisms in that perspective), and as *reflectivity* (e.g., has a capacity for reflecting on other organisms from that perspective). While a notation system that distinguishes between these different uses would add extra clarity to the text, we will allow the context to clarify the usage—generally using no hyphen for the number of individuals (e.g., "a 2nd person walked into the room") and a hyphen for the perspective on individual takes (e.g., "she took a 2nd-person perspective on the situation").

Since this is an important but complex area of the Integral Model, we would like to further unpack a few distinctions here for the advanced students of Integral Theory. A fuller discussion of "integral math" (e.g., representing a 1st-person perspective of 1st person realities as 1-p x 1p) can be found in appendix B of Wilber, "Excerpt C." Development involves the increased capacity to take perspectives. For example, an individual at the red altitude of psychological development (see chap. 4 for a discussion of development and altitude), which is associated with egocentrism (1-p) can experience feelings (1p) and mutual resonance (2p), and play a video game (3p): Red 1-p = 123p

It is not until this individual transforms to amber altitude, which is associated with ethnocentrism (2-p), that they can begin to take a 1st, 2nd, and 3rd person perspective on 1st, 2nd, and 3rd person perspectives. For example, he can take a 1st person perspective (1p) on another person's emotions (1p), their relationship (2p), and their behavior (3p): Amber 2-p = 123p x 123p.

When an individual transforms to the orange altitude, which is associated with worldcentrism (3-p), they can take a 1st, 2nd, and 3rd person perspective on 1st, 2nd, and 3rd person perspectives on 1st, 2nd, and 3rd person perspectives. For example, they can take a 2nd person perspective (2p) on another person (2p) taking a 3rd person perspective (3p): 3-p = 123p x 123p x 123p. So at the red altitude an individual would say, "I think that . . ."; at the amber altitude an individual would be more inclined to say, "What do you think about...?"; and at the orange altitude an individual would likely say, "What do you think about what I think about . . . ?" In other words, at red there is only one perspective, at amber there is a perspective on a perspective, and at orange there is a perspective on a perspective on a perspective.

Thus development is x-p perspective increasing: Red is 1-p, amber is 2-p, orange is 3-p, green is 4-p, teal or turquoise is 5-p. However, keep in mind that each altitude has a 1st-person awareness (1-p), but this awareness is characterized by the ability to take (or not to take) multiple perspectives (on perspectives). Thus my 1-p ("in") can be a 5-p ("from"). Understandably this can be confusing. In the following table, the first variable (e.g., 1-p) refers to how many perspectives you can take (i.e., the perspectives *from* which you encounter the other perspectives *in*). In other words, x-p is the perspective from which 123p occurs in.

Red	1-p = 123p
Amber	2-p = 123p x 123p
Orange	3-p = 123p x 123p x 123p
Green	4-p = 123p x 123p x 123p x 123p
Teal+	5-p = 123p x 123p x 123p x 123p x 123p

Each additional perspective is not a linear increase of just another perspective but is basically an exponential capacity to take a 1st, 2nd, and 3rd person perspective on all the previous perspectives. Thus, 3-p often is used to represent both 4th-person and 5th-person perspectives. Basically, to be able to recognize that there are 1st, 2nd, and 3rd persons is a 4th-person perspective. A 5th-person perspective is the ability to take all three perspectives on all three perspectives. To date, researchers like Jane Loevinger and Susanne Cook-Greuter have demonstrated that this process of perspective-taking goes up to a 5th-person perspective. See Wilber, "Excerpt C," 140–41 for a discussion of 1st- through 7th-person perspectives, wherein Wilber claims that Indra's Net as presented in the *Avatamsaka Sutra* only goes to a 4th-person perspective.

So imagine a group of three people sitting around having coffee. If everyone in that group can only take a 1st-person perspective, then they can basically only each say: "This is how I feel" (1p), "This is how I understand you" (2p), "This is what I see" (3p). If everyone in that group can also take a 2nd-person perspective, then they can also say: "This is how you feel" (1p), "This is how you understand him" (2p), "This is what you see" (3p). If everyone in that group can also take a 3rd-person perspective, then they can also say: "This is how you feel about his feeling that way" (1p), "This is how you understand his being understood by you" (2p), "This is what you see him seeing" (3p). In short:

A 1st person perspective is: subjectivity or direct awareness.

A 1st-person perspective is: I see the world in this way.

A 2nd person perspective is: intersubjectivity, or mutual resonance/understanding.

A 2nd-person perspective is: I see that you see the world in a particular way. At this level, I can share my perspective with you, but I cannot take your perspective—that comes next!

A 3rd person perspective is: objectivity, or sensorial awareness.

A 3rd-person perspective is: I see that you see that he sees the world in a particular way. Thus, "I can see an object" (3p) is not the same as "I can take an objective perspective" (3-p).

Now, in terms of nonhuman organisms, 1p, 2p, and 3p go all the way down (as far down as you want to push them). Likewise 1-p goes all the way down, and 2-p goes all the way down but is limited to shared depth. Thus, the 2-p between an atom and a human is quite different from the 2-p of a dog and a human. However, 3-p doesn't go all the way down. In fact, it appears to be restricted to humans and some primates. "If a wolf is signaling another wolf about prey they are hunting, that wolf is necessarily in a first-person stance to the other or second-person wolf about the prey (or third person). If a bacterium is signaling another bacterium using chemical messengers, that *already* is a first-, second-, and third-person situation" (Wilber, "Excerpt C," 133–34, italics in original).

Thus a dog, as a 1st-person (1-p), can take 1st, 2nd, and 3rd person perspectives. It perceives the world through subjective (experiences emotions), intersubjective (communicates with other dogs and its owner), and objective perspectives (sees toys, cars, and other dogs). In other words, the dog has an awareness that recognizes itself (1p), knows its owner is different from itself (2p), and knows that the Frisbee is different (3p) from either itself (1p) or its owner (2p). However, the dog can't take a 2nd-person perspective. For example, it doesn't have the capacity to think that your view on Frisbee rules is right or wrong. Nor can it take a 3rd-person perspective and think that Jane's perspective of your view on Frisbee rules is right or wrong. However, a dog can see a Frisbee, which is a 3-p object (because that is our description of it), but it can't take the 3-p perspective.

Here is another example. Imagine a dog (1-p) has a thorn in its leg. As a result it experiences pain (1p). It comes to you (2p), recognizing you as its owner. The two of you have shared depth due to your mammalian heritage (i.e., a limbic system). As a result you two can feel each other's feelings (2p). The dog then begins communicating with you by barking, whimpering, and nudging you. This communication is an exchange of objective signifiers (3-p). While the dog uses and sees 3-p objects, it can't take 3-p perspectives. A dog recognizes a 3-p but not from a 3-p (universally reflective stance) to arrive at what an object is: object perspective but not from an objective perspective.

So while a dog is a 1-p that can take a 1p, it is not self-reflective (1-p x 1-p). If I come over to the dog, at first I'm simply a 2nd person (2p). But as soon as we resonate with each other we become a 2-p (i.e., a 1st-person plural: 1-p*pl). But the dog is not taking a 2nd-person (2-p) perspective on my situation. So the dog recognizes itself as an I and you and the two of you as a we, and is talking about an it (thorn). The dog's awareness recognizes that all these objects and subjects are all different. Otherwise it wouldn't bark at me about the thorn.

In integral math this process is often represented by three variables as: quadrant x quadrivium x quadrant. In other words as: an organism's I/We/It awareness x how it is perceiving x some object (see note 26 in chap. 8).

10. On animal awareness, see Bekoff, *If You Tame Me* and *Animal Passions and Beastly Virtues;* Bekoff and Byers, *Animal Play;* Griffin, *The Question of Animal Awareness* and *Animal Minds;* and Ristau, *Cognitive Ethology*.

11. Wilber, "Excerpt D," part 1.

12. Many ecotheorists identify the development of abstract language as a major cause of our fall from ecological harmony (see chap. 9). Of course, we maintain that such thoughts are misguided. For example, David Abram's widely popular *The Spell of the Sensuous* identifies the shift from an oral language tradition to an alphabetic written one as the primary cultural break with the natural world: "Indeed, it was only then, under the slowly spreading influence of alphabetic technology, that 'language' was beginning to separate itself from the animate flux of the world" (p. 107); "Only with the emergence of the phonetic alphabet, and its appropriation by the ancient Greeks, . . . did the written images lose all evident ties to the larger field of expressive beings. Each image now came to have a strictly *human* referent: each letter was now associated purely with a gesture or sound of the human mouth. Such images could no longer function as windows opening on to a more-than-human field of powers, but solely as mirrors reflecting the human form back upon itself. . . . Only thus, with the advent and spread of phonetic writing, did the rest of nature begin to lose its voice" (p. 138). While Abram's book has much to recommend it, we differ with this position. In some important respects, to claim that the artifact of the written alphabet served to remove us from nature is analogous to claiming that an anthill separates ants from nature. Imbedded in this thinking is the very dualism that the holistic approaches are trying to overcome. Language is in fact a development within Nature. For another attempt to criminalize alphabetic literacy, see Shlain, *The Alphabet versus the Goddess*: "The decline of the Goddess began

when some clever Sumerian first pressed a sharp stick into wet clay and invented writing. The relentless spread of the alphabet two thousand years later spelled her demise. The introduction of the written word, and then the alphabet, into the social intercourse of humans initiated a fundamental change in the way newly literate cultures understood their reality. It was this dramatic change in mind-set, I propose, that was primarily responsible for fostering patriarchy" (p. 7). For us, such major shifts in society are complex tetra-occasions and defy simple causal explanations. And language, a brilliant expression of our depth, cannot be only a crime. For a sophisticated examination of the positive role that language both oral and written (i.e., maps with place names) has in relationship to the natural and social world, we highly recommend Basso, *Wisdom Sits in Places*.

At the same time, we also find both of their presentations partially valuable! For example, Abram's work shares our efforts to honor the more-than-human world and is an exemplar of eco-phenomenology.

13. In view of these remarks, we wrote this book from within a set of perspectives influenced and structured by our own AQAL constellation of quadrants, levels, lines, states, and types: the quadrants we orient to and from, our center of gravity, our various lines of development, the states of consciousness we are familiar with and have access to, and our personality types. We ask you to help us expand the perspectives, include what we omit, and correct blind spots and distortions. Also, much of this book makes assertions from the 3rd-person perspective. Acting as an Integral Ecologist, however, requires adopting not only a 3rd-person perspective but also 1st- and 2nd-person perspectives. Thus, Integral Ecology provides complex 3rd-person maps of environmental phenomena in their interior and exterior manifestations, shared 2nd-person concepts, distinctions, and terms that provide a language that individuals separated by expertise can speak in pursuit of integral solutions. Integral Ecology also provides 1st-person practices for exploring and transforming your own interior space. Practicing Integral Ecology involves acknowledging that ecosystems consist of plants and animals that also take 1st-, 2nd-, and 3rd-person perspectives. An integral ecology must explore the best practices available for taking these perspectives into account. Memorizing diagrams, no matter how complex, cannot do this. Also see note 25.

14. For a concise presentation of the five elements of the Integral Model, see chap. 1 of Wilber, *Integral Spirituality*. For an overview of the history of Ken Wilber's thought, see Frank Visser, *Ken Wilber: Thought as Passion*. Also see Brad Reynolds, *Embracing Reality*, a chapter-by-chapter summary of Wilber's major works, and *Where's Wilber At?*

15. It is worth noting, as Visser does in *Ken Wilber* (pp. 193–94), that Wilber's quadrants are foreshadowed by the economist E. F. Schumacher in his book *A Guide for the Perplexed* (1977), where he identifies four fields of knowledge: the invisible inner experiences of individuals and groups as well as their outer appearances.

16. Similarly the Integral Model can support the development of such fields as integral biology, integral botany, integral zoology, and so on. It is our hope that this book will serve as a template and inspiration for others to make other scientific endeavors integral. See Koller's three articles ("An Introduction to Integral

Science," "Architecture of an Integral Science," and "The Data and Methodologies of Integral Science") for the outlines of a general integral science.

17. Ideally, Integral Ecology should also call on people to reconstruct—using imagination or empathic resonance, which in its most radical expression could include altered states—the experience of nonhuman organisms affected by environmental problems.

18. Currently there is not a discipline solely devoted to the study of animal cultures. However, we suspect that such a field (e.g., zooethnology, zoopology, or bioethnography) will emerge in the near future. In the last 20 years there has been a good deal of groundbreaking research into the study of animal culture. Most of this research is associated with primatologists like Jane Goodall and Frans B. M. de Waal, who have demonstrated the complex cultural differences among various groups of chimps in their natural habitats. See Goodall, *The Chimpanzees of Gombe*; also consult de Waal, *The Ape and the Sushi Master*. In 1999, *Nature* ran a special issue devoted to cultures in chimpanzees. For one of the first treatments of animal culture, see Bonner, *The Evolution of Culture in Animals*. The most sophisticated overview of animal culture to date is de Waal and Tyack, *Animal Social Complexity*. This 600-page anthology presents the latest research and proof of the existence of culture in primates, elephants, dolphins, whales, hyenas, sea lions, birds, and bats.

19. Wilber, *A Brief History of Everything*, 114.

20. Wilber, "Excerpt D," 3:1.

21. Note that there is widespread disagreement as to what a species is or even whether "species" is a useful construct. See Wheeler and Meier, *Species Concepts and Phylogenetic Theory*, and Parvis, Jones, and Mace, "Extinction."

22. Levels can be construed in two distinct ways: as a general level of altitude or as a specific level of development associated with a particular line or capacity. In this book, unless otherwise noted, we will be using "level" to refer to levels of lines.

23. Another example is levels of ecological complexity in the LR. See Salthe, "Summary of the Principles of Hierarchy Theory." Salthe's work has considerable affinity with Wilber's all-quadrant model, despite certain differences. An evolutionary theorist, Salthe makes a distinction between *scalar hierarchies* of nested extensions (e.g., [ecosystem [population [organism]]]) and *specification hierarchies* of ordered intensional complexity, presented as integrative levels (e.g., { physical world { chemical world { biological world { social world { mental world }}}}}). In terms of Integral Theory, scalar hierarchies occur with inside-outside boundaries whereas specification hierarchies occur with internal-external distinctions. Also, the former deals with span and the latter with depth. Just as Wilber cautions against the conflation of individual and social holarchies, Salthe cautions against conflating scalar and specification hierarchies. For a list of 9 principles of scalar and 14 of specification hierarchies, see chap. 2 of Salthe, *Development and Evolution*. For additional discussions see Salthe, "Summary of the Principles of Hierarchy Theory," *Evolving Hierarchical Systems*; Van de Vijver, Salthe, and Delpos, *Evolutionary Systems*.

24. This 3rd-person description actually requires a 5-p perspective. See note 9 above.

25. The adage "The map is not the territory" comes from Alfred Korzybski, the founder of general semantics. He introduced this dictum in 1931 through a paper he presented at a gathering of the American Association for the Advancement of Science. It appeared in print several years later (Korzybski, *Science and Sanity*, 747–61.) In general this dictum is a postmodern critique of the representational paradigm, which often commits what is called the "myth of the given." This phrase was first used by Sellars in the 1950s (Sellars, "Empiricism and the Philosophy of Mind"). In response to the extreme social constructionist views of postmodern theorists attacking science with the myth of the given, Karl Popper retorted by exposing the "myth of the framework" (Popper, "The Myth of the Framework"). For an erudite discussion of this dualism of map and framework represented by these two myths, see Ferrer, *Revisioning Transpersonal Theory*. These modern and postmodern myths are important reminders that the definition of and relationship between the "map" and the "territory" is understood differently within various worldviews. Integral Theory avoids both myths by highlighting that the map and the territory constitute each other in a recursive enactive tetra-arising fashion. In other words, while it is important to keep in mind that the "map is not the territory," in some ways the map *is* the territory, and at the same time the territory *is also* the map.

26. Following Jürgen Habermas and Max Weber, Wilber notes that a hallmark of modernity is the separation of domains that are not separated in premodern societies. The Big Three correspond to Kant's three major critiques: Kant examined the domain of Its in *The Critique of Pure Reason*, the domain of We/You in *The Critique of Practical Reason*, and the domain of I (personal subjectivity) in *The Critique of Judgment*. So although Wilber uses the Big Three to simplify the model, he also uses it to refer to a philosophical discourse that began with Plato.

27. With regard to validity claims made in any of the quadrants, we can ask for three judgments: Is it true? Is it good? Is it beautiful? For instance, an equation in physics may be judged not only true but also beautiful and an instance of a good of its kind. Likewise, a hermeneutic judgment may be assessed not only as good but as beautiful and true by a community of other competent hermeneutic inquirers. Thus, there are validity claims connected with each of the 4 quadrants, and there are judgments associated with the quadrants, but these judgments can also be used to assess realities in other quadrants. For example, in the LL there is the validity claim of justness and the associated judgment of the Good: "What should we do?" But just because this judgment starts in the LL doesn't mean it is exclusively tied to it—you can employ this judgment in the context of the other quadrants: a shovel is not just (validity claim) but it can be put to good use (judgment). Thus, validity claims are quadrants and judgments are quadriviums. This distinction between validity claims and judgments is a unique contribution of Integral Theory. Also, Wilber notes that using AQAL we can extend Habermas's validity claims into nature. Because Habermas himself fails to do this, his framework has been limited to creating an environmental ethic based solely on communicative exchange between humans. "Using AQAL, we can see the validity claims (or if you prefer, their precursors) extending

all the way down into "lower forms" of nature, and thus the communicative accord reached between humans is but the tip of an *inter-holonic network* found in atoms, ants, and apes" (Wilber, "Excerpt C," 183).

28. According to Wilber, the act of interpreting something should always give rise to three interrelated questions: (1) what do we want to know? (2) what are the specific paradigms, injunctions, and methods needed to generate knowledge about what we want to know? and (3) what is the developmental level of the knower? Put briefly: *What* part of reality are we looking at? *How* is the looking being done? and *Who* is looking? We take up these three aspects of interpreting in great detail in chap. 6.

29. Wilber, "Excerpt C," 137, italics in original.

30. Ibid.

31. Ibid.

32. For example, at the amber altitude the Integral Model is typically viewed in a formulaic fashion, with one approach being right and others wrong; at the orange altitude, the Integral Model provides a complex way to categorize phenomena and identify causal relationships; at the green altitude the Integral Model supports the inclusion of many different and even contradictory approaches; at the teal altitude the Integral Model is dynamic representation of multiple realities that influence each other at various scales; at the turquoise altitude the Integral Model is a perspectival dance where epistemological and ontological realities are enacted through methodological injunctions.

33. Vision-logic, the cognitive stage necessary to support Integral awareness, is typically subdivided into early, middle, and late stages. Early vision-logic differentiates reality into relativistic systems, while middle and late vision-logic add up and integrate those perspectives into systems of systems. Vision-logic is often referred to as the first "postformal" stage of cognitive development since it is immediately beyond or "after" formal operational cognition. It is not yet "transrational," but rather the end-limits of rational thought. Vision-logic is, in a sense, the bridge between the mental and the transmental. Vision-logic enables you to adopt the perspectives offered by many different ways of disclosing a phenomenon, in order to develop a comprehensive understanding of that phenomenon. (Vision-logic includes the capacity to take five perspectives simultaneously: a 5th-person perspective. It is this capacity that allows for worldcentric and planetcentric awareness. For a fuller discussion of vision-logic, see Wilber, *Sex, Ecology, Spirituality*.

Let's back up: *Egocentrism* is the result of an exclusive identification with your own 1st-person perspective. *Ethnocentrism* is the result of an ability to submit your perspective to an exclusive identity with a group's 2nd-person perspective (i.e., your family, tribe, political affiliation). *Sociocentrism* is the increasing capacity to release your exclusive identity with your 1st- and 2nd-person perspectives and consider outside views, or a 3rd-person perspective. Consequently, you can hold your view, our view, and their view all with equal consideration. *Worldcentrism* builds on this ability to genuinely consider two different perspectives in addition to your own view and your primary group's view. This is a 4th-person perspective. *Planetcentrism* then is the increased ability to take on additional perspectives such as those held by plants and animals. The quadrants merely outline four perspectival sets from which to interpret complex phenomena. To engage in specific kinds of interpretation, you must engage in the *disclosive practices* of the specific discipline, say the practices of a biochemist, a sociologist, an environmental ethicist, or a psychologist. In one sense, integral knowledge of a phenomenon is the totality of interpretative perspectives taken on it by investigators using reliable methods. Can such methods be improved? Can communities be misguided? Certainly, but Integral Theory and Integral Ecology make no claim to infallibility. Instead we seek to make judgments based on the best available knowledge from a host of different perspectives.

34. Important methods of knowing and modes of experience, such as meditation, contemplative prayer, shamanic rituals, and numerous everyday life practices, are *not* included in this diagram!

35. Seeking to halt what Habermas called the "colonization of the life-world" by the natural sciences, some deconstructive postmodernists aligned themselves with environmentalists who likewise criticized the modern megamachine. Environmentalists soon discovered, however, that postmodernism undermined not only the foundations of Western capitalism and science, including systems science, but also the foundation for environmentalism itself, which often called upon natural science to buttress politically influential and publicly appealing claims about the integrity, harmony, and balance of nature. For postmodernists, all too many environmentalists depicted "nature" as an ultimate source, origin, or foundation, whereas in important respects nature is a social and cultural construct, not a "thing-in-itself" to which individuals and groups have privileged access. (The issue of the social/cultural construction of nature has generated a vast and growing literature. For one of the best defenses of a social constructionist approach, see Vogel, *Against Nature*. See also notes 43 and 44 in chap. 1 and note 2 in chap. 9.)

36. For an important account of how scientific truth claims are being challenged by alternative perspectives, see Tainter, Allen, and Hoekstra, "Energy Transformations and Post-Normal Science."

37. For an overview of the holon concept and how Wilber's use of it differs from that of Arthur Koestler, the originator of the term, see Edwards, "A Brief History of Holons."

38. Wilber speaks of this shift from wilber-4 AQAL to wilber-5 AQAL + post-metaphysics. We discuss post-metaphysics in chap. 6, but for now suffice it to say that by post-metaphysics Wilber means refusing to posit preexisting domains of reality, as did such great premodern thinkers as Plotinus and modern thinkers as Aurobindo, to whose thought Wilber appealed frequently in *Sex, Ecology, Spirituality*. Post-metaphysicians argue that the stages of consciousness to which the perennial philosophy refers evolved or developed over thousands of years as Kosmic habits, rather than being set in place from the beginning.

39. For a great overview of action inquiry—an approach to using 1st-, 2nd-, and 3rd-person perspectives in real time—see Torbert, *Action Inquiry*.

40. Luhmann, *Ecological Communication*, 5–6.

41. Ibid., 6l.

42. Uexküll (1864–1944) was a Baltic German biologist who developed the concept of umwelt to explore in a scientific way how organisms subjectively perceive their environment. Much of his writing and research was devoted to describing the various subjective worlds of animals. He is considered the founder of biosemiotics. See T. von Uexküll, "Introduction"; J. von Uexküll, *The Theory of Meaning* and *A Stroll through the Worlds of Animals and Men*; and Kull, *Jakob von Uexküll*.

43. Uexkull's notion of functional cycle describes how an organism enacts its umwelt or surrounding environment through cognitively registering phenomena and then responding. As a result, it has many similarities with the descriptions of structural coupling in autopoeisis theory. While the functional cycle is typically understood in autopoeitic terms (zone 5), it can be understood in structural terms (zone 2). For example, Jean Piaget recognized this twofold process in cognition. He labeled these two functions *organization* and *adaptation*. Organization refers to the many interrelationships between cognitive activities and adaptation points to an organism's interaction with the environment. In effect, organization is the inner process and adaptation is the outer process. Like Uexküll, Piaget saw these two aspects of cognition as inseparable: "They are two complementary processes of a single mechanism, the first being the internal aspect of the cycle of which adaptation constitutes the external aspect. . . . These two aspects of thought are indissociable: it is by adapting to things that thought organizes itself and it is by organizing itself that it structures things" (Miller, 2002, 64).

Piaget's language is very similar to a description of the functional cycle used by biosemioticians. The value in pointing this out is that it highlights why some uses of "umwelt" refer to Left-Hand structural-phenomenological-interpretive realities and some to Right-Hand autopoeitic-behavioral-informational realities: the functional cycle can be used to describe zone 2 structures as associated with Piaget) and zone 5 structural couplings as associated with Maturana and Varela. Adding to the confusion, Maturana and Varela refer to their autopoeitic approach as "biophenomenology," which is misleading in that they are not using "phenomenology" to refer to subjective realities associated with zone 1 or 2. Rather they are describing the cognitive or phenomenal world of the organism—the world that is enacted by what the organism is capable of registering through its senses and reacting to through its behavior—zones 5 and 6.

One of the powerful insights that autopoiesis makes is that organisms do not have a systems view as part of their overall cognition. Thus, Maturana and Varela avoid using systems theory to explain the principles of biological phenomenology. Only humans, and a select few at that, have the developmental capacity to cognize the web of life. Another important insight of autopoiesis is that an organism isn't simply a strand in the great pregiven web of life but is an active agent bringing forth and enacting its world. See Wilber, "Excerpt C," 9–17.

44. See Sharov, "Pragmatism and Umwelt-Theory." The translation of "umwelt" as both "environment" and as "an organism's subjective universe" has created confusion because biologists think of environment as something external and independent of an organism's perception of it. However, what Uexküll was highlighting in his use of "umwelt" is that an organism enacts the "outside" world. In other words, Uexküll was carefully avoiding the myth of the given by emphasizing the perspectival nature of the environment. Thus, "umwelt" is not referring to the UL phenomenology of an organism (1p: 1-p) but rather to how organisms (1p) perceive the UR objective world around themselves through their senses (3-p). In other words, the umwelt of an organism is not subjective in the sense of a subject perceiving its own embodiment (1p: 1-p) but rather in the sense of recognizing that there is a subject perceiving an object and enacting its world (1p: 3-p). Thus AQAL helps us clarify what is meant by "umwelt." But what umwelt theory does is acknowledge the subjectivity of an organism, thereby paving the way to expanding umwelt (1p: 3-p) to include subjective and intersubjective perspectives enacted by the organism (1p: 1-p, 2-p, 3-p). Thus, umwelt theory recognizes an organism's subjective (1p) universe (3-p) and is often used by biosemioticians to explore an organism's subjective (1p) social world (2-p) and sometimes used to explore an organisms subjective (1p) inner world (1-p). Integral Ecology makes explicit from the outset that an organism has quadrants (1p) and therefore has subjective 1-p, intersubjective 2-p, and (inter)objective 3-p perspectives.

Likewise, "umwelten," the plural of umwelt, generally refers to the shared or overlapping of an umwelt between two or more organisms. For an exploration of organisms and their translation between partially shared worldspaces, either intraspecies or interspecies, see Kull and Torop, "Biotranslation." The most extensive discussion of umwelt occurs in Kull, *Jakob von Uexküll*, which has over 20 articles on it, including a 150-page section devoted to umwelt. For an interesting article that links umwelt theory with the deep ecology platform, see Tonnessen, "Umwelt Ethics." For a valuable and lucid overview of four approaches to how subjects perceive objects and their environment, see Susi and Ziemke, "On the Subject of Objects." This article contains one of the best summaries of umwelt theory that we have come across.

45. Integral Theory acknowledges that hunter-gather societies also had a large impact on their natural surroundings. In fact, most organisms alter their environments and in the process produce artifacts and systems of artifacts. For example, birds create nests, but this is "natural." This is not getting "away" from nature. Organisms are constantly intervening in their environments, sometimes in detrimental ways. Thus, when environmentalists condemn human artifacts, they are often relying on a dualism that separates humans from the natural world and other organisms and depicts us as transcending nature.

46. Luhmann, *Ecological Communication*, xviii.

47. Ibid., 28–29. Italics added.

48. Ibid., 29.

49. See Wilber, "Excerpt C," 64–76.

50. For a discussion of this point and the difference between hermeneutics and ecology in the context of ecological awareness, see Wilber, "Excerpt C," 8–32. Wilber points out that one of the insights of recognizing that humans (and other organisms) are members and not simply parts of an ecosystem is that ecological awareness is not the result of ecosystems, nor can it be accounted for, let alone explained by, systems theory. "Ecological awareness is not living in accord with all sensorimotor its, but living in solidarity with all sentient I's, an awareness itself not found in any ecosystem as such" (Wilber, "Excerpt C," 31).

51. Wilber, "Excerpt C," 89. Italics in original.

52. Ibid. 176–77. Italics in original.

53. In addition, organisms are members of multiple levels of an ecosystem. Since this is such an important though complex point, it is worth quoting Wilber at length:

> Quick example: A wolf, hunting in a pack, lets out a warning call to the members of the pack. That vocal, physical vibration is part of a physical social system—in this case, the social system of communication among member wolves—and thus those particular wolves have dimensions of being-in-the-world that are both inside and internal to that specific social system of wolf hunting. Those physical sounds also fall on several surrounding trees, but have no discernible or significant impact on them, nor are they registered as communicative sounds by the trees, which are therefore not part of (i.e., not members of) the small, local, wolf-pack social system itself. However, the wolves and the trees are participating in exchanges involving biochemical life functions, vegetative physiology, cellular and molecular interactions, and so on—the wolves and trees are members of various local social systems *at those levels*, but not at the level of evolutionary complexity of vocal communication. Thus, the trees are actually *external* to several ecosystems that the wolves are members of. Both the wolves and the trees—and all sentient beings—exist in holarchical levels of ecosystems and social systems (or holarchical levels of relational exchange), based largely on the levels of evolutionary complexity of the organisms themselves (which determine the levels of the interactions with other same-depth holons). This allows us to construct holarchies based on complexity or depth, not merely on size or span. . . . (Wilber, "Excerpt D," 164; italics in original)

54. Luhmann, *Ecological Communication*, 14.

55. See chap. 11 for a discussion of climate change from an Integral perspective.

56. Luhmann, *Ecological Communication*, 14.

57. Far from being a typical systems theory reductionist, Luhmann is a sophisticated theorician familiar with traditions of inquiry into consciousness and culture. However, in the context of his social theory he focused on the inside of systems. See Brier's "Luhmann Semioticized" for an integration of experiential and cultural realities with Luhmann's work.

Chapter 3. A Developing Kosmos

1. Puhakka, "Restoring Connectedness in the Kosmos," 11.

2. See the phenomenologically driven interdisciplinary work of Stefanovic: "Interdisciplinarity and Wholeness"; *Safeguarding Our Common Future*; "Phenomenological Reflections on Ecosystem Health"; "Phenomenological Encounters with Place"; "An Integrative Framework for Sustainability: The Case of the Hamilton Harbour Ecosystem." We discuss Stefanovic's work in chap. 8.

The work of the historical ecologist Carole Crumley also recognizes the need for and works toward an integrative framework. She writes: "Although earlier deterministic, mechanistic, and dualistic characterizations have been rejected, there remains a great challenge to contemporary researchers: they must identify and employ an integrated framework, which must accommodate spatially and temporally specific natural and social scientific information and include evidence of changing values, perceptions, and awareness. Construction of this integrated framework is not yet in place" (in Balee, *Advances to Historical Ecology*, x).

We think that Integral Ecology provides this integrated framework! Ted Gragson, another historical ecologist, discusses an integrative approach to ecology: "History by comparison to ecology is an old discipline, but historical ecology emerges from the relatively recent transformation of 'old' history into 'new' history and 'reductionistic' ecology into 'integrative' ecology. . . . Historical ecology as the convergence of new history and integrative ecology is increasingly practiced across a diverse set of fields from zoology through anthropology" (Gragson, "Time in Service to Historical Ecology," 2–3).

See also the integrative work of Saunders and Myers, "The Emerging Field of Conservation Psychology." Saunders explains that conservation psychology is modeled on conservation biology in that both are "synthetic fields that mobilize contributions from other fields and subdisciplines toward conservation-related missions" (p. 139). She lists 18 distinct disciplines (ibid., fig. 1) that play a part in conservation psychology, which includes a developmental dimension.

In the same article Saunders presents a 4-quadrant chart of "one way to organize possible Conservation Psychology research areas" (p. 141). This chart is divided along the same axis of Integral Theory's 4-quadrant chart: Conservation Behaviors and Caring About/Valuing Nature at both the Individual and Group Level, which creates the following four cells: individual behaviors; collective action; personal connections to animals, places, ecosystems, etc.; and social norms and discourses. Saunders's chart also points out that these four areas can be approached from three different types of research: theoretical (developing conceptual models), applied (identifying effective strategies), and evaluative (measuring success).

3. See the section "Compound Individuals and Compound Networks" in Wilber, "Excerpt C," 61–64.

4. Lyle, *Design for Human Ecosystems*, 17.

5. Wilber explains: "The one thing I do know, and that I would like to emphasize, is that any integral theory is just that—a mere theory. . . . [All my books] are simply

maps of a territory, shadows of a reality, gray symbols dragging their bellies across the dead page, suffocated signs full of muffled sound and faded glory, signifying absolutely nothing. . . . There follows a book of maps; hopefully more comprehensive maps, but maps nonetheless. Please use them only as a reminder to take up dancing itself, to inquire into this Self of yours, this Self that holds this page and this Kosmos all in a single glance. And then express that glory in integral maps . . ." (Wilber, "Foreword to *Integral Medicine*," xiv–xv).

6. The reason some details are omitted is that the stages presented are ones that the average reader would be familiar with and as such are highlights from more detailed developmental sequences. All the Right-Hand stages are based on Eric Jantsch's diagram used by Wilber in *Sex, Ecology, Spirituality*. Thus, the stages used in the 4-quadrant diagram are generally accepted sequences of development, which have a developmental logic of transcend and include. That is, holarchy is built into space-time. The controversy is not the sequence, which is largely agreed upon; the controversy is over what is driving the process. Integral Theory posits that it is Eros: the drive toward greater complexity in the service of including more of reality.

7. In chap. 7 we discuss the various views on the ecological falls from grace. Consider the novel *Ishmael*, in which a talking ape condemns the invention of agriculture for driving humanity down a long road of decline toward alienation from and destruction of nature. The irony of this book, of course, is that the ape speaks like a human. The human capacity for complex speech enables us to form these very opinions and make the distinctions needed for new domains to arise, such as agriculture, wheels, cities, books, and computers. There is a vast literature on the considerable language capacity of nonhuman primates, especially apes and chimpanzees, and there is little evidence that primates can make the complex and disclosive distinctions associated with human speech. See, for example, Roger Fouts with Stephen Tukel Mills, *Next of Kin: My Conversations with Chimpanzees*; S. Savage-Rumbaugh, S. G. Shanker, and T. J. Taylor, *Apes, Language, and the Human Mind*; and D. L. Cheney and R. M. Seyfarth, *How Monkeys See the World*.

8. For a comprehensive overview of the history of the emergence of evolutionary thought during Darwin's lifetime, see Richards, *Darwin and the Emergence of Evolutionary Theories*. Also see Ruse, *The Darwinian Revolution*.

9. And Integral evolutionary theory involves not only sexual selection (UL) and environmental pressures (LR) but organisms interpreting their environment and one another (LL) as well as experiencing themselves (UL). So not only is it survival of the strongest (UR) and fittest (LR) but survival of the best interpreter (LL) and experiencer (UL). Biosemiotics provides a powerful critique of neo-Darwinism's Right-Hand emphasis by claiming that interpretation is the primary drive of evolution. For example, in addition to genetic fitness, Hoffmeyer discusses "semiotic fitness" or an "increasing depth of meaning," which, he explains, "results in the continuing growth of depth of interpretative patterns accessible to life" (Hoffmeyer, "The Unfolding Semiosphere," 291). Of course, many biosemioticians are exchanging one quadrant absolutism for another, favoring cultural (LL) selection over natural (UR

and LR) selection—which is why an integral evolutionary theory would include all 4 quadrants as aspects of evolutionary selection. Wilber states, "Thus, each holon must be able to register the external it-world accurately enough (*truth*); each holon must be able to register its internal I-world accurately enough (*truthfulness*); it must be able to fit with its communal or social system of its (*functional fit*); and it must be able to adequately negotiate its cultural milieu of we (*meaning*)" (Wilber, "ExcerptA," 34; italics in original).

In a recent book, *Evolution in Four Dimensions*, Eva Jablonka and Marion Lamb take a step in this direction by including four inheritance systems in their presentation of evolution: genetic, epigenetic, behavioral, and symbolic. However, their approach still is overly reliant on Right-Hand factors. Even their discussion of language and symbolic communication is largely couched in LR terms and concepts.

For a recent discussion of Jakob von Uexküll's own evolutionary position, see Kull, "Uexküll and the Post-modern Evolutionism." Kull argues that Uexküll's views on evolution are often seen as being emblematic of a premodern understanding when in fact they are more representative of a postmodern perspective alongside approaches such as autopoiesis. As such, biosemiotics is viewed as ushering in post-Darwinian biology (in contrast to the long-standing neo-Darwinian period), which naturalizes interiority by recognizing through the functional cycle that subject and object (i.e., the organism and the environment) enact and co-constitute each other. For an interesting integration of Darwinian thought and the hermeneutics of Heidegger, see Markos, Grygar, Kleisner, and Neubauer, "Towards a Darwinian Biosemiotics." And for an exploration of the role biosemiotics can play integrating Darwinian and creationist views of evolution, see Rothschild, *Creation and Evolution*. In fact, Rothschild is credited with coining the term "biosemiotics" (see Kull, "On the History of Joining *bio* with *semio*").

10. See note 61 in chap. 1. See Kull, "Biosemiotics in the Twentieth Century" and "Outlines for a Post-Darwinian Biology" for a discussion of the relationship between neo- and post-Darwinism in the context of biosemiotics.

11. See Flew and Habermas, "My Pilgrimage from Atheism to Theism."

12. At the end of the 19th century, many scientists irrationally feared the biological decline of the human species. Some social Darwinists even argued that the white, northern European race was losing its competitive advantage over the Slavic, Asian, and African races. According to these alarmists (and racists), whose ideas promoted the U.S. eugenics program that later influenced Nazi racial purification, European vitality was being drained by three factors: (1) urban industrialism, which enervated males by cutting them off from their instinctual roots and natural setting, (2) racial mixing (miscegenation, mongrelization), and (3) misplaced pity that supported and thus encouraged the survival of degenerate and otherwise deficient people. That the German biologist Ernst Haeckel, the man who first defined ecology, contributed to social Darwinism should give every environmentalist pause.

13. Schrödinger, *What Is Life?* For a wonderful tour of many influential definitions of life, see the appendix in Barbieri, *The Organic Codes*.

14. See Salthe, "An Exercise in the Natural Philosophy of Ecology," "The Natural Philosophy of Ecology," and "Natural Philosophy"; Kay, "Ecosystems as Self-Organizing Holarchic Open Systems" and "A Non-Equilibrium Thermodynamic Framework." See also the work of Robert E. Ulanowicz, including *Growth and Development* and *Ecology*. Finally, see related work on ecosystem resilience: Holling, "Resilience and Stability of Ecological Systems" and "The Resilience of Terrestrial Ecosystems," and several essays in Gunderson and Holling, *Panarchy*.

15. See Sheldrake, *Rebirth of Nature*, for an overview of the ways in which the theory of evolution has yet to be integrated into many contemporary disciplines.

16. Although Darwin's hypothesis that species descended from prior life forms flew in the face of biblical claims, the Big Bang hypothesis was in many ways consistent with the biblical creation story told in Genesis. The scientist who devised the original version of the Big Bang hypothesis in 1927 was a Belgian Catholic priest, Georges Lemaître.

17. Historical ecology draws on a Marxist analysis to explore and document the ongoing dialectical relationship within the landscape between humans and nature.

18. Although Marx's postcapitalist vision has been largely discredited, many moderns—including the classical economists whom he criticized—retain faith in some version of a progressive conception of history. Moreover, because social scientists agree that technological innovation and economic factors play the primary role in development, social scientists typically downplay or ignore the powerful ways that personal and cultural factors influence development. For overviews of green Marxism, see Benton's edited volume *The Greening of Marxism* and O'Conner's *Natural Causes: Essays in Ecological Marxism*.

19. As we've mentioned, practitioners of the physical and social sciences concluded that there was no need to consider subjective and intersubjective insights, because subjectivity and intersubjectivity were emotion-laden and thus lacked truth and value. Indeed, terms like "interior," "subjectivity," or "consciousness," associated with the eye of mind or Spirit, referred to nothing, because these did not manifest to the material eye of flesh. See Wilber, *Eye to Eye*, *The Eye of Spirit*, and *The Marriage of Sense and Soul*, for a discussion of the three eyes of knowing.

20. Recognition of the universality of a developing interior, an evolving consciousness, which we address in the next chapter, is a major step toward reenchanting the Kosmos, but such reenchantment must occur without either personal or social regression. Carl Sagan once wrote that premodern peoples lived in a "demon-haunted" world that was also repressive, violent, and intolerant. Discovery that interiority is present in the Kosmos from top to bottom, and that mind is at work from beginning to end, is not an invitation to recall the Inquisition nor an excuse to reduce science to superstitious, religiously dogmatic assertions. Instead, it should invite curiosity, wonder, and a deeper desire to understand the order of this universe.

21. For postmodern theorists who believe that German National Socialism, or something like it, is an outcome of the dialectic of Enlightenment, we maintain that Na-

tional Socialism was an example of reactionary modernism, or a premodern, tribal social organization that used technology for its militaristic and genocidal aims; see Herf, *Reactionary Modernism*. Likewise, Soviet Marxism's dream of a worker's paradise turned into a social and ecological disaster. Soviet Marxism attained and held power by devastating the bourgeoisie, a group that included not only capitalists but also the middle class and educated groups committed to worldcentric moral values. By attempting to eliminate a major stage in personal and cultural development, Soviet Marxism reverted to the preceding level of development—an authoritarian social organization. (On this topic, see Chantrill, "Ken Wilber.")

22. For additional criticism of developmental models, see Miller, *Toward a Feminist Developmental Psychology*. For a cross-cultural exploration of human development, see Gardiner and Kosmitzki, *Lives across Cultures*. For an insightful discussion of philosophical issues of developmental models within an education context, see van Haaften, *Philosophy of Development*.

23. For an interesting anthology of writings dealing with the relationship between the biosphere and the noosphere, see Samson, *The Biosphere and Noosphere Reader*.

24. Rowe, "Transcending This Poor Earth."

25. Recently, Tim Quick published a response to the debate between Zimmerman and Rowe that took place in the pages of *The Trumpeter*. See Quick, "Holons or Gestalts?" We appreciate the charitable and constructive tone of Quick's interesting article. We both got a lot out of his presentation of Arne Naess's use of the concept of gestalts. We hope that this chapter in particular and this book in general will further clarify the issues, especially around holons, that Quick is addressing and provide a more nuanced presentation of the value of Integral Theory to ecology than Ken Wilber has provided.

26. Ibid., 7. The internet text lacks page numbers. Page numbers refer to pagination of the printed document.

27. Wilber has since postulated that perspectives are even more basic than holons, but holons partly embody perspectives. In commenting on this chapter, Stan Salthe objected that holons cannot be understood in developmental terms, because holons belong to what Salthe calls the *scalar* hierarchy, in which one holon subsumes those under it, but is in turn subsumed by the holon senior to it, in Chinese-box fashion. In contrast to this synchronic, spatial containment hierarchy stands what Salthe calls the diachronic specification hierarchy, which is diachronic because it involves development of ever greater complexity over long periods of time. We ask: Can the same phenomenon, for example, an organism, be viewed in terms of two different frameworks? Viewed as a holon according to the logic of spatial containment, an organism is a "part" of the holons that contain it. Viewed as an organism according to the logic of development, the organism has attained a level of complexity (e.g., significant interiority) that cannot be understood as a "part" of a more encompassing system, especially if that system lacks the focused interiority attained by the organism. Obviously, the issues here are both important and difficult to resolve. Some important texts on hierarchy theory include Salthe, *Evolving Hierarchical Systems*;

R. V. O'Neill, D. L. DeAngelis, J. B. Waide, and T. F. H. Allen, *A Hierarchical Concept of Ecosystems*; T. F. H. Allen and V. Ahl, *Hierarchy Theory*.

28. Originally, in *Sex, Ecology, Spirituality*, Wilber referred to this self-immanence as "self-dissolution" but later changed it when he realized that "self-dissolution" was the negative expression of this capacity.

29. Wilber says this of the 20 tenets on p. xxii in the preface to the 2nd ed. of *Sex, Ecology, Spirituality*:

> Chapter 2 outlines "twenty tenets" that are common to evolving or growing systems wherever we find them. Many people counted them up and didn't get twenty, and they wanted to know if they had missed something. This simply depends on what you count as a tenet. I give twelve numbered tenets. Number 2 contains four tenets, and number 12 contains five. That's nineteen altogether. Throughout the book, I give three additions. That's twenty-two. But one or two of the tenets are not really characteristics, just simple word definitions (e.g., tenet 7 and possibly 9). So that leaves around twenty actual tenets, or actual characteristics of evolution. But there is nothing sacred about the number twenty; these are just some of the more noticeable trends, tropisms, or tendencies of evolution.

30. Koestler, *The Ghost in the Machine*. This book remains a very important contribution to many different domains of thought. For an interesting comparison of Wilber and Koestler, see Edwards, "A Brief History of Holons."

31. Rowe, "Transcending This Poor Earth," 2. We do not believe that reductionism is an issue of size, as he has argued here. Reductionism is that which denies interiority no matter the size, or tries to explain data from one injunction through the concepts of another injunction.

32. Ibid., 2.

33. Ibid.

34. Ibid.

35. Allen and Starr, *Hierarchy*, cited in Rowe, "Transcending This Poor Earth," 3. Rowe provides no page number for the quotation, which begins "prefer to . . ." and which we were unable to locate in a reasonable amount of time. Moreover, Rowe claims that Allen and Starr argue "that all hierarchies (holarchies) are congruent because their common denominator is *information*." In our reading of Allen and Starr's book we did not encounter this concept, nor is the term "information" included in the glossary or index.

36. Rowe, "Transcending This Poor Earth," 3.

37. Feibleman, "Theory of Integrative Levels." This remains a noteworthy essay.

38. Rowe, "Transcending This Poor Earth," 4.

39. Ibid., 4.

40. Ibid.

41. Ibid.

42. Some critics charge that in domains such as community, cultural cohesiveness, and individual meaningfulness, Western social systems may compare unfavorably with some premodern systems. (For an excellent discussion of these issues, see Owen, *Between Reason and History*.) These charges must be evaluated on a case-by-case basis. Consider the enormous scope of "meaningfulness" and "relationship" available to many 21st-century individuals. Moreover, many premodern societies included practices and attitudes that a very large percentage of modern people would find deeply problematic. Indeed, the very activity of comparing and evaluating many different cultures presupposes a cognitive perspective and conceptual resources available to very few individuals in premodern societies, which are unapologetically ethnocentric.

43. David Owen recently remarked that Habermas conceives the rationalization process of social evolution "as progressive changes in structures of consciousness which determine the range of possible variations a society can embody. Thus, the institutions of two empirical societies may appear significantly different, while they are both conditioned by the same deep structure of consciousness" (Owen, *Between Reason and History*, 175).

44. Rowe, "Transcending This Poor Earth," 6.

45. In commenting on an earlier version of this essay, Stanley N. Salthe maintains that there *is* a necessary pattern in virtually all forms of evolution involving dissipative structures: from exuberant immaturity, to maturity, and finally to senescence as the phenomenon (e.g., organism, ecosystem) becomes so overloaded with information that it can no longer cope with perturbations and heads toward destruction (and subsequent releasing/recycling of its components). On this universal developmental pattern, which Salthe theorizes in terms of "infodynamics," see Salthe, *Development and Evolution*.

46. As mentioned in note 23 in chap. 2, Wilber's holarchy has much in common with Stanley N. Salthe's distinction between scalar and specification hierarchy. An early version of this hierarchy is found in Salthe, *Evolving Hierarchical Systems*, 167, fig. 16. See also Salthe, *Development and Evolution*, and his essay, "Summary of the Principles of Hierarchy Theory." According to Salthe, scalar hierarchy involves parts nested within wholes, as in the case of Koestler and Wilber's individual holons from subatomic structures to the human organism. Salthe's cosmic "specification hierarchy" parallels Wilber's cosmic holarchy. *Salthe*: {material world {biological world {psychological world {human psychological world}}}}. *Wilber*: Physiosphere gives rise to biosphere gives rise to noosphere gives rise to human interiority. Salthe's specification hierarchy is based on "qualitative differences," greater "intensional complexity," and "emergence" of new properties elicited by a "final cause." Salthe suggests (personal communication) that Wilber's famous quadrant diagram tends to conflate scalar and specification hierarchies. In our view Wilber's distinction between individual and social holons helps avoid such conflation. The biosphere is a developmental classification that includes both individual and social holons. Individual holons are a nested hierarchy of parts and wholes, whereas—at least in important respects—social holons are communities constituted by members. The

worldviews of Salthe and Wilber have a great deal in common, despite certain differences.

47. Rowe gives no page number for this citation from Wilber, *A Brief History of Everything*. However, the actual quote can be found on p. 185 and reads, "And you might awake one morning and find that nature is a part of you, literally internal to your being."

48. Rowe, "Transcending This Poor Earth," 6.

49. Ibid.

50. Ibid.

51. Ibid.

52. Wilber, *Sex, Ecology, Spirituality*, 97.

53. Ibid., 383.

54. Capra, *The Web of Life*.

55. Wilber, *A Brief History of Everything*, 186.

56. For suggesting that we indicate that greater depth includes greater individuation, we thank Stan Salthe. Salthe emphasizes that emergent levels are in important respects increasingly constrained by the prior levels; hence, he is leery of our talk of transcending and including those levels. Nevertheless, he remarks: "While there are fewer individuals as we ascend a hierarchy, importantly these are increasingly *individuated*, and this might be a basis for transcendence in the [specification or developmental hierarchy]" (Salthe, personal communication).

57. Following suggestions made by Fred Kofman, Wilber describes two other kinds of wholes—heaps and artifacts—that have whole-part relations that are quite different from individual and social holons. So Rowe is incorrect in contending that Wilber bases everything on an invalid extrapolation of one particular kind of part-whole relationship, that of the organism and its parts. See Kofman, "Holons, Heaps and Artifacts." Wilber discusses Kofman's ideas in part 2 of "On Critics, Integral Institute, My Recent Writing." For a comprehensive overview of nine distinct whole-part relationships, see Morin's article "From the Concept of System to the Paradigm of Complexity."

58. For a discussion of the unique whole-part relationships of these holons as well as artifacts and heaps, see *Integral Psychology*.

59. Wilber, *Sex, Ecology, Spirituality*, 71–72.

60. Ibid., 72.

61. Ibid., 62–63.

62. Ibid., 72. In the first edition of *Sex, Ecology, Spirituality*, immediately following this sentence, we read: "Its [that is, the individual holon's] environment we will call the *social holon*."

63. Ibid.

64. Ibid.

65. Jantsch, *The Self-Organizing Universe*. Wilber provides no page reference for the quotation. Cited in Wilber, *Sex, Ecology, Spirituality*, 73.

66. Wilber, *Sex, Ecology, Spirituality*, 79.

67. Wilber, "Excerpt D," 4:2.

68. Wilber, *Sex, Ecology, Spirituality*, 73.

69. Ibid., 77–78.

70. Ibid., 90.

71. Ibid., 91.

72. Wilber, "Excerpt D," part 4.

73. For a defense of emergence as an ontological phenomenon, and thus not merely an epistemological phenomenon, see Silberstein and McGeever, "The Search for Ontological Emergence."

74. Wilber, *A Brief History of Everything*, 65.

75. Rowe, "Transcending This Poor Earth," 7.

76. Ibid.

77. David Ray Griffin, in his essay "Whitehead's Deeply Ecological Worldview," makes this point when he explains that Deep Ecologists use the term "intrinsic value" to actually refer to what he calls "inherent value" (i.e., "ecological value" or Whiteheadian extrinsic value), and then they go on to confuse extrinsic value with ground value by only emphasizing extrinsic value (what is most fundamental to ecosystems) but talking as if everything had equal value. Thus, Deep Ecologists use "intrinsic value" in the exact opposite sense than Whiteheadians and we do. Griffin explains: "That is, those species whose (individual) members have the *least intrinsic value*, such as bacteria, worms, trees, and the plankton, have the *greatest ecological value*: without them, the whole ecosystem would collapse. By contrast, those species whose members have the *greatest intrinsic value* (meaning the richest experience and thereby the most value for themselves), such as whales, dolphins, and primates, have the *least ecological value*" (pp. 202–3). Griffin then goes on to point out that inherent value (i.e., what we call ground value) includes both intrinsic and extrinsic values.

78. Rolston, *Environmental Ethics*. See also Birch and Cobb, *The Liberation of Life*.

79. As quoted in Rowe, "Transcending This Poor Earth," 7.

80. Ibid.

81. Wilber, *A Brief History of Everything*, 184.

82. Ibid.

83. Ibid., 185.

84. In *Toward a Unified Ecology*, Allen and Hoekstra argue persuasively that there are six different criteria for defining ecology: the landscape, the ecosystem, the community, the organism, the population, and the biome/biosphere. These criteria demarcate different research domains. In *Ecology*, Dodson describes these six versions of ecological science.

85. Callicott, "The Metaphysical Implications of Ecology." More so than Callicott, Odum acknowledged the relative autonomy of hierarchal structures in ecosystems, and he viewed ecosystems as inclusive of the biosphere. For a highly informed discussion of these difficult issues, see Salthe, *Development and Evolution* and "Summary of the Principles of Hierarchy Theory."

86. Warren and Cheney, "Ecological Feminism and Ecosystem Ecology." See also Warren and Cheney, "Ecosystem Ecology and Metaphysical Ecology." For an insightful analysis of the problems involved in attempting to make ecological science the foundation for environmental philosophy in its metaphysical, epistemological, and ethical dimensions, see de Laplante, "Environmental Alchemy." De Laplante discusses some very important issues that we cannot address within the limits of this chapter.

87. Hierarchy theory has grown enormously since the initial work done by Feibleman and Koestler. See Pattee, *Hierarchy Theory*; Eldredge and Salthe, "Hierarchy and Evolution"; Salthe, *Evolving Hierarchical Systems*; O'Neill et al., *A Hierarchical Concept of Ecosystems*; R. V. O'Neill, "Perspectives in Hierarchy and Scale"; Blitz, *Emergent Evolution*; Ahl and Allen, *Hierarchy Theory*; Gilbert and Sarkar, "Embracing Complexity"; Mitchell, *Biological Complexity and Integrative Pluralism*.

88. Here we gloss over the extremely difficult issue of the ontological status of "species"! Is "species" a term of classification? Is it a reference to a population of similar organism that endures over time? Is it a higher-order, more fundamental, and even more "valuable" aspect of reality than the individuals that instantiate them? To the last question, Callicott would answer yes, because what is real and important are species and systems of which they are parts, whereas the individuals are constantly arising and vanishing, merely temporary instantiations of the enduring DNA code constituting the "essence" of the species. For a detailed overview of the debate around the status of species involving five distinct concepts, see Wheeler and Meier, *Species Concepts and Phylogenetic Theory*.

89. Buege, "An Ecologically-Informed Ontology for Environmental Ethics."

90. Ibid, 12.

91. Rolston, *Environmental Ethics*, 175.

92. Ibid., 174.

93. Rowe, "Transcending This Poor Earth," 8.

94. Ibid., 8.

95. Rowe, "Transcending This Poor Earth," 9.

96. Ibid., 4–5.

97. Ibid., 4.

98. Ibid., 6.

99. Ibid.

100. Rowe, "Transcending This Poor Earth," 9.

101. Ibid.

102. Ibid.

103. Ibid.

104. Rowe, "Transcending This Poor Earth," 5.

105. Ibid., 8.

106. Rowe, "From Shallow to Deep Ecological Thinking," 6.

107. Rowe, "Transcending This Poor Earth," 5.

108. Salthe reviews objections to the individuality of ecosystems in *Evolving Hierarchical Systems*, 205–9.

109. Ibid., 209–14.

110. For a valuable presentation of "nexus-agency," see Wilber, "Excerpt C," 95–99.

111. Ibid., 108, italics in original. Elsewhere Wilber makes the important point that within Integral Theory "an organism is an 'I,' an ecology is a 'we,' and whenever ecology is called an organism, there is a hidden 'I,' often that of the theorist." Wilber, "Excerpt D," 117.

112. Salthe, "Infodynamics."

113. Salthe, *Development and Evolution*, 174.

114. Salthe, "Infodynamics," 8. On the idea of organisms as superorganisms, see Depew and Weber, *Darwinism Evolving*, 474–75.

115. See Zimmerman, "Ecofascism."

116. Koestler, *The Ghost in the Machine*, 237–42.

117. Ibid., 233.

118. Rowe, "The Integration of Ecological Studies."

119. Ibid., 117.

120. Ibid.

121. Ibid.

122. Ibid., 118.

123. Salthe, *Development and Evolution*, 224.

124. Allen and Hoekstra, "Comment on Rowe's Article."

125. Ibid., 118.

126. Ibid.

127. Ibid.

128. Ibid.

129. Ibid.

130. Wilber, *Sex, Ecology, Spirituality*, 5–6.

131. Zimmerman, *Contesting Earth's Future*.

Chapter 4. Developing Interiors

1. Charles S. Peirce, "The Basis of Pragmaticism," 258.

2. Developmental psychology has come a long way since it became a household term through the pioneering work of Jean Piaget and his student Lawrence Kohlberg. During the last 20 years a variety of critiques have been leveled at developmental psychology and its various stage models. Prominent critiques include pointing out how stage models have been and can be androcentric, cognitive-centric, Eurocentric, and so on. These critiques, far from dislodging the findings of researchers such as Piaget, have helped contemporary developmentalists situate their findings into more sophisticated understandings of infants, children, and adult development and have in our opinion sufficiently addressed these critiques.

In spite of the many inadequacies that have been exposed in early developmental models and theories, two general findings continually surface. First, individuals throughout their lifespan have the capacity to develop more and more complex ways of being in the world, and this spectrum of development can at the very least be heuristically be broken into stages. Second, cross-cultural evidence continues to support the position that development is shared by humans the world over. While rates of development differ from culture to culture and various developmental skills or capacities are emphasized differently between cultures—as far as the existing evidence is concerned—people across cultures accomplish the main developmental tasks in the same sequential order. See Gardiner and Kosmitzki, *Lives Across Cultures*. This is not to deny the wealth of diversity between individuals and cultures but rather to point out that we all share roughly the same contours of psychological development.

While we use an explicitly developmental framework (based on Cook-Greuter's developmental model of ego development), we are very aware—and our approach points out—that developmental psychology is only one of eight major methods of understanding the transformational dynamics of humans. Our emphasis on development is the result of a number of reasons: the insights of developmental psychology have been largely overlooked in the context of ecology and environmental studies; models of developmental psychology have a massive amount of empirical evidence supporting them; developmental psychology provides one of the most effective ways to include interiors in ecology—in that psychological structures create reliable patterns of non/ecological awareness; the inclusion of a framework of developmental psychology provides powerful pragmatic real-world and real-time application; and there is much explanatory power in interpreting our current ecological situations through a developmental lens.

3. Eco-psychologists, for instance, maintain that personal and cultural pathologies arise when urban children cannot form proper relationships with nature due to socioeconomic and technological realities.

For a comprehensive study of the ways the cultural development of both scientific and religious thought is illustrative of the stages of cognitive development associated with the research of psychologists like Jean Piaget, James Fowler, and Robert Bellah, see Barnes, *Stages of Thought*.

4. For example, in *Mother Earth*, Gill reexamines the role that European influence has had on the American Indian concept of "mother earth." Additional examples include the ways human project their own fear of death onto the biosphere and rally behind apocalyptic messages of environmental crises (see, e.g., Bluehdorn, *Post-ecological Politics*).

5. In fact, Wilber soberly points out:

> We are nowhere near the [New Age] Millennium. In fact, at this point in history, the most radical, pervasive, and earth-shaking transformation would occur simply if everybody truly evolved to a mature, rational, and responsible ego, capable of freely participating in the open exchange of mutual self-esteem (and even better, to centauric self-actualization). *There* is the "edge of history." There would be a *real* New Age. We are nowhere near the stage "beyond reason," simply because we are nowhere yet near universal reason itself. Thus, the single greatest service that trans-personalists, as well as humanists, could now perform is to champion, not just trans-reason, but an honest embrace of simple reason itself. (Wilber, *Up from Eden*, 328) (italics in original).

6. The rungs represent basic or enduring structures (e.g., Piaget's stages of cognitive development—once you have the capacity for concepts, you are using capacities associated with lower levels such as symbol formation), and the view represents the transitional structures (e.g., Kohlberg's stages of moral development—once you are at moral stage 5 you cannot see the world only from moral stage 1).

7. Wilber, "Excerpt D," part 3.

8. For example, the absence of your biological mother may lead you to try to reconnect with "mother" through Gaia imagery or you may disregard any notion of the earth as nurturing and motherlike. Or, consider Tom's development. Tom is a

man with unresolved issues with his father. When Tom confronts authority figures, (such as CEOs responsible for clear cutting in a national forest area), he projects his unresolved issues with his father onto this authority figure and often regresses to a childish emotional state. Quite possibly, the authority figure dislikes this, and then projects and regresses, too. As you can imagine, this produces a (non)dialogue. Integral Ecologists must identify and integrate those dissociated, neglected, unconscious psychic contents that arise at the least propitious moments.

The importance of understanding object-relations theory (e.g., as developed by psychoanalysts such as Melanie Klein and Donald Winnicott and the transpersonal psychologist A. H. Almaas) in this context cannot be overemphasized. We are constantly playing out the internalized relationships from our childhood with nature and with people who are "for or against" nature. For when we define nature in a particular way based on these relationships, we define ourselves: if nature is dangerous, we are fearful; if nature is a commodity, we are consumers; if nature is home, we are protective, and so on.

9. Theorists disagree about how many levels are involved in moral development. Indeed, the literature on human psychological and moral development is vast. Some researchers claim that there are no such levels, or that they are not transcultural. See Rest, Narvaez, Bebeau, and Thoma, *Postconventional Moral Thinking*; and Lind, "The Cross-Cultural Validity of the Moral Judgment Test: Findings from 27 Cross-Cultural Studies" and "The Meaning and Measurement of Moral Judgment Competence Revisited."

10. In chap. 9 we discuss his more recent addition of a "nature" line.

11. Wilber, "Waves, Streams, and Self," n. 6.

12. This movement is a 3-2-1 process: where content moves from an abstract 3rd-person position (see it) to a more interactive 2nd-person position (talk to it) and finally becomes an embodied 1st-person position (be it). We have observed a similar process with regard to students of the Integral Model. It starts out as an abstract map, then over several years of working with it becomes a shared language used with others in the community, and then with additional years of application becomes a lived reality of embodied moment-to-moment awareness.

13. These terms for the transpersonal levels are based on Sri Aurobindo's research and are associated with the cognitive line.

14. Robert Kegan estimates in *In Over Our Heads* that it takes at least five years on average for an individual to move from one order of consciousness to the next. For a more detailed presentation of the pathologies that occur all long the spectrum, see Wilber's "Spectrum of Pathology" in Wilber, Engler, and Brown, *Transformations of Consciousness*.

15. Wilber's first published use of the spectrum of light as altitude is found in *Integral Spirituality*; however, a similar use of the rainbow occurred in his first book, *Spectrum of Consciousness*. Wilber's colors are taken from chakra psychology and its use of the natural rainbow as a spectrum of consciousness moving from the denser dark

colors to lighter ones, and should not be confused with the Spiral Dynamics model of development, which also uses colors to signify levels of development.

16. Note that there is some overlap between these colors and those associated with the Spiral Dynamics model of vMeme development. While this can be confusing, it forces writers to specify if they are using a color to refer to altitude or to values, which is ultimately more clarifying.

17. Accepting people where they are is one of the distinguishing characteristics of Integral practitioners as opposed to people operating at green altitude, who want to transform others according to their postmodern, multicultural, environmental values. People operating at green sincerely believe that the world would be better if traditional and modern individuals adopted *their* postmodern values, which unrecognized by green would require vertical (hierarchical) transformation. While we agree that the world would benefit from more individuals being informed by postmodern considerations (especially leaders), we would rather see healthy expressions of traditional, modern, and postmodern individuals and systems. Besides, vertical transformation occurs more naturally when there is healthy expression all along the spiral of development. New Paradigm rhetoric usually suggests that the world would be revolutionized in a short period of time *if only* a critical mass of people were to adopt the values of the speaker. Certainly these people have good intentions, but we strongly caution against such proclamations.

18. For research on population percentages of each level, see Kegan, *In Over Our Heads;* Cook-Greuter, "Postautonomous Ego Development"; Ray and Anderson, *The Cultural Creatives*. All these sources are largely in agreement that the center of gravity of at least half the population is amber.

19. See Cook-Greuter, "Postautonomous Ego Development"; Miller and Cook-Greuter, *Creativity, Spirituality, and Transcendence;* Miller and Cook-Greuter, *Transcendence and Mature Thought in Adulthood;* Torbert, *The Power of Balance;* Torbert, *Action Inquiry*. Cook-Greuter and Torbert's research is based in large part on Jane Loevinger's 1998 Sentence Completion Test, which has had over 10,000 tests performed, and their Leadership Development Profile, which has had over 6,000 tests performed. See Loevinger, "History of the Sentence Completion Test (SCT)."

20. Other important developmental researchers that specialize in postformal operations include Kurt Fischer, Francis Richards, Michael Commons, Charles "Skip" Alexander, Juan Pascual-Leone, and Michael A. Basseches. See Commons, Richards, and Armon, *Beyond Formal Operation;* Commons, Armon, Kohlberg, Richards, Grotzer, and Sinnott, *Adult Development: Volume 2;* and Alexander and Langer, *Higher Stages of Human Development*. For an exploration of the correlations between behavioral complexity and interior complexity, see the work of Elliott Jaques (e.g., Jaques and Cason, *Human Capability*). For a comprehensive treatment of cognitive development, see Flavell, Miller, and Miller, *Cognitive Development*. For a comprehensive survey of the major and minor theories of development, see Miller, *Theories of Developmental Psychology*. For an interesting exploration of

the philosophical underpinnings of developmental approaches, see Van Haaften, Korthals, and Wren, *Philosophy of Development*.

21. Graves's research has served as the basis for Spiral Dynamics developed by two of his students, Don Beck and Christopher Cowan. In their book *Spiral Dynamics*, Beck and Cowan use the term "Value Memes" (vMemes) to describe the unspoken contexts (centers of gravity) that organize people's lives as they pass through progressively more inclusive stages of development.

22. For a succinct and accessible article that provides a lot of detail for each self, see Cook-Greuter, "A Detailed Description of the Development of Nine Action Logics."

23. One of the things that makes Cook-Greuter's model of development more appealing to us than the Spiral Dynamics model concerns this level of awareness. Spiral Dynamics views this level as being expressed through primitive and indigenous peoples who have a magical sensibility and take sanctuary in rituals and living in harmony with nature. The impression that one gets from Beck and Cowan's *Spiral Dynamics* is that large numbers of contemporary adults are living at this stage. Our sense is that research does not support this claim beyond a superficial association. This is a very low level of development and is in effect pre-egoic. By associating indigenous people with this level of development much confusion results. After all, indigenous people of yesteryear and today exist at all levels. While Spiral Dynamics acknowledges this, we have not found Spiral Dynamics' description of this level to be of practical service in dealing with the complex issues around indigenous worldviews, the environment, and development. For these reasons we feel Cook-Greuter's model and research provide a much cleaner and clearer foundation from which to make sense of these delicate issues.

24. Nash, *The Rights of Nature*. For an insightful exploration of the historical relationship between Muir's preservationism and Pinchot's conservationism, see Nash, *Wilderness and the American Mind*, and Oelschlaeger, *The Idea of Wilderness*.

25. The tyranny of the discount rate contributes to short-term strategies. The discount rate is an economic concept according to which something worth $1,000 today is worth much less 30 or 40 years from now. In discounting the future, it emphasizes the value in the present. So when concluding the value of a species, say a half-century from now, they consider that species as virtually without economic value, due to the discount rate. To combat the tyranny of the discount rate would require a near-revolution in how neoclassical economics depicts the value of commodities (including species, if they can be priced).

26. Ken Wilber has written extensively about what the calls the "Mean Green Meme." See *Boomeritis*. For a valuable critique of Wilber's characterization of green see Gary P. Hampson, "Integral Re-views Postmodernism."

27. For this description of the teal/yellow wave, see Beck and Cowan, *Spiral Dynamics*, chap. 15.

28. Linscott, "Sustainable Development Strategy." Note that for ease of reading and to avoid confusion we have used Wilber's altitude schema in place of some of the Spiral Dynamics terms that Linscott used in the original quote.

29. James D. Proctor also discusses the difficulties posed by the clash of multiple perspectives in the spotted owl controversy in the Pacific Northwest. See Proctor, "Whose Nature?" in Cronon, *Uncommon Ground*. The latter includes several essays consistent with the teal-holistic wave's commitment to honoring multiple perspectives on nature.

30. Note: This level is also called "construct-aware."

31. We are deeply indebted to Susanne Cook-Greuter for her critical suggestions that greatly improved the quality of this section of the book. Any remaining problems, omissions, or inaccuracies are our responsibility.

32. In terms of the third tier, Susanne Cook-Greuter (personal communication) explains:

> I see the ego-aware stage as the last step on the route to ego-transcendence. In my view it comes after the earlier and more general construct-aware insights. I am in the process of figuring out whether it is more useful to see it as a subset (special case) of the construct-aware stage (late vision-logic), whether it is an alternate expression of it, or whether one could consider it a separate stage. Currently, I see it as the first. The unitive stage is the only stage defined in [my model] that is ego-transcendent or transpersonal. All the other stages belong to the personal, storied with the ego as the central processing unit.

33. Due to constructive criticism, and Wilber's recognition of the value of postmodern thought, he converted to an explicit post-metaphysical approach rather than relying on preordained, metaphysical systems. We discuss this approach in detail in chap. 5.

34. Charles Saunders Peirce (1839–1914) is a leading American philosopher who made important contributions to the fields of mathematics, epistemology, metaphysics, logic, and research. He is a founder of the fields of pragmatism and semiotics. Peirce's semiotics has a triadic structure: a sign, the object it represents, and the intepretant. His own pansemiotic position extends this triadic structure throughout the universe as universal categories. The field of biosemiotics draws heavily on his work. Wilber's post-metaphysical material is also influenced by Peirce, especially his integral semiotics. For an overview of Peirce's contribution to semiotics, see Lidov, *Elements of Semiotics;* and to biosemiotics, see Vehkavaara, "From the Logic of Science to the Logic of the Living."

35. Wilber, "Excerpt D," part 4.

36. Intensive modern agriculture in the American Midwest, for instance, allowed New Englanders to abandon their farms, which had deforested much of that part of the United States. In the past century, much of this land has been reforested.

Similarly William Denevan in "The Pristine Myth" points out that when Europeans first made contact with the Americas, the Europeans discovered a highly humanized landscape (i.e., it had been massively altered from agriculture, logging, burning, and so on). During that initial contact, diseases were passed to the indigenous people. Their populations were decimated over the next hundred years. Consequently, by the time the original colonies were established, the landscape had "recovered." Unaware of this, settlers depicted the land as a "pristine," "virgin," "Eden-like" paradise. In effect, the image of an unspoiled wilderness in the early American mind was largely the result of early European contact having killed of millions of individuals before the first settlers and towns began to carve their livelihood out of the wilds.

37. For a review of the debate over modern technology, see Zimmerman, *Heidegger's Confrontation with Modernity* and *Contesting Earth's Future*.

38. Albert Borgmann, *Technology and the Character of Contemporary Life*.

39. Beck, *The Risk Society*.

40. For a useful overview, see O'Riordan and Cameron, *Interpreting the Precautionary Principle*. See also Manson, "Formulating the Precautionary Principle."

41. Sunstein, *Laws of Fear*. See also Slovic, *The Perception of Risk*.

42. Wilber, *A Brief History of Everything*, 64.

43. See Spaargaren, *The Ecological Modernization of Production and Consumption*; Spaargaren, Mol, and Buttel, *Environment and Global Modernity*; Buttel, "Ecological Modernization as Social Theory"; York and Rosa, "Key Challenges to Ecological Modernization Theory."

44. Schnaiberg, *The Environment;* Gould, Pellow, and Schnaiberg, "Everything You Wanted to Know about the Treadmill." The literature on this topic is extensive and important.

45. See Foster, "The Treadmill of Accumulation."

46. Wilber, *A Brief History of Everything*, 242.

47. These discussions (a combination of ecopsychology and eco-social theory) are useful, however limited. We don't want to suggest that there is one modern self. Everyone deviates from the norm; there is no one singular modern subject. Still, there are interesting and important commonalities.

48. On the topic of the deep connection between the Romantic and the modern self, when compared with the fragmented postmodern self, see Gergen, *The Saturated Self*.

49. See, for instance, Sessions, "Reinventing Nature, the End of Wilderness?"

50. Strand, "The Whole Package," 85. Also see note 56 in this chapter. Similarly, in *The Specter of Speciesism*, Waldau provides a convincing case that Buddhism is, in important ways, as speciesist as Christianity. Waldau effectively challenges the

romantic notion that Buddhism is inherently pro-animals due to its emphasis on benefiting all sentient beings.

51. According to Stan Salthe in a personal communication, "It's impossible to say whether Gaia is the same type as a single organism because we at our scale (like blood cells in our bodies) cannot apprehend it in such a way as to judge the issues. So, if that view [Gaia is like a cell] 'fits' theoretically, one can provisionally say that it is the same type." While acknowledging the fact that the issue at hand has not yet been decided, we contest the claim that humans are akin to blood cells in the human body. Blood cells lack the capacity to represent the body in which they are contained, whereas human beings can launch satellites that can take photographs of the planet. The cognitive capacities of the observers in questions—blood cells versus adult humans—differs greatly.

52. Interestingly, turquoise awareness is capable of perceiving both, which is partly why this confusion is so prevalent. See Wilber, *Integral Spirituality*, 263. Wilber calls this mistake a "referential falsehood," which refers to the fact that the referent exists (e.g., Gaia is a single unified consciousness), but it is mistakenly thought to exist in a place where it does not (e.g., the Right-Hand empirical world) when in fact that referent only exists in the Left-Hand interpretive world. In other words, they have an accurate description of the phenomena, but they have the wrong Kosmic address. Wilber also describes "phenomenal falsehoods," which occur when individuals assume that what they can see is true for everyone.

53. Prior to the rise of modern science, most people assumed that animals had feelings and even intentions, and that plants had some sort of inner life, however difficult this was to define. So for thousands of years many adhered to some version of panpsychism, or a belief that everything was in some measure ensouled, alive, and vital (see Skrbina, *Panpsychism in the West*). When Descartes bifurcated creation into extended things (matter) and thinking things (human minds), he concluded that animals lacked mind and hence had *no interior life*. This is a classic form of dissociation. This led to terrible treatment of animals. A good deal of groundbreaking research has occurred in the past decade, which has led to the establishment of the field of emotional ethology. As we stated in chap. 1, we agree with cognitive and emotional ethologists, biosemioticians, and animal rights activists that interiority is not restricted to human beings. (See Grandin, *Animals in Translation*; Bekoff, *The Smile of a Dolphin* and *The Emotional Lives of Animals*; and the widely popular book by Masson and McCarthy, *When Elephants Weep*.)

Within the context of early modern science, this mind-body dualism (more accurately described as mind-body dissociation) affected attitudes not only toward animals but toward humans as well. If the material universe is nothing but a wholly predictable clockwork, and if human bodies are merely cogs in this machine, then humans lack freedom. Kant's attempt to include moral freedom by limiting the reach of natural science to physical phenomena failed to take hold. According to the reasoning of the natural sciences, to understand the human being, we need only to study objective and interobjective perspectives. It became increasingly difficult to take moral freedom seriously.

We think that differentiating mind from body, without dissociating them, is crucial to understanding a person's relation to their own body, and humanity's relation with living nature.

54. It remains to be seen whether the emerging field of biosemiotics will adequately model the interiority of individuals that are members of complex systems, all of which involve "signaling" by organisms at virtually all of their various hierarchical levels. Currently the field is divided between interpretative and systemic approaches.

55. Wilber, *A Brief History of Everything*, 105–6.

56. For example, Joanna Macy's *Mutual Causality in Buddhism and General Systems Theory* suffers from not more clearly distinguishing between UL and LR realities and their associated methods and practices. For instance, in her introduction Macy explains, "Systems concepts provide explanations and analogies which can illuminate Buddhist ideas that are less accessible from a linear causal point of view" (p. 1). However, this mixing and matching of concepts derived from LR analysis with ideas arrived at through UL contemplation runs the risk of conflating these importantly different domains as well as reducing and distorting one through the terms of the other. Such important efforts as Macy's would be more successful if they took extra time to methodologically differentiate phenomenological insights from systemic ones before they set about integrating them. As discussed in note 52 above, confusing phenomenological interconnection with systemic interconnection is called a referential falsehood.

57. Wilber, "Excerpt D," part 2.

58. There are growing number of postmodern systems theorists, including M. Ceruti, E. Morin, H. Von Foerster, E. Von Glasersfeld, F. Varela, H. Maturana, and N. Luhmann.

59. Wilber, *A Brief History of Everything*, 106.

60. Integral Theory's IMP highlights that we can explore each quadrant from the inside or the outside through established methodological families.

61. See Aaran E. Gare's excellent book *Postmodernism and the Environmental Crisis*, which he describes as "the outcome of a continuing search for the roots of the global ecological crisis, and the cultural conditions required to overcome it" (p. vii).

62. Silos, "The Politics of Consciousness."

63. Michael Shellenberger and Ted Nordhaus wrote a controversial memorandum entitled "The Death of Environmentalism: Global Warming Politics in a Post-environmental World." Originally it was presented at the October 2004 meeting of the Environmental Grantmakers Association. Later it was posted on the World Wide Web. Recently, they have expanded on their analysis in their book *Break Through*. Their work has much to recommend it. In many ways, their approach contains aspects that overlap with our own (e.g., they are critical of the eco-apocalyptic narrative, draw on Maslow's developmental model, highlight the importance of socio-

cultural realities, emphasize worldviews and values, and point to the creativity of human beings in addressing our ecological problems). However, there are some important differences that, in our opinion, prevent their otherwise promising approach from being integral. In particular, they don't include post-personal stages of development, they have a narrow view of spirituality, they overemphasize sociopolitical realities, and they often fail to include the dignity of postmodernism. In short, in spite of their integrative style they still favor—at the expense of others—certain levels (orange and teal) and particular quadrants (LR).

64. Of course, we understand that it is an all-quadrant affair. Many people have developed a conscience in regard to the environment but are bound by LR systems to act against their conscience. They usually work in publicly held corporations and are legally required to maximize profits. People working in government agencies often face similar moral dilemmas. Making changes in environmentally destructive but legal corporate practices would require significant pressure from many different angles, including moral, political, and economic. Requiring corporations to live up to the conditions of their state charters, which typically demand that their activity serve the public good, would be one place to start. Encouraging corporations to follow the best green practices as a way of maximizing profit is another approach that has had some successful outcomes. The "natural capitalism" advocated by Paul Hawken, Amory Lovins, and L. Hunter Lovins comes to mind. Integral sustainable development is taking important strides, particularly in convincing major organizations—including leaders within agencies at the U.N. and World Bank—to include the interior domains in development programs.

Chapter 5. Defining, Honoring, and Integrating the Multiple Approaches to Ecology

1. Dodson, *Ecology*, 3.

2. The need for integration within the field of ecology has been recognized by many theorists and practicing ecologists; see chap. 1, note 46.

3. Our approach also includes *when* nature is perceived, either in terms of the developmental stage of the perceiver (the Who) or evolutionary complexity of the perceived (the What), and *where* nature is perceived (i.e., the Kosmic address). Thus, our use of the What, Who, and How of nature serves as a shorthand that includes the When and Where of nature. See also note 10 in chap. 6.

4. For an extended discussion of the myth of the given, see chap. 4 of this book and Wilber, *Integral Spirituality*.

5. For an overview of the development of the discipline of ecology and 18th–19th-century science, consult any number of the following sources: Bowler, *The Earth Encompassed*; Bramwell, *Ecology in the 20th Century*; Golley, *A History of the Ecosystem Concept in Ecology*; Worster, "The Ecology of Order and Chaos" and *Nature's Economy*; McIntosh, *The Background of Ecology*; Keller and Golley, *The Philosophy of Ecology*; Grove, *Green Imperialism*; Merchant, *The Death of Nature*; Porter, *The Cambridge History of Science*; Mitman, *The State of Nature*. For an

in-depth philosophical analysis of the evolution of theoretical positions within scientific ecology, see Cuddington and Beisner, *Ecological Paradigms Lost.*

6. Throughout this book we will often point to just 100 (of the 200) perspectives presented in the appendix. This is because we are typically highlighting a subset within the appendix (i.e., schools of ecology proper), whereas our use of 200 refers to ecology schools plus environmental studies schools and schools of ecological thought.

7. For Aristotle's own works, consult Aristotle, *The Complete Works*. For a short overview of Aristotle's relationship to ecology, consult Palmer, *Fifty Key Thinkers on the Environment*. In addition, there are a number of great anthologies that explore Aristotle's and other Greek philosophers' contributions to ecology; consult Westra and Robinson, *The Greeks and the Environment* and *Thinking About the Environment;* Boudouris and Kalimtzis, *Philosophy and Ecology;* Roberts, *Approaches to Nature in the Middle Ages*. A number of other articles exploring Aristotle's contribution have also appeared recently: Foster, "Aristotle and the Environment"; Garrett, "Aristotle, Ecology and Politics"; Peden and Hudson, *Communitarianism, Liberalism, and Social Responsibility;* Glazebrook, "Art or Nature?" Lastly, an interesting dissertation explores the relationship between Aristotle, Heidegger, and nature: Monti, "Origin and Ordering."

8. There are a number of excellent surveys of attitudes toward nature over the last 2,000 years. Consult Glacken, *Traces on the Rhodian Shore;* Torrance, *Encompassing Nature;* Coates, *Nature;* and Marshall, *Nature's Web;* as well as a number of books that focus on notions of nature: Collingwood, *The Idea of Nature;* Huth, *Nature and the American;* Teich, Porter, and Gustafsson, *Nature and Society in Historical Context;* Nash, *Wilderness and the American Mind;* Oelschlaeger, *The Idea of Wilderness;* Williams, *Wilderness and Paradise in Christian Thought*.

9. Haeckel, *Generelle Morphologie der Organismen* (translated into English as *General Morphology*). See Merchant, "The Rise of Ecology, 1890–1990," for an informative overview of the origins of "ecology" in relationship to Haeckel.

10. Including *The History of Creation* (1873), a popular version of *General Morphology*, and *The Evolution of Man* (1879). It was through the translation of the former that the word "ecology" was first used in English.

11. Merchant, "The Rise of Ecology, 1890–1990," 160.

12. As quoted in Dodson, *Ecology*, 2. Note that this definition first appeared in 1949 in English via a frontispiece in Allee et al., *Principles of Animal Ecology*.

13. That organisms were the preferred unit of study reflects the macroscopic view of the mid-19th century. This view neglected both the microscopic dimensions (lower scales) and wider dimensions, including ecosystem at larger scales of organization.

14. For more information on the Darwinian synthesis, consult Mayr and Provine, *The Evolutionary Synthesis*.

15. These four categories are informed in part by Merchant, "The Rise of Ecology, 1890–1990." We will not cover the many psychological, cultural, social, and behav-

ioral dimensions that have contributed to the formation of each definition, but we affirm the importance of such tetra-analysis. Ecological historians such as Donald Worster and Carolyn Merchant have provided us with many such details.

16. It is worth noting that before plant ecology emerged around 1916, Ellen Swallow Richards, a graduate of the Massachusetts Institute of Technology, established Human Ecology in the 1890s. "She argued that fresh air and clean water free of pollutants from factories were as necessary to human health as was good nutrition. Swallow's approach was quite different from that of scientific ecology, which was developing about the same time. She viewed humans as part of nature, while scientific ecology, as it later developed, separated humans from nature in order to study the environment prior to human influence." Merchant, "The Rise of Ecology, 1890–1990," 163. For additional information on Richards, see her book *Sanitation in Daily Life*; McIntosh, *The Background of Ecology*; and Clarke, *Ellen Swallow*.

17. As cited in Worster, *Nature's Economy*, 211.

18. This approach was influenced by the misguided agricultural practices that contributed to the Dust Bowl of the 1930s ("Drought follows the plow," as was often said at the time).

19. This approach developed toward the end of the 19th century and focused on abundance because animals were easier to study where they were more densely populated.

20. In 1954, H. G. Andrewartha and L. C. Birch merged the concepts of distribution (plant ecology) and abundance (animal ecology) to form another influential definition of ecology. They did this by defining the limits of distribution as the place where abundance falls to zero. Thus, in the now classic text *The Distribution and Abundance of Animals*, they defined ecology as "the scientific study of interactions affecting the distribution and abundance of organisms." This definition, like Haeckel's, emphasizes organisms as the keystone to ecology but focuses instead on *groups* of organisms (their distribution and abundance), thereby influencing community and population-centered ecologists. While offering precision by virtue of its clear parameters, this definition omits many important dimensions of ecology (e.g., abiotic components). Andrewartha and Birch also introduced a classification of four general environmental factors: weather (physical and chemical factors); food (other organisms, nutrients, inorganic compounds); other organisms, both of different species (competitors, predators, pathogens, symbionts) and same species (family members, social groups, mates); place in which to live (nest sites, shelter, niche). Andrewartha and Birch's definition of ecology has had a lasting effect on the field. By the 1960s, experimental studies of animal distributions were well established, and by the 1970s, studies in plant demography (i.e., abundance) were fully established.

21. See Gleason, "The Individualistic Concept of the Plant Association." For more information on the historical dynamics between Clements's holistic approach and Gleason's reductionistic approach, consult Barbour, "Ecological Fragmentation in the Fifties."

22. Elton was also known for defining ecology as "scientific natural history."

23. Elton, *Animal Ecology*, vii–viii.

24. The term "ecosystem" was first introduced in 1935 by the English ecologist Arthur Tansley, a critic of Clement's organismic approach and an important figure in establishing economic ecology. This usage appeared in Tansley, "The Use and Abuse of Vegetational Concepts." Tansley's original formulation of "ecosystem" provided the foundation for a whole new approach to ecology:

> But the more fundamental conception is, as it seems to me, the whole *system* (in the sense of physics, including not only the organism-complex, but also the whole complex of physical factors forming what we call the environment of the biome—the habitat factors in the widest sense.) It is the systems so formed which, from the point of view of the ecologist, are the basic units of nature on the face of the earth. These *ecosystems*, as we may call them, are of the most various kinds and sizes. They form one category of the multitudinous physical systems of the universe, which range from the universe as a whole down to the atom. (Tansley, "The Use and Abuse of Vegetational Concepts," 299)

25. Odum, *Fundamentals of Ecology*, 9.

26. During the 1950s, the Darwinian synthesis was complete, and the role of evolution began to be more thoroughly considered by ecologists. V. C. Wynne-Edwards, a British zoologist, proposed, like Allee before him, that selection could occur at the level of the group and not just through individuals. Because the emerging new field of ecosystem ecology did not examine natural systems with any kind of time-space relations, this presented a major challenge. Two distinct schools of ecology emerged: *ecosystem ecology* and *evolutionary ecology*. New ecologists were trained in the latter and slowly gained influence in journals and the academy. One reason evolutionary ecology gained prominence is that ecosystem ecology did not fit into the university system very well, since its techniques of research required a team of scientists and the academy prized individual scholarship. Thus, students could not easily take ecosystems as a topic of papers and dissertations. In spite of these pressures, ecosystem ecology has maintained itself as a distinct approach in North America, though it has not been able to establish itself in other countries.

27. Worster, *Nature's Economy*, 408.

28. Chaotic ecology is also referred to as the new ecology (e.g., Partridge, "Reconstructing Ecology"), disequilibrium ecology (Pickett, Parker, and Fiedler, "The New Paradigm in Ecology"), and disturbance ecology (Woods, "Upsetting the Balance of Nature"). For a great overview of the sociological and political dimensions to an outdated adherence to a "balance of nature" perspective, see Chase, *In a Dark Wood*. Note that "new ecology" was also used in the 40s to describe economic approaches to ecology.

29. For two other sources, see William Drury and John Anderson, *Chance and Change: Ecology for Conservationists*, and J. M. Cusing, Robert Costantino, Brian

Dennis, Robert Desharnais, and Shandelle Henson, *Chaos in Ecology: Experimental Nonlinear Dynamics*.

30. Botkin, *Discordant Harmonies*, 62.

31. The above discussion on the three definitions and their relationships is based on an article by the Institute of Ecosystem Studies entitled "Defining Ecology"; Golley, *A History of the Ecosystem Concept in Ecology*; McIntosh, *The Background of Ecology*; Bowler, *The Earth Encompassed*; Merchant, "The Rise of Ecology, 1890–1990"; Botkin, *Discordant Harmonies*, chap. 1.

32. In the context of biosemiotics, Kull discusses three metaphors of biology as associated with different historical ages: the premodern ladder (*scala naturae*), the modern tree of life, and the postmodern web of life (Kull, "Ladder, Tree, Web").

33. It is important to keep in mind that the notion of exploitation is dependent on where the perceiver is standing in relation to nature. Also, what constitutes exploitation will vary depending on such integral factors as the altitude of the perceiver and the quadrant under consideration.

34. Beginning in the 1950s the concept of an ecosystem emerged for the first time. This view saw nature as competitive parts that fit together into a functional whole. In the 1960s and 1970s, as the environmental movement began, ecosystems were still viewed in mechanistic (i.e., cybernetic) terms, but in contrast to the competitive understanding of economic ecology, the early environmentalists emphasized balance and cooperation. They still viewed ecological phenomena predominantly as complex machines, yet now many imagined themselves as one with this great, complex, interlocking Web. This reemergence of the concept of the super-organism, albeit in mechanistic garb, gave way to an understanding in the late 20th century of ecosystems as dynamic and chaotic. This most recent contender has discredited many notions of a "balance of nature." Much of the environmental legislation passed in the 1960s and 1970s, including the Endangered Species Act, was informed by an ecological science that spoke of the stability, harmony, balance, and fragility of ecosystems. The rise of chaos theory allowed scientists to conceive of natural phenomena as always in a state of near-constant dynamic change, with order emerging out of and returning to chaos. The startling idea that nature not only is subject to, but *also depends for its vitality on*, violent perturbations such as earthquakes, hurricanes, floods, and fires pulled the rug out from under old-time environmentalism. Unfortunately, conservative legislators, armed with scientific claims that nature is not harmonious and balanced but is always on the edge of chaos and in flux, were quick to conclude; since the future of ecological systems is uncertain, why should we worry so much about our incursions? Environmentalists countered that human interventions often occur on a compressed time scale and in a larger area than "natural" perturbations. In light of our ever-evolving definitions of nature, it is crucial that we do not naively superimpose our current understanding onto earlier times; nor should we assume that nature, ecology, and ecosystems are all signifiers for the same signified or referent.

35. In their essay "What Is Ecosystem Health and Why Should We Worry About It?" Haskell, Norton, and Costanza inadvertently spell out the need for an integral approach to environmental protection. Ecosystem health, they write, "cannot be defined or understood simply in biological [UR] or ethical [LL] or aesthetic [UL] or historical [LL and LR] terms. Many approaches must be used in clarifying the goals of environmental protection" (p. 3).

36. For example, you can place the term "ecology" after every plant, insect, animal or type of habitat and quickly arrive at over 1,000 distinct ecologies (e.g., beetle ecology, oak ecology, tundra ecology).

37. Dodson, *Ecology*, 4. See also the companion volume of primary source readings, Dodson, *Readings in Ecology*.

38. Each definition is from Dodson, *Ecology*, 6; the representative question each approach would ask about a rural North American landscape is from p. 8; the tools and techniques used by each approach to answer such questions is from p. 15.

39. Ibid., 3.

40. Much of the work toward a unified ecology has been done in the context of hierarchy theory, which has developed an increasingly complex epistemological position around perspectives. Consult Allen and Starr, *Hierarchy*; Salthe, *Evolving Hierarchical Systems*; O'Neill et al., *A Hierarchical Concept of Ecosystems*; Allen and Hoekstra, *Toward a Unified Ecology*; Ahl and Allen, *Hierarchy Theory* (see note 46, chap. 1).

41. In general our classification of schools is based on the What of the approach. In other words, we group schools primarily according to what they focus on or take as object. A more detailed classification system would also consider the Who and the How. But for our purposes it is sufficient to just use the What, in large part because the Who and the How often, though not always, follow the What. For example, because practitioners of behavioral ecology focus on the exterior of individuals (the What), they tend to come from an orange altitude (the Who) and use empirical techniques (the How). Thus the What, Who, and How of behavioral ecology all have a scientific-rational-empirical emphasis.

42. In "Excerpt C," note 19, Wilber outlines five critiques of typical Web of Life approaches, all of which we agree with and have pointed out over the last five chapters: (1) they commit the myth of the given, (2) they impose "anthropic cognition" on other organisms, (3) they commit subtle reductionism by reducing interiors to complex Its, (4) they fail to grasp that the some total of exteriors as enacted, (5) they continually confuse individual and social holons making Gaia a super-organism.

43. Notice how the different schools within each terrain could be further categorized based on their zone (the How). For example, in the terrain of culture, ecological hermeneutics is primarily a zone 3 approach, whereas spiritual ecology is primarily a zone 4 approach. Likewise, some schools combine two zones, like architectural phenomenology, which utilizes both zone 1 and zone 8. See chap. 8 for a fuller discussion of the 8 zones as ecological modes of investigation.

Alongside its commitment to categorize different schools of ecology, Integral Ecology recognizes that the complexity of ecosystems is always elusive: a splendid display of emptiness and form that will forever transcend our conceptual frameworks. After all, definitions of ecology are simply abstractions from the multidimensionality of the natural world. Each definition inevitably reduces the complexity of ecosystems in some important way.

44. Wilber, *Integral Spirituality*, 255.

45. See Wilber, *Integral Spirituality*, appendix II. We feel Wilber's concept of Kosmic address is original and provides an important integrative function in the context of organizing and making sense of the hundreds of perspectives on ecology and the environment.

46. Additional elements of the AQAL matrix, such as lines, states, and types, can also help specify the Kosmic address of the perceiver, or Who is doing the looking, as well as How it is being investigated and What is focused on. We use only two variables (altitude and quadrant/quadrivium) here for simplicity and because these two variables represent the bare minimum needed to establish the Kosmic address.

47. In fact, different phenomena exist (i.e., are enacted) only in specific levels of developmental complexity. For example: atoms are first enacted in the orange world, human and animal rights are first enacted in the green world, vacuum potentials are first enacted in the teal world. For an illustrative list of representative phenomena that are enacted within each altitude, see Wilber, *Integral Spirituality*, 258–59, in particular and appendix II in general. For an important discussion of how Integral Theory uses the concept of enactment and situates Varela's enactive paradigm in the AQAL model, see Wilber, *Sex, Ecology, Spirituality*, 2nd ed., 560–62, 568–70, 590–95, 734–41, and Wilber, "Excerpt C" and "Excerpt D."

48. It is with this altitude that scientific ecology comes into existence as well as the first expressions of environmental ethics, where rights are extended through rational logic to nonhuman beings.

49. At this altitude the intrinsic value of all life (i.e., biodiversity) and even landscapes is emphasized.

50. The reason we are stopping at turquoise is that it represents the current leading edge of consciousness. However, this is not to deny that there are higher and more complex ways to perceive nature.

51. Wilber, *Integral Spirituality*, 260. Likewise, the Integral Model only exists in its fullness at a turquoise level. Wilber explains:

> Thus, the **referents** (or real objects) of the **signifiers** in [the AQAL map] exist in a turquoise worldspace in a 3rd-person dimension/perspective. And they can be seen (brought forth or enacted) only by subjects at a turquoise altitude in a 3rd-person perspective—which is to say, only subjects with that Kosmic address will be able to bring forth the correct **signifieds** that correspond with the signifiers

in [the AQAL map], and thus will be able to see and understand the real **referents** of those **signifiers**. (That is, they will be able to enter into communities of knowing who work together to decide on the contours of reality enacted at those Kosmic addresses, and whether something does or does not, in fact, exist at those addresses). (Wilber, *Integral Spirituality*, 256–57; bold in original)

52. See ibid., 263–66, for additional ways to represent a Kosmic address. Wilber refers to the confusion of a signifier with the wrong referent as a *referential falsehood*. In other words, the referent of the signifier, although it exists somewhere, does not exist where the person believes it does. This is an example of getting the wrong Kosmic address of a phenomenon. As Wilber points out, this often occurs with Gaia: the signifier "Gaia is a planetary organism," which has a subjective referent, is thought to actually have an interobjective referent (p. 263). (See note 52, chap. 4.)

53. Ibid., 250–51.

54. Ibid.

55. This illustration of there being no single tree has much in common with the famous "meadow" example from Uexküll, "Theory of Meaning," 29–30, which was a delight to discover after this section was written and is worth quoting at length:

> Are we not taught . . . that the forest, for instance, which the poets praise as the most beautiful place of sojourn for human beings, is in no way grasped in its full meaning if we relate it only to ourselves?
>
> Before we follow this thought further, a sentence from the Umwelt chapter of Sombart's book *About the Human* may be cited: "No 'forest' exists as an objectively prescribed environment. There exists only a forester-, hunter-, botanist-, walker-, nature-enthusiast-, wood-gatherer-, berry-picker- and a fairytale-forest in which Hansel and Gretel lose their way." The meaning of the forest is multiplied a thousandfold if its relationships are extended to animals, and not only limited to human beings:
>
> There is, however, no point in becoming intoxicated with the enormous number of Umwelts (subjective universes) that exist in the forest. It is much more instructive to pick out a typical case in order to take a look into the relationship-network of the Umwelts.
>
> Let us consider, for example, the stem of a blooming meadow-flower and ask ourselves which roles are assigned to it in the following four Umwelts: (1) In the Umwelt of a girl picking flowers, who gathers herself a bunch of colorful flowers that she uses to adorn her bodice; (2) In the Umwelt of an ant, which uses the regular design of the stem-surface as the ideal path in order to reach its food-area in the flower-petals; (3) In the Umwelt of a cicada-larva, which bores into the sap-paths of the stem and uses it to extract the sap in order to construct the liquid walls of its airy house; (4) In the Umwelt of a cow, which grasps the stems and the flowers in order to push them into its wide mouth and utilizes them as fodder. According to the Umwelt-stage on which it appears, the identical flower stem at times plays the role of an ornament, sometimes the role of a path, some-

times the role of an extraction-point, and finally the role of a morsel of food. . . . Each Umwelt forms a closed unit in itself, which is governed, in all its parts, by the meaning it has for the subject.

Uexküll goes on to explain that the space of meaning is generated by the animal's sensory capacities to perceive its world. While Uexküll tends to emphasize an organisms capacity for 3rd-person perception via its senses, he situates these perceptions within a context of meanings that are shared with similar organisms (i.e., 2nd-person perception) and highlights that the organism has a "subjective universe" (i.e., 1st-person perception)

While Uexküll did not fully develop the implications of his umwelt theory for animal phenomenology and intersubjectivity, his theory and its recent developments, via the field of biosemiotics, does, once situated in the AQAL framework, provide us with a solid foundation upon which to build an integral ecology that recognizes the interior lives of organisms.

Another worthwhile illustration of the perspectival nature of trees occurs in Don Inde's *Experimental Phenomenology* (pp. 37–38), which provides a thought experiment of two seers viewing a tree: one who takes a 3rd-person view of the tree at noon and one who takes more of a 1st- and 2nd-person view at dusk or dawn. Because each seer is approaching the tree with a worldview of what the best conditions are to reveal the tree, they end up encountering different trees. Inde explains that "this simplified example shows that each seer sees what he already believes is 'out there'; his seeing confirms him in his metaphysics. Phenomenology holds that reality belief must be suspended in order to allow the full range of appearances to show themselves" (p. 38).

56. Wilber, "Excerpt C," 136.

57. This example is taken from Wilber, "Excerpt C," 150.

58. Wilber, "Excerpt D," 7.

Chapter 6. Ecological Terrains

1. R. Bruce Hull, *Infinite Nature*, 215.

2. In addition to the What, the 4 terrains can be understood in the context of both the Who and How. Thus, we give a brief example of the Who, How, and What of a lake from each of the quadrants/terrains. The Who involves *looking from* aspects of your own awareness that can perceive any phenomenon. This is the realm of epistemology: what is real. So you can examine the lake using your senses (UR) to see it, feel it, taste it, smell it, and listen to it; using phenomenological inquiry (UL) to notice your felt experience of the water; using intersubjective connections (LL) to recall memories of the lake and local legends about its dark depths; using systemic analysis (LR) to notice and measure patterns of nutritional flow, watershed dynamics, and vegetative distributions.

The How involves *looking with* methods associated with each quadrant to examine any phenomenon. This is the realm of methodology. Notice that the methods were required to explore, disclose, and generate the What. Additional methods that could be used to examine the lake include bracketing your experience (UL), testing the water quality (UR), reading local folktales about the lake (LL), using satellite mapping (LR).

The What involves *looking at* any phenomena that can be understood from all 4 quadrants. This is the realm of ontology. In Integral Theory this is also called a quadrivia. When examining the lake, you could choose to focus on how the interiors of fish are impacted by motorboats or how different people experience water: fear, joy, excitement, aprehension (UL), the chemical and nutritional composition of the water as well as its flow and evaporation rates (UR), the complexity of turtle culture among red-eared sliders or the ways different groups of people in the area connect to and make meaning related to the lake: symbolic, religious, historical (LL), and/or the role and function of the lake in the watershed and the various social systems connected to the water (economic, political, and educational) (LR). Clearly, the Who, How, and What framework honors many aspects of the lake in all 4 terrains that would not otherwise be accounted for or included in most ecological approaches but nonetheless are part of the phenomena of the lake.

3. Guattari, *The Three Ecologies*. Not surprisingly, others have come to a similar conclusion and have independently proposed the need for three ecologies. The post-structuralist French philosopher Félix Guattari in his *Les Trois Écologies*, calls for a mental ecology (focused on human subjectivity), a social ecology (concerned with social relations), and a natural ecology (observing environmental processes). Similarly, the voluntary-simplicity proponent Duane Elgin, in a report to the Fetzer Institute ("Global Consciousness Change: Indicators of an Emerging Paradigm"), states: "The 'ecological' challenges we face are not even purely physical. Many are social and spiritual as well. It is difficult to imagine a positive future that does not value, integrate, and balance three major ecologies" (p. 3). He describes a "physical ecology that is sustainable," a "social ecology that is satisfying," and a "spiritual ecology that is soulful."

While both Guattari and Elgin share an understanding of the need to approach the environment recognizing the three value spheres of subjectivity, intersubjectivity, and objectivity—they highlight slightly different aspects within each sphere. This difference in emphasis occurs in part because they are concerned with different aspects or levels within each sphere. For example, in regard to the domain of subjectivity, Guattari emphasizes the mental while Elgin highlights the spiritual.

The different approaches to the three ecologies as represented by Guattari and Elgin further clarify the need for an even larger integrative framework: one that not only recognizes the importance of understanding the ecological correlate of each value sphere, but one that can also identify the various levels and aspects (qualities) that exist within each sphere. This is exactly what Integral Ecology provides: a capacity to identify, situate, and interconnect the complexity of each value sphere. As a result, this more accurate map helps to ensure the inclusion and consideration of all aspects of a given environmental situation. This in turn allows all ecosystems (wild, urban, rural) and their members to be honored as having objective, interobjective, subjective, and intersubjective dimensions and results in a truly integral approach to ecology.

4. While the individual interiors of nonhuman beings are elusive, they are not inaccessible. As we discuss in chap. 8, Integral Methodological Pluralism allows the Integral Ecologist to consider and include these important realities even if they remain partially hidden or distinct from human experience. Combining various methodological approaches (e.g., phenomenology, structuralism, hermeneutics, empiricism) serves to "triangulate" data that provide a bearing on the phenomenological realities of other species in a meaningful way. The interior worlds (the so-called black box) of other species become increasingly discernible with each additional methodology. For example, in the field of biosemiotics, much work is being done to research and build upon Jakob von Uexküll's concept of umwelt. Consult Kull, "Jakob von Uexkull," in the 800+-page special issue of *Semiotica*.

5. Just as the individual interiors of nonhuman beings are elusive, so are the collective interiors (cultures) of insects, birds, and mammals. Nevertheless, this ambiguity does not mean they do not exist or that they do not in some way overlap or "touch" our own interiors. Much work is being done in environmental semiotics (ecosemiotics, biosemiotics, zoosemiotics), interspecial communication, and the study of animal culture (mostly bird and mammal) to shed light on these realities. For some articles on ecosemiotics, see Noth and Kull, *Semiotics of Nature*; Noth, "Ecosemiotics"; Kull, "Semiotic Ecology."

When these research efforts are set within an Integral Methodological Pluralism, Integral Ecologists can begin to access, understand, and include nonhuman cultural realities alongside and linked to ecosystems analysis.

6. "Tetra-hension" is an expanded version of Whitehead's notion of "prehension." Wilber also refers to it as "quadratic prehension" and "com-prehension" ("Excerpt A," 87). It describes the basic unconscious unit of perception in the UL. See Wilber, "Excerpt C" and "Excerpt D."

7. Facts adapted from Walker, *The Woman's Dictionary of Symbols and Sacred Objects*, 468. Keep in mind that shared meaning is not just how we reflect on objects but also how we enact them.

8. Facts adapted from Whitney, *Western Forests*, 383.

9. Facts adapted from Stuckey and Palmer, *Western Trees*, 81.

10. A fourth term (x Where) can be added to indicate either an inside or outside location (e.g., the 8 zones). In "Integral Geography" Integral theorist Brian Eddy has proposed including both "x When" and "x Where."

11. Hoffmeyer, *Signs of Meaning in the Universe*, 58.

12. The frog's senses can be understood either from a physiological dimension, such as its nervous system and neurochemistry, or from a phenomenological dimension. In the former, we as humans would be examining via a quadrivium the UR

quadradic dimension of the frog. In the latter, the frog would be examining via a quadrivium the UR quadradic dimension of its environment through the phenomenological experiences (umwelt) it has through registering signals with its senses. If we try to understand that process of the frog in an objective way, we get Varela's biological phenomenology. A quadradic perspective of the behavior terrain of a frog emphasizes how the frog itself perceives its exterior world—just as the terrain of experience highlights how a frog perceives its interior world.

13. The point we are making is twofold. First, an Integral approach to organisms (members of an ecosystem) and to frogs in particular would not just focus on the exterior quadrivia of a frog, its behavior and its ecosystemic role, but would also include a recognition of its interior quadrivia, namely, that it has subjective experiences and intersubjective resonance. Second, an Integral approach would *also* honor a frog as a perceiving being with four distinct dimensions-perspectives. In other words, it is not enough to expand our two quadrivium of the frog, how we hold it as an object of investigation, to include two more quadrivium looking at its subjective and intersubjective dimensions. In addition to the fact that we can view the frog in four ways, we must also recognize that the frog itself views its world in four ways—that it actually *has or even exists as* four perspectives. Thus, ecological science examines only two of the four quadrivia of a frog and denies that it has any perspectives of its own (i.e., the frog's quadrants), thereby investigating only two of the eight dimension-perspectives. To grant it any quadrant perspectives would be to reorganize it as a subject in its own right. Integral Ecology recognizes all four quadrivia and all 4 quadrants of all organisms (with IMP this becomes 8 and 8!).

Thus, modern biology and ecology are based on taking a quadrivium of an objective organism using zone 6 (empiricism) and zone 8 (systems theory), whereas semiotic biology (e.g., biosemiotics) is based on complementing those quadrivia with a quadratic understanding that recognizes in various ways that the organism is a subject that tetra-hends its environment. Thus, biosemiotics takes a quadratic view of a subjective organism using primarily the inner zones of 5 (autopoiesis) and 3 (hermeneutics) and to some extent 1 (phenomenology) and 7 (social autopoiesis).

14. As autopoiesis points out, the frog does not see the ecosystem; in fact it does not even exist for the frog—only what could be called an "eco-patch" exists in the frog's cognization of its environment. In other words, frogs do not see systems, they see bugs (and not in the way we see them!). This is what Varela referred to as the biological phenomenology (i.e., the study of what phenomena exist within various worldspaces of biological organisms) or the view from the inside. This "insider's view" of the frog is not a view of the frog's 1st-person experience or "I-ness," but rather is a conceptual-scientific view of how the frog cognizes and registers various phenomena within its biological world. See the classic paper by Lettvin, Maturana, McCulloch, and Pitts, "What the Frog's Eye Tells the Frog's Brain," and Maturana and Varela's *Autopoiesis and Cognition* and their more accessible overview of autopoiesis, *The Tree of Knowledge*. For a great overview of autopoiesis applied across multiple domains and its relationship to the phenomenological approaches of Husserl and Heidegger, see Mingers's *Self-Producing Systems*. An important book that highlights

the zone 5 aspects of autopoiesis is Winograd and Flores's *Understanding Computers and Cognition*, which draws many parallels between computer design and cognitive processes. Weber's "The 'Surplus of Meaning': Biosemiotic Aspects in Francisco J. Varela's Philosophy of Cognition" provides an important look at how Varela's work can serve biosemiotics.

15. Hoffmeyer developed the idea of a semiotic niche: "The semiosphere imposes limitations on the *umwelt* of its resident population in the sense that to hold its own in the semiosphere a population must occupy a 'semiotic niche'" (Hoffmeyer, *Signs of Meaning in the Universe*, 59).

16. The concept of "semiosphere" was developed by the Estonian semiotician Jurdi Lotman ("O Semiosfere"). For articles on Lotman's notion of the semiosphere, see Merrell, "Lotman's Semiosphere, Peirce's Categories, and Cultural Forms of Life," and Chang, "Is Language a Primary Modeling System? On Juri Lotman's Concept of Semiosphere." For other discussions of the semiosphere, see the special issue "Semiotics in the Biosphere," *Semiotica* 120 (nos. 3/4), which is dedicated to reviewing Hoffmeyer's *Signs of Meaning in the Universe*. For an interesting article that explores the development of semiotic space (i.e., environment, umwelt, semiosphere) in relationship to traditional, modern, and postmodern worldviews, see Lotman, "Umwelt and Semiosphere."

17. "Semiosphere is the set of all interconnected Umwelts. Any two Umwelts, when communicating, are a part of the same semiosphere" (Kull, "On Semiosis, Umwelt, and Semiosphere," 305). Kull has a slightly different opinion of the semiosphere than Hoffmeyer. Hoffmeyer suggests that the semiosphere may be partially independent of the organisms' umwelt. Kull asserts that the semiosphere is "entirely created by the organisms' Umwelts. Organisms are themselves creating signs, which become the constituent parts of the semiosphere. This is not an adaptation to environment, but the creation of a new environment" (ibid.). We conceive of the semiosphere as intersubjective space generated by all the phenomenological spaces of various organisms in communication both in terms of zone 3 hermeneutics and zone 7 social autopoiesis. As a result, we often use the semiotic niche to highlight the LL hermeneutic dimension of an organism. However, we recognize that some uses of "semiotic niche" are more closely aligned with zone 7. When using concepts like "semiotic niche" or "umwelt," you have to be clear about whether you are talking about them in the context of quadrants or quadrivia. For example, we use "umwelt" to refer sometimes to the UL dimension of an organism (e.g., when taking a quadrivium on an organism and acknowledging it is a subject) and sometimes to the quadrants of an organism (i.e., their capacity to take perspectives: tetra-hension). Either usage is accurate as long as you are clear which one you are utilizing.

For a discussion of subjectivity in the context of organisms and the semiosphere, see Hoffmeyer's discussion of "What it is to be a subject?" and "semiotic freedom" (Hoffmeyer, "The Unfolding Semiosphere"). While we commend Hoffmeyer's inclusion of subjectivity, we note that he all too often discusses the exterior aspects of signaling in the semiosphere: "sounds, odours, movements, colours, electric fields, waves of any kind, chemical signals, touch, etc." (p. 290). Though he is explicit that "semiotic freedom does not simply refer to the wealth or quantity of semiotic processes around but rather to the quality of such processes: the depth of meaning a culture, an individual or a species is capable of communicating" (p. 291). Thus, Hoffmeyer includes information (span) and knowledge (depth) in his understanding of semiotic processes within and between organisms. Also see Hoffmeyer, "Evolutionary Intentionality."

For a presentation of 17 different but complementary definitions of semiosphere, see Kull, "Semiosphere and a Dual Ecology." Interestingly, in this article Kull argues that biosemiotics is a qualitative ecology: "There is an ecology that has been developed as a natural science, according to the Modernist model of science—a field of quantitative research of environment with organic systems in it, without any intrinsic value or meaning in itself. And there is an ecology that includes meaning and value. The latter would include ecophilosophy, biosemiotics, semiotic ecology. . . . Thus, semiosphere is a concept of [a] fundamentally postmodern approach . . ." (p. 184). He goes on to define the semiosphere as a "heterogeneous space (or communicative medium) enabling qualitative diversity to emerge, to fuse, and to sustain" (p. 185). Similarly, in his article "Does a Robot Have an Umwelt?" Emmeche refers to biosemiotics as a "qualitative organiscism."

18. See Wilber, "Excerpt C" and "Excerpt D," for a discussion of Integral semiotics. Stan Salthe points out that biosemiotics typically uses the Peircean formulation of object, sign, and interpretant, which is triadic, not quadratic (personal communication). Our quadrants, however, can be interpreted in ways that are largely consistent with Peirce's scheme. Thus an integral approach to biosemiotics (LL) would include the study of bio-syntax (LR), bio-signifieds (UL), and bio-signifiers (UR).

19. Hoffmeyer echoes this in his emphasis on *semiotic freedom*: "The most pronounced feature of organic evolution is not the creation of a multiplicity of amazing morphological structures, but the general expansion of 'semiotic freedom,' that is to say the increase in richness or 'depth' of meaning that can be communicated: from pheromones to birdsong and from antibodies to Japanese ceremonies of welcome" (Hoffmeyer, *Signs of Meaning in the Universe*, 61).

Hoffmeyer is very clear that what he has in mind with the use of semiotic "depth" and "freedom" is to be contrasted with "information," which in our framework would be associated with the Right-Hand quadrants: "Semiotic freedom refers not only to the quantitative mass of semiotic processes involved but even more so to the quality of these processes. We could perhaps define it as the 'depth of meaning'" that an individual or a species is capable of communicating" (ibid., 62). In fact, Hoffmeyer prefers semiotic freedom over semiotic depth due to an association of depth with "logical depth." However for our purposes we will use "semiotic depth" since depth is already a term used in the Integral Model to refer to the complexity of interiors.

20. Ibid., 59. Biosemiotics emphasizes that sign production and interpretation are fundamental to organisms. Consequently, many biosemioticians take the position

counter to neo-Darwinism and claim that the mechanism of evolution is not survival of the fittest but rather an organism's ability to interpret its environment. See Hoffmeyer's discussion of intentionality in "Origin of Species by Natural Translation." This amounts to nothing less than a revolution within ecological sciences because interiors become more important than exteriors. An Integral approach emphasizes the tetra-enactment of evolution where experience, behaviors, interpretations, and environmental pressures all play an important role (see note 9, chap. 3).

21. This pansemiotics serves as the LL correlate to our panpsychist (pan-interiority) position associated with the UL. Peirce believed that everything in the universe was perfused with signs. Claus Emmeche outlines the pansemiotic thesis:

> The universe is perfused with signs, semiosis is not only a process found in all living nature among beings which are organic, functional wholes (organisms as interpreters, or interpretants). The sign, its object and its interpretant are universal categories, which existed (eventually in degenerate form) even before the origin of life. The pansemiotic thesis may be read as a version of panpsychism: the idea that matter is effete mind, or that the qualities of experience, sensation, pain or feeling come in degrees, and that even inorganic systems may have, eventually to very small degrees, such qualities. If one does not like the idea of emergence (as a sudden appearance of qualitative new irreducible properties, cf. Baas and Emmeche 1997), and embraces a continuity thesis (that mind is continuous with matter, and that systems with meaning-attributing capacities have originated from, or are a certain organization of, material systems), one is more inclined to such a view of nature, according to which mental phenomena are not simply found in the brain (and presuppose the body of a whole multicellular organism) because also single cells of any kind, not only complexes of nerve cells, have "mind", "feeling", "consciousness" (or semiotic capacity)—at least to a tiny degree. Of course, a problem with this idea is that it is painfully difficult to give precise scientific meaning to the claim that single cells or even non-cellular systems have feeling, even to a very tiny degree, if one by meaning demands clear and fulfilled conditions of verification (or assertability or falsifiability) and not just appeals to special intuitions that seem to differ among semioticians as well as metaphysicists. This demand of clarity may be perceived by general semiotics as unnecessarily restrictive, but its fulfilment should facilitate communication between scientists and semioticians. (Emmeche, "The Biosemiotics of Emergent Properties in a Pluralist Ontology," 91).

22. Wilber, "Excerpt A," 80–81. For another example of hermeneutics occurring between a human and a nonhuman organism, in this case gorillas, see Wilber, "Excerpt D," 62–64.

23. For example, in the context of this more commonly used narrow sense, Emmeche claims that the umwelt of an animal is not the same thing as the mind of an animal: "The mind is a broader notion than the Umwelt, so, for instance, there can be a lot of activity in a living organism which is of a mental, or semiotic, character, but which does not figure as a part of the animal's phenomenal world" (Emmeche,

"Does a Robot Have an Umwelt?" 654). Even though for him umwelt is just a particular aspect of mind, Emmeche goes on to explain that "the Umwelt notion is of central importance to the development of a coherent theory of the qualitative experiential world of the organism, a task present day biology must face, instead of continuing to ignore a huge phenomenal realm of the living world—the experiential world of animal appetites, desires, feelings, sensations, etc." (p. 660).

24. For another extensive example of the quadrants of an organism, see Wilber's discussion of a goose, "Excerpt C," 49–50 and 100–104; and of a bacterium, "Excerpt C," 56–60, which he concludes by explaining that "each bacterium has an interior sensation (or prehension), an exterior registration (or rudimentary cognition of its enacted world), an inter-exterior system of communication (which forms part of its social system or ecosystem), and therefore an inter-interior harmonic resonance with other bacteria (and other sentient beings)" (p. 60).

25. Note that in figure 6.4 we are emphasizing the subjective and intersubjective dimensions of umwelt and semiosphere in contrast to typical objects of study of biology and ecology, which cannot recognize these interior dimensions of organisms due to their methodological approach. However, it is important that both the umwelt and semiosphere of an organism can be approached as quadrants or quadrivia.

In some uses, "umwelt" and "semiosphere" refer to the realities revealed by zones 5 and 7 respectively. In this context, then, traditional biology deals with the outside zones of an organism and its environment (i.e., biosphere), and semiotic biology deals with the inside zones. However, because biosemiotics also uses Left-Hand methodologies, it explores the qualitative aspects of organisms as well. Note how this contrasts with autopoiesis (e.g., cognitive biology), which often acknowledges animal interiors but is not able to say anything about them because their methodology only gives them access to the inside of exteriors (zone 5 and 7 realities). In other words, what distinguishes umwelt theory from being just autopoiesis is not the recognition that an organism is a subject—both do that; but only biosemiotics uses Left-Hand methodologies that allow it to discuss the phenomenological, interpretative, motivational, and intentional aspects of an organism. Of course, there are biosemioticians who reduce their field to autopoietic methods, but in general biosemiotics is a multiple-zone endeavor spanning across Left- and Right-Hand methodologies.

26. We generated this example at the first Integral Ecology meeting held at the Integral Institute.

27. Here we are using the 12 niches from a quadrivial perspective. They can also be used to understand the quadratic dimensions of any ecological occasion.

28. Consult Wilber, "Excerpt G."

29. See Wilber, *Sex, Ecology, Spirituality*, 517–20, for a discussion of these three values in relationship to environmental ethics.

30. For a deeper discussion of "ontology" in the context of a post-metaphysical approach, consult Wilber, "Excerpt A," "Excerpt B," "Excerpt C," and "Excerpt D."

31. Merleau-Ponty (*The Primacy of Perception* and *The Visible and the Invisible*); Abram, *The Spell of the Sensuous*; Brown and Toadvine, *Eco-Phenomenology*; Fisher, *Radical Ecopsychology*; Hiss, *The Experience of Place*; Levin, *The Body's Recollection of Being*; Seamon, *Dwelling, Seeing, and Designing*; and Cataldi and Hamrick, *Merleau-Ponty and Environmental Philosophy*.

32. Conrad, "Life on Land"; Dufrechou, *Coming Home to Nature through the Body*; Endredy, *Earthwalks for Body and Spirit*; Olson, *Body and Earth*; Sheets-Jonstone, *The Roots of Thinking*.

33. Dodds and Tavernor, *Body and Building*; Norberg-Schulz, *Intentions in Architecture* and *Genius Loci*; and Pile, *The Body and the City*.

34. Kahn, *The Human Relationship with Nature*; Forest, "'Ought' and 'Can' in Environmental Ethics"; Greenwald, "Environmental Attitudes"; Latonick-Flores, "Awakening to the Ecotragedy"; Robbins and Greenwald, "Environmental Attitudes Conceptualized through Developmental Theory"; Swearingen, "Moral Development and Environmental Ethics"; Wilber, *Sex, Ecology, Spirituality*.

35. Cobb, *The Ecology of Imagination in Childhood*; Hart, *Children's Experience of Place*; Kahn and Kellert, *Children and Nature*; Melson, *Why the Wild Things Are*; Myers, *Children and Animals*; Nabhan and Trimble, *The Geography of Childhood*.

36. Bechtel and Churchman, *Handbook of Environmental Psychology*; Bell, Greene, Fisher, and Baum, *Environmental Psychology*.

37. Gibson and Pick, *An Ecological Approach to Perceptual Learning and Development*; Gibson, *The Ecological Approach to Visual Perception*; Reed, *Encountering the World*.

38. Clinebell, *Ecotherapy*; Cohen, *Reconnecting with Nature*; Fischer, *Radical Ecopsychology*; Roszak, *The Voice of the Earth*; Roszak, Gomes, and Kanner, *Ecopsychology*; Sewall, *Sight and Sensibility*. It is worth noting:

> The apparent richness and variety of the contributions to *Ecopsychology* are somewhat deceptive, however. Of the 26 papers in *Ecopsychology*, only one seems to have been written by an academic psychologist and only one by a scientific ecologist. With a few exceptions, the rest emphasize only two views of psychology—psychodynamic and transpersonal. . . . Besides representing a somewhat narrow view of the new field, the collection in *Ecopsychology* lacks geographical and cultural diversity: All the writers are North American and more than 60% live in California. (Scull, "Let a Thousand Flowers Bloom," 1)

Our experience is that this assessment is representative of much of the current field of ecopsychology.

39. Clayton and Opotow, *Identity and the Natural Environment*; Jolma, *Attitudes toward the Outdoors*; Kidner, *Nature and Psyche*; Milton, *Loving Nature*; Stern, "Psychological Dimensions of Global Environmental Change"; Thomashow, *Ecological Identity*; Tuan, *Topophilia*; and Winter and Koger, *The Psychology of Environmental Problems*.

40. Bekoff and Jamieson, *Readings in Animal Cognition;* Cheney and Seyfarth, *How Monkeys See the World;* Dawkins, *Through Our Eyes Only?;* Griffin, *Animal Minds.*

41. Fox, *Towards a Transpersonal Ecology;* Naess, *Ecology, Community and Lifestyle;* Macy, "The Ecological Self"; Wilber, *Sex, Ecology, Spirituality.*

42. Oliver, *New and Selected Poems;* Snyder, *No Nature.*

43. Gibson, *Multiple Chemical Sensitivity.*

44. Merleau-Ponty's notion of "flesh" (*The Primacy of Perception* and *The Visible and the Invisible*), which is also the basis of Abram's eco-poetic *Spell of the Sensuous;* Whitehead's philosophy (*Process and Reality*); as well as the somatic work of Anna Halprin's Environmental Dance (*Moving toward Life, Returning Home,* and *Embracing Earth*), Emily Conrad Da'oud's Continuum (Conrad, "Life on Land"), Bonnie Bainbridge's Body-Mind Centering (Hartley, *Wisdom of the Body Moving*); and Andrea Olsen's earth-based Authentic Movement (Stromsted, "Dancing Body Earth Body").

45. Bigwood, *Earth Muse;* Casey, *The Fate of Place;* Halprin, *Moving toward Life;* Holler, "Thinking with the Weight of the Earth"; Keller, *From a Broken Web;* and Mazis, *Earthbodies.*

46. Bowler, *The Earth Encompassed;* Callicott and Nelson, *The Great New Wilderness Debate;* Collingwood, *The Idea of Nature;* Nash, *Wilderness and the American Mind;* Oelschlaeger, *The Idea of Wilderness;* Worster, *Nature's Economy.*

47. Callicott and Ames, *Nature in Asian Traditions of Thought;* Glacken, *Traces on the Rhodian Shore;* Kerr, *Dogs and Demons;* McLuhan, *The Way of the Earth;* Preece, *Animals and Nature;* Tetsuro, *Climate and Culture;* Torrance, *Encompassing Nature—A Sourcebook.*

48. Elgin's *Awakening Earth,* which can be seen as a hermeneutical engagement of a genealogy of "ecology," provides a developmentally based understanding of worldviews and how they relate to human-nature relationships. See Kealey, *Revisioning Environmental Ethics,* and Wilber, *Up from Eden.*

49. Cronon, *Uncommon Ground;* Evernden and Leslie, *The Social Creation of Nature;* Oelschlaeger, *Postmodern Environmental Ethics;* Simmons, *Interpreting Nature;* Soule and Lease, *Reinventing Nature?;* Zimmerman, *Contesting Earth's Future.*

50. Ferry, *The New Ecological Order;* Guattari, *The Three Ecologies;* Hallman, "Nietzsche's Environmental Ethics"; Hayden, "Gilles Deleuze and Naturalism"; Miller, *The Vegetative Soul;* Vogel, "Nature as Origin and Difference"; Zimmerman, "What Can Continental Philosophy Contribute to Environmentalism?" In addition, there is a whole subsection of environmental thought informed by continental thought that can be called "animal philosophy." See Atterton and Calarco, *Animal Philosophy;* Wolch and Emel, *Animal Geographies;* Wolfe, *Animal Rites* and *Zoontologies.*

51. Brulle, *Agency, Democracy, and Nature;* Buren, "Critical Environmental Hermeneutics"; Dryzek, "Green Reason"; Vogel, *Against Nature;* Whitebook, "The Problem of Nature in Habermas."

52. Casey, *Getting Back into Place, The Fate of Place, Representing Place,* and *Earth Mapping;* Jackson, *Becoming Native to This Place;* Gallagher, *Power of Place;* Hogan, *Dwellings;* Malpas, *Place and Experience;* Snyder, *A Place in Space;* Sobel, *Children's Special Places;* Spretnak, *The Resurgence of the Real;* Stine, *The Earth at Our Doorstep.*

53. Adams, Hoelscher, and Till, *Textures of Place;* Tilley, *A Phenomenology of Landscape;* Tuan, *Space and Place* and *Topophilia.*

54. Groth and Bressi, *Understanding Ordinary Landscapes;* Jackson, *The Necessity for Ruins, and Other Topics* and *A Sense of Place, a Sense of Time;* Kunstler, *The Geography of Nowhere;* Wilson, *The Culture of Nature.*

55. Andruss et al., *Home! A Bioregional Reader;* Berry, *The Unsettling of America* and *Sex, Economy, Freedom, and Community;* Mills, *In Service of the Wild;* Sale, *Dwellers in the Land;* Snyder, *The Practice of the Wild;* Vitek and Jackson, *Rooted in the Land.*

56. Abbey, *Desert Solitaire;* Dillard, *Pilgrim at Tinker Creek;* Emerson, "On Nature"; Glotfely and Fromm, *The Ecocriticism Reader;* Jensen, *A Language Older Than Words;* Muir, *The Wilderness World of John Muir;* Tempest, *Red.*

57. Basso, *Wisdom Sits in Places;* Maffi, *On Biocultural Diversity;* Nabhan, *Cultures of Habitat;* Nettle and Romaine, *Vanishing Voices;* Spirn, *The Language of Landscape.*

58. Cuomo, *Feminism and Ecological Communities;* Daly, *Gyn/ecology;* Diamond and Orenstein, *Reweaving the World;* Plumwood, *Feminism and the Mastery of Nature;* Warren, *Ecofeminism* and *Ecofeminist Philosophy.*

59. C. Keller, *From a Broken Web;* E. F. Keller, *Reflections on Gender and Science;* Griffin, *Woman and Nature;* Harding, *The Science Question in Feminism;* Harroway, *Simians, Cyborgs and Women;* Merchant, *The Death of Nature;* Schiebinger, *Nature's Body;* Shiva, *Staying Alive.*

60. Berkofer, *The White Man's Indian;* Buege, "The Ecological Noble Savage Revisited"; Coward, *The Newspaper Indian;* P. Deloria, *Playing Indian;* V. Deloria, *Red Earth, White Lies;* Dilworth, *Imagining Indians in the Southwest;* Grande, "Beyond the Ecological Noble Savage"; Martin, "The American Indian as Miscast Ecologist"; Taylor, "Earthen Spirituality or Cultural Genocide?"

61. Ellen, Parkes, and Bicker, *Indigenous Environmental Knowledge and Its Transformations;* Laird, *Biodiversity and Traditional Knowledge.*

62. Berkes, *Sacred Ecology;* Callicott, *Earth's Insights;* Hughes, *North American Indian Ecology;* Metzner, *The Well of Remembrance;* Mosley, *African Philosophy;*

Piacentini, *Story Earth;* Taylor, "Earth and Nature-Based Spirituality (Part 1)" and "Earth and Nature-Based Spirituality (Part 2)."

63. Frohoff and Peterson, *Between Species;* Lauck, *The Voice of the Infinite in the Small;* Kaza, *The Attentive Heart;* Midgely, *Animals and Why They Matter;* Shepard, *The Others;* Wilson, *Biophilia.*

64. Bonner, *The Evolution of Culture in Animals;* de Waal and Tyack, *Animal Social Complexity;* Stanford, *Significant Others;* Wrangham et al., *Chimpanzee Cultures.*

65. Whitehead, *Process and Reality.*

66. Birch and Cobb, *The Liberation of Life;* Birch, Eakin, and McDaniel, *Liberating Life;* Griffin, "Whitehead's Deeply Ecological Worldview"; C. Keller, *From a Broken Web.*

67. Des Jardins, *Environmental Ethics;* Zimmerman et al., *Environmental Philosophy.*

68. Schweitzer, *Civilization and Ethics;* Taylor, *Respect for Nature.*

69. Regan, *The Case for Animal Rights;* Rollin, *The Unheeded Cry;* Singer, *Animal Liberation;* and Sorabji, *Animal Minds and Human Morals.*

70. Callicott, *Companion to a Sand County Almanac;* Callicott, *In Defense of the Land Ethic;* Callicott, *Beyond the Land Ethic;* Callicott and Freyfogle, *For the Health of the Land;* Leopold, *A Sand County Almanac and Sketches Here and There.*

71. Rolston, *Philosophy Gone Wild* and *Environmental Ethics.*

72. Amrine, *Goethe in the History of Science;* Bortoft, *The Wholeness of Nature;* Keller and Freeman, *A Feeling for the Organism;* Schad, *Man and Mammals;* Seamon and Zajonc, *Goethe's Way of Science;* Schwenk, *Sensitive Chaos;* Suchantke, *Eco-Geography;* and Wilkes, *Flowforms.*

73. Adams, *Ecofeminism and the Sacred;* Howell, *A Feminist Cosmology;* MacKinnon and McIntyre, *Readings in Ecology and Feminist Theology;* McFague, *The Body of God;* Ruether, *Gaia and God.*

74. Boff, *Ecologia* and *Cry of the Earth, Cry of the Poor;* Gutierrez, *A Theology of Liberation.*

75. Fox, *The Coming of the Cosmic Christ* and *Creation Spirituality.*

76. Berry, *The Dream of the Earth* and *The Great Work;* Swimme, *The Universe Is a Green Dragon* and *The Hidden Heart of the Cosmos;* Swimme and Berry, *The Universe Story.* For an overview of Teilhard de Chardin's eco-theology, see Fabel and St. John's *Teilhard in the Twenty-first Century.*

77. Gottlieb, *This Sacred Earth;* Kaza and Kraft, *Dharma Rain;* Nasr, *Man and Nature;* Prime, *Vedic Ecology;* Spretnak, *States of Grace;* Tucker, *Worldly Wonder;* Tucker and Grim, *Worldviews and Ecology;* and Waskow, *Torah of the Earth.*

78. Tucker and Grim, *Religions of the World and Ecology.*

79. For a comprehensive and erudite examination, see Waldau and Patton, *A Communion of Subjects.*

80. McClellan, "Nondual Ecology."

81. Abrams, *The Mirror and the Lamp* and *Natural Supernaturalism;* Bate, *Romantic Ecology;* and Myerson, *The Transcendentalists.*

82. Goodwin, *How the Leopard Changed Its Spots;* Harman and Sahtouris, *Biology Revisioned.*

83. Brown, *Tom Brown's Field Guide to Nature Observation and Tracking* and *Tom Brown's Field Guide to Wilderness Survival;* McNab, *Living off the Land;* Rezendes, *Tracking and the Art of Seeing.*

84. Carson, *Silent Spring;* Markowitz and Rosner, *Deceit and Denial;* McCully, *Life Support;* Radetsky, *Allergic to the Twentieth Century;* Wargo, *Our Children's Toxic Legacy.*

85. Davis, *When Smoke Ran Like Water;* Garrett, *The Coming Plague.*

86. Berthold-Bond, *Better Basics for the Home;* Brower and Leon, *The Consumer's Guide to Effective Environmental Choices;* D. L. Dadd, *Home Safe Home;* Earth Works Group, *50 Simple Things You Can Do to Save the Earth;* Hayes, *The Official Earth Day Guide to Planet Repair;* Robbins, *Diet for a New America.*

87. Brown, *Organic Living;* Elgin, *Awakening Earth;* Luhrs, *The Simple Living Guide;* Mate, *A Reasonable Life;* Pierce, *Choosing Simplicity.*

88. Bullard, *Dumping in Dixie;* Camacho, *Environmental Injustices, Political Struggle;* Cole and Foster, *From the Ground Up;* Johansen and Grinde, *Ecocide of Native America;* Weaver, *Defending Mother Earth.*

89. Butterfly Hill, *The Legacy of Luna;* Roddick, *Take It Personally;* Switzer, *Environmental Activism.*

90. Barlow, *Sacred Sites of the West;* Devereauz, *The Sacred Place;* Milne, *Sacred Places in North America;* Olsen, *Sacred Places.*

91. Jones, *The Social Face of Buddhism.*

92. Dodson, *Ecology.*

93. Bateson, *Steps to an Ecology of Mind* and *Mind and Nature;* Botkin, *Discordant Harmonies;* Capra, *The Web of Life;* Laszlo, *The Systems View of the World;* Oyama, *Evolution's Eye;* Worster, "The Ecology of Order and Chaos"; Taylor, *The Moment of Complexity;* Von Bartalanffy, *General Systems Theory.*

94. Lovelock, *The Ages of Gaia* and *Gaia;* Margulis, *Symbiotic Planet.*

95. Clark, *Averting Extinction;* Czech and Krausman, *The Endangered Species Act;* Grumbine, *Ghost Bears;* Leakey and Lewin, *The Sixth Extinction;* Reakarkudla,

Wilson, and Wilson, *Biodiversity II*; Tudge, *Last Animals at the Zoo!*; Wilson, *The Diversity of Life*.

96. Hemenway and Todd, *Gaia's Garden*; Holmgren *Permaculture*; and Mollison, *Permaculture*. Not surprisingly, many permaculturists consider their approach to be quite integral. Our view, on the other hand, sees a lot of integral potential within the permaculture movement, but as of now it seems to be predominately defined by LR realities and considerations (see, e.g., fig. 1, "The Permaculture Flower," in Holmgren, *Permaculture*, xx). Permaculture is by and large a LR design framework, but because it is issued from the green altitude, its proponents often have an affinity for indigenous knowledge, holism, and "home-grown" spirituality. So while many permaculturists talk about UL spirituality and LL ethics and see their design work as an expression of both, permaculture lacks a framework that explicitly includes individual and collective interiors. Not to mention that many permaculture proponents are guilty of Romanticism, the pre/trans fallacy, flatland holism, and antimodernity sentiments.

Tim Winton, an Integral practitioner in Australia (see chap. 10) with a strong background in permaculture, provides a helpful Integral analysis of permaculture that is worth quoting at length:

> I think the time may be right for permaculture, as a humanist ecological design science with a strong ethical basis, to be transcended and included into an 'integral permaculture.' . . . I am beginning to see the impact integral understandings and practice are having on my students in the Permaforest Trust permaculture training program. . . . The potential of integral permaculture is that it could help round out some of the LR lopsidedness of permaculture as it stands now, and help permaculture fulfill its early promise as an effective cultural alternative to the industrial model it wanted so desperately to change. Thirty years later, it is obvious that this promise has not yet been realized.
>
> A thumbnail critique of permaculture as an integral pursuit should include quadrants, levels, lines, states, and types. For now I will just focus mostly on quadrants and levels.
>
> *Quadrants*: Permaculture is most relevant to and most highly developed in "It space," particularly the LR in dealing with ecological systems and the application of ecological principles to design for sustainable human habitation. Ethics are also a core aspect of permaculture and give credible weight to permaculture practice in "We space," although it is my experience that beyond the deep recognition of permaculture ethical principles their application is not as well developed as design practice in "It space." "I space" is almost completely undeveloped in permaculture beyond the rational level—anything more is most often actively discouraged, although interestingly pattern literacy, which is a strong theme in permaculture, is essentially at the Teal altitude. The integral impulse is there but only one leg of the Big Three is well developed—the other two are shaky at best and often lead, ironically, to unsustainable permaculture projects.
>
> *Levels:* Permaculture at its best attracts people at healthy Green verging into Teal and in this instance permaculture can be practiced at this level to good effect. My experience is that permaculture also attracts a lot of people in unhealthy

Green where Green is infected with unhealthy Red. In this case, permaculture is practiced as an angry critique of industrial society—this is the impression a lot of "mainstream" people have of permaculture, and in many cases it is well deserved. Also, there is a general lack of understanding regarding developmental unfolding in both personal and cultural contexts. While agricultural systems are successfully adapted from Red and [Amber] cultures, these cultures are often then elevated beyond higher levels of cultural unfolding.

Summary: In general there is very little recognition in permaculture literature and practice that genuine post-personal development in the UL is important; practices in the LL are not well developed or integrated with well developed practices in LR or UR; Levels or Stages of cultural and personal development are poorly understood, while the same evolutionary unfolding in the Right Hand quadrants is well understood as ecological succession; Spirit is acknowledged only uneasily; and an understanding of Lines of development, which would aid in the resolution of interpersonal conflict due to the high percentage of individualistic Types, is not in evidence.

While permaculture shows some signs of an impulse and capacity for integral practice, for all of the reasons above as well as others, it is reasonable to understand permaculture as a discipline which is only beginning its journey to integral modes of practice. (Personal communication, August 10, 2006.)

97. Kimbrell, *Fatal Harvest;* Nader and Teitel, *Genetically Engineered Food;* Shiva, *Stolen Harvest.*

98. Daly and Cobb, *For the Common Good;* Hawken, *The Ecology of Commerce;* Hawken, Lovins, and Lovins, *Natural Capitalism;* Ricklefs, *The Economy of Nature;* Sagoff, *The Economy of the Earth;* Schumacher, *Small Is Beautiful.*

99. Berkes and Folke, *Linking Social and Ecological Systems;* Norberg-Hodge, *Ancient Futures;* Sachs, *The Development Dictionary;* Shiva, *Close to Home.*

100. Alexander, *A Pattern Language* and *The Timeless Way of Building;* Grange, *Nature* and *The City;* Miller, *The Lewis Mumford Reader;* Mumford, *The City in History;* Register, *Ecocity Berkeley.*

101. Welter, *Biopolis.* Geddes was a Scottish urbanist and biologist who is often considered the grandfather of urban planning. He used a triadic concept of place, work, and folk that wove together the geographical (It/s), historical (We), and Spiritual (I) aspects of a city. Geddes's "biopolis" recognized interiors and exteriors within individuals and collectives and serves as a fruitful basis for an integral approach to urban development. Also, see Geddes's seminal work, *Cities in Evolution.*

102. Boris, *Art and Labor;* Cumming and Kaplan, *The Arts and Crafts Movement;* Wheeler, *Ruskin and Environment.*

103. Bookchin, *Ecology and Revolutionary Thought;* Bookchin, *Remaking Society* and *The Philosophy of Social Ecology;* Watson, *Beyond Bookchin.*

104. Parson, *Marx and Engels on Ecology;* Schmidt, *The Concept of Nature in Marx.*

105. Keulartz, *The Struggle for Nature*; Darier, *Discourses of the Environment*; McGee, "The Relevance of Foucault to Whiteheadian Environmental Ethics."

106. Eckersley, *Environmentalism and Political Theory*; Goodin, *Green Political Theory*; Graham, *Environmental Politics and Policy, 1960s to 1990s*; Griffin and Falk, *Postmodern Politics for a Planet in Crisis*; Meyer, *Political Nature*; Morris, *The Political Writings of William Morris*; Spretnak, *Spiritual Dimensions of Green Politics*; Wall, *Green History*; Whiteside, *Divided Natures*; Zimmerman, "Possible Political Problems of Earth-Based Religiosity."

107. Hutter, *A Reader in Environmental Law*; Hoban and Brooks, *Green Justice*.

108. Korten, *When Corporations Rule the World*; Mander and Goldsmith, *The Case Against the Global Economy*; Nader, *The Case Against "Free Trade"*; Shiva, *Water Wars*.

109. Bowers, *Education, Cultural Myths, and the Ecological Crisis* and *The Culture of Denial*; Orr, *Ecological Literacy* and *Earth in Mind*; and Stone and Barlow, *Ecological Literacy*.

110. Benyus, *Biomimicry*; McDonough and Braungart, *Cradle to Cradle*; Todd and Todd, *Bioshelters, Ocean Arks, City Farming*; Van der Ryn and Cowan, *Ecological Design*; Watkinson, *William Morris as Designer*; Willis, *The Sand Dollar and the Slide Rule*.

111. Hoffman, *Frank Lloyd Wright*; Jencks, *The Architecture of the Jumping Universe*; Pearson, *Earth to Spirit*; Ruskin, *Seven Lamps of Architecture*.

112. King, *Buildings of Earth and Straw*; Lacinski and Bergeron, *Serious Strawbale*; Pearson, *The Natural House Book*; Stiles and Stiles, *Rustic Retreats*.

113. Mander, *Four Arguments for the Elimination of Television* and *In the Absence of the Sacred*; Rothenberg, *Hand's End*; Sale, *Rebels Against the Future*; Shiva, *Monocultures of the Mind*.

114. Brown, Gardner, and Halweil, *Beyond Malthus*; Cohen, *How Many People Can the Earth Support?*; Ehrlich, *The Population Bomb*; Livi-Bacci, *A Concise History of World Population*.

115. Anderson, *You Have to Admit It's Getting Better*; Easterbrook, *A Moment on the Earth*; Lomborg, *The Skeptical Environmentalist*; Pimm, *The World According to Pimm*; Worldwatch Institute, *State of the World 2003*. Also see chap. 9.

116. Adams and McShane, *The Myth of Wild Africa*; Atkins, Roberts, and Simmons, *People, Land, and Time*; Balee, *Advances in Historical Ecology*; Barry, *Environment and Social Theory*; Bonnicksen, *America's Ancient Forests*; Boyd, *Indians, Fire, and the Land in the Pacific Northwest*; Butzer, "The Americas Before and After 1492"; Crumley, *Historical Ecology*; Denevan, "The Pristine Myth"; Delcourt and Delcourt, *Prehistoric Native Americans and Ecological Change*; Harms, *Games Against Nature*; Isenberg, *The Destruction of the Bison*; Jackson and Jackson, *Environmental Science*; Krech, *The Ecological Indian*; Markham, *A Brief History of Pollution*;

Moran, *People and Nature*; Redman, *Human Impact on Ancient Environments*; Russell, *People and the Land Through Time*; Simmons, *Changing the Face of the Earth*; Vecsey and Venables, *American Indian Environments*; White, *Land Use, Environment, and Social Change*; Whitney, *From Coastal Wilderness to Fruited Plain*; Williams, *Deforesting the Earth*.

117. Diamond, *Guns, Germs, and Steel*; Fernandez-Armesto, *Civilizations*; Griffiths and Robin, *Ecology and Empire*; Hillel, *Out of the Earth*; Montgomery, *Dirt*; Ponting, *A Green History of the World*.

118. Callicott, *Companion to a Sand County Almanac*; Campbell, *The Masks of God*; Carrasco and Kleit, *City of Sacrifice*; Clendinnen, *Aztecs*; Davies, *Human Sacrifice in History and Today*; Drescher and Engerman, *A Historical Guide to World Slavery*; Edgerton, *Sick Societies*; Girard, *Violence and the Sacred*; Goldman, *The Anthropology of Cannibalism*; Green, *The Role of Human Sacrifice in the Ancient Near East*; Hughes, *Human Sacrifice in Ancient Greece*; Keeley, *War Before Civilization*; LeBlanc, *Prehistoric Warfare in the American Southwest*; McDowell, *Hamatsa*; Nolan, Lenski, and Lenski, *Human Societies*; Petrinovich, *The Cannibal Within*; Sartore, *Humans Eating Humans*; Schele and Miller, *The Blood of Kings*; Tierney, *The Highest Altar*; Schwarz, "Indian Rights and Environmental Ethics"; Tsosie, "Tribal Environmental Policy in an Era of Self-Determination"; Turner and Turner, *Man Corn*; White, *Prehistoric Cannibalism at Mancos 5Mtumr-2346*; Windschatle, *The Killing of History*; and for an interesting look at cannibalism throughout the animal kingdom, see Elgar and Crespi, *Cannibalism*.

119. Salthe, "The Natural Philosophy of Ecology."

120. Crowley and Crowley, *Wabi Sabi Style*; Hale, *The Practical Encyclopedia of Feng Shui*; Koren, *Wabi Sabi*; Roberst, *Fast Feng Shui*; Wong, *A Master Course in Feng Shui*.

121. Cowan, *Ley Line and Earth Energies*; Devereux and Pennick, *Lines on the Landscape*; Devereux, *Places of Power*; Lane, *Landscapes of the Sacred*; Pogačnik, *Sacred Geography*; Sheldrake, *A New Science of Life* and *The Presence of the Past*; Sullivan, *Ley Lines*; Swan, *Sacred Places* and *The Power of Place*; Webster, *The Art of Dowsing*.

122. An important approach to the subtle energies of nature that is not based on nonordinary states of consciousness is the research and innovative designs of Viktor Schauberger. Much of Schauberger's efforts dealt with water (e.g., implosion technology and fluidic vortices) and renewable energy. In particular, he viewed water as living and as becoming fully mature as it flows toward the ocean and is enriched with various minerals, trace elements, and salts. When water is stripped of these qualities, it becomes sterile, lifeless, and even dangerous as it leaches out of anything it comes into contact with (e.g., when drunk by an organism) the biochemical composition associated with mature water. For a popular introduction to Schauberger's work, see Alexandersson, *Living Water*, or Bartholomew, *Hidden Nature*. For a more detailed presentation, see Coats, *Living Energies*. Coats also has

edited much of Schauberger's work into the 4-volume Eco-Technology Series (vol.1, *The Water Wizard;* vol.2, *Nature as Teacher;* vol.3, *The Fertile Earth;* vol.4, *The Energy Revolution*). For a biographical overview of Schauberger's life and intellectual landscape, see Cobbald's recent *Victor Schauberger.* Schauberger's discoveries and designs share much in common with Goethean science's flowforms (see Schwenk, *Sensitive Chaos;* Wilkes, *Flowforms*).

123. Halifax, *Shamanic Voices* and *The Fruitful Darkness;* Harner, *The Way of the Shaman;* Holler, *The Black Elk Reader;* Irwin, *Native American Spirituality;* Neihardt, *Black Elk Speaks;* Walsh, *The Spirit of Shamanism* and *The World of Shamanism.*

124. Bache, *Dark Night, Early Dawn;* McKenna, *Food of the Gods;* Schultes, Hoffman, and Ratsch, *Plants of the Gods.*

125. Bockemuhl, *Toward a Phenomenology of the Etheric World;* Hodson, *Fairies at Work and at Play* and *The Kingdom of Faerie;* Metzner, *Ayahuasca;* Nagel, "Are Plants Conscious?"; Nahmad, *Fairy Spells;* Pogačnik, *Nature Spirits and Elemental Beings;* Steiner *Nature Spirits* and *Harmony of the Creative World;* and Stewart, *Earth Light* and *Power within the Land.*

An Integral approach to nature spirits needs to distinguish between magenta magic, amber mythic, and indigo psychic interpretations and experiences of them. What is needed is an Integral approach to beings experienced in the subtle realm and to nature mysticism in general (see chap. 9). In addition, such an approach would not only distinguish between prepersonal and authentic transpersonal UL experiences of nature spirits, but would also consider LL cultural fairy traditions and folklore (see note 128), the UR occult sciences and practices developed to observe and study them (see note 132), and their influence on ecological relationships and dynamics (see note 133). Unfortunately, the notion of nature spirits is so thoroughly associated with magical thinking and mythic beliefs that it makes any serious attempt to explore this fascinating aspect of ecology seem laughable. Integral Ecology provides a sophisticated framework from which to begin exploring the subtle dimensions of ecology, while avoiding the pitfalls of New Age understandings associated with nature spirits. It is important to not just reduce NATURE to the gross realm but make room to begin exploring NATURE'S subtle realm in a rigorous and scientific (i.e., broad empiricist) fashion.

126. Grof, *The Holotropic Mind* and *Psychology of the Future.*

127. Brunke, *Animal Voices;* Cabarga, *Talks with Trees;* Hiby, *Conversations with Animals;* Meyer, *The Animal Connection;* Myers, *Communicating with Animals;* Sheldrake, *Dogs That Know When Their Owners Are Coming Home* and *The Sense of Being Stared At;* Smith, *Animal Talk* and *When Animals Speak.*

128. Anderson, *Green Man;* Doel and Doel, *The Green Man in Britain;* Matthews, *The Green Man.* Also see Narvaez, *The Good People,* for a recent collection of essays exploring the cultural aspect of fairy lore.

129. Moura, *Green Witchcraft;* Brodle, *Earthdance;* Cunningham, *Wicca;* Orr, *Ritual;* Starhawk, *The Spiral Dance;* Starhawk and Macha NightMare, *The Pagan*

Book of Living and Dying; West, *The Real Witches' Handbook*; Wood, *Sisters of the Dark Moon*.

130. Cruden, *The Spirit of Place*; Lake-Thom, *Spirits of the Earth*; Mails, *Dancing in the Paths of the Ancestors*; McCarthy, *Celebrating the Earth*.

131. Anderson, *Peyote*; Dugan, *The Vision Quest of the Plains Indians*; Foster, *The Book of the Vision Quest*; Jastrab, *Sacred Manhood Sacred Earth*; Linn and Linn, *Quest*; Maley, *Desert Shamanism*.

132. Steiner, *Nature Spirits*; Wright, *Co-creative Science*; Tompkins, *The Secret Life of Nature*.

133. Findhorn Community, *The Findhorn Garden*; McCoy, *A Witch's Guide to Faery Folk*; Pogačnik, *Nature Spirits and Elemental Beings* and *Healing the Heart of the Earth*; Riddell, *The Findhorn Community*; Roads, *Talking with Nature*; Telesco and Telesco, *Dancing with Devas*; Tompkins and Bird, *Secrets of the Soil*; Wright, *Perelandra Garden Workbook*.

134. Schilthuis, *Biodynamic Agriculture*; Steiner, *Agriculture* and *What Is Biodynamics?*

135. Fukuoka, *The One-Straw Revolution* and *The Natural Way of Farming*; Fukuoka and Metreaud, *The Road Back to Nature*.

136. Warren, *Ecofeminist Philosophy*.

137. Here we draw on Edgar Morin, *Homeland Earth*, and his notion of recursivity.

Chapter 7. Ecological Selves

1. As quoted in T. von Uexküll's "Introduction," 9.

2. Goethe, *Scientific Studies*, 39.

3. Even healthy versions of magenta and red altitude worldviews tend to be ecologically destructive since they are both egocentric.

4. Kegan, *The Evolving Self* and *In Over Our Heads*.

5. Gebser, *The Ever-Present Origin*.

6. Habermas, *The Theory of Communicative Action*.

7. Vision-logic is the first wave of postrational consciousness. Wilber sometimes refers to this as the centaur or existential level. Unlike rational consciousness, which is linear, vision-logic is a mode of consciousness that is systemic in its patterns. In its more complex forms it consists of working within systems of systems. It is the mode of consciousness that integrates body and mind. See Wilber, *Sex, Ecology, Spirituality*. Also see note 33 in chap. 2.

8. It is not enough, for example, for individuals to identify with the biosphere-as-self (as in Deep Ecology's emphasis on an expanded self-identity). After all, if individuals are "one with Gaia" but lack the capacity to transcend their differences with

fellow citizens at local, regional, national, and global levels, little if any progress will occur in securing effective solutions. Integral Ecology recognizes that for an *eco*centric approach to manifest in ourselves, and our communities, individuals have to work together to stabilize *world*centric patterns of relationship. Otherwise, ecologically concerned individuals who are ostensibly "one" with the Earth are likely to propagate dynamics of "othering" against their human "neighbors." For example, many environmentalists unnecessarily demonize business owners and politicians in an effort to highlight ecological degradation.

9. We recognize that there are many other important psychological dimensions and perspectives that must be included in an integral approach to ecology. Our decision to focus on developmental psychology in this chapter is based on two considerations: the stages of psychological development have a profound impact on how the environment is perceived and acted upon, and there is relatively little scholarship and research exploring the relationship between structural-psychological development and nature. For additional sources on psychology and nature, see Bonnes, Lee, and Bonaiutom, *Psychological Theories for Environmental Issues;* Schmuck and Schultz, *Psychology of Sustainable Development*; Stern, "Psychological Dimensions of Global Environmental Change"; Nickerson, *Psychology and Environmental Change;* Kidner, *Nature and Psyche;* Winter and Koger, *The Psychology of Environmental Problems.* Two important volumes on psychology that don't focus on ecology or the environment but that contain important implications for the dynamics of environmental decision-making include: Gilovitch, Griffin, and Kahneman, *Heuristics and Biases,* and Kahneman and Tversky, *Choices, Values and Frames.*

10. For an extensive discussion of the relationship between ecology and structuralism, see Wilber, "Excerpt D," 30–35.

11. Searles writes, in *The Nonhuman Environment in Normal Development and in Schizophrenia,* p. 55:

> I believe that every human being, however emotionally healthy, has known, at one time or another in his life, the following feelings which . . . hold sway in psychotic, and to some degree in neurotic, patients: feelings of regard for certain elements in his nonhuman environment as being integral parts of himself—and upon the loss of such objects, feelings of having lost a part of himself; a resentful conviction that some animal or inanimate object is being accorded more consideration and more love that he himself is receiving; anxiety lest he become, or be revealed as, nonhuman; desires to become nonhuman; and experiences of his own reacting to another human being as if the latter were an animal or an inanimate object. Further, I think it could readily be shown that normal, adult human beings frequently undergo "phylogenetic regression," in waking life as well as in dreams, as a means of gaining release from the demands of interpersonal living and as a means of gaining a restoration of emotional energy so that, refreshed now, they can participate in more strictly human interpersonal relatedness with new freshness and vigour.

12. Also see Gardiner and Kosmitzki, *Lives Across Cultures.*

13. See Bronfenbrenner, "Environments in Developmental Perspective"; Bronfenbrenner and Morris, "The Ecology of Developmental Processes."

14. While Bronfenbrenner's model successfully explores the multiple contexts of cultural and social influence on individual development, his model confuses individual and social holons by placing the individual in the middle of the four concentric circles and confuses Lower-Left and Lower-Right relationships by making the LL the outermost circle (macrosystem) and describing the remaining circles predominantly in LR terms.

15. His more popular books include *The Tender Carnivore and the Sacred Game* (1973), *Nature and Madness* (1982), and *Coming Home to the Pleistocene* (1998).

16. This romanticization of pre-agricultural times is often called primitivism and has been with us since the birth of the Western mind. For a historical overview of the prevalence of primitivism, see Lovejoy and Boas, *Primitivism and Related Ideas in Antiquity*, and Boas, *Primitivism and Related Ideas in the Middle Ages*.

17. See Naess, *Ecology, Community and Lifestyle*.

18. One of the problems facing approaches that create a binary between *ego*centric and *ego*centric or posit a triadic movement from self-centered to human-centered (i.e., anthropocentric) to nature-centered (i.e., bio- or ecocentric) is that all too often the bio- or ecocentric values are articulated in a way that exclude human beings. Consequently, Integral Ecology avoids the use of "biocentric" or "ecocentric" to refer to its approach and instead uses "planetcentric," which explicitly includes humans and the planet (plants, animals, and biospheric processes) in moral responsibility. Thus, the movement of self-identity is from egocentric (me) to ethnocentric (me + *my group*) to sociocentric (me + my group + *my nation*) to worldcentric (me + my group + my nation+ *all peoples*) to planetcentric (me + my group + my nation + all peoples + *all beings*).

19. See Fox, "Transpersonal Ecology" and *Toward a Transpersonal Ecology*.

20. Fox, "Transpersonal Ecology," 68.

21. See Kellert, *The Value of Life*, 46–53.

22. Ibid., 47.

23. Ibid.

24. Eagles and Muffitt, "An Analysis of Children's Attitudes toward Animals"; Kellert and Westervelt, "Children's Attitudes, Knowledge and Behaviors toward Animals: Phase V"; Kellert, "Attitudes toward Animals: Age-Related Development among Children."

25. Gilligan, *In a Different Voice*.

26. See Kahn, *The Human Relationship with Nature*, 184–87, for some additional discussion of gender differences in environmental perception and values. For

a comprehensive review of recent research on gender and the environment, see Zelenzny, Chua, and Aldrich, "New Ways of Thinking about Environmentalism."

27. In this section we have focused on the vertical component of Kellert's work because he is one of the few precursors to the structural-development approach of Integral Ecology who is interested in how different worldviews of nature develop holarchically. In addition to this initial "all-level" approach, Kellert also provides an "all-quadrant" framework for understanding the "multidimensional nature" of wildlife policy:

> As this diagram indicates, effective conservation policy, whether for endangered species or any other wildlife, requires the consideration of biophysical [UR], socioeconomic [LR], institutional-regulatory [LR], and valuation forces [UL and LL]. Moreover, these dimensions must be understood in relation to the competitive interactions of various stakeholders. And all these forces and interactions tend to change with time. Endangered species and wildlife policy should thus be regarded as a complex multidimensional, dynamic, and dialectical process, difficult to understand, and even harder to control and render more efficient and effective. (Kellert, *The Value of Life*, 155–56)

For Kellert, loss of biodiversity does not merely represent a loss of species and ecosystems, but also signals a loss of human experience and opportunity for meaning-making. For him, loss of biodiversity is an all-quadrants affair. We feel that Kellert's framework is attempting to highlight the ways all 4 quadrants tetra-arise and tetra-mesh. Kellert's framework was developed with Tim Clark and, in addition to chap. 7 in *The Value of Life*, it has been described in detail in their coauthored "'The Theory and Application of a Wildlife Policy Framework." Clearly Kellert has intuited the value of an all-quadrant, all-level approach to ecology, and as such we honor and recognize his contribution to the work we ourselves are doing.

28. See Myers, *Children and Animals*.

29. Myers, Saunders, and Birjulin, "Emotional Dimensions of Watching Zoo Animals"; Myers and Russell, "Human Identity in Relation to Wild Black Bears"; Myers, Saunders, and Garrett, "What Do Children Think Animals Need? Developmental Trends"; Myers, Saunders, and Garrett, "What Do Children Think Animals Need? Aesthetic and Psycho-Social Conceptions"; Bott, Cantrill, and Myers, "Place and the Promise of Conservation Psychology"; Myers, "Symbolic Animals and the Developing Self"; Myers and Saunders, "Animals as Links to Developing Caring Relationships with the Natural World"; Myers, "Human Development as Transcendence of the Animal Body and the Child-Animal Association in Psychological Thought."

30. Conservation psychology recognizes the importance of including developmental perspectives of psychology in the understanding of ecological relationships and identity within a 4-quadrant perspective. Thus, it is our sense that conservation psychology holds much promise in helping to make environmentalists and ecologists more aware of the value that developmental psychology has for their endeavors, and thereby helps address an area that has been largely ignored.

31. In addition to the published work of Myers and Kahn, a number of doctoral researchers have drawn on Robert Kegan's model of development, which uses five orders of consciousness, to explore the development of environmental attitudes and values. In 1992 Jill Greenwald completed her PhD dissertation at the University of Massachusetts, "Environmental Attitudes: A Structural Developmental Model." Using Kegan's model, she researched women's attitudes concerning their understanding of and relationship to the environment. Her sample base was of 27 Caucasian women between the ages of 40 and 49 from a single Massachusetts town who all scored between 2nd and 4th order through the subject-object interview. Her research demonstrated that Kegan's model can be effectively generalized to environmental attitudes. She also found that the younger women were significantly higher in ego development than the older women. She attributed this finding to historical factors. She also found a strong statistical correlation between the ways women expressed concern for the environment and how they reportedly took care of their own health, which suggests that self-identity (i.e., personal concerns) was reflective of environmental concerns. In conclusion, she points out that "developmental stage does not predict whether or not the woman is going to be concerned about the environment. Rather, stage provides information as to the reasons for which the woman may be concerned" (Greenwald, "Environmental Attitudes," 110). For an article-length summary of her research, see Robbins and Greenwald, "Environmental Attitudes Conceptualized through Developmental Theory."

In 2004, Jill Flores completed her doctoral research in psychology at Saybrook Graduate School and Research Center. For her dissertation, "Awakening to the Eco-Tragedy," she worked with a sample of 52 adult Christians who completed an environmental rating scale and a behavioral checklist. From these assessments a subgroup of 12 individuals were identified who held pro-environmental attitudes. This subgroup was given Kegan's subject-object interview to assess their level of psychological development. Flores's findings suggest that

> In spite of their initial expression of pro-environmental beliefs, participants with less complex-meaning-making were not able to epistemologically construct descriptions of human interference in the natural world. Nor could they discern in their ideology the implicit functions of social and political systems in forming their environmental worldviews. These limitations left those participants particularly vulnerable to broad systematic efforts that willfully distort environmental information in the service of ideology. (Flores, "Awakening to the Eco-Tragedy," ii–iii)

In addition to this important insight, she echoes Greenwald when she concludes that Kegan's 4th order of consciousness is an important part of "pro-environmental awareness" but that it alone does not ensure ecological beliefs, attitudes, or behaviors.

Currently, Keith Johnston, a doctoral student at Australian National University, is focusing on the ego development requirements of leaders managing the environment. His research is drawing on the adult development theories and research of individuals like Robert Kegan, Bill Torbert, Elliott Jaques, and Susanne Cook-Greuter. He is combining this with environmental management theory to create a developmental framework for environmental decisions and management. He proposes using this framework to assess levels of development within individuals involved in conservation in New Zealand. He is primarily interested in researching the kind of developmental levels necessary to make decisions on complex environmental issues. Regardless of his findings, the results of Johnston's doctoral research will be an important contribution to the issues explored in this chapter. Our hope is that more and more researchers will build on these initial findings and explore the important relationships between developmental structures and environmental attitudes.

32. Important articles that preceded this groundbreaking book and include some of its original research include Kahn and Friedman, "On Nature and Environmental Education"; Kahn, "Bayous and Jungle Rivers"; Kahn, "Children's Moral and Ecological Reasoning about the Prince William Sound Oil Spill"; Kahn, "Developmental Psychology and the Biophilia Hypothesis"; Howe, Kahn, and Friedman, "Along the Rio Negro"; Kahn and Weld, "Environmental Education"; Kahn and Friedman, "Environmental Views and Values of Children in an Inner-City Black Community."

33. See Kahn, "Nature and Moral Development"; Friedman, Kahn, and Borning, "Value Sensitive Design and Information Systems"; Kahn, "The Development of Environmental Moral Identity"; Kahn, "Ape Cognition and Why It Matters for the Field of Psychology"; Kahn and Lourenço, "Water, Air, Fire, and Earth"; Kahn, "Children's Affiliations with Nature."

34. Kahn and Kellert, *Children and Nature*.

35. Based on Kahn's "Children's Affiliations with Nature":

> In general [in our published scientific papers on how children value nature], we found a comparatively large use of anthropocentric reasoning (roughly 95 percent) and a small use of biocentric reasoning (roughly 5 percent). We also found this pattern to occur in the Amazonia Study—which included a population of children who lived in a small village along the Rio Negro that is inaccessible except by boat. This finding was surprising because it could reasonably be expected that children who live intimately with nature would have a greater biocentric affiliation with the land and animals. Instead, only in the Lisboa Study—which included an adolescent and college-age population—did we find that certain questions pulled more biocentric responses than anthropocentric responses. (p. 97)

36. Here we are using "ecocentric" interchangeably with "biocentric" and referring to the common use of this among environmentalists to mean being able to take nonhuman aspects of the planet into moral consideration.

37. Anthony, "Ecopsychology and the Deconstruction of Whiteness."

38. Kahn, "Children's Affiliations with Nature," 96.

39. Ibid. 97.

40. Kahn, *The Human Relationship with Nature*, 165.

41. We want to thank our colleague at the Integral Institute Jon Geselle for pointing this out. See Geselle, "The Development of Ecological Concern."

42. See Louv, *Last Child in the Woods*, 71, and Durie, "An Interview with Howard Gardner." The other seven intelligences are linguistic, logical-mathematical, spatial, bodily-kinesthetic, musical, interpersonal, and intrapersonal intelligence. All of Gardner's lines are basically talent lines and as such are not necessary for any other lines to develop. In other words, they are not associated with subjective identity or connected to the self-related lines.

43. To be clear, Kahn is not making these correlations. However, we find his research to be suggestive/supportive of such a schema. In an exploratory spirit we offer figure 7.2 as a way of conceptualizing the possible development of bio- and ecocentrism. Before any hard conclusions are reached, we will have to wait for Kahn and others to do more research on biocentric reasoning and its various expressions along the spiral of development.

44. Kellert, "Experiencing Nature."

45. For Kellert's sources of cognitive development, see Bloom, Engelhart, Furst, Hill, and Krathwohl, *Taxonomy of Educational Objectives, Handbook I*; Maker, *Teaching Models of the Gifted*; and Piaget, *The Child's Conception of the World*. It is not always clear how Kellert distinguishes intellectual development from the cognitive development described by Piaget and neo-Piagetians, which emphasizes perspective-taking over intellectual mastery of knowledge, though Kellert does draw on Piaget's research both in this article and in his work in general, as does Kahn.

For emotional development, see Salovey and Mayer, "Emotional Intelligence"; Mayer, Caruso, and Salovey, "Emotional Intelligence Meets Traditional Standards for an Intelligence"; Mayer, Salovey, and Caruso, "Models of Emotional Intelligence"; Mayer, Caruso, and Salovey, "Selecting a Measure of Emotional Intelligence"; Mayer, Salovey, and Caruso, "Emotional Intelligence as Zeitgeist, as Personality, and as a Mental Ability." Daniel Goleman is popularly associated with emotional intelligence due to his successful book *Emotional Intelligence* (1995), which remained on the *New York Times* best-seller list for almost a year. However, there has been criticism of how he has made use of the concept of EQ. See Hein, "Critical Review of Daniel Goleman."

Additional existing research on various lines of development that would be good starting points for exploring ecological lines includes Robert Selman's research on interpersonal development (Selman, *The Growth of Interpersonal Understanding*; Schultz and Selman, "Ego Development and Interpersonal Development in Young Adulthood"; Selman, *The Promotion of Social Awareness*); Clare Graves's research on values development (Graves, *Graves: Levels of Human Existence* and *The Never Ending Quest*); Beck and Cowan, *Spiral Dynamics*; and Lawrence Kohlberg's research on moral development (Kohlberg, *The Meaning and Measurement of Moral Development*; *The Philosophy of Moral Development*; and *The Psychology of Moral Development*). Both Kellert and Kahn are informed by Kohlberg (and Piaget). As implied in the text, Kahn and Kellert are the only researchers we are aware of who have developed their own stage conceptions of ecological lines (biocentric reasoning and ecological values respectively) based on their original research. Abigail Housen's research on aesthetic development is explored in the text in connection to Jon Geselle's work, and Susanne Cook-Greuter's research on ego-development serves as the basis for the self-identities of the ecological selves discussed in this chapter.

46. Housen, "The Eye of the Beholder"; DeSantis and Housen, "A Brief Guide to Developmental Theory and Aesthetic Development"; Kellert, *The Value of Life* and *Kinship to Mastery;* and Stables, "Environmental Literacy."

47. Consult Cook-Greuter, "Postautonomous Ego Development"; Torbert, *The Power of Balance* and *Action Inquiry*. Their research represents the most sophisticated and extensive full-spectrum (prepersonal, personal, post-personal, and post-postpersonal) research available. Their levels match very closely the levels of Spiral Dynamics. For a graphic comparison between Spiral Dynamics and the Leadership Development Framework created by Paul Landraitis from Integral Development Associates, visit www.harthillusa.com.

48. For a helpful overview of each level of ego development, see Clayton and Opotow's important anthology, *Identity and the Natural Environment;* Naess's classic *Ecology, Community and Lifestyle;* Thomashow, *Ecological Identity;* and Weigert *Self, Interaction, and Natural Environment*.

49. For a succinct and accessible article that provides a lot of detail for each self, see Cook-Greuter, "A Detailed Description of the Development of Nine Action Logics."

50. In addition each eco-self can be viewed through typologies such as the nine personalities of the enneagram system. For a valuable presentation of how all nine types are influenced by Loevinger's stages correlated with amber, orange, and green levels of psychological development, see Empereur, *The Enneagram and Spiritual Direction*. In fact, we feel that enneagram types can be considered more fundamental than levels of development. Just as states of consciousness are interpreted by levels of development; levels of development are contextualized by enneagram type. As a result, knowing someone's enneagram type can be more important than knowing their level of development since their personality type will be expressed through any level they have stabilized.

Also, masculine and feminine orientations as discussed by Gareth Hill, which have both a static and a dynamic expression, can be found within each eco-self (Hill, *Masculine and Feminine*).

51. Starhawk, *The Spiral Dance;* Sjoo, *The Great Cosmic Mother*.

52. Albanese, *Nature Religion in America* and *Reconsidering Nature Religion*.

53. Lévi-Strauss, *Totemism Today*.

54. Brodle, *Earthdance;* Foster, *The Book of the Vision Quest*.

55. Moura, *Green Witchcraft*; Lipp, *The Elements of Ritual.*

56. Tobias, *A Vision of Nature*; Diamond, *In Search of the Primitive*; Horon, *The Pure State of Nature.*

57. Huhndorf, *Going Native*; Taylor, "Earthen Spirituality or Cultural Genocide?"; Deloria, *Playing Indian*; Martin, "The American Indian as Miscast Ecologist."

58. Shepard, *Nature and Madness*; Goldsmith, *The Way*; Berman, *The Reenchantment of the World.*

59. Adams, *Ecofeminism and the Sacred*; Low and Tremayne, *Sacred Custodians of the Earth?*

60. Zakin, *Coyotes and Town Dogs.*

61. Abby, *The Monkey Wrench Gang*; Forman, *Ecodefense.*

62. Love, *Ecotage*; Manes, *Green Rage.*

63. Arnold, *Ecoterror*; Scarce, *Eco Warriors.*

64. Bell, *Seven Summits.*

65. Olsen, *Women Who Risk*; Soares and Powers, *Extreme Sea Kayaking.*

66. Chapman, *What a Hunter Brings Home*; Etling, *Hunting Superbucks.*

67. Nash, *Wilderness and the American Mind*; Fussell, *Frontier.*

68. McPherson and McPherson, *Primitive Wilderness Living and Survival Skills.*

69. Stein, *When Technology Fails*; Bingham, *The NEW Passport to Survival.*

70. Hofstadter, *Social Darwinism in American Thought.*

71. Fox, *Towards a Transpersonal Ecology.*

72. Merchant, *Reinventing Eden*; Nash, *Wilderness and the American Mind.*

73. Nash, *Wilderness and the American Mind*; Carroll, *Puritanism and the Wilderness.*

74. Schwartz and Schwartz, *The Scouting Way*; Baden-Powell, *Scouting for Boys*; Mechling, *On My Honor.*

75. Landy, Roberts, and Thomas, *The Environmental Protection Agency.*

76. Kubasek and Silverman, *Environmental Law*; Hutter, *A Reader in Environmental Law.*

77. Hodges, *Sworn to Protect*; Grosz, *Wildlife Wars.*

78. Spence, *Dispossessing the Wilderness.*

79. Anderson, *Managing Our Wildlife Resources*; Butcher, *American's National Wildlife Refuges.*

80. Czech and Krausman, *The Endangered Species Act*; Noss, O'Connell, and Murphy, *The Science of Conservation Planning.*

81. Tennyson, *A Singleness of Purpose.*

82. Graham and Buchheister, *The Audubon Ark.*

83. Hawken, Lovins, and Lovins, *Natural Capitalism.*

84. Hays, *Conservation and the Gospel of Efficiency.*

85. Oelschlaeger, *The Idea of Wilderness.*

86. Nash, *Wilderness and the American Mind.*

87. Dodson, *Ecology*; Brown and Real, *Foundations of Ecology*; Ricklefs and Miller, *Ecology.*

88. Des Jardins, *Environmental Ethics.*

89. Alexander, *A Pattern Language*; Register, *Ecocity Berkeley.*

90. Des Jardins, *Environmental Ethics.*

91. Light and Katz, *Environmental Pragmatism.*

92. Bell et al., *Environmental Psychology*; Bechtel and Churchman, *Handbook of Environmental Psychology.*

93. Brower and Leon, *The Consumer's Guide to Effective Environmental Choices.*

94. Kimbrell, *The Fatal Harvest Reader.*

95. Fox, *Towards a Transpersonal Ecology.*

96. Sessions, *Deep Ecology for the 21st Century*; Drengson and Inoue, *The Deep Ecology Movement.*

97. Warren, *Ecofeminism* and *Ecofeminist Philosophy*; Mies and Shiva, *Ecofeminism*; Griffin, *Woman and Nature*; Diamond and Orenstein, *Reweaving the World*; Keller, *Reflections on Gender and Science*; Plumwood, *Feminism and the Mastery of Nature.*

98. Bookchin, *The Philosophy of Social Ecology*; Watson, *Beyond Bookchin.*

99. Regan, *The Case for Animal Rights*; Rollin, *The Unheeded Cry.*

100. Taylor, *Respect for Nature*; Schweitzer, *Civilization and Ethics.*

101. Rolston, *Philosophy Gone Wild* and *Environmental Ethics.*

102. Roszak, Kanner, and Gomes, *Ecopsychology*; Sewall, *Sight and Sensibility*; Fisher, *Radical Ecopsychology.* For an Integral critique of ecopsychology, see Wilber, *Sex, Ecology, Spirituality*, 710–16.

103. Cole and Foster, *From the Ground Up*; Adamson, Evans, and Stein, *The Environmental Justice Reader.*

104. Goodin, *Green Political Theory*; Whiteside, *Divided Natures*.

105. Abram, *The Spell of the Sensuous*.

106. Worster, *Nature's Economy*; Nash, *Wilderness and the American Mind*; Golley, *A History of the Ecosystem Concept in Ecology*; Merchant, *The Death of Nature*.

107. Vitek and Jackson, *Rooted in the Land*; Sale, *Dwellers in the Land*.

108. Pimm, *The World According to Pimm*; McKibben, *The End of Nature*.

109. Evernden and Leslie, *The Social Creation of Nature*; Simmons, *Interpreting Nature*.

110. Guattari, *The Three Ecologies*.

111. Swimme and Berry, *The Universe Story*; Berry, *The Dream of the Earth*.

112. Berry, *Teilhard in the Ecological Age*; O'Brien, "Teilhard's View of Nature and Some Implications for Environmental Ethics."

113. Lovelock, *Gaia*.

114. Bateson, *Steps to an Ecology of Mind*.

115. Capra, *The Web of Life*; Laszlo, *The Systems View of the World*.

116. Spretnak, *The Resurgence of the Real*.

117. Leopold, *A Sand County Almanac and Sketches Here and There*.

118. Berkes and Folke, *Linking Social and Ecological Systems*; Gunderson and Holling, *Panarchy*.

119. Morin and Kern, *Homeland Earth*.

120. Schilthuis, *Biodynamic Agriculture*; Steiner, *Agriculture*.

121. Elgin, *Awakening Earth*.

122. Kealey, *Revisioning Environmental Ethics*.

123. Birch and Cobb, *The Liberation of Life*; Griffin, *Whitehead's Deeply Ecological Worldview*; Palmer, *Environmental Ethics and Process Thinking*.

124. Boff, *Ecology and Liberation*; Boff et al., *Ecology and Poverty*.

125. Dodson, *Ecology*.

126. Fox, *Towards a Transpersonal Ecology*.

127. Hargens, "Integral Development."

128. Tissot, "Integral Marine Ecology."

129. Zimmerman, *Contesting Earth's Future*; "Possible Political Problems of Earth-Based Religiosity"; "Ken Wilber's Critique of Ecological Spirituality"; "Integral Ecology."

130. Riddell, "Evolving Approaches to Conservation."

131. Eddy, "An Integral Approach to Sustainable Development"; "A Comparative Review of Ecosystem Modeling and 'Integral Theory'"; "Some First Principles of an 'Integral Geography'"; "Integral Geography."

132. Owens, "An Integral Approach to Sustainable Consumption and Waste Reduction."

133. Kreisberg, "The Twelve Niches of Plant Medicine."

134. Snorf, "Integral Eco-Design."

135. Hochachka, "Integral Community Development in San Juan del Gozo, El Salvador" and "Integrating Interiority in Community Development."

136. Prpich, "A Critical Analysis of the National Standard of Canada for Organic Agriculture."

137. Johnston, "Case Study: Passing Green Building Codes in Alameda County, California, USA."

138. Wahl, "Design for Human and Planetary Health," and "Scale-Linking Design for Systemic Health."

139. Wight, "Integrating 'It' and 'We' with the 'I' of the Beholder"; "Rethinking Regions as Holons in Holarchies"; "Place, Placemaking and Planning, Part 1"; "Placemaking as Applied Integral Ecology."

140. Salthe, "The Natural Philosophy of Ecology."

141. Cook-Greuter, "A Detailed Description of the Development of Nine Action Logics in the Leadership Development Framework."

142. Emerson, *On Nature*.

143. Bortoft, *The Wholeness of Nature*.

144. Armstrong, *Francis and Clare*.

145. Wilber, *Sex, Ecology, Spirituality*.

146. Macy, "The Ecological Self."

147. Bache, "The Eco-Crisis and Species Ego-Death."

148. Cunningham, *Wicca;* diZerega, Frew, and Wilber, "Neopaganism and the Mystical Tradition"; Farrar and Farrar, *The Witches' Way*; Frew, "Harran."

149. Jones, *The Social Face of Buddhism*.

150. McClellan, "Nondual Ecology."

151. Fox, *Towards a Transpersonal Ecology*.

152. Neihardt, *Black Elk Speaks;* Holler, *The Black Elk Reader*.

153. Fox, *The Coming of the Cosmic Christ* and *Creation Spirituality*.

154. Fukuoka, *The One-Straw Revolution* and *The Natural Way of Farming*.

155. Metzner, *Green Psychology*.

156. Metzner, *Ayahuasca*.

157. Halifax, *Shamanic Voices*; Walsh, *The Spirit of Shamanism* and *The World of Shamanism*.

158. Dugan, *The Vision Quest of the Plains Indians*; Jastrab, *Sacred Manhood, Sacred Earth*.

159. Wright, *Perelandra Garden Workbook*; Telesco and Telesco, *Dancing with Devas*.

Chapter 8. Ecological Research

1. Dodson, *Ecology*, 15.

2. Wilber, "Excerpt A," 4:1.

3. For a great article on the barriers to interdisciplinarity within an ecological context, see Lele and Norgaard, "Practicing Interdisciplinarity." All of the considerations that this article raises are worth exploring in the development of Integral research. Also see Mitchell's "Why Integrative Pluralism?" for an exploration of the importance of integrating multiple "scientific theories, models, and explanations of complex biological phenomena" (p. 81). What is so interesting about Mitchell's article is how, without any knowledge of Wilber's Integral Methodological Pluralism, it recognizes the need for "a multidimensional framework in which knowledge claims may be located and to use this more complex framework to explore the variety of epistemic practices that constitute science" (p. 82).

4. Integrative Ecology Group, "Research," http://ieg.ebd.csic.es/ieg_res.html.

5. UC Davis Graduate Group in Ecology, "Integrative Ecology," http://ecology.ucdavis.edu/AOE/inteco/Inteco_home.htm.

6. Similarly, the Integrative Ecology Group at the University of Southampton explains that their aim is "to better understand the means by which the demands of needing and being [sic] resources affect the interdependent levels of ecological organisation. The group has been formed to bring together individual areas of expertise in ecological techniques and to actively promote interaction and collaboration between individuals and research programmes." They go on to list the methodologies they are using: "recombinant DNA technology, large-scale simulation/modelling, computer/digital imaging, microsatellite markers, molecular genetics (differential display, microarrays, antisense/overexpression), GIS." See www.sbs.soton.ac.uk/research/groups/ieg. Also, the University of Florida School of Natural Resources and Environment has an Interdisciplinary Ecology Graduate Program that provides "interdisciplinary coursework in the basic and applied science of ecology and

the social, political, and economic sciences, leading to MS and PhD degrees. . . . Research areas of ecology graduate students range across natural resource ecology, environmental policy, and sustainable development." See http://snre.ufl.edu/graduate. Once again, "integrative" and "interdisciplinary" are used to refer solely to integration within and between the Right-Hand terrains.

7. Ecology and Society, "Focus and Scope," www.ecologyandsociety.org.

8. Ibid.

9. See Kurt Koller's work on Integral science (e.g., "An Introduction to Integral Science," "Architecture of an Integral Science," and "The Data and Methodologies of Integral Science").

10. For example: Hierarchy theory identifies various scales of complexity within ecosystems that different approaches take as primary. Once these scales are acknowledged, their respective approaches can be unified into a single ecological framework (consult Allen and Hoekstra, *Toward a Unified Ecology*). However, it is important to note that scientific understanding generally decreases as the scale increases. Also, the field of ecological modeling has pioneered efforts to identify patterns across ecological approaches in order to provide a comprehensive ecosystem theory; consult Jørgensen and Muller, *Handbook of Ecosystem Theories and Management*; Jørgensen, *Integration of Ecosystem Theories*.

11. For an overview of the contours of this dispute, see Howe, "Against the Quantitative-Qualitative Incompatibility Thesis, or Dogmas Die Hard," and Reichardt and Rallis, *The Qualitative-Quantitative Debate*.

12. Johnson and Onwuegbuzie, "Mixed Methods Research," 17.

13. Ibid., 24.

14. Creswell, *Research Design*, 18–19.

15. Johnson and Turner, "Data Collection Strategies in Mixed Methods Research."

16. For textbooks, see Brewer, *Foundations of Multimethod Research*; Creswell, *Research Design* and *Designing and Conducting Mixed Methods Research*; Johnson and Christensen, *Educational Research*; Tashakkori, *Mixed Methodology* and *Handbook of Mixed Methods in Social and Behavioral Research*. For articles, see Caracelli and Greene, "Data Analysis Strategies for Mixed-Method Evaluation Designs"; Greene, Caracelli, and Graham, "Toward a Conceptual Framework of Mixed-Method Evaluation Designs"; Li, Marquart, and Zercher, "Conceptual Issues and Analytic Strategies in Mixed-Method Studies of Preschool Inclusion." Journals such as *Field Methods* and *Educational Evaluation and Policy Analysis* have been important outlets for explorations and presentations of mixed methods approaches. In January 2007 the first journal solely devoted to mixed methods began publishing quarterly: *Journal of Mixed Methods Research*. The editors of this journal, John Creswell and Abbas Tashakkori, are both leaders in the field making important contributions to the development of this exciting new approach to research.

17. Johnson and Onwuegbuzie, "Mixed Methods Research," 15.

18. Integral math can be performed with up to five terms. For this chapter, as throughout this book, we will use three.

19. The families are named in a particular way, based on the definition provided and examples given; this usage differs in minor ways from how these terms are used in other contexts (e.g., philosophy). For additional explanation on the "technical" use of these terms in Integral Theory, consult Wilber's Excerpts A through D.

20. For an overview of this emerging field see the "Integral Research" special issues of *Journal of Integral Theory and Practice* 3 nos. 1–2 (2008). Also see the Integral Research Center, www.integralresearchcenter.org.

21. Tobert, *The Power of Balance*; "Transforming Social Science to Integrate Quantitative, Qualitative, and Action Research"; "A Developmental Approach to Social Science"; "The Practice of Action Inquiry"; *Action Inquiry*.

22. Bortoft, *The Wholeness of Nature*; Seamon and Zajonc, *Goethe's Way of Science*.

23. Braud, "Integral Inquiry."

24. See the case study by Owens, "An Integral Approach to Sustainable Consumption and Waste Reduction," for a 12-niche analysis of the barriers to sustainable consumption and waste management in Calgary, Canada.

25. For an example of IMP in another context, see Esbjörn-Hargens and Wilber, "Towards a Comprehensive Integration of Science and Religion."

26. Note that for each of the 4 terrains in this section, we are using the first three variables of a four-variable integral math equation to create a sentence to summarize each method's combination of perspectives. These three variables represent the Who x How x What: Who is looking (i.e., the perspective looking); How they are looking (i.e., the informal or formal method used to look); and What is being looked at (i.e., the dimension looked at). In this context the Who is a quadrant (Q), the How is a quadrivium (Qv), and the What is a quadrant (Q): Q x Qv x Q or space x mode x dimension. In other words, Who x How x What represents the phenomenological space or awareness of the perceiving subject doing the enacting x the mode of perception that is enacting the phenomenon/dimension x the dimension (perspective or aspect) of an occasion being enacted.

27. Abram, *The Spell of the Sensuous*.

28. Fisher, *Radical Ecopsychology*.

29. Brown and Toadvine, *Eco-Phenomenology*.

30. O'Connell, *The Elephant's Secret Sense*; Hughes, *Sensory Exotica*; Ford, "Plants Have Senses" in *The Secret Language of Life*. This research also supports Z5 and Z6 explorations.

31. Kahn, *The Human Relationship with Nature*; Kahn and Kellert, *Children and Nature*.

32. Bekoff and Jamieson, *Readings in Animal Cognition*. This research also supports Z1, Z5, and Z6 explorations.

33. Kegan, *The Evolving Self* and *In Over Our Heads*.

34. Latonick-Flores, "Broadening Environmental Awareness"; Greenwald, "Environmental Attitudes"; Swearingen, "Moral Development and Environmental Ethics."

35. Nash, *Wilderness and the American Mind*.

36. Casey, *Getting Back into Place* and *The Fate of Place*.

37. Steeves, *Animal Others*.

38. Keller, "Ecological Hermeneutics."

39. Van Buren, "Critical Environmental Hermeneutics."

40. Mugerauer, "Language and the Emergence of the Environment"; Basso, *Wisdom Sits in Places*.

41. Tucker and Grim, *Religions of the World and Ecology*.

42. Laird, *Biodiversity and Traditional Knowledge*; Ellen, Parkes, and Bicker, *Indigenous Environmental Knowledge and Its Transformations*.

43. For a sampling of the vast literature on environmental ethics, see the contents of the two leadings journals in the field: *Environmental Ethics* and *Environmental Values*.

44. Bortoft, *The Wholeness of Nature*; Seamon and Zajonc, *Goethe's Way of Science*.

45. Elgin, *Awakening Earth*.

46. Balee, *Advances in Historical Ecology*; Crumley, *New Directions in Anthropology and Environment*.

47. Merchant, *The Death of Nature*; Sturgeon, *Ecofeminist Natures*.

48. Maturana and Varela, *Autopoiesis and Cognition*. The term "biophenomenology" is a bit misleading since Varela acknowledged that his method does not allow one to know the inside of the interior of biological phenomena but rather the inside of the exterior. In addition to the biological phenomenology of Varela and Maturana, which accounts for how organisms enact phenomena through 3rd-person perspectives (i.e., autopoiesis), we need a *phenomenological biology* that accounts for how organisms enact phenomena through 1st-person perspectives. Such a phenomenological biology would obviously combine Z1 phenomenology with Z6 biological empiricism. In addition, it would also highlight how these two zones are conascent with Z2 structuralism and Z5 cognitive biology (i.e., autopoiesis). There are a number of biologists and philosophers with something to contribute to the development of a phenomenological biology, including Goethe's participatory biology, Hans Jonas's hermeneutical biology, Emanuel Radl's interpretative biology, and Jakob von Uexküll's theoretical biology. In fact, Uexküll's approach has been

referred to as "phenomenological biology." This is not surprising given that not only does biosemiotics recognize the role that structurally coupling (e.g., Uexküll's functional cycle) plays in an organism's enacting its phenomenal world (umwelt), but it also recognizes (via Peircean pansemiotics) that organisms have a 1st-person perspective.

Evan Thompson's compelling *Mind in Life* lays some important groundwork by linking autopoiesis and phenomenology. However, we feel his analysis doesn't go far enough (down the evolutionary spectrum). Much of his analysis is focused on subjectivity within human organisms, and his efforts would be bolstered by more differentiation and integration between Z1 and Z5. Our sense is that there is a subtle tendency in Thompson's work to privilege Z5 and interpret some phenomenological concepts and realities in predominantly autopoietic terms. For example, his notion of "sensorimotor subjectivity" seems to draw more on cognitive science than it does on phenomenology.

49. Petitot et al., *Naturalizing Phenomenology*.

50. Luhmann, *Ecological Communication* and *Social Systems*.

51. Geyer and Zouwen, *Sociocybernetics*.

52. Dodson, *Ecology*.

53. Von Bertalanffy, *General Systems Theory*.

54. Oyama, *Evolution's Eye*.

55. He uses the phrase "minding animals" to refer to both respecting animals and acknowledging that they have rich mental lives of their own (i.e., giving them a mind).

56. Bekoff, *Animal Passions and Beastly Virtues*, 1.

57. Bekoff, "The Public Lives of Animals," 128.

58. See Bekoff's many books and edited volumes: Bekoff and Jamieson, *Interpretation and Explanation in the Study of Animal Behavior*, vols. 1 and 2; Bekoff and Jamieson, *Readings in Animal Cognition*; Allen and Bekoff, *Species of Mind*; Bekoff, *Encyclopedia of Animal Rights and Animal Welfare*; Bekoff and Byers, *Animal Play*; Bekoff, *The Smile of a Dolphin*; Goodall and Bekoff, *The Ten Trusts*; Bekoff, Allen, and Burghardt, *The Cognitive Animal*; Bekoff, *Minding Animals*; Bekoff, *Encyclopedia of Animal Behavior*; Bekoff, *Animal Passions and Beastly Virtues*; Bekoff, *The Emotional Life of Animals*.

59. Bekoff, *Animal Passions and Beastly Virtues*, 23.

60. For a fascinating article on Darwin's view on worm subjectivity, see Crist, "The Inner Life of Earthworms."

61. Bekoff, *Animal Passions and Beastly Virtues*, 3. We would also add to this list morality or moral agency. For a solid case of animal morality see de Waal's many books such as *Peacemaking among Primates* and *Good Natured*. See Rottschaefer,

The Biology and Psychology of Moral Agency for an evolutionary model of moral agency. For a discussion of primate tool making and use see de Waal and Lanting, *Bonobo*. See endnote 76 for an overview of sources on animal culture. For an extensive review of the research on self-awareness in animals see Parker, Mitchell, and Boccia, *Self-Awareness in Animals and Humans*. Also see Mitchell, *Pretending and Imagination in Animals and Children*. For an thorough exploration of rationality in animals see Hurley and Nudds, *Rational Animals?* Also see the article on meta-cognition (i.e., reason) by Foote and Crystal, J., "Metacognition in the Rat." For a fascinating look at art created by primates see Lenain, *Monkey Painting*. For an overview of the ways animals self-medicate (zoopharmacosnosy) see Engel, *Wild Health*. For a detailed presentation of reflective communication in grey parrots see Pepperberg, *The Alex Studies*; and for animal communication in general see Hauser and Konishi, *The Design of Animal Communication*; Friend, *Animal Talk*; and Morton and Page's volume with the same name, *Animal Talk*.

62. One of the last lines in the sand that makes humans a unique animal is religion and spirituality. However, even this line is being wind blown with new scholarship on the biological basis of religion, psychedelic use by animals, and displays of awe and excitement by animals in response to natural phenomena. See King, *Evolving God: A Provocative View of the Origins of Religion*; Samorini, *Animals and Psychedelics: The Natural World and the Instinct to Alter Consciousness*; Verbeek and de Waal, "The Primate Relationship with Nature: Biophilia as a General Pattern"; and Wallauer, "Do Chimpanzees Feel Reverence for Nature?"

63. Bekoff, *Animal Passions and Beastly Virtues*, 7.

64. In particular see Allen and Bekoff, *Species of Mind*, and Bekoff, *Animal Passions and Beastly Virtues* and *The Emotional Life of Animals*.

65. Bekoff, *Animal Passions and Beastly Virtues*, 41.

66. Ibid., 47.

67. Griffin, *The Question of Animal Awareness*. Also see Griffin's other influential books: *Animal Thinking* and *Animal Minds*.

68. For a brief exploration of phenomenological ethology, see Churchill's two-part article "Intercorporeality Gestural Communication and the Voices of Silence," inspired by an encounter with a primate and based on the work of Merleau-Ponty and Heidegger. Another exploration of animal phenomenology that uses Merleau-Ponty is Powley, *An Exploration of a Merleau-Pontyan Approach to Cognitive Ethology*. Similarly, the work of key figures in phenomenology has been linked and extended into animality. See Lotz, "Psyche or Person? Husserl's Phenomenology of Animals," and Skocz, "Wilderness: A Zoocentric Phenomenology—From Hediger to Heidegger."

Also Kenneth Shapiro has developed a bodily based 1st-person method for conducting research on animal phenomenology. See Shapiro, "Understanding Dogs through Kinesthetic Empathy, Social Construction, and History" and "A Phenomenological Approach to the Study of Nonhuman Animals." One of the first explorations of animal phenomenology can be found in F. J. J. Buytendijk's *The Mind of a Dog* (1935).

Arguably the most important contribution to date for the emerging field of phenomenological ethology is Lotz and Painter's recent *Phenomenology and the Non-human Animal*. In addition, two noteworthy examples of phenomenological ethology include the biology professor Bernd Heinrich's study of raven awareness (*Mind of a Raven*) and the ethology professor Vilmos Csanyi's study of canine awareness (*If Dogs Could Talk*). While both Heinrich and Csanyi rely heavily on behaviors (Z6), they both take seriously animal interiors (Z1) and use their own shared depth and resonance with animals (Z3) to explore them.

69. Zoophenomenology is the label we give to the efforts to study animal interiors (e.g., bodily sensations, emotions, and awareness) using zone 1 methods. We also use the phrase "phenomenological ethology" (taken from Scott Churchill) interchangeably with "zoophenomenology." See Churchill's two-part article "Intercorporeality Gestural Communication and the Voices of Silence."

The study of animal emotions ranges from Charles Darwin's classic *The Expression of the Emotions in Man and Animals* to the widely popular books by Jeffrey Moussaieff Masson on animal emotions: *When Elephants Weep; Dogs Never Lie about Love; The Pig Who Sang to the Moon; The Nine Emotional Lives of Cats*. While Masson is not a biologist or ethologist, he is psychoanalytically trained, which does give him expertise in the world of emotions and makes his books worthwhile for scholars even though they are written for a general audience.

70. For this reason we think biosemiotics is a strong candidate for becoming an Integral science. In other words, the Integral framework could actually help clarify the various methodological approaches in biosemiotics: an Integral biosemiotics could be a subset of the Integral semiotics that Wilber is currently developing. Also see chap. 6, note 13, for a discussion of how biosemiotics is a quadradic science (as opposed to biology, which is a quadrivial science), which uses all four inside zones.

For a history of biosemiotics and an overview of its various schools, see Favareau, "The Evolutionary History of Biosemiotics." Not surprisingly, even biosemiotics is divided into Right-Hand (autopoietic/cognitive and ethology) approaches (e.g., Marcello Barbieri's "organic codes") and Left-Hand (interpretive/hermeneutic) approaches (e.g., Hoffmeyer's "signs of meaning" and Markos's biohermeneutics). Currently, biosemiotics is most influenced by the Copenhagen-Tartu school (Søren Brier, Claus Emmeche, Jesper Hoffmeyer, Kalevi Kull, and Thomas Sebeok), which is grounded in the Peircean-Uexküllian approach that emphasizes the interpretive (e.g., interior) dimensions of cells and organisms. In contrast, Marcello Barbieri, an embryologist, is not willing to push interpretation that far down the evolutionary spectrum and emphasizes that cells are "code makers," not "interpreters." Barbieri takes a more mechanistic and quantitative approach to semiosis. Since Barbieri is the editor of the first introductory book of biosemiotics (*Introduction to Biosemiotics*) and is editor-in-chief of the *Journal of Biosemiotics*, which was launched in 2006, it remains to be seen to what extent he might influence the emergence of Right-Hand approaches to semiosis. This Left-Hand versus Right-Hand tension in biosemiotics is explored by Artmann, "Computing Codes versus Interpreting Life."

71. Comparative psychology, also referred to as animal psychology, is often concerned with animal intelligence and animal cognition. As such, it draws heavily on zone 5 and 6 methods to understand how learning occurs across different species. The first textbook of comparative psychology was *The Animal Mind* by Margaret Floy Washburn (1908). Washburn was the first woman to receive a PhD in psychology. Even though she had zone 6 behavioralist leanings, she argued for the existence of animal consciousness (Z1), which was quite a radical position at the time. Furthermore, she claimed that animal minds were structurally similar to human minds and drew from behavioralism, structuralism, functionalism, and gestalt psychology. Washburn uses the term "structure" to refer to psychological (Z2), cognitive (Z5), and physiological (Z6) structures. While comparative psychology is not strictly or even primarily a zone 2 approach, it does provide a good starting point for a more specific zone 2 approach to animal psychology.

One of the most important and exciting examples of recent work being done in comparative psychology is the research of Daniel Povinelli on the folk psychology and folk physics of primates. Povinelli is the Director of the Cognitive Evolution Group (www.cognitiveevolutiongroup.org) at the University of Louisiana at Lafayette. According to his website, he is particularly interested in "using rigorous, systematic behavioral/psychological studies of chimpanzees and human adults and children to help identify those aspects of the mind that both unify and distinguish human and chimpanzee cognition." In particular, the CEG focuses on self-understanding (self concepts), theories of mind (social cognition), and theories of world (causal reasoning). Note that essentially this triadic commitment is none other than the Big Three of self, culture, and nature. In all three foci there is an explicit acknowledgment of the 1st-, 2nd-, and 3rd-person perspectives that primates can take. In fact, much of Povinelli's research is devoted to understanding the kinds of perspectives that nonhuman minds can and cannot inhabit. See Povinelli, *Folk Physics for Apes: The Chimpanzee's Theory of How the World Works*; Bering and Povinelli, "Comparing Cognitive Development"; Povinelli, "Behind the Ape's Appearance"; Povinelli and Vonk, "We Don't Need a Microscope to Explore the Chimpanzee's Mind"; Penn and Povinelli, "On the Lack of Evidence That Chimpanzees Possess Anything Remotely Resembling a 'Theory of Mind'"; and Penn and Povinelli, "Causal Cognition in Human and Nonhuman Animals."

72. For an important and in many ways groundbreaking approach to this in the context of animal cognition and intelligence, see Commons, "Measuring an Approximate g in Animals and People." One thing that makes Commons's work foundational to any Z2 approach to animals is his use of a model of hierarchical complexity for tasks performed.

73. For an introduction to zoohermeneutics, see Scheers, "Hermeneutics and Animal Being." Scheers explores, among others, Heidegger's lectures on animals' world experience given in 1929–1930 where he explicitly discusses animal being for over 200 pages (*The Fundamental Concepts of Metaphysics*, part 2). Heidegger himself suggests that his concept of being-in-the-world (*In-der-Welt-sein*) was inspired in part by Uexküll's notion that each animal exists in a worldspace within which things show

up that are pertinent to that animal (ibid., 261–264; also see Agamben, *The Open*). Another central contribution to zoohermeneutics, Hans Jonas's groundbreaking *The Phenomena of Life,* draws on Heidegger and Whitehead in the context of the biological sciences. For a book-length treatment of Heidegger's relationship to animals, see David Krell's critical *Daimon Life* and Schalow's positive *The Incarnality of Being*. Scheers also draws heavily on Uexküll's umwelt theory. Merleau-Ponty also drew on Uexküll's umwelt theory for his own explorations of the relationship between the phenomenological body and the world (Merleau-Ponty, *Nature*, 167–78). For an overview of Uexküll's influence on continental thinkers such as Heidegger, Merleau-Ponty, and Deleunze, see Buchanan's *Onto-Ethologies*.

As can be expected, J. Uexküll figures prominently in the emerging field of zoohermeneutics. See also Chang, "Semiotician or Hermeneutician? Jacob von Uexkull Revisited" and "The 'Philological Understanding' of Jacob von Uexkull." A related field is biohermeneutics. See the work of S. V. Chebanov: "Biology and Humanitarian Culture"; "Man as Participant to Natural Creation"; "Enlogue as Quasipersonal Interaction"; "The Role of Hermeneutics in Biology"; "Bio-Hermeneutics and Hermeneutics of Biology."

Another biohermeneut worth noting is Anton Markoš's *Readers of the Book of Life*. Zoohermeneutics and biohermeneutics should not be confused with eco(logical) hermeneutics. Whereas the first two focus on how nonhuman organisms interpret their environment, the latter is primarily concerned with how humans use language to disclose nature. Even though ecohermeneutics focuses on language, it does not make use of semiotics, as the other two do.

74. See Levinson, *Pets and Human Development;* Toynbee, *Animals in Roman Life and Art;* Wilson, *Biophilia;* Serpell, *In the Company of Animals;* Willis, *Signifying Animals;* Kellert, *The Value of Life;* Podberscek, Paul, and Serpell, *Companion Animals and Us*.

75. Thomas Sebeok has written extensively on zoosemiotics: *Animal Communication; How Animals Communicate; Perspectives in Zoosemiotics; Essays in Zoosemiotics*.

76. See the groundbreaking research of the primatologist Frans de Waal: Wrangham et al., *Chimpanzee Cultures;* de Waal, *Chimpanzee Politics;* de Waal and Tyack, *Animal Social Complexity;* de Waal, *Bonobo;* de Waal, *Good Natured;* de Waal, *Peacemaking among Primates*. Also see McGrew, *Chimpanzee Material Cultures*, and Cheney and Seyfarth, *Baboon Metaphysics*. For an in-depth examination of and lively debate on whale and dolphin culture, see the special issue of *Behavioral and Brain Sciences* (vol. 24, no. 2), which provides a target article by Luke Rendell and Hal Whitehead ("Culture in Whales and Dolphins") along with 37 scholarly peer responses from professionals in many related fields and the authors' responses to those comments. Also see Mann et al., *Cetacean Societies*. Two noteworthy volumes are Fragaszy and Perry, *The Biology of Traditions*, and Avital and Jablonka, *Animal Traditions*. One of the first books on the topic is Bonner, *The Evolution of Culture in Animals*.

77. For more information on Bradshaw's important research, see her website, www.kerulos.org, which discusses the current research she is involved in: post-traumatic stress disorder in animals, an elephant sanctuary for trauma recovery, parrot relational trauma, body therapy for elephants, and a community and elephant restoration project in Uganda. Also see Bradshaw and Schore, "How Elephants Are Opening Doors: Neuroethology, Attachment, and Social Context"; Bradshaw, Schore, Brown, Poole, and Moss, "Elephant Breakdown"; Bradshaw and Finlay, "Natural Symmetry"; Bradshaw and Watkins, "Trans-species Psychology: Theory and Praxis"; Bradshaw, Linden, and Schore, "Behavioral and Physiological Effects of Trauma on Psittacine Birds"; Bradshaw and Sapolsky, "Mirror, Mirror"; Bradshaw, "Not by Bread Alone: Symbolic Loss, Trauma, and Recovery in Elephant Communities"; Bradshaw, "No Longer a Mind of Our Own" and *Elephant Breakdown.*

78. Brier, *Cybersemiotics*; "Cybersemiotics: A Suggestion for a Transdisciplinary Framework"; "Biosemiotics and the Foundation of Cybersemiotics"; "The Cybersemiotic Explanation of the Emergence of Cognition"; "What Is a Possible Ontological and Epistemological Framework for a True Universal 'Information Science'?"; "Cybersemiotics: A New Interdisciplinary Development"; "From Second-Order Cybernetics to Cybersemiotics"; "Cyber-Semiotics: On Autopoiesis, Code-Duality and Sign Games in Biosemiotics."

79. See Henzi and Barrett, "The Historical Socioecology of Savanna Baboons"; Wittemyer, Douglas-Hamilton, and Getz, "The Socioecology of Elephants"; Andersen, Armitage, and Hoffman, "Socioecology of Marmots"; Anthony, "Socioecology of a Terrestrial Salamander."

80. The 8 zones can also lend themselves to a new meta-discipline of ethology: Integral ethology. Ideally Integral ethology would have a single ethological discipline per zone. For example: Z1, phenomenological ethology; Z2, developmental ethology; Z3, intersubjective ethology; Z4, cultural ethology; Z5, cognitive ethology; Z6, classical ethology; Z7, communication ethology; and Z8, social ethology. In addition, Integral ethology would also study the various lines of development within animals as well as the spectrum of states they can experience and their various personality types. For information on personalities in animals, see Dall, "Behavioural Biology," and Gosling, "From Mice to Men." For an extensive bibliography of research on personality in animals such as dogs, primates, birds, hyenas, and fish, visit the Animal Personality Institute online at www.animalpersonality.org.

81. In addition to Bekoff's interdisciplinary approach to animal minds, various criteria have been advanced by ethologists for gaining access to animal interiors. For example, on the more simplistic end of the spectrum there is Donald Griffin, who relies on analogy, novel behavior, and animal communication (Bekoff, *Animal Passions and Beastly Virtues*, 55). A more sophisticated approach can be found in Jakob von Uexküll's umwelt research methodology (*Umwelt-Forschung*), which is designed to "research into phenomenal worlds, self-worlds or subjective universes, i.e., the worlds around animals as they perceive them" (T. von Uexküll, "Introduction," 1). This methodology is built on the assumption that reality is not independent of its

observer but rather is enacted through signs and consists of reconstructing the way nature is enacted. In short, the biosemiotics methodology has post-metaphysical leanings. This post-metaphysical potential of biosemiotics (especially if it is situated within the Integral Approach) is revealed by Torsten Ruting:

> Uexküll focused on meaningful responses which enable every organism, humans included, to actively realize its own life-world—its unique *Umwelt*. Consequently, scientists were subjects interpreting and constructing their objects. Besides this refutation of scientific objectivism, Uexküll's concept of the universe as the creation of countless individual *Umwelten* challenged the idea of one universal objective world. Refuting reproaches of solipsism, Uexküll did not deny the existence of a physical world, but rejected the claims of its universally equal intersubjective significance and labeled them "metaphysical." However, Uexküll emphasized that intersubjective (interspecies) understanding is the central aim of biological investigation. (Ruting, "History and Significance of Jakob von Uexküll and of His Institute in Hamburg," 49)

Thus, biosemiotics accomplishes its methodology through what Thure von Uexküll terms "participatory observation," which involves documenting what signs are registered by the organism through a detailed understanding of the sensory organs of the observer. With this understanding in place, it is possible to "observe how the living being decodes the signs it receives in the course of its behavioral activity" (ibid., 4). Thus, participation refers to the ethologist's capacity to reconstruct the umwelt of an organism based on what signs it can receive and how it interprets them. Thure von Uexküll is clear that this process is not one of "sympathetic understanding" and should not be confused with empathy. This umwelt-research method aims to create a theory of what nature means to the vast number of umwelts that are enacting the world around them and overlapping with one another. There are two primary tasks identified by Thure von Uexküll that umwelt research must accomplish (ibid., 10). First, it must provide a description of how sign processes at the cellular level give rise to more complicated umwelts and in particular the umwelt of the researcher. Second, it must provide a description of how the outside objective world that is perceived by an organism arises out of its subjective umwelt. The solution to this second task involves integrating epistemology and biology. Only when these two tasks are accomplished, T. von Uexküll claims, "can the true task of Umwelt-research be tackled: to construct from the discoveries we have made regarding the construction of our human subjective universe a model for the construction of the subjective universes of other living beings (their Umwelts)" (ibid.). The field of biosemiotics is devoted to accomplishing these tasks and has made much progress in these areas. For more information on biosemiotics, see chap. 6. We feel that our proposal toward other minds includes both the insights and techniques of Griffin and Uexküll and adds to them, resulting in a very effective way of accessing and describing animal subjectivity. For a fascinating account of how biosemiotic research can shed light on the inner world (i.e., qualitative and subjective) of animal minds, see the four case studies presented in Pain, "Inner Representations and Signs in Animals."

82. Bekoff, *Animal Passions and Beastly Virtues*, 25.

83. In *Sex, Ecology, Spirituality*, Wilber explains what he means by mutual understanding: "The depth in me ('lived experience') must empathetically align itself, intuitively feel into, the corresponding depth (or lived experience) that I seek to understand in others, and not simply blankly register an empirical patch. *Mutual understanding* is a type of interior harmonic resonance of depth: 'I know what you mean!'" (Wilber, *Collected Works*, 6:133; italics in original).

84. See the endnote "Solidarity and Post-Kantian Internality" in Wilber, "Excerpt C," 183–188.

85. Wilber, "Excerpt C," 184–85.

86. A great example of the value of shared depth for understanding the 1st-person perspective of animals can be found in the work of Temple Grandin, who has used her autism to bridge the interior world of animals, particularly livestock (she is an animal scientist). See Grandin and Johnson, *Animals in Translation*. Also see the field research of the psychologist/anthropologist Barbara Smuts (e.g., "Encounters with Animal Minds").

87. Wilber, *Sex, Ecology, Spirituality*, 570. Elsewhere in *Sex, Ecology, Spirituality*, he explains that "a worldspace is not simply pregiven and then merely *represented* via a *correspondence* of agency with its allegedly separable communions (other agencies). Rather, the coherency of its agency (autonomy), structurally coupled with other communing agencies, enacts a worldspace mutually codetermined" (*Collected Works*, 6:569; italics in original).

88. Wilber, *A Brief History of Everything*, 99–100.

89. Wilber refers to the direct prehension of another individual's interior (i.e., the direct experience of another's feelings) as "tele-prehension." He gives three examples: psychic phenomena, nonduality, and harmonic resonance. Wilber, "Excerpt C," 48–49. Elsewhere, Wilber discusses the possibility of doing hermeneutics with sentient beings either through a phenomenological resonance (Z1) or through communicative exchange (Z3). The former is a direct route, whereas the latter is a reconstructive route and assumes a certain amount of overlap between the interiors of yourself and the other organism. "In short, through an exchange of third-person tokens (or signifiers), I attempt to understand second-person realities ('you') not as an *object* or 'it,' but as a *subject* or first-person 'I'—a bearer of consciousness, meaning, and intentionality. . ." Wilber, "Excerpt C," 181. Italics in original.

90. Wilber, *A Brief History of Everything*, 78. Wilber gives a similar example using a deer-human interaction:

> A deer sees me approach. It *sees* my exterior form, my shape, and registers all the appropriate physical stimuli coming from my form to the deer. But what do they *mean*? Am I the friendly fellow with the food, or the hunter with the rifle? The deer must *interpret* its stimuli in the *context* of its own *worldspace* and how I might *affect* it. And it is not just a matter of seeing: the deer sees just fine. But

it might be *mistaken* in its *interpretation*; I might actually have the rifle, not the food. All the physical stimuli are hitting the deer fully (that's not the problem); the problem is, what do they actually *mean*? The surfaces are *given*, but what is lurking in the depths? What are the intentions lying behind the surfaces? What is transmitted empirically but not merely given empirically? (*Collected Works*, 6:132–33; italics in original)

91. Sources on language: Bekoff reports that "in the prestigious journal *Science*, researchers in Germany report that a dog named Rico has a vocabulary of about two hundred words and is able to figure out that an unfamiliar sound referred to an unfamiliar toy" (*Animal Passions and Beastly Virtues*, 13). For documentation on the language abilities of Kanzi, a bonobo chimp, see Savage-Rumbaugh and Lewin, *Kanzi*; Savage-Rumbaugh, Shanker, and Taylor, *Apes, Language, and the Human Mind*. For vervet monkeys see Cheney and Seyfarth, *How Monkeys See the World*. For Alex the grey parrot see Pepperberg, *The Alex Studies*; "Cognitive and Communicative Abilities of Grey Parrots"; and "An Avian Perspective on Language Evolution."

92. For instance, the field biologist and environmental scientist turned animal communicator Marta Williams provides insight into how animals self disclose to reveal their interiors. See Williams, *Learning Their Language* and *Beyond Words*.

93. Wilber, *Integral Psychology*, 161.

94. With regard to nonordinary states, we feel that altered states and meditative states can both provide invaluable information about animal interiors. Rupert Sheldrake has done much to document the occurrence of telepathy between animals and humans (e.g., *Dogs That Know When Their Owners Are Coming Home* and *The Sense of Being Started At*). Most impressive is the research he is doing with the African grey parrot N'Kisi. See www.sheldrake.org/nkisi. Other important work on animal telepathy includes William Long's classic *How Animals Talk*; Penelope Smith's provocative "New Agey" *Animal Talk* and *When Animals Speak*; and John Lilly's controversial research with dolphins: *Man and Dolphin*, *The Mind of the Dolphin*, and *Communication between Man and Dolphin*. The transpersonal psychologist Stan Grof has documented the occurrence of individuals identifying with animals in the 1st-person and even obtaining information about their biology that they could not have known otherwise. See Grof, *Psychology of the Future*, 62–64 and 271–73 and *The Holotropic Mind*, 98–103. Also of interest in this context are the many ways various species seek out consciousness-altering substances. Thus it would seem that altered states of awareness play a role in consciousness all along the evolutionary spectrum: see Samorini, *Animals and Psychedelics*.

95. Bekoff builds on Burghardt's notion and calls his version "biocentric anthropocentrism" (Bekoff, *Animal Passions and Beastly Virtues*, 26–27). While we like Bekoff's version, we prefer Burghardt's phrase. See Burghardt, "Cognitive Ethology and Critical Anthropomorphism." For additional discussion on anthropomorphism and its value, see Mitchell, Thompson, and Miles, *Anthropomorphism, Anecdote, and Animals*; Fisher, "The Myth of Anthropomorphism"; and Daston and Mitman,

Thinking with Animals. For an example of the scientific value of attributing "semiotic competence" (i.e., interpretive capacities) to animals via analogy to humans, see Roepstorff, "Thinking with Animals."

96. Bekoff, *Animal Passions and Beastly Virtues*, 26.

97. Ibid., 27.

98. Stefanovic, *Safeguarding Our Common Future*. Other relevant writings of Stefanovic include "Phenomenological Reflections on Ecosystem Health"; "Phenomenological Encounters with Place"; "Interdisciplinarity and Wholeness"; "An Integrative Framework for Sustainability: The Hamilton Harbour Ecosystem."

99. The listing of the various disciplines associated with each focus group is based on Sproule-Jones, "Comments on ECOWISE," 2.

100. Stefanovic, *Safeguarding Our Common Future*, 154–55.

101. Ibid., 155.

102. For a detailed description of the matrix used, see Stefanovic, "An Integrative Framework for Sustainability: The Hamilton Harbour Ecosystem." For a philosophical overview of the project, see Stefanovic, "Interdisciplinarity and Wholeness."

103. Stefanovic, *Safeguarding Our Common Future*, 157.

104. Ibid.

105. "Interdisciplinarity and Wholeness," 81.

106. Keep in mind that this analysis is based on the published materials on the project and might not reflect accurately a more detailed evaluation of the project and its focus groups. Nevertheless, we feel the illustration is true to the general contours of the Ecowise project and as such is a valuable analysis for highlighting the additional disciplinary clarity that the IMP framework provides.

107. For some ideas of what this might look like, see Winter and Koger, *The Psychology of Environmental Problems*, chap. 6.

108. See Morton's *Ecology without Nature* for a similarly argued position that we need a "denaturalized" nature that dissolves the Romantic notion of pristine nature. Like Wilber, Morton argues that one of the biggest roadblocks to the environmental movement is the idea of nature. Thus, to save the environment, one has to realize that "nature" actually doesn't exist. Also see Costanza, Graumlich, and Steffen's *Sustainability or Collapse?* for an integrated history of people (cultural history) and the earth (environmental history). This pioneering text does much to illustrate that culture and nature come together and recursively inform each other.

Chapter 9. Ecological Harmony and Environmental Crisis in a Post-Natural World

1. Cronon, *Uncommon Ground*, 89.

2. In 1980 Carolyn Merchant identified the "death of nature" as occurring with the scientific revolution. (See Merchant, *The Death of Nature*.) This 17th-century movement marked the shift from the premodern organic worldview where women were favorably identified with living ("Mother") nature to a modern mechanistic, dead nature where men's association with culture and scientific investigation was prioritized. Thus, to make room for a postmodern worldview of nature, Merchant provides an ecofeminist critique of the hierarchical categories of lower feminine nature and higher masculine culture. This hierarchical polarization, Merchant points out, lies at the heart of Western metaphysics and social organization, and is driving culture's destruction of nature. By deconstructing the nature-culture dualism, she calls for a rebirth of "organic" nature and the rise of an egalitarian ecological holism where "each part contributes equal value to the healthy functioning of the whole" (ibid., 293). Almost a decade later Bill McKibben echoed Merchant by announcing the "end of nature." (See McKibben, *The End of Nature*.) His assessment highlights that we have reached a point in human history where the entire biosphere and pristine wilderness in particular is now irreparably altered due to climate changes caused by our unimpeded cultural activities. Thus, both Merchant and McKibben in various ways charge culture with the crime of killing nature.

With all due respect to what Merchant and McKibben have achieved, we must ask, "Which nature had *died*? Which nature has *ended*?" and "What *is* the relationship between nature and culture?" In our view, Merchant commits a pre/post fallacy by reinstating a premodern version of organic nature in postmodern times. Moreover, her reliance on ecological holism leaves her open to the charge of subtle reductionism. Similarly, McKibben's position is compromised because he relies on a notion of pristine nature or untouched wilderness. This view fails to acknowledge the insights of historical ecology and environmental anthropology, which demonstrate that throughout history humans in various ways have continually impacted their local surroundings and the global biosphere. By building his argument on the myth of pristine nature, McKibben inadvertently reinforces the nature versus culture schism that cripples so much of contemporary environmental thought and ecological action.

In our reading, Merchant and McKibben interpret nature as older and more foundational, whereas culture is newer and more destructive. While both Merchant and McKibben identify culture as having a problematic relationship with nature, they emphasize different sides of the culture-nature dichotomy in their proposed solutions. Merchant stresses culture. Hence, after providing a helpful historical discussion of how different worldviews (organismic, mechanical, holistic) circumscribe nature, she proposes that we adopt a new cultural worldview: holism. In contrast, McKibben emphasizes nature, by arguing that we have crossed a point of no return in which the nature that once existed independently of humans is no longer pure. He proposes that we take steps to protect nature from being further altered by human activity. So while Merchant highlights culturally inflected nature, McKibben is committed to a culturally free nature. These differences of emphasis underscore two of the main current positions regarding the nature-culture relationship: the social construction of nature versus the mirror of nature. In other words, although culture creates nature (myth of the framework), nature must be protected from culture (myth of the given).

Another common view in the nature versus culture debate is to declare that one pole is good and the other bad. All too often, environmentalists and ecologists are guilty of psychological splitting: making nature good and culture bad. While the opposite (nature is bad, culture is good) is also of interest to us, we will just focus on the former here. You can see this in the ways that people naively think that if something is natural or organic it is unquestionably good for you, whereas if it is cultural or synthetic it is viewed with suspicion or outright condemnation. Such simplistic thinking also leads many environmentalists to fear that all human activity is justified insofar as humans are animals and human culture is an expression of nature. This position attempts to collapse culture into nature. Similarly, such a naive position often makes modernists nervous, because they view it as threatening human uniqueness by naturalizing culture. In short, if we make human culture natural, there is nothing to stop culture from killing nature. On the other hand, if we acknowledge that nature is cultural (i.e., all organisms have some sort of intersubjectivity and thus culture), there is nothing "in kind" to distinguish human uniqueness from the rest of the animal kingdom. Using the Integral framework, we can recognize human culture as a development in nature without giving carte blanche to the destruction of the biosphere, and we can also acknowledge the ways organisms have culture without losing sight of the unique linguistic, religious, ethical, and philosophical contributions of humans.

Lastly, for an interesting multi-zone analysis of how culture and nature are implicated in each other see Atran and Medin's *The Native Mind and the Cultural Construction of Nature.* In this well researched exploration of the development of biological cognition Atran and Medin draw on cognitive psychology (Z5) and cultural anthropology (Z4) to study folk taxonomies of indigenous people. They do a great job of combining the methodological strengths of each discipline as a way of overcoming their singular limits.

3. For a discussion of the Wilber-Combs Lattice, see chap. 4, "States and Stages," in Wilber, *Integral Spirituality*.

4. See Kahn, "Children's Affiliations with Nature," 100–101, and *The Human Relationship with Nature.*

5. Kahn found in his Lisboa Study ("Children's Affiliations with Nature," 101) that of all these types of harmony with Nature, only compositional reasoning increased with age, from 3% with 5th-graders to 31% with 8th-graders to 52% with 11th-graders, and 71% with college students. This strongly suggests that this form of harmony with nature has a developmental dimension.

6. There are a variety of important texts on nature spirituality and mysticism within each of these traditions. For Native American spirituality, see the following: Ake

Hultkrantz's *The Religions of the American Indians* and *Native Religions of North America* are classic erudite descriptions of the diversity and similarity of Native North American religion and spirituality; Lawrence Sullivan's edited volume *Native Religions and Cultures of North America* examines culturally significant spiritual moments in eight Native American cultures; Sam Gill's *Native American Religions* provides a general overview, while his *Mother Earth* critically examines the concept of "Mother Earth" in Native religion; Elisabeth Tooker's *Native North American Spirituality of the Eastern Woodlands* and Miguel Leon-Portilla's *Native Meso-American Spirituality* make available rare translations of Native American spiritual texts, prayers, and rituals; Joel Martin's *The Land Looks After Us* is an important history of Native American religion; Lee Irwin's *Native American Spirituality* provides a much-needed critical exploration of many themes related to Native religion; Dennis McPherson and J. Douglas Rabb's *Indian from the Inside: A Study in Ethno-Metaphysics* provides an interesting look at Native philosophy and spiritual cosmology; Ronald Niezen's *Spirit Wars* is a sobering examination of the cultural and religious genocide of Native peoples; the Sioux activist and theologian Vine Deloria, Jr.'s *God Is Red* is a classic; the Sioux author Ed McGaa's *Mother Earth Spirituality* and his recent *Nature's Way* provide invaluable 1st-person accounts of Native beliefs and practices; Clyde Holler's *The Black Elk Reader* raises many important issues around the relationship between Christianity and Native spirituality while at the same time providing a critical evaluation of John Neihardt's classic *Black Elk Speaks*; Catherine Albanese's *Nature Religion in America* and *Reconsidering Nature Religion* both offer a scholarly exploration of the role of nature in American spirituality from pre-European contact to the present; Bruce Wilshire's *The Primal Roots of American Philosophy* offers an interesting exploration of the role Native spirituality has played in American philosophy; John Gatta's *Making Nature Sacred* provides insight into the role of nature in contemporary North American spirituality.

For Australian aboriginal spirituality, see A. P. Elkin's classic *Aboriginal Men of High Degree*; Max Charlesworth's scholarly volume *Religious Business*; Lynne Hume's *Ancestral Power*, focusing on the dreamtime from a multidisciplinary approach to consciousness; James Cowan's accessible overview *Aborigine Dreaming*; Harvey Arden's *Dreamkeepers*; Bruce Chatwin's travelogue, *The Songlines*; and Bill Stanner's influential collection of essays, *White Man Got No Dreaming*. Also noteworthy is Tiffany Behringer's research on the dreamtime using cognitive anthropology and bodylore.

For neo-paganism, see Sarah Pike's illuminating ethnography, *Earthly Bodies, Magical Selves*; Sabina Magliocco's ethnography, *Witching Culture*, which explores the role anthropology and folklore have played in the development of this religious movement; Michael York's *Pagan Theology*, an important discussion of the theological dimensions of paganism as a world religion; Jordan Paper's introduction to polytheism, *The Deities Are Many*, providing important connections to contemporary pagans; Jenny Blain's *Nine Worlds of Seid-Magic*, a case study of the role ancient Norse and Icelandic beliefs play in contemporary neo-pagan groups; Sabina Magliocco's weaving of personal experience, folklore, and history in *Witching Culture*; Joanne Pearson's edited volume, *Nature Religion Today*; Chas Clifton and

Graham Harvey's invaluable guide, *The Paganism Reader*, which covers historical roots and contemporary issues and debates.

The religious scholar Graham Harvey has written a number of noteworthy books, including *Contemporary Paganism*, which provides a solid introduction to the tradition of paganism; *Animism*, the single best recent volume on this all too often misunderstood perspective on nature; and *Shamanism*, one of the best collections of essays exploring, among other things, the differences between traditional practitioners and current appropriators such as Michael Harner.

For an introduction to the contributions of alchemy to nature mysticism see Peter Wilson, Christopher Bamford, and Kevin Townley's *Green Hermeticism*. This text serves the important role of exploring the relationship between the Western esoteric tradition and ecology. This is the only book we know of that does.

For New England transcendentalism, see Catherine Albanese's historical *Corresponding Motion;* Joel Myerson's comprehensive reader of writings between 1836 and 1844 from well-known and lesser figures, *The Transcendentalists;* and Arthur Versluis's encyclopedic *American Transcendentalism and Asian Religions*.

For African spirituality, see Jacob Olupona's edited volumes *African Spirituality*, *African Traditional Religions in Contemporary Society*, and *Religious Plurality in Africa;* E. Thomas Lawson's *Religions of Africa*, a general overview that focuses on two representative religious groups, the Yoruba of Nigeria and the Zulu of South Africa; Dominique Zahan's *The Religion, Spirituality, and Thought of Traditional Africa;* Albert Mosley's *African Philosophy*, providing a number of articles that explore epistemological and metaphysical issues related to African spirituality; and the books of Malidoma Patrice Somé, arguably the most well-known practitioner of African spirituality, popularized through his *Ritual*, *Of Water and the Spirit*, and *Healing Wisdom of Africa*.

7. Important exceptions are Roger Walsh's books, *The Spirit of Shamanism* and *The World of Shamanism*, which we highly recommend. See also Walsh, "Shamanic Experiences." One of the most striking implications of Walsh's analysis is that while traditional shamans can and do have on rare occasions causal and nondual state experiences, they predominantly have subtle state experiences. These state experiences are then often interpreted through primarily amber structures, or, in the case of neoshamanism, through green structures.

8. This chart is based on the Wilber-Combs Lattice. See Wilber, *Integral Spirituality*.

9. Here we refer to gross, subtle, causal as actual realms and not just as physical energies in the UR.

10. See Wilber, "Excerpt G," for an overview of subtle energies and these realms.

11. For an encyclopedic text on nature religions see Frazer's 700-page classic, *The Worship of Nature*. For a recent detailed overview of the relationship between ecology and nature religions, see Taylor's scholarly "Ecology and Religion: Ecology and Nature Religions." Albanese's scholarship in this area (e.g., *Nature Religion in America* and her more recent *Reconsidering Nature Religion*) has done much

to show the dark side (i.e., ethnocentrism and the desire to control nature) of nature religions, which are all too often thought of as positively promoting ecological and social harmony. For two interesting contemporary approaches to nature religion, see Corrington's *Nature's Religion* and Crosby's *A Religion of Nature*. For a recent anthology of essays on nature and the sacred see McDonald's *Seeing God Everywhere*.

12. In this context, people often equate "nature mysticism" and "shamanism." For our purposes, we avoid using the word "shamanism" in this way, for several reasons. First of all, shamanism is typically associated with indigenous people, whereas nature mysticism can occur among individuals from any culture. Thus, in this sense shamanism is a subset of nature mysticism. Second, "shamanism" is used in different ways by the anthropological and New Age communities. Cultural anthropologists use "shamanism" in the broad sense to generally describe tribal elders who serve a particular healing and leadership role in their communities. These individuals do not necessarily experience altered states associated with nature mysticism. New Age environmentalists, on the other hand, usually use "shamanism" to describe practices associated with nature mysticism by individuals who can experience altered states of union with the natural world, often through the use of entheogens. In addition, some New Age approaches to shamanism are basically just programs of guided meditation with suggestive imagery and trance beats. These are often called "neoshamanism" or more derogatorily "plastic shamanism."

One of the key issues that applies to both anthropological and New Age uses is the question of what are the stabilized levels of psychological development for shamans, whether they are traditional tribal elders or urban environmentalists; and what kinds of states of consciousness do they have access to? We feel that these two questions are best handled by the Integral nature mysticism matrix discussed in the text.

For some books on neoshamanism from a New Age environmental perspective, see Michael Harner's classic *The Way of the Shaman* and Sandra Ingerman's recent *Soul Retrieval*. Harner's particular synthesis of shamanic beliefs and practices is referred to as "core shamanism." The medical anthropologist, Buddhist, and deep ecologist Joan Halifax has done much to popularize shamanism among environmentalists. Her *Shamanic Voices* provided one of the first widely available collections of 1st-person narratives of shamans from around the world and is a rich collection of descriptions of many aspects of shamanic life. Jeremy Narby and Francis Huxley's *Shamans Through Time* provides a collection of essays that explore the 500-year history of Western culture's attempts to interpret shamans. Tom Crockett has created what he calls "integral shamanism," which draws on Wilber's work a little bit and distinguishes between pre- and postrational worldviews; see Crockett, "Integral Shamanic Dreamwork" and *Stone Age Wisdom*.

For some general books on shamanism from an anthropological perspective, see Mircea Eliade's classic 1951 text, *Shamanism*, which effectively introduced shamanism to Western culture, and Ake Hultkrantz's exploration of Native American shamanism in *Shamanic Healing and Ritual Drama*. Since 2000 a number of valuable critical perspectives on shamanism have emerged. For example, Alice Kehoe argues in *Shamans and Religion* that the term "shaman" is being widely misused, and proceeds to deconstruct much of the naive understandings of shamanism. Robert Wallis's *Shamans/Neo-Shamans* critically examines both ancient and New Age expressions of shamanism and discusses such important issues as cultural appropriation. Similarly, Philip Jenkins's *Dream Catchers* focuses on the long history of European American appropriation of Native American spirituality.

Graham Harvey's edited volume *Shamanism* provides one of the best coverages of a wide range of topics and multiple perspectives. Andrei Znamenski's edited three-volume set (1,200 pages!), *Shamanism,* brings together key articles and texts on shamanism from anthropologists, religious scholars, psychologists, and historians, providing a comprehensive survey of the many issues involved with the study of shamanism. For an extensive bibliography on shamanism that covers Siberian, Central Asian, Finno-Uralic, Celtic, African, and nontraditional contemporary shamanism; shamanism among Native Americans in North and South America and in South and East Asia; and shamanism and ethnobotany, see http://deoxy.org/shaover.htm. Shelley Osterreich's annotated bibliography, *Native North American Shamanism,* is a valuable resource for materials published up to 1998.

13. Recently, Whitney Hibbard completed his PhD dissertation at Saybrook, which researched the transpersonal experiences of non-Native practitioners of the Native American sweat lodge ceremony. He identified 32 distinct transpersonal experiences and then organized them into 18 categories and 18 subcategories. In addition, Hibbard compared his findings with Stan Grof's taxonomy of 42 types of transpersonal experience. He found that less than half of his findings fit into Grof's taxonomy and that only 10 of Grof's 42 types were experienced by Hibbard's interviewees. It would be very interesting to use the Integral nature mysticism framework to analyze both Hibbard's and Grof's taxonomies. Clearly, given the variety of experiences of nature mysticism, something like our Integral Approach is needed to honor the complexity and span of experiences that people have. See Hibbard, "Whitney Hibbard Reports of Transpersonal Experiences of Non-Native Practitioners of the Native American Sweat Lodge Ceremony." Also see Dan Merkur's discussion of 24 types of psychedelic experience (Merkur, *The Ecstatic Imagination*). Merkur also provides an interesting argument that unitive experiences occur in patterns isomorphic with cognitive development (Merkur, *Mystical Moments and Unitive Thinking*).

14. For some classic studies of nature mysticism, see Spinoza's *Tractatus Theologico-Politicus* (1670) and his *Ethics* (1676); William Ralph Inge's chapter "Nature Mysticism and Symbolism" in his *Christian Mysticism* (1899); John Edward Mercer's *Nature Mysticism* (1913), which to this day is the only book in English that bears this title; Edward Gall's chapter "Nature Mysticism" in his *Mysticism throughout the Ages* (1934); Richard Charles Zaehner's *Mysticism Sacred and Profane* (1957); and F. C. Happold's *Mysticism* (1962).

Surprisingly, nothing much has been written in the last 50 years on nature mysticism. One notable exception is Paul Marshall's recent *Mystical Encounters with the Natural World.* However, much of what Marshall is exploring is mystical experience in general or mystical experiences that take place in the context of nature but where nature is not necessarily the content. Nevertheless, his wide-ranging exploration is an important contribution toward an Integral understanding of nature mysticism. Also see Carl von Essen's worthwhile *The Hunter's Trance: Nature, Spirit and Ecology*, which explores nature mysticism in a Romantic vein through archetypal figures such as "the explorer" "the hunter" and "the poet." Another valuable source of contemporary writings on nature mysticism can be found in recent dissertations. For example, Dufrechou, "Coming Home to Nature Through the Body," and Hill, "Mountains and Mysticism." Dufrechou has summarized his research in his article "We Are One."

There are a number of sages and mystics from various traditions who are not "nature mystics" per se but who have had profound experiences of the natural world and who have discussed nature in the context of their spiritual insight: Carl Jung (Meredith Sabini's *The Earth Has Soul*), His Holiness the Dalai Lama (Dalai Lama, *On the Environment*), J. Krishnamurti (*On Nature and the Environment*), and Thomas Merton (*When the Trees Say Nothing*).

In addition, many American writers and poets have been viewed as nature mystics, such as Walt Whitman, Ralph Waldo Emerson, Henry David Thoreau, Richard Jefferies, John Muir, Annie Dillard, Mary Oliver, and Gary Snyder. For an exploration of some of these authors as nature mystics, see King, "Nature Mysticism in the Writings of Traherne, Whitman, Jefferies and Krishnamurti," and Elliott, "The Mysticism of Annie Dillard's *Pilgrim at Tinker Creek*." Not to mention the Romantic poets (Wordsworth, Blake, Keats, Tennyson, and Browning).

15. For an extensive exploration of altered states of union with nature in general and whales and dolphins in particular, see DeMares, *Peak Experiences with Cetaceans*. DeMares generalizes the discussion further in his article with Krycka, "Wild-Animal-Triggered Peak Experiences." It would be interesting to take DeMares's phenomenological descriptions and interpret them using the Integral nature mysticism lattice.

16. Often in environmental work, *ego*centric drives employ *eco*centric rhetoric. Be on the lookout for how this manifests in yourself and others.

17. All too often, various ecological theorists insert their own perspective in the place of Gaia's. In other words, a theorist's ego becomes the dominant monad for Gaia. Given that Gaia is a collective and does not have a dominant monad, it is practically speaking impossible for any individual to know what Gaia wants or needs. Once you begin speaking on behalf of Gaia, you are likely projecting your own needs and values. The scale to which this occurs among even the most respected and sophisticated thinkers and activists should give rise to caution. What is needed is a much more humble and self-reflective approach to the many perspectives that inhabit the Earth.

18. Both of us trace our abiding appreciation of the natural world in part to childhood experiences. Sean grew up in the wilds of the Olympic Peninsula and on the rugged southern coast of Oregon, where he spent most of his time outside hiking, fishing and hunting, photographing nature, and learning natural history. Michael spent countless hours playing in creeks, climbing glacial cliffs, and meandering through fields and streams on the outskirts of a small town in Ohio.

19. Kahn, as discussed in chap. 7 and mentioned above, has carried out cross-cultural empirical research on children in order to test E. O. Wilson's biophilia hypothesis. According to Wilson, biophilia—love of all life—is a universal trait because it proved adaptive for humans many thousands of years ago. As a psychologist, Kahn wanted to know whether biophilia emerges in the course of moral development, rather than being genetically encoded. As we discussed, Kahn found remarkably similar, stage-specific moral attitudes toward nature on the part of poor African-American children growing up in heavily polluted Houston neighborhoods and children growing up in a Brazilian village with easy access to the pristine rainforest. In both settings, children at the same level of moral development exhibited moral concern about the well-being of plants and animals. The most interesting aspect of this study is that both groups of children, those living in the concrete jungle and those living in the most luscious rainforest on Earth, demonstrated predominantly anthropocentric moral reasoning, with some children in each group occasionally providing a biocentric justification. These findings totally contradict what most environmentalists would assume. The findings indicate that the inner-city Houston youth demonstrate just as much biocentric reasoning as children living in one of the most unspoiled and wild environments in the world, and that those children in the Amazon exhibit anthropocentric reasoning most of the time. The fact that the city youth have any biocentric reasoning is just as interesting as the fact that the indigenous youth have so little.

Other researchers have concluded that once children become adolescents, they lose interest in exploring the outdoors and turn their attention to developing relationships with their peers. For instance, Rachel Kaplan and Stephen Kaplan conclude: "adolescents, compared to younger and older groups, have lower preference for natural settings and greater appreciation for certain kinds of developed areas. The latter tend to be places that suggest action and activity" (Kaplan and Kaplan, "Adolescents and the Natural Environment," 233–36). As Mergen points out, the question becomes: Have adolescents typically become more absorbed in their peer relationships, perhaps as a way of preparing for adult roles, including parenting? Or do modern adolescents live in significantly changed circumstances? Many adolescents are so captivated by digital entertainment and communication media designed to keep them constantly in touch with one another that little time remains for exploring natural or seminatural places. Even defining such places is difficult, as Peter Kahn has observed, because of what he calls "environmental amnesia." People grow up in natural environments that are often significantly degraded compared with what the previous generation knew. What appears "natural" to the current generation in a given locale might be regarded very differently by parents or grandparents. While this is a valuable insight, Kahn does not note the possibility that "environmental amnesia" can work in the other direction: for instance, an area can become increasingly free of pollution, as in the case of American cities, without this fact being noticed by current generations, who all too often think pollution is worse than it has ever been.

20. For research on the developmental value of exposure to nature and animal others, see Melson, *Why the Wild Things Are*; Kahn and Kellert, *Children and Nature*; and Sobel, *Beyond Ecophobia*.

21. See Louv, *Last Child in the Woods*.

22. Roszak, "Awakening the Ecological Unconscious," 2.

23. Ibid.

24. Not all ecopsychologists accept Roszak's view of the ecological unconscious. For example, see Andy Fisher's *Radical Ecopsychology*. Interestingly, Fisher's four principles of ecopsychology correspond loosely with the 4 terrains: the psychological task, "To acknowledge and better understand the human-nature relationship as a relationship" (p. 7) [terrain of culture]; the philosophical task, "To place psyche (soul, anima, mind) back into the (natural) world" (p. 9) [terrain of experience]; the practical task, "to develop therapeutic and recollective practices toward an ecological society" (p. 12) [terrain of behavior]; the critical task, "to engage in ecopsychologically based criticism" of society and its systems (p. 16) [terrain of systems].

25. Shepard, *Nature and Madness*, 124.

26. Ibid.

27. Ibid., 126.

28. Ibid., 123.

29. Shepard, *The Tender Carnivore and the Sacred Game*.

30. See Carolyn Merchant's *Reinventing Eden* for a scholarly historical discussion of the Garden of Eden myth and its relationship to ecology. For an overview of the various cross-cultural versions of the Eden myth, see Evan Eisenberg's *The Ecology of Eden*. Also see Alston Chase's *In a Dark Wood* for a sophisticated rebuttal to the various myths of nature that result from an Edenic view. In fact, there has been a growing effort to save the environment from the green altitude environmentalists, who all to often adhere to naive views of nature and science. See Martin Lewis's *Green Delusions*, which echoes much of Chase's critique, and Peter Huber's *Hard Green*, which proposes an orange altitude "manifesto" for the environment.

A number of erudite examinations of the structure and history of apocalyptic rhetoric and belief, especially in American culture, were published around (not surprisingly) the turn of the new millennium. Of particular interest are Catherine Keller's ecofeminist *Apocalypse Now and Then*, the American folklorist analysis of Daniel Wojcik's *End of the World As We Know It*, the rhetorical model advanced by Stephen O'Leary's *Arguing the Apocalypse*, and Frederic Baumgartner's *Longing for the End*, which provides an exposé of millennialism throughout the history of Christianity.

Many doomsday environmentalists would do well to familiarize themselves with the deep structure and historical prevalence of the "end of the world" predictions, especially given the resurgence of New Age prophecies related to the year 2012 (based on the Mayan calendar, which is interpreted to end on December 21, 2012). As a result of the above-cited millennial scholarship, books such as Daniel Pinchbeck's *2012* and Lawrence Joseph's *Apocalypse 2012*, which combine the ecological crisis with religious sensationalism, are more predictable than they are predictive. Not surprisingly, the authors are largely unaware that they are the most recent expressions of an age-old tale. For a refreshing approach to a posthuman world, see Alan Weisman's provocative thought experiment, *The World without Us*.

31. See, for example, Nordhaus and Shellenberger, *Break Through* (chap. 1 and p. 149).

32. The following list is an expanded version of Scull, "The Separation from More-Than-Human Nature." These are just a representative sampling, and many authors will cite other reasons in various contexts. We want to thank Nick Hedlund for researching the various root etiologies of the ecological crisis and identifying a number of quotes. Sources for the list are as follows: Abram, *The Spell of the Sensuous*; Allman, *The Stone Age Present*; T. Berry, *The Great Work*; W. Berry, *The Unsettling of America*; Bookchin, *The Ecology of Freedom*; Caldicott, *If You Love This Planet*; Cohen, *Reconnecting with Nature*; Durning, "Are We Happy Yet?"; Elgin, *Voluntary Simplicity*; Ehrlich, *The Population Bomb*; Glendinning, *My Name Is Chellis and I'm in Recovery from Western Civilization*; Gore, *Earth in the Balance*; Greenway, "The Wilderness Effect and Ecopsychology"; Hardin, "The Tragedy of the Commons"; Hillman, "A Psyche the Size of the Earth"; Mander, *In the Absence of the Sacred*; Meadows, Randers, and Meadows, *Limits to Growth*; Metzner, "The Psychopathology of the Human-Nature Relationship"; Orr, *Earth in Mind*; Quinn, *Ishmael*; Roszak, *The Voice of the Earth*; Russell, *Waking Up in Time*; Scitovsky, *The Joyless Economy*; Sewall, "The Skill of Ecological Perception"; Shepard, *Nature and Madness*; Shiva, *Earth Democracy*; Suzuki, *The Sacred Balance*; White, "The Historical Roots of Our Ecological Crisis"; Wilson, *Sociobiology*; Winter, *Ecological Psychology*.

33. The postmodern attack on and vilification of these scientists, the scientific revolution, and science in general has been forcefully challenged by a number of recent books and articles. For example, in her book *Francis Bacon* (especially chaps. 24 and 33) and her more recent article "Francis Bacon, Slave-Driver or Servant of Nature? Is Bacon to Blame for the Evils of Our Polluted Age?" Mathews demonstrates through extensive scholarship that Bacon has been grossly misinterpreted by intellectuals and postmodernists such as Karl Popper, Herbert Marcuse, and Martin Heidegger. This mishandling in turn set the stage for an ongoing dismissal of Bacon by green altitude environmentalists for his purported championing of science to "torture nature's secrets out of her." As a result she points out that "when we turn to what Bacon *actually said*, however, these various dismissals prove groundless. Absorbed in their respective ideologies, their authors had failed to read Bacon in the original—and to read him in context—thus making a travesty of his model of science, and reducing his philosophy to a handful of slogans" (p. 1, italics in original). Likewise, Radner and Radner in *Animal Consciousness* challenge the prevailing view of Descartes and show how his theory of mind actually can be placed in service of the study of animal consciousness. Not to mention that Newton was more an alchemist and theologian than he was a rationalist and a scientist—even though his contributions to math (e.g., the invention of calculus) and physics (e.g., the discovery of gravity) were foundational to the scientific revolution.

For more general critiques of how green altitude intellectuals, academics, philosophers, and environmentalists have dismissed science, see Sokal and Bricmont, *Fashionable Nonsense: Postmodern Intellectuals' Abuse of Science*; Koertge, *A House Built on Sand: Exposing Postmodernist Myths about Science*; and Gross and Levitt, *Higher Superstition: The Academic Left and Its Quarrels with Science.*

34. In terms of Wilber's developmentalism, he claims that the "Fall" occurred with involution and therefore is the result of moving from the timeless to the timebound. Consequently, he finds the attempts to recontact our early states misguided in that they are confusing timebound states with timeless ones.

35. See chap. 2, note 12, for a discussion of the relationship of language and our alleged separation from nature.

36. See Zimmerman, *Heidegger's Confrontation with Modernity.*

37. For a devastating critique of this position, see Eller, *The Myth of Matriarchal Prehistory: Why an Invented Past Won't Give Women a Future.* In 2004 she received a $40,000 National Endowment for the Humanities grant to further develop the themes of this book in her next: *From Motherright to Gylany: The Myth of Matriarchal Prehistory, 1861–2000.* For a response to Eller by a Marija Gimbutas proponent, see Marler, "The Myth of Universal Patriarchy."

38. See Edgerton, *Sick Societies*; LeBlanc, *Constant Battles*; Keeley, *War Before Civilization.*

39. Krech, *The Ecological Indian*; Vecsey and Venables, *American Indian Environments.* In fact, much of the rhetoric around the current mass extinction is ideologically driven. It is quite commonplace to see various estimates claiming that anywhere from 50% to 80% of all species will be extinct in the next 20–100 years. However, such claims seem suspect since scientists don't even have a clear definition of species; don't have a reliable estimate of current numbers of species on the planet; have notoriously been bad at making similar environmental predictions; don't yet have reliable models of planetary systems, climate change, extinction, etc.; and have only recently (since the 1980s) begun to understand and study extinction.

We are not questioning that the planet is currently in the midst of the sixth mass extinction; it is clear that humans are impacting biodiversity in an unprecedented fashion through over-harvesting/fishing/hunting, inadvertently introducing new species into areas, habitat destruction, and climate change. The scientific community is largely unified in that assessment. What we are questioning is how very few in the general discussion seem to question how many species are going extinct and how fast are they going extinct. The formulas and math behind such estimates are rarely presented or examined critically. Seldom do proponents of the current mass extinction list current losses of species—e.g., Yangtze River Dolphin (*Lipotes vexillifer*) in 2006; Miss Waldron's Red Colobus Monkey (*Procolobus badius waldronae*) in 2001; Golden Toad (*Bufo periglenes*) in 1989. Thus, you often hear about the massive extinction crisis currently underway, but you hardly ever see a list of 5 to 10 species that have either gone extinct in the last year or that are now on the endangered species list. (For a list of the top most endangered species in the world according to the World Wildlife Fund, see http://tiger.towson.edu/users/agoodw1/top_ten_endangered_species.htm.) Clearly, since we are in a major period of extinction, we should become more aware of the examples of species lost that are occurring each year and should examine more closely the analysis of the contributing factors leading toward extinction for each species. Interestingly, the 2006 IUCN (International Union for Conservation of Nature and Natural Resources) Red List of Threatened Species (www.iucnredlist.org) highlights that only 735 species have been documented as going extinct since 1500 C.E. The IUCN lists around 59,000 species of mammals, birds, reptiles, amphibians, and fishes. Of these 59,000 less than half were evaluated in 2006 for threatened status. Of those evaluated, 5,624 were identified as threatened, almost 2,000 of them being amphibians. Thus, the total percentage of threatened species in these categories ranges between 10% and 23%, depending on how you are looking at the data (www.iucnredlist.org/info/tables/table1). The IUCN reports that the number of endangered species in the above-mentioned categories more than doubled from 774 in 1996 to 1776 in 2006 (www.iucnredlist.org/info/tables/table2). Clearly there is a huge discrepancy between the number of species documented as going extinct and the number often estimated to have become extinct in the last 100 years.

In short, we feel the popular discussion of the sixth extinction lacks a sufficient amount of detail and examples, which renders a fruitful dialogue difficult. We also feel that embedded in the standard postmodern view of the sixth mass extinction are dualisms that separate humans from nature, divide culture from the natural world, and view mass extinctions as inherently bad. After all, the first five mass extinctions were "natural." We propose that such metaphysical positions need to be examined from an Integral perspective. Such a perspective would include some of the following considerations.

In their "Extinction" essay, Purvis, Jones, and Mace document over 20 common species concepts in use by scientists. Wheeler and Meier explore five of these competing theories in their informative *Species Concepts and Phylogenetic Theory*. For an interesting article on the species issue within the context of biosemiotics, see Schult, "Species, Signs, and Intentionality." For a great introduction to the complex issues of taxonomy and systematic naming as well as an extensive overview of thousands of organisms (mostly alive) at various scales of classification (kingdom, phylum, class, order, family, genus, and species), see Tudge's *The Variety of Life*. As presented by Tudge, life is believed to have emerged on this planet around 4,000 million years ago, about 500 million years after the earth formed. In the last 600 million years there have been five mass extinctions. There are currently an estimated 1.7 million documented species alive today, most of which are insects, the majority being beetles (~350,000 beetles have been described, in contrast to ~250,000 plants and ~4,000 mammals). Various biologists estimate the total number of species to be anywhere between 8 and 100 million (with 30–50 million being the most commonly cited estimate). Keep in mind that these estimates do not include the variety of bacteria now being catalogued, which could double the above figures. It is often suggested that the total number of species on the planet only represents 1% of all species that have ever lived. However, Tudge feels this 1% is actually an underrepresentative figure. He calculates that in contrast to the nearly 2 million species today there have been around 4,000 *billion* species in total (Tudge, *The Variety of Life*, 6–9). Similarly, Newman and Palmer in *Modeling Extinction* suggest that for every species alive today there are around 1,000 that are extinct (Raup also makes this claim in *Extinction*, 3–6, as does Ellis in *No Turning Back*, 20). Consequently, it would appear that extinction is more common than is often recognized, has occurred for many non-anthropogenic reasons, and likely has served an important evolutionary function.

In fact, Newman and Palmer go on to argue that most species have become extinct within 10 million years of their first appearance and that it is this admittedly high rate of extinction that has contributed to the current level of biodiversity on this planet. The reason for this is that when ecological niches are repopulated after extinctions, a wider range of adaptation strategies are developed by organisms than through the gradual process of phyletic transformation. They point out that if previous trends are any indication, then most of the current species alive on the planet will all be extinct within the next 10 million years.

Thus, with regard to the mass extinction of species that is occurring around the globe, one can point out that this could actually be in service of biodiversity, since research shows that after each of the five previous mass extinctions that have occurred there has been a large increase in biodiversity—often an exponential leap. So if environmentalists really want to promote biodiversity, they should consider helping to eliminate as many current species as possible. . . . Of course we say this tongue in cheek and see the limits of such a stance, but what we want to highlight is that it is often hard to tell whether our actions are beneficial or harmful in the long run. It is not unreasonable—though this might be undesirable—to imagine a proliferation of biodiversity on this planet as the result of another mass extinction. Of course, humans don't typically plan in terms of millions of years. We feel it is important to consider such large time scales when discussing mass extinctions.

The causes of both extinctions of individual species ("micro extinctions") and mass extinctions ("macro extinctions") are varied and can be intrinsic (evolutionary changes) and extrinsic (environmental changes). For a great review of the various possible causes (e.g., impact by comets, sea-level changes, volcanic activity, climate change, human activity), see Hallam's *Catastrophes and Lesser Calamities* and Hallam and Wignall's *Mass Extinctions and Their Aftermath*. For a concise overview of human-caused extinctions from premodern to contemporary society, with lots of informative charts and tables, see Broswimmer's *Ecocide*. See Ellis's *No Turning Back* for a worthwhile overview of extinction dynamics and considerations.

An important practice of grounding the abstract numbers of species loss is to view books that contain drawings and pictures of extinct species. Becoming familiar with the actual visual appearance of hundreds of extinct species can go a long way in putting faces on those organisms that once roamed the Earth and bring to the surface of our awareness the urgency of the current situation. For a beautifully illustrated presentation of over 100 species from every continent that have gone extinct since the European discovery of North America, see Flannery and Schouten's

A *Gap in Nature*. For a similar text that covers over 300 extinct species and provides a seven-page list of over 600 human-caused extinctions since prehistoric times, see Balouet's *Extinct Species of the World*. For an impressive presentation of over 500 species of now extinct prehistoric animals, accompanied by color plates, see Palmer's *The Marshall Illustrated Encyclopedia of Dinosaurs and Prehistoric Animals*.

40. Keep in mind that authentic respect requires taking the role of the other, and the extent to which prehistoric and even contemporary indigenous peoples could or can do this is debatable. This is not a negative reflection on indigenous peoples, as this is the case for non-indigenous peoples as well. The vast majority of individuals of yesteryear and today are ethnocentric.

41. Preece's heavily researched *Animals and Nature* does a great job of showing how "the much-maligned Western tradition has far more to commend it than is customarily recognized, and that the much-valued Oriental and Aboriginal orientations to animals and nature have habitually been described in a misleadingly rosy hue" (inside cover flap).

42. The original speech was given by Chief Seattle in the coastal Salish language and delivered in 1854 during treaty negotiations. Thirty-three years later, Dr. Henry Smith, who was purportedly present at the speech, translated his impressions of the speech into English and published a version of it in the *Seattle Sunday Star* on October 29, 1887. Smith's version of the speech served as the basis for Ted Perry's movie script rendition written in 1972.

Given the 33 years that transpired between Chief Seattle's famous speech and the writing an English translation, there is no way to verify what was spoken. Scull points to the *Seattle Sunday Star's* version of Seattle's speech (www.suquamish.nsn.us/seattle.html) as the most credible, albeit approximate, version (Scull, "Chief Seattle, er, Professor Perry Speaks," 1).

43. See Scull, "Chief Seattle, er, Professor Perry Speaks," and Rothenberg and Ulvaeus, "Will the Real Chief Seattle Please Stand Up? An Interview with Ted Perry."

44. See Wilber, "Excerpt D," for a discussion of a sociograph.

45. Redman, *Human Impact on Ancient Environments*, xi.

46. Buell, *From Apocalypse to Way of Life*, xii.

47. From the jacket cover of Edith Cobb's *The Ecology of Imagination in Childhood*. Another fitting quote for this eco-self is "Gaia needs us. She is calling us, asking us to accept this ministry as our sacred duty. In the spirit of the Celtic Druids and the Priestesses at Delphi of long ago, I invite all . . . —whatever path or name you choose—to hear her voice, to step forward and make magic with Gaia" (Howell, *Making Magic with Gaia*, 5–6).

48. Earth First! "About Earth First!" www.earthfirst.org/about.htm.

49. John Paul II, *The Ecological Crisis*, nos. 1, 15.

50. Gifford Pinchot, *An American Fable*; Microsoft Encarta Reference Library 2003.

51. Baskin, *The Work of Nature*, 9.

52. Morin and Kern, *Homeland Earth*, 74.

53. Wilber, *Brief History of Everything*, 311. Emphasis in original.

54. McClellan, "Nondual Ecology," 5–6.

55. Also, *the* crisis is always from a human perspective. How might the ecological crisis look to a woodpecker who is able to increase its habitat range because competing birds have been forced out of the area due to climate change, or a virus that is able to propagate like crazy due to transcontinental flights, or to a plant species that that is capable of colonizing new areas due to toxic pollution? It is important to keep in mind that if Gaia had a dominant monad, she might not actually consider herself to be in a crisis.

56. Keep in mind that a sociocybernetic view is only one of the 8 eco-modes. The above analysis of the eco-selves is a structural analysis: the outside view of individual interiors.

57. Connell, "Multiple Constructions of the Environmental Crisis."

58. Ibid., 1

59. Ibid., 2

60. Ibid., 4

61. Ibid., 6.

62. Ibid.

63. Ibid., 9.

64. Ibid., 7.

65. Examples of the environment getting worse include the familiar litany of increased loss of habitat and biodiversity, species extinction, deforestation, climate change, desertification, air and water pollution, increased use of unrenewable resources, poverty, world hunger, occurrence of new diseases, poaching, soil erosion, use of pesticides, etc.

66. Examples of the environment getting better include the Koyoto accords as representative of one of the first transnational environmental agreements, decreases of air and water pollution in many parts of the word, the rebounding of many near-extinct and threatened species (e.g., as a result of the Endangered Species Act), over 1,800 species protected under the Endangered Species Act, more environmental legislation being passed and enforced across the globe, breakthroughs in solar and alternative energy, an increasing number of nonprofit and for-profit organizations dedicated to the global commons, more and more open space being established

as well as more national parks being created, an increase in environmental awareness and action among many worldviews and value systems, more grassroots and internet-based efforts to protect the environment, etc. For an overview of how everyday people are protecting the planet, see Suzuki, Dressel, and Saunders, *Good News for a Change*.

A field of environmental studies that promotes the notion that things are getting better is *environmental optimism (aka environmental skepticism)*. This approach began in 1987 with the Brundtland report, *Our Common Future*, which went against the more common dire tone of environmental reports and suggested an optimistic outlook on environmental and sustainability issues. Environmental optimism challenges the "doomsday" perspective associated with environmentalists by appealing to the scientific method to demonstrate that things are either not as bad as has been reported (e.g., in the media) or that they are actually getting better. Key figures include Gregg Easterbrook, Bjørn Lomborg, Ronald Bailey, and Richard North. Foundational publications include Ron Bailey's *Eco-Scam: The False Prophets of Ecological Apocalypse* (1993), Gregg Easterbrook's *A Moment on the Earth: The Coming Age of Environmental Optimism* (1995), and Ron Bailey's edited volume *The True State of the Planet: Ten of the World's Environmental Researchers in a Major Challenge to the Environmental Movement* (1995). Current publications include Bjørn Lomborg's *The Skeptical Environmentalist: Measuring the Real State of the World* (2001), Patrick Michaels's *Meltdown: The Predictable Distortion of Global Warming by Scientists, Politicians, and the Media* (2002), and Terry Anderson's edited volume *You Have to Admit It's Getting Better: From Economic Prosperity to Environmental Quality* (2004).

While environmental optimism provides a needed corrective to environmental pessimism, we feel that the above-mentioned authors and books tend to overstate (as do their counterparts) their case at the expense of a more nuanced understanding. For example, see Paul and Anne Ehrlich's *Betrayal of Science and Reason*, which provides a presentation of the dangers of overtly optimistic rhetoric, which they label "brownwash." Thus, Integral Ecology tempers their views by embracing all three sides of its slogan: things are getting worse, things are getting better, things are Always Already perfect.

It is worth noting that the "facts" of things getting worse tend to be more contested across worldviews (e.g., traditional, modern, postmodern) than the "facts" associated with things getting better. So while we can often agree on the good news, we can't find the same consensus with regard to the bad news. For example, even though there is ostensibly widespread agreement on global climate change, there is much disagreement as to what exactly is happening, why, and what we should do about it. In other words, it is really only at a highly abstract level that many people agree on the bad news. Once you dig down into their specific perspectives, you find much variance.

67. Many environmentalists and scientists consider the controversial author of *The Skeptical Environmentalist* (TSE) and Danish statistician Bjørn Lomborg and his colleagues to be in this latter camp. However, we feel that is a misrepresentation. Lomborg ignited a fierce emotional debate within the scientific and environmental

communities with the publication of TSE in 2001. Famous scientists and global leaders denounced TSE as unscientific, fraudulent, and flawed. Others of equal stature hailed it as refreshing, well researched, and important. One reason this debate became and has remained so charged is that it represents a worldview clash between orange economic optimists and green environmental pessimists. Interestingly, *The Economist* defended Lomborg while *Scientific America* attacked him. There are obviously other important contributing factors, but this one alone should be a reminder to transcend the pro-Lomborg and anti-Lomborg rhetoric and look for the Integral patterns.

Official complaints against TSE were brought by a few scientists to the Danish Committees on Scientific Dishonesty, which led to an investigation of the contents of TSE. On January 6, 2003, the body ruled that the book was in fact scientifically dishonest but that Lomborg himself wasn't guilty since he lacked the prerequisite expertise in the fields of science involved. In short, the committee found him guilty of fabrication, selective citation, misuse of statistical methods, distorted conclusions, plagiarism, and deliberate misinterpretation of data. Not surprisingly, this ruling gave the anti-Lomborg camp all it needed to discard the entire book as naive at best and scientifically dishonest at worst. However, the story doesn't end there, though most green altitude environmentalists have conveniently not followed the more recent developments that exonerate Lomborg.

A month after the above-mentioned ruling, Lomborg filed a complaint against the decision. Ten months later the Ministry found that the investigation had been based on a number of procedural errors and invalidated the committee's previous findings. The committee decided to drop the case. As a result, a pro-Lomborg petition emerged that was signed by over 300 Danish scientists (many of them social scientists) criticizing the committee's investigation and claiming that 25 of the 27 accusations were either unsubstantiated or off the mark. In response to this petition, an anti-Lomborg one emerged that supported the committee's findings and was signed by over 600 scientists (most of them natural and medical scientists).

In 2003 a group of four international scientists (Rorsch, Frello, Soper, and de Lange) got together and penned "An Analysis of the Nature of the Opposition Raised against the Book *The Skeptical Environmentalist* by B. Lomborg." In this seemingly balanced document, where they are also critical of Lomborg, they conclude: "From the scholarly point of view the exchange of criticism and opinions between Lomborg and his opponents has been unsatisfactory (p. 6)"; "The critiques and the accusations are written in emotional style and in a—for scientists—very unprofessional, insufficiently matter-of-fact way. It appears that the opponents just refused to open their minds to views alternative to their own or to grasp the scope of the book as a whole" (ibid.); and "In our opinion, even when Lomborg had it all wrong, the opponents are guilty of (a) false, at least imprudent accusations, (b) misquotations and selective quotations and (c) ad hominem attacks" (p. 8).

Clearly, smart, rational, and environmentally concerned professionals are divided over Lomborg's findings/presentation, which suggests that something is at stake beyond mere "facts." In many ways, Lomborg's analysis strikes at the heart of a worldview that many environmentalists hold dear, and as a result its conclusions are very threatening. As Integral Ecologists it is important to approach TSE with an open mind while remaining aware of the limits of its perspective. The Lomborg case is fascinating in terms of worldview dynamics as well as epistemological issues around scientific truth and the role of emotions in scientific debate. Much is to be gained by reading TSE; much is to be lost by rejecting it out of hand due to its alleged missteps.

Since the controversy over TSE, Lomborg has continued to be involved in tackling global environmental issues. In 2004 he edited and published *Global Crises, Global Solutions*, which uses a panel of experts in economics to examine almost a dozen of the most serious problems facing the planet and offer next-step solutions. In 2006 an abridged version of this book was published: *How to Spend $50 Billion to Make the World a Better Place*. He currently has published a book on climate change—*Cool It: The Skeptical Environmentalist's Guide to Global Warming*—which will likely be as controversial as TSE given the heated debates around the subject. At the time of our writing, this new book has 83 customer reviews on amazon.com; eleven 1-star reviews and 41 5-star reviews. This echoes the emotionally charged reviews found at amazon.com with TSE, which currently has 89 1-star reviews and 165 5-star reviews.

68. Speth, *Red Sky at Morning*, xiii.

69. Amory Lovins, personal communication.

70. McClellan, "Nondual Ecology," 8–9.

71. Snyder, *Turtle Island*, 102.

72. Because Integral Ecology embraces all three positions (things are getting worse, are getting better, and are perfect just the way they are), it is an approach that is not dependent on the existence of a crisis. Many current approaches to ecology and the environment have largely organized themselves around the existence of *the* ecological crisis. Given that there has always basically been an eco-crisis and there always likely will be one, these approaches will always have something valuable to contribute to solution generation. However, we feel that such entrenched commitments (on either side of the polarity: denial and apocalyptism) ultimately are limiting and work against the kind of nuanced solutions we need.

Chapter 10. Practices for Cultivating Integral Ecological Awareness

1. Macy, *World as Lover, World as Self*, 183.

2. For a comprehensive treatment of the connection between indigenous religions and ecology, see John Grim's *Indigenous Traditions and Ecology*, which is one of 10 books in the Harvard Forum on Religion and Ecology; we highly recommend the others as well. Another valuable resource is Bron Taylor's erudite and massive (nearly 2,000 pages) *Encyclopedia of Religion and Nature*.

3. See Paul Allen and Joan deRis Allen's *Francis of Assisi's Canticle of the Creatures*, which explores the historical context for this famous prayer and presents how it

is being used contemporaneously. Also see the naturalist and theologian Edward Armstrong's *Saint Francis—Nature Mystic*.

4. A number of sources are available for psychological and spiritual practices that cultivate a profound relationship to nature. The former environmental studies professor and current meditation teacher John Milton's *Sky Above Earth Below* is a contemporary approach to cultivating nature mysticism that integrates aspects of Tibetan Buddhism, indigenous shamanism, Hindu Tantra, and esoteric Christianity. Similarly, in *Awake in the Wild* the wilderness meditation instructor Mark Coleman draws on the *vipassana* tradition in Buddhism to present a multitude of practices aimed at using the natural world as a teacher. The psychologist and wilderness guide Bill Plotkin's wonderful *Soulcraft* details nearly 50 nature-based practices that provide "a trail guide for the mystical descent into the underworld of the soul" (p. 2). Similarly, in *Well Mind, Well Earth* the environmental educator Michael J. Cohen has developed and identified over 100 practices and sensory awareness exercises for (re)connecting with nature. These practices serve as the foundation for Cohen's "Integrated Ecology" a form of applied ecopsychology. Andrea Olsen's experiential guide, *Body and Earth*, provides one of the few substantial explorations of our somatically based connection to nature. Jim Merkel's *Radical Simplicity* offers three tools for individuals to accomplish personal sustainability goals and reduce their ecological footprint. Ed O'Sullivan and Marilyn Taylor's edited volume, *Learning toward an Ecological Consciousness*, provides a wide-ranging selection of practices from various educators.

In *Coming Back to Life*, the Buddhist author and activist Joanna Macy has expanded her *Despair and Personal Empowerment in the Nuclear Age* to create a new collection of practices for "reconnecting" our lives with nature. Many of these practices are designed for small-group work. A Buddhist located in New Zealand, Tarchin Hearn has described himself as a "yogi for the natural world" and provides teachings and practices informed by the science of ecology. He is the author of *Natural Awakening*, which contains over 20 practices. Susan Murphy is a lyrical Zen master attuned to ecological awareness and aboriginal dreamtime whose *Upside-Down Zen* offers a refreshing series of daily practices to transform our lives into wakeful living.

The emerging field of eco- or green yoga has done much to connect the path of yoga to ecology through spiritual practices. For example, see Georg Feuerstein's "The Practice of Eco-Yoga," Henryk Skolimowsky's *EcoYoga*, David Frawley's *Yoga and the Sacred Fire*, and Thia Luby's *Yoga of Nature*. Also visit www.greenyoga.org and www.yogagaia.com. For practices working with nature's subtle energy and nature intelligences, see Marko Pogačnik's *Healing the Heart of the Earth, Sacred Geography*, and *Turned Upside Down*; and Michaelle Small Wright's *Perelandra Garden Workbook* and *Perelandra Garden Workbook II*.

The world-renowned tracker and survivalist Tom Brown, Jr., has written a number of important books on his experiences of nature mysticism, including *The Tracker, The Vision*, and *Awakening Spirits*. His critically acclaimed series of wilderness field guides is highly recommended, especially *Tom Brown's Field Guide to Nature Observation and Tracking, Tom Brown's Field Guide to the Forgotten*

Wilderness, and *Tom Brown's Field Guide to Living with the Earth*. Also consult his *The Science and Art of Tracking*. All of these guides provide important interior and exterior practices.

In addition, Jon Young, who was a student of Tom Brown, has founded the Wilderness Awareness School in Washington and has created two invaluable audio courses, *Seeing Through Native Eyes* and *Advanced Bird Language*. Both of these present important and effective practices for cultivating what Young now is calling "Integral Awareness," which combines Native American awareness skills, wildlife science and ecology, animal perspective taking, and inner reflection. These courses are especially effective in demonstrating the possibility of profound intersubjective connection and understanding between humans and animals.

5. If you look at this quadrant figure and follow the movement from Resonating to Observing to Patterning to Experiencing, you will notice it creates an infinity symbol, which is a helpful reminder of the elusive mystery of the natural world that is revealed when you perform this practice.

6. See Ken Wilber, Terry Patten, Adam Leonard, and Marco Morelli's *Integral Life Practice*. Integral Life Practice is similar to the Integral Transformative Practice associated with George Leonard and Mike Murphy's *The Life We Are Given*. The main difference is that ILP uses the AQAL framework to organize practices and as a result it is, among other things, modular and scalable. ITP began in 1992 when Leonard and Murphy ran a two-year course in human transformation. ILP began in 2003 when the Integral Institute began to develop seminars inspired by ITP. Building on the general goals of ITP, the Integral Institute and its lead trainers were able to quickly develop a unique approach to transformative practice using the Integral model.

7. For a comprehensive overview of Integral Life Practice and the various modules, see the boxed set *Integral Life Practice Starter Kit: The Simplest Practice You Can Do to Wake Up*.

8. We would like to thank Jon Geselle for his thoughts on this process that helped us develop this practice.

9. See Wilber, *Sex, Ecology, Spirituality* (pp. 285–93, 390, 607–12) and *A Brief History of Everything* (pp. 202–7) for a discussion of the Eco-Noetic Self and the World Soul.

10. The I-WISE in Boulder, Colorado, March 10–12, 2006.

Chapter 11. Integral Ecology in Action

1. Lewin, *Field Theory in Social Science*, 169.

2. Kazantzakis, *Saviors of God*, 99.

3. In addition to his fictional novel *State of Fear* that challenges the status of global warming, Michael Crichton has delivered a number of controversial speeches over the last few years, including a lecture at Caltech in 2003, "Aliens Cause Global

Warming," and one at the Commonwealth Club of California in the same year, "Environmentalism as a Religion," and at the National Press Club in 2005, "The Impossibility of Prediction." For some of the criticism of Crichton's positions on the environment and climate change, see the environment scholar David Sandalow, "Michael Crichton and Global Warming"; the meteorologist Jeffrey Masters, "Review of Michael Crichton's *State of Fear*"; Union of Concerned Scientists, "Crichton's Thriller *State of Fear*: Separating Fact from Fiction."

4. For a book version of Gore's documentary, see Gore, *An Inconvenient Truth*. Gore is also the author of the 1992 bestseller *Earth in Balance: Ecology and the Human Spirit* and as such has been a longtime environmental politician and activist.

5. The literature on global warming and climate change is vast and growing. A few of the more accessible books that acknowledge and try to respond to climate change include McKibben's now-classic *The End of Nature*, one of the first books to discuss global warming in detail; Houghton's solid overview *Global Warming*; and Burroughs's *Climate Change*. For a well-stated defense of why we should act in spite of the difficulty of prediction, see Philander, *Is the Temperature Rising? The Uncertain Science of Global Warming*. For a persuasive historical perspective, see Flannery, *The Weather Makers*. Another worthwhile historical overview is the environmental journalist Eugene Linden's book on paleoclimatology, *The Winds of Change*. Elizabeth Kolbert's *Field Notes from a Catastrophe* is a recent volume also worth consulting. A good general overview of the symptoms, science, and solutions of climate change can be found in Henson's *The Rough Guide to Climate Change*.

There are also a number of skeptical voices in the conversation. See the environmental journalist and skeptic Ronald Bailey's *Global Warming and Other Eco Myths* for some of the more polarized anti-global warming rhetoric out there. However, he has changed his position recently, accepting the reality of global warming (see Bailey, "We're All Global Warmers Now"), though he still is critical of much of the discussion of climate change. The climatologist Patrick Michaels is one of the most famous climate change skeptics. His book *Meltdown* presents many of the positions that skeptics take to challenge conventional wisdom on global warming. His edited volume *Shattered Consensus* features chapters by 10 scientists focusing on various dimensions of climate change. Also see the environmental scientist Fred Singer's *Hot Talk, Cold Science*. For a list of over 50 well-known global warming skeptics, see http://en.wikipedia.org/wiki/Category:Global_warming_skeptics; many of these are scientists with respectable credentials such as Richard Lindzen, MIT meteorology professor and member of the National Academy of Sciences; Robert Balling, director of the Office of Climatology and professor of geography at Arizona State University; and paleoclimatologist Tim Patterson. In fact, some scientists, such as Sherwood Idso, president of the Center for the Study of Carbon Dioxide and Global Change, claim that global warming will actually be a good thing for humanity and the planet.

6. For a succinct overview of the global warming controversy for a general audience, see http://en.wikipedia.org/wiki/Global_warming_controversy. This article is also the source of many of the questions formulated in this section.

7. For an extensive analysis of climate change response using the Integral Model, see Riedy, "The Eye of the Storm." Riedy's Integral work is showcased later in the chapter as well.

8. See, for example, Riedy and Brown, "Use of the Integral Framework to Design Developmentally-Appropriate Sustainability Communications." In this article, they draw on *cultural theory*—a social theory of environmental decision-making and consumption that identifies five worldviews, each with a "myth of nature": *fatalists*, who view nature as unpredictable; *hierarchists*, who view nature as in need of management; *individualists*, who see nature as robust and benign; *egalitarians*, who see nature as fragile; and *hermits*, who see nature as resilient (see Thompson, Ellis, and Wildawsky, *Cultural Theory*, and Thompson and Rayner, "Cultural Discourses"). While cultural theory doesn't posit a developmental relationship between these cultural types Riedy and Brown make linkages between a number of these cultural types and the eco-selves. In our interpretation of cultural theory, we see all five types as having a developmental relationship, essentially moving from the Eco-Warrior to the Eco-Holist. Admittedly, their presentation of the fatalist is more of a passive Eco-Warrior than our characterization, but there still remains much overlap. Likewise, their discussion of the hermit often could apply to any developmental level, but much of their presentation does seem to suggest a second-tier worldview. Proponents of cultural theory often focus on the three most common "pro-active" cultures (hierarchical, individualistic, and egalitarian) and argue for an equal inclusion of their partial insights into policy decisions, "claiming that they balance each other for the greater good" (Seyfang, "Consuming Values and Contested Cultures," 335).

9. For well-done presentations on why scientists cannot predict the environmental future, see Orrell, *The Future of Everything*; and Pilkey and Pilkey-Jarvis, *Useless Arithmetic*.

10. See Owens, "An Integral Approach to Sustainable Consumption and Waste Reduction."

11. Ibid., 97.

12. Ibid., 107.

13. Ibid.

14. Prpich, "An Integral Analysis of the National Standard of Canada for Organic Agriculture."

15. Ibid., 145.

16. Ibid., 147–48.

17. Ibid., 150.

18. See Feinstein, "The Pepperweed Issue in Sebastopol, CA."

19. Ibid., 100.

20. Ibid., 86.

21. He can be contacted at brad.arkell@dpiw.tas.gov.au.

22. Hay, *Main Currents in Western Environmental Thought*.

23. Eckersley, *The Green State*. Also see Barry and Eckersley, *The State and the Global Ecological Crisis*.

24. Eddy, "Integral Geography." Also see Eddy, "A Comparative Review of Ecosystem Modeling and 'Integral Theory'" and "Some First Principles of an 'Integral Geography'"; Eddy and Taylor, "Exploring the Concept of Cybercartography Using the Holonic Tenets of Integral Theory." See also Eddy's PhD thesis, "The Use of Maps and Map Metaphors for Integration in Geography" (Ottawa: Carleton University, 2006).

25. Eddy, "Integral Geography," 151–52.

26. Ibid., 152.

27. Ibid., 152–53.

28. The anthroposphere can be understood as the higher levels of the noosphere.

29. Ibid., 153.

30. Ibid., 155.

31. Ibid., 156.

32. Other organizations that have been identified as using the Integral Model include Educate Girls Globally (an Integral international development approach to girls' education), Brandt21 Forum (an upgrading of the Brandt Report to include an Integral perspective), the Centre for Global Negotiations, *Kosmos Journal* (a publication that promotes an Integral approach to international affairs and global policy issues), Caribbean Institute (a provider of Integral social and environmental sustainability initiatives), Avastone Consulting (offering training in Integral leadership for sustainability to Fortune 500 clients), and Integral Africa (an Integral Approach to training African leaders). See Brown, "Major Initiatives Which Have Used, or Are Using, the Integral Framework for Social, Environmental, and Economic Development."

33. See Brown, "The Use of an Integral Approach by UNDP's HIV/AIDS Group as Part of Their Global Response to the HIV/AIDS Epidemic."

34. Gueye et al., *Community Capacity Enhancement Handbook*, 33.

35. See Sharma, *Responding to HIV/AIDS*, 13, 25–30.

36. For example, Robertson Work, the principal advisor in the Bureau for Development Policy at UNDP headquarters in New York, uses the Integral framework as part of his program to train national and local leaders about decentralized governance; June Kunugi, the senior UNICEF representative to Oman, uses the model to develop all their assessments, advocacy programs, communication techniques, and speeches; leaders at UNICEF Bangladesh have used the model in a variety of

ways: as part of their learning program, to identify and assess interventions, to support education and cultivate multiple intelligences, and to organize conferences; leadership at the UNICEF Regional Office for South Asia has used the model for staff development and to design their Women's Right to Life and Health project; and leadership at UNICEF Bhutan has used it for their Safe Motherhood project. See Brown, "The Use of an Integral Approach by UNDP's HIV/AIDS Group as Part of Their Global Response to the HIV/AIDS Epidemic," appendix C. Also, the Integral framework is used by iSchaik Development Associates, who serve as consultants to various UN agencies and international development groups.

37. See http://multiplex.integralinstitute.org/Public/cs/files/43/sustainability/default.aspx.

38. Brown, personal communication, August 8, 2006.

39. Brown's writings and publications include "The Four Worlds of Sustainability"; "The Use of an Integral Approach by UNDP's HIV/AIDS Group as Part of Their Global Response to the HIV/AIDS Epidemic"; "Integrating the Major Research Methodologies Used in Sustainable Development"; "Theory and Practice of Integral Sustainable Development: Part 1"; "Theory and Practice of Integral Sustainable Development: Part 2"; "Integral Communication for Sustainability"; Brown and Riedy, "Use of the Integral Framework to Design Developmentally Appropriate Sustainability Communications."

40. See www.hbr.com.au.

41. Will Varey, personal communication, July 12, 2006.

42. Ibid. Varey has also used the Integral Model to create strategy frameworks, which have been adopted by state governments for water sustainability and waste management. He has also created Integral programs to address crimes and substance abuse in rural indigenous communities.

43. Ibid.

44. He can be contacted at mdekay@utk.edu; also see his website www.ecodesignresources.net.

45. Also involved are Kevin Snorf (Integral ecological design), Steven Nadel (Integral architecture), and Laura Curley (Integral landscape architecture).

46. DeKay and Guzowski, "A Model for Integral Sustainable Design Explored through Daylighting," 3.

47. Ibid.

48. Ibid., 4. Note that the third level of the 12 niches in figures 11.1 and 11.4 is based on an earlier version than the one presented in this book.

49. See DeKay, "Integral Sustainable Design."

50. Ibid.

51. Ibid.

52. For an overview of the project, see http://www.tdot.state.tn.us/sr475/library/bcgitdot.pdf.

53. DeKay, personal communication, July 7, 2006.

54. He can be contacted at jwight@cc.umanitoba.ca.

55. Wight, "Placemaking as Applied Integral Ecology." For some of his other work, see "Integrating 'It' and 'We' with the 'I' of the Beholder"; "Rethinking Regions as Holons in Holarchies"; "Place, Placemaking and Planning."

56. Wight, "Placemaking as Applied Integral Ecology," 128.

57. Ibid., 131.

58. See http://datasearch.uts.edu.au/isf/about/details.cfm?StaffId=2442. Dr. Riedy can be contacted at Christopher.Riedy@uts.edu.au.

59. Riedy, "The Eye of the Storm." Also see Riedy, "Two Social Practices to Support Emergence of a Global Collective" and "Integrating Exterior and Interior Knowledge in Sustainable Development Policy."

60. See www.drishti.ca. Originally Drishti was called "Centre for Integral Ecology," but Gail Hochachka changed it to "Integral Action" to reflect a wider context of Integral practice.

61. See www.drishti.ca/homepage.htm.

62. Ibid.

63. See www.drishti.ca/projects.htm.

64. See www.drishti.ca/integral_applications.htm.

65. See www.drishti.ca/resources.htm.

66. Hochachka, *Developing Sustainability, Developing the Self.* This book has also been translated into Spanish. Other important writings by Hochachka include: "Introductory Paper on Integral International Development"; "Integral International Development"; "Ecological Governance in Nigeria"; and "Integral Assessment Methodology" (with Sandra Thomson). All of these resources and more can be downloaded at www.drishti.ca/resources.htm.

67. This project was funded by Canada's International Development Research Centre, with Gail Hochachka as project coordinator and primary researcher.

68. The seven organizations as described by Hochachka included "*UNDP Bureau for Development Policy's Leadership for Results Programme*—a leadership program carried out in 40 countries, with curriculum that includes elements of an Integral approach, that aims to catalyze leadership from within individuals, to create break-through initiatives for addressing the HIV/AIDS epidemic; *Educate Girls Globally*—a US-based NGO that works in India to increase girls' education using

a methodology influenced by Integral Theory that mobilizes parents and other community members to consciously engage in the process of girls' education, and thus also community wellbeing; *Integral Africa*—a pan-African leadership development Master's program, informed by an Integral approach, being pioneered by Ethiopian-born Yene Assegid, to develop the incredible potentials and insights of leaders for development across the African continent; *Institute for Action and Progress (INAPRO)*—a grassroots rural organization that works in the Andes of Peru to facilitate healthy psychological and emotional development of Quechua children and youth, in service of improved community resilience in a post-war context; *Centro Bartolomé de las Casas (CBC)*—a popular education organization in San Salvador that works to foster community wellbeing using an integrative approach (both folk integral and integrally informed). The work seeks to address concrete needs, such as local economy, yet also enables people to draw upon their own subjectivity to heal from post-conflict trauma and to make new meanings of society, regarding faith, gender dynamics, social roles, violence, and more; *EveryONE-Ethiopia*—a non-governmental organization that works with the interior dimension of personal development with the exterior dimension of economic development as an integrated approach to addressing the HIV/AIDS epidemic with several different target groups in 5 sub-cities in Addis Ababa; and *Integrated Service for AIDS Prevention and Support Organization (ISAPSO)*—a non-governmental organization in Addis Ababa that works with various target groups to address HIV/AIDS, particularly low income women, with a methodology that creates peer education support groups as platforms for learning about HIV/AIDS, for professional development and income generation, as well as empowerment and personal change" (Hochachka, "An Integral Approach to Well Being," 1–2).

69. For a detailed description of these findings, see Hochachka, "An Integral Approach to Well Being," appendix 1.

70. See www.onesky.ca.

71. Community Conversations is an excellent set of methodologies used by the UNDP HIV/AIDS Group in their Community Capacity Enhancement (see www.undp.org/hiv/docs/prog_guides/cce_handbook.pdf). Negash Shiferaw at negashiferaw@yahoo.com is the person to follow up with regarding this work and its impacts.

72. Hochachka, personal communication, July 15, 2006.

73. Ibid.

74. See www.permaforesttrust.org.au.

75. Winton, personal communication, July 10, 2006.

76. See www.nickwilding.com.

77. See www.community.nickwilding.com.

78. She can be contacted at marilyn@integralcity.com.

79. Also, Hamilton's doctoral research used Integral frameworks and methodologies to study learning and leadership in self-organizing online community systems.

80. This table is based on her website at www.integralcity.com, which also contains additional information for each of the quadrants of the integral city.

81. For example, see her dissertation: Hamilton, "The Berkana Community of Conversations"; or some of her recent articles, "Four Questions That Release the Potential of Your City"; "Integral Community"; "Integral Framework for Sustainable Planning."

82. Kevin Snorf can be contacted at Kevin@warrioryoga.com.

83. Snorf, personal communication, August 2, 2006.

84. Ibid.

85. See www.teleosis.org.

86. Kreisberg uses the Integral Model in a number of his courses within this program. This program is part of JFKU's Integral Studies Department, which is based on the Integral AQAL framework. The other programs in the department are the Integral Psychology Program, the Integral Theory Program, and the Consciousness and Transformative Studies Program. See www.jfku.edu.

87. See www.teleosis.org/symbiosis-current.php.

88. Currently Prescott College in Arizona (www.prescott.edu) is one of the best places to study Integral Ecology. While they don't have a specific degree in Integral Ecology, a number of students have created an individualized degree that has allowed them to study environmental and ecological issues from an Integral perspective. There are numerous faculty there who are familiar with the Integral Model and capable of supporting students in getting an Integral education.

The California Institute of Integral Studies now offers a four-course Integral Ecology track in the Philosophy, Cosmology, and Consciousness (PCC) program (www.ciis.edu/pcc/integralecology.php). However, this track uses "integral" in a very broad nonspecific sense, which in our opinion makes its approach almost synonymous with the deep ecology movement. Consequently, the view of Integral Ecology presented in this volume is quite different in many respects from the one presented in this track. In fact, many of the critiques we advance throughout this volume apply in various degrees to the assumptions, beliefs, and views held by some of the faculty and many of the students within the PCC program. Nevertheless, it is exciting to see a variety of integral ecologies emerging, and we trust that this track will make an important contribution to the exploration and application of an integral approach to ecology.

It is worth noting that Sean earned his doctorate from the PCC program in 2005. During his six years within the program, he was the only one talking about and writing on Integral Ecology. As noted in the preface, much of this volume is based on his dissertation research. Soon after he graduated, individuals within the program began to use "integral ecology" to refer to an approach quite different from the one he researched and developed. Sean currently teaches online courses in Integral Ecology through JFKU's Master of Arts in Integral Theory. These courses are open to graduate students at other institutions. Contact Sean for information.

89. Kreisberg, "Toward an Integral Ecology of Medicinal Plants," 5.

90. Kreisberg, "Spiral Dynamics, *Sequoia Sempervirens* and the Redwood Forests," 3.

91. Kreisberg, "An Integral Ecology of Sudden Oak Death," 5–6.

92. Ibid., 67.

93. See www.teleosis.org/mission.php.

94. See www.teleosis.org/programs.php.

95. Teleosis Institute, "The Green Health Care Program," 13.

96. See www.openskywilderness.com.

97. Dupuy, personal communication, July 28, 2006. Also see Dupuy, "Integral Recovery."

98. Ibid.

99. Ibid.

100. Ibid.

101. Dupuy has also rewritten the Twelve Steps of AA as the Integral Twelve Steps, updating and bringing forward the many gifts of the Alcoholics Anonymous tradition, but at the same time getting rid of some of the outdated material.

102. See www.greenlighttrust.org.

103. Personal communication, August 20, 2006.

104. See www.whatsworking.com.

105. See Johnston and Master, *Green Remodeling*; Johnston, *Building Green in a Black and White World*.

106. Integral Naked, "Who Is David Johnston?" http://in.integralinstitute.org/contributor.aspx?id=46.

107. Ibid.

108. As quoted on the homepage of Next Step Integral: www.nextstepintegral.org.

109. For details on all their foci and programs, see www.nextstepintegral.org.

110. Martineau, "Humanity, Forest Ecology, and the Future in a British Columbia Valley: A Case Study," 36–40.

111. For an engaging transcript of an interview recorded in February 2007, in which Martineau discusses the three-year effort of SIFCo, see http://integralecology.ca/articles.php. Also see his recent article, "Humanity, Forest Ecology, and the Future in a British Columbia Valley: A Case Study," where he discusses the entire project in detail.

112. See http://integralecology.ca/index.php.

113. Personal communication, January 7, 2007.

114. Martineau, "Humanity, Forest Ecology, and the Future in a British Columbia Valley," 29.

115. For four additional Integral Ecology case studies, see the special double issue of *World Futures: The Journal of General Evolution* 61, nos. 1–2 (2005), devoted to Integral Ecology and guest-edited by Sean Esbjörn-Hargens.

Case Studies

Notes for case studies can be found at the end of each case study.

Conclusion. The Integral Ecology Advantage

1. Clark, "How Wide is Deep Ecology?" 3–4.

REFERENCES

Abbey, E. 1968. *Desert Solitaire*. Tucson: University of Arizona Press.

———. 1975/2000. *The Monkey Wrench Gang*. New York: Perennial Classics.

Abe, Joseph M., P. E. Dempsey, and D. A. Bassett. 1998. *Business Ecology: Giving Your Organization the Natural Edge*. Boston: Butterworth-Heinemann.

Abram, D. 1996. *The Spell of the Sensuous: Perception and Language in a More-Than-Human World*. New York: Pantheon Books.

Abrams, M. H. 1965. *The Mirror and the Lamp: Romantic Theory and the Critical Tradition*. Rev. ed. Oxford: Oxford University Press.

———. 1973. *Natural Supernaturalism: Tradition and Revolution in Romantic Literature*. New York: W. W. Norton.

Acampora, R. R. 1994. "Using and Abusing Nietzsche for Environmental Ethics." *Environmental Ethics* 16: 87–194.

Adams, C. J., ed. 1993. *Ecofeminism and the Sacred*. New York: Continuum.

Adams, J. S., and T. O. McShane. 1996. *The Myth of Wild Africa: Conservation without Illusion*. Berkeley: University of California Press.

Adams, P. C., S. Hoelscher, and K. E. Till, eds. 2001. *Textures of Place: Exploring Humanist Geographies*. Minneapolis: University of Minnesota Press.

Adamson, J., M. M. Evans, and R. Stein, eds. 2002. *The Environmental Justice Reader: Politics, Poetics, and Pedagogy*. Tucson: University of Arizona Press.

Agamben, G. 2004. *The Open: Man and Animal*. Stanford, CA: Stanford University Press.

Ahl, V., and T. F. H. Allen. 1996. *Hierarchy Theory: A Vision, Vocabulary, and Epistemology*. New York: Columbia University Press.

Albanese, C. L. 1977. *Corresponding Motion: Transcendental Religion and the New America*. Philadelphia: Temple University Press.

———. 1991. *Nature Religion in America: From the Algonkian Indians to the New Age*. Chicago: University of Chicago Press.

———. 2002. *Reconsidering Nature Religion*. Valley Forge, PA: Trinity Press International.

Albanese, C. L., ed. 1988. *The Spirituality of the American Transcendentalists*. Macon, GA: Mercer University Press.

Alexander, C. 1977. *A Pattern Language: Towns, Buildings, Construction*. New York: Oxford University Press.

———. 1979. *The Timeless Way of Building*. New York: Oxford University Press.

———. 2002–2003. *The Order of Nature: An Essay on the Art and the Nature of the Universe*. Vols. 1–4. Berkeley: Center for Environmental Structure.

Alexander, C., and E. Langer, eds. 1990. *Higher Stages of Human Development: Perspectives on Adult Growth*. New York: Oxford University Press.

Alexanderson, O. 1984. *Living Water: Viktor Schauberger and the Secrets of Natural Energy*. Bath, UK: Gateway Books.

Allee, W. C., A. Emerson, O. Park, T. Park, and K. Schmidt. 1949. *Principles of Animal Ecology*. Philadelphia: W. B. Saunders.

Allen, C., and M. Bekoff. 1997. *Species of Mind: The Philosophy and Biology of Cognitive Ethology*. Cambridge, MA: MIT Press.

Allen, P. M., and J. D. Allen. 1996. *Francis of Assisi's Canticle of the Creatures: A Modern Spiritual Path*. New York: Continuum.

Allen, T. F. H., and T. W. Hoekstra. 1992a. "Comment on Rowe's Article." *Functional Ecology* 6 (1): 119.

———. 1992b. *Toward a Unified Ecology*. New York: Columbia University Press.

Allen, T. F. H., and T. B. Starr. 1982. *Hierarchy: Perspectives for Ecological Complexity*. Chicago: University of Chicago Press.

Allman, W. F. 1994. *The Stone Age Present: How Evolution Has Shaped Modern Life—From Sex, Violence, and Language to Emotions, Morals, and Communities*. New York: Simon & Schuster.

Allred, B. W., and E. S. Clements, eds. 1949. *The Dynamics of Vegetation: Selections from the Writings of F. E. Clements*. New York: H. W. Wilson Company.

Almaas, A. H. 2004. *The Inner Journey Home: Soul's Realization of the Unity of Reality*. Boston: Shambhala Publications.

Altieri, M. A. 1995. *Agroecology: The Science of Sustainable Agriculture*. 2nd ed. Boulder, CO: Westview Press.

Amrine, F., ed. 1996. *Goethe in the History of Science*. 2 vols. New York: Peter Lang.

Andersen, D. C., K. B. Armitage, and R. S. Hoffman. 1976. "Socioecology of Marmots: Female Reproductive Strategies." *Ecology* 57 (3): 552–60.

Anderson, E. F. 1996. *Peyote: The Divine Cactus*. Tucson: University of Arizona Press.

Anderson, S. 2001. *Managing Our Wildlife Resources*. 4th ed. Upper Saddle River, NJ: Prentice Hall.

Anderson, T. L., ed. 2004. *You Have to Admit It's Getting Better: From Economic Prosperity to Environmental Quality*. Stanford, CA: Hoover Press.

Anderson, W. 1991. *Green Man: The Archetype of Our Oneness with the Earth*. San Francisco: Harper.

Andrewartha, H. G., and L. C. Birch. 1954. *The Distribution and Abundance of Animals*. Chicago: University of Chicago Press.

Andruss, V., C. Plant, J. Plant, and S. Wright, eds. 1990. *Home! A Bioregional Reader*. Philadelphia: New Society.

Anthony, C. 1995. "Ecopsychology and the Deconstruction of Whiteness." In *Ecopsychology: Restoring the Earth, Healing the Mind,* edited by T. Roszak, M. E. Gomes, and A. D. Kanner. San Francisco: Sierra Club Books, 263–78.

Anthony, C. D. 1995. "Socioecology of a Terrestrial Salamander: Juveniles Enter Adult Territories During Stressful Foraging Periods." *Ecology* 76 (2): 533–43.

Aquirre, A. A., R. S. Ostfeld, G. M. Tabor, C. House, and M. C. Pearl, eds. 2002. *Conservation Medicine: Ecological Health in Practice*. Oxford: Oxford University Press.

Araya, D. 2002. "Integral Religion: Uniting Eros and Logos." *Integral World: Exploring Theories of Everything*. September. www.integralworld.net/araya.html (accessed August 16, 2003).

Arden, H. 1994. *Dreamkeepers: A Spirit-Journey into Aboriginal Australia*. New York: HarperCollins.

Aristotle. 1984. *The Complete Works of Aristotle: The Revised Oxford Translation*. Edited by J. Barnes. 2 vols. Bollingen Series. Princeton, NJ: Princeton University Press.

Armstrong, E. A. 1973. *Saint Francis: Nature Mystic*. Berkeley: University of California Press.

Armstrong, R. J., ed. 1988. *Francis and Clare: The Complete Works*. New York: Paulist Press.

Armstrong-Buck, S. 1986. "Whitehead's Metaphysical System as a Foundation for Environmental Ethics." *Environmental Ethics* 8 (Fall): 241–59.

———. 1991. "What Process Philosophy Can Contribute to the Land Ethic and Deep Ecology." *Trumpeter* 8 (1): 29–34.

Arnold, R. 1987. *Ecology Wars: Environmentalism as if People Mattered*. Bellevue, WA: Free Enterprise Press.

———. 1997. *Ecoterror: The Violent Agenda to Save Nature; The World of the Unabomber*. Bellevue, WA: Free Enterprise Press.

Arnol'd, V. 1992. *Catastrophe Theory*. 3rd ed. Berlin: Springer.

Artmann, S. 2007. "Computing Codes versus Interpreting Life." In *Introduction to Biosemiotics*, edited by M. Barbieri. Dordrecht, Netherlands: Springer, 209–33.

Ashton, P., S. Harris, and A. Lowe. 2004. *Ecological Genetics: Design, Analysis, and Application*. Malden, MA: Blackwell.

Astin, A. W. 2000. "Conceptualizing Service-Learning Research Using Ken Wilber's Integral Framework." *Michigan Journal of Community Service Learning* 7 (Fall): 98–104.

Astin, J. A., and A. W. Astin. 2002. "An Integral Approach to Medicine." *Alternative Therapies in Health and Medicine* 8 (2): 70–75.

Atkins, P., I. Simmons, and B. Roberts. 1998. *People, Land, and Time: An Historical Introduction to the Relations between Landscape, Culture and Environment*. New York: Arnold.

Atran, S., and D. Medin. 2008. *The Native Mind and the Cultural Construction of Nature*. Cambridge, MA: MIT Press.

Atterton, P., and M. Calarco, eds. 2004. *Animal Philosophy: Essential Readings in Continental Thought*. New York: Continuum.

Aurobindo, S. 1973. *On Nature*. Auroville, India: All India Press.

Ausubel, K. 2004. *Ecological Medicine: Healing the Earth, Healing Ourselves*. San Francisco: Sierra Club Books.

Avital, E., and E. Jablonka. 2005. *Animal Traditions: Behavioral Inheritance in Evolution*. Cambridge: Cambridge University Press.

Bache, C. M. 2000a. *Dark Night, Early Dawn: Steps to a Deep Ecology of Mind*. Albany, NY: SUNY Press.

———. 2000b. "The Eco-Crisis and Species Ego-Death: Speculations on the Future." *Journal of Transpersonal Psychology* 32 (1): 89–94.

Backster, C. 2003. *Primary Perception: Biocommunication with Plants, Living Foods, and Human Cells*. Anza, CA: White Rose Millennium Press.

Baden-Powell, R. 1908. *Scouting for Boys: A Handbook for Instruction in Good Citizenship*. Astoria, IL: Stevens.

Bailey, B. 1993. *Eco-Scam: The False Prophets of Ecological Apocalypse*. New York: St. Martin's Press.

Bailey, R., ed. 1995. *The True State of the Planet: Ten of the World's Environmental Researchers in a Major Challenge to the Environmental Movement*. New York: Free Press.

———. 2002. *Global Warming and Other Eco Myths: How the Environmental Movement Uses False Science to Scare Us to Death*. Roseville, CA: Prima Lifestyles.

———. 2005. "We're All Global Warmers Now: Reconciling Temperature Trends That Are All Over the Place." *Reason*, August 11. www.reason.com/links/links 081105.shtml.

Balee, W., ed. 1998. *Advances in Historical Ecology*. New York: Columbia University Press.

Balouet, J. 1990. *Extinct Species of the World*. New York: Barrons.

Baluska, F., S. Mancuso, and D. Volkmann, eds. 2006. *Communication in Plants: Neuronal Aspects of Plant Life*. Berlin: Springer.

Barbieri, M. 2003. *The Organic Codes: An Introduction to Semantic Biology*. Cambridge: Cambridge University Press.

———. 2006. *Introduction to Biosemiotics: The New Biological Synthesis*. Dordrecht, Netherlands: Springer.

Barbour, M. 1996. "Ecological Fragmentation in the Fifties." In *Uncommon Ground: Rethinking the Human Place in Nature*, edited by W. Cronon. New York: W. W. Norton, 233–55.

Barfield, O. 1957/1988. *Saving the Appearances: A Study in Idolatry*. Rev. ed. Hanover, NH: Wesleyan University Press.

Barkow, J., L. Cosmides, and J. Tooby, eds. 1992. *The Adapted Mind: Evolutionary Psychology and the Generation of Culture*. Oxford: Oxford University Press.

Barlow, B. 1996. *Sacred Sites of the West*. St. Paul, MN: Llewellyn.

Barnes, M. 2002. *Stages of Thought: The Co-evolution of Religious Thought and Science*. New York: Oxford University Press.

Barry, J. 1999. *Environment and Social Theory*. London: Routledge.

Barry, J., and R. Eckersley. 2005. *The State and the Global Ecological Crisis*. Cambridge, MA: MIT Press.

Bartholomew, A. 2003. *Hidden Nature: The Startling Insights of Viktor Schauberger*. Edinburgh, UK: Floris Books.

Baskin, Y. 1998. *The Work of Nature: How the Diversity of Life Sustains Us*. Washington, DC: Island Press.

Basso, K. H. 1996. *Wisdom Sits in Places: Landscape and Language among the Western Apache*. Albuquerque: University of New Mexico Press.

Bate, J. 1991. *Romantic Ecology: Wordsworth and the Environmental Tradition*. London: Routledge.

Bateson, G. 1972/2000. *Steps to an Ecology of Mind*. Chicago: University of Chicago Press.

———. 1979/2002. *Mind and Nature: A Necessary Unity*. Cresskill, NJ: Hampton Press.

Baumgartner, F. J. 2001. *Longing for the End: A History of Millennialism in Western Civilization*. New York: St. Martin's Press.

Bauwens, M. 2003. "Three Challenges for Global Religion in the 21st Century." *Integral World: Exploring Theories of Everything*. August. www.integralworld.net/bauwens.html (accessed August 16, 2003).

Bawden, G., and R. Reycraft, eds. 2001. *Environmental Disaster and the Archaeology of Human Response*. Albuquerque: University of New Mexico Press.

Bechtel, R. B., and A. Churchman, eds. 2002. *Handbook of Environmental Psychology*. Indianapolis: John Wiley & Sons.

Beck, D., and G. Linscott. 1991. *The Crucible: Forging South Africa's Future*. Denton, TX: New Paradigm Press.

Beck, D. E., and C. C. Cowan. 1996. *Spiral Dynamics: Mastering Values, Leadership, and Change*. Oxford: Blackwell.

Beck, U. 1992. *The Risk Society: Toward a New Modernity*. Thousand Oaks, CA: SAGE.

Behringer, T. 2005. "Embodied Dreaming." Unpublished manuscript.

Bekoff, M. 1998. *Encyclopedia of Animal Rights and Animal Welfare*. Westport, CT: Greenwood.

———. 2000. *The Smile of a Dolphin: Remarkable Accounts of Animal Emotions*. New York: Discovery Books.

———. 2002. *Minding Animals: Awareness, Emotions, and Heart*. New York: Oxford University Press.

———. 2004a. *Encyclopedia of Animal Behavior*. 3 vols. Westport, CT: Greenwood.

———. 2004b. *If You Tame Me: Understanding Our Connection with Animals*. Philadelphia: Temple University Press.

———. 2006a. "Animal Passions and Beastly Virtues: Cognitive Ethology as the Unifying Science for Understanding the Subjective, Emotional, Empathic, and Moral Lives of Animals." *Zygon* 41: 71–104.

———. 2006b. *Animal Passions and Beastly Virtues: Reflections on Redecorating Nature*. Philadelphia: Temple University Press.

———. 2006c. "The Public Lives of Animals: A Troubled Scientist, Pissy Baboons, Angry Elephants, and Happy Hounds." *Journal of Consciousness Studies* 13 (5): 115–31.

———. 2007. *The Emotional Lives of Animals: A Leading Scientist Explores Animal Joy, Sorrow, and Empathy—and Why They Matter.* Novato, CA: New World Library.

Bekoff, M., C. Allen, and G. M. Burghardt, eds. 2002. *The Cognitive Animal.* Cambridge, MA: MIT Press.

Bekoff, M., and J. A. Byers, eds. 1998. *Animal Play: Evolutionary, Comparative and Ecological Perspectives.* New York: Cambridge University Press.

Bekoff, M., and D. Jamieson, eds. 1990a. *Interpretation and Explanation in the Study of Animal Behavior.* Vol. 1, *Interpretation, Intentionality, and Communication.* Boulder, CO: Westview Press.

———, ed. 1990b. *Interpretation and Explanation in the Study of Animal Behavior.* Vol. 2, *Explanation, Evolution, and Adaptation.* Boulder, CO: Westview Press.

———, ed. 1996. *Readings in Animal Cognition.* Cambridge, MA: MIT Press.

Bell, P., T. Greene, J. Fisher, and A. Baum. 1996. *Environmental Psychology.* Orlando, FL: Harcourt Brace.

Bell, S., ed. 2000. *Seven Summits: The Quest to Reach the Highest Point on Every Continent.* Boston: Bulfinch Press.

Bentham, J. 1823/1963. *Introduction to the Principles of Morals and Legislation.* New York: Free Press.

Benton, T., ed. 1996. *The Greening of Marxism.* New York: Guilford Press.

Benyus, J. 2002. *Biomimicry: Innovation Inspired by Nature.* New York: Harper Perennial Library.

Bering, J. M. and D. J. Povinelli. 2003. "Comparing Cognitive Development." In *Primate Psychology: Bridging the Gap between the Mind and Behavior of Human and Nonhuman Primates,* edited by D. Maestripieri. Cambridge, MA: Harvard University Press.

Berkes, F. 1999. *Sacred Ecology: Traditional Ecological Knowledge and Resource Management.* Philadelphia: Taylor and Francis.

Berkes, F., and C. Folke, eds. 2002. *Linking Social and Ecological Systems.* Cambridge: Cambridge University Press.

Berkofer, R. Jr. 1979. *The White Man's Indian: Images of the American Indian from Columbus to the Present.* New York: Vintage Books.

Berkowitz, A., C. Nilon, and K. Hollweq, eds. 2002. *Understanding Urban Ecosystems.* New York: Springer.

Berleant, A. 1992. *The Aesthetics of Environment*. Philadelphia: Temple University Press.

Berman, M. 1981. *The Reenchantment of the World*. Ithaca, NY: Cornell University Press.

Berry, T. 1982. *Teilhard in the Ecological Age*. Teilhard Studies 7. Chambersburg, PA: Anima Books.

———. 1988. *The Dream of the Earth*. San Francisco: Sierra Club Books.

———. 1999. *The Great Work: Our Way into the Future*. New York: Bell Tower.

Berry, W. 1977/1986. *The Unsettling of America: Culture and Agriculture*. San Francisco: Sierra Club Books.

———. 1993. *Sex, Economy, Freedom, and Community*. New York: Pantheon.

Bertalanffy, L. 1968. *General Systems Theory*. New York: Braziller.

Berthold-Bond, A. 2001. *Better Basics for the Home: Simple Solutions for Less Toxic Living*. New York: Three Rivers Press.

Biehl, J., and P. Staudenmaier. 1995. *Ecofascism: Lessons from the German Experience*. Oakland, CA: AK Press.

Bigwood, C. 1993. *Earth Muse: Feminism, Nature, and Art*. Philadelphia: Temple University Press.

Bilsky, L. 1980. *Historical Ecology: Essays on Environment and Social Change*. Port Washington, NY: Kennikat Press.

Bingham, R., and C. Bingham. 1999. *The NEW Passport to Survival: 12 Steps to Self-Sufficient Living*. Sandy, UT: Natural Meals.

Birch, C., and J. Cobb, Jr. 1990. *The Liberation of Life: From the Cell to the Community*. Denton, TX: Environmental Ethics Books.

Birch, L. C., W. Eakin, and J. McDaniel, eds. 1990. *Liberating Life: Contemporary Approaches to Ecological Theology*. Maryknoll, NY: Orbis Books.

Blain, J. 2002. *Nine Worlds of Seid-Magic: Ecstasy and Neo-Shamanism in North European Paganism*. London: Routledge.

Blitz, D. 1992. *Emergent Evolution: Qualitative Novelty and the Levels of Reality*. Dordrecht, Netherlands: Kluwer.

Bloom, B. S., M. B. Engelhart, E. J. Furst, W. H. Hill, and D. R. Krathwohl. 1956. *Taxonomy of Educational Objectives, Handbook I: The Classification of Educational Goals—Cognitive Domain*. New York: Longman.

Blühdorn, I. 2000. *Post-ecological Politics: Social Theory and the Abdication of the Ecologist Paradigm*. New York: Routledge.

Boas, G. 1997. *Primitivism and Related Ideas in the Middle Ages*. Baltimore, MD: Johns Hopkins University Press.

Bockemuhl, J., ed. 1985. *Toward a Phenomenology of the Etheric World*. Spring Valley, NY: Anthroposophic Press.

Boff, L. 1995a. *Ecologia: Grito da Terra, Grito dos Pobres*. São Paulo: Editora Atica.

———. 1995b. *Ecology and Liberation: A New Paradigm*. Maryknoll, NY: Orbis Books.

———. 1997. *Cry of the Earth, Cry of the Poor*. Maryknoll, NY: Orbis Books.

Boff, L., and V. Elizondo, eds. 1995. *Ecology and Poverty: Cry of the Earth, Cry of the Poor*. London: SCM Press.

Bonde, L. O. 2001. "Steps toward a Meta-theory of Music Therapy: An Introduction to Ken Wilber's Integral Psychology and a Discussion of Its Relevance for Music Therapy." *Nordic Journal of Music Therapy* 10 (2): 176–87.

Bonner, J. T. 1980. *The Evolution of Culture in Animals*. Princeton, NJ: Princeton University Press.

Bonnes, M., T. Lee, and M. Bonaiutom, eds. 2003. *Psychological Theories for Environmental Issues*. Aldershot, UK: Ashgate.

Bonnicksen, T. 2000. *America's Ancient Forests: From the Ice Age to the Age of Discovery*. New York: John Wiley & Sons.

Bookchin, M. 1970. *Ecology and Revolutionary Thought*. New York: Times Change Press.

———. 1989. *Remaking Society*. Montreal: Black Rose Books.

———. 1990. *The Philosophy of Social Ecology: Essays on Dialectical Naturalism*. Montreal: Black Rose Books.

———. 1991. *The Ecology of Freedom: The Emergence and Dissolution of Hierarchy*. Rev. ed. Montreal: Black Rose Books.

Borgmann, A. 1987. *Technology and the Character of Contemporary Life*. Chicago: University of Chicago Press.

Boris, E. 1986. *Art and Labor: Ruskin, Morris, and the Craftsman Ideal in America*. Philadelphia: Temple University Press.

Bortoft, H. 1996. *The Wholeness of Nature: Goethe's Way toward a Science of Conscious Participation in Nature*. Hudson, NY: Lindisfarne Press.

Botkin, D. 2000. *Discordant Harmonies: A New Ecology for the Twenty-first Century*. Oxford: Oxford University Press.

Bott, S., J. G. Cantrill, and O. E. Myers, Jr. 2003. "Place and the Promise of Conservation Psychology." *Human Ecology Review* 10 (2): 100–112.

Boudouris, K., and K. Kalimtzis, eds. 1999. *Philosophy and Ecology*. Vol. 1. Athens: IONIA.

Bourg, D., and S. Erkman. 2003. *Perspectives on Industrial Ecology*. Sheffield, UK: Greenleaf Publishing.

Bowers, C. A. 1993. *Education, Cultural Myths, and the Ecological Crisis: Toward Deep Changes*. Albany, NY: SUNY Press.

———. 1997. *The Culture of Denial: Why the Environmental Movement Needs a Strategy for Reforming Universities and Public Schools*. Albany, NY: SUNY Press.

Bowler, P. J. 1992. *The Earth Encompassed: A History of the Environmental Sciences*. New York: W. W. Norton.

Boyd, R., ed. 1999. *Indians, Fire, and the Land in the Pacific Northwest*. Corvallis: Oregon State University Press.

Bradshaw, G. A. 2004. "Not by Bread Alone: Symbolic Loss, Trauma, and Recovery in Elephant Communities." *Society and Animals* 12 (2): 143–58.

———. 2006. "No Longer a Mind of Our Own." *Seed*. June/July. www.seedmagazine.com/news/2006/06/no_longer_a_mind_of_our_own.php. (accessed May 14, 2007).

———. Forthcoming. *Elephant Breakdown: The Psychological Study of Animal Cultures in Crisis*.

Bradshaw, G. A., and B. L. Finlay. 2005. "Natural Symmetry." *Nature* 435: 149.

Bradshaw, G. A, P. Greene Linden, and A. N. Schore. 2005. "Behavioral and Physiological Effects of Trauma on Psittacine Birds." Proceedings of American Avian Veterinarian Conference, August.

Bradshaw, G. A., and R. M. Sapolsky. 2006. "Mirror, Mirror." *American Scientist* 94 (6). November-December www.kerulos.org. (accessed May 14, 2007).

Bradshaw, G. A., and A. N. Schore. 2007. "How Elephants Are Opening Doors: Neuroethology, Attachment, and Social Context." *Ethology* 113 (5): 426–36.

Bradshaw, G. A, A. N. Schore, J. L. Brown, J. H. Poole, and C. J. Moss. 2005. "Elephant Breakdown." *Nature* 433: 807.

Bradshaw, G. A., and M. Watkins. 2006. "Trans-species Psychology: Theory and Praxis." *Spring* 75: 1–2.

Bramwell, A. 1989. *Ecology in the Twentieth Century: A History*. New Haven, CT: Yale University Press.

Braud, W. 1998. "Integral Inquiry: Complementary Ways of Knowing, Being, and Expression." In *Transpersonal Research Methods for the Social Sciences*, edited by W. Braud and R. Anderson. Thousand Oaks, CA: SAGE, 35–67.

Brewer, J. 2005. *Foundations of Multimethod Research: Synthesizing Styles*. Newbury Park, CA: SAGE.

Brier, S. 1995. “Cyber-Semiotics: On Autopoiesis, Code-Duality and Sign Games in Biosemiotics.” *Cybernetics and Human Knowledge* 3 (1): 3–14.

———. 1996a. “Cybersemiotics: A New Interdisciplinary Development Applied to the Problems of Knowledge Organisation and Document Retrieval in Information Science.” *Journal of Documentation* 52 (3): 296–344.

———. 1996b. “From Second-Order Cybernetics to Cybersemiotics: A Semiotic Re-entry into the Second-Order Cybernetics of Heinz von Foerster.” *Systems Research* 13: 229–44.

———. 1997. “What Is a Possible Ontological and Epistemological Framework for a True Universal ‘Information Science’? The Suggestion of a Cybersemiotics.” *World Futures: The Journal of General Evolution* 49: 297–308.

———. 1998a. “The Cybersemiotic Explanation of the Emergence of Cognition: The Explanation of Cognition, Signification and Communication in a Non-Cartesian Cognitive Biology.” *Evolution and Cognition* 4 (1): 90–102.

———. 1998b. “Cybersemiotics: A Suggestion for a Transdisciplinary Framework for Description of Observing, Anticipatory and Meaning Producing Systems.” In *Computing Anticipatory Systems*, edited by D. M. Dubois. Woodbury, NY: American Institute of Physics, 182–93.

———. 1999. “Biosemiotics and the Foundation of Cybersemiotics.” *Semiotica* 127 (1/4): 169–98.

———. 2002. “Luhmann Semioticized.” *Journal of Sociocybernetics* 3 (2): 12–22.

———. 2003. “The Cybersemiotic Model of Communication.” *TripleC* 1 (1): 71–94.

———. 2008. *Cybersemiotics: Why Information Is Not Enough*. Toronto: University of Toronto Press.

Brodle, J. 1995. *Earthdance: A Year of Pagan Rituals*. Chieveley, UK: Capall Bann.

Bromley, D. 1995. *Handbook of Environmental Economics*. Malden, MA: Blackwell.

Bronfenbrenner, U. 1979. *The Ecology of Human Development: Experiments by Nature and Design*. Cambridge, MA: Harvard University Press.

———. 1999. “Environments in Developmental Perspective: Theoretical and Operational Models.” In *Measuring Environment across the Life Span*, edited by S. L. Friedman and T. D. Wachs. Washington, DC: American Psychological Association Press, 3–28.

Bronfenbrenner, U., and P. A. Morris. 1998. “The Ecology of Developmental Processes.” In *Handbook of Child Psychology*: Vol. 1, *Theoretical Models of Human Development*, edited by R. M. Lerner. New York: Wiley, 993–1028.

Broswimmer, F. J. 2002. *Ecocide: A Short History of the Mass Extinction of Species.* London: Pluto Press.

Brower, M., and W. Leon. 1999. *The Consumer's Guide to Effective Environmental Choices: Practical Advice from the Union of Concerned Scientists.* New York: Three Rivers Press.

Brown, B. C. 2003. "What Really Matters? A Youth's Quest for Keys to Effecting Global Change." *Spirituality and Reality: New Perspectives on Global Issues* 2 (2): 10–12.

———. 2005. "Integral Communication for Sustainability." *Kosmos: An Integral Approach to Global Awakening* 4 (2): 17–20.

———. 2006a. "Theory and Practice of Integral Sustainable Development: Part 1—Quadrants and the Practitioner." *AQAL: Journal of Integral Theory and Practice* 1 (2): 366–405.

———. 2006b. "Theory and Practice of Integral Sustainable Development: Part 2—Values, Developmental Levels, and Natural Design." *AQAL: Journal of Integral Theory and Practice* 1 (2): 406–77.

———. Forthcoming. "The Four Worlds of Sustainability: Drawing upon Four Universal Perspectives to Support Sustainability Initiatives." *AQAL: Journal of Integral Theory and Practice.*

———. Forthcoming. "Integrating the Major Research Methodologies Used in Sustainable Development." *AQAL: Journal of Integral Theory and Practice.*

———. "Major Initiatives Which Have Used, or Are Using, the Integral Framework for Social, Environmental, and Economic Development." Integral Sustainability Center. www.multiplex.integralinstitute.org/Public/cs/files/43/sustainability/default.aspx.

———. "The Use of an Integral Approach by UNDP's HIV/AIDS Group as Part of Their Global Response to the HIV/AIDS Epidemic." Integral Sustainability Center. www.multiplex.integralinstitute.org/Public/cs/files/43/sustainabilitydefault.aspx.

Brown, B. C., and C. Riedy. 2006. "Use of the Integral Framework to Design Developmentally Appropriate Sustainability Communication." In chapter 34 of *Innovation, Education, and Communication for Sustainable Development,* edited by W. Leal Filho. New York: Peter Lang.

Brown, C. S., and T. Toadvine. 2003. *Eco-Phenomenology: Back to the Earth Itself.* Albany, NY: SUNY Press.

Brown, J. H. 2000. *Macroecology.* Chicago: University of Chicago Press.

Brown, J. H., and L. Real. 1991. *Foundations of Ecology: Classic Papers with Commentaries.* Chicago: University of Chicago Press.

Brown, L. 2001. *Organic Living: Simple Solutions for a Better Life*. London: DK.

Brown, L., G. Gardner, and B. Halweil. 1999. *Beyond Malthus: Nineteen Dimensions of the Population Challenge*. New York: W. W. Norton.

Brown, L. R. 1995. "Ecopsychology and the Environmental Revolution: An Environmental Foreword." In *Ecopsychology: Restoring the Earth, Healing the Mind*, edited by T. Roszak, M. E. Gomes, and A. D. Kanner. San Francisco: Sierra Club Books, xiii–xvi.

Brown, T. Jr. 1983. *Tom Brown's Field Guide to Nature Observation and Tracking*. New York: Berkley Books.

———. 1983. *Tom Brown's Field Guide to Wilderness Survival*. New York: Berkley Books.

———. 1986. *The Tracker*. New York: Berkley Books.

———. 1987. *Tom Brown's Field Guide to the Forgotten Wilderness*. New York: Berkley Books.

———. 1988. *The Vision*. New York: Berkley Books.

———. 1994. *Awakening Spirits*. New York: Berkley Books.

———. 1999. *The Science and Art of Tracking*. New York: Berkley Books.

Bruggemeier, F., M. Cioc, and T. Zeller, eds. 2005. *How Green Were the Nazis? Nature, Environment, and Nation in the Third Reich*. Athens: Ohio University Press.

Brulle, R. 2000. *Agency, Democracy, and Nature*. Cambridge, MA: MIT Press.

Brunke, D. B. 2002. *Animal Voices: Telepathic Communication in the Web of Life*. Rochester, VT: Inner Traditions.

Bryson, S. 2002. *Ecopoetry: Critical Introduction*. Salt Lake City: University of Utah Press.

Buchanan, B. 2008. *Onto-Ethologies: The Animal Environments of Uexküll, Heidegger, Merleam-Ponty, and Deleuze*. Albany, NY: SUNY Press.

Buege, D. 1996. "The Ecological Noble Savage Revisited." *Environmental Ethics* 18 (Spring): 71–88.

———. 1997. "An Ecologically-Informed Ontology for Environmental Ethics." *Biology and Philosophy* 12: 1–20.

Buell, F. 2003. *From Apocalypse to Way of Life*. New York: Routledge.

Buell, L. 2005. *The Future of Environmental Criticism: Environmental Crisis and Literary Imagination*. New York: Blackwell.

Bullard, R. 2000. *Dumping in Dixie: Race, Class, and Environmental Quality*. Boulder, CO: Westview Press.

Buren, J. V. 1991. "Cognitive Ethology and Critical Anthropomorphism: A Snake with Two Heads and Hognose Snakes That Play Dead." In *Cognitive Ethology: The Minds of Other Animals—Essays in Honor of Donald R. Griffin*, edited by C. A. Ristau. Hillsdale, NJ: Lawrence Erlbaum, 53–90.

———. 1995. "Critical Environmental Hermeneutics." *Environmental Ethics* 17 (Fall): 259–75.

Burkhardt, R. W. Jr. 2005. *Patterns of Behavior: Konrad Lorenz, Niko Tinbergen, and the Founding of Ethology*. Chicago: University of Chicago Press.

Burroughs, W. J. 2001. *Climate Change: A Multidisciplinary Approach*. Cambridge: Cambridge University Press.

Buss, D. 2007. *Evolutionary Psychology: The New Science of the Mind*. 3rd ed. New York: Allyn & Bacon.

Buss, L. 1987. *The Evolution of Individuality*. Princeton, NJ: Princeton University Press.

Butcher, R. 2003. *American's National Wildlife Refuges: A Complete Guide*. Boulder, CO: Roberts Rinehart.

Butlin, R. A., and N. Roberts, eds. 1995. *Ecological Relations in Historical Times: Human Impact and Adaptation*. Oxford: Blackwell Scientific.

Buttel, F. H. 2000. "Ecological Modernization as a Social Theory." *Geoforum* 31: 57–65.

Butterfly Hill, J. 2001. *The Legacy of Luna: The Story of a Tree, a Woman and the Struggle to Save the Redwoods*. San Francisco: Harper.

Butzer, K. W., ed. 1992. "The Americas Before and After 1492: An Introduction to Current Geographical Research." *Annals of the Association of American Geographers* 82 (3): 345–68.

Buytendijk, F. J. J. 1935. *The Mind of a Dog*. London: George Allen & Unwin.

Cabarga, L. 1997. *Talks with Trees*. Los Angeles: Iconoclassics.

Caldicott, H. 1992. *If You Love This Planet: A Plan to Heal the Earth*. New York: W. W. Norton.

Caldwell, D. 1985. *Planetary Ecology*. New York: Van Nostrand Reinhold.

Callicott, J. B. 1986. "The Metaphysical Implications of Ecology." *Environmental Ethics* 8 (4): 301–16.

———. 1987. *Companion to a Sand County Almanac: Interpretive and Critical Essays*. Madison: University of Wisconsin Press.

———. 1989. *In Defense of the Land Ethic: Essays in Environmental Philosophy*. Albany, NY: SUNY Press.

———. 1994. *Earth's Insights: A Multicultural Survey of Ecological Ethics from the Mediterranean Basin to the Australian Outback*. Berkeley: University of California Press.

———. 1999. *Beyond the Land Ethic: More Essays in Environmental Philosophy*. Albany, NY: SUNY Press.

Callicott, J. B., and R. T. Ames. 1989. *Nature in Asian Traditions of Thought: Essays in Environmental Philosophy*. Albany, NY: SUNY Press.

Callicott, J. B., and E. T. Freyfogle, eds. 1999. *For the Health of the Land: Previously Unpublished Essays and Other Writings by Aldo Leopold*. Washington, DC: Island Press.

Callicott, J. B., and Nelson, M. P., eds. 1998. *The Great New Wilderness Debate: An Expansive Collection of Writings Defining Wilderness from John Muir to Gary Snyder*. Athens: University of Georgia Press.

Camacho, D. E., ed. 1998. *Environmental Injustices, Political Struggle: Race, Class, and the Environment*. Durham, NC: Duke University Press.

Campbell, J. 1962/1976. *The Masks of God: Oriental Mythology*. New York: Penguin Books.

Capra, F. 1996. *The Web of Life: A New Scientific Understanding of Living Systems*. New York: Anchor Books.

———. 2000. *The Tao of Physics: An Exploration of the Parallels between Modern Physics and Eastern Mysticism*. 4th ed. Boston: Shambhala Publications.

Caracelli, V. W., and J. C. Greene. 1993. "Data Analysis Strategies for Mixed-Method Evaluation Designs." *Educational Evaluation and Policy Analysis* 15 (2): 195–207.

Carlson, A. 1984. "Nature and Positive Aesthetics." *Environmental Ethics* 6: 5–34.

Carlson, A., and A. Berleant, eds. 2004. *The Aesthetics of Natural Environments*. Toronto: Broadview Press.

Carrasco, D., and M. Kleit. 1999. *City of Sacrifice: The Aztec Empire and the Role of Violence in Civilization*. Boston: Beacon Press.

Carroll, G. R., and M. T. Hannan. 2000. *The Demography of Corporations and Industries*. Princeton, NJ: Princeton University Press.

Carroll, P. 1969. *Puritanism and the Wilderness: The Intellectual Significance of the New England Frontier, 1629–1700*. New York: Columbia University Press.

Carson, R. 1962. *Silent Spring*. Boston: Houghton Mifflin.

Casagrande, D. 1999. "Information as Verb: Re-conceptualizing Information for Cognitive and Ecological Models." *Georgia Journal of Ecological Anthropology* 3: 4–14.

Casey, E. 1993. *Getting Back into Place: Toward a Renewed Understanding of the Place-World.* Bloomington: Indiana University Press.

———. 1997. *The Fate of Place: A Philosophical History*. Berkeley: University of California Press.

———. 2002. *Representing Place: Landscape Painting and Maps*. Minneapolis: University of Minnesota Press.

———. 2005. *Earth-Mapping: Artists Reshaping Landscape*. Minneapolis: University of Minnesota Press.

Cataldi, S. L., and W. S. Hamrick, eds. 2007. *Merleau-Ponty and Environmental Philosophy: Dwelling on the Landscapes of Thought.* Albany, NY: SUNY Press.

Catton, W. R. Jr. 1980. *Overshoot: The Ecological Basis of Revolutionary Change.* Urbana: University of Illinois Press.

Chalmers, D. J. 1996. *The Conscious Mind: In Search of a Fundamental Theory.* New York: Oxford University Press.

Chalquist, C. 2007. *Terrapsychology: Reengaging the Soul of Place*. New Orleans: Spring Journal Books.

Chang, H. 2003. "Is Language a Primary Modeling System? On Juri Lotman's concept of Semiosphere." *Sign Systems Studies* 31 (1): 9–23.

———. 2004a. "The 'Philological Understanding' of Jacob von Uexküll." Paper presented at the 4th Gatherings in Biosemiotics Prague, July 1–5.

———. 2004b. "Semiotician or Hermeneutician? Jacob von Uexküll Revisited." *Sign Systems Studies* 32 (1): 115–38.

Chantrill, C. "Ken Wilber." *Road to the Middle Class*. www.roadtothemiddleclass.com/chappies.php?id=85 (accessed November 30, 2005).

Chapin, F. S. III, P. Matson, and H. Mooney. 2004. *Principles of Terrestrial Ecosystem Ecology*. 2nd ed. New York: Springer.

Chapman, S. 2001. *What a Hunter Brings Home: Pursuing the Trophies That Matter Most.* Eugene, OR: Harvest House.

Charlesworth, M., ed. 1998. *Religious Business: Essays on Australian Aboriginal Spirituality*. Cambridge: Cambridge University Press.

Chase, A. 2000. *In a Dark Wood: The Fight over Forests and the Myths of Nature.* New ed. London: Transaction Publishers.

Chatwin, B. 1987. *The Songlines*. New York: Penguin.

Chebanov, S. V. 1993. "Biology and Humanitarian Culture: The Problem of Interpretation in Biohermeneutics and Hermeneutics of Biology." In *Lectures in Theoretical Biology: The Second Stage*, edited by K. Kull and T. Tiivel. Tallinn: Estonian Academy of Sciences, 219–48.

———. 1994. "Man as Participant to Natural Creation. Enlogue and Ideas of Hermeneutics in Biology." *Biology Forum* 87 (1): 39–48.

———. 1995. "Enlogue as Quasipersonal Interaction: Biohermeneutic Issues." *European Journal of Semiotic Studies* 7 (3–4): 439–66.

———. 1998. "The Role of Hermeneutics in Biology." In *Sociobiology and Bioeconomics: The Theory of Evolution in Biological and Economic Theory*, edited by P. Koslowski. New York: Springer, 141–72.

———. 1999. "Bio-Hermeneutics and Hermeneutics of Biology." *Semiotica* 127 (1/4): 215–26.

Cheney, D. L., and R. M. Seyfarth. 1990. *How Monkeys See the World: Inside the Mind of Another Species.* Chicago: University of Chicago Press.

———. 2007. *Baboon Metaphysics: The Evolution of a Social Mind.* Chicago: University of Chicago Press.

Churchill, S. D. 2000. "Intercorporeality Gestural Communication and the Voices of Silence: Toward a Phenomenological Ethology: Part 1." *Somatics* 11 (1): 28–32.

———. 2001. "Intercorporeality Gestural Communication and the Voices of Silence: Toward a Phenomenological Ethology: Part 2." *Somatics* 11 (2): 40–45.

Clark, J. 2000. "How Wide Is Deep Ecology?" In *Beneath the Surface: Critical Essays in the Philosophy of Deep Ecology*, edited by E. Katz, A. Light, and D. Rothenberg. Cambridge, MA: MIT Press, 3–16.

Clark, T. 1997. *Averting Extinction: Reconstructing Endangered Species Recovery.* New Haven, CT: Yale University Press.

Clarke, R. 1973. *Ellen Swallow: The Woman Who Founded Ecology.* Chicago: Follett.

Clayton, P. and A. Peacocke, eds. 2004. *In Whom We Live and Move and Have Our Being: Panentheistic Reflections on God's Presence in a Scientific World.* Cambridge, UK: William B. Eerdmans Publishing Company.

Clayton, S., and S. Opotow. 2003. *Identity and the Natural Environment: The Psychological Significance of Nature.* Cambridge, MA: MIT Press.

Clements, F. E., and V. E. Shelford. 1939. *Bio-Ecology.* New York: Wiley.

Clendinnen, I. 1991/1997. *Aztecs: An Interpretation.* Cambridge: Cambridge University Press.

Clifton, C., and G. Harvey, eds. 2003. *The Paganism Reader.* London: Routledge.

Clinebell, H. 1996. *Ecotherapy: Healing Ourselves, Healing the Earth: Guide to Ecologically Grounded Personality Theory.* Binghamton, NY: Haworth Press.

Coates, P. 1998. *Nature: Western Attitudes Since Ancient Times*. Berkeley: University of California Press.

Coats, C. 2002. *Living Energies*. 2nd ed. Bath, UK: Gateway Books.

Cobb, E. 1993. *The Ecology of Imagination in Childhood*. Dallas, TX: Spring.

Cobbald, J. 2007. *Viktor Schauberger: A Life of Learning from Nature*. Edinburgh, UK: Floris Books.

Cohen, J. E. 1995. *How Many People Can the Earth Support?* New York: W. W. Norton.

Cohen, M. J. 1993. "Integrated Ecology: The Process of Counseling with Nature." *The Humanistic Psychologist* 21: 277–95.

———. 1995. *Well Mind, Well Earth: 109 Environmentally Sensitive Activities for Stress Management, Spirit and Self-Esteem*. 5th ed. Friday Harbor, WA: World Peace University Press.

———. 1997. *Reconnecting with Nature: Finding Wellness through Restoring Your Bond with the Earth*. Corvallis, OR: Ecopress.

Cole, L. W., and S. Foster. 2000. *From the Ground Up: Environmental Racism and the Rise of the Environmental Justice Movement*. New York: New York University Press.

Coleman, M. 2006. *Awake in the Wild: Mindfulness in Nature as a Path of Self-Discovery*. Makowao, HI: Inner Ocean Publishing.

Collingwood, R. G. 1945/1960. *The Idea of Nature*. Oxford: Oxford University Press.

Common, M, and S. Stagl. 2005. *Ecological Economics: An Introduction*. New York: Cambridge University Press.

Commons, M. L. 2006. "Measuring an Approximate g in Animals and People." *Integral Review* 3: 82–99.

Commons, M. L., C. Armon, L. Kohlberg, F. A. Richards, T. A. Grotzer, and J. D. Sinnott, eds. 1990. *Adult Development*. Vol. 2, *Models and Methods in the Study of Adolescent and Adult Thought*. New York: Praeger.

Commons, M. L., F. A. Richards, and C. Armon. 1984. *Beyond Formal Operations: Late Adolescent and Adult Cognitive Development*. New York: Praeger.

Connell, D. 2002. "Multiple Constructions of the Environmental Crisis: A Sociocybernetic View." *Journal of Sociocybernetics* 3 (2): 1–12.

Connell, R. W. 1995. *Masculinities*. Berkeley and Los Angeles: University of California Press.

Conrad, E. 1995. "Life on Land." In *Bone, Breath, and Gesture: Practices of Embodiment*, edited by D. Johnson. Berkeley, CA: North Atlantic Books, 297–312.

Cook-Greuter, S. 1999. "Postautonomous Ego Development: A Study of Its Nature and Measurement." PhD diss., Harvard University Graduate School of Education.

———. 2002. "A Detailed Description of the Development of Nine Action Logics in the Leadership Development Framework: Adapted from Ego Development Theory." www.harthillusa.com/Detailed%20descrip.%20of%20ego%20development%20stages.pdf (accessed January 10, 2004).

Cooper, J. W. 2006. *Panentheism: The Other God of the Philosophers—from Plato to the Present.* Grand Rapids, MI: Baker Academic.

Cooper, M. 1986. "The Ecology of Writing." *College English* 48: 364–75.

Corrington, R. S. 1997. *Nature's Religion*. Lanham, MD: Rowman & Littlefield.

Costanza, R., L. Graumlich, and W. Steffen, eds. 2007. *Sustainability or Collapse? An Integrated History and Future of People on Earth.* Cambridge, MA: MIT Press.

Costanza, R., B. G. Norton, and B. D. Haskell, eds. 1992. *Ecosystem Health: New Goals for Environmental Management.* Washington, DC: Island Press.

Cowan, D. 2003. *Ley Line and Earth Energies: An Extraordinary Journey into the Earth's Natural Energy System*. Kempton, IL: Adventures Unlimited Press.

Cowan, J. 2002. *Aborigine Dreaming: An Introduction to the Wisdom and Magic of the Aboriginal Traditions*. London: Thorsons.

Coward, J. 1999. *The Newspaper Indian: Native American Identity in the Press, 1820–1890*. Chicago: University of Illinois Press.

Cox, G. W. 1993. *Conservation Ecology: Biosphere and Biosurvival.* Dubuque, IA: William C. Brown.

Cox, R. 2005. *Environmental Communication and the Public Sphere*. Thousand Oaks, CA: SAGE.

Creswell, J. 2002. *Research Design: Qualitative, Quantitative, and Mixed Methods Approaches*. 2nd ed. Thousand Oaks, CA: SAGE.

Creswell, J., and V. Clark. 2006. *Designing and Conducting Mixed Methods Research*. Newbury Park, CA: SAGE.

Crichton, M. 2004. *State of Fear*. New York: HarperCollins.

Crist, E. 2002. "The Inner Life of Earthworms: Darwin's Argument and Its Implications." In *The Cognitive Animal*, edited by M. Bekoff, C. Allen, and G. M. Burghardt. Cambridge, MA: MIT Press, 3–8.

Crockett, T. 2003. *Stone Age Wisdom: The Healing Principles of Shamanism*. Gloucester, MA: Fair Winds Press.

———. 2004. "Integral Shamanic Dreamwork: A Proposal." *Dream Time* 1 (1): 14–17.

Cronon, W. 1983/1989. *Changes in the Land: Indians, Colonists, and the Ecology of New England*. New York: Hill & Wang.

———. 1996. *Uncommon Ground: Toward Reinventing Nature*. New York: W. W. Norton.

Crosby, D. A. 2002. *A Religion of Nature*. Albany, NY: SUNY Press.

Crowley, J., and S. Crowley. 2001. *Wabi Sabi Style*. Layton, UT: Gibbs Smith.

Cruden, L. 1995. *The Spirit of Place: A Workbook for Sacred Alignment, Ceremonies and Visualizations for Cultivating Your Relationship with the Earth*. Rochester, VT: Inner Traditions.

Crumley, C., ed. 1994. *Historical Ecology: Cultural Knowledge and Changing Landscapes*. Santa Fe, NM: School of American Research Press.

———. 2001. *New Directions in Anthropology and Environment: Intersections*. New York: Rowman & Littlefield.

Csanyi, V. 2005. *If Dogs Could Talk: Exploring the Canine Mind*. New York: North Point Press.

Cuddington, K., and B. Beisner. 2005. *Ecological Paradigms Lost: Routes of Theory Change*. San Diego, CA: Elsevier Academic Press.

Cumming, E., and W. Kaplan. 1991. *The Arts and Crafts Movement*. London: Thames & Hudson.

Cunningham, S. 1990. *Wicca: A Guide for the Solitary Practitioner*. St. Paul, MN: Llewellyn.

Cuomo, C. J. 1998. *Feminism and Ecological Communities: An Ethic of Flourishing*. New York: Routledge.

Cusing, J. M., R. Costantino, B. Dennis, R. Desharnais, and S. Henson. 2002. *Chaos in Ecology: Experimental Nonlinear Dynamics*. San Diego, CA: Academic Press.

Czech, B., and P. Krausman. 2001. *The Endangered Species Act: History, Conservation Biology, and Public Policy*. Baltimore: Johns Hopkins University Press.

Dadd, D. L. 1997. *Home Safe Home: Protecting Yourself and Your Family from Everyday Toxics and Harmful Household Products*. New York: J. P. Tarcher.

Dalai Lama, His Holiness. 1995. *On the Environment: Collected Statements*. Dharamsala, India: DIIR Publications.

Dall, S. R. 2004. "Behavioural Biology: Fortune Favors Bold and Shy Personalities." *Current Biology* 14: R470–72.

Daly, H. 1991. *Steady-State Economics*. 2nd ed. Washington, DC: Island Press.

Daly, H. E., and J. B. Cobb. 1989. *For the Common Good: Redirecting the Economy toward Community, the Environment and a Sustainable Future*. Boston: Beacon Press.

Daly, M. 1978. *Gyn/ecology*. Boston: Beacon Press.

Darier, E., ed. 1999. *Discourses of the Environment*. Oxford: Blackwell.

Daston, L., and G. Mitman, eds. 2005. *Thinking with Animals: New Perspectives on Anthropomorphism*. New York: Columbia University Press.

Daston, L., and F. Vidal, eds. 2004. *The Moral Authority of Nature*. Chicago: University of Chicago Press.

Davenport, T., and L. Prusak. 1997. *Information Ecology: Mastering the Information and Knowledge Environment*. New York: Oxford University Press.

Davies, N. 1981. *Human Sacrifice in History and Today*. London: Macmillan.

Davis, D. L. 2002. *When Smoke Ran Like Water: Tales of Environmental Deception and the Battle Against Pollution*. New York: Basic Books.

Davis, S. 1997. *Kid Mystic* (CD). Dharma Pop Records.

Dawkins, M. S. 1993. *Through Our Eyes Only? The Search for Animal Consciousness*. San Francisco: W. H. Freeman.

Dawkins, R. 2006. *The Selfish Gene*. 3rd ed. New York: Oxford University Press.

de Groot, S. R., and P. Mazur. 1984. *Non-equilibrium Thermodynamics*. New York: Dover Publications.

DeKay, M. "Integral Sustainable Design: Re-integrating What Modernism Differentiated and Post-modernism Dissociated." Society of Building Science Educators. www.sbseretreat.lsu.edu/theme.html.

DeKay, M., and M. Guzowski. 2006. "A Model for Integral Sustainable Design Explored through Daylighting." Presented at the American Solar Energy Society Conference, Denver, CO, July 7–13.

de Laplante, K. 2004. "Environmental Alchemy: How to Turn Ecological Science into Ecological Philosophy." *Environmental Ethics* 26, no. 4: 361–80.

Delcourt, P. A., and H. R. Delcourt. 2004. *Prehistoric Native Americans and Ecological Change: Human Ecosystems in Eastern North America Since the Pleistocene*. New York: Cambridge University Press.

Deloria, P. 1998. *Playing Indian*. New Haven, CT: Yale University Press.

Deloria, V. Jr. 1972/1994. *God Is Red: A Native View of Religion*. 2nd ed. Golden, CO: Fulcrum.

———. 1995. *Red Earth, White Lies*. New York: Scribner.

DeMares, R. 1998. "Peak Experiences with Cetaceans: A Phenomenological Study." PhD diss., Union Institute Graduate School.

DeMares, R., and K. Krycka. 1998. "Wild-Animal-Triggered Peak Experiences: Transpersonal Aspects." *Journal of Transpersonal Psychology* 30 (2): 161–77.

Denevan, W. 1992. "The Pristine Myth: The Landscape of the Americas in 1492." *Annals of the Association of American Geographers* 82 (3): 369–85.

Depew, D. J., and B. H. Weber. 1995. *Darwinism Evolving: Systems Dynamics and the Genealogy of Natural Selection*. Cambridge, MA: MIT Press.

Derr, P. 2003. *Case Studies in Environmental Ethics*. New York: Rowman & Littlefield.

DeSantis, K., and A. Housen. 2000. "A Brief Guide to Developmental Theory and Aesthetic Development." *Visual Understanding in Education*. www.vue.org/download.html.

Des Jardins, J. R. 1993. *Environmental Ethics: An Introduction to Environmental Philosophy*. Belmont, CA: Wadsworth.

Devereux, P. 1999. *Places of Power: Measuring the Secret Energy of Ancient Sites*. London: Blandford Press.

———. 2001. *The Sacred Place: The Ancient Origin of Holy and Mystical Sites*. London: Cassell Academic.

Devereux, P., and N. Pennick. 1989. *Lines on the Landscape: Leys and Other Linear Enigmas*. London: Hale.

de Waal, F. B. M. 1989. *Peacemaking among Primates*. Cambridge, MA: Harvard University Press.

———. 1996. *Good Natured: The Origins of Right and Wrong in Humans and Other Animals*. Cambridge, MA: Harvard University Press.

———. 1997. *Bonobo: The Forgotten Ape*. Berkeley: University of California Press.

———. 2000. *Chimpanzee Politics: Power and Sex among Apes*. Baltimore: Johns Hopkins University Press.

———. 2001. *The Ape and the Sushi Master*. New York: Basic Books.

de Waal, F. B. M., and P. L. Tyack. 2003. *Animal Social Complexity: Intelligence, Culture, and Individualized Societies*. Cambridge, MA: Harvard University Press.

Diamond, I., and G. Orenstein, eds. 1990. *Reweaving the World: The Emergence of Ecofeminism*. San Francisco: Sierra Club Books.

Diamond, J. 1997. *Guns, Germs, and Steel: The Fates of Human Societies*. New York: W. W. Norton.

———. 2004. *Collapse: How Societies Choose to Fail or Succeed*. New York: Viking Press.

Diamond, S. 1981. *In Search of the Primitive*. New York: Transaction Books.

Dicke, M., and W. Takken. 2006. *Chemical Ecology: From Gene to Ecosystem*. London: Springer.

Dillard, A. 1974. *Pilgrim at Tinker Creek*. New York: HarperCollins.

Dilworth, L. 1992. *Imagining Indians in the Southwest: Persistent Visions of a Primitive Past*. Washington, DC: Smithsonian Institution Press.

diZerega, G., D. Frew, and K. Wilber. 1999. "Neopaganism and the Mystical Tradition." *Integral World: Exploring Theories of Everything*. November. www.integralworld.net/dizerega.html (accessed November 20, 2003).

Dobrin, S., and C. Weisser. 2002. *Natural Discourses: Toward Ecocomposition*. Albany, NY: SUNY Press.

Dodd, J., and R. Stanton. 1990. *Paleoecology: Concepts and Applications*. 2nd ed. New York: Wiley-Interscience.

Dodds, G., and R. Tavernor. 2002. *Body and Building: Essays on the Changing Relation of Body and Architecture*. Cambridge, MA: MIT Press.

Dodson, S. 1998. *Ecology*. New York: Oxford University Press.

———. 1999. *Readings in Ecology*. New York: Oxford University Press.

Doel, F., and G. Doel. 2001. *The Green Man in Britain*. Stroud, UK: Tempus.

Doolittle, W. 2002. *Cultivated Landscapes of Native North America*. Oxford: Oxford University Press.

Drengson, A., and Y. Inoue. 1995. *The Deep Ecology Movement: An Introductory Anthology*. Berkeley, CA: North Atlantic Books.

Drescher, S., and S. Engerman, eds. 1998. *A Historical Guide to World Slavery*. New York: Oxford University Press.

Drexler, E. 1993. *Unbounding the Future: The Nanotechnology Revolution*. New York: Quill.

Drury, W., and J. Anderson. *Chance and Change: Ecology for Conservationists*. Berkeley: CA: University of California Press.

Dryzek, J. 1990. "Green Reason: Communicative Ethics for the Biosphere." *Environmental Ethics* 12 (Fall): 195–210.

Dufrechou, J. 2002. "Coming Home to Nature through the Body: An Intuitive Inquiry into Experience of Grief, Weeping and Other Deep Emotions in Response to Nature." PhD diss., Institute of Transpersonal Psychology.

———. 2004. "We Are One: Grief, Weeping, and Other Deep Emotions in Response to Nature as a Path toward Wholeness." *The Humanistic Psychologist* 32 (4): 357–78.

Dugan, K. M. 1985. *The Vision Quest of the Plains Indians: Its Spiritual Significance*. Lewiston, NY: Edwin Mellen Press.

Dukas, R., ed. 1998. *Cognitive Ecology: The Evolutionary Ecology of Information Processing and Decision Making*. Chicago: University of Chicago Press.

Dunlap, R. E., and W. Michelson, eds. 2001. *Handbook of Environmental Sociology*. Westport, CT: Greenwood Press.

Dupuy, J. 2007. "Toward an Integral Recovery Model for Drug and Alcohol Addiction." *AQAL: Journal of Integral Theory and Practice* 2 (3): 26–42.

Durie, R. 1997. "An Interview with Howard Gardner." *Mindshift Connection*, Spring.

Durning, A. T. 1995. "Are We Happy Yet?" In *Ecopsychology: Restoring the Earth, Healing the Mind*, edited by T. Roszak, M. E. Gomes, and A. D. Kanner. San Francisco: Sierra Club Books.

Dusenbery, D. 1992. *Sensory Ecology: How Organisms Acquire and Respond to Information*. New York: Freeman.

Eagles, P. F. J., and S. Muffitt. 1990. "An Analysis of Children's Attitudes toward Animals." *Journal of Environmental Education* 21 (3): 41–44.

Earth Works Group. 1989. 50 *Simple Things You Can Do to Save the Earth*. Berkeley, CA: Earthworks Press.

Easterbrook, G. 1995. A *Moment on the Earth: The Coming Age of Environmental Optimism*. New York: Viking.

Eckersley, R. 1992. *Environmentalism and Political Theory: Toward an Ecocentric Approach*. Albany, NY: SUNY Press.

———. 2004. *The Green State: Rethinking Democracy and Sovereignty*. Cambridge, MA: MIT Press.

Ecological Visions Committee. 2004. "Ecological Science and Sustainability for a Crowded Planet." Report presented to the Ecological Society of America, April. http://esa.org/ecovisions.

Eddy, B. 2001. "An Integral Approach to Sustainable Development." Working paper, Carleton University, Ottawa.

———. 2002a. "A Comparative Review of Ecosystem Modeling and 'Integral Theory': A Theoretical Basis for Modeling Human-Environment Interaction in Geography." Working paper, Carleton University, Ottawa.

———. 2002b. "Some First Principles of an 'Integral Geography': An Approach to Modeling Human-Environment Problem Spaces." Working paper, Carleton University, Ottawa.

———. 2005. "Integral Geography: Space, Place and Perspective." *World Futures: The Journal of General Evolution* 61 (1–2): 151–63.

Eddy. B., and D. R. F. Taylor. 2006. "Exploring the Concept of Cybercartography

Using the Holonic Tenets of Integral Theory." In *Cybercartography: Theory and Practice*, edited by D. R. F. Taylor. Amsterdam: Elsevier, 35–61.

Edgerton, R. 1992. *Sick Societies: Challenging the Myth of Primitive Harmony*. New York: Free Press.

Edwards, M. 2002. "Through AQAL Eyes, Part 2: Integrating Holon Theory and the AQAL Framework." *Integral World: Exploring Theories of Everything* (November). www.integralworld.net/edwards6.html (accessed August 16, 2003).

———. 2003. "A Brief History of Holons." *Integral World: Exploring Theories of Everything* (October). www.integralworld.net/edwards13.html (accessed May 9, 2006).

Ehrlich, P. 1968. *The Population Bomb*. New York: Ballantine.

Ehrlich, P. R., and A. H. Ehrlich. 1996. *Betrayal of Science and Reason: How Anti-environmental Rhetoric Threatens Our Future*. Washington, DC: Island Press.

Einarsen, J. 2000. "Satish Kumar Takes Deep Ecology a Step Further." *Kyoto Journal* 43. www.reverentialecology.org/kyoto_interview.htm (accessed August 14, 2007).

Eisenberg, E. 1998. *The Ecology of Eden: An Inquiry into the Dream of Paradise and a New Vision of Our Role in Nature*. New York: Random House.

Eldredge, N., and S. N. Salthe. 1984. "Hierarchy and Evolution." *Oxford Surveys in Evolutionary Biology* 1: 184–208.

Elgar, M. A., and B. J. Crespi, eds. 1992. *Cannibalism: Ecology and Evolution among Diverse Taxa*. New York: Oxford University Press.

Elgin, D. 1993a. *Awakening Earth: Exploring the Evolution of Human Consciousness*. New York: William Morrow.

———. 1993b. *Voluntary Simplicity: Toward a Way of Life That Is Outwardly Simple, Inwardly Rich*. Rev. ed. New York: Quill.

———. 1997. "Global Consciousness Change: Indicators of an Emerging Paradigm." www.simpleliving.net/awakeningearth/reports.asp#03 (accessed December 14, 1997).

Eliade, M. 1958. *Patterns in Comparative Religion*. New York: Harper.

———. 1959. *The Sacred and the Profane: The Nature of Religion*. New York: Harper.

———. 1964/2004. *Shamanism: Archaic Techniques of Ecstasy*. Princeton, NJ: Princeton University Press.

Elkin, A. P. 1977/1993. *Aboriginal Men of High Degree: Initiation and Sorcery in the World's Oldest Tradition*. Rochester, VT: Inner Traditions.

Ellen, R. F., P. Parkes, and A. Bicker, eds. 2000. *Indigenous Environmental Knowledge and Its Transformations: Critical Anthropological Perspectives*. New York: Routledge.

Eller, C. 2000. *The Myth of Matriarchal Prehistory: Why an Invented Past Won't Give Women a Future*. Boston: Beacon Press.

Ellingson, T. 2001. *The Myth of the Noble Savage*. Berkeley: University of California Press.

Elliott, S. S. 1994. "The Mysticism of Annie Dillard's *Pilgrim at Tinker Creek*." www.well.com/user/elliott/smse_dillard.html (accessed August 10, 2005).

Ellis, R. 2004. *No Turning Back: The Life and Death of Animal Species*. New York: HarperCollins.

Elton, C. 1949. *Animal Ecology*. Chicago: University of Chicago Press.

———. 1958/2000. *The Ecology of Invasion by Animals and Plants*. Chicago: University of Chicago Press.

Emerson, R. W. 1969. "On Nature." In *The Best of Ralph Waldo Emerson: Essays, Poems, Addresses*, edited by G. Haight. Roslyn, NY: Walter J. Black, 73–116.

Emmeche, C. 1999. "The Biosemiotics of Emergent Properties in a Pluralist Ontology." In *Semiosis. Evolution. Energy: Towards a Reconceptualization of the Sign*, edited by E. Taborsky. Aachen, Germany: Shaker Verlag, 89–108.

———. 2001. "Does a Robot Have an Umwelt?" In *Jakob von Uexküll: A Paradigm for Biology and Semiotics*, edited by K. Kull. Special volume, *Semiotica* 134 (1–4): 653–93.

Emmeche, C., J. Hoffmeyer, and K. Kull, eds. 2002. *Biosemiotics*. Special issue, *Sign Systems Studies* 30: 1.

Emmeche, C., K. Kull, and F. Stjernfelt. 2002. *Reading Hoffmeyer, Rethinking Biology*. Tartu, Estonia: Tartu University Press.

Empereur, J. 1997. *The Enneagram and Spiritual Direction: Nine Paths to Spiritual Guidance*. New York: Continuum.

Endredy, J. 2002. *Earthwalks for Body and Spirit: Exercises to Restore Our Sacred Bond with the Earth*. Rochester, VT: Inner Traditions.

Engel, C. 2002. *Wild Health*. New York: Houghton Mifflin.

Esbjörn-Hargens, S. 2005a. "Integral Ecology: The What, Who, and How of Environmental Phenomena." *World Futures: The Journal of General Evolution* 61 (1–2): 5–49.

———. 2005b. "Integral Education by Design: How Integral Theory Informs Teaching, Learning, and Curriculum in a Graduate Program." *ReVision* 28 (3): 21–29.

———. 2006. "Integral Research: A Multi-method Approach to Investigating Phenomena." *Constructivism and the Human Sciences* 11 (1–2): 79–107.

———. 2007. "Integral Teacher, Integral Students, Integral Classroom: Applying Integral Theory to Education" *AQAL: Journal of Integral Theory and Practice* 2 (2): 72–103.

Esbjörn-Hargens, S., ed. 2005. *Integral Ecology*. Special double issue. *World Futures: The Journal of General Evolution* 61 (1–2).

Esbjörn-Hargens, S., and K. Wilber. 2006. "Towards a Comprehensive Integration of Science and Religion: A Post-metaphysical Approach." In *The Oxford Handbook of Science and Religion*. Oxford: Oxford University Press, 523–46.

Etling, K. 2001. *Hunting Superbucks: How to Find and Hunt Today's Trophy Mule and Whitetail Deer*. New York: Lyons Press.

Evernden, N., and L. Leslie. 1992. *The Social Creation of Nature*. Baltimore: Johns Hopkins University Press.

Ewert, J. P. 1980. *Neuroethology: An Introduction to the Neurophysiological Fundamentals of Behaviour*. New York: Springer-Verlag.

Fabel, A., and D. St. John, eds. 2003. *Teilhard in the Twenty-first Century. The Emerging Spirit of Earth*. Maryknoll, NY: Orbis Books.

Falk, D., M. Palmer, and J. Zedler, eds. 2006. *Foundations of Restoration Ecology: The Science and Practice of Ecological Restoration*. Washington, DC: Island Press.

Farrar, S., and J. Farrar. 1984. *The Witches' Way*. London: Robert Hale.

Favareau, D. 2007. "The Evolutionary History of Biosemiotics." In *Introduction to Biosemiotics*, edited by M. Barbieri. Dordrecht, Netherlands: Springer, 209–33.

Feibleman, J. K. 1959. "Theory of Integrative Levels." *British Journal for the Philosophy of Science* 5 (17): 59–66.

Feinstein, K. 2006. "The Pepperwood Issue in Sebastopol, CA: A Case Study." Master's thesis, John F. Kennedy University.

Fernandez-Armesto, F. 2001. *Civilizations: Culture, Ambition, and the Transformation of Nature*. New York: Touchstone.

Ferrer, J. 2002. *Revisioning Transpersonal Theory: A Participatory Vision of Human Spirituality*. Albany, NY: SUNY Press.

Ferry, L. 1992. *The New Ecological Order*. Translated by C. Volk. Chicago: University of Chicago Press.

Feuerstein, G. 2003. "The Practice of Eco-Yoga." In *The Deeper Dimension of Yoga*, 210–14. Boston: Shambhala Publications.

Feuerstein, G., and B. Feuerstein. 2007. *Green Yoga*. New York: Traditional Yoga Studies.

Fill, A., and P. Muhlhausler, eds. 2001. *The Ecolinguistics Reader: Language, Ecology and Environment*. New York: Continuum.

Findhorn Community. 1976. *The Findhorn Garden*. New York: HarperCollins.

Fisher, A. 2002. *Radical Ecopsychology: Psychology in the Service of Life*. Albany, NY: SUNY Press.

Fisher, J. A. 1996. "The Myth of Anthropomorphism." In *Readings in Animal Cognition*, edited by M. Bekoff and D. Jamieson. Cambridge, MA: MIT Press, 3–16.

Fisher, R. M. 2003. "'Lighting Up' the Integral: A Critical Review of Ken Wilber's Philosophy and Theories Related to Education." Working paper.

Flader, S. L., and J. B. Callicott, eds. 1992. *The River of the Mother of God and Other Essays by Aldo Leopold*. Madison: University of Wisconsin Press.

Flannery, T. 1994. *The Future Eaters: An Ecological History of the Australasian Lands and People*. New York: George Braziller.

———. 2001. *The Eternal Frontier: An Ecological History of North America and Its People*. New York: Grove Press.

———. 2005. *The Weather Makers*. New York: Atlantic Monthly Press.

Flannery, T., and P. Schouten. 2001. *A Gap in Nature: Discovering the World's Extinct Species*. New York: Atlantic Monthly Press.

Flavell, J. H., P. H. Miller, and S. A. Miller. 2002. *Cognitive Development*. 4th ed. Upper Saddle River, NJ: Prentice Hall.

Flew, A., and G. Habermas. 2004. "My Pilgrimage from Atheism to Theism: A Discussion between Antony Flew and Gary Habermas." *Philosophia Christi* 6 (Winter): 197–211.

Flores, J. 2004. "Awakening to the Eco-Tragedy: An Ideological and Epistemological Inquiry into Consistencies and Contradictions in Christian Environmental Awareness." PhD diss., Saybrook Graduate School and Research Center.

Foote, A., and J. Crystal. 2007. "Metacognition in the Rat." *Current Biology* 17 (6): 551–55.

Ford, B. 2000. "Plants Have Senses." In *The Secret Language of Life: How Animals and Plants Feel and Communicate*. New York: Fromm International, chap. 5.

Ford, D. H., and R. M. Lerner. 1992. *Developmental Systems Theory: An Integrative Approach*. Thousand Oaks, CA: SAGE.

Ford, E. B. 1975. *Ecological Genetics*. 4th ed. London: Chapman and Hall.

Foreman, D., and Haywood, B. 1993. *Ecodefense: A Field Guide to Monkey Wrenching*. 3rd ed. Chico, CA: Abbzug.

Forest, M. S. E. 1992. "'Ought' and 'Can' in Environmental Ethics: Ethical Extensionism and Moral Development." Master's thesis, Colorado State University, Fort Collins.

Fortin, M., and M. R. T. Dale. 2005. *Spatial Analysis: A Guide for Ecologists*. New York: Cambridge University Press.

Foster, J. B. 2005. "The Treadmill of Accumulation: Schnaiberg's Environment and Marxian Political Economy." *Organization and Environment* 18 (1): 7–18.

Foster, S., with M. Little. 1992. *The Book of the Vision Quest: Personal Transformation in the Wilderness*. New York: Simon & Schuster.

Foster, S. E. 2002. "Aristotle and the Environment." *Environmental Ethics* 24 (Winter): 409–28.

Foucault, M. 1984. *The Use of Pleasure*. Vol. 2, *The History of Sexuality*. Translated by R. Hurley. New York: Vintage Books.

Fouts, R., and S. T. Mills. 1997. *Next of Kin: My Conversations with Chimpanzees*. New York: Avon Books.

Fox, C., D. Roff, and D. Fairbairn, eds. 2001. *Evolutionary Ecology: Concepts and Case Studies*. New York: Oxford University Press.

Fox, M. 1988. *The Coming of the Cosmic Christ: The Healing of Mother Earth and the Birth of a Global Renaissance*. San Francisco: Harper.

———. 1991. *Creation Spirituality: Liberating Gifts for the Peoples of the Earth*. San Francisco: Harper.

Fox, W. 1990. "Transpersonal Ecology: 'Psychologizing' Ecophilosophy." *Journal of Transpersonal Psychology* 22 (1): 59–96.

———. 1990. *Toward a Transpersonal Ecology: Developing New Foundations for Environmentalism*. Albany, NY: SUNY Press.

Fragaszy, D. M., and S. Perry, eds. 2003. *The Biology of Traditions: Models and Evidence*. Cambridge: Cambridge University Press.

Frankel, C. 2004. *Out of the Labyrinth*. Rhinebeck, NY: Monkfish Book Publishing Co.

Frawley, D. 2004. *Yoga and the Sacred Fire: Self-Realization and Planetary Transformation*. Twin Lakes, WI: Lotus Press.

Frazer, J. G. 1926. *The Worship of Nature*. London: Macmillan.

———. 1994. *The Golden Bough: A History of Myth and Religion*. London: Macmillan.

Freeman, A., ed. 2006. *Consciousness and Its Place in Nature: Does Physicalism Entail Panpsychism?* Exeter, UK: Imprint Academic.

Frew, D. 1999. "Harran: Last Refuge of Classical Paganism." *The Pomegranate: A New Journal of Neopagan Thought* 3 (9). http://chass.colostate-pueblo.edu/natrel/pom/old/POM9a1.html (accessed January 15, 2002).

Friedman, B., P. H. Kahn, Jr., and A. Borning. 2006. "Value Sensitive Design and Information Systems." In *Human-Computer Interaction in Management Information Systems*, edited by D. Galletta and P. Zhang. Armonk, NY: Sharpe.

Friedman, T. L. 2005. "Geo-Greening by Example." *New York Times*, www.nytimes.com/2005/03/27/opinion/27friedman.html (accessed March 27, 2005).

Friend, T. 2004. *Animal Talk: Breaking the Codes of Animal Language.* New York: Free Press.

Frohoff, T., and B. Peterson, eds. 2003. *Between Species: Celebrating the Dolphin-Human Bond.* San Francisco: Sierra Club Books.

Frumkin, H. 2005. *Environmental Health: From Global to Local.* San Francisco: Jossey-Bass.

Fukuoka, M. 1978. *The One-Straw Revolution: An Introduction to Natural Farming.* Emmaus, PA: Rodale Press.

———. 1985. *The Natural Way of Farming: The Theory and Practice of Green Philosophy.* Tokyo: Japan Publications.

Fukuoka, M., M. Fukuoka, and F. Metreaud. 1987. *The Road Back to Nature: Regaining the Paradise Lost.* Tokyo: Japan Publications.

Fuller, M. 2007. *Media Ecologies: Materialist Energies in Art and Technoculture.* Cambridge, MA: MIT Press.

Fussell, E. 1965. *Frontier: American Literature and the American West.* Princeton, NJ: Princeton University Press.

Gall, E. 1934. *Mysticism Throughout the Ages.* London: Rider.

Gallagher, W. 1994. *Power of Place: How Our Surroundings Shape Our Thoughts, Emotions, and Actions.* New York: Harper Perennial.

Gallopin, G., and P. Raskin. 2002. *Global Sustainability: Bending the Curve.* London: Routledge.

Gardell, M. 2003. *Gods of the Blood: The Pagan Revival and White Separatism.* Durham, NC: Duke University Press.

Gardiner, H. W., and C. Kosmitzki. 2004. *Lives Across Cultures: Cross-cultural Human Development.* Boston: Allyn & Bacon.

Gare, A. 1995. *Postmodernism and the Environmental Crisis.* London: Routledge.

Garrett, J. 1991. "Aristotle, Ecology and Politics: Theoria and Praxis for the Twenty-first Century." In *Communitarianism, Liberalism, and Social Responsibility*, edited by C. Peden and Y. Hudson, 122–46. Lewiston, NY: Edwin Mellen Press.

Garrett, L. 1994. *The Coming Plague: Newly Emerging Diseases in a World Out of Balance*. New York: Farrar, Straus & Giroux.

Gaston, K. J., and T. M. Blackburn. 2000. *Pattern and Process in Macroecology*. Malden, MA: Blackwell Science.

Gatta, J. 2004. *Making Nature Sacred: Literature, Religion, and Environment in America from the Puritans to the Present*. Oxford: Oxford University Press.

Gebser, J. 1949/1953/1985. *The Ever-Present Origin*. Translated by N. Barstad and A. Mickunas. Athens: Ohio University Press.

Geddes, P. 1915. *Cities in Evolution: An Introduction to the Town Planning Movement and the Study of Civics*. London: Williams & Norgate.

Geldard, R., ed. 2005. *The Essential Transcendentalists*. New York: Penguin.

George, P. A. n.d. "*Mandala*: Buddhist Tantric Diagrams." http://ccat.sas.upenn.edu/george/mandala.html (accessed February 16, 2006).

Gergen, K. 2000. *The Saturated Self: Dilemmas of Identity in Contemporary Life*. New York: Basic Books.

Geselle, J. 2005. "The Development of Ecological Concern: An Integral Approach." Working paper.

Geyer, F., and J. V. D. Zouwen. 2001. *Sociocybernetics: Complexity, Autopoiesis, and Observation of Social Systems*. Westport, CT: Greenwood.

Gibbs, J. J., D. Giever, and K. A. Pober. 2000. "Criminology and the Eye of the Spirit: An Introduction and Application of the Thoughts of Ken Wilber." *Journal of Contemporary Criminal Justice* 16 (1): 99–127.

Giblett, R. 1996. *Postmodern Wetlands: Culture, History, Ecology*. Edinburgh, UK: Edinburgh University Press.

———. 2004. *Living with the Earth: Mastery to Mutuality*. London: Salt Publishing.

Gibson, E., and A. Pick. 2000. *An Ecological Approach to Perceptual Learning and Development*. New York: Oxford University Press.

Gibson, J. 1986. *The Ecological Approach to Visual Perception*. Hillsdale, NJ: Lawrence Erlbaum.

Gibson, P. R. 2000. *Multiple Chemical Sensitivity: A Survival Guide*. Oakland, CA: New Harbinger.

Gigerenzer, G., and R. Selten, eds. 2001. *Bounded Rationality: The Adaptive Toolbox*. Cambridge, MA: MIT Press.

Gilbert, S. F., and S. Sarkar. 2000. "Embracing Complexity: Organicism for the Twenty-first Century." *Developmental Dynamics* 219: 1–9.

Gill, S. 1982. *Native American Religions: An Introduction*. Belmont, CA: Wadsworth.

———. 1991. *Mother Earth: An American Story*. Chicago: Chicago University Press.

Gilligan, C. 1982. *In a Different Voice: Psychological Theory and Women's Development*. Cambridge, MA: Harvard University Press.

Gilmore, R. *Catastrophe Theory for Scientists and Engineers*. New York: Dover Publications.

Gilovitch, T., D. Griffin, and D. Kahneman, eds. 2002. *Heuristics and Biases: The Psychology of Intuitive Judgment*. Cambridge: Cambridge University.

Girard, R. 1972. *Violence and the Sacred*. Translated by P. Gregory. Baltimore: Johns Hopkins University Press.

Glacken, C. 1973. *Traces on the Rhodian Shore: Nature and Culture in Western Thought from Ancient Times to the End of the Eighteenth Century*. Berkeley: University of California Press.

Glazebrook, T. 2003. "Art or Nature? Aristotle, Restoration Ecology, and Flowforms." *Ethics and the Environment* 8 (1): 23–36.

Gleason, H. A. 1926. "The Individualistic Concept of the Plant Association." *Bulletin of the Torrey Botanical Club* 53: 7–26.

Glendinning, C. 1994. *My Name Is Chellis and I'm in Recovery from Western Civilization*. Boston: Shambhala Publications.

Glotfely, C., and H. Fromm, eds. 1996. *The Ecocriticism Reader: Landmarks in Literary Ecology*. Athens: University of Georgia Press.

Gobster, P., and R. B. Hull, eds. 2000. *Restoring Nature: Perspectives from Humanities and Social Sciences*. Washington DC: Island Press.

Goethe, J. W. 1995. *Scientific Studies* (*Goethe: The Collected Works*, vol. 12). Princeton, NJ: Princeton University Press. Translated by D. Miller.

Goldman, L. 1999. *The Anthropology of Cannibalism*. Westport, CT: Bergin & Garvey.

Goldsmith, E. 1993. *The Way: An Ecological Worldview*. Boston: Shambhala Publications.

Goleman, D. 1995. *Emotional Intelligence: Why It Can Matter More Than IQ*. New York: Bantam.

Golinski, J. 2005. *Making Natural Knowledge: Constructivism and the History of Science*. Chicago: University of Chicago Press.

Golley, F. B. 1996. *A History of the Ecosystem Concept in Ecology: More Than the Sum of the Parts*. New Haven, CT: Yale University Press.

Goodall, J. 1986. *The Chimpanzees of Gombe: Patterns of Behavior*. Boston: Belknap Press of Harvard University Press.

Goodall, J., and M. Bekoff. 2002. *The Ten Trusts: What We Must Do to Care for the Animals We Love*. San Francisco: HarperCollins.

Goodin, R. E. 1992. *Green Political Theory*. Cambridge, MA: Polity Press.

Goodman, A. H., and T. L. Leatherman, eds. 1999. *Building a New Biocultural Synthesis: Political-Economic Perspectives on Human Biology*. Ann Arbor: University of Michigan.

Goodwin, B. 1994. *How the Leopard Changed Its Spots: The Evolution of Complexity*. London: Weidenfeld & Nicolson.

Gordon, G., and S. Esbjörn-Hargens, 2007. "Are We Having Fun Yet? An Exploration of the Transformative Power of Play." *Journal of Humanistic Psychology* 35 (1): 198–222.

Gore, A. 1993. *Earth in Balance: Ecology and the Human Spirit*. New York: Plume.

———. 2006. *An Inconvenient Truth*. Emmaus, PA: Rodale.

Gosling, S. D. 2001. "From Mice to Men: What Can We Learn about Personality from Animal Research?" *Psychological Bulletin* 127: 45–86.

Gottlieb, R. S. 1996. *This Sacred Earth: Religion, Nature, Environment*. New York: Routledge.

Gould, J. 1982. *Ethology: The Mechanisms and Evolution of Behavior*. New York: W. W. Norton.

Gould, K. A., D. N. Pellow, and A. Schnaiberg. 2004. "Everything You Wanted to Know about the Treadmill but Were Afraid to Ask." *Organization and Environment* 17 (3): 296–316.

Graedel, T. E., and B. R. Allenby. 2002. *Industrial Ecology*. 2nd ed. Englewood Cliffs, NJ: Prentice Hall.

Gragson, T. L. 2005. "Time in Service to Historical Ecology." *Ecological and Environmental Anthropology* 1 (1) (February): 2–9. www.uga.edu/eea/01_2005/gragson_2005.pdf.

Graham, F., and C. Buchheister. 1992. *The Audubon Ark: A History of the National Audubon Society*. Austin: University of Texas Press.

Graham, O. L. Jr., ed. 2000. *Environmental Politics and Policy, 1960s to 1990s*. Issues in Policy History 9. University Park: Pennsylvania State University Press.

Grande, S. M. A. 1999. "Beyond the Ecological Noble Savage: Deconstructing the White Man's Indian." *Environmental Ethics* 21 (Fall): 307–20.

Grandin, T., and C. Johnson. 2006. *Animals in Translation*. London: Bloomsbury.

Grange, J. 1997. *Nature: An Environmental Cosmology*. Albany, NY: SUNY Press.

———. 1999. *The City: An Urban Cosmology*. Albany, NY: SUNY Press.

Graves, C. 2002. *Graves: Levels of Human Existence*. Edited by W. Lee, C. C. Cowan, and N. Todorovic. Santa Barbara, CA: ECLET.

———. 2005. *The Never Ending Quest: Dr. Clare W. Graves Explores Human Nature*. Edited by C. C. Cowan and N. Todorovic. Santa Barbara, CA: ECLET.

Green, A. 1975. *The Role of Human Sacrifice in the Ancient Near East*. Missoula, MT: Scholars Press for the American Schools of Oriental Research.

Greenberg, G. 1998. *Comparative Psychology: A Handbook*. London: Routledge.

Greene, J. C. 1999. *Debating Darwin*. Claremont, CA: Regina Books.

Greene, J. C., V. J. Caracelli, and W. F. Graham. 1989. "Toward a Conceptual Framework for Mixed-Method Evaluation Designs." *Educational Evaluation and Policy Analysis* 11: 255–74.

Greenwald, J. 1992. "Environmental Attitudes: A Structural Developmental Model." PhD diss., University of Massachusetts.

Greenway, R. 1995. "The Wilderness Effect and Ecopsychology." In *Ecopsychology: Restoring the Earth, Healing the Mind*, edited by T. Roszak, M. E. Gomes, and A. D. Kanner. San Francisco: Sierra Club Books, 122–35.

Gregory, N. 2005. *Physiology and Behavior of Animal Suffering*. Cambridge, MA: Blackwell.

Grey, A. 1991. *Sacred Mirrors: The Visionary Art of Alex Grey*. Rochester, NY: Inner Traditions.

———. 1998. *The Mission of Art*. Boston: Shambhala Publications.

———. 2001. *Transfigurations*. Rochester, NY: Inner Traditions.

Griffin, David R. 1994. "Whitehead's Deeply Ecological Worldview." In *Worldviews and Ecology*, edited by M. E. Tucker and J. Grim, 190–206. Maryknoll, NY: Orbis Books.

Griffin, D. R., and R. Falk, eds. 1993. *Postmodern Politics for a Planet in Crisis*. Albany, NY: SUNY Press.

Griffin, Donald R. 1976. *The Question of Animal Awareness: Evolutionary Continuity of Mental Experience*. New York: Rockefeller University Press.

———. 1984. *Animal Thinking*. Cambridge, MA: Harvard University Press.

———. 1992. *Animal Minds*. Chicago: University of Chicago Press.

Griffin, S. 1978. *Woman and Nature: The Roaring Inside Her*. New York: Harper & Row.

Griffiths, T., and L. Robin, eds. 1997. *Ecology and Empire: Environmental History of Settler Societies*. Seattle: University of Washington Press.

Grim, J., ed. 2001. *Indigenous Traditions and Ecology: The Interbeing of Cosmology and Community*. Cambridge, MA: Harvard University Press.

Grof, S. 1992. *The Holotropic Mind: The Three Levels of Human Consciousness and How They Shape Our Lives*. San Francisco: HarperCollins.

———. 2000. *Psychology of the Future: Lessons from Modern Consciousness Research*. Albany: SUNY Press.

Gross, P. R., and N. Levitt. 1997. *Higher Superstition: The Academic Left and Its Quarrels with Science*. Baltimore: Johns Hopkins University Press.

Grosz, T. 1999. *Wildlife Wars: The Life and Times of a Fish and Game Warden*. Boulder, CO: Johnson Books.

Groth, P., and T. Bressi. 1997. *Understanding Ordinary Landscapes*. New Haven, CT: Yale University Press.

Grove, R. H. 1995. *Green Imperialism: Colonial Expansion, Tropical Island Edens, and the Origins of Environmentalism, 1600–1860*. New York: Cambridge University Press.

Grubb, T. Jr. 2003. *The Mind of the Trout: A Cognitive Ecology for Biologists and Anglers*. Madison: University of Wisconsin Press.

Grumbine, R. E. 1992. *Ghost Bears: Exploring the Biodiversity Crisis*. Washington, DC: Island Press.

Guattari, F. 1989/2000. *The Three Ecologies*. Translated by I. Pindar and P. Sutton. London: Alhlone Press.

Gueye, M., D. Diouf, T. Chaava, and D. Tiomkin. 2005. *Community Capacity Enhancement Handbook*. New York: UNDP.

Gunderson, L. H., and C. S. Holling. 2001. *Panarchy: Understanding Transformations in Systems of Humans and Nature*. Washington, DC: Island Press.

Gussow, J. D. 1978. *The Feeding Web: Issues in Nutritional Ecology*. Palo Alto, CA: Full Publishing Co.

Gutierrez, G. 1988. *A Theology of Liberation: History, Politics and Salvation*. Maryknoll, NY: Orbis Books.

Habermas, J. 1984–1985. *The Theory of Communicative Action*. 2 vols. Translated by T. McCarthy. Boston: Beacon Press.

———. 1985/1995. *The Philosophical Discourse of Modernity: Twelve Lectures.* Translated by F. G. Lawrence. Cambridge, MA: MIT Press.

———. 1994. "The Critique of Reason as an Unmasking of the Human Sciences: Michel Foucault." In *Critique and Power: Recasting the Foucault/Habermas Debate,* edited by M. Kelly. Cambridge, MA: MIT Press.

Haeckel, E. 1866. *Generelle Morphologie der Organismen.* Berlin: Georg Reimer.

———. 1873. *The History of Creation* London: D. Appleton and Company.

———. 1879. *The Evolution of Man.* London: C. Kegan Paul & Co.

Hale, G. 2002. *The Practical Encyclopedia of Feng Shui.* New York: Lorenz Books.

Halifax, J. 1979/1991. *Shamanic Voices: A Survey of Visionary Narratives.* New York: E. P. Dutton.

———. 1993. *The Fruitful Darkness: Reconnecting with the Body of the Earth.* San Francisco: Harper.

Hallam, A., and P. B. Wignall. 1997. *Mass Extinctions and Their Aftermath.* New York: Oxford University Press.

Hallam, T. 2004. *Catastrophes and Lesser Calamities: The Causes of Mass Extinctions.* New York: Oxford University Press.

Haller, R., and C. Kramer, eds. 2007. *Horticulture Therapy Methods: Making Connections in Health Care, Human Service, and Community Programs.* New York: Haworth Press.

Hallman, M. 1991. "Nietzsche's Environmental Ethics." *Environmental Ethics* 13 (Summer): 99–125.

Halprin, A. 1995a. *Moving toward Life: Five Decades of Transformational Dance.* London: Wesleyan University Press.

———. 1995b. *Embracing Earth: Dances with Nature* (VHS, 23 min.). Sausalito, CA: Open Eye Pictures.

———. 2003. *Returning Home* (VHS, 43 min.). Sausalito, CA: Open Eye Pictures.

Hamilton, M. 1999. "The Berkana Community of Conversations: A Study of Leadership Skill Development and Organizational Leadership Practices in a Self-Organizing Online Microworld." PhD diss., Columbia Pacific University.

———. 2005. "Four Questions That Release the Potential of Your City." Paper presented to the Canadian Institute of Planners.

———. 2006a. "Integral Community: A Common Meta-Language for Planning and Managing Change." *Journal of Change Management* 19 (3).

———. 2006b. "Integral Framework for Sustainable Planning: A Prototype for Emergent Well Being." Paper presented at World Planners Congress.

Hampsom, G. P. 2007. "Integral Re-views Postmodernism: The Way Out Is Through." *Integral Review* 4: 110–73.

Hannan, M. T., and J. H. Freeman. 1989. *Organizational Ecology*. Cambridge, MA: Harvard University Press.

Happold, F. C. 1962. *Mysticism: A Study and an Anthology*. New York: Penguin.

Haraway, D. 1991. *Simians, Cyborgs and Women: The Reinvention of Nature*. New York: Routledge.

Hardin, G. 1968. "The Tragedy of the Commons." *Science* 162 (3859): 1243–48.

Harding, S. 1986. *The Science Question in Feminism*. Ithaca, NY: Cornell University Press.

Hargens, S. 2001. "Intersubjective Musings: A Response to Christian de Quincey's 'The Promise of Integralism.'" *Journal of Consciousness Studies* 8 (12): 35–78.

———. 2002. "Integral Development: Taking the Middle Path towards Gross National Happiness." *Journal of Bhutan Studies* 6 (Summer): 24–87.

Harguindey, S. 2003. "Spirit and Politics for the XXI Century." Working paper.

Harman, W. W., and E. Sahtouris. 1998. *Biology Revisioned*. Berkeley, CA: North Atlantic Books.

Harmon, D. 2002. *In Light of Our Differences: How Diversity in Nature and Culture Makes Us Human*. Washington, DC: Smithsonian Institution Press.

Harms, R. 1987. *Games Against Nature: An Eco-Cultural History of the Nunu of Equatorial Africa*. Cambridge: Cambridge University Press.

Harner, M. 1990. *The Way of the Shaman*. 2nd ed. San Francisco: Harper.

Harris, P. 2001. *Frequently Asked Questions about Christian Meditation: The Path of Contemplative Prayer*. Ottawa, ON: Novalis.

Harrod, H. 2000. *The Animals Came Dancing: Native American Sacred Ecology and Animal Kinship*. Tucson, AZ: University of Arizona Press.

Hart, R. 1979. *Children's Experience of Place*. New York: Irvington.

Hartley, L. 1989. *Wisdom of the Body Moving: An Introduction to Body-Mind Centering*. Berkeley, CA: North Atlantic Books.

Harvey, G. 2000. *Contemporary Paganism: Listening People, Speaking Earth*. New York: New York University Press.

———. 2006. *Animism: Respecting the Living World*. New York: Columbia University Press.

Harvey, G., ed. 2002. *Shamanism: A Reader*. London: Routledge.

Haskell, B. D., B. G. Norton, and R. Costanza. 1992. "What Is Ecosystem Health and Why Should We Worry About It?" In *Ecosystem Health: New Goals for*

Environmental Management, edited by R. Costanza, B. G. Norton, and B. D. Haskell. Washington, DC: Island Press.

Hauser, M. D., and M. Konishi, eds. 2003. *The Design of Animal Communication*. Cambridge, MA: MIT Press.

Hawken, P. 1993. *The Ecology of Commerce*. New York: Harper Business.

Hawken, P., A. Lovins, and L. H. Lovins. 1999. *Natural Capitalism*. New York: Brown.

Hawley, A. 1986. *Human Ecology: A Theoretical Essay*. Chicago: University of Chicago Press.

Hay, P. 2002. *Main Currents in Western Environmental Thought*. Bloomington: Indiana University Press.

Hayden, P. 1997. "Gilles Deleuze and Naturalism: A Convergence with Ecological Theory and Politics." *Environmental Ethics* 19 (Summer): 185–204.

Hayes, D. 2000. *The Official Earth Day Guide to Planet Repair*. Covelo, CA: Island Press.

Hays, S. 1959/1999. *Conservation and the Gospel of Efficiency: The Progressive Conservation Movement, 1890–1920*. Pittsburgh, PA: University of Pittsburgh Press.

Hearn, T. 1995. *Natural Awakening: The Way of the Heart*. Auckland, New Zealand: Wangapeka Books.

Heft, H. 2005. *Ecological Psychology in Context: James Gibson, Roger Barker, and the Legacy of William James*. Hillsdale, NJ: Lawrence Erlbaum.

Heidegger, M. 2001. *The Fundamental Concepts of Metaphysics: World, Finitude, Solitude*. Bloomington: Indiana University Press.

Hein, S. "Critical Review of Daniel Goleman." www.eqi.org/gole.htm.

Heinrich, B. 1999. *Mind of a Raven: Investigations and Adventures with Wolf-Birds*. New York: HarperCollins.

Helvarg, D. 2004. *The War Against Greens: The "Wise-Use" Movement, the New Right, and the Browning of America*. 2nd ed. San Francisco: Sierra Club Books.

Hemenway, T. 2001. *Gaia's Garden: A Guide to Homescale Permaculture*. White River Junction, VT: Chelsea Green.

Henson, R. 2006. *The Rough Guide to Climate Change*. New York: Penguin Books.

Henzi, S. P., and L. Barrett. 2005. "The Historical Socioecology of Savanna Baboons." *Journal of the Zoological Society of London* 265: 215–26.

Herf, J. 1984. *Reactionary Modernism: Technology, Culture, and Politics in Weimar and the Third Reich*. Cambridge: Cambridge University Press.

Hibbard, W. "Reports of Transpersonal Experiences of Non-native Practitioners of the Native American Sweat Lodge Ceremony." PhD diss., Saybrook Graduate School and Research Center.

Hiby, L. 1998. *Conversations with Animals: Cherished Messages and Memories as Told by an Animal Communicator*. Troutdale, OR: New Sage Press.

Hill, G. 1992. *Masculine and Feminine: The Natural Flow of Opposites in the Psyche*. Boston: Shambhala Publications.

Hill, R. 2003. "Mountains and Mysticism: Observing Transformation When Climbing in Thin Air." PhD diss. proposal, Institute of Transpersonal Psychology, Palo Alto, CA.

Hill, R. L. 2004. "Complexity 'Humbles' Environmental Chiefs." *Oregonian*, sec. C, August 5.

Hillel, D. 1992. *Out of the Earth: Civilization and the Life of the Soil*. Berkeley: University of California Press.

Hillman, J. 1995. "A Psyche the Size of the Earth: A Psychological Foreword." In *Ecopsychology: Restoring the Earth, Healing the Mind*, edited by T. Roszak, M. E. Gomes, and A. D. Kanner. San Francisco: Sierra Club Books, xvii–xxiii..

———. 2007. *Animal Presences*. Putnam, CT: Spring Publications.

Hiss, T. 1990. *The Experience of Place*. New York: Knopf.

Hoban, T. M., and R. O. Brooks. 1987/1996. *Green Justice: The Environment and the Courts*. Boulder, CO: Westview Press.

Hochachka, G. 2001. "Integral Community Development in San Juan del Gozo, El Salvador: Including Communities, Ecosystems and 'Interiority' in the Developmental Process." Master's thesis, University of Victoria, BC.

———. 2005a. *Developing Sustainability, Developing the Self*. Victoria, BC: POLIS Project on Ecological Governance.

———. 2005b. "Integrating Interiority in Community Development." *World Futures: The Journal of General Evolution* 61 (1–2): 110–26.

———. 2007. "An Integral Approach to Well Being." Working paper.

Hodges, T. 1999. *Sworn to Protect*. Oroville, CA: T & C Books.

Hodson, G. 1922/1982. *Fairies at Work and at Play*. London: Theosophical Publishing House.

———. 1924. *The Kingdom of Faerie*. London: Theosophical Publishing House.

Hoffman, D. 1995. *Frank Lloyd Wright: Architecture and Nature*. New York: Dover.

Hoffmeyer, J. 1996a. *Signs of Meaning in the Universe*. Bloomington: Indiana University Press.

———. 1996b. "Evolutionary Intentionality." In *Proceedings from the Third European Conference on Systems Science*, edited by E. Pessa, A. Montesanto, and M. P. Penna. Rome: Edzioni Kappa, 699–703.

———. 1998. "The Unfolding Semiosphere." *Evolutionary Systems: Biological and Epistemological Perspectives on Selection and Self-Organization*, edited by G. Van de Vijver, S. N. Salthe, and M. Delpos. Dordrecht, Netherlands: Kluwer, 281–93.

———. 2001. "Origin of Species by Natural Translation." *Athanor* 12 (4): 240–55.

Hofstadter, R. 1992. *Social Darwinism in American Thought*. Boston: Beacon Press.

Hogan, L. 1995. *Dwellings: A Spiritual History of the Living World*. New York: W. W. Norton.

Holler, C., ed. 2000. *The Black Elk Reader*. Syracuse, NY: Syracuse University Press.

Holler, L. 1990. "Thinking with the Weight of the Earth: Feminist Contributions to an Epistemology of Concreteness." *Hypatia* 5 (1): 2–23.

Holling, C. S. 1973. "Resilience and Stability of Ecological Systems." *Annual Review of Systematics and Ecology* 4: 1–23.

———. 1986. "The Resilience of Terrestrial Ecosystems." In *Sustainable Development of the Biosphere*, edited by W. C. Clark and R. E. Munn. Cambridge: Cambridge University Press, 292–317.

Holmgren, D. 2002. *Permaculture: Principles and Pathways Beyond Sustainability*. Victoria, Australia: Holmgren Design Services.

Hoopes, J. 1991. *Peirce on Signs: Writings on Semiotic by Charles Sanders Peirce*. Chapel Hill, NC: University of North Carolina Press.

Horkheimer, M., and T. W. Adorno. 2002. *Dialectic of Modernity: Philosophical Fragments*. Translated by E. Jephcott. Stanford, CA: Stanford University Press.

Horon, D. 2000. *The Pure State of Nature: Sacred Cows, Destructive Myths and the Environment*. Crows Nest, Australia: Allen & Unwin.

Houghton, J. 1997. *Global Warming: The Complete Briefing*. Cambridge: Cambridge University Press.

Housen, A. 1983. "The Eye of the Beholder: Measuring Aesthetic Development." PhD diss., Harvard University.

Howe, D., P. H. Kahn, Jr., and B. Friedman. 1996. "Along the Rio Negro: Brazilian Children's Environmental Views and Values." *Developmental Psychology* 32: 979–87.

Howe, K. R. 1988. "Against the Quantitative-Qualitative Incompatibility Thesis or Dogmas Die Hard." *Educational Researcher* 17: 10–16.

Howell, F. C. 2002. *Making Magic with Gaia: Practices to Heal Ourselves and Our Planet*. Boston: Red Wheel/Weiser.

Howell, N. 2000. *A Feminist Cosmology: Ecology, Solidarity, and Metaphysics*. Amherst, NY: Humanity Press.

Huber, P. 1999. *Hard Green: Saving the Environment from the Environmentalists, A Conservative Manifesto*. New York: Basic Books.

Hughes, D. D. 1991. *Human Sacrifice in Ancient Greece*. New York: Routledge.

Hughes, H. 2001. *Sensory Exotica: A World beyond Human Experience*. Cambridge, MA: Bradford Books.

Hughes, J. D. 1975. *Ecology in Ancient Civilizations*. Albuquerque: University of New Mexico Press.

———. 1983/1996. *North American Indian Ecology*. El Paso, TX: Western Press.

———. 1994. *Pan's Travail: Environmental Problems of the Ancient Greeks and Romans*. Baltimore: Johns Hopkins University Press.

Huhndorf, S. M. 2001. *Going Native: Indians in the American Cultural Imagination*. Ithaca, NY: Cornell University Press.

Hull, R. B. 2006. *Infinite Nature*. Chicago: University of Chicago Press.

Hultkrantz, A. 1967/1980. *The Religions of the American Indians*. Translated by M. Setterwall. Berkeley: University of California Press.

———. 1987. *Native Religions of North America: The Power of Visions and Fertility*. San Francisco: HarperCollins.

———. 1992. *Shamanic Healing and Ritual Drama: Health and Medicine in Native North American Religious Traditions*. New York: Crossroad.

Hume, L. 2002. *Ancestral Power: The Dreaming, Consciousness and Aboriginal Australians*. Melbourne: Melbourne University Press.

Hunter, B., ed. 1999. *A Reader in Environmental Law*. New York: Oxford University Press.

Hurley, S., and M. Nudds. 2006. *Rational Animals?* New York: Oxford University Press.

Huth, H. 1957/1991. *Nature and the American: Three Centuries of Changing Attitudes*. New ed. Lincoln: University of Nebraska Press.

Hutter, B. M., ed. 1999. *A Reader in Environmental Law*. New York: Oxford University Press.

Ihde, D. 1986. *Experimental Phenomenology: An Introduction*. Albany, NY: SUNY Press.

Inge, W. R. 1899. *Christian Mysticism*. London: Methuen.

Ingerman, S. 2006. *Soul Retrieval: Mending the Fragmented Self.* San Francisco: Harper.

Ingersoll, R. E., and C. F. Rak. 2006. *Psychopharmacology for Helping Professionals: An Integral Exploration.* Toronto: Thomson Brooks/Cole.

Institute of Ecosystem Studies. n.d. "Defining Ecology." www.ecostudies.org/definition_ecology.html (accessed January 8, 2005).

Integral Institute. 2005. *Integral Life Practice Starter Kit: Version 1.0.* Boulder, CO: Integral Institute.

Irwin, L., ed. 2000. *Native American Spirituality: A Critical Reader.* Lincoln: University of Nebraska Press.

Isenberg, A. C. 2000. *The Destruction of the Bison: An Environmental History, 1750–1920.* Cambridge: Cambridge University Press.

Jablonka, E., and M. J. Lamb. 2005. *Evolution in Four Dimensions: Genetic, Epigenetic, Behavioral, and Symbolic Variation in the History of Life.* Cambridge, MA: MIT Press.

Jackson, A., and J. Jackson. 1996. *Environmental Science: The Natural Environment and Human Impact.* Harlow, UK: Longman.

Jackson, J. B. 1980. *The Necessity for Ruins, and Other Topics.* Amherst: University of Massachusetts Press.

———. 1996. *A Sense of Place, a Sense of Time.* New Haven, CT: Yale University Press.

Jackson, W. 1994. *Becoming Native to This Place.* Lexington: University Press of Kentucky.

Jacobsen, J., and J. Firor. 1993. *Human Impact on the Environment: Ancient Roots, Current Challenges.* Boulder, CO: Westview Press.

Jamieson, D. 2003. *A Companion to Environmental Philosophy.* Malden, MA: Blackwell.

Jamison, A. 2001. *The Making of Green Knowledge: Environmental Politics and Cultural Transformation.* Cambridge: Cambridge University Press.

Jantsch, E. 1980. *The Self-Organizing Universe.* Oxford and New York: Pergamon Press.

Jaques, E., and K. Cason. 1994. *Human Capability: A Study of Individual Potential and Its Application.* Falls Church, VA: Cason Hall.

Jastrab, J. 1995. *Sacred Manhood, Sacred Earth: A Vision Quest into the Wilderness of a Man's Heart.* New York: Harper Perennial Library.

Jauregui, J. 2002. "Jorge Mario Jauregui on Favela's Urbanization." Interview by E. Blum and P. Neitzke. www.jauregui.arq.br/favelas_interview.html (accessed October 10, 2002).

Jellicoe, G. A. 1995. *The Landscape of Man: Shaping the Environment from Prehistory to the Present Day*. 3rd ed. New York: Thames & Hudson.

Jencks, C. 1995. *The Architecture of the Jumping Universe: How Complexity Science Is Changing Architecture and Culture*. London: Academy Editions.

Jenkins, P. 2004. *Dream Catchers: How Mainstream America Discovered Native Spirituality*. New York: Oxford University Press.

Jensen, D. 2000. *A Language Older Than Words*. New York: Context Books.

———. 2006. *End-Game*. Vols. 1–2. New York: Seven Stories Press.

Johansen, B., and D. Grinde, Jr. 1995. *Ecocide of Native America: Environmental Destruction of Indian Lands and Peoples*. Santa Fe, NM: Clear Light.

John Paul II. 1989. "The Ecological Crisis: A Common Responsibility." 1990 World Day of Peace message, January 1.

Johnson, D. H. 1994. *Body, Spirit and Democracy*. Berkeley, CA: North Atlantic Books.

Johnson, R. B., and L. B. Christensen. 2004. *Educational Research: Quantitative, Qualitative, and Mixed Approaches*. Boston: Allyn & Bacon.

Johnson, R. B., and A. J. Onwuegbuzie. 2004. "Mixed Methods Research: A Research Paradigm Whose Time Has Come." *Educational Researcher* 33 (7): 14–26.

Johnson, R. B., and L. A. Turner. 2003. "Data Collection Strategies in Mixed Methods Research." In *Handbook of Mixed Methods in Social and Behavioral Research*, edited by A. Tashakkori and C. Teddlie. Thousand Oaks, CA: SAGE, 297–319.

Johnston, D. 2000. *Building Green in a Black and White World*. Washington, DC: Home Builder Press.

———. 2005. "Case Study: Passing Green Building Codes in Alameda County, California, USA." Working paper.

Johnston, D., and K. Master. 2004. *Green Remodeling: Changing the World One Room at a Time*. Gabriola Island, BC: New Society.

Jolma, D. J. 1994. *Attitudes toward the Outdoors: An Annotated Bibliography of U.S. Survey and Poll Research Concerning the Environment, Wildlife and Recreation*. Jefferson, NC: McFarland & Company.

Jonas, H. 1966. *The Phenomena of Life: Toward a Philosophical Biology*. Chicago: University of Chicago Press.

———.1984. *The Imperative of Responsibility: In Search of an Ethics for the Technological Age*. Chicago: University of Chicago Press.

Jones, K. 1989. *The Social Face of Buddhism: An Approach to Political and Social Activism*. Boston: Wisdom.

Jørgensen, S. E. 1997. *Integration of Ecosystem Theories: A Pattern*. 2nd rev. ed. Boston: Kluwer.

Jørgensen, S. E., and G. Bendoricchio, eds. 2001. *Fundamentals of Ecological Modeling*. 3rd ed. Kidlington, UK: Elsevier Science.

Jørgensen, S. E., B. Halling-Sorensen, and S. N. Nielsen, eds. 1995. *Handbook of Environmental and Ecological Modeling*. Boca Raton, FL: CRC Press.

Jørgensen, S. E., and F. Muller, eds. 2000. *Handbook of Ecosystem Theories and Management*. London: CRC Press.

Jung, C. G. 2002. *The Earth Has a Soul: The Nature Writings of C. G. Jung*. Berkeley, CA: North Atlantic Books.

Kahn, P. H. Jr. 1997a. "Bayous and Jungle Rivers: Cross-cultural Perspectives on Children's Environmental Moral Reasoning." In *Culture as a Context for Moral Development: New Perspectives on the Particular and the Universal*, edited by H. Saltzstein. San Francisco: Jossey-Bass, 23–36.

———. 1997b. "Children's Moral and Ecological Reasoning about the Prince William Sound Oil Spill." *Developmental Psychology* 33: 1091–96.

———. 1997c. "Developmental Psychology and the Biophilia Hypothesis: Children's Affiliation with Nature." *Developmental Review* 17: 1–61.

———. 1999. *The Human Relationship with Nature: Development and Culture*. Cambridge, MA: MIT Press.

———. 2002. "Children's Affiliations with Nature: Structure, Development, and the Problem of Environmental Generational Amnesia." In *Children and Nature: Psychological, Sociocultural, and Evolutionary Investigations*, edited by P. H. Kahn, Jr., and S. R. Kellert. Cambridge, MA: MIT Press, 93–116.

———. 2003a. "Ape Cognition and Why It Matters for the Field of Psychology." *Human Development* 46: 161–68.

———. 2003b. "The Development of Environmental Moral Identity." In *Identity and the Natural Environment*, edited by S. Clayton and S. Opotow. Cambridge, MA: MIT Press, 113–14.

———. 2006. "Nature and Moral Development." In *Handbook of Moral Development*, edited by M. Killen and J. G. Smetana. Mahwah, NJ: Lawrence Erlbaum, 461–82.

Kahn, P. H. Jr., and B. Friedman. 1995. "Environmental Views and Values of Children in an Inner-City Black Community." *Child Development* 66: 1403–17.

———. 1998. "On Nature and Environmental Education: Black Parents Speak from the Inner City." *Environmental Education Research* 4: 25–39.

Kahn, P. H. Jr., and S. R. Kellert, eds. 2002. *Children and Nature: Psychological, Sociocultural, and Evolutionary Investigations*. Cambridge, MA: MIT Press.

Kahn, P. H. Jr., and O. Lourenço. 2002. "Water, Air, Fire, and Earth—A Developmental Study in Portugal of Environmental Moral Reasoning." *Environment and Behavior* 34: 405–30.

Kahn, P. H. Jr., and A. Weld. 1996. "Environmental Education: Toward an Intimacy with Nature." *Interdisciplinary Studies in Literature and Environment* 3 (2): 165–68.

Kahneman, D., and A. Tversky, eds. 2000. *Choices, Values and Frames.* Cambridge: Cambridge University Press.

Kaldahl, E. 2000. *Environmental Archaeology: Principles and Practice.* Cambridge: Cambridge University Press.

Kaplan R., and S. Kaplan. 2002. "Adolescents and the Natural Environment: A Time Out?" In *Children and Nature: Psychological, Sociocultural, and Evolutionary Investigations*, edited by P. H. Kahn, Jr., and S. R. Kellert. Cambridge, MA: MIT Press, 227–57.

Karasov, W. H., and C. Martinez del Rio. 2007. *Physiological Ecology: How Animals Process Energy, Nutrients, and Toxins*. Princeton, NJ: Princeton University Press.

Kaufman, F. 2002. *Foundations of Environmental Philosophy: A Text with Readings.* New York: McGraw-Hill.

Kay, C., and R. T. Simmons, eds. 2002. *Wilderness and Political Ecology: Aboriginal Influences and the Original State of Nature*. Salt Lake City: University of Utah Press.

Kay, J. J. 1991. "A Non-equilibrium Thermodynamic Framework for Discussing Ecosystem Integrity." *Environmental Management* 15 (4): 483–595.

———. 2000. "Ecosystems as Self-Organizing Holarchic Open Systems: Narratives and the Second Law of Thermodynamics." In *Handbook of Ecosystem Theories and Management,* edited by S. E. Jørgensen and F. Muller. London: CRC Press, 135–60.

Kaza, S. 1993. *The Attentive Heart.* New York: Ballantine.

Kaza, S., and K. Kraft. 2000. *Dharma Rain: Sources of Buddhist Environmentalism.* Boston: Shambhala Publications.

Kazantzakis, N. 1960. *The Saviors of God: Spritual Exercises.* New York: Simon & Schuster.

Kealey, D. 1990. *Revisioning Environmental Ethics*. Albany, NY: SUNY Press.

Keeley, L. 1997. *War Before Civilization: The Myth of the Peaceful Savage*. New York: Oxford University Press.

Kegan, R. 1982. *The Evolving Self: Problem and Process in Human Development.* Cambridge, MA: Harvard University Press.

———. 1994. *In Over Our Heads: The Mental Demands of Modern Life*. Cambridge, MA: Harvard University Press.

Kehoe, A. 2000. *Shamans and Religion: An Anthropological Exploration in Critical Thinking*. London: Waveland Press.

Keller, C. 1986. *From a Broken Web: Separation, Sexism, and Self*. Boston: Beacon Press.

———. 1996. *Apocalypse Now and Then: A Feminist Guide to the End of the World*. Boston: Beacon Press.

———. 2003. *Face of the Deep: A Theology of Becoming*. New York: Routledge.

Keller, D. R. 1998. "Ecological Hermeneutics." Paper presented at the Philosophy and the Environment Conference, Boston, August 10–15. www.bu.edu/wcp/Papers/Envi/EnviKell.htm (accessed September 15, 2003).

Keller, D. R., and F. B. Gollev, eds. 2000. *The Philosophy of Ecology: From Science to Synthesis*. Athens: University of Georgia Press.

Keller, E. F. 1995. *Reflections on Gender and Science*. New Haven, CT: Yale University Press.

Keller, E. F., and W. H. Freeman. 1993. A *Feeling for the Organism: The Life and Work of Barbara McClintock*. New York: W. H. Freeman.

Kellert, S. R. 1985. "Attitudes toward Animals: Age-related Development Among Children." *Journal of Environmental Education* 16: 29–39.

———. 1996. *The Value of Life: Biological Diversity and Human Society*. Washington, DC: Island Press.

———. 1997. *Kinship to Mastery: Biophilia in Human Evolution and Development*. Washington, DC: Island Press.

———. 2002. "Experiencing Nature: Affective, Cognitive, and Evaluative Development in Children." In *Children and Nature: Psychological, Sociocultural, and Evolutionary Investigations*, edited by P. H. Kahn, Jr. and S. R. Kellert. Cambridge, MA: MIT Press, 117–51.

Kellert, S. R., and T. Clark. 1991. "The Theory and Application of a Wildlife Policy Framework." In *Public Policy and Wildlife Conservation*, edited by W. R. Mangun and S. S. Nagel. New York: Greenwood Press, 17–35.

Kellert, S. R., and M. O. Westervelt. 1983. *Children's Attitudes, Knowledge and Behaviors toward Animals:* Phase V. United States Department of the Interior Fish and Wildlife Service. Washington, DC: Superintendent of Documents, U.S. Government Printing Office.

Kelly, S. 1998. "Revisioning the Mandala of Consciousness: A Critical Appraisal of Wilber's Holarchical Paradigm." In *Ken Wilber in Dialogue: Conversations with Leading Transpersonal Thinkers*, edited by D. Rothberg and S. Kelly. Wheaton, IL: Quest Books, 117–30.

———. 1999. "The Complexity of Consciousness and the Consciousness of Complexity." In *Proceedings of the 43rd Annual Conference of the International Society for the Systems Sciences*, edited by J. K. Allen, M. L. W. Hall, and J. Wilby. Asilomar, CA: ISSS.

———. 2000. "Transpersonal Psychology and the Paradigm of Complexity." *Integralis: Journal of Integral Consciousness, Culture and Science* 1 (0).

———. 2003. "Space, Time, and Spirit: The Analogical Imagination and the Evolution of Transpersonal Theory." *Journal of Transpersonal Psychology* 34 (2): 73–100.

———. Forthcoming. "Integral Ecology and the Paradigm of Complexity." *Futures*.

Kempton, W. M., J. S. Boster, and J. A. Hartley. 1996. *Environmental Values in American Culture*. Cambridge, MA: MIT Press.

Kerr, A. 2001. *Dogs and Demons: Tales from the Dark Side of Japan*. New York: Hill & Wang.

Keulartz, J. 1995. *The Struggle for Nature: A Critique of Radical Ecology*. Translated by R. Kuitenbrouwer. New York: Routledge.

Kibert, C. J., J. Sendzimir, and G. B. Guy, eds. 2001. *Construction Ecology: Nature as a Basis for Green Buildings*. New York: Routledge.

Kidner, D. 2001. *Nature and Psyche: Radical Environmentalism and the Politics of Subjectivity*. Albany, NY: SUNY Press.

Killingsworth, J. 2004. *Walt Whitman and the Earth: A Study of Ecopoetics*. Iowa City: University of Iowa Press.

Killingsworth, M. J., and J. Palmer. 1992. *Ecospeak: Rhetoric and Environmental Politics in America*. Carbondale: Southern Illinois University Press.

Kimbrell, A., ed. 2002a. *Fatal Harvest: The Tragedy of Industrial Agriculture*. Washington, DC: Island Press.

———. 2002b. *The Fatal Harvest Reader*. Washington, DC: Island Press.

King, B. 1997. *Buildings of Earth and Straw: Structural Design for Rammed Earth and Straw Bale Architecture*. Sausalito, CA: Ecological Design Press.

King, B. J. 2007. *Evolving God: A Provocative View of the Origins of Religion*. New York: Doubleday.

King, M. 1995. "Nature Mysticism in the Writings of Traherne, Whitman, Jefferies and Krishnamurti." http://web.ukonline.co.uk/mr.king/writings/essays/essaysukc/natmysta.html (accessed September 15, 2000).

Kingsland, S. 1995. *Modeling Nature: Episodes in the History of Population Ecology*. 2nd ed. Chicago: Chicago University Press.

Kinsley, D. 1995. *Ecology and Religion: Ecological Spirituality in Cross-cultural Perspective*. Englewood Cliffs, NJ: Prentice Hall.

Kitching, R. L. 1983. *Systems Ecology*. Brisbane, Australia: University of Queensland Press.

Klein, J. T. 1990. *Interdisciplinarity: History, Theory, and Practice*. Detroit, MI: Wayne State University Press.

———. 1996. *Crossing Boundaries: Knowledge, Disciplinarities, and Interdisciplinarities*. Charlottesville: University Press of Virginia.

Koberstein, P. 2003. "Journey to the Heart of the Great Bear Rainforest." *Cascadia Times*. Summer. www.times.org/archives/2003/bcforests.htm (accessed March 27, 2006).

Koertge, N., ed. 2000. *A House Built on Sand: Exposing Postmodernist Myths about Science*. New York: Oxford University Press.

Koestler, A. 1967/1976. *The Ghost in the Machine*. New York: Random House.

Kofman, F. 2001. "Holons, Heaps and Artifacts." *Integral World: Exploring Theories of Everything*. January. www.integralworld.net/kofman.html (accessed August 16, 2003).

Kohlberg, L. 1981a. *The Meaning and Measurement of Moral Development*. Worcester, MA: Clark University Press.

———. 1981b. *The Philosophy of Moral Development: Moral Stages and the Idea of Justice*. Vol. 1 of *Essays in Moral Development*. San Francisco: Harper & Row.

———. 1984. *The Psychology of Moral Development: The Nature and Validity of Moral Stages*. Vol. 2 of *Essays in Moral Development*. San Francisco: Harper & Row.

Kolbert, E. 2006. *Field Notes from a Catastrophe*. New York: Bloomsbury.

Koller, K. 2006a. "An Introduction to Integral Science." *AQAL: Journal of Integral Theory and Practice* 1 (2): 237–49.

———. 2006b. "Architecture of an Integral Science." *AQAL: Journal of Integral Theory and Practice* 1 (2): 250–75.

———. 2006c. "The Data and Methodologies of Integral Science." *AQAL: Journal of Integral Theory and Practice* 1 (3): 158–83.

Kolstad, C. 1999. *Environmental Economics*. New York: Oxford University Press.

Koren, L. 1994. *Wabi Sabi: For Artists, Designers, Poets and Philosophers*. Berkeley, CA: Stone Bridge Press.

Korten, D. C. 1995. *When Corporations Rule the World*. West Hartford, CT: Kumarian Press & Berrett-Koehler.

Korzybski, A. 1933. *Science and Sanity: An Introduction to Non-Aristotelean Systems and General Semantics*. San Francisco: Institute of General Semantics.

Kot, M. 2001. *Elements of Mathematical Ecology*. Cambridge, MA: Cambridge University Press.

Kovel, J. 2002. *The Enemy of Nature: The End of Capitalism or the End of the World?* New York: Zed Books.

Krebs, C. 1994. *Ecology: The Experimental Analysis of Abundance and Distributions*. Reading, MA: Addison-Wesley.

Krebs, J. R., and N. B. Davis, eds. 1997. *Behavioural Ecology: An Evolutionary Approach*. 4th ed. Oxford: Blackwell.

Krech, S. III. 1999. *The Ecological Indian: Myth and History*. New York: W. W. Norton.

Krech, S., J. R. McNeill, and C. Merchant, eds. 2004. *Encyclopedia of World Environmental History*. London: Routledge.

Kreisberg, J. 2002. "Spiral Dynamics, *Sequoia Sempervirens* and the Redwood Forests: An Integral Ecology." Working paper.

———. 2003a. "Toward an Integral Ecology of Medicinal Plants." Working paper.

———. 2003b. "The Twelve Niches of Plant Medicine." Working paper.

———. 2005. "An Integral Ecology of Sudden Oak Death." Master's thesis, Prescott College.

Krell, D. 1992. *Daimon Life: Heidegger and Life-Philosophy*. Bloomington: Indiana University Press.

Krishna, T. 2001. *The Vaastu Workbook: Using the Subtle Energies of the Indian Art of Placement*. Rochester, VT: Destiny Books.

Krishnamurti, J. 1991. *On Nature and the Environment*. San Francisco: HarperSanFrancisco.

Kubasek, N., and G. Silverman. 2001. *Environmental Law*. 4th ed. Upper Saddle River, NJ: Prentice Hall.

Kull, K. 1998a. "On Semiosis, Umwelt, and Semiosphere." *Semiotica* 120 (3/4): 299–310.

———. 1998b. "Semiotic Ecology: Different Natures in the Semiosphere." *Sign Systems Studies* 26: 344–68.

———. 1999a. "Biosemiotics in the Twentieth Century: A View from Biology." *Semiotica* 127 (1/4): 385–414.

———. 1999b. "Outlines for a Post-Darwinian Biology." *Folia Baeriana* 7: 129–42.

———. 1999c. “On the History of Joining *Bio* with *Semio*: F. S. Rothschild and the Biosemiotic Rules” *Sign Systems Studies* 27: 128–38.

———. 2000. “An Introduction to Phytosemiotics: Semiotic Botany and Vegetative Sign Systems.” *Sign Systems Studies* 28: 326–50.

———. 2003. “Ladder, Tree, Web.” *Sign Systems Studies* 31 (2): 589–603.

———. 2004. “Uexküll and the Post-modern Evolutionism.” *Sign Systems Studies* 32 (1/2): 99–114.

———. 2005. “Semiosphere and a Dual Ecology: Paradoxes of Communication.” *Sign Systems Studies* 33 (1): 175–89.

Kull, K., ed. 2001. *Jakob von Uexküll: A Paradigm for Biology and Semiotics*. Special volume, *Semiotica* 134 (1–4).

Kull, K., and P. Torop. 2000. “Biotranslation: Translation between Umwelten.” In *Translation Translation*, edited by Susan Petrilli. Amsterdam: Rodopi, 313–28.

Kull, K., and W. Nöth, eds. 2001. *Semiotics of Nature*. Kassel, Germany: Kassel University Press.

Kunstler, J. H. 1993. *The Geography of Nowhere: The Rise and Decline of America's Man-Made Landscape*. New York: Simon & Schuster.

Kusbasek, N., and G. Silverman. 2007. *Environmental Law*. 6th ed. Upper Saddle River, NJ: Prentice Hall.

Lacinski, P., and M. Bergeron. 2000. *Serious Strawbale: A Home Construction Guide for All Climates*. White River Junction, VT: Chelsea Green.

Laird, S. A., ed. 2002. *Biodiversity and Traditional Knowledge*. London: Earthscan.

Lake-Thom, B. 1997. *Spirits of the Earth: A Guide to Native American Nature Symbols, Stories, and Ceremonies*. New York: Plume.

Landy, M., M. Roberts, and S. Thomas. 1994. *The Environmental Protection Agency: Asking the Wrong Questions; From Nixon to Clinton*. New York: Oxford University Press.

Lane, B. 2001. *Landscapes of the Sacred: Geography and Narrative in American Tradition*. Baltimore: Johns Hopkins University Press.

Laszlo, E. 1996. *The Systems View of the World: A Holistic Vision for Our Time*. Cresskill, NJ: Hampton Press.

Latonick-Flores, J. “Awakening to the Ecotragedy: An Examination of Kegan's ‘In Over Our Heads’ Thesis and the Transformation of Ecological Consciousness.” (working paper, Saybrook Graduate School, 2003).

———. 2004. “Broadening Environmental Awareness: Another Perspective on the New Environmental Paradigm/Dominant Social Paradigm Theory.” PhD diss., Saybrook Graduate School, San Francisco.

Lauck, J. E. 2002. *The Voice of the Infinite in the Small: Re-visioning the Insect-Human Connection*. Boston: Shambhala Publications.

Lauzon, A. 1998. "Adult Education and the Human Journey: An Evolutionary Perspective." *International Journal of Lifelong Education* 17 (2): 131–45.

Lawrence, J. E. 2007. *Apocalypse 2012: A Scientific Investigation into Civilization's End*. New York: Broadway Books.

Lawson, T. E. 1985. *Religions of Africa: Traditions in Transformation*. Prospect Heights, IL: Waveland Press.

Leakey, R., and R. Lewin. 1995. *The Sixth Extinction: Patterns of Life and the Future of Humankind*. New York: Doubleday.

LeBlanc, S. 1999. *Prehistoric Warfare in the American Southwest*. Salt Lake City: University of Utah Press.

———. 2004. *Constant Battles: The Myth of the Peaceful, Noble Savage*. New York: St. Martin's Griffin.

Lee, P. C., ed. 2001. *Comparative Primate Socioecology*. Cambridge: Cambridge University Press.

Lele, S., and R. B. Norgaard. 2005. "Practicing Interdisciplinarity." *BioScience* 55 (11) 967–76.

Lenain, T. 1991. *Monkey Painting*. London: Reaktion Books.

Lentz, D., ed. 2000. *Imperfect Balance: Landscape Transformations in the Pre-Columbian Americas*. New York: Columbia University Press.

Leonard, G., and M. Murphy. 2005. *The Life We Are Given*. New ed. New York: Tarcher.

Leonard, W. 2002. *Human Nutritional Ecology and Evolution*. Boulder, CO: Westview Press.

Leon-Portilla, M., ed. 1980. *Native Meso-American Spirituality*. Mahwah, NJ: Paulist Press.

Leopold, A. 1949. *A Sand County Almanac and Sketches Here and There*. New York: Oxford University Press.

Lettvin, J. Y., H. R. Maturana, W. S. McCulloch, and W. H. Pitts. 1965. "What the Frog's Eye Tells the Frog's Brain." In *Embodiments of Mind* by W. S. McCulloch. Cambridge, MA: MIT Press, 230–55.

Levin, D. M. 1985. *The Body's Recollection of Being: Phenomenological Psychology and the Deconstruction of Nihilism*. London: Routledge & Kegan Paul.

Levin, S., and T. Hallam, 1990. *Applied Mathematical Ecology*. New York: Springer.

Levinson, B. 1972. *Pets and Human Development*. London: Wiedenfeld & Nicholson.

Lévi-Strauss, C. 1962. *Totemism Today*. Translated by R. Needham. Boston: Beacon Press.

Lewin, K. 1951. *Field Theory in Social Science: Selected Theoretical Papers*. Edited by D. Cartwright. New York: Harper & Rowe.

Lewis, B. 1983. *Bioacoustics: A Comparative Approach*. New York: Academic Press.

Lewis, C. S. 1960. *Studies in Words*. Cambridge: Cambridge University Press.

Lewis, M. 1992. *Green Delusions: An Environmentalist Critique of Radical Environmentalism*. Durham, NC: Duke University Press.

Li, S., J. M. Marquart, and C. Zercher. 2000. "Conceptual Issues and Analytic Strategies in Mixed-Method Studies of Preschool Inclusion." *Journal of Early Intervention* 23: 116–32.

Liddick, D. 2006. *Eco-Terrorism: Radical Environmental and Animal Liberation Movements*. Westport, CT: Praeger.

Lidov, D. 1999. *Elements of Semiotics*. New York: St. Martin's Press.

Light, A., and E. Katz, eds. 1996. *Environmental Pragmatism*. New York: Routledge.

Lilly, J. 1961. *Man and Dolphin*. Garden City, NY: Doubleday.

———. 1967. *The Mind of the Dolphin*. Garden City, NY: Doubleday.

———. 1978. *Communication between Man and Dolphin*. New York: Crown.

Lind, G. 2005. "The Cross-cultural Validity of the Moral Judgment Test: Findings from 27 Cross-cultural Studies." Presentation at the Conference of the American Psychological Association, Washington, DC, August 18–21.

———. 2008. The Meaning and Measurement of Moral Judgment Competence Revisited: A Dual-Aspect Model. In *Contemporary Philosophical and Psychological Perspectives on Moral Development and Education*, edited by D. Fasko and W. Willis. Cresskill, NJ: Hampton Press.

Linden, E. 2006. *The Winds of Change: Climate, Weather, and the Destruction of Civilizations*. New York: Simon & Schuster.

Linn, D., and M. Linn. 1999. *Quest: A Guide for Creating Your Own Vision Quest*. New York: Ballantine Wellspring.

Linscott, G. n.d. "Sustainable Development Strategy in a Country of Partial Development—Viewed from the Values Perspective of Spiral Dynamics." *Global Values Network*. www.globalvaluesnetwork.com/00ArticleContent.asp?aid=70 (accessed March 3, 2006).

Lipp, D. 2003. *The Elements of Ritual: Air, Fire, Water and Earth in the Wiccan Circle*. St. Paul, MN: Llewellyn.

Livi-Bacci, M. 2001. *A Concise History of World Population: An Introduction to Population Processes.* 3rd ed. Oxford: Blackwell.

Lockwood, J., M. Hoopes, and M. Marchetti. 2006. *Invasion Ecology.* Malden, MA: Blackwell.

Loevinger, J. 1998. "History of the Sentence Completion Test (SCT) for Ego Development." In *Technical Foundations for Measuring Ego Development: The Washington University Sentence Completion Test,* edited by J. Loevinger. Mahwah, NJ: Lawrence Erlbaum, 1–10.

Lomborg, B. 1998/2001. *The Skeptical Environmentalist: Measuring the Real State of the World.* Cambridge: Cambridge University Press.

———. 2008. *Cool It: The Skeptical Environmentalist's Guide to Global Warming.* New York: Knopf.

Lomborg, B., ed. 2004. *Global Crises, Global Solutions.* Cambridge: Cambridge University Press.

———. 2006. *How to Spend $50 Billion to Make the World a Better Place.* 2nd rev. ed. Cambridge: Cambridge University Press.

Lomolino, M. V., B. R. Riddle, and J. H. Brown. 2005. *Biogeography.* 3rd ed. Sunderland, MA: Sinauer Associates.

Lomolino, M. V., D. F. Sax, and J. H. Brown, eds. 2004. *Foundations of Biogeography: Classic Papers with Commentaries.* Chicago: University of Chicago Press.

Long, W. 1919. *How Animals Talk.* New York: Harper.

Lorbiecki, M. 2005. *Aldo Leopold: A Fierce Green Fire.* Helena, MT: Falcon.

Lorenz, E. 1993. *The Essence of Chaos.* Seattle: University of Washington Press.

Lotman, J. 1984. "O Semiosfere." *Sign Systems Studies* 17: 5–23.

Lotman, M. 2002. "Umwelt and Semiosphere" *Sign Systems Studies* 30 (1): 33–39.

Lotz, C. 2006. "Psyche or Person? Husserl's Phenomenology of Animals." In *Interdisziplinäre Perspektiven der Phänomenologie,* edited by D. Lohmar and D. Fonfara. Phänomenologica 177. Dordrecht, Netherlands: Springer, 190–204.

Louv, R. 2006. *Last Child in the Woods: Saving Our Children from Nature-Deficit Disorder.* Chapel Hill, NC: Algonquin Books.

Love, S. 1972. *Ecotage.* New York: Pocketbooks.

Lovejoy, A. O., and G. Boas. 1935/1997. *Primitivism and Related Ideas in Antiquity.* Baltimore: Johns Hopkins University Press.

Lovelock, J. 1988. *The Ages of Gaia: A Biography of Our Living Earth.* New York: W. W. Norton.

———. 2000. *Gaia: A New Look at Life on Earth.* 3rd ed. New York: Oxford University Press.

Low, A., and S. Tremayne. 2002. *Sacred Custodians of the Earth? Women, Spirituality and the Environment*. Oxford: Berghahn Books.

Lubbock, J. 1870/1889. *The Origin of Civilization and the Primitive Condition of Man*. London: Macmillan.

Luby, T. 2004. *Yoga of Nature: Union with Fire, Earth, Air, and Water.* Santa Fe, NM: Clear Light.

Luhmann, N. 1984/1995. *Social Systems*. Translated by J. Bednarz and D. Baecker. Stanford, CA: Stanford University Press.

———. 1986. *Ecological Communication*. Translated by J. Bednarz. Chicago: University of Chicago Press.

Luhrs, J. 1997. *The Simple Living Guide: A Sourcebook for Less Stressful, More Joyful Living*. New York: Broadway Books.

Luure, A. 2001. "Lessons from Uexküll's Antireductionism and Reductionism: A Pansemiotic View." *Semiotica* 134 (1): 311–22.

Lyle, J. T. 1985. *Design for Human Ecosystems: Landscape, Land Use, and Natural Resources*. New York: Van Nostrand Reinhold.

Lythgoe, J. 1979. *Ecology of Vision*. Oxford: Clarendon Press.

MacKinnon, M. H., and M. McIntyre. 1995. *Readings in Ecology and Feminist Theology*. Kansas City, KS: Sheed & Ward.

Macy, J. 1983. *Despair and Personal Empowerment in the Nuclear Age*. Gabriola Island, BC: New Society.

———. 1990. "The Ecological Self: Postmodern Ground for Right Action." In *Sacred Interconnections: Postmodern Spirituality, Political Economy, and Art*, edited by D. Griffin. Albany, NY: SUNY Press, 35–48.

———. 1991a. *Mutual Causality in Buddhism and General Systems Theory*. Albany, NY: SUNY Press.

———. 1991b. *World As Lover, World As Self.* Berkeley, CA: Parallax Press.

Macy, J., and M. Brown. 1998. *Coming Back to Life*. Gabriola Island, BC: New Society.

Maffi, L., ed. 2001. *On Biocultural Diversity: Linking Language, Knowledge, and the Environment.* Washington, DC: Smithsonian Institution Press.

Magliocco, S. 2004. *Witching Culture: Folklore and Neo-Paganism in America*. Philadelphia: University of Pennsylvania Press.

Mahoney, M. J., and A. Marquis. 2002. "Integral Constructivism and Dynamic Systems in Psychotherapy Processes." *Psychoanalytic Inquiry* 22 (5): 794–813.

Mails, T. 1999. *Dancing in the Paths of the Ancestors: The Culture, Crafts, and Ceremonies of the Hopi, Zuni, Acoma, Laguna, and Rio Grande Pueblo Indians of Yesterday and Today*. New York: Marlowe.

Main, J. 1985. *The Way of Unknowing*. London: Darton, Longman & Todd.

Maker, C. J. 1982. *Teaching Models of the Gifted*. Austin: Pro-ed.

Maley, K. 2002. *Desert Shamanism: A Catholic Bishop's Vision Quest*. Bloomington, IN: First Books Library.

Malpas. J. 1999. *Place and Experience: A Philosophical Topography*. Cambridge: Cambridge University Press.

Mander, J. 1977. *Four Arguments for the Elimination of Television*. New York: Morrow.

———. 1992. *In the Absence of the Sacred: The Failure of Technology and the Survival of the Indian Nations*. San Francisco: Sierra Club Books.

Mander, J., and E. Goldsmith, eds. 1996. *The Case Against the Global Economy: And for a Turn toward the Local*. San Francisco: Sierra Club Books.

Manes, C. 1990. *Green Rage*. Boston: Little, Brown.

Mann, C. C. 1995. *1491: New Revelations of the Americas Before Columbus*. New York: Knopf.

Mann, J., R. C. Connor, P. L. Tyack, and H. Whitehead. 2000. *Cetacean Societies: Field Studies of Dolphins and Whales*. Chicago: University of Chicago Press.

Manson, N. 2002. "Formulating the Precautionary Principle." *Environmental Ethics* 24 (Fall): 263–74.

Margulis, L. 2000. *Symbiotic Planet: A New Look at Evolution*. New York: Basic Books.

Marineau, S. 2007. "Humanity, Forest Ecology, and the Future in a British Columbia Valley: A Case Study." *Integral Review* 4: 26–43.

Markham, A. 1994. *A Brief History of Pollution*. New York: St. Martin's Press.

Markos, A. 2002. *Readers of the Book of Life: Contextualizing Developmental Evolutionary Biology*. Oxford: Oxford University Press.

Markos, A., F. Grygar, K. Kleisner, and Z. Neubauer. 2007. "Towards a Darwinian Biosemiotics: Life as Mutual Understanding." In *Introduction to Biosemiotics*, edited by M. Barbieri. Dordrecht, Netherlands: Springer, 235–55.

Markowitz, G., and D. Rosner. 2002. *Deceit and Denial: The Deadly Politics of Industrial Pollution*. Berkeley: University of California Press.

Marquis, A., and E. S. Warren. 2004. "Integral Counseling: Prepersonal, Personal, and Transpersonal in Self, Culture, and Nature." *Constructivism in the Human Sciences* 9 (1): 111–32.

Marquis, A., and K. Wilber. "Unification beyond Eclecticism: Integral Psychotherapy." (working paper, *Journal of Psychotherapy Integration*, 2004).

Marshall, P. 2005. *Mystical Encounters with the Natural World: Experiences and Explanations*. New York: Oxford University Press.

Marshall, P. H. 1992. *Nature's Web: An Exploration of Ecological Thinking*. London: Simon & Schuster.

Martin, C. 1981. "The American Indian as Miscast Ecologist." In *Ecological Consciousness*, edited by R. Schutz and J. Hughes. Washington, DC: University Press of America, 137–48.

Martin, J. W. 1999. *The Land Looks After Us: A History of Native American Religion*. New York: Oxford University Press.

Martin, P., and R. Klein, eds. 1984. *Quaternary Extinctions: A Prehistoric Revolution*. Tucson: University of Arizona Press.

Masson, J. M. 1998. *Dogs Never Lie About Love: Reflections on the Emotional World of Dogs*. New York: Three Rivers Press.

———. 2002. *The Nine Emotional Lives of Cats: A Journey into the Feline Heart*. New York: Ballantine Books.

———. 2003. *The Pig Who Sang to the Moon: The Emotional World of Farm Animals*. New York: Ballantine Books.

Masson, J. M., and S. McCarthy. 1996. *When Elephants Weep: The Emotional Lives of Animals*. New York: Delta.

Masters, J. 2005. "Review of Michael Crichton's *State of Fear*." The Weather Underground. www.wunderground.com/education/stateoffear.asp (accessed January 15, 2006).

Mate, F. 2000. *A Reasonable Life: Toward a Simpler, Secure, More Humane Existence*. 2nd ed. Pflugerville, TX: Albatross.

Mathews, F. 2003. *For Love of Matter: A Contemporary Panpsychism*. Albany, NY: SUNY Press.

Mathews, N. 1996. *Francis Bacon: The History of a Character Assassination*. New Haven, CT: Yale University Press.

———. 2003. "Francis Bacon, Slave-driver or Servant of Nature? Is Bacon to Blame for the Evils of Our Polluted Age?" www.sirbacon.org/mathewsessay.htm (accessed March 10, 2007).

Matthews, J. 2002. *The Green Man: Spirit of Nature*. Boston: Red Wheel / Weiser.

Maturana, H. R., and F. J. Varela. 1987. *The Tree of Knowledge: The Biological Roots of Human Understanding*. Boston: New Science Library / Shambhala Publications.

———. 1991. *Autopoiesis and Cognition: The Realization of the Living*. Boston: Reidel.

May, R. 2007. *Theoretical Ecology: Principles and Applications*. 3rd ed. New York: Oxford University Press.

Mayer, J. D., D. R. Caruso, and P. Salovey. 1999. "Emotional Intelligence Meets Traditional Standards for an Intelligence." *Intelligence* 27: 267–98.

———. 2000. "Selecting a Measure of Emotional Intelligence: The Case for Ability Scales." In *The Handbook of Emotional Intelligence*, edited by R. Bar-On and J. D. A. Parker. New York: Jossey-Bass, 320–42.

Mayer, J. D., P. Salovey, and D. R. Caruso. 2000a. "Emotional Intelligence as Zeitgeist, as Personality, and as a Mental Ability." In *The Handbook of Emotional Intelligence*, edited by R. Bar-On and J. D. A. Parker. New York: Jossey-Bass, 92–117.

———. 2000b. "Models of Emotional Intelligence." In *Handbook of Human Intelligence*, edited by R. J. Sternberg. New York: Cambridge University Press, 396–420.

Mayr, E., and W. Provine, eds. 1998. *The Evolutionary Synthesis: Perspectives on the Unification of Biology*. Cambridge, MA: Harvard University Press.

Mazis, G. A. 2002. *Earthbodies: Rediscovering Our Planetary Senses*. Albany, NY: SUNY Press.

McCarthy, D., and L. King. 2005. *Environmental Sociology: From Analysis to Action*. Lanham, MD: Rowman & Littlefield.

McCarthy, S. 1991. *Celebrating the Earth: An Earth-Centered Theology of Worship with Blessings, Prayers and Rituals*. San Jose, CA: Resource.

McClellan, J. 1993. "Nondual Ecology: In Praise of Wildness and in Search of Harmony with Everything That Moves." http://spot.colorado.edu/~mcclelr/NondualEcologyLight.htm (accessed August 16, 2000). This article also appeared in the Winter 1993 issue of *Tricycle: The Buddhist Review*.

McCoy, E. 1994. *A Witch's Guide to Faery Folk: Reclaiming Our Working Relationship with Invisible Helpers*. St. Paul, MN: Llewellyn.

McCully, M., ed. 2002. *Life Support: The Environment and Human Health*. Cumberland, RI: MIT Press.

McDonald, B., ed. 2003. *Seeing God Everywhere: Essays on Nature and the Sacred*. Bloomington, IN: World Wisdom.

McDonough, W., and M. Braungart. 2002. *Cradle to Cradle: Remaking the Way We Make Things*. New York: North Point Press.

McDowell, J. 1997. *Hamatsa: The Enigma of Cannibalism on the Pacific Northwest Coast*. Vancouver, BC: Ronsdale Press.

McFague, S. 1993. *The Body of God: An Ecological Theology*. Minneapolis, MN: Fortress Press.

McGaa, E. 1990. *Mother Earth Spirituality: Native American Paths to Healing Ourselves and Our World*. San Francisco: HarperCollins.

———. 2004. *Nature's Way: Native Wisdom for Living in Balance with Earth*. San Francisco: HarperCollins.

McGee, G. 1994. "The Relevance of Foucault to Whiteheadian Environmental Ethics." *Environmental Ethics* 16 (Winter): 419–24.

McGrew, W. C. 1992. *Chimpanzee Material Culture*. Cambridge: Cambridge University Press.

McIntosh, R. P. 1985. *The Background of Ecology: Concept and Theory*. New York: Cambridge University Press.

McKenna, T. 1993. *Food of the Gods: The Search for the Original Tree of Knowledge, A Radical History of Plants, Drugs, and Human Evolution*. New York: Bantam Doubleday Dell.

McKibben, B. 1999. *The End of Nature*. New York: Anchor.

McKusick, J. 2000. *Green Writing: Romanticism and Ecology*. New York: Palgrave Macmillan.

———. 2005. "Ecology." In *Romanticism: An Oxford Guide*, edited by N. Roe. New York: Oxford University Press.

McLuhan, M. 1964. *Understanding Media: The Extensions of Man*. New York: McGraw-Hill.

McLuhan, T. C. 1994. *The Way of the Earth: Encounters with Nature in Ancient and Contemporary Thought*. New York: Simon & Schuster.

McNab, C. 2002. *Living off the Land: An Illustrated Guide to Tracking, Building Traps, Constructing Shelters, Tool Making, Finding Water, Foraging for Food and Much More*. Guilford, CT: Lyons Press.

McPherson, D., and D. Rabb. 1993. *Indian from the Inside: A Study in Ethno-Metaphysics*. Thunder Bay, ON: Lakehead University.

McPherson, J., and G. McPherson. 1993. *Primitive Wilderness Living and Survival Skills: Naked into the Wilderness*. New York: John McPherson.

Meadows, D., J. Randers, and D. Meadows. 2004. *Limits to Growth: The Thirty-Year Update*. White River Junction, VT: Chelsea Green.

Mechling, J. 2001. *On My Honor: Boy Scouts and the Making of American Youth*. Chicago: University of Chicago Press.

Meine, C. 1988. *Aldo Leopold: His Life and Work*. Madison: University of Wisconsin Press.

Melson, G. 2001. *Why the Wild Things Are: Animals in the Lives of Children*. Cambridge, MA: Harvard University Press.

Mercer, J. E. 1913. *Nature Mysticism*. London: George Allen.

Merchant, C. 1980. *The Death of Nature: Women, Ecology and the Scientific Revolution*. San Francisco: HarperCollins.

———. 2002. "The Rise of Ecology, 1890–1990." In *The Columbia Guide to American Environmental History*. New York: Columbia University Press, chap. 9.

———. 2003. *Reinventing Eden: The Fate of Nature in Western Culture*. New York: Routledge.

Merideth, R. 1993. *The Environmentalist's Bookshelf: A Guide to the Best Books*. New York: G. K. Hall & Co.

Merkel, J. 2003. *Radical Simplicity: Small Footprints on a Finite Earth*. Gabriola Island, BC: New Society.

Merkur, D. 1998. *The Ecstatic Imagination: Psychedelic Experiences and the Psychoanalysis of Self-Actualization*. Albany, NY: SUNY Press.

———. 1999. *Mystical Moments and Unitive Thinking*. Albany, NY: SUNY Press.

Merleau-Ponty, M. 1964/2000. *The Primacy of Perception*. Translated by J. E. Edie. Evanston, IL: Northwestern University Press.

———. 1968/2000. *The Visible and the Invisible*. Translated by C. Lefort. Evanston, IL: Northwestern University Press.

———. 2003. *Nature: Course Notes from the Collège de France*. Translated by Robert Vallier. Evanston, IL: Northwestern University Press.

Merrell, F. 2001. "Lotman's Semisophere, Peirce's categories, and Cultural Forms of Life." *Sign Systems Studies* 29 (2): 385–414.

Merton, T. 2003. *When the Trees Say Nothing*. Notre Dame, IN: Sorin Books.

Messer, E., and M. Lambek. 2001. *Ecology and the Sacred: Engaging the Anthropology of Roy A. Rappaport*. Ann Arbor: University of Michigan Press.

Metzner, R. 1994. *The Well of Remembrance: Rediscovering the Earth Wisdom Myths of Northern Europe*. Boston: Shambhala Publications.

———. 1995. "The Psychopathology of the Human-Nature Relationship." In *Ecopsychology: Restoring the Earth, Healing the Mind*, edited by T. Roszak, M. E. Gomes, and A. D. Kanner. San Francisco: Sierra Club Books, 55–67.

———, ed. 1999a. *Ayahuasca: Human Consciousness and the Spirits of Nature*. New York: Thunder's Mouth Press.

———. 1999b. *Green Psychology. Transforming Our Relationship to the Earth*. Rochester, VT: Inner Traditions.

Meyer, J. 2000. *The Animal Connection: A Guide to Intuitive Communication with Your Pet.* New York: Plume.

Meyer, J. M. 2001. *Political Nature: Environmentalism and the Interpretation of Western Thought*. Cambridge, MA: MIT Press.

Michaels, P. J. 2004. *Meltdown: The Predictable Distortion of Global Warming by Scientists, Politicians, and the Media*. Washington, DC: Cato Institute.

———. 2005. *Shattered Consensus: The True State of Global Warming*. Lanham, MD: Rowman & Littlefield.

Midgley, M. 1983. *Animals and Why They Matter.* Atlanta: University of Georgia Press.

Mies, M., and V. Shiva. 1993. *Ecofeminism.* London: Zed Books.

Mikulas, W. 2001. *The Integrative Helper: Convergence of Eastern and Western Traditions*. Belmont, CA: Wadsworth.

Milani, B. 2000. *Designing the Green Economy: The Postindustrial Alternative to Corporate Globalization*. Lanham, MD: Rowman & Littlefield.

Miller, D., ed. 1986. *The Lewis Mumford Reader*. New York: Pantheon Books.

Miller, E. P. 2002. *The Vegetative Soul: From Philosophy of Nature to Subjectivity in the Feminine.* Albany, NY: SUNY Press.

Miller, J. G. 1978. *Living Systems*. New York: McGraw-Hill.

Miller, M. E., and S. R. Cook-Greuter. 1994. *Transcendence and Mature Thought in Adulthood*. Lanham, MD: Rowman & Littlefield.

———. 2000. *Creativity, Spirituality, and Transcendence: Paths to Integrity and Wisdom in the Adult Self.* Stamford, CT: Ablex.

Miller, P. 2000. *Toward a Feminist Developmental Psychology*. New York: Routledge.

———. 2002. *Theories of Developmental Psychology*. 4th ed. New York: Worth.

Mills, S. 1995. *In Service of the Wild: Restoring and Reinhabiting Damaged Land.* Boston: Beacon Press.

Milne, C. 1999. *Sacred Places in North America: A Journey into the Medicine Wheel.* New York: Stewart, Tabori & Chang.

Milton, J. 2005. *Sky Above, Earth Below: Spiritual Practice in Nature*. Boulder, CO: Sentient.

Milton, K. 2002. *Loving Nature: Towards an Ecology of Emotion*. New York: Routledge.

Mingers, J. 1995. *Self-Producing Systems: Implications and Applications of Autopoiesis*. New York: Plenum Press.

Mitchell, R. W., ed. 2002. *Pretending and Imagination in Animals and Children.* New York: Cambridge University Press.

Mitchell, R. W., N. Thompson, and L. Miles, eds. 1997. *Anthropomorphism, Anecdote, and Animals: The Emperor's New Clothes?* Albany, NY: SUNY Press.

Mitchell, S. D. 2003. *Biological Complexity and Integrative Pluralism.* New York: Cambridge University Press.

———. 2004. "Why Integrative Pluralism?" Special double issue. *Emergence: Complexity and Organization* (E:CO) 6 (1–2): 81–91.

Mitman, G. 1992. *The State of Nature: Ecology, Community, and American Social Thought, 1900–1950.* Chicago: University of Chicago Press.

Mollison, B. C. 1997. *Permaculture: A Designer's Manual.* Tyalgum, Australia: Tagari.

Mongomery, D. R. 2007. *Dirt: The Erosion of Civilizations.* Berkeley: University of California Press.

Monti, M. J. 1997. "Origin and Ordering: Aristotle, Heidegger, and the Production of Nature." PhD diss., State University of New York, Binghamton.

Moran, E. F. 2006. *People and Nature: An Introduction to Human Ecological Relations.* Malden, MA: Blackwell.

Moran, E. F., and R. Gillett-Netting, eds. 2000. *Human Adaptability: An Introduction to Ecological Anthropology.* 2nd ed. Boulder, CO: Westview Press.

Moran, J. 2002. *Interdisciplinarity.* London: Routledge.

Morin, E. 1992. "From the Concept of System to the Paradigm of Complexity." Translated by Sean Kelly. *Journal of Social and Evolutionary Systems* 15 (4): 371–85.

———. 2005. "RE: From Prefix to Paradigm." Translated by Frank Poletti and Sean Kelly. *World Futures: The Journal of General Evolution* 61 (4): 254–67.

Morin, E., and A. B. Kern. 1999. *Homeland Earth: A Manifesto for the New Millennium.* Cresskill, NJ: Hampton Press.

Morin, P. J. 1999. *Community Ecology.* Malden, MA: Blackwell Science.

Morris, W. 1979. *The Political Writings of William Morris.* London: Lawrence Wishart.

Morton, T. 2005. "Ecology." In *Romanticism: An Oxford Guide,* edited by N. Roe. New York: Oxford University Press.

———. 2007. *Ecology without Nature: Rethinking Environmental Aesthetics.* Boston: Harvard University Press.

Mosley, A., ed. 1995. *African Philosophy: Selected Readings.* Englewood Cliffs, NJ: Prentice Hall.

Moura, A. [Aoumiel, pseud.]. 1996. *Green Witchcraft: Folk Magic, Fairy Lore and Herb Craft*. St. Paul, MN: Llewellyn.

Moyer, B. 2001. *Doing Democracy: The MAP Model for Organizing Social Movements*. Gabriola Island, BC: New Society.

Mugerauer, R. 1985. "Language and the Emergence of the Environment." In *Dwelling, Place and Environment*, edited by D. Seamon and R. Mugerauer. New York: Columbia University Press, 51–70.

———. 1995. *Interpreting Environments: Tradition, Deconstruction, Hermeneutics*. Houston: University of Texas Press.

Muhlhausler, P. 2003. *Language of Environment, Environment of Languages: A Course in Ecolinguistics*. London: Battlebridge Publications.

Muir, J. 1954. *The Wilderness World of John Muir*. Edited by E. W. Teale. New York: Houghton Mifflin.

Mulhall, D. 2002. *Our Molecular Future: How Nanotechnology, Biotechnology, Robotics, Genetics and Artificial Intelligence Will Transform Our World*. Amherst, NY: Prometheus Books.

Müller, F. M. 1888. *Natural Religion*. London: Longmans.

Mumford, L. 1961. *The City in History: Its Origins, Its Transformations and Its Prospects*. New York: Harcourt, Brace & World.

Murphy, S. 2004. *Upside-Down Zen: A Direct Path into Reality*. Victoria, Australia: Lothian Books.

Myers, A. 1997. *Communicating with Animals: Unleashing the Spiritual Connection between People and Animals*. New York: McGraw-Hill.

Myers, G. 1998. *Children and Animals: Social Development and Our Connections to Other Species*. Boulder, CO: Westview Press.

Myers, O. E. Jr. 1999. "Human Development as Transcendence of the Animal Body and the Child-Animal Association in Psychological Thought." *Society and Animals* 7 (1): 121–40.

———. 2002. "Symbolic Animals and the Developing Self." *Anthrozoös: A Multidisciplinary Journal of the Interactions of People and Animals* 15 (1): 19–36.

Myers, O. E. Jr., and A. Russell. 2004. "Human Identity in Relation to Wild Black Bears: A Natural-Social Ecology of Subjective Creatures." In *Identity and the Natural Environment*, edited by S. Clayton and S. Opotow. Cambridge, MA: MIT Press, 67–90.

Myers, O. E. Jr., and C. Saunders. 2002. "Animals as Links to Developing Caring Relationships with the Natural World." In *Children and Nature: Psychological, Sociocultural and Evolutionary Investigations*, edited by P. H. Kahn, Jr., and S. R. Kellert. Cambridge, MA: MIT Press, 153–78.

Myers, O. E. Jr., C. Saunders, and A. Birjulin. 2004. "Emotional Dimensions of Watching Zoo Animals: An Experience Sampling Study Building on Insights from Psychology." *Curator* 47 (3): 299–321.

Myers, O. E. Jr., C. D. Saunders, and E. Garrett. 2003. "What Do Children Think Animals Need? Aesthetic and Psycho-Social Conceptions." *Environmental Education Research* 9 (2): 305–25.

———. 2004. "What Do Children Think Animals Need? Developmental Trends." *Environmental Education Research* 10 (4): 545–62.

Myerson, J., ed. 1984. *The Transcendentalists: A Review of Research and Criticism*. New York: Modern Language Association.

Nabhan, G. P. 1997. *Cultures of Habitat: On Nature, Culture, and Story*. Washington, DC: Counterpoint.

Nabhan, G. P., and S. Trimble. 1994. *The Geography of Childhood: Why Children Need Wild Places*. Boston: Beacon Press.

Nader, R. 1993. *The Case Against "Free Trade": GATT, NAFTA, and the Globalization of Corporate Power*. San Francisco: Earth Island Institute.

Nader, R., and M. Teitel. 2001. *Genetically Engineered Food: Changing the Nature of Nature*. Rochester, VT: Inner Traditions.

Naess, A. 1989. *Ecology, Community and Lifestyle*. Translated and edited by D. Rothenberg. New York: Cambridge University Press.

Nagel, A. 1997. "Are Plants Conscious?" *Journal of Consciousness Studies* 4 (3): 215–30.

Nahmad, C. 1998. *Fairy Spells: Seeing and Communicating with the Fairies*. London: Souvenir Press.

Narby, J. 2005. *Intelligence in Nature: An Inquiry into Knowledge*. New York: Tarcher.

Narby, J., and F. Huxley, eds. 2001. *Shamans Through Time: 500 Years on the Path to Knowledge*. New York: Tarcher.

Narvaez, P., ed. 1997. *The Good People: New Fairylore Essays*. Lexington: University Press of Kentucky.

Nash, R. 1967/1982. *Wilderness and the American Mind*. New Haven, CT: Yale University Press.

———. 1989. *The Rights of Nature*. Madison: University of Wisconsin Press.

Nasr, S. H. 1990. *Man and Nature: The Spiritual Crisis of Modern Man*. Boston: Unwin Paperbacks.

Nazarea, V., ed. 2003. *Ethnoecology: Situated Knowledge/Located Lives*. Tucson: University of Arizona Press.

Neihardt, J. G. 1932/1979. *Black Elk Speaks: Being the Life Story of a Holy Man of the Oglala Sioux*. Lincoln: University of Nebraska Press.

Netting, R. 1986. *Cultural Ecology*. 2nd ed. Long Grove, IL: Waveland Press.

Nettle, D., and S. Romaine. 2000. *Vanishing Voices: The Extinction of the World's Languages*. Oxford: Oxford University Press.

Newman, E. I. 2001. *Applied Ecology and Environmental Management*. 2nd ed. Blackwell.

Newman, M. C., and W. H. Clements, 2007. *Ecotoxicology: A Comprehensive Treatment*. Boca Raton, FL: CRC Press.

Newman, M. E. J., and R. G. Palmer. 2003. *Modeling Extinction*. New York: Oxford University Press.

Nickerson, R. 2002. *Psychology and Environmental Change*. London: Lawrence Erlbaum.

Nicolescu, B. 2002. *Manifesto of Transdisciplinarity*. Albany, NY: SUNY Press.

Niezen, R. 2000. *Spirit Wars: Native North American Religions in the Age of Nation Building*. Berkeley: University of California Press.

Noel, D. C. 1990. "Soul and Earth: Traveling with Jung toward an Archetypal Ecology" (Part 1). *Quadrant* (Fall-Winter).

———. 1991. "Soul and Earth: Traveling with Jung toward an Archetypal Ecology" (Part 2). *Quadrant* (Spring-Summer).

Nolan, P., G. Lenski, and J. Lenski. 1994. *Human Societies: An Introduction to Macrosociology*. New York: McGraw-Hill.

Nolt, J. 2006. "The Move from *Good* to *Ought* in Environmental Ethics." *Environmental Ethics* 28 (4): 355–74.

Norberg-Hodge, H. 1991. *Ancient Futures: Learning from Ladakh*. San Francisco: Sierra Club Books.

Norberg-Schulz, C. 1965. *Intentions in Architecture*. Cambridge, MA: MIT Press.

———. 1980. *Genius Loci: Towards a Phenomenology of Architecture*. New York: Rizzoli.

Nordhaus, T., and M. Shellenberger. 2007. *Break Through: From the Death of Environmentalism to the Politics of Possibility*. New York: Houghton Mifflin.

Norton, B. G. 1991. *Toward Unity among Environmentalists*. New York: Oxford University Press.

Noss, R., M. O'Connell, and D. Murphy. 1997. *The Science of Conservation Planning: Habitat Conservation under the Endangered Species Act*. Washington, DC: Island Press.

Nöth, W. 1998. "Ecosemiotics." *Sign Systems Studies* 26: 332–42.

Nöth, W., and K. Kull, eds. 2001. *Semiotics of Nature*. Special issue. *Sign Systems Studies* 29 (1).

O' Brien, J. F. 1988. "Teilhard's View of Nature and Some Implications for Environmental Ethics." *Environmental Ethics* 10 (Winter): 329–46.

O'Connell, C. 2007. *The Elephant's Secret Sense: The Hidden Life of the Wild Herds of Africa*. New York: Free Press.

O'Conner, J. 1998. *Natural Causes: Essays in Ecological Marxism*. New York: Guilford Press.

Odum, E. P. 1953. *Fundamentals of Ecology*. Philadelphia: W. B. Saunders.

Odum, E., and G. Barrett. 2004. *Fundamentals of Ecology*. 5th ed. New York: Brooks Cole.

Odum, H. T. 1994. *Ecological and General Systems: An Introduction to Systems Ecology*. Niwot: University Press of Colorado.

Oelschlaeger, M. 1991. *The Idea of Wilderness: From Prehistory to the Age of Ecology*. New Haven, CT: Yale University Press.

———. 1995. *Postmodern Environmental Ethics*. Albany, NY: SUNY Press.

O'Leary, S. D. 1998. *Arguing the Apocalypse: A Theory of Millennial Rhetoric*. New ed. New York: Oxford University Press.

Oliver, M. 1993. *New and Selected Poems*. Boston: Beacon Press.

Olsen, B. 2000. *Sacred Places: 101 Spiritual Sites around the World*. San Francisco: Consortium of Collective Consciousness.

Olsen, M. 2003. *Women Who Risk: Profiles of Women in Extreme Sports*. Long Island City, NY: Hatherleigh Press.

Olson, A. 2002. *Body and Earth: An Experiential Guide*. Hanover, CT: Middlebury College Press.

Olupona, J., ed. 1991. *African Traditional Religions in Contemporary Society*. New York: New Era Books.

———. 1995. *Religious Plurality in Africa*. New York: Mouton De Gruyter.

———. 2001. *African Spirituality: Forms, Meanings and Expressions*. New York: Herder & Herder.

O'Neill, R. V. 1989. "Perspectives in Hierarchy and Scale." In *Perspectives in Ecological Theory*, edited by J. Roughgarden, R. M. May, and S. A. Levin. Princeton, NJ: Princeton University Press, 140–56.

O' Neill, R. V., D. De Angelis, J. Waide, and T. F. H. Allen. 1986. *A Hierarchical Concept of Ecosystems*. Princeton, NJ: Princeton University Press.

O'Riordan, T., and J. Cameron, eds. 1994. *Interpreting the Precautionary Principle.* London: Earthscan.

Orr, D. 1992. *Ecological Literacy: Education and the Transition to a Postmodern World.* Albany, NY: SUNY Press.

———. 1994. *Earth in Mind: On Education, Environment, and the Human Prospect.* Washington, DC: Island Press.

Orr, E. 2001. *Ritual.* London: Thorsons.

Orrell, D. 2006. *The Future of Everything: The Science of Prediction.* New York: Thunder's Mouth Press.

Osterreich, S. 1998. *Native North American Shamanism: An Annotated Bibliography.* Westport, CT: Greenwood Press.

O'Sullivan, E., and M. Taylor, eds. 2004. *Learning toward an Ecological Consciousness: Selected Transformative Practices.* New York: Palgrave.

Ott, E. 2002. *Chaos in Dynamical Systems.* 2nd ed. New York: Cambridge University Press.

Owen, D. S. 2002. *Between Reason and History: Habermas and the Idea of Progress.* Albany, NY: SUNY Press.

Owens, C. 2005. "An Integral Approach to Sustainable Consumption and Waste Reduction." *World Futures: The Journal of General Evolution* 61 (1–2): 96–109.

Oyama, S. 2000. *Evolution's Eye: A Systems View of the Biology-Culture Divide.* Durham, NC: Duke University Press.

Pain, S. P. 2007. "Inner Representations and Signs in Animals." In *Introduction to Biosemiotics*, edited by M. Barbieri. Dordrecht, Netherlands: Springer, 409–55.

Painter, C., and C. Lotz, eds. 2007. *Phenomenology and the Non-human Animal.* New York: Springer.

Pallasmaa, J. 2005. *The Eyes of the Skin. Architecture and the Senses.* New York: Wiley.

Palmer, C. 1998. *Environmental Ethics and Process Thinking.* Oxford: Clarendon Press.

Palmer, D., ed. 2003. *The Marshall Illustrated Encyclopedia of Dinosaurs and Prehistoric Animals.* New York: New Line Books.

Palmer, J. A. 1998. *Environmental Education in the Twenty-first Century: Theory, Practice, Progress and Promise.* London: Routledge.

Palmer, J. A., ed. 2001. *Fifty Key Thinkers on the Environment.* London: Routledge.

Paper, J. 2005. *The Deities Are Many: A Polytheistic Theology.* Albany, NY: SUNY Press.

Parker, S. T., Mitchell, R. W., and Boccia, M. L., eds. 1994. *Self-Awareness in Animals and Humans: Developmental Perspectives*. Cambridge: Cambridge University Press.

Parkes, G. 1994. *Composing the Soul: Reaches of Nietzsche's Psychology*. Chicago: University of Chicago Press.

———. 1998. "Staying Loyal to the Earth: Nietzsche as an Ecological Thinker." In *Nietzsche's Futures*, edited by J. Lippitt. London: Macmillan, 167–88.

———. 2005. "Nietzsche's Environmental Philosophy: A Trans-European Perspective." *Environmental Ethics* 27 (Spring): 77–91.

Parson, H. L., ed. 1977. *Marx and Engels on Ecology*. Westport, CT: Greenwood Press.

Partridge, E. 2000. "Reconstructing Ecology." In *Ecological Integrity: Integrating Environment, Conservation, and Health*, edited by D. Pimentel, L. Westra, and R. F. Noss. Washington, DC: Island Press, 79–98.

Passino, K. M. 2004. *Biomimicry for Optimization, Control, and Automation*. London: Springer.

Pattee, H. H., ed. 1973. *Hierarchy Theory: The Challenge of Complex Systems*. New York: George Braziller.

Paulson, D. 1999a. "The Near-Death Experience: An Integration of Cultural, Spiritual, and Physical Perspectives." *Journal of Near Death Studies* 18 (1): 13–45.

———. 1999b. "Successfully Marketing Skin Moisturizing Products." *Soap/Cosmetics/Chemical Specialties* 3 (8): 24–27.

———. 1999c. *Topical Antimicrobial Testing and Evaluation*. New York: Marcel Dekker.

———. 2002. *Competitive Business Caring Business: An Integral Business Perspective for the 21st Century*. New York: Paraview Press.

Pearson, D. 1994. *Earth to Spirit: In Search of Natural Architecture*. San Francisco: Chronicle Books.

———. 1998. *The Natural House Book: Creating a Healthy, Harmonious, and Ecologically Sound Home*. 2nd ed. New York: Fireside.

Pearson, J., ed. 1998. *Nature Religion Today: Paganism in the Modern World*. Edinburgh, UK: Edinburgh University Press.

Peden, C., and Y. Hudson, eds. 1991. *Communitarianism, Liberalism, and Social Responsibility*. Lewiston, NY: Edwin Mellen Press.

Peet, R., and M. Watts, eds. 2004. *Liberation Ecologies*. 2nd ed. London: Routledge.

Peirce, C. S. 1906/1991. "The Basis of Pragmaticism." In *Peirce on Signs: Writings on Semiotic*, edited by J. Hoopes. Chapel Hill, NC: UNC Press, 253–59.

Penn, D. C. and D. J. Povinelli. 2007a. "Causal Cognition in Human and Nonhuman Animals: A Comparative, Critical Review." *Annual Review of Psychology*, 58: 97–118.

———. 2007b. "On the Lack of Evidence That Chimpanzees Possess Anything Remotely Resembling a 'Theory of Mind.'" *Philosophical Transactions of the Royal Society*, B 362: 731–44.

Pepperberg, I. M. 1999. *The Alex Studies: Cognitive and Communicative Abilities of Grey Parrots*. Cambridge, MA: Harvard University Press.

———. 2004. "Cognitive and Communicative Abilities of Grey Parrots: Implications for the Enrichment of Many Species." *Animal Welfare* 13: S203–8.

———. 2005. "An Avian Perspective on Language Evolution: Implications of Simultaneous Development of Vocal and Physical Object Combinations by a Grey Parrot (*Psittacus Erithacus*)." In *Language Origins: Perspectives on Evolution*, edited by M. Tallerman. Oxford: Oxford University Press, chap. 11.

Perlman, F. 2002. *Against His-Story, Against Leviathan*. Detroit, MI: Black & Red Books.

Persinger, M. A. 1987. "Geopsychology and Geopsychopathology: Mental Processes and Disorders Associated with Geochemical and Geophysical Factors." *Cellular and Molecular Life Sciences* 43 (1): 92–104.

Peschek, M. "Integral Gender." Working paper, Universität Bremen.

Petitot, J., F. Varela, B. Pachoud, and J. Roy, eds. 2000. *Naturalizing Phenomenology: Issues in Contemporary Phenomenology and Cognitive Science*. Stanford, CA: Stanford University Press.

Petrinovich, L. 2000. *The Cannibal Within*. Evolutionary Foundations of Human Behavior. New York: Aldine De Gruyter.

Philander, S. G. 2000. *Is the Temperature Rising? The Uncertain Science of Global Warming*. Princeton, NJ: Princeton University Press.

Phillips, D. 2003. *The Truth of Ecology: Nature, Culture, and Literature in America*. New York: Oxford University Press.

Piacentini, P., ed. 1993. *Story Earth: Native Voices on the Environment*. San Francisco: Mercury House.

Piaget, J. 1969. *The Child's Conception of the World*. Totowa, NJ: Littlefield, Adams.

Pianka, E. 1999. *Evolutionary Ecology*. 6th ed. San Francisco: Benjamin Cummings.

Pickett, S. T. A., V. T. Parker, and P. L. Fiedler. 1992. "The New Paradigm in Ecology: Implications for Conservation above the Species Level." In *Conservation Biology*, edited by P. L. Fiedler and S. K. Jain. New York: Chapman & Hall, 66–88.

Pierce, L. B. 2000. *Choosing Simplicity: Real People Finding Peace and Fulfillment in a Complex World*. Carmel, CA: Gallagher Press.

Pike, S. M. 2001. *Earthly Bodies, Magical Selves: Contemporary Pagans and the Search for Community*. Berkeley: University of California Press.

Pile, S. 1996. *The Body and the City: Psychoanalysis, Space, and Subjectivity*. London: Routledge.

Pilkey, O. H., and L. Pilkey-Jarvis. 2006. *Useless Arithmetic: Why Environmental Scientists Can't Predict the Future*. New York: Columbia University Press.

Pimm, S. L. 2001. *The World According to Pimm: A Scientist Audits the Earth*. Blacklick, OH: McGraw-Hill.

Pinchbeck, D. 2006. *2012: The Return of Quetzalcoati*. New York: Tarcher.

Plumwood, V. 1993. *Feminism and the Mastery of Nature*. New York: Routledge.

Podberscek, A. L., E. S. Paul, and J. A. Serpell, eds. 2005. *Companion Animals and Us: Exploring the Relationship between People and Pets*. Cambridge: Cambridge University Press.

Pogačnik, M. 1995. *Nature Spirits and Elemental Beings: Working with the Intelligence of Nature*. Translated and edited by K. Werner and G. McNamara. Tallahassee, FL: Findhorn Press.

———. 1998. *Healing the Heart of the Earth: Restoring the Subtle Levels of Life*. Tallahassee, FL: Findhorn Press.

———. 2004. *Turned Upside Down: A Workbook on Earth Changes and Personal Transformation*. Great Barrington, MA: Lindisfarne Books.

———. 2007. *Sacred Geography: Co-creating the Earth Cosmos*. Herndon, VA: Lindisfarne Books.

Pojman, L., and P. Pojman, eds. 2007. *Environmental Ethics: Readings in Theory and Application*. New York: Wadsworth Publishing.

Pollack, R. 1994. *Signs of Life: The Language and Meanings of DNA*. Boston: Houghton Mifflin.

Ponting, C. 1991. *A Green History of the World: The Environment and the Collapse of Great Civilizations*. New York: Penguin Books.

Popper, K. 1994. "The Myth of the Framework" In *The Myth of the Framework: In Defense of Science and Rationality*, edited by M. A. Notturno. New York: Routledge, 33–64.

Porter, R., ed. 2003. *The Cambridge History of Science*. Vol. 4, *Eighteenth-Century Science*. New York: Cambridge University Press.

Povinelli, D. J. 2003. *Folk Physics for Apes: The Chimpanzee's Theory of How the World Works*. Rev. ed. Oxford: Oxford University Press.

———. 2004. "Behind the Ape's Appearance: Escaping Anthropomorphism in the Study of Other Minds. *Daedalus: Journal of the American Academy of Arts and Sciences* (Winter): 29–41.

——— and J. Vonk. 2006. "We Don't Need a Microscope to Explore the Chimpanzee's Mind." *In Rational Animals*, edited by S. Hurley. Oxford: Oxford University Press, 385–412.

Powley, C. J. 1999. "An Exploration of a Merleau-Pontyan Approach to Cognitive Ethology." PhD diss., University of Colorado.

Preece, R. 1999. *Animals and Nature: Cultural Myths, Cultural Realities*. Vancouver, BC: UBC Press.

Prime, R. 2002. *Vedic Ecology: Practical Wisdom for Surviving the Twenty-first* Century. Novato, CA: Mandala.

Proctor, J. D. 1995. "Whose Nature? The Contested Moral Terrain of Ancient Forests." In *Uncommon Ground: Rethinking the Human Place in Nature*, edited by W. Cronon. New York: W. W. Norton, 269–97.

Prpich, W. 2005. "An Integral Analysis of the National Standard of Canada for Organic Agriculture." *World Futures: The Journal of General Evolution* 61 (1–2): 138–50.

Puhakka, K. 1995. "Restoring Connectedness in the Kosmos: A Healing Tale of a Deeper Order." *Integral World: Exploring Theories of Everything*. www.integralworld.net/rev/rev_ses_puhakka.html (accessed February 17, 2006).

Purvis, A., K. E. Jones, and G. M. Mace. 2000. "Extinction." *BioEssays* 22: 1123–33.

Quick, T. 2006. "Holons or Gestalts? A Response to Wilber and Zimmerman." *The Trumpeter* 22 (2): 9–25.

Quinn, D. 1992. *Ishmael*. New York: Bantam/Turner Books.

Radetsky, P. 1997. *Allergic to the Twentieth Century: The Explosion in Environmental Allergies—From Sick Buildings to Multiple Chemical Sensitivity*. Edited by B. Phillips. New York: Little, Brown.

Radner, D., and M. Radner. 1989. *Animal Consciousness*. Buffalo, NY: Prometheus Books.

Rainforest Solutions Project. 2006. "British Columbia's Great Bear Rainforest Review." May. www.savethegreatbear.org/Greatbear%20Review%20May%20 2006.pdf (accessed June 1, 2006).

Randolph, T. G. 1990. *An Alternative Approach to Allergies: The New Field of Clinical Ecology Unravels the Environmental Causes of Mental and Physical Ills*. New York: Harper Perennial.

Rappaport, R. A. 1968. *Pigs for the Ancestors: Ritual in the Ecology of a New Guinea People*. New Haven, CT: Yale University Press.

———. 1979. *Ecology, Meaning and Religion*. Richmond, CA: North Atlantic Books.

Raskin, P., T. Banuri, G. Gallopin, P. Gutman, and A. Hammond. 2002. *Great Transition: The Promise and Lure of the Times Ahead*. Boston: Stockholm Environment Institute/Tellus Institute.

Raup, D. M. 1991. *Extinction: Bad Genes or Bad Luck?* New York: W.W. Norton.

Ray, P. H., and S. R. Anderson. 2001. *The Cultural Creatives*. New York: Three Rivers Press.

Reakarkudla, M., D. Wilson, and E. O. Wilson, eds. 1996. *Biodiversity II: Understanding and Protecting Our Biological Resources*. Washington, DC: Joseph Henry Press.

Real, L. 1993. "Toward a Cognitive Ecology." *Trends in Ecological Evolution*. 8: 413–17

Redford, K. 1991. "The Ecological Noble Savage." *Cultural Survival Quarterly* 15 (1): 46–48.

Redman, C. 1999. *Human Impact on Ancient Environments*. Tucson: University of Arizona Press.

Reed, E. S. 1996. *Encountering the World: Toward an Ecological Psychology*. New York: Oxford University Press.

Regan, T. 1983. *The Case for Animal Rights*. Berkeley: University of California Press.

Register, R. 1987. *Ecocity Berkeley: Building Cities for a Healthy Future*. Berkeley, CA: North Atlantic Press.

Reichardt, C. S., and S. F. Rallis, eds. 1994. *The Qualitative-Quantitative Debate: New Perspectives*. New Directions for Program Evaluation 61. San Francisco: Jossey-Bass.

Rendell, L., and H. Whitehead. 2001. "Culture in Whales and Dolphins." *Behavioral Brain Sciences* 24 (2): 309–82.

Rentschler, M. 2006a. "Introducing Integral Art." *AQAL: Journal of Integral Theory and Practice* 1 (1): 34–41.

———. 2006b. "Understanding Integral Art." *AQAL: Journal of Integral Theory and Practice* 1 (1): 42–59.

Rest, J., D. Narvaez, M. Bebeau, and S. Thoma. 1999. *Postconventional Moral Thinking: A Neo-Kohlbergian Approach*. Mahwah, NJ: Lawrence Erlbaum.

Reynolds. B. 2004. *Embracing Reality: The Integral Vision of Ken Wilber, A Historical Survey and Chapter-by-Chapter Guide to Wilber's Work*. New York: Jeremy P. Tarcher/Penguin.

———. 2006. *Where's Wilber At? Ken Wilber's Integral Vision in the New Millennium*. St. Paul, MN: Paragon House.

Rezendes, P. 1999. *Tracking and the Art of Seeing: How to Read Animal Tracks and Sign*. New York: HarperCollins.

Richards, E. S. 1907. *Sanitation in Daily Life*. Boston: Whitcomb & Borrows.

Richards, R. 1987. *Darwin and the Emergence of Evolutionary Theories of Mind and Behavior*. Chicago: University of Chicago Press.

Ricklefs, R. E. 1993. *The Economy of Nature*. 3rd ed. New York: W. H. Freeman.

Ricklefs, R. E., and G. Miller. 1999. *Ecology*. 4th ed. New York: W. H. Freeman.

Riddell, C. 1997. *The Findhorn Community: Creating a Human Identity for the Twenty-first Century*. Tallahassee, FL: Findhorn Press.

Riddell, D. 2005. "Evolving Approaches to Conservation: Integral Ecology and British Columbia's Great Bear Rainforest." *World Futures: The Journal of General Evolution* 61 (1–2): 63–78.

Riedy, C. 2005a. "The Eye of the Storm: An Integral Perspective on Sustainable Development and Climate Change Response." PhD diss., Institute for Sustainable Futures, University of Technology, Sydney.

———. 2005b. "Integrating Exterior and Interior Knowledge in Sustainable Development Policy." In *Ecopolitics XVI: Transforming Environment Governance for the Twenty-first Century*, edited by C. Star. Brisbane, Australia: Ecopolitics Association of Australasia/Centre for Governance and Public Policy.

———. 2006. "Two Social Practices to Support Emergence of a Global Collective." *Journal of Future Studies* 10 (4): 45–60.

Riedy, C., and B. Brown. 2006. "Use of the Integral Framework to Design Developmentally-Appropriate Sustainability Communications." In *Innovation, Education and Communication for Sustainable Development*, edited by W. Leal Filho. Hamburg, Germany: TuTech Innovation, chap. 34.

Ristau, C., ed. 1991. *Cognitive Ethology: The Minds of Other Animals—Essays in Honor of Donald R. Griffin*. Hillsdale, NJ: Lawrence Erlbaum.

Riversong, M. 1993. *Design Ecology*. Bayside, CA: Borderland Sciences.

Roads, M. J. 1987. *Talking with Nature: Sharing the Energies and Spirits of Trees, Plants, Birds, and Earth*. Tiburon, CA: H. J. Kramer.

Robbins, J. 1987. *Diet for a New America: How Your Food Choices Affect Your Health, Happiness, and the Future of Life on Earth*. Walpole, NH: Stillpoint.

Robbins, J. G., and R. Greenwald. 1994. "Environmental Attitudes Conceptualized Through Developmental Theory: A Qualitative Analysis." *Journal of Social Issues* 50 (3): 29–47.

Robbins, P. 2004. *Political Ecology: A Critical Introduction*. Malden, MA: Blackwell.

Roberts, L. D., ed. 1982. *Approaches to Nature in the Middle Ages*. Binghamton, NY: MRTS.

Roberts, S. 2001. *Fast Feng Shui: Nine Simple Principles for Transforming Your Home*. Kahalui, HI: Lotus Pond Press.

Robertson, D. 2002. "Public Ecology: Linking People, Science, and the Environment." PhD diss., Virginia Tech, Blacksburg.

Rockwood, L., and G. Bertola. 2006. *Introduction to Population Ecology*. Malden, MA: Blackwell.

Roddick, A., ed. 2001. *Take It Personally: How to Make Conscious Choices to Change the World*. Berkeley, CA: Conari Press.

Rodes, B. K., and R. Odell. 1992. *A Dictionary of Environmental Quotes*. Baltimore: Johns Hopkins University Press.

Roe, N., ed. 2005. *Romanticism: An Oxford Guide*. New York: Oxford University Press.

Roepstorff, A. 2001. "Thinking with Animals." In *Semiotics of Nature*, edited by W. Nöth and K. Kull. Special issue. *Sign Systems Studies* 29 (1): 203–17.

Rollin, B. 1989. *The Unheeded Cry: Animal Consciousness, Animal Pain, and Science*. Oxford: Oxford University Press.

Rolston, H., III. 1986. *Philosophy Gone Wild: Essays in Environmental Ethics*. Buffalo, NY: Prometheus.

———. 1988. *Environmental Ethics: Duties to and Values in the Natural World*. Philadelphia: Temple University Press.

Roof, N. 2003. "Integral Approaches That Transform Us and the World." *Spirituality and Reality: New Perspectives on Global Issues* 3 (2): 7–10.

Rorsch, A., T. Frello, R. Soper, and A. de Lange. 2003. "An Analysis of the Nature of the Opposition Raised Against the Book *The Skeptical Environmentalist* by B. Lomborg." www.stichting-han.nl/lomborg.htm (accessed May 5, 2006).

Rose, D. B., and L. Robin. 2004. "The Ecological Humanities in Action: An Invitation." *Australian Humanities Review* 31–32. www.lib.latrobe.edu.au/AHR/archive/Issue-April-2004/rose.html (accessed March 10, 2007).

Rosenweig, M. 2003. *Win-Win Ecology: How the Earth's Species Can Survive in the Midst of Human Enterprise*. Oxford: Oxford University Press.

Roszak, T. 1992. *The Voice of the Earth*. New York: Simon & Schuster.

———. 1993. "Awakening the Ecological Unconscious." *In Context* 34 (Winter). www.context.org/ICLIB/IC34/Roszak.htm (accessed March 5, 2006).

Roszak, T., M. E. Gomes, and A. D. Kanner, eds. 1995. *Ecopsychology: Restoring the Earth, Healing the Mind*. San Francisco: Sierra Club Books.

Rothberg, D., and S. Kelly, eds. 1998. *Ken Wilber in Dialogue: Conversations with Leading Transpersonal Thinkers*. Wheaton, IL: Quest Books.

Rothenberg, D. 1993. *Hand's End: Technology and the Limits of Nature*. Berkeley: University of California Press.

Rothenberg, D. 1999. "Will the Real Chief Seattle Please Stand Up? An Interview with Ted Perry." In *The New Earth Reader*, edited by M. Ulveaus. Cambridge, MA: MIT Press, 36–51.

Rothschild, F. S. 2000. *Creation and Evolution: A Biosemiotic Approach*. Translated by Jozef Hes. New Brunswick, NJ: Transaction Publishers.

Rothschild, M. 1995. *Bionomics: Economy as Ecosystem*. New York: Henry Holt & Company.

Rottschaefer, W. 1998. *The Biology and Psychology of Moral Agency*. New York: Cambridge University Press.

Rowe, J. S. 1992. "The Integration of Ecological Studies." *Functional Ecology* 6 (1): 115–19.

Rowe, S. 1996. "From Shallow to Deep Ecological Philosophy." *The Trumpeter* 13 (1). http://trumpeter.athabascau.ca/content/v13.1/8-rowe.html (accessed April 10, 2005).

———. 2001. "Transcending This Poor Earth—à la Ken Wilber." *The Trumpeter* 17 (1). http://trumpeter.athabascau.ca/content/v17.1/rowe.html (accessed April 10, 2005).

Ruether, R. 1992. *Gaia and God: An Ecofeminist Theology of Earth Healing*. San Francisco: Harper.

Ruse, M. 1979. *The Darwinian Revolution: Science Red in Tooth and Claw*. Chicago: University of Chicago Press.

———. 2007. *Darwinism and Its Discontents*. Cambridge: Cambridge University Press.

Ruskin, J. 1995. *Seven Lamps of Architecture*. New York: Dover.

Russell, E. 1997. *People and the Land Through Time: Linking Ecology and History*. New Haven, CT: Yale University Press.

Russell, P. 1992. *Waking Up in Time: Finding Inner Peace in Times of Accelerating Change*. Novato, CA: Origin Press.

Ruting, T. 2004. "History and Significance of Jakob von Uexküll and of His Institute in Hamburg." *Sign Systems Studies* 31 (1/2): 35–72.

Ryder, R. 1975. *Victims of Science: The Use of Animals in Research*. New York: Open Gate Press.

Sachs, W., ed. 1992. *The Development Dictionary: A Guide to Knowledge as Power*. London: Zed Books.

Sack, R. 2003. *A Geographical Guide to the Real and the Good*. London: Routledge.

Sagoff, M. 1989. *The Economy of the Earth: Philosophy, Law and the Environment*. Cambridge: Cambridge University Press.

———. 1992. "Has Nature a Good of Its Own?" In *Ecosystem Health: New Goals for Environmental Management*, edited by R. Costanza, B. G. Norton, and B. D. Haskell. Washington, DC: Island Press.

———. 2000. "What's Wrong with Exotic Species?" In *Philosophical Dimensions of Public Policy*, edited by V. Gehring and W. Galston. New Brunswick: Transaction Publishers, 327–40.

Sale, K. 1985. *Dwellers in the Land: The Bioregional Vision*. San Francisco: Sierra Club Books.

———. 1995. *Rebels Against the Future: The Luddites and Their War on the Industrial Revolution*. New York: Addison-Wesley.

Salovey, P., and J. D. Mayer. 1990. "Emotional Intelligence." *Imagination, Cognition, and Personality* 9: 185–211.

Salt, H. 1892. *Animals' Rights: Considered in Relation to Social Progress*. New York: Centaur Press.

Salthe, S. N. 1985. *Evolving Hierarchical Systems: Their Structure and Representation*. New York: Columbia University Press.

———. 1993. *Development and Evolution: Complexity and Change in Biology*. Cambridge, MA: MIT Press.

———. 2001. "Summary of the Principles of Hierarchy Theory" www.nbi.dk/~natphil/salthe/Hierarchy_th.html (accessed August 14, 2006).

———. 2002a. "An Exercise in the Natural Philosophy of Ecology." *Ecological Modeling* 158 (3): 167–79.

———. 2002b. "Summary of the Principles of Hierarchy Theory." *General Systems Bulletin* 31: 13–17.

———. 2003. "Infodynamics, a Developmental Framework for Ecology/Economics." *Conservation Ecology* 7 (3). www.consecol.org/vol7/iss3/art3.

———. 2004. "The Natural Philosophy of Ecology: Developmental Systems Ecology." *Ecological Complexity* 2: 1–19.

———. 2006. "Natural Philosophy: Developmental Systems in the Thermodynamic Perspective." Unpublished manuscript.

Samorini, G. 2002. *Animals and Psychedelics: The Natural World and the Instinct to Alter Consciousness*. Rochester, VT: Park Street Press.

Samson, P. R., ed. 1999. *The Biosphere and Noosphere Reader*. London: Routledge.

Sandalow, D. 2005. "Michael Crichton and Global Warming." *Environment and Energy*, January 28. www.brookings.edu/views/op-ed/fellows/sandalow20050128.htm (accessed March 10, 2006).

Sartore, R. 1994. *Humans Eating Humans: The Dark Shadow of Cannibalism*. Notre Dame, IN: Cross Cultural.

Saunders, C. D., and O. E. Myers, Jr. 2003. "The Emerging Field of Conservation Psychology." *Human Ecology Review* 10 (2): 137–49.

Saussure, F. 1916/1983. *Course in General Linguistics*. Translated by R. Harris. London: Duckworth.

Savage-Rumbaugh, S., and R. Lewin. 1994. *Kanzi: The Ape at the Brink of the Human Mind*. New York: Wiley.

Savage-Rumbaugh, S., S. G. Shanker, and T. J. Taylor. 1998. *Apes, Language, and the Human Mind*. New York: Oxford University Press.

Scarce, R. 1990. *Eco Warriors: Understanding the Radical Environmental Movement*. Chicago: Noble Press.

Schad, W. 1977. *Man and Mammals: Toward a Biology of Form*. Garden City, NY: Waldorf Press.

Schalow. F. 2006. *The Incarnality of Being: The Earth, Animals, and the Body in Heidegger's Thought*. Albany, NY: SUNY Press.

Schauberger, V. 1997. *The Water Wizard*. Bath, UK: Gateway Books.

———. 1998. *Nature as Teacher*. Bath, UK: Gateway Books.

———. 2000. *The Fertile Earth*. Bath, UK: Gateway Books.

———. 2001. *Energy Evolution*. Bath, UK: Gateway Books.

Scheers, P. 2001."Hermeneutics and Animal Being: The Question of Animal Interpretation." www.lancs.ac.uk/depts/philosophy/awaymave/onlineresources/peter%20scheers.rtf (accessed April 24, 2006).

Schele, L., and M. A. Miller. 1986. *The Blood of Kings: Dynasty and Ritual in Maya Art*. New York: George Braziller.

Schiebinger, L. 1993. *Nature's Body: Gender in the Making of Modern Science*. Boston: Beacon Press.

Schilthuis, W. 1994. *Biodynamic Agriculture*. Hudson, NY: Anthroposophic Press.

Schlitz, M., T. Amorok, and M. Micozzi, eds. 2004. *Consciousness and Healing: Integral Approaches to Mind-Body Medicine*. St Louis, MO: C. V. Mosby.

Schmidt, A. 1971. *The Concept of Nature in Marx*. London: New Left Books.

Schmitz-Guenther, T. 1999. *Living Spaces: Ecological Building and Design*. Cologne: Konemann.

Schmuck, P., and W. Schultz, eds. 2002. *Psychology of Sustainable Development*. Norwell, MA: Kluwer.

Schnaiberg, A. 1980. *The Environment: From Surplus to Scarcity*. New York: Oxford University Press.

Schneider, S. H., J. R. Miller, E. Crist, and P. J. Boston, eds. 2004. *Scientists Debate Gaia: The Next Century*. Cambridge, MA: MIT Press.

Schneider, E., and D. Sagan. 2006. *Into the Cool: Energy, Flow, Thermodynamics, and Life*. New ed. Chicago: Chicago University Press.

Schrödinger, K. 1947. *What Is Life? The Physical Aspect of the Living Cell*. New York: Macmillan.

Schult, J. 1992. "Species, Signs, and Intentionality." In *Biosemiotics: The Semiotic Web 1991*, edited by T. Sebeok and J. Umiker-Sebeok. New York: Mouton de Gruyter, 317–32.

Schultes, R. E., A. Hoffman, and C. Ratsch. 2002. *Plants of the Gods: Their Sacred Healing and Hallucinogenic Powers*. Rochester, VT: Healing Arts Press.

Schultz, L. H., and R. L. Selman. 1998. "Ego Development and Interpersonal Development in Young Adulthood: A Between-Model Comparison." In *Personality Development*, edited by P. M. Westenberg, A. Blasi, and L. D. Cohn. Mahwah, NJ: Lawrence Erlbaum.

Schumacher, E. F. 1973. *Small Is Beautiful: Economics as if People Mattered*. New York: Harper.

Schwartz, S., and J. Schwartz. 2001. *The Scouting Way*. San Clemente, CA: Scouting Way Press.

Schwartz, T. 1995. *What Really Matters: Searching for Wisdom in America*. New York. Bantam Books.

Schwarz, D. O. 1987. "Indian Rights and Environmental Ethics." *Environmental Ethics* 9 (Winter): 291–302.

Schweitzer, A. 1946. *Civilization and Ethics*. London: A. & C. Black.

Schwenk, T. 1996. *Sensitive Chaos: The Creation of Flowing Forms in Water and Air*. 2nd ed. London: Rudolf Steiner Press.

Scitovsky, T. 1992. *The Joyless Economy: The Psychology of Human Satisfaction*. New York: Oxford University Press.

Scott Cato, M., and M. Kennet. 1999. *Green Economics: Beyond Supply and Demand to Meeting People's Needs*. Aberystwyth, Wales: Green Audit Books.

Scull, J. 1999. "The Separation from More-Than-Human Nature." http://members.shaw.ca/jscull/SEPARATE.pdf (accessed March 20, 2004).

———. 1999–2000. "Let a Thousand Flowers Bloom: A History of Ecopsychology." In *Gatherings: Seeking Ecopsychology*, Winter. www.ecopsychology.org/journal/gatherings/index.html (accessed November 18, 2006).

———. 2003. "Chief Seattle, er, Professor Perry speaks: Inventing Indigenous Solutions to the Environmental Problem." www.ecopsychology.org/journal/gatherings2/scull.htm (accessed March 20, 2007).

Seamon, D., ed. 1993. *Dwelling, Seeing, and Designing: Toward a Phenomenological Ecology*. Albany, NY: SUNY Press.

Seamon, D., and A. Zajonc, eds. 1998. *Goethe's Way of Science: A Phenomenology of Nature*. Albany, NY: SUNY Press.

Searles, H. 1960. *The Nonhuman Environment in Normal Development and in Schizophrenia*. Madison, CT: International Universities Press.

Sebeok, T. 2001. *Global Semiotics*. Bloomington: Indiana University Press.

Sebeok, T., ed. 1968. *Animal Communication: Techniques of Study and Results of Research*. Bloomington: Indiana University Press.

———. 1972. *Perspectives in Zoosemiotics*. The Hague: Mouton.

———. 1977. *How Animals Communicate*. Bloomington: Indiana University Press.

———. 1990. *Essays in Zoosemiotics*. Toronto: University of Toronto.

———. 1998. *Semiotics in the Biosphere*. Special issue. *Semiotica* 120 (3/4).

Sebeok, T., J. Hoffmeyer, and C. Emmeche, eds. 1999. *Biosemiotica*. Special issue. *Semiotica* 127 (1/4).

Sebeok, T., and J. Umiker-Sebeok, eds. 1992. *Biosemiotics: The Semiotic Web 1991*. New York: Mouton de Gruyter.

Sellars, W. 1956. "Empiricism and the Philosophy of Mind." In Minnesota Studies in the Philosophy of Science. Vol. 1, *Foundations of Science and the Concepts of Psychology and Psychoanalysis*, edited by H. Feigl and M. Scriven. Minneapolis: University of Minnesota Press, 253–329.

Selman, R. L. 1980. *The Growth of Interpersonal Understanding*. San Diego, CA: Academic Press.

———. 2003. *The Promotion of Social Awareness: Powerful Lessons from the Partnership of Developmental Theory and Classroom Practice*. New York: Russell Sage Foundation.

Serpell, J. A. 1986. *In the Company of Animals: A Study of Human-Animal Relationships*. London: Basil Blackwell.

Sessions, G., ed. 1995. *Deep Ecology for the Twenty-first Century*. Boston: Shambhala Publications.

———. 1996. "Reinventing Nature, the End of Wilderness? A Response to William Cronon's *Uncommon Ground*." *The Trumpeter* 13 (1): 34–38.

Sewall, L. 1995. "The Skill of Ecological Perception." In *Ecopsychology: Restoring the Earth, Healing the Mind*, edited by T. Roszak, M. E. Gomes, and A. D. Kanner. San Francisco: Sierra Club Books, 201–15.

———. 1999. *Sight and Sensibility: The Ecopsychology of Perception*. New York: Tarcher & Putnam.

Seyfang, G. 2004. "Consuming Values and Contested Cultures: A Critical Analysis of the UK Strategy for Sustainable Consumption and Production." *Review of Social Economy* 62 (3): 323–38.

Shapiro, K. J. 1990. "Understanding Dogs through Kinesthetic Empathy, Social Construction, and History. *Anthrozoös* 3: 184–95.

———. 1997. "A Phenomenological Approach to the Study of Animals." In *Anthropomorphism, Anecdote, and Animals*, edited by R. W. Mitchell, N. Thompson, and L. Miles. Albany, NY: SUNY Press, 277–95.

Sharma, M. 2005. *Responding to HIV/AIDS: Measuring Results*. New York: UNDP.

Sharov, A. 2001. "Pragmatism and Umwelt-theory." In *Jakob von Uexküll: A Paradigm for Biology and Semiotics*, edited by K. Kull. Special volume. *Semiotica* 134 (1–4): 211–28.

Sheets-Johnstone, M. 1990. *The Roots of Thinking*. Philadelphia: Temple University Press.

Sheldrake, R. 1981. *A New Science of Life: The Hypothesis of Morphic Resonance*. Rochester, VT: Park Street Press.

———. 1988. *The Presence of the Past: Morphic Resonance and the Habits of Nature*. Rochester, VT: Park Street Press.

———. 1994. *Rebirth of Nature: The Greening of God*. New York: Park Street Press.

———. 1999. *Dogs That Know When Their Owners Are Coming Home*. New York: Crown.

———. 2003. *The Sense of Being Stared At*. New York: Crown.

Shellenberger, M., and T. Nordhaus. 2004. "The Death of Environmentalism: Global Warming Politics in a Post-environmental World." Paper presented to the Environmental Grantmakers Association, Miami, FL, October.

Shepard, P. 1973. *The Tender Carnivore and the Sacred Game*. New York: Scribners.

———. 1982. *Nature and Madness*. San Francisco: Sierra Club Books.

———. 1996. *The Others: How Animals Made Us Human*. Washington, DC: Island Press.

———. 1998. *Coming Home to the Pleistocene*. Washington, DC: Island Press.

———. n.d. "Nature and Madness." www.primitivism.com/nature-madness.htm (accessed March 5, 2006).

Shiva, V., ed. 1994. *Close to Home: Women Reconnect Ecology, Health and Development Worldwide*. Philadelphia: New Society.

———. 2005. *Earth Democracy: Justice, Sustainability, and Peace*. Cambridge, MA: South End Press.

Shlain, L. 1988. *Staying Alive: Women, Ecology, and Development*. London: Zed Books.

———. 1993. *Monocultures of the Mind: Perspectives on Biodiversity and Biotechnology*. London: Zed Books.

———. 1998. *The Alphabet versus the Goddess: The Conflict between Word and Image*. New York: Penguin.

———. 1999. *Stolen Harvest: The Hijacking of the Global Food Supply*. Cambridge, MA: South End Press.

———. 2002. *Water Wars: Privatization, Pollution, and Profit*. Cambridge, MA: South End Press.

Silberstein, M., and J. McGeever. 1999. "The Search for Ontological Emergence." *Philosophical Quarterly* 49 (195): 182–200.

Silos, M. 2002. "The Politics of Consciousness." *International Journal of Humanities and Peace* 18 (1): 44–51.

Simmons, A., A. N. Popper, and R. R. Fay, eds. 2002. *Acoustic Communication*. New York: Springer.

Simmons, I. G. 1989. *Changing the Face of the Earth: Culture, Environment, History*. Oxford: Basil Blackwell.

———. 1993. *Interpreting Nature: Cultural Constructions of the Environment*. New York: Routledge.

Simonds, J. O., and B. Starke. 2006. *Landscape Architecture*. 4th ed. New York: McGraw-Hill.

Simson, S. P., and M. C. Straus, eds. 2003. *Horticulture as Therapy: Principles and Practice*. New York: Haworth Press.

Singer, F. 1998. *Hot Talk, Cold Science: Global Warming's Unfinished Debate*. Oakland, CA: Independent Institute.

Singer, P. 1990. *Animal Liberation*. 2nd ed. New York: Random House.

Sjoo, M. 1991. *The Great Cosmic Mother: Rediscovering the Religion of the Earth*. 2nd ed. San Francisco: Harper.

Skocz, D. E. 2004. "Wilderness: A Zoocentric Phenomenology—From Hediger to Heidegger." In *Imaginatio Creatrix*, edited by A. T. Tymieniecka. Boston: Springer, 217–44.

Skolimowsky, H. 1994. *EcoYoga: Practice and Meditations for Walking in Beauty on the Earth*. London: Gaia Books.

Skrbina, D. 2005. *Panpsychism in the West*. Cambridge, MA: MIT Press.

Skyttner, L. 2006. *General Systems Theory: Perspectives, Problems, Practice*. 2nd ed. London: World Scientific Publishing Company.

Slaughter, R. A. 1997. "Ken Wilber's Path to Transformational Futures." *New Renaissance* 7 (3): 23–25.

———. 1998. "Transcending Flatland: Implications of Ken Wilber's Metanarrative for Futures Studies." *Futures* 30 (6): 519–33.

———. 1999a. "A New Framework for Environmental Scanning." *Foresight: The Journal of Future Studies, Strategic Thinking and Policy* 1 (5): 387–97.

———. 1999b. "An Outline of Critical Futures Studies." In *Futures for the Third Millennium: Enabling the Forward View*. Sydney: Prospect Media.

———. 2001. "Knowledge Creation, Futures Methodologies and the Integral Agenda." *Foresight: The Journal of Future Studies, Strategic Thinking and Policy* 3 (5): 407–18.

———. 2002. "Beyond the Mundane: Reconciling Breadth and Depth in Futures Work." *Futures* 34 (5): 493–507.

Slovic, P. 2000. *The Perception of Risk*. London: Earthscan.

Smith, B. 2001. "Objects and Their Environments: From Aristotle to Ecological Ontology." In *The Life and Motion of Socio-Economic Units* (GISDATA 8), edited by A. Frank, J. Raper, and J. Cheylan. London: Taylor and Francis, 79–97.

Smith, B., and D. Mark. 2003. "Do Mountains Exist? Towards an Ontology of Landforms." *Environment & Planning* 30 (3): 411 27.

Smith, B., and A. Varzi. 2002. "Surrounding Space: The Ontology of Organism-Environment Relations." *Theory in Biosciences* 121 (2): 139–62.

Smith, D. 2007. *Urban Ecology*. London: Routledge.

Smith, P. 1999a. *Animal Talk: Interspecies Telepathic Communication*. Hillsboro, OR: Beyond Words.

———. 1999b. *When Animals Speak: Advanced Interspecies Communication*. Hillsboro, OR: Beyond Words.

Smuts, B. 2001. "Encounters with Animal Minds." *Journal of Consciousness Studies* 8 (5–7): 293–309.

Snorf, K. 2003. "Integral Eco-Design." Working paper.

Snyder, G. 1974. *Turtle Island*. New York: New Directions.

———. 1992. *No Nature: New and Selected Poems*. New York: Pantheon Books.

———. 1995a. *A Place in Space: Ethics, Aesthetics, and Watersheds*. Washington, DC: Counterpoint.

———. 1995b. *The Practice of the Wild*. New York: North Point Press.

Soares, E., and M. Powers. 1999. *Extreme Sea Kayaking: A Survival Guide*. Camden, ME: International Marine/Ragged Mountain Press.

Sobel, D. 1993. *Children's Special Places: Exploring the Role of Forts, Dens, and Bush Houses in Middle Childhood*. Tucson, AZ: Zephyr Press.

———. 2005. *Beyond Ecophobia: Reclaiming the Heart in Nature Education*. Great Barrington, MA: Orion Society.

Sokal, A., and J. Bricmont. 1999. *Fashionable Nonsense: Postmodern Intellectuals' Abuse of Science*. New York: Picador.

Somé, M. P. 1993. *Ritual: Power, Healing and Community*. New York: Penguin Books.

———. 1994. *Of Water and the Spirit: Ritual, Magic and Initiation in the Life of an African Shaman*. New York: Penguin Books.

———. 1998. *Healing Wisdom of Africa*. New York: Tarcher.

Soper, K. 1995. *What Is Nature?* Cambridge, MA: Blackwell.

Sorabji, R. 1993. *Animal Minds and Human Morals: The Origins of the Western Debate*. Ithaca, NY: Cornell University Press.

Soulé, M., and G. Lease, eds. 1995. *Reinventing Nature? Responses to Postmodern Deconstruction*. Washington, DC: Island Press.

Southwick, C. 1996. *Global Ecology in Human Perspective*. New York: Oxford University Press.

Spaargaren, G. 1997. *The Ecological Modernization of Production and Consumption*. Wageningen, Netherlands: Wageningen University.

Spaargaren, G., A. P. J. Mol, and F. H. Buttel. 2000. *Environment and Global Modernity*. London: SAGE.

Spence, M. D. 2000. *Dispossessing the Wilderness: Indian Removal and the Making of the National Parks*. New York: Oxford University Press.

Speth, J. G. 2005. *Red Sky at Morning*. New Haven, CT: Yale University Press.

Spinoza, B. 1910. *Ethics*. London: Orion.

———. 1991. *Tractatus Theologico-Politicus*. Leiden, Netherlands: Brill Academic Publishers.

Spirn, A. W. 1998. *The Language of Landscape*. New Haven, CT: Yale University Press.

Spretnak, C. 1987. *Spiritual Dimensions of Green Politics*. Rochester, VT: Bear & Company.

———. 1991. *States of Grace: The Recovery of Meaning in the Postmodern Age*. San Francisco: Harper.

———. 1999. *The Resurgence of the Real*. New York: Routledge.

Sproule-Jones, M. 1995. "Comments on ECOWISE, a Multidisciplinary Research Program on the Hamilton Harbour Watershed, Lake Ontario." *Polycentric Circles* 2 (1): 2.

Stables, A. 1998. "Environmental Literacy: Functional, Cultural, Critical; The Case of the SCAA Guidelines." *Environmental Education Research* 4 (2): 155–64.

Stanford, C. 2001. *Significant Others: The Ape-Human Continuum and the Quest for Human Nature*. New York: Basic Books.

Stanner, W. E. H. 1979. *White Man Got No Dreaming: Essays 1938–1973*. Canberra: Australian National University Press.

Starhawk. 1995. *The Spiral Dance: A Rebirth of the Ancient Religion of the Great Goddess*. San Francisco: Harper.

Starhawk and M. Macha NightMare. 1995. *The Pagan Book of Living and Dying: Practical Rituals, Prayers, Blessings, and Meditations on Crossing Over*. San Francisco: Harper.

Steele, J. 2005. *Ecological Architecture: A Critical History*. New York: Thames & Hudson.

Steeves, H. P. 1999. *Animal Others: On Ethics, Ontology, and Animal Life*. Albany, NY: SUNY Press.

Stefanovic, I. 1995. "An Integrative Framework for Sustainability: The Case of the Hamilton Harbour Ecosystem." Paper presented at the Synthesis of Tradition and Modernity for a Sustainable Society Conference, Bombay, September 23–30.

———. 1996. "Interdisciplinarity and Wholeness: Lessons from Eco-research on the Hamilton Harbor Ecosystem." *Environments: Journal of Interdisciplinary Studies* 23 (3): 74–94.

———. 1997. "An Integrative Framework for Sustainability: The Hamilton Harbour Ecosystem." *Ekistics* 63 (382): 83–91.

———. 1998. "Phenomenological Encounters with Place: Cavtat to Square One." *Journal of Environmental Psychology* 18 (1): 31–44.

———. 2000a. "Phenomenological Reflections on Ecosystem Health." *Ethics and the Environment* 5 (2): 253–69.

———. 2000b. *Safeguarding Our Common Future: Rethinking Sustainable Development*. Albany, NY: SUNY Press.

Steffen, A., ed. 2006. *Worldchanging: A User's Guide for the Twenty-first Century*. New York: Harry N. Abrams.

Stein, M. 2000. *When Technology Fails: A Manual for Self-Reliance and Planetary Survival*. Santa Fe, NM: Clear Light.

Stein, Z. 2007. "Modeling the Demands of Interdisciplinarity: Toward a Framework for Evaluating Interdisciplinary Endeavors." *Integral Review* 4: 91–107.

Steiner, F. 2002. *Human Ecology: Following Nature's Lead*. Washington, DC: Island Press.

Steiner, R. 1993. *Agriculture: Spiritual Foundations for the Renewal of Agriculture*. Kimberton, PA: Bio-Dynamic Farming & Gardening Association.

———. 1995. *Nature Spirits: Selected Lectures*. London: Rudolf Steiner Press.

———. 2001. *Harmony of the Creative Word: The Human Being and the Elemental, Animal, Plant and Mineral Kingdoms*. London: Rudolf Steiner Press.

———. 2005. *What Is Biodynamics? A Way to Heal and Revitalize the Earth*. Great Barrington, MA: SteinerBooks.

Stewart, O. 2002. *Forgotten Fires: Native Americans and the Transient Wilderness*. Norman: University of Oklahoma Press.

Stewart, R. J. 1992. *Earth Light: The Ancient Path to Transformation—Rediscovering the Wisdom of the Celtic and Faery Lore*. Lake Toxaway, NC: Mercury Publishing.

———. 1998. *Power within the Land: The Roots of Celtic and Underworld Traditions Awakening the Sleepers and Regenerating the Earth*. Lake Toxaway, NC: Mercury Publishing.

Stiles, D. R., and J. T. Stiles. 1998. *Rustic Retreats: A Build-It-Yourself Guide*. Pownal, VT: Storey Books.

Stine, A., ed. 1996. *The Earth at Our Doorstep: Contemporary Writers Celebrate the Landscape of Home*. San Francisco: Sierra Club Books.

Stjernfelt, F. 2002. "*Tractatus Hoffmeyerensis*: Biosemiotics as Expressed in Twenty-two Basic Hypotheses." *Sign Systems Studies* 30 (1): 337–44.

Stone, M. K., and Z. Barlow, eds. 2005. *Ecological Literacy: Educating Our Children for a Sustainable World*. San Francisco: Sierra Club Books.

Strand, C. 2005. "The Whole Package." *Tricycle* (Summer): 82–87.

Stromsted, T. 2002. "Dancing Body Earth Body: Andrea Olsen's Story." *Somatics* (Spring/Summer): 10–20.

Stuckey, M., and G. Palmer. 1998. *Western Trees: A Field Guide for Weekend Naturalists*. Helena, MT: Falcon.

Sturgeon, N. 1997. *Ecofeminist Natures: Race, Gender, Feminist Theory, and Political Action*. London: Routledge.

Suchantke, A. 2001. *Eco-Geography: What We See When We Look at Landscapes*. Great Barrington, MA: Lindisfarne Books.

Sullivan, D. 1990. *Sacred Places*. Santa Fe, NM: Bear.

———. 2000. *Ley Lines: A Comprehensive Guide to Alignments*. London: Piatkus Books.

Sullivan, L., ed. 2003. *Native Religions and Cultures of North America: Anthropology of the Sacred*. New ed. New York: Continuum International.

Sunstein, C. R. 2005. *Laws of Fear: Beyond the Precautionary Principle*. New York: Cambridge University Press.

Sunstein, C. R., and M. C. Nussbaum. 2004. *Animal Rights: Current Debates and New Directions*. New York: Oxford University Press.

Susi, T., and T. Ziemke. 2005. "On the Subject of Objects: Four Views on Object Perception and Tool Use." *TripleC* 3 (2): 6–19.

Sutton, M. O., and E. N. Anderson. 2004. *An Introduction to Cultural Ecology*. Lanham, MD: AltaMira Press.

Suzuki, D. 1997. *The Sacred Balance: Rediscovering Our Place in Nature*. Vancouver, BC: Greystone Books.

Suzuki, D. T., H. Dressel, and G. L. Saunders. 2003. *Good News for a Change: How Everyday People Are Helping the Planet*. Vancouver, BC: Greystone Books.

Swan, J. A. 1991. *The Power of Place: Sacred Ground in Natural and Human Environments*. Wheaton, IL: Quest.

Swearingen, T. 1996. *The Hidden Heart of the Cosmos: Humanity and the New Story*. Maryknoll, NY: Orbis Books.

Swimme, B. 1985. *The Universe Is a Green Dragon: A Cosmic Creation Story*. Santa Fe, NM: Bear.

———. 1989. "Moral Development and Environmental Ethics." PhD diss., University of Washington, Seattle.

Swimme, B., and T. Berry. 1992. *The Universe Story: From the Primordial Flaring Forth to the Ecozoic Era, A Celebration of the Unfolding of the Cosmos*. San Francisco: HarperCollins.

Switzer, J. V. 2003. *Environmental Activism: Reference Handbook*. Santa Barbara, CA: ABC-Clio.

Taborsky E., ed. 1999. *Semiosis, Evolution, Energy: Towards a Reconceptualization of the Sign*. Aachen, Germany: Shaker Verlag.

Tainter, J. A., T. F. H. Allen, and T. W. Hoekstra. 2006. "Energy Transformations and Post-Normal Science." *Energy* 31 (1): 44–58.

Tansley, A. 1935. "The Use and Abuse of Vegetational Concepts and Terms." *Ecology* 16 (3): 284–307.

Tarnas, R. 1991. *The Passion of the Western Mind*. New York: Ballantine Books.

Tashakkori, A. 1998. *Mixed Methodology: Combining Qualitative and Quantitative Approaches*. Newbury Park, CA: SAGE.

———. 2002. *Handbook of Mixed Methods in Social and Behavioral Research*. Newbury Park, CA: SAGE.

Taylor, B. 1997. "Earthen Spirituality or Cultural Genocide? Radical Environmentalism's Appropriation of Native American Spirituality." *Religion* 27 (2): 183–215.

———. 2001a. "Earth and Nature-Based Spirituality (part 1): From Deep Ecology to Radical Environmentalism." *Religion* 31 (2): 175–93.

———. 2001b. "Earth and Nature-Based Spirituality (part 2): From Earth First! and Bioregionalism to Scientific Paganism and the New Age." *Religion* 31 (3): 225–45.

———. 2005a. *The Encyclopedia of Religion and Nature*. New York: Thoemmes Continuum.

———. 2005b. "Ecology and Religion: Ecology and Nature Religions," *Encyclopedia of Religion*, vol. 4. 2nd ed., edited by Lindsay Jones. New York: Macmillan Reference, 2661–68.

Taylor, M. 2001. *The Moment of Complexity: Emerging Network Culture*. Chicago: University of Chicago Press.

Taylor, P. 1986. *Respect for Nature*. Princeton, NJ: Princeton University Press.

Teich, M., R. Porter, and B. Gustafsson, eds. 1997. *Nature and Society in Historical Context*. New York: Cambridge University Press.

Teleosis Institute. 2005. "The Green Health Care Program: Good for People and the Environment; Executive Summary." *Green Health Care*. www.teleosis.org/ghcp-exec.ph. (accessed August 7, 2006).

Telesco, T., and P. Telesco. 1999. *Dancing with Devas: Connecting with the Spirits and Elements of Nature*. Laceyville, PA: Todd Hall.

Tempest, T. 2001. *Red: Passion and Patience in the Desert*. New York: Pantheon Books.

Tennyson, J. 1977. *A Singleness of Purpose: The Story of Ducks Unlimited*. Chicago: Ducks Unlimited.

Tetsuro, W. 1961/1988. *Climate and Culture: A Philosophical Study*. Translated by G. Bowans. Westport, CT: Greenwood Press.

Thayer, R. 2003. *LifePlace: Bioregional Thought and Practice*. Berkeley: University of California Press.

Theodoropoulos, D. 2003. *Invasion Biology: Critique of a Pseudoscience*. Blythe, CA: Avvar Books.

Thomashow, M. 1995. *Ecological Identity: Becoming a Reflective Environmentalist*. Cambridge, MA: MIT Press.

Thompson, A. *Homes That Heal (and Those That Don't): How Your Home Could be Harming Your Family's Health*. Gabriola Island, BC: New Society Publishers.

Thompson, E. 2007. *Mind in Life: Biology, Phenomenology, and the Sciences of Mind*. Cambridge, MA: Belknap Press.

Thompson, M., R. Ellis, and A. Wildawsky. 1990. *Cultural Theory*. Boulder, CO: Westview Press.

Thompson, M., and S. Rayner. 1998. "Cultural Discourses." In *Human Choice and Climate Change*. Vol. 1, *The Societal Framework*, edited by S. Rayner and E. L. Malone. Columbus, OH: Battelle Press, 265–343.

Tierney, P. 1989. *The Highest Altar: The Story of Human Sacrifice*. New York: Viking Press.

Tilley, C. 1997. *A Phenomenology of Landscape: Places, Paths and Monuments*. Oxford, UK: Berg Publishers.

Tilman, D., and P. Kareiva, eds. 1997. *Spatial Ecology: The Role of Space in Population Dynamics and Interspecific Interactions*. Princeton, NJ: Princeton University Press.

Tissot, B. 2005. "Integral Marine Ecology: Community-Based Fishery Management in Hawai'i." *World Futures: The Journal of General Evolution* 61 (1–2): 79–95.

Tobias, M. 1995. *A Vision of Nature: Traces of the Original World*. Kent, OH: Kent State University Press.

Todd, N. J., and J. Todd. 1984. *Bioshelters, Ocean Arks, City Farming: Ecology as the Basis of Design*. San Francisco: Sierra Club Books.

Tompkins, P. 1997. *The Secret Life of Nature: Living in Harmony with the Hidden World of Nature Spirits, from Fairies to Quarks*. London: HarperCollins.

Tompkins, P., and C. Bird. 1998. *Secrets of the Soil: New Solutions for Restoring Our Planet*. Eagle River, AK: Earthpulse Press.

Tonnessen, M. 2003. "Umwelt Ethics." *Sign Systems Studies* 31 (1): 281–99.

Tooker, E., ed. 1979. *Native North American Spirituality of the Eastern Woodlands: Sacred Myths, Dreams, Visions, Speeches, Healing Formulas, Rituals, and Ceremonials*. Mahwah, NJ: Paulist Press.

Torbert, W. 1991. *The Power of Balance: Transforming Self, Society, and Scientific Inquiry*. Newbury Park, CA: SAGE.

———. 2000a. "A Developmental Approach to Social Science: A Model for Analyzing Charles Alexander's Scientific Contributions." *Journal of Adult Development* 7 (4): 255–67.

———. 2000b. "Transforming Social Science to Integrate Quantitative, Qualitative, and Action Research." In *Transforming Social Inquiry, Transforming Social Action*, edited by F. Sherman and W. R. Torbert. Norwell, MA: Kluwer, 67–92.

———. 2001. "The Practice of Action Inquiry." In *Handbook of Action Research*, edited by P. Reason and H. Bradbury. London: SAGE, 250–60.

———. 2004. *Action Inquiry: The Secret of Timely and Transforming Leadership*. San Francisco: Berrett-Koehler.

Torrance, R. M. 1998. *Encompassing Nature: A Sourcebook; Nature and Culture from Ancient Times to the Modern World*. Washington, DC: Counterpoint.

Townsend, A. K. 2006, *Green Business: A Five-Part Model for Creating an Environmentally Responsible Company*. Atglen, PA: Schiffer Publishing.

Toynbee, J. M. C. 1973. *Animals in Roman Life and Art*. London: Thames & Hudson.

Trewavas, A. 2003. "Aspects of Plant Intelligence." *Annals of Botany* 92: 1–20.

Truax, B. 2000. *Acoustic Communication*. 2nd ed. Westport, CT: Ablex Publishing.

Truax, B., ed. 1978. *Handbook for Acoustic Ecology*. Burnaby, BC: Aesthetic Research Centre.

Tsosie, R. 1996. "Tribal Environmental Policy in an Era of Self-Determination." *Vermont Law Review* 21 (1): 225–39.

Tuan, Y. F. 1977/2002. *Space and Place: The Perspective of Experience*. Minneapolis: University of Minnesota Press.

———. 1990. *Topophilia: A Study of Environmental Perception, Attitudes, and Values*. New York: Columbia University Press.

Tucker, M. E. 2003. *Worldly Wonder: Religions Enter Their Ecological Phase*. Chicago: Open Court.

Tucker, M. E., and J. Grim, eds. 1994. *Worldviews and Ecology: Religion, Philosophy, and the Environment*. Maryknoll, NY: Orbis Books.

———. 1994–2002. *Religions of the World and Ecology*. 10 vols. Cambridge, MA: Harvard University Press.

Tudge, C. 1992. *Last Animals at the Zoo! How Mass Extinction Can Be Stopped*. Washington, DC: Island Press.

———. 2000. *The Variety of Life: A Survey and a Celebration of All the Creatures That Have Ever Lived*. New York: Oxford University Press.

Turner, C. II, and J. Turner. 1999. *Man Corn: Cannibalism and Violence in the Prehistoric American Southwest*. Salt Lake City: University of Utah Press.

Tylor, E. B. 1871. *Primitive Culture: Researches into the Development of Mythology, Philosophy, Religion, Art and Custom*. London: Murray.

Tyssen, R. 1996. "Ecosystem Management: Is It the Answer to Environmental Crisis?" Master's thesis, York University, Toronto.

Uexküll, J. von. 1982. "The Theory of Meaning." Special issue. *Semiotica* 42 (1).

———. 1992. *A Stroll through the Worlds of Animals and Men*. Special issue. *Semiotica* 89 (4).

Uexküll, T. von. 1982. "Introduction: Meaning and Science in Jacob von Uexküll's Concept of Biology." *Semiotica* 42 (1): 1–24.

Ulanowicz, R. E. 1986. *Growth and Development: Ecosystems Phenomenology*. New York: Springer-Verlag.

———. 1997. *Ecology: The Ascendant Perspective*. New York: Columbia University Press.

Union of Concerned Scientists. 2004. "Crichton's Thriller *State of Fear*: Separating Fact from Fiction." *Global Environment*. http://go.ucsusa.org/global_environment/global_warming/page.cfm?pageID=1670 (accessed August 2006).

Vale, T., ed. 2002. *Fire, Native Peoples, and the Natural Landscape*. Washington, DC: Island Press.

van Andel, J., and J. Aronson, eds. 2005. *Restoration Ecology: The New Frontier*. New York: Wiley.

Van Buren, J. 1995. "Critical Environmental Hermeneutics." *Environmental Ethics* 17 (Fall): 259–75.

Van der Ryn, S., and S. Cowan. 1996. *Ecological Design*. Washington, DC: Island Press.

Van de Vijver, G., S. N. Salthe, and M. Delpos, eds. 1998. *Evolutionary Systems: Biological and Epistemological Perspectives on Selection and Self-Organization*. Dordrecht, Netherlands: Kluwer.

Van Haaften, A. W., M. Korthals, and T. E. Wren, eds. 1996. *Philosophy of Development: Reconstructing the Foundations of Human Development and Education*. Norwell, MA: Kluwer.

Van Wyck, P. C. 1997. *Primitives in the Wilderness: Deep Ecology and the Missing Human Subject*. Albany, NY: SUNY Press.

Vecsey, C., and R. Venables, eds. 1980. *American Indian Environments: Ecological Issues in Native American History*. Syracuse, New York: Syracuse University Press.

Vehkavaara, T. 2006. "From the Logic of Science to the Logic of the Living: The Relevance of Charles Peirce to Biosemiotics." In *Introduction to Biosemiotics: The New Biological Synthesis*, edited by M. Barbieri. Dordrecht, Netherlands: Springer, chap. 11.

Verbeek, P., and F. B. M. de Waal. 2002. "The Primate Relationship with Nature: Biophilia as a General Pattern." In *Children and Nature: Psychological, Sociocultural, and Evolutionary Investigations*, edited by P. H. Kahn, Jr., and S. R. Kellert. Cambridge, MA: MIT Press, 93–116.

Vernadsky, V. 1926/1998. *The Biosphere*. New York: Copernicus Books.

Versluis, A. 1993. *American Transcendentalism and Asian Religions*. New York: Oxford University Press.

Visser, F. 2003. *Ken Wilber: Thought as Passion*. Albany, NY: SUNY Press.

———. 2005. "Translations." *Integral World: Exploring Theories of Everything*. www.integralworld.net/translations.html (accessed January 16, 2005).

Vitek, W., and W. Jackson, eds. 1996. *Rooted in the Land*. New Haven, CT: Yale University Press.

Vogel, S. 1996. *Against Nature: The Concept of Nature in Critical Theory*. Albany, NY: SUNY Press.

———. 1998. "Nature as Origin and Difference: On Environmental Philosophy and Continental Thought." *Philosophy Today* 24: 169–81.

Von Bertalanffy, L. 1968. *General Systems Theory*. London: Allen Lane.

Von Essen, C. 2007. *The Hunter's Trance: Nature, Spirit, and Ecology*. Great Barrington, MA: Lindisfarne Books.

Voros, J. 2001. "Reframing Environmental Scanning: An Integral Approach." *Foresight: The Journal of Future Studies, Strategic Thinking, and Policy* 3 (6): 533–52.

Wahl, D. C. 2006. "Design for Human and Planetary Health." In *Management of Natural Resources, Sustainable Development and Ecological Hazards*, edited by C. A. Brebba, M. E. Conti, and E. Tiezzi. Boston: WIT Press, 285–96.

———. 2007. "Scale-Linking Design for Systemic Health." *International Journal of Ecodynamics* 2 (1): 1–16.

Waldau, P. 2001. *The Specter of Speciesism: Buddhist and Christian Views of Animals*. New York: Oxford University Press.

Waldau, P., and K. Patton, eds. 2006. *A Communion of Subjects: Animals in Religion, Science, and Ethics*. New York: Columbia University Press.

Walker, B. 1988. *The Woman's Dictionary of Symbols and Sacred Objects*. San Francisco: Harper.

Wall, D. 1994. *Green History: A Reader in Environmental Literature, Philosophy and Politics*. New York: Routledge.

Wallauer, B. 2002. "Do Chimpanzees Feel Reverence for Nature?" www.janegoodall.org/chimp_central/chimpanzees/behavior/rain_dance.asp (accessed May 30, 2007).

Wallis, R. J. 2003. *Shamans/Neo-Shamans: Ecstasy, Alternative Archaeologies and Contemporary Pagans*. London: Routledge.

Walsh, R. 1992. *The Spirit of Shamanism*. Los Angeles: Tarcher.

———. 2001. "Shamanic Experiences: A Developmental Analysis." *Journal of Humanistic Psychology* 41 (3): 31–52.

———. 2002. "Terrorism and Other Global Terrors: An Integral Analysis." *Journal of Transpersonal Psychology* 34 (1): 13–22.

———. 2007. *The World of Shamanism: New Views of an Ancient Tradition*. Woodbury, MN: Llewellyn.

Waltner-Toews, D., J. J. Kay, C. Neudoerffer, and T. Gitau. 2003. "Perspective Changes Everything: Managing Ecosystems from the Inside Out." *Frontiers in Ecology and the Environment* 1 (1): 23–30.

Wargo, J. 1996. *Our Children's Toxic Legacy: How Science and Law Fail to Protect Us from Pesticides*. New Haven, CT: Yale University Press.

Warren, K., ed. 1997. *Ecofeminism: Women, Culture, Nature*. Bloomington: Indiana University Press.

———. 2000. *Ecofeminist Philosophy*. New York: Rowman & Littlefield.

Warren, K. J., and J. Cheney. 1991. "Ecological Feminism and Ecosystem Ecology." *Hypatia* 6 (1): 179–97.

———. 1993. "Ecosystem Ecology and Metaphysical Ecology: A Case Study." *Environmental Ethics* 15 (2): 99–116.

Washburn, M. 1994. *Transpersonal Psychology in Psychoanalytic Perspective*. Albany, NY: SUNY Press.

———. 1995. *The Ego and the Dynamic Ground: A Transpersonal Theory of Human Development*. 2nd ed. Albany, NY: SUNY Press.

———. 1998. "The Pre/Trans Fallacy Reconsidered." In *Ken Wilber in Dialogue: Conversations with Leading Transpersonal Thinkers*, edited by D. Rothberg and S. Kelly. Wheaton, IL: Quest Books, 62–87.

———. 2003a. *Embodied Spirituality in a Sacred World*. Albany, NY: SUNY Press.

———. 2003b. "Transpersonal Dialogue: A New Direction." *Journal of Transpersonal Psychology* 34 (2): 1–20.

Washburn, M. F. 1908. *The Animal Mind: A Textbook of Comparative Psychology*. New York: Macmillan.

Waskow, A., ed. 2000. *Torah of the Earth: Exploring Four Thousand Years of Ecology in Jewish Thought*. 2 vols. Woodstock, VT: Jewish Lights.

Watkinson, R. 1990. *William Morris as Designer*. London: Trefoil.

Watson, D. 1996. *Beyond Bookchin: Preface for a Future Social Ecology*. Detroit, MI: Black & Red Books.

Weaver, J., ed. 1996. *Defending Mother Earth: Native American Perspectives on Environmental Justice*. New York: Orbis Books.

Weber, A. 2002. "The 'Surplus of Meaning.' Biosemiotic Aspects in Francisco J. Varela's Philosophy of Cognition." *Cybernetics & Human Knowing* 9 (2): 11–29.

Webster, R. 2001. *The Art of Dowsing*. Edinburgh, UK: Castle Books.

Weigert, A. J. 1997. *Self, Interaction, and Natural Environment*. Albany, NY: SUNY Press.

Weil, P. 2002. *The Art of Living in Peace: Guide to Education for a Culture of Peace*. Paris: UNESCO.

Weisman, A. 2007. *The World Without Us*. New York: St. Martin's Press.

Weissmann, D. 2007. *The Cage: Must, Should, and Ought from Is*. Albany, NY: SUNY Press.

Welter, V. M. 2002. *Biopolis: Patrick Geddes and the City of Life*. Cambridge, MA: MIT Press.

West, K. 2001. *The Real Witches' Handbook: A Complete Introduction to the Craft*. London: Thorsons.

Westley, F., S. R. Carpenter, W. A. Brock, C. S. Holling, and L. H. Gunderson. 2001. "Why Systems of People and Nature Are Not Just Ecological Systems." In *Panarchy: Understanding Transformations in Systems of Humans and Nature*, edited by L. H. Gunderson and C. S. Holling. Washington, DC: Island Press, 103–19.

Westra, L. 1997. "Aristotelian Roots of Ecology: Causality, Complex Systems Theory, and Integrity." In *The Greeks and the Environment*, edited by L. Westra and T. M. Robinson. Lanham, MD: Rowman & Littlefield, 83–98.

Westra, L., and T. M. Robinson, eds. 1997. *The Greeks and the Environment*. Lanham, MD: Rowman & Littlefield.

———. 2002. *Thinking about the Environment: Our Debt to the Classical and Medieval Past*. Lanham, MD: Lexington.

Wheeler, M., ed. 1995. *Ruskin and Environment: The Storm-Cloud of the Nineteenth Century*. Manchester, UK: Manchester University Press.

Wheeler, Q., and R. Meier, eds. 2000. *Species Concepts and Phylogenetic Theory: A Debate*. New York: Columbia University Press.

Wheeler, W. 2006. *The Whole Creature: Complexity, Biosemiotics and the Evolution of Culture*. London: Lawrence & Wishart.

White, L. Jr. 1967. "The Historical Roots of Our Ecological Crisis." *Science* 155 (3767): 1203–7.

White, R. 1980. *Land Use, Environment, and Social Change: The Shaping of Island County, Washington*. Seattle: University of Washington Press.

White, T. 1992. *Prehistoric Cannibalism at Mancos 5Mtumr-2346*. Princeton, NJ: Princeton University Press.

Whitebook, J. 1996. "The Problem of Nature in Habermas." In *Minding Nature: The Philosophers of Ecology*, edited by D. Macauley. New York: Guilford, 283–317.

Whitehead, A. N. 1979. *Process and Reality: An Essay in Cosmology*. Edited by A. Griffin, D. Ray, and D. Sherburne. New York: Free Press.

Whiteside, K. H. 2002. *Divided Natures: French Contributions to Political Ecology*. Cambridge, MA: MIT Press.

Whitman, W. 1990. *Leaves of Grass*. New York: Oxford University Press.

Whitney, G. 1994. *From Coastal Wilderness to Fruited Plain: A History of Environmental Change in Temperate North America, 1500 to Present*. New York: Cambridge University Press.

Whitney, S. 1997. *Western Forests*. New York: Knopf.

Wiener, N. 1965. *Cybernetics: Or the Control and Communication in the Animal and the Machine*. 2nd ed. Cambridge, MA: MIT Press.

Wiens, J., M. R. Moss, M. G. Turner, and D. Mladenoff, eds. 2006. *Foundation Papers in Landscape Ecology*. New York: Columbia University Press.

Wiersma, G. B., ed. 2004. *Environmental Monitoring*. Boca Raton, FL: CRC Press.

Wight, I. 1999. "Integrating 'It' and 'We' with the 'I' of the Beholder: From Planning the City's Region to Making Our Collective Home-Place." Paper presented at the CIP National Conference, Montreal, June.

———. 2000. "Rethinking Regions as Holons in Holarchies: Toward a More Integral City-Region Planning Practice." Paper presented to the Association of Collegiate Schools of Planning, Atlanta, GA, November.

———. 2002. "Place, Placemaking and Planning, Part 1: Wilber's Integral Theory." Paper presented to the Association of Collegiate Schools of Planning, Baltimore, November.

———. 2005. "Placemaking as Applied Integral Ecology: Evolving an Ecologically-Wise Planning Ethic." *World Futures: The Journal of General Evolution* 61 (1–2): 127–37.

Wilber, K. 1977. *The Spectrum of Consciousness*. Wheaton, IL: Theosophical Publishing House.

———. 1979. *No Boundary: Eastern and Western Approaches to Personal Growth*. Los Angeles: Center.

———. 1980. *The Atman Project: A Transpersonal View of Human Development*. Wheaton, IL: Quest.

———. 1981. *Up from Eden*. Garden City, NY: Anchor Books/Doubleday.

———, ed. 1982. *The Holographic Paradigm and Other Paradoxes: Exploring the Leading Edge of Science*. Boston: Shambhala Publications.

———. 1983a. *Eye to Eye: The Quest for the New Paradigm*. New York: Anchor Books / Doubleday.

———. 1983b. *A Sociable God: A Brief Introduction to a Transcendental Sociology*. New York: McGraw-Hill.

———, ed. 1984. *Quantum Questions: Mystical Writings of the World's Great Physicists*. Boulder, CO: Shambhala Publications.

———. 1991. *Grace and Grit: Spirituality and Healing in the Life of Treya Killam Wilber*. Boston: Shambhala Publications.

———. 1995. *Sex, Ecology, Spirituality: The Spirit of Evolution*. Boston: Shambhala Publications.

———. 1996. *A Brief History of Everything*. Boston: Shambhala Publications.

———. 1997. *The Eye of Spirit: An Integral Vision for a World Gone Slightly Mad*. Boston: Shambhala Publications.

———. 1998. *The Marriage of Sense and Soul: Integrating Science and Religion*. New York: Random House.

———. 1999. *One Taste: The Journals of Ken Wilber.* Boston: Shambhala Publications.

———. 1999–2000. *The Collected Works of Ken Wilber.* 8 vols. Boston: Shambhala Publications.

———. 2000. *Integral Psychology: Consciousness, Spirit, Psychology, Therapy.* Boston: Shambhala Publications.

———. 2001a. "On the Nature of a Post-metaphysical Spirituality." 2 parts. Ken Wilber Online, August 6. http://wilber.shambhala.com/html/misc/habermas/index.cfm (accessed November 18, 2003).

———. 2001b. *A Theory of Everything: An Integral Vision for Business, Politics, Science, and Spirituality.* Boston: Shambhala Publications.

———. 2001c. "On Critics, Integral Institute, My Recent Writing, and Other Matters of Little Consequence: A Shambhala Interview with Ken Wilber." 3 parts. Ken Wilber Online. http://wilber.shambhala.com/html/interviews/interview1220.cfm (accessed August 3, 2003).

———. 2001d. "Waves, Streams, and Self—A Summary of My Psychological Model." Ken Wilber Online. http://wilber.shambhala.com/html/books/psych_model/psych_model1.cfm (accessed November 18, 2003).

———. 2002a. *Boomeritis: A Novel That Will Set You Free.* Boston: Shambhala Publications.

———. 2002b. "Introduction to Excerpts, from vol. 2 of the Kosmos Trilogy." Ken Wilber Online. http://wilber.shambhala.com/html/books/kosmos/index.cfm (accessed November 18, 2004).

———. 2002c. "Excerpt A: An Integral Age at the Leading Edge." 5 parts. Ken Wilber Online. http://wilber.shambhala.com/html/books/kosmos/excerptA/part1.cfm (accessed November 18, 2004).

———. 2002d. "Excerpt B: The Many Ways We Touch: Three Principles Helpful for Any Integrative Approach." 3 parts. Ken Wilber Online. http://wilber.shambhala.com/html/books/kosmos/excerptB/part1.cfm (accessed November 18, 2004).

———. 2002e. "Excerpt C: The Ways We Are in This Together: Intersubjectivity and Interobjectivity in the Holonic Kosmos." 4 parts. Ken Wilber Online. http://wilber.shambhala.com/html/books/kosmos/excerptD/part1.cfm (accessed November 18, 2004).

———. 2002f. "Excerpt D: The Look of a Feeling; The Importance of Post/Structuralism." 4 parts. Ken Wilber Online. http://wilber.shambhala.com/html/books/kosmos/excerptD/part1.cfm (accessed November 18, 2004).

———. 2002g. "Excerpt G: Toward a Comprehensive Theory of Subtle Energies." 4 parts. Ken Wilber Online. http://wilber.shambhala.com/html/books/kosmos/excerptG/part1.cfm (accessed November 18, 2004).

———. 2003a. "Foreword." In *Ken Wilber: Thought as Passion* by Frank Visser. Albany, NY: SUNY Press, xi–xv.

———. 2003b. "Foreword to *Integral Medicine: A Noetic Reader*." Ken Wilber Online, May 16. http://wilber.shambhala.com/html/misc/integral-med-1.cfm (accessed February 16, 2006).

———. 2004. "The Integral Vision of Healing." In *Consciousness and Healing: Integral Approaches to Mind-Body Medicine*, edited by M. Schlitz, T. Amorok, and M. Micozzi. St. Louis, MO: C. V. Mosby, i–xx.

———. 2006. *Integral Spirituality: A Startling New Role for Religion in the Modern and Postmodern World*. Boston: Shambhala Publications.

Wilber, K., D. Anthony, and B. Ecker, eds. 1987. *Spiritual Choices: The Problem of Recognizing Authentic Paths to Inner Transformation*. New York: Paragon House.

Wilber, K., J. Engler, and D. Brown, eds. 1986. *Transformations of Consciousness*. Boston: Shambhala Publications.

Wilber, K., T. Patten, A. Leonard, and M. Morelli. 2008. *Integral Life Practice*. Boston: Integral Books.

Wiley, A. S. 2004. *An Ecology of High-Altitude Infancy: A Biocultural Perspective*. New York: Cambridge University Press.

Wilkes, J. 2003. *Flowforms: The Rhythmic Power of Water*. Edinburgh, UK: Floris Books.

Williams, G. H. 1962. *Wilderness and Paradise in Christian Thought*. New York: Harper & Brothers.

Williams, M. 2002. *Deforesting the Earth: From Prehistory to Global Crisis*. Chicago: University of Chicago.

———. 2003. *Learning Their Language: Intuitive Communication with Animals and Nature*. Novato, CA: New World Library.

———. 2005. *Beyond Words: Talking with Animals and Nature*. Novato, CA: New World Library.

Willis, D. 1995. *The Sand Dollar and the Slide Rule: Drawing Blueprints from Nature*. Reading, MA: Addison-Wesley.

Willis, R., ed. 1994. *Signifying Animals: Human Meaning in the Natural World*. London: Routledge.

Wilpert, G. 2001. "Integral Politics: A Spiritual Third Way." *Tikkun* 16 (4): 44–49.

Wilshire, B. 2000. *The Primal Roots of American Philosophy: Pragmatism, Phenomenology and Native American Thought.* University Park: Pennsylvania State University Press.

Wilson, A. 1992. *The Culture of Nature: North American Landscape from Disney to the Exxon Valdez.* Cambridge, MA: Blackwell.

Wilson, E. O. 1975. *Sociobiology.* Cambridge, MA: Harvard University Press.

———. 1984. *Biophilia: The Human Bond with Other Species.* Cambridge, MA: Harvard University Press.

———. 1992. *The Diversity of Life.* Cambridge, MA: Harvard University Press.

———. 1998. *Consilience: The Unity of Knowledge.* New York: Knopf.

Wilson, P. L., C. Bamford, and K. Townley. 2007. *Green Hermeticism: Alchemy and Ecology.* Great Barrington, MA: Lindisfarne Books.

Windschatle, K. 1996. *The Killing of History: How Literary Critics and Social Theorists Are Murdering Our Past.* New York: Free Press.

Winograd, T., and F. Flores. 1986. *Understanding Computers and Cognition: A New Foundation to Design.* Norwood, NJ: Ablex Corporation.

Winter, D. 1995. *Ecological Psychology: Healing the Split between Planet and Self.* San Francisco: HarperCollins.

Winter, D., and S. Koger. 2004. *The Psychology of Environmental Problems.* 2nd ed. Mahwah, NJ: Lawrence Erlbaum.

Wittemyer, G., I. Douglas-Hamilton, and W. M. Getz. 2005. "The Socioecology of Elephants: Analysis of the Processes Creating Multitiered Social Structures." *Animal Behavior* 69: 1357–71.

Wojcik, D. 1999. *The End of the World As We Know It: Faith, Fatalism, and Apocalypse in America.* New York: New York University Press.

Wolch, J., and J. Emel, eds. 1998. *Animal Geographies: Place, Politics, and Identity in the Nature-Culture Borderlands.* New York: Verso.

Wolfe, C. 2003. *Animal Rites: American Culture, the Discourse of Species, and Posthumanist Theory.* Chicago: University of Chicago Press.

Wolfe, C., ed. 2003. *Zoontologies: The Question of the Animal.* Minneapolis: University of Minnesota Press.

Wong, E. 2001. *A Master Course in Feng Shui.* Boston: Shambhala Publications.

Wood, G. 2001. *Sisters of the Dark Moon: Thirteen Rituals of the Dark Goddess.* St. Paul, MN: Llewellyn.

Woods, M. 1998. "Upsetting the Balance of Nature: Can Wilderness Preservation Survive the New Ecology?" Paper presented at the meeting of the International Society for Environmental Ethics, Los Angeles, March.

World Commission on Environment and Development (WCED). 1987. *Our Common Future*. Oxford: Oxford University Press.

Worldwatch Institute. 2003. *State of the World 2003*. New York: W. W. Norton.

Worster, D. 1993. "The Ecology of Order and Chaos." In *Environmental Ethics*, edited by S. Armstrong and R. Botzler. New York: McGraw-Hill, 39–43.

———. 1994. *Nature's Economy: A History of Ecological Ideas*. 2nd ed. Cambridge: Cambridge University Press.

Wrangham, R. W., W. C. McGrew, F. De Waal, and P. Heltne, eds. 1996. *Chimpanzee Cultures*. Cambridge, MA: Harvard University Press.

Wright, M. S. 1990. *Perelandra Garden Workbook II: Co-creative Energy Process for Gardening, Agriculture and Life*. Warrenton, VA: Perelandra.

———. 1993. *Perelandra Garden Workbook: A Complete Guide to Gardening with Nature Intelligences*. 2nd ed. Warrenton, VA: Perelandra.

———. 1997. *Co-creative Science: A Revolution in Science Providing Real Solutions for Today's Health and Environment*. Jefferson, VA: Perelandra.

Wu, J., and R. J. Hobbs, eds. 2007. *Key Topics in Landscape Ecology*. Cambridge: Cambridge University Press.

Wynne-Edwards, V. C. 1986. *Evolution Through Group Selection*. Oxford: Blackwell.

York, M. 2005. *Pagan Theology: Paganism as a World Religion*. New ed. New York: New York University Press.

York, R., and E. A. Rosa. 2003. "Key Challenges to Ecological Modernization Theory." *Organization and Environment* 16 (3): 273–88.

Young, J. 1996. *Seeing Through Native Eyes: Understanding the Language of Nature* (audio). Duvall, WA: Wilderness Awareness School Owlink Media. 6 cassettes.

———. 1999. *Advanced Bird Language: Reading the Concentric Rings of Nature* (audio). Duvall, WA: Wilderness Awareness School Owlink Media. 6 cassettes.

Young, S., ed. 2001. *The Emergence of Ecological Modernisation: Integrating the Environment and the Economy?* New York: Routledge.

Zaehner, R. C. 1957. *Mysticism Sacred and Profane: An Inquiry into Some Varieties of Praeternatural Experience*. Oxford: Clarendon Press.

Zahan, D. 1983. *The Religion, Spirituality, and Thought of Traditional Africa*. Translated by K. Martin and L. Martin. Chicago: University of Chicago Press.

Zakin, S. 1995. *Coyotes and Town Dogs: Earth First! and the Environmental Movement*. New York: Penguin USA.

Zelenzny, L. C., P. P. Chua, and C. Aldrich. 2000. "New Ways of Thinking about Environmentalism: Elaborating on Gender Differences in Environmentalism." *Journal of Social Issues* 56 (3): 443–57.

Zerzan, J., ed. 2005. *Against Civilization: Readings and Reflections*. Los Angeles: Feral House.

Zimmerer, K. S., and T. J. Bassett. 2003. *Political Ecology: An Integrative Approach to Geography and Environment-Development Studies*. New York: Guilford Press.

Zimmerman, M. 1985. "Anthropocentric Humanism and the Arms Race." In *Nuclear War: Philosophical Perspectives*, edited by M. Fox and L. Groarke. New York: Peter Lang, 26–43.

———. 1990. *Heidegger's Confrontation with Modernity*. Bloomington: Indiana University Press.

———. 1994. *Contesting Earth's Future: Radical Ecology and Postmodernity*. Berkeley: University of California Press.

———. 1996. "A Transpersonal Diagnosis of the Ecological Crisis." *ReVision: A Journal of Consciousness and Transformation* 18 (4): 38–48.

———. 1997. "Ecofascism: A Threat to American Environmentalism?" In *The Ecological Community*, edited by R. S. Gottlieb. New York: Routledge, 229–54.

———. 2000. "Possible Political Problems of Earth-Based Religiosity." In *Beneath the Surface: Critical Essays in the Philosophy of Deep Ecology*, edited by E. Katz, A. Light, and D. Rothenberg. Cambridge, MA: MIT Press, 169–94.

———. 2001. "Ken Wilber's Critique of Ecological Spirituality." In *Deep Ecology and World Religions*, edited by D. Barnhill and R. Gottlieb. Albany, NY: SUNY Press, 243–69.

———. 2004. "What Can Continental Philosophy Contribute to Environmentalism?" In *Rethinking Nature: Essays in Environmental Philosophy*, edited by B. Foltz and R. Frodeman. Bloomington: Indiana University Press, 207–30.

———. 2005. "Integral Ecology: A Perspectival, Developmental, and Coordinating Approach to Environmental Problems." *World Futures: The Journal of General Evolution* 61 (1–2): 50–62.

———. 2008. "Nietzsche and Ecology: A Critical Inquiry." In *Reading Nietzsche in the Margins*, edited by S. V. Hicks and A. Rosenberg. Lafayette, IN: Purdue University Press, 165–85.

Zimmerman, M., J. Callicott, J. Clark, K. Warren, and I. Klaver. 2004. *Environmental Philosophy: From Animal Rights to Radical Ecology*. 4th ed. Upper Saddle River, NJ: Prentice Hall.

Znamenski, A., ed. 2004. *Shamanism: Critical Concepts*. 3 vols. London: Routledge.

Zupanc, G. K. H. 2004. *Behavioral Neurobiology: An Integrative Approach*. Oxford: Oxford University Press.

CREDITS

Chapter 2, "It's All About Perspectives" was previously published as "Defending the Importance of the Holarchical-Developmental Scheme for Environmentalism" in *AQAL: Journal of Integral Theory and Practice*, 1 (3), 40–100. © 2006 Michael E. Zimmerman.

Chapter 5, "Defining, Honoring, and Integrating the Multiple Approaches to Ecology," was previously published as "Integral Ecology: An Ecology of Perspectives" in *AQAL: Journal of Integral Theory and Practice*, 1 (1), 267–304. © 2006 Sean Esbjörn-Hargens.

Chapters 6, 7, and 11, "Ecological Terrains," "Ecological Selves," and "Integral Ecology in Action," were adapted from "Integral Ecology: A Post-Metaphysical Approach to Environmental Phenomena" in *AQAL: Journal of Integral Theory and Practice*, 1 (1), 305–378. © 2006 Sean Esbjörn-Hargens.

Chapter 8, "Ecological Research," was previously published as "Integral Ecological Research: Using IMP to Explore Animal Consciousness and Sustainability" in *Journal of Integral Theory and Practice*, 3 (1), 35–72. © 2008 Sean Esbjörn-Hargens.

The following illustrations are used with permission from Ken Wilber:

Figure 1.2: The ego and the eco before the collapse

Figure 3.1: The AQAL map

Figure 4.1: The 4 quadrants in human development

Figure 4.3: Wilber's color spectrum of consciousness

Figure 7.4: Important lines of development

Color Plate 2: The 8 ecological selves

Figure 8.1: The 8 fundamental perspectives

Figure 8.2: The 8 methodological zones

Figure 9.2: Nature mysticism lattice

INDEX